CONTROL SYSTEMS ENGINEERING

Modelling and Simulation, Control Theory and Microprocessor Implementation

STEPHEN P. BANKS
Department of Control Engineering
University of Sheffield

Prentice-Hall International

Englewood Cliffs, New Jersey
London Mexico New Delhi
Rio de Janeiro Singapore Sydney
Tokyo Toronto Wellington

To David and Yvonne

Library of Congress Cataloging in Publication Data

Banks, Stephen P.
Control systems engineering.

Bibliography: p.
Includes index.
1. Automatic control. 2. Control theory.
3. Microprocessors. I. Title.
TJ213.B224 1986 629.8 85–20491
ISBN 0–13–171794–4

British Library Cataloguing in Publication Data

Banks, Stephen P.
Control systems, engineering: modelling and simulation, control theory and microprocessor implementation. – (Prentice-Hall International series in Systems and control engineering)
1. Control theory
I. Title
629.8 QA402.3

ISBN 0–13–171794–4
ISBN 0–13–171778–2 Pbk

Prentice-Hall Inc., *Englewood Cliffs, New Jersey*
Prentice-Hall International (UK) Ltd, *London*
Prentice-Hall of Australia Pty Ltd, *Sydney*
Prentice-Hall Canada Inc., *Toronto*
Prentice-Hall Hispanoamericana S.A., *Mexico*
Prentice-Hall of India Private Ltd, *New Delhi*
Prentice-Hall of Japan Inc., *Tokyo*
Prentice-Hall of Southeast Asia Pte Ltd, *Singapore*
Editora Prentice-Hall do Brasil Ltda, *Rio de Janeiro*
Whitehall Books Ltd, *Wellington, New Zealand*

Printed and bound in Great Britain for
Prentice-Hall International (UK) Ltd,
66 Wood Land End, Hemel Hempstead,
Hertfordshire, HP2 4RG
by A. Wheaton & Co. Ltd, Exeter

1 2 3 4 5 90 89 88 87 86

ISBN 0-13-171794-4
ISBN 0-13-171778-2 PBK

CONTENTS

Preface vii
Introduction What is Control Systems Engineering? ix

PART ONE
Modelling and Simulation

1 Modelling of Physical Systems 3
1.1 Introduction 3
1.2 Modelling of mechanical systems 4
1.3 Modelling of lumped electrical systems 22
1.4 Modelling of electromechanical systems 29
1.5 Finite-dimensional systems 33
1.6 Distributed parameter systems 35
1.7 Stochastic systems 58
1.8 System elements and signal flow graphs 60
1.9 Identification methods 66
1.10 Sensitivity theory 69
1.11 Exercises 71
References 75

2 Analogue Simulation 77
2.1 Introduction 77
2.2 Linear analogue systems 77
2.3 Non-linear systems 88
2.4 Design of computer amplifiers 106
2.5 Exercises 119
References 121

3 Digital Simulation 123
3.1 Introduction 123
3.2 Digital computers 124

3.3 Numerical techniques for ordinary differential equations 133
3.4 Algebraic equations 143
3.5 Interpolation 145
3.6 Simulation languages 148
3.7 Simulation of distributed parameter systems 149
3.8 Simulation of stochastic systems 178
3.9 Exercises 180
References 183

PART TWO
Control Theory

4 Linear Systems Theory **187**
4.1 Introduction 187
4.2 Laplace transformation 187
4.3 Linear system stability 196
4.4 Frequency response analysis 204
4.5 Root locus theory 207
4.6 Classical compensation 212
4.7 Discrete systems 219
4.8 State variable approach 233
4.9 Multivariable systems 240
4.10 Exercises 254
References 256

5 Non-linear Systems **258**
5.1 Introduction 258
5.2 Linearization of non-linear systems 259
5.3 Phase plane analysis and classification of linear systems 263
5.4 Non-linear system stability 274
5.5 Approximate methods for non-linear oscillations 300
5.6 Bilinear systems 312
5.7 Exercises 317
References 319

6 Optimal Control Theory **320**
6.1 Introduction 321
6.2 Classical optimization theory 322
6.3 Classical calculus of variations 332
6.4 The Pontryagin maximum principle 342
6.5 Dynamic programming 355

6.6 The linear regulator and generalizations 361
6.7 Exercises 369
References 371

7 Stochastic Systems 373
7.1 Introduction 373
7.2 Discrete estimation and filtering 374
7.3 Continuous estimation and filtering 391
7.4 Linear stochastic control 402
7.5 Non-linear estimation 407
7.6 Computational aspects of linear filtering 411
7.7 System identification 413
7.8 Exercises 420
References 422

8 Adaptive Control, Self-tuning Control and Variable-structure Systems 423
8.1 Introduction 423
8.2 Adaptive control 424
8.3 Self-tuning control 442
8.4 Variable-structure systems 451
8.5 Exercises 465
References 466

PART THREE
Microprocessor Implementation

9 Introduction to the Z80 Microprocessor and Assembly Programming 471
9.1 Introduction 471
9.2 The Z80 microprocessor 472
9.3 The Z80 instruction set 488
9.4 Assembly programming 513
9.5 Other microprocessors 535
9.6 Exercises 535
References 536

10 Microprocessor Interfacing Techniques 537
10.1 Introduction 537
10.2 Three bus architecture and timing 538
10.3 Interfacing memory 543
10.4 Input–output ports, interrupts and the PIO chip 549
10.5 Interfacing peripherals 564

10.6 D/A, A/D conversion and data acquisition 572
10.7 Examples of real system interfaces 577
10.8 Conclusion 587
10.9 Exercises 587
References 588

Appendix 1 Summary of mathematical ideas used in the book 589
Appendix 2 Elementary probability theory 601
Index 610

PREFACE

This book evolved from courses at Sheffield University given to undergraduate and postgraduate students in control engineering. The intention is to present, in a single volume, the three main aspects of control systems engineering – namely, modelling and simulation, control theory, and microprocessor implementation. It is aimed at the undergraduate and postgraduate levels, but some more specialists sections have been included which can be omitted on first reading.

Different chapters may serve as a basis for various courses on modelling, linear systems theory, stochastic systems etc., although, because of the limitations of space, most will need supplementing from the cited literature at the end of each chapter. However, it is hoped that each chapter will provide at least an introduction to that particular field. At the end of each chapter there is also a set of problems which illustrate the theory and which the serious student should at least read and attempt as many as possible.

Prerequisites to reading this book are elementary courses on mathematics, physics, probability and some simple electronics, although the main results in mathematics and probability which are used here are presented in two appendices. The three parts of the book may be read independently and can serve as introductions to their respective topics.

Part One is concerned with modelling and simulation and considers mainly those systems which can be modelled by classical physics. Both analogue and digital simulation of (ordinary and partial) differential equations, delay systems etc. are discussed and an introduction to the finite element technique is also given. Part Two deals with control theory and consists of chapters on linear and non-linear systems theory, optimal control theory, filtering, self-tuning and adaptive control, and variable structure systems. Finally, Part Three discusses the Z80 μP (microprocesor) in both its hardware and software aspects and describes its application in control.

In a single book of this length, which attempts to cover the whole field of control systems engineering, I have had to make difficult decisions on which parts of the theory to include. The contents therefore reflect my own prejudices on what control systems engineering is about. In particular, a chapter on analogue simulation is included even though most real-time simulation is now done digitally, using one of the real-time languages currently available. However, analogue machines are still widely used and I feel that it is important for students at least to be aware of the analogue approach to systems design. Secondly, in Part Three of the book, the reader may ask why I have chosen the Z80 microprocessor to develop digital controllers, rather than a general purpose processor or a 16- or 32-bit chip. I have chosen the Z80, in particular, since I wish to illustrate specific software and interface circuits, which would be more difficult with a general purpose microprocessor, and I have not chosen a 16- or 32-bit chip because, in my experience, students tend to understand the operation of an 8-bit chip more easily than 16- or 32-bit chips.

The appendices are designed only as a collection of results which the reader is expected to know (or at least have met before). In particular, the reader is assumed to have a knowledge of elementary probability theory – Appendix 2 is just to provide a list of the results of advanced probability theory necessary for understanding continuous filtering. The only other prerequisite is some elementary electronics.

The author wishes to express his gratitude to colleagues for many useful discussions on various aspects of the book and in particular Dr S. A. Billings for the generous use of his lecture notes on self-tuning and adaptive control.

S. P. B.

INTRODUCTION

WHAT IS CONTROL SYSTEMS ENGINEERING?

Control systems engineering, unlike most other engineering disciplines, is not concerned initially with any particular system or structure, although the theories developed are intended for application to many types of real systems – aircraft control, nuclear reactors, chemical plants, even economic modelling and control. The main object of control systems theory is to abstract the salient properties from as large a class of systems as possible and then derive results which will be useful in the control of all the systems in the class. The most important class, as far as control engineering is concerned, is that of linear systems, mainly because the linear class is mathematically tractable and can be used to model a large variety of systems, at least in the neighbourhood of some operating condition. Indeed the whole purpose of classical feedback control is to keep the system outputs close to some nominal trajectory so that the linear model of the system is valid.

This does not imply, of course, that non-linear systems are not important – many systems have essential non-linearities which cannot be 'linearized' away. The 'theory' of non-linear systems is by no means as well developed as that of linear systems and, as yet, no coherent theory of control design exists for any reasonable class of non-linear systems. One promising line of attack seems to be the bilinear class which will be discussed briefly later in the book. In fact, it can be shown that many reasonable types of non-linear systems may be approximated by bilinear ones. However, many types of non-linear systems including discontinuities such as switches, hysteresis or deadzones still require special treatment.

In view of the fact that all real systems are non-linear, the first major task of control systems engineering is to obtain a good model of the system, before any simplifications, such as linearizations, can be made. The modelling exercise often requires the application of many different types of expert knowledge and skill – aerodynamics, nuclear physics, theory of chemical equilibria and economic theory for the systems mentioned above.

The main purpose of modelling is to reduce the physical system to a set of mathematical equations. These may be non-linear ordinary or partial differential equations, difference equations, delay equations etc., and many types of physical systems with apparently differing structure can be subsumed under a common form of equation. Systems which can be modelled by considering the equations of classical physics are studied in Part One.

Once a good model of any particular system has been developed, it is usually necessary to validate the model against real input–output data from the system. This involves the process of simulation which is considered in detail in Part One. We may use analogue amplifiers to simulate differential equations which will result in an analogue simulation, or we may use digital simulation by discretizing the equations and solving the resulting difference equations on a digital computer. Although most simulations (including real-time studies) are now carried out digitally, it is still important to understand analogue systems since most real signals require some form of analogue processing before being input to a digital system, as we shall see in chapter 10 when we consider microprocessor interfacing.

Having obtained a validated model of the system, we can now apply the theories of control which are developed independently of any particular system, as stated before. This is the real power of control – the techniques which have been discovered for general linear systems will apply to any system of this form. An extensive collection of results have been developed for general systems control and the main ones, which we consider in Part 2, are classical compensator design, multivariable theory, non-linear stability and design, noise filtering and optimal control, and we give an introduction to self-tuning and adaptive control and variable structure systems. The control systems engineer must be familiar with a large class of methods since some will be more appropriate in certain situations than others.

The task of the control systems engineer is only two-thirds complete, however, when the appropriate control strategy has been chosen for the given system. The final part of any real control project is to connect the system to the controller. This can often be done using analogue filters, which is merely an application of the ideas of analogue computing. However, because of their greater flexibility and reliability, many feedback control systems are now digitally controlled. The software and hardware interfacing of a microprocessor to a real system is therefore an essential element of the control systems engineer's job. This task will be discussed in detail in the last section of the book and is based on the Z80 microprocessor.

Here, then, are the three major aspects of control systems engineering: modelling and simulation, control theory and microcomputer implementation, and it is these with which this book is concerned. One of the intentions of the book is to stimulate a wider interest in this challenging but rewarding discipline, but the reader should also be aware of the modern 'spin-offs' from control, such as image processing, information technology, robotics etc., which are not covered here.

PART ONE

Modelling and Simulation

1 MODELLING OF PHYSICAL SYSTEMS

1.1 INTRODUCTION

In order to control a physical system it is necessary to obtain a good mathematical 'model' of the system. When models are difficult or impossible to obtain we must consider other methods of control which will be discussed later in the book. However, for the present we shall assume that it is desirable to obtain the best possible model for the system. A mathematical model of a system S which depends on some input or forcing function u, producing an output $y(u)$, is a set of mathematical equations S_M which also depend on an input u_M, producing an output $y_M(u_M)$. Ideally, if the model S_M is a perfect representation of S, then for any (physically reasonable) input u to S we have

$$y(u) = y_M(u) \tag{1.1.1}$$

i.e. if we input the same function u to S and S_M we obtain the same output. Of course, mathematical idealizations never model real physical processes exactly and so (1.1.1) will only be an approximate equality, so that

$$y(u) = y_M(u) + \varepsilon(u) \tag{1.1.2}$$

where ε is the modelling error, which may depend on the input u.

A good example of the input-dependent error ε is in mechanical systems which are modelled very accurately by Newton's equations. However, for sufficiently large input functions, which may be related to particle velocities, the mathematical model may require substantial relativistic corrections $\varepsilon(u)$. This is, of course, an extreme example and the physical processes which we shall study in this chapter may all be modelled accurately using classical physics.

We shall begin by studying dynamics from the generalized Lagrangian approach, since this method allows one to obtain equations for complex systems fairly simply. We shall then discuss electrical network theory and introduce the physical modelling necessary for the electric motor, which is extremely important for control implementations. Physical systems which are not described by ordinary differential equations are then considered. These include systems with delay giving rise to differential delay equations, and diffusion processes and fluid motion which require partial differential equations for their accurate representation. Finally we shall mention the important concept of noise which is present in most real systems.

1.2 MODELLING OF MECHANICAL SYSTEMS

1.2.1 Newton's Second Law and Inertial Reference Frames

Setting up dynamical equations of motion for a mechanical system requires a choice of a particular reference frame in which one can define a co-ordinate system. Any frame of reference which is moving with uniform velocity relative to the fixed stars is called an **inertial reference frame.** Of course, the stars are not truly fixed and so we can only obtain approximations to such a reference frame. However, even though the Earth is rotating, for many applications a frame fixed in Earth axes is approximately inertial.

Relative to an inertial reference frame (with rectangular Cartesian co-ordinates), Newton discovered that the equation

$$\begin{aligned} \boldsymbol{F} &= \frac{\mathrm{d}(m\boldsymbol{v})}{\mathrm{d}t} \\ &= m\frac{\mathrm{d}\boldsymbol{v}}{\mathrm{d}t}, \quad \text{if } m \text{ is time independent} \end{aligned} \tag{1.2.1}$$

is valid, where $\boldsymbol{F}$ is the real (physical) force acting on a particle of mass m, producing a velocity $\boldsymbol{v}$. In fact, one could define an inertial reference frame as one in which (1.2.1) holds. It is, of course, possible to write down dynamical equations in any reference frame, but the force $\boldsymbol{F}$ will then include terms which are 'fictitious' forces introduced by virtue of the acceleration of the general frame relative to the inertial one. For example, if a two-dimensional frame (x_2, y_2) is rotating counter-clockwise with angular velocity ω relative to an inertial frame (x_1, y_1), it is easy to show that the equations of motion of a particle in (x_2, y_2) coordinates are

$$\left.\begin{aligned} F_{x_2} &= m\ddot{x}_2 - mx_2\omega^2 - 2m\omega\dot{y}_2 \\ F_{y_2} &= m\ddot{y}_2 - my_2\omega^2 + 2m\omega\dot{x}_2 \end{aligned}\right\} \tag{1.2.2}$$

where $\boldsymbol{F} = (F_{x_2}, F_{y_2})$ is the 'real' force acting on the particle resolved into x_2, y_2 components.

In any frame of reference we can use an infinite variety of different coordinate systems, which may be curvilinear and non-orthogonal. For a mechanical system, the number of **degrees of freedom** is the number of independent coordinates (excluding time) required to specify the position of each component part of the system. For example, each free particle has three degrees of freedom and each free solid body has six degrees of freedom (three translational and three rotational degrees of freedom). Hence a system of p free particles has $3p$ degrees of freedom, but if there are any constraints the number of degrees of freedom will be reduced. For example, if a bead is constrained to move on a straight rod rotating in a known way about the origin then the distance of the particle from the origin is a sufficient coordinate (fig. 1.1), and the particle has only one degree of freedom. In this case the coordinates x and y are related by

$$y = \tan(\omega t)x \qquad (1.2.3a)$$

Also

$$z = 0 \qquad (1.2.3b)$$

since the particle is constrained to move in the (x, y)-plane. Equations (1.2.3) are called the equations of constraint. Note that they may depend explicitly on time. In general, a system of p particles has

$$n = 3p\text{-(number of equations of constraint)} \qquad (1.2.4)$$

degrees of freedom.

In a system of p particles we denote by $q_1, q_2, \ldots, q_{3p}$ their **generalized coordinates**, i.e. any coordinates required to specify the positions of p free particles. It is often convenient to relate these coordinates to inertial rectangular coordinates x_i, y_i, z_i $(1 \leqslant i \leqslant p)$. Then we can write

$$x_i = x_i\,(q_1, q_2, \ldots, q_{3p}, t), \text{ etc.} \qquad (1.2.5)$$

If there are $3p$-n constraints then we can write the constraint equations as

$$c_i(q_1, q_2, \ldots, q_{3p}, t) = 0, \quad 1 \leqslant i \leqslant 3p\text{-}n \qquad (1.2.6)$$

which may be complicated non-linear equations. We can say that $3p$-n of the coordinates are **superfluous** and, if we can solve the equations (1.2.6),

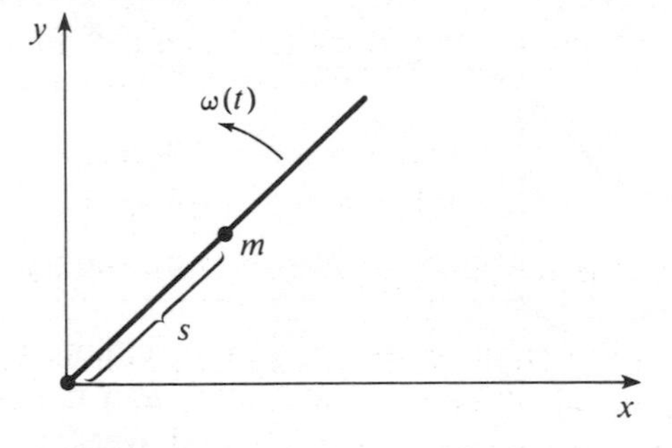

Fig. 1.1 Mass on a rotating wire.

we can replace (1.2.5) by

$$x_i = x_i(q_1, q_2, \ldots, q_n, t), \text{ etc.} \tag{1.2.7}$$

(or by any other set of n independent coordinates of the set $q_1, q_2, \ldots, q_{3_p}$), and we say that the system is **holonomic**.

1.2.2 Work and Kinetic Energy

Suppose that a force $\boldsymbol{F}$ acts on a particle which moves on a path joining the points a and b (fig. 1.2). Then the work done by $\boldsymbol{F}$ along the path is, by definition,

$$\begin{aligned} W &= \int_a^b \boldsymbol{F} \cdot \mathrm{d}\boldsymbol{s} \\ &= \int_a^b (F_x\,\mathrm{d}x + F_y\,\mathrm{d}y + F_z\,\mathrm{d}z) \end{aligned} \tag{1.2.8}$$

However, if the frame (x, y, z) is inertial, then, by Newton's second law, $\boldsymbol{F} = m(\ddot{x}, \ddot{y}, \ddot{z})$ and, since $\ddot{x}\,\mathrm{d}x = \dot{x}\,\mathrm{d}\dot{x}$, we have

$$\begin{aligned} W &= \int_a^b m(\dot{x}\,\mathrm{d}\dot{x} + \dot{y}\,\mathrm{d}\dot{y} + \dot{z}\,\mathrm{d}\dot{z}) \\ &= \frac{m}{2}(\dot{x}^2 + \dot{y}^2 + \dot{z}^2)\Big|_a^b \end{aligned}$$

If the velocity at a equals 0, then

$$W = \tfrac{1}{2} m v_b^2 \tag{1.2.9}$$

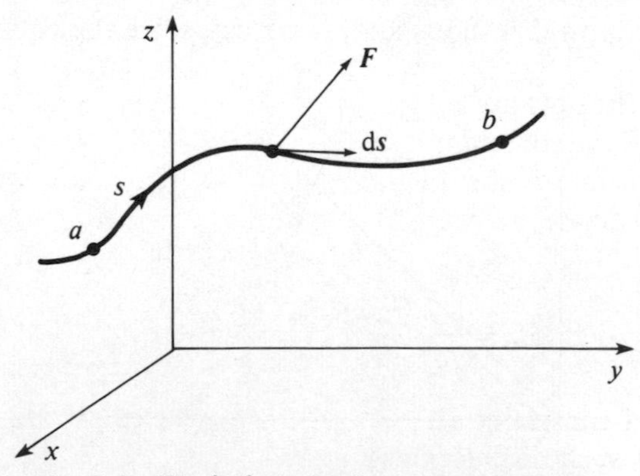

Fig. 1.2 Work done by force F along a line.

where v_b is the velocity of the particle at b. The quantity in (1.2.9) is called the **kinetic energy** of the particle and is the work required to give the particle a velocity v, in an inertial frame.

The total kinetic energy of p particles is

$$T = \tfrac{1}{2}\sum_{i=1}^{p} m_i v_i^2$$

(in inertial coordinates). Differentiating (1.2.7), we can express T in non-inertial coordinates. (See section 1.11, exercise 2.)

1.2.3 Virtual Work and Lagrange's Equations

Solving Newton's equations of motion gives rise to a trajectory for each particle in the system. Any displacement along the motion is consistent with the applied forces and constraining forces (such as reactions between different parts of the system and tensions in rigid rods) and is called a **real displacement**. Although not consistent with the real motion of the system, we can conceive of arbitrary displacements of the generalized coordinates, which are therefore called **virtual displacements**. Generally, from (1.2.5) we have

$$\delta x_i = \frac{\partial x_i}{\partial q_1}\,\delta q_i + \frac{\partial x_i}{\partial q_2}\,\delta q_2 + \ldots + \frac{\partial x_i}{\partial q_{3p}}\,\delta q_{3p} + \frac{\partial x_i}{\partial t}\,\delta t \qquad (1.2.10)$$

However, such a displacement implies that the equations of constraint are violated and hence the forces of constraint do work during the displacement. If we remove the superfluous coordinates but still allow moving frames and constraints, we obtain

$$\delta x_i = \frac{\partial x_i}{\partial q_1}\,\delta q_1 + \frac{\partial x_i}{\partial q_2}\,\delta q_2 + \ldots + \frac{\partial x_i}{\partial q_n}\,\delta q_n + \frac{\partial x_i}{\partial t}\,\delta t \qquad (1.2.11)$$

and now the constraints are not violated but since any moving constraints change in time δt, the constraining forces still do work. The most important virtual displacement for the **Lagrange method** is to allow only the non-superfluous coordinates to vary. Then we have the instantaneous virtual displacement

$$\delta x_i = \frac{\partial x_i}{\partial q_1}\,\delta q_1 + \ldots + \frac{\partial x_i}{\partial q_n}\,\delta q_n \qquad (1.2.12)$$

and the forces of constraint do no work; hence we can ignore them in the consideration of such displacements.

The work done in a virtual displacement of type (1.2.12) can be found easily. In fact

$$\delta W = \sum_{i=1}^{p} \boldsymbol{F} \cdot \delta \boldsymbol{s}_i$$

$$= \sum_{i=1}^{p} (F_{x_i}\,\delta x_i + F_{y_i}\,\delta y_i + F_{z_i}\,\delta z_i)$$

(for a system of p particles) and so, using (1.2.12),

$$\delta W = \sum_{j=1}^{n} \sum_{i=1}^{p} \left(F_{x_i} \frac{\partial x_i}{\partial q_j} + F_{y_i} \frac{\partial y_i}{\partial q_j} + F_{z_i} \frac{\partial z_i}{\partial q_j} \right) \delta q_j \tag{1.2.13}$$

Now, in an inertial frame, Newton's second law implies that

$$\delta W = \sum_{j=1}^{n} \sum_{i=1}^{p} m_i \left(\ddot{x}_i \frac{\partial x_i}{\partial q_j} + \ddot{y}_i \frac{\partial y_i}{\partial q_j} + \ddot{z}_i \frac{\partial z_i}{\partial q_j} \right) \delta q_j$$

However,

$$\ddot{x}_i \frac{\partial x_i}{\partial q_j} = \frac{\mathrm{d}}{\mathrm{d}t}\left(\dot{x}_i \frac{\partial x_i}{\partial q_j} \right) - \dot{x}_i \frac{\mathrm{d}}{\mathrm{d}t}\left(\frac{\partial x_i}{\partial q_j} \right)$$

$$= \frac{\mathrm{d}}{\mathrm{d}t}\left(\dot{x}_i \frac{\partial x_i}{\partial q_j} \right) - \dot{x}_i \frac{\partial \dot{x}_i}{\partial q_j}$$

since

$$\dot{x}_i = \sum_{k=1}^{n} \frac{\partial x_i}{\partial q_k} \dot{q}_k \quad \text{(by (1.2.7) with } \delta t = 0)$$

whence

$$\frac{\partial \dot{x}_i}{\partial \dot{q}_j} = \frac{\partial x_i}{\partial q_j}$$

and

$$\frac{\partial \dot{x}_i}{\partial q_j} = \sum_{k=1}^{n} \frac{\partial}{\partial q_j} \left(\frac{\partial x_i}{\partial q_k} \dot{q}_k \right)$$

$$= \frac{\mathrm{d}}{\mathrm{d}t} \left(\frac{\partial x_i}{\partial q_j} \right)$$

by the chain rule. Therefore,

$$\delta W = \sum_{j=1}^{n} \left[\frac{\mathrm{d}}{\mathrm{d}t} \frac{\partial}{\partial \dot{q}_j} \sum_{i=1}^{p} m_i \frac{(\dot{x}_i^2 + \dot{y}_i^2 + \dot{z}_i^2)}{2} - \frac{\partial}{\partial q_j} \sum_{i=1}^{p} m_i \frac{(\dot{x}_i^2 + \dot{y}_i^2 + z_i^2)}{2} \right] \delta q_j$$

$$= \sum_{j=1}^{n} \left[\frac{\mathrm{d}}{\mathrm{d}t} \left(\frac{\partial T}{\partial \dot{q}_j} \right) - \frac{\partial T}{\partial q_j} \right] \delta q_j$$

where T is the total kinetic energy of the system. Comparing this with (1.2.13) we have

$$\frac{\mathrm{d}}{\mathrm{d}t}\left(\frac{\partial T}{\partial \dot{q}_j}\right) - \frac{\partial T}{\partial q_j} = F_{q_j}, \quad 1 \leqslant j \leqslant n \tag{1.2.14}$$

where

$$F_{q_j} \triangleq \sum_{i=1}^{p} \left(F_{x_i} \frac{\partial x_i}{\partial q_j} + F_{y_i} \frac{\partial y_i}{\partial q_j} + F_{z_i} \frac{\partial z_i}{\partial q_j} \right) \tag{1.2.15}$$

is the **generalized force** related to coordinate q_j. (We have used the fact that the arbitrary displacements δq_j are independent.) Note that, by (1.2.13), since the δq_k are independent, F_{q_j} may be found by taking a virtual displacement of q_j only, while fixing all other generalized coordinates. Then

$$\delta W_{q_j} = F_{q_j}\, \delta q_j$$

The equations (1.2.14) are **Lagrange's equations of motion**, and require the total kinetic energy T and the generalized forces F_{q_j} expressed in non-superfluous coordinates.

We have defined the work done by a force $\boldsymbol{F}$ along a path γ joining two points a and b as

$$W = \int_\gamma \boldsymbol{F} \cdot \mathrm{d}\boldsymbol{s} = \int_\gamma \sum_{i=1}^{p} (F_{x_i}\, \mathrm{d}x_i + F_{y_i}\, \mathrm{d}y_i + F_{z_i}\, \mathrm{d}z_i)$$

If the work is independent of the particular path joining a and b then we say that $\boldsymbol{F}$ is **conservative**. In this case it is wellknown that $\boldsymbol{F}$ can be expressed as the gradient of a potential V, i.e.

$$F_{x_i} = -\frac{\partial V}{\partial x_i}, \quad F_{y_i} = -\frac{\partial V}{\partial y_i}, \quad F_{z_i} = -\frac{\partial V}{\partial z_i} \tag{1.2.16}$$

A necessary and sufficient condition for (1.2.16) to hold is

$$\frac{\partial F_{x_i}}{\partial y_j} = \frac{\partial F_{y_j}}{\partial x_i}, \quad \frac{\partial F_{x_i}}{\partial z_j} = \frac{\partial F_{z_j}}{\partial x_i}, \quad \text{etc.} \tag{1.2.17}$$

Now, using (1.2.15)

$$F_{q_j} = -\sum_{i=1}^{p} \left(\frac{\partial V}{\partial x_i} \frac{\partial x_i}{\partial q_j} + \frac{\partial V}{\partial y_i} \frac{\partial y_i}{\partial q_j} + \frac{\partial V}{\partial z_i} \frac{\partial z_i}{\partial q_j} \right)$$

and so

$$\boxed{F_{q_j} = -\frac{\partial V}{\partial q_j}} \tag{1.2.18}$$

Hence, by Lagrange's equations,

$$\boxed{\frac{\mathrm{d}}{\mathrm{d}t}\left(\frac{\partial L}{\partial \dot{q}_j}\right) - \frac{\partial L}{\partial q_j} = F^{\mathrm{d}}_{q_j}} \tag{1.2.19}$$

where

$$L = T - V$$

is called the **Lagrangian function**, and $F^{\mathrm{d}}_{q_j}$ contains only dissipative forces.

Note that if L does not depend on time explicitly and all forces are conservative, then it is easy to see that

$$\boxed{T + V = \text{constant}} \tag{1.2.20}$$

(see section 1.11 exercise 3), i.e. the total energy is constant.

The most common type of dissipative (i.e. non-conservative) forces are those due to friction and viscous drag. Such a force may often be written in the form

$$f = av^n, \quad \text{for some constant } a \tag{1.2.21}$$

where $n = 0$ for a frictional force and $n = 1$ for a simple low velocity viscous drag. Since $\dot{x}/v$ is the cosine of the angle between the x-axis and v, we have

$$\begin{aligned} f_x &= -av^n\frac{\dot{x}}{v} \\ &= -a\dot{x}v^{n-1} \end{aligned}$$

and similarly for f_y, f_z. Hence

$$\delta W = -\sum_{i=1}^{p} a_i(\dot{x}_i\,\delta x_i + \dot{y}_i\,\delta y_i + \dot{z}_i\,\delta z_i)v_i^{n-1}$$

which can be written in generalized coordinates by using (1.2.7). We obtain

$$\delta W = \sum_{j=1}^{n} F^{\mathrm{d}}_j\,\delta q_j \tag{1.2.22}$$

where $F^{\mathrm{d}}_j = F^{\mathrm{d}}_j(q_1, \dot{q}_1, \ldots, q_n, \dot{q}_n)$ are the generalized dissipative forces.

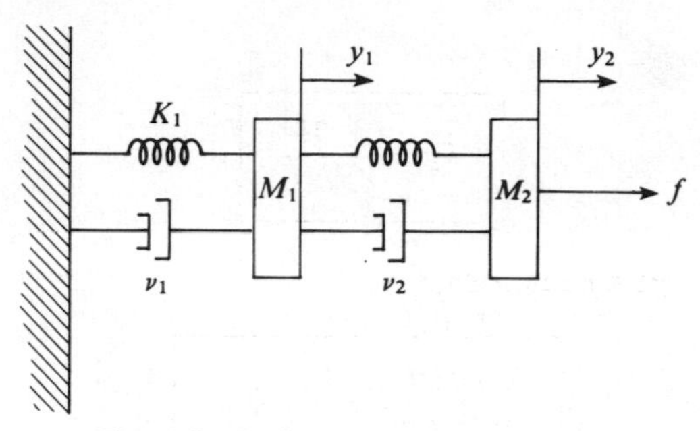

Fig. 1.3 Spring-mass-damper system.

Example 1.2.1

As an example consider the mass–spring–damper system in fig. 1.3. Choose y_1, y_2 as generalized coordinates. Then the kinetic energy is

$$T = \tfrac{1}{2}M_1\dot{y}_1^2 + \tfrac{1}{2}M_2\dot{y}_2^2$$

the potential energy is

$$V = \tfrac{1}{2}K_1y_1^2 + \tfrac{1}{2}K_2(y_2 - y_1)^2$$

and the generalized forces are

$$F_{y_1} = -\nu_1\dot{y}_1 + \nu_2(\dot{y}_2 - \dot{y}_1)$$
$$F_{y_2} = -\nu_2(\dot{y}_2 - \dot{y}_1) + f$$

Hence, by (1.2.19) we obtain the equations

$$M_1\ddot{y}_1 + K_1y_1 - K_2(y_2 - y_1) = -\nu_1\dot{y}_1 + \nu_2(\dot{y}_2 - \dot{y}_1)$$
$$M_2\ddot{y}_2 + K_2(y_2 - y_1) = f - \nu_2(\dot{y}_2 - \dot{y}_1)$$

Example 1.2.2

As a simple example of the power of the Lagrangian method, consider a dumb-bell free to move in a vertical plane (X, Y) shown in fig. 1.4. The system has three degrees of freedom and the kinetic energy is

$$T = m_1\frac{\dot{x}_1^2}{2} + m_1\frac{\dot{y}_1^2}{2} + m_2\frac{\dot{x}_2^2}{2} + m_2\frac{\dot{y}_2^2}{2}$$

$$= \frac{m_1 + m_2}{2}(\dot{x}^2 + \dot{y}^2) + \tfrac{1}{2}(m_1l_1^2 + m_2l_2^2)\,\dot{\theta}^2$$

$$= \frac{m_1 + m_2}{2}(\dot{x}^2 + \dot{y}^2) + \frac{I}{2}\,\dot{\theta}^2$$

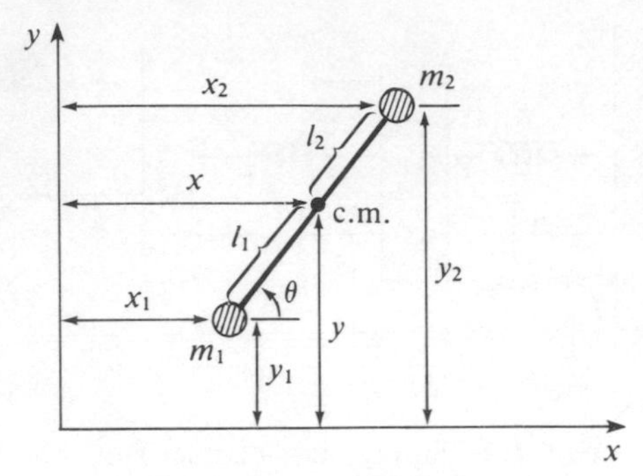

Fig. 1.4 Freely moving dumbell.

where I is the moment of inertia of the rod about the centre of mass (c.m.). Thus, by (1.2.14), Lagrange's equations are

$$(m_1 + m_2)\ddot{x} = F_x, \quad (m_1 + m_2)\ddot{y} = F_y, \quad I\ddot{\theta} = F_\theta$$

To find the generalized forces F_x, F_y, F_θ we can consider virtual displacements of x, y, independently. Hence, if x changes by δx and y, θ are fixed, gravity does no work, and so $F_x = 0$. Similarly, $F_y = -(m_1 + m_2)g$ and since (x, y) is the c.m., $F_\theta = 0$. Hence

$$\ddot{x} = 0, \quad \ddot{y} = -g, \quad \ddot{\theta} = 0$$

and so the c.m. moves as a free particle under gravity and the rod rotates with constant angular velocity. Note that, as stated above the forces of constraint can be ignored, which means that there was no need to introduce the tension in the rod.

1.2.4 Moments of Inertia

Consider the continuous body B shown in fig. 1.5. The moment of inertia of B about the axis OP is defined as

$$I_{OP} = \int_B h^2 \, d\mu$$

$$= \int_B [(x^2 + y^2 + z^2)(l^2 + m^2 + n^2) - (lx + my + nz)^2] \, d\mu$$

where (l, m, n) are the direction cosines of OP relative to (x, y, z). Hence

$$\boxed{I_{OP} = I_x l^2 + I_y m^2 + I_z n^2 - 2I_{xy} lm - 2I_{xz} ln - 2I_{yz} mn} \quad (1.2.23)$$

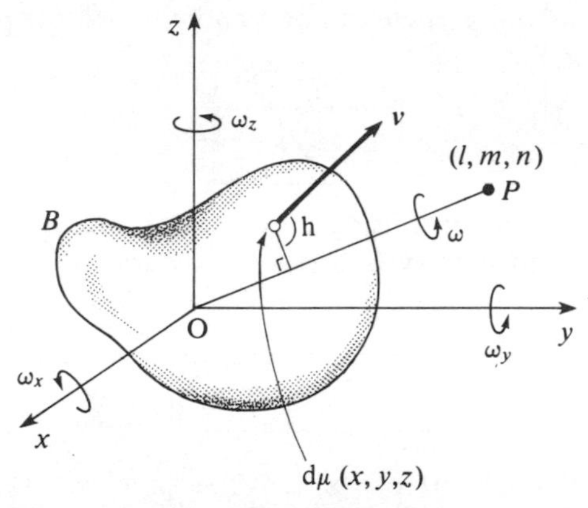

Fig. 1.5 Rotating body.

where I_x, I_y, I_z are the moments of inertia about the x, y, z axes and

$$I_{xy} = \int xy \, d\mu,$$

etc. are the **products of inertia.** The relation (1.2.23) is true for any line OP. Consider the points (x, y, z) on each such line at a distance $\pm 1/(I_{OP})^{1/2}$ from the origin. Then, by (1.2.23), these points satisfy the equation

$$\boxed{I_x x^2 + I_y y^2 + I_z z^2 - 2I_{xy}xy - 2I_{xz}xz - 2I_{yz}yz = 1} \tag{1.2.24}$$

and form the **ellipsoid of inertia** about O.

The **principal moments of inertia** are solutions I^p of the equation

$$\begin{vmatrix} I^p - I_x & I_{xy} & I_{xz} \\ I_{xy} & I^p - I_y & I_{yz} \\ I_{xz} & I_{yz} & I^p - I_z \end{vmatrix} = 0 \tag{1.2.25}$$

and the corresponding normalized eigenvectors define their direction cosines. (See appendix 1 for more discussion of eigenvalues and eigenvectors of a matrix.)

1.2.5 Lagrange's Equations for a Rigid Body

Referring again to fig. 1.5, if the body rotates with an angular speed ω about OP and we define the vector $\boldsymbol{\omega}$ along OP by

$$\begin{aligned} \boldsymbol{\omega} &= \omega(l, m, n) \\ &= (\omega_x, \omega_y, \omega_z) \end{aligned}$$

then the velocity $\boldsymbol{v}$ of the particle dm perpendicular to the plane containing OP and the particle is

$$\boxed{\boldsymbol{v} = \boldsymbol{\omega} \wedge \boldsymbol{r}} \qquad (r = (x, y, z))$$

Since no conditions have been imposed on the frame in fig. 1.5, if $X_I Y_I Z_I$ is an inertial frame and a body B is translating and rotating as in fig. 1.6, then

$$\boxed{\boldsymbol{v} = \boldsymbol{v}_O + \boldsymbol{\omega} \wedge \boldsymbol{r}} \tag{1.2.26}$$

where $\boldsymbol{v}$ is the inertial velocity of dμ, $\boldsymbol{\omega}$ is the angular velocity of the body and $\boldsymbol{v}_O$ is the velocity of an arbitrary point O fixed in the body, relative to the inertial frame. Expressed in terms of any frame XYZ with origin at O, (1.2.26) becomes

$$\boxed{\begin{aligned} v_x &= v_{Ox} + \omega_y z - \omega_z y \\ v_y &= v_{Oy} + \omega_z x - \omega_x z \\ v_z &= v_{Oz} + \omega_x y - \omega_y x \end{aligned}} \tag{1.2.27}$$

Note that, although the XYZ frame can be rotating in any way relative to the body, it is often taken to be body fixed, i.e. rotating with the body.

Now, the kinetic energy of the body is

$$T = \tfrac{1}{2} \int (v_x^2 + v_y^2 + v_z^2)\, \mathrm{d}\mu$$

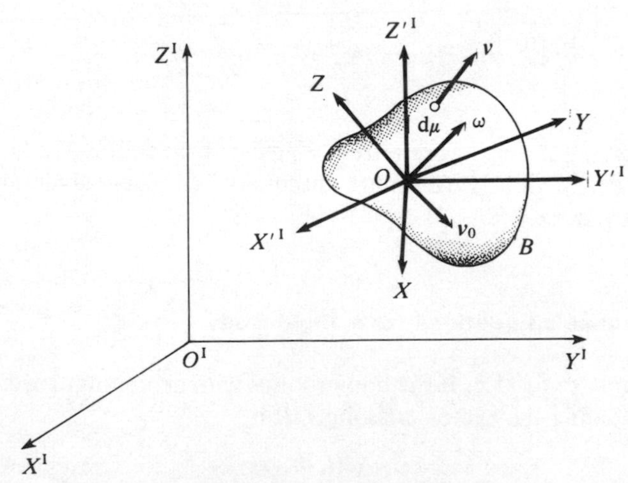

Fig. 1.6 Freely moving solid body.

from which we see that

$$T=\tfrac{1}{2}Mv_O^2+\tfrac{1}{2}[I_x\omega_x^2+I_y\omega_y^2+I_z\omega_z^2-2I_{xy}\omega_x\omega_y-2I_{xz}\omega_x\omega_z-2I_{yz}\omega_y\omega_z]$$
$$+M[v_{Ox}(\omega_y\bar{z}-\omega_z\bar{y})+v_{Oy}(\omega_z\bar{x}-\omega_x\bar{z})+v_{Oz}(\omega_x\bar{y}-\omega_y\bar{x})] \tag{1.2.28}$$

where all quantities are measured relative to the XYZ axes and $\bar{x}, \bar{y}, \bar{z}$ are the centre of mass coordinates (again relative to XYZ). Lagrange's equations of motion can now be found from (1.2.28) as before and the generalized forces are found in exactly the same way as in the case of particles. However, it is often useful to express the equations in terms of special angles relating the inertial frame to the body fixed axes. These angles are called Euler's angles, denoted by (ψ, θ, ϕ) and defined as in fig. 1.7. Starting from an inertial frame $X^I Y^I Z^I$ we rotate the initially coincident frame XYZ through ψ about OZ^I, then about ON by the angle θ, and finally about OZ through ϕ. The direction cosines of the axes OX, OY, OZ relative to OX^I, OY^I, OZ^I are

$$\begin{array}{c|ccc} & OX & OY & OZ \\ OX^I & \begin{pmatrix}\cos\phi\cos\psi \\ -\sin\phi\sin\psi\cos\theta\end{pmatrix} & \begin{pmatrix}-\sin\phi\cos\psi \\ -\cos\phi\sin\psi\cos\theta\end{pmatrix} & \sin\theta\sin\psi \\ OY^I & \begin{pmatrix}\cos\phi\sin\psi \\ +\sin\phi\cos\psi\cos\theta\end{pmatrix} & \begin{pmatrix}-\sin\phi\sin\psi \\ +\cos\phi\cos\psi\cos\theta\end{pmatrix} & -\sin\theta\cos\psi \\ OZ^I & \sin\theta\sin\phi & \sin\theta\cos\phi & \cos\theta \end{array} \tag{1.2.29}$$

By setting $\psi=0$ it can be seen that the direction cosines between OX, OY, OZ and OZ^I, ON are given by

$$\begin{array}{c|ccc} & OX & OY & OZ \\ OZ^I & \sin\theta\sin\phi & \sin\theta\cos\phi & \cos\theta \\ ON & \cos\phi & -\sin\phi & 0 \end{array} \tag{1.2.30}$$

Hence the components ω_x, ω_y, ω_z of the angular velocity of the body in XYZ axes are

$$\begin{aligned} \omega_x &= \dot{\psi}\sin\theta\sin\phi+\dot{\theta}\cos\phi \\ \omega_y &= \dot{\psi}\sin\theta\cos\phi-\dot{\theta}\sin\phi \\ \omega_z &= \dot{\phi}+\dot{\psi}\cos\theta \end{aligned} \tag{1.2.31}$$

Suppose that the origin O of the XYZ axes is also translating arbitrarily in

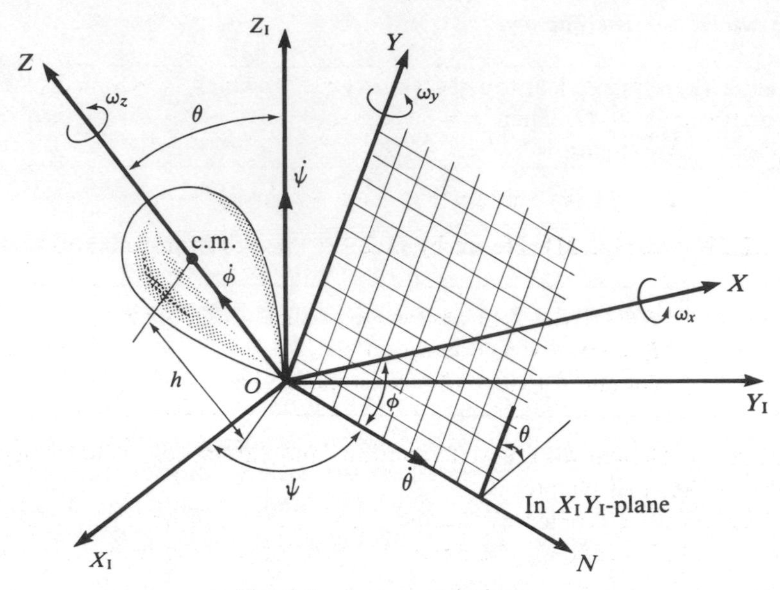

Fig. 1.7 A symmetrical top.

inertial axes with inertial components x_1, y_1, z_1. Then the speed v_O of the translating origin relative to inertial space is just

$$v_O = (\dot{x}_1^2 + \dot{y}_1^2 + \dot{z}_1^2)$$

Now, to apply (1.2.28) recall that v_O must be expressed in instantaneous XYZ axes. However, this is just given by

$$v_O = E^{\mathrm{T}}(\dot{x}_1, \dot{y}_1, \dot{z}_1)^{\mathrm{T}}$$

where E is the matrix in (1.2.29). Hence T can be expressed in terms of the six coordinates $x_1, y_1, z_1, \theta, \psi, \phi$ and their derivatives. To find the generalized forces $F_{x_1}, F_{y_1}, F_{z_1}, F_\theta, F_\psi, F_\phi$ note that if the components of all forces acting on the body are known in the inertial axes, then $F_{x_1}, F_{y_1}, F_{z_1}$ are just the total resolved forces in the X^{I}, Y^{I}, Z^{I} directions. Also, if f is any force acting on the body, then the torque τ equals $f \wedge r$ in body fixed XYZ axes. However, as above $f = E^{\mathrm{T}} f_{\mathrm{I}}^{\mathrm{T}}$, where f_{I} is the known inertial space force. Hence,

$$\boxed{\begin{aligned} F_\theta &= \tau_x \cos\phi - \tau_y \sin\phi \\ F_\phi &= \tau_z \\ F_\psi &= \tau_x \sin\theta \sin\phi + \tau_y \sin\theta \cos\phi + \tau_z \cos\theta \end{aligned}} \tag{1.2.32}$$

by (1.2.30).

Example 1.2.3 – The Top and the Gyroscope

Consider the symmetrical top shown in fig. 1.7, which is assumed to have stationary apex at O. Then $I_x = I_y$ and the products of inertia are zero. Hence the Lagrangian is

$$L = \tfrac{1}{2}[I_x(\dot{\theta}^2 + \dot{\psi}^2 \sin^2\theta) + I_z(\dot{\phi} + \dot{\psi}\cos\theta)^2] - Mgr\cos\theta$$

by (1.2.28) and (1.2.31). Hence by (1.2.19) the equations of motion are

$$\boxed{\begin{aligned} &I_x\ddot{\theta} + [(I_z - I_x)\dot{\psi}\cos\theta + I_z\dot{\phi}]\dot{\psi}\sin\theta = Mgr\sin\theta \\ &I_z(\dot{\psi}\cos\theta + \dot{\phi}) = \text{constant} + c_\phi \\ &I_x\dot{\psi}\sin^2\theta + c_\phi\cos\theta = \text{constant} \end{aligned}} \tag{1.2.33}$$

These are non-linear differential equations and can be solved numerically, by the methods of chapter 3.

The double gimbaled gyroscope shown in fig. 1.8 may be considered equally easily by defining the Euler angles relative to an inertial frame as shown. Then the kinetic energy is just

$$T = \tfrac{1}{2}I_x(\dot{\psi}^2\sin^2\theta + \dot{\theta}^2) + \tfrac{1}{2}I_z(\dot{\psi}\cos\theta + \dot{\phi})^2$$

and if the bearing frictions are small and the gimbals are weightless, the generalized forces are zero and equations of motion similar to (1.2.33) follow immediately (without the $Mgr\sin\theta$ term).

1.2.6 Euler's Equations of Motion

Suppose that the axes XYZ in fig. 1.6 have angular velocity $\boldsymbol{\omega}$ relative to the

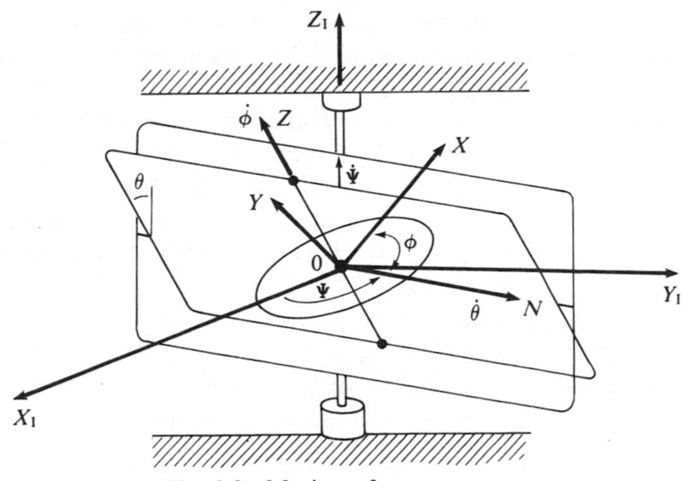

Fig. 1.8 Motion of a gyroscope.

inertial axes and suppose that the particle $d\mu$† shown is now given an arbitrary acceleration $a^I = (a_{x^I}^I, a_{y^I}^I, a_{z^I}^I)$ relative to inertial axes. (Suppose for the moment that the particle is 'free', i.e. not part of a solid body.) Then if E is the matrix in (1.2.29), the inertial space acceleration of the particle relative to the XYZ axes is

$$\begin{aligned} a^I &= E^T(\ddot{x}_I, \ddot{y}_I, \ddot{z}_I)^T \\ &= E^T\ddot{r}^I \end{aligned}$$

Also, if r_0^I is the vector joining O^I to O, then

$$r^I = r_0^I + Er$$

Hence,

$$\ddot{r}^I = \ddot{r}_O^I + \ddot{E}r + E\ddot{r} + 2\dot{E}\dot{r}$$

It follows that, by using (1.2.29) and (1.2.31), we obtain the equation

$$\boxed{\mathbf{a}^I = \ddot{r}_O + a + 2\omega \wedge v + \dot{\omega} \wedge r + \omega \wedge (\omega \wedge r)} \tag{1.2.34}$$

where all quantities on the right are measured relative to XYZ axes. Now, if the particle is again assumed to be part of a body B and the axes XYZ are body fized, then ω is the angular velocity of the body and (1.2.34) reduces to the equations

$$\left.\begin{aligned} a_x^I &= a_{Ox} - x(\omega_y^2 + \omega_z^2) + y(\omega_y\omega_x - \dot{\omega}_z) + z(\omega_x\omega_z + \dot{\omega}_y) \\ a_y^I &= a_{Oy} + x(\omega_x\omega_y + \dot{\omega}_z) - y(\omega_x^2 + \omega_z^2) + z(\omega_y\omega_z - \dot{\omega}_x) \\ a_z^I &= a_{Oz} + x(\omega_x\omega_z - \dot{\omega}_y) + y(\omega_y\omega_z + \dot{\omega}_x) - z(\omega_x^2 + \omega_y^2) \end{aligned}\right\} \tag{1.2.35}$$

where a_x^I, a_y^I, a_z^I are the components of a^I in XYZ axes.

Now the particle $d\mu$ will experience some force $\mathbf{f}^I$ per unit volume due to all externally applied forces acting on the body.

Hence

$$d\mu\, a^I = f^I\, dV$$

Resolving along XYZ axes gives

$$d\mu\, a_x^I = f_x^I, \text{ etc.}$$

and so

$$\int (a_z^I y + a_y^I z)\, d\mu = \int (f_z^I y - f_y^I z)\, dV = \tau_x \tag{1.2.36}$$

where τ_x is the torque about the X axis, together with similar expressions

---

†$d\mu$ has coordinates (x^I, y^I, z^I) with respect to $X^IY^IZ^I$ axes and (x, y, z) relative to XYZ.

for τ_y and τ_z. Hence, by (1.2.35) and (1.2.36),

$$\begin{aligned}
&M(a_{Oz}\bar{y} - a_{Oy}\bar{z}) + I_x\dot{\omega}_x + (I_z - I_y)\,\omega_y\omega_z + I_{xy}(\omega_x\omega_z - \dot{\omega}_y)\\
&\quad - I_{xz}(\omega_x\omega_y + \dot{\omega}_z) + I_{yz}(\omega_z^2 - \omega_y^2) = \tau_x\\
&M(a_{Ox}\bar{z} - a_{Oz}\bar{x}) + I_y\dot{\omega}_y + (I_x - I_z)\,\omega_x\omega_z + I_{yz}(\omega_y\omega_x - \dot{\omega}_z)\\
&\quad - I_{xy}(\omega_y\omega_z + \dot{\omega}_x) + I_{xz}(\omega_x^2 - \omega_z^2) = \tau_y\\
&M(a_{Oy}\bar{x} - a_{Ox}\bar{y}) + I_z\dot{\omega}_z + (I_y - I_x)\omega_x\omega_y + I_{xz}(\omega_y\omega_z - \dot{\omega}_x)\\
&\quad - I_{yz}(\omega_x\omega_z + \dot{\omega}_y) + I_{xy}(\omega_y^2 - \omega_x^2) = \tau_z
\end{aligned} \tag{1.2.37}$$

where $\bar{x}, \bar{y}, \bar{z}$ are the coordinates of the centre of mass relative to XYZ. Note that, in this formulation, all external forces, including forces of constraint must be included and so it is more difficult to apply than Lagrange's method. If O is at the centre of mass and the principal axes of inertia are along X, Y, Z, then (1.2.37) reduces to

$$\begin{aligned}
I_x^{\mathrm{p}}\dot{\omega}_x + (I_z^{\mathrm{p}} - I_y^{\mathrm{p}})\,\omega_y\omega_z &= \tau_x\\
I_y^{\mathrm{p}}\dot{\omega}_y + (I_x^{\mathrm{p}} - I_z^{\mathrm{p}})\,\omega_x\omega_z &= \tau_y\\
I_z^{\mathrm{p}}\dot{\omega}_z + (I_y^{\mathrm{p}} - I_x^{\mathrm{p}})\,\omega_x\omega_y &= \tau_z
\end{aligned} \tag{1.2.38}$$

If the particle $d\mu$ has a force per unit volume $\boldsymbol{f}$ acting on it, then

$$d\mu\,\ddot{\boldsymbol{r}}^{\mathrm{I}} = \boldsymbol{f}\,dV$$

and so

$$\int \ddot{\boldsymbol{r}}^{\mathrm{I}}\,d\mu = \boldsymbol{f}\,dV = \boldsymbol{F}$$

where $\boldsymbol{F}$ is the total force on the body. Hence,

$$M\ddot{\tilde{\boldsymbol{r}}}^{\mathrm{I}} = \boldsymbol{F}$$

where $\tilde{\boldsymbol{r}}^{\mathrm{I}}$ is the centre of mass in inertial coordinates, and so

$$M\tilde{\boldsymbol{a}} = \boldsymbol{F} \tag{1.2.39}$$

where $\tilde{\boldsymbol{a}}$ is the acceleration of the centre of mass. The equations (1.2.37) and (1.2.39) consitute Euler's six equations of motion of a rigid body, and consist of three rotational equations and three describing the motion of the centre of mass. For a more extensive discussion of classical dynamics, see, for example, Stephenson (1960).

1.2.7 Application of Euler's Equations -- Aircraft Dynamics

An extremely important area of modelling and control is in the aerospace industry; indeed much of classical and optimal control was developed to provide adequate theory for early flight programmes and for the more recent spacecraft flight controllers. It seems appropriate, therefore, to give an outline of the main modelling tasks involved in setting up the equations of motion of an aircraft. Consider therefore a relatively slow-moving aircraft as shown in fig. 1.9. (This assumption is merely to ensure that the Earth axes form an inertial frame. If the aircraft is very fast then the gravity vector will not remain fixed in an inertial frame.) The axes in fig. 1.9 should be compared with those in fig. 1.7. Since the products of inertia I_{yz} and I_{xy} are zero, it follows from (1.2.39) and (1.2.37) that we have

$$\boxed{\begin{aligned}
&M(\dot{u} + \omega_y w - \omega_z v) = f_x \\
&M(\dot{v} + \omega_z u - \omega_x w) = f_y \\
&M(\dot{w} + \omega_x v - \omega_y u) = f_z \\
&I_x\dot{\omega}_x + (I_z - I_y)\,\omega_y\omega_z - I_{xz}(\omega_x\omega_y + \dot{\omega}_z) = \tau_x \\
&I_y\dot{\omega}_y + (I_x - I_z)\,\omega_x\omega_z + I_{xz}(\omega_x^2 - \omega_z^2) = \tau_y \\
&I_z\dot{\omega}_z + (I_y - I_x)\,\omega_x\omega_y + I_{xz}(\omega_y\omega_z - \dot{\omega}_x) = \tau_z
\end{aligned}} \tag{1.2.40}$$

where u, v, w are the components of the inertial space velocity of the c.m. relative to the XYZ axes. The forces and torques are produced by gravity, the thrust (for a jet aircraft) and the aerodynamic forces. Using (1.2.30) it is clear that the gravity forces along X, Y, Z are

$$-mg\sin\theta\sin\phi, \quad -mg\sin\theta\cos\phi, \quad -mg\cos\theta \tag{1.2.41}$$

respectively. The thrust will generally produce components T_x, T_z along x and z and a pitching moment m_T about the c.m. The aerodynamic forces can be derived by noting that it can be shown (Milne-Thomson, 1968; see

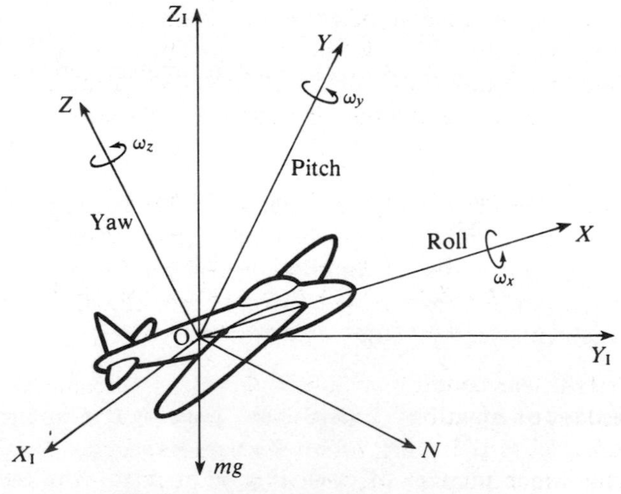

Fig. 1.9 Relative axes for an aircraft.

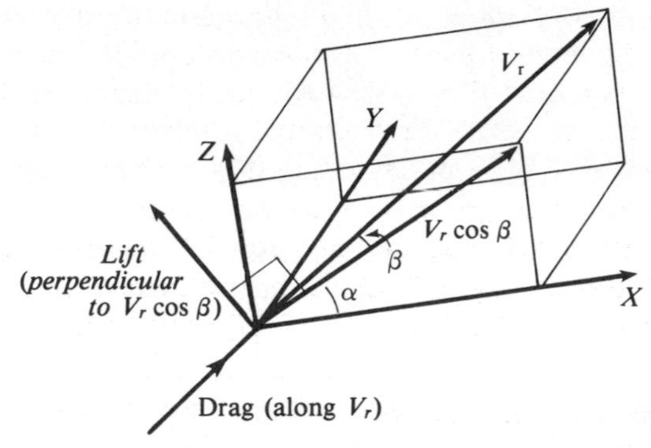

Fig. 1.10 Lift and drag vectors in aircraft dynamics.

also section 1.6.3) that the force acting on a solid moving through a liquid is of the form

$$F = C_F \tfrac{1}{2}\rho v^2 A \tag{1.2.42}$$

where C_F is a dimensionless constant, ρ is the fluid density, v is the relative velocity of the solid and A is a characteristic area of the solid. Suppose that the relative wind is oriented as shown in fig. 1.10. Then we may express the steady aerodynamic forces along the body axes in terms of the lift L and drag D as follows:

$$\left.\begin{aligned} f_x^a &= -L\sin\alpha + D\cos\beta\cos\alpha \\ f_y^a &= D\sin\beta \\ f_z^a &= L\cos\alpha + D\cos\beta\sin\alpha \end{aligned}\right\} \tag{1.2.43}$$

L and D can be written in the form of (1.2.42). Control surfaces, such as flaps and rudders, produce aerodynamic moments and perturbed forces. The non-linear equations (1.2.40) are usually linearized about a motion which contains no translational or rotational accelerations, i.e. along the axes of the vehicle. The equations of such a motion are

$$\left.\begin{aligned} M(\omega_{yO}w_O - \omega_{zO}v_O + g\sin\theta_O\sin\phi_O) &= f_{xO} \\ M(\omega_{zO}u_O - \omega_{xO}w_O + g\sin\theta_O\cos\phi_O) &= f_{yO} \\ M(\omega_{xO}v_O - \omega_{yO}u_O + g\cos\theta_O) &= f_{zO} \\ (I_z - I_y)\omega_{yO}\omega_{zO} - I_{xz}\omega_{xO}\omega_{yO} &= \tau_{xO} \\ (I_x - I_z)\omega_{xO}\omega_{zO} + I_{xz}(\omega_{xO}^2 - \omega_{zO}^2) &= \tau_{yO} \\ (I_y - I_x)\omega_{xO}\omega_{yO} + I_{xz}\omega_{yO}\omega_{zO} &= \tau_{zO} \end{aligned}\right\} \tag{1.2.44}$$

and are called the **trim conditions**. Subtracting (1.2.44) from (1.2.40) results in a linearized set of equations (for small angles θ, ϕ) with perturbed forces and torques δf_x, δf_y, δf_z, $\delta\tau_x$, $\delta\tau_y$, $\delta\tau_z$ on the right-hand side. These normally depend on the Mach number M, the angle of attack α, the relative wind velocity V_r and the control surface deflections Δ_i. Each perturbed force is determined by writing, for example,

$$\delta f_x = \frac{\partial f_x}{\partial M}\delta M + \frac{\partial f_x}{\partial \alpha}\delta\alpha + \frac{\partial f_x}{\partial V_r}\delta V_r + \sum \frac{\partial f_x}{\partial \Delta_i}\delta\Delta_i + \dots$$

by Taylor's theorem (appendix 1). The terms $\partial f_x/\partial M$ etc. are called **aerodynamic derivatives**, and are often determined experimentally. A complete discussion of the resulting linear equations is given by McRuler *et al.* (1973).

1.3 MODELLING OF LUMPED ELECTRICAL SYSTEMS

1.3.1 Maxwell's Equations and Lumped Systems

We shall begin this section by recalling Maxwell's equations for a general electromagnetic field. These equations are

$$\boxed{\begin{aligned} \operatorname{div} \boldsymbol{D} &= \rho \\ \operatorname{div} \boldsymbol{B} &= 0 \\ \operatorname{curl} \boldsymbol{H} &= \boldsymbol{J} + \frac{\partial \boldsymbol{D}}{\partial t} \\ \operatorname{curl} \boldsymbol{E} &= -\frac{\partial \boldsymbol{B}}{\partial t} \end{aligned}} \tag{1.3.1}$$

where $\boldsymbol{E}$ is the electric field intensity, $\boldsymbol{D} = \varepsilon\varepsilon_0\boldsymbol{E}$ is the displacement vector (for homogeneous isotropic media), $\boldsymbol{B}$ is the magnetic induction vector, $\boldsymbol{H} = \boldsymbol{B}/\mu\mu_0$ is the magnetic intensity (for homogeneous isotropic media), ρ is the charge density and $\varepsilon, \varepsilon_0, \mu, \mu_0$ are constants (Kraus and Carver, 1973).

Moreover, we have Ohm's law

$$\boldsymbol{J} = \sigma\boldsymbol{E} \tag{1.3.2}$$

where σ is the conductivity and the equation of charge conservation

$$\operatorname{div} \boldsymbol{J} + \frac{\partial \rho}{\partial t} = 0 \tag{1.3.3}$$

In a conductor, Ohm's law may be written

$$\boldsymbol{J} = \sigma(\boldsymbol{E} + \boldsymbol{E}')$$

where $\boldsymbol{E}$ is the conservative field component in the conductor and $\boldsymbol{E}'$ is the non-conservative field associated with seats of emf or induced emf's.

If we assume that the conductor has constant cross-sectional area A, then taking the line integral along the axis of the conductor between some points a and b we have

$$\int_a^b \sigma(\boldsymbol{E}+\boldsymbol{E}')\cdot \mathrm{d}\boldsymbol{l} = \int_a^b \boldsymbol{J}\cdot \mathrm{d}\boldsymbol{l} \tag{1.3.4}$$

The emf between a and b is defined by

$$E_{ab} = \int_a^b \boldsymbol{E}'\cdot \mathrm{d}\boldsymbol{l}$$

The conservative term leads to a potential difference

$$-V_b + V_a = \int_a^b \boldsymbol{E}\cdot \mathrm{d}\boldsymbol{l}$$

Now if we consider lumped circuits, then $\boldsymbol{J}$ and $\mathrm{d}\boldsymbol{l}$ are parallel and so

$$A\boldsymbol{J}\cdot \mathrm{d}\boldsymbol{l} = \mathrm{d}li$$

where i is the current; hence, by (1.3.4)

$$\begin{aligned} -V_b + V_a + E_{ab} &= \frac{1}{\sigma A}\int_a^b A\boldsymbol{J}\cdot \mathrm{d}\boldsymbol{l} \\ &= \frac{l_{ab}i}{\sigma A} \\ &= R_{ab}i \end{aligned}$$

where l_{ab} is the length of the conductor and R_{ab} is, by definition, its resistance.

For homogeneous and isotropic media it follows from (1.3.1) that

$$\begin{aligned} \operatorname{div}(\boldsymbol{E}\wedge\boldsymbol{H}) &= \boldsymbol{H}\cdot \operatorname{curl}\boldsymbol{E} - \boldsymbol{E}\cdot\operatorname{curl}\boldsymbol{H} \\ &= -\boldsymbol{E}\cdot\boldsymbol{J} - \frac{\partial u}{\partial t} \end{aligned}$$

where $u = \frac{1}{2}(\boldsymbol{E}\cdot\boldsymbol{D} + \boldsymbol{B}\cdot\boldsymbol{H})$. Hence, integrating over the total volume V_c occupied by the circuit, we have

$$-\int_{V_c} \boldsymbol{E}\cdot \boldsymbol{J}\mathrm{d}V = \int_{V_c} \operatorname{div}(\boldsymbol{E}\wedge\boldsymbol{H})\,\mathrm{d}V + \int_{V_c}\frac{\partial u}{\partial t}\mathrm{d}V$$

However,

$$\int_{V_c} \operatorname{div}(\boldsymbol{E}\wedge\boldsymbol{H})\,\mathrm{d}V = \int_{S_c}(\boldsymbol{E}\wedge\boldsymbol{H})\cdot \mathrm{d}\boldsymbol{S} \quad \text{(by Gauss' theorem)}$$

where S_c is the surface containing the circuit. Since $\boldsymbol{E}\wedge\boldsymbol{H}$ represents the radiation, this is small over S_c for most circuits and so we shall neglect this term. Hence we obtain

$$\int_{V_c} \boldsymbol{E}'\cdot\boldsymbol{J}\,\mathrm{d}V = \frac{1}{\sigma}\int_{V_c} J^2\,\mathrm{d}V + \int_{V_c}\frac{\partial u}{\partial t}\mathrm{d}V \tag{1.3.5}$$

using $\boldsymbol{J} = \sigma(\boldsymbol{E}' + \boldsymbol{E})$. But

$$\frac{1}{\sigma}\int_{V_c} J^2\,\mathrm{d}V = \frac{l_{ab}i^2}{\sigma A} = Ri^2 \tag{1.3.6}$$

for each resistor in the system,

$$\frac{1}{2}\int_{V_c}\frac{\partial}{\partial t}(\boldsymbol{B}\cdot\boldsymbol{H})\,\mathrm{d}V = i_1L_{12}\frac{\partial i_2}{\partial t} + i_2L_{12}\frac{\partial i_1}{\partial t} \tag{1.3.7}$$

for each pair of inductors with mutual inductance L_{12}, carrying currents i_1, i_2 ($i_1 = i_2$ for a single coil with self-inductance L_{11}) and

$$\frac{1}{2}\frac{\partial}{\partial t}\int_V (\boldsymbol{E}\cdot\boldsymbol{D})\,\mathrm{d}\tau = \frac{iq}{C} \tag{1.3.8}$$

for each capacitor C carrying current i and storing charge q. Hence, if we have N separate loops in a system then, by (1.3.5–8),

$$\sum_{j=1}^{N} E_j i_j = \sum_{j=1}^{N}\left\{i_j\sum_{k=1}^{N} L_{jk}\frac{\partial i_k}{\partial t} + i_j^2 R_j + \frac{i_j q_j}{C_j}\right\}$$

where i_j is the current in the jth loop, E_j is the emf in loop j, L_{jk} is the mutual inductance between loops j and k, R_j is the total resistance, C_j the capacitance and q_j its charge in the jth loop. Since the loops are independent,

$$E_j = \sum_{k=1}^{N} L_{jk}\frac{\partial i_k}{\partial t} + i_jR_j + \frac{q_j}{C_j} \tag{1.3.9}$$

where

$$\frac{\mathrm{d}q_j}{\mathrm{d}t} = i_j(t)$$

(1.3.9) is **Kirchhoff's first law**, i.e. the instantaneous emf in any loop is equal to the voltage drop across the passive elements. **Kirchhoff's second law** merely states that charge is conserved at a node, i.e. the algebraic sum of the currents converging at a node is zero (fig. 1.11).

We can also obtain an equation for loop m in terms of branch elements

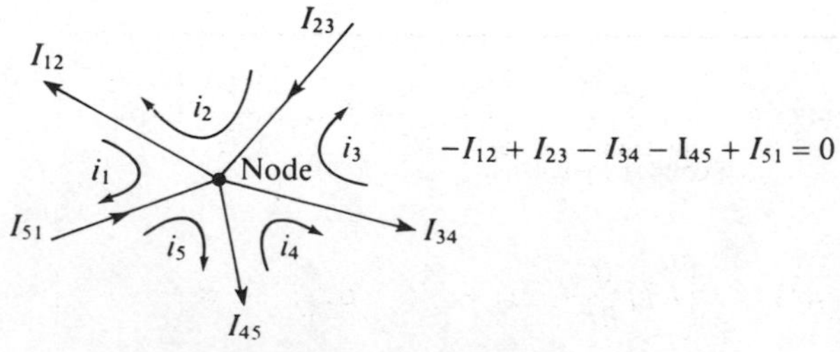

Fig. 1.11 Current node equation.

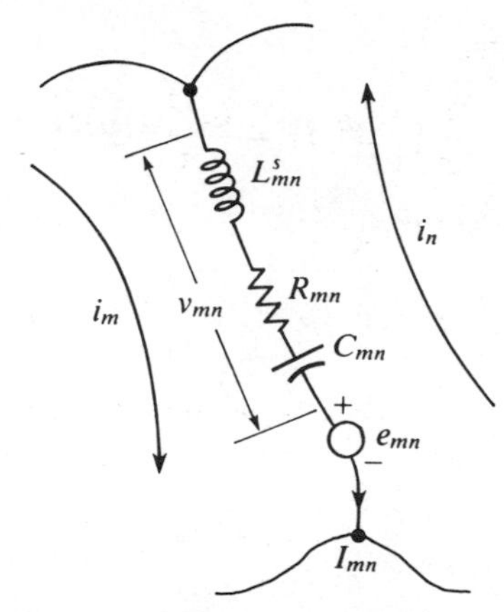

Fig. 1.12 Passive circuit branch.

as follows. Consider the branch which is common to loops m and n as in fig. 1.12, where we now use a double suffix notation for the passive elements and we shall ignore mutual inductances. Note that L^s_{mn} is a **self**-inductance as indicated by the superscript s. By Kirchhoff's first law applied to the mth loop we have

$$\sum_{j=1}^{N} e_{mj}(t) = \sum_{j=1}^{N} v_{mj}(t)$$

where

$$v_{mj} = L^s_{mj} \frac{\mathrm{d}}{\mathrm{d}t}(i_m - i_n) + R_{mn}(i_m - i_n) + \frac{1}{C_{mn}} \int (i_m - i_n)\,\mathrm{d}t \quad \text{if } m \neq j$$

and

$$v_{mm} = L^s_{mm} \frac{\mathrm{d}i_m}{\mathrm{d}t} + R_{mm} i_m + \frac{1}{C_{mm}} \int i_m \,\mathrm{d}t$$

(Note that e_{mm} is an algebraic sum of all emfs in loop m which occur in branches which do not adjoin any other loop and similarly, v_{mm}, L^s_{mm}, R_{mm}, $1/C_{mm}$ may be sums over all such branches.) Hence if we define the operators

$$Z_{kk}(t) = \sum_{\substack{\text{all loops } n \\ \text{adjoining loop } k}} \left\{ L^s_{kn} \frac{\partial}{\partial t} + R_{kn} + \frac{1}{C_{kn}} \int \mathrm{d}t \right\}$$

$$Z_{kj}(t) = -\left\{ L^s_{kj} \frac{\partial}{\partial t} + R_{kj} + \frac{1}{C_{kj}} \int \mathrm{d}t \right\}, \quad k \neq j \tag{1.3.10}$$

then we have

$$\sum_{n=1}^{N} e_{mn}(t) = \sum_{n=1}^{N} Z_{mn}(t) i_n(t)$$

If e_m is the total emf in loop m, then

$$\boxed{\boldsymbol{e} = Z\boldsymbol{i}} \tag{1.3.11}$$

where $\boldsymbol{e} = (e_1, \ldots, e_m)^{\mathrm{T}}$, $\boldsymbol{i} = (i_1, \ldots, i_n)^{\mathrm{T}}$ and $Z = (Z_{ij})$.

1.3.2 Example

Consider the application of (1.3.11) to the simple two loop system in fig. 1.13. Then

$$e_1 = e_{11}, \quad e_2 = 0$$

$$z_{11} = L_{11}^{s} \frac{\mathrm{d}}{\mathrm{d}t} + \frac{1}{C_{12}} \int \mathrm{d}t$$

$$Z_{12} = Z_{21} = -\frac{1}{C_{12}} \frac{\mathrm{d}}{\mathrm{d}t}$$

$$Z_{22} = R_{22} + \frac{1}{C_{12}} \int \mathrm{d}t$$

Hence, by (1.3.11),

$$e_{11} = L_{11}^{s} \frac{\mathrm{d}i_1}{\mathrm{d}t} + \frac{1}{C_{12}} \int i_1 \, \mathrm{d}t - \frac{1}{C_{12}} \int i_2 \, \mathrm{d}t$$

$$0 = -\frac{1}{C_{12}} \int i_1 \, \mathrm{d}t + \frac{1}{C_{12}} \int i_2 \, \mathrm{d}t + R_{22} i_2$$

1.3.3 Thévenin's and Norton's Theorems

It is convenient to replace the operators $\partial/\partial t$ and $\int \mathrm{d}t$ by the symbols s and $1/s$ in (1.3.10). For the present this can be regarded merely as a notational

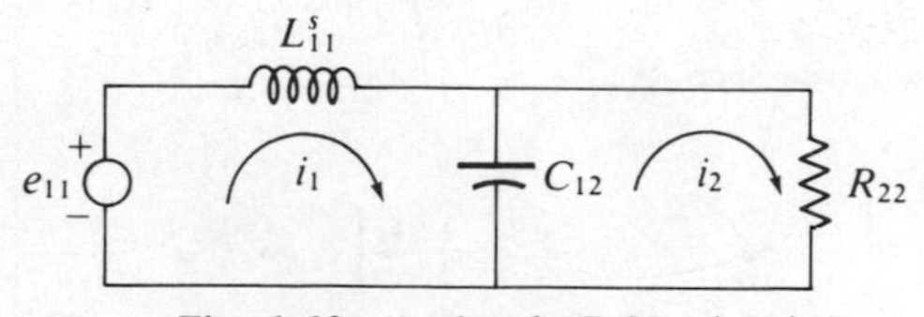

Fig. 1.13 A simple RCL circuit.

change, and will be made precise later when we discuss the Laplace transform. If we put $s = j\omega$, the resulting expressions for Z_{kk} and Z_{kj} are related to the familiar phasor representation of passive elements. A typical term

$$z(s) = Ls + R + \frac{1}{sC}$$

is called the impedance due to L, R and C and the matrix Z in (1.3.11) is called the **impedance matrix** of the system.

Consider now the general circuit in fig. 1.14 which has only voltage sources and passive elements. The total branch impedances are denoted $z_{ij}(s)$. Then by (1.3.10) we have

$$\begin{pmatrix} Z_{11}(s) & Z_{12}(s) & \dots & Z_{1N}(s) \\ \vdots & \vdots & & \vdots \\ Z_{N1}(s) & Z_{N2}(s) & \dots & \{Z_{NN}(s)/ + Z_L\} \end{pmatrix} \begin{pmatrix} I_1(s) \\ \vdots \\ I_N(s) \end{pmatrix} = \begin{pmatrix} E_1(s) \\ \vdots \\ E_N(s) \end{pmatrix} \quad (1.3.12)$$

where

$$Z_{ii}(s) = \sum_{\text{all } k} z_{ik}(s)$$

and

$$Z_{ij}(s) = -z_{ij}(s), \quad i \neq j$$

(Note that, in the s respresentation, currents and voltages are represented by capital letters rather than lower case as in the time domain.) Let

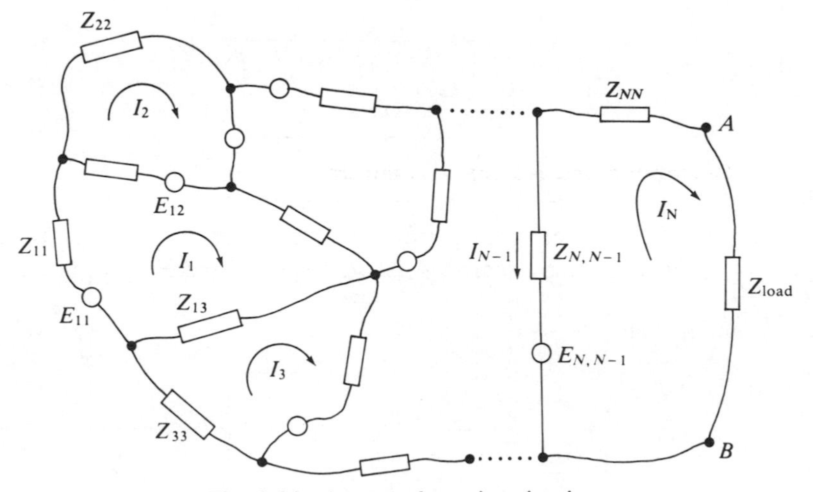

Fig. 1.14 A general passive circuit.

$Z' = (Z_{ij})$ and $P = (p_{ij})$ where $p_{NN} = 1$ and $p_{ij} = 0$ for other i, j. Then (1.3.12) becomes

$$(Z' + PZ_L)I(s) = E(s)$$

Suppose that $Z'(s)$ is not identically singular. Then if $Y(s) = (Z'(s))^{-1}$ (so that Y is an **admittance matrix**), we have

$$[I + Y(s)PZ_L]I(s) = Y(s)E(s)$$

and if $(I + Y(s)PZ_L)^{-1}$ exists,

$$I(s) = [I + Y(s)PZ_L]^{-1}Y(s)E(s)$$

It follows that

$$I_N(s) = \sum_{k=1}^{N} Y_{Nk}E_k(s)/(1 + Z_L Y_{NN}) \tag{1.3.13}$$

Consider the open-circuit voltage E_{oc} between A and B (setting $Z_L = \infty$ in $Z_L I_N$) and the short circuit current I_{sc} when $Z_L = 0$. Then

$$E_{oc} = \sum_{k=1}^{N} Y_{Nk}E_k / Y_{NN}$$

$$I_{sc} = \sum_{k=1}^{N} Y_{Nk}E_k$$

and so by (1.3.13)

$$\boxed{[Z_{int}(s) + Z_L(s)]\, I_N(s) = E_{oc}(s)} \tag{1.3.14}$$

where

$$\boxed{Z_{int}(s) = \frac{1}{Y_{NN}} = \frac{E_{oc}}{I_{sc}}} \tag{1.3.15}$$

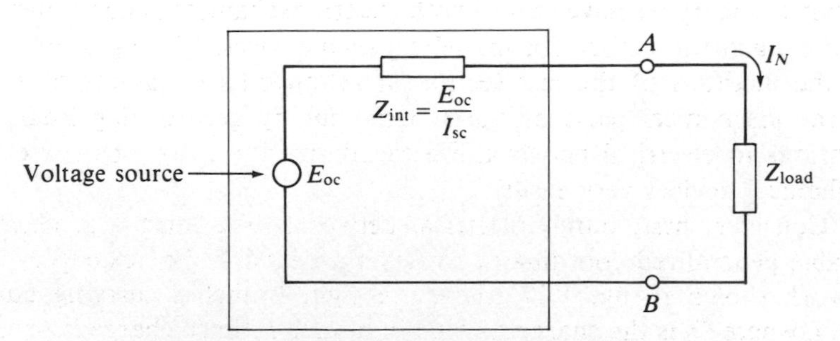

Fig. 1.15 Illustration of Thévenin's theorem.

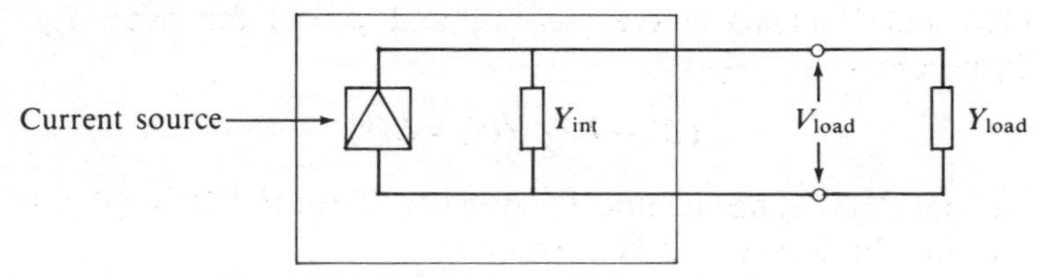

Fig. 1.16 Illustration of Norton's theorem.

The equations (1.3.14) and (1.3.15) constitute Thévenin's theorem and show that the general system in fig. 1.14 may be replaced by the equivalent circuit in fig. 1.15.

Again by (1.3.13) we have

$$I_N(s) + Y_{NN}Z_L I_N = I_{sc}(s)$$

However,

$$I_N(s) = Y_L(s)V_N(s)$$

where $Y_L = 1/Z_L$ and V_N is the voltage across the load. Hence,

$$\boxed{[Y_{\text{int}}(s) + Y_L(s)]\,V_N(s) = I_{sc}(s)} \tag{1.3.16}$$

where $Y_{\text{int}} = 1/Z_{\text{int}}$. This is Norton's theorem and is the dual of Thévenin's theorem (fig. 1.16). For further discussion of networks which only have voltage or current sources and passive elements see Guillemin (1957).

1.4 MODELLING OF ELECTROMECHANICAL SYSTEMS

1.4.1 Lagrange's Equations for Electrical Systems

In the last section we have considered a method for writing down the equations describing a general lumped passive electrical network. In many cases in control theory we have to deal with electromechanical systems and so we require a general method for modelling such devices. We could, of course, use the methods of the last section and Newton's equations of motion for the mechanical parts of the system, but by generalizing Lagrange's equations to electrical networks we can derive the equations for electromechanical devices very easily.

Consider, first, purely electrical networks. We must first decide on suitable generalized coordinates for such a network. For example, in the network shown in fig. 1.17, there are eight branches carrying currents $\dot{Q}_i = I_i$ where Q_i is the charge flowing in branch i. Since charge is produced by electrons (or holes in transistors) it seems reasonable to choose the

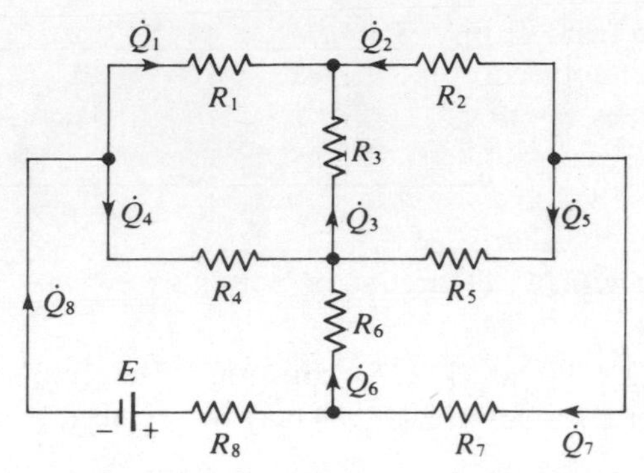

Fig. 1.17 A simple four-loop circuit.

charges Q_i as the generalized coordinates. Then $\dot{Q}_i$ and $\ddot{Q}_i$ are the generalized velocities and accelerations respectively. Note that not all eight charges in fig. 1.17 are independent, however. In fact at each node we have a constraint due to the conservation of charge. For the circuit in fig. 1.17 we have

$$\begin{aligned}
&\dot{Q}_1 + \dot{Q}_2 + \dot{Q}_3 = 0\\
&\dot{Q}_2 + \dot{Q}_7 - \dot{Q}_5 = 0\\
&\dot{Q}_1 + \dot{Q}_4 - \dot{Q}_8 = 0\\
&\dot{Q}_3 + \dot{Q}_5 - \dot{Q}_4 - \dot{Q}_6 = 0\\
&\dot{Q}_8 + \dot{Q}_6 - \dot{Q}_7 = 0
\end{aligned}$$

These equations are not independent, however; in fact the last equation is a linear combination of the others, which are independent. Hence the system has four degrees of freedom and we can choose, for example, Q_1, Q_2, Q_5, Q_8 as independent coordinates. Then

$$\begin{aligned}
&Q_3 = -(Q_1 + Q_2)\\
&Q_4 = Q_8 - Q_1\\
&Q_6 = Q_7 - Q_8 = Q_5 - Q_2 - Q_8\\
&Q_7 = Q_5 - Q_2
\end{aligned}$$

are four superfluous coordinates, and just as for mechanical systems, must be removed from the expression for the Lagrangian.

For an electrical system the magnetic energy of a system of s coils can be shown to be

$$T_e = \tfrac{1}{2}\sum_{\substack{i=1\\j=1}}^{s} L_{ij}\dot{Q}_i\dot{Q}_j \tag{1.4.1}$$

where L_{ij} are the mutual inductances of the coils. Superfluous coordinates should be removed from T_e just as in the mechanical case. We shall assume that the non-superfluous coordinates are Q_i, $1 \leqslant i \leqslant n$.

The potential energy of the system consists of the part due to energy sources (batteries, generators, etc.) and that due to charge stored on capacitors. If E is a voltage source then the energy supplied to the system is $-EQ$ where Q is the charge produced by the source, assumed to flow in the positive E direction. (The negative sign merely represents the fact that the zero potential has been referred to $Q=0$.) Since the energy in an isolated capacitor C is just $\frac{1}{2}Q^2/C$, the total potential energy of the system is

$$V_e = \frac{1}{2}\sum_{i=1}^{l}\frac{Q_i^2}{C_i} - \sum_{j=1}^{m} E_j Q_j \tag{1.4.2}$$

Again, superfluous coordinates must be removed from V_e.

Finally, we must identify the generalized forces F_{Q_i}. These are made up of the conservative 'forces' $-\partial V_e/\partial Q_i$ from (1.4.2) and the dissipative 'forces' $(F_{Q_i})_r$ produced by the resistors, which dissipate energy in the form of heat. The virtual work done when charge Q_i flows through the resistance R_i is $-R_i\dot{Q}_i\,\delta Q_i$ and so the total virtual work is

$$\delta W = -\sum_{i=1}^{r} R_i\dot{Q}_i\,\delta Q_i$$

Eliminating superfluous coordinates given $(F_{Q_i})_r$. Hence

$$F_{Q_i} = -\frac{\partial V}{\partial Q_i} - (F_{Q_i})_r$$

and by Lagrange's equations we obtain the equations

$$\frac{\mathrm{d}}{\mathrm{d}t}\left(\frac{\partial L_e}{\partial \dot{Q}_i}\right) - \frac{\partial L_e}{\partial Q_i} = (F_{Q_i})_r \tag{1.4.3}$$

Consider the system in fig. 1.18 as an example of the application of (1.4.3). There are two degrees of freedom since

$$Q_1 = Q_2 + Q_3$$

The kinetic energy is

$$T_e = \tfrac{1}{2}(L_{11}\dot{Q}_1^2 + L_{22}\dot{Q}_2^2 + L_{33}\dot{Q}_3^2 + 2L_{12}\dot{Q}_1\dot{Q}_2 + 2L_{13}\dot{Q}_1\dot{Q}_3 + 2L_{23}\dot{Q}_2\dot{Q}_3)$$

assuming mutual inductances between the coils. Eliminating $\dot{Q}_3$ we have

$$T_e = \tfrac{1}{2}[L_{11}\dot{Q}_1^2 + L_{22}\dot{Q}_2^2 + L_{33}(\dot{Q}_1 - \dot{Q}_2)^2 + 2L_{12}\dot{Q}_1\dot{Q}_2 + 2L_{13}\dot{Q}_1(\dot{Q}_1 - \dot{Q}_2) + 2L_{23}\dot{Q}_2(\dot{Q}_1 - \dot{Q}_2)]$$

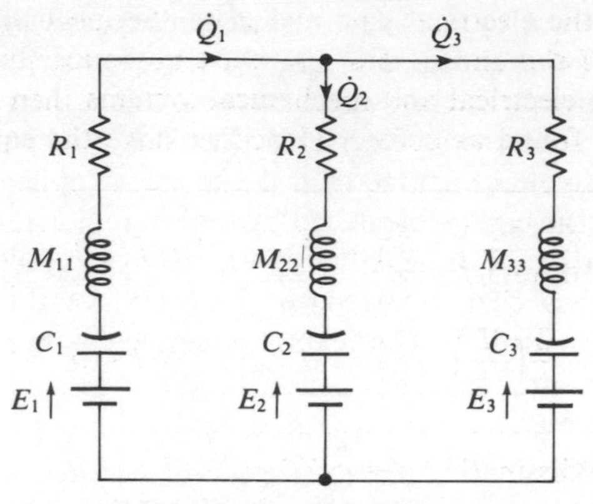

Fig. 1.18 An LRC circuit.

Also, V_e is clearly given by

$$V_e = \frac{1}{2}\left[\frac{Q_1^2}{C_1} + \frac{Q_2^2}{C_2} + \frac{(Q_1 - Q_2)^2}{C_3}\right] - E_1Q_1 + E_2Q_2 + E_3(Q_1 - Q_2)$$

after eliminating Q_3. Finally,

$$\begin{aligned}\delta W &= -R_1\dot{Q}_1\,\delta Q_1 - R_2\dot{Q}_2\,\delta Q_2 - R_3\dot{Q}_3\,\delta Q_3 \\ &= [R_3\dot{Q}_2 - (R_1 + R_3)\dot{Q}_1]\,\delta Q_1 + [R_3\dot{Q}_1 - (R_2 + R_3)\dot{Q}_2]\,\delta Q_2\end{aligned}$$

and so

$$(F_{Q_1})_r = R_3\dot{Q}_2 - (R_1 + R_3)\dot{Q}_1$$
$$(F_{Q_2})_r = R_3\dot{Q}_1 - (R_2 + R_3)\dot{Q}_2$$

The differential equations for Q_i now follow easily from (1.4.3).

We may now consider electromechanical systems by combining the above Lagrangian equations for mechanical and electrical systems. The beauty of this method can now be seen, for the Lagrangian of the combined electrical and mechanical system is just the sum of the separate Lagrangians. Hence,

$$\boxed{L = T_e - V_e + T_m - V_m = L_e + L_m} \qquad (1.4.4)$$

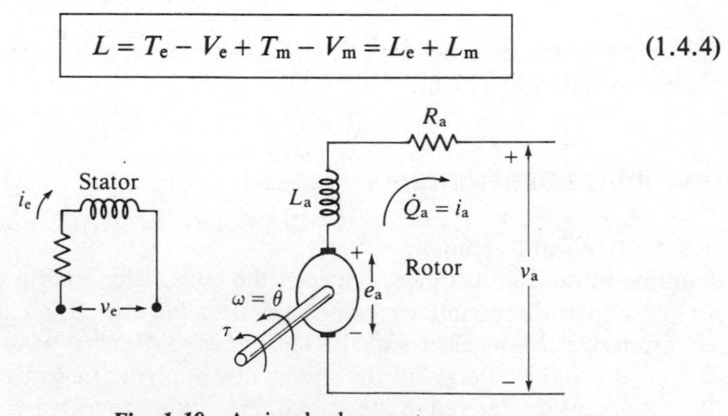

Fig. 1.19 A simple d.c. motor.

where e refers to the electrical part and m to the mechanical part of the system. If Q_i, $1 \leqslant i \leqslant n$ and q_j, $1 \leqslant j \leqslant m$ are non-superfluous generalized coordinates of the electrical and mechanical systems then the generalized forces F_{Q_i}, F_{q_j} are found as before. Hence we have the equations

$$\boxed{\begin{aligned} \frac{\mathrm{d}}{\mathrm{d}t}\left(\frac{\partial L}{\partial \dot{Q}_i}\right) - \frac{\partial L}{\partial Q_i} &= (F_{Q_i})_{\mathrm{r}}, \quad 1 \leqslant i \leqslant n \\ \frac{\mathrm{d}}{\mathrm{d}t}\left(\frac{\partial L}{\partial \dot{q}_j}\right) - \frac{\partial L}{\partial q_j} &= (F_{q_j})_{\mathrm{d}}, \quad 1 \leqslant j \leqslant m \end{aligned}} \tag{1.4.5}$$

where $(F_{q_j})_{\mathrm{d}}$ is the dissipative part of F_{q_j}.

Consider the d.c. motor shown in fig. 1.19. The Lagrangian for the rotor system is

$$L = \tfrac{1}{2}J\dot{\theta}^2 + \tfrac{1}{2}L_{\mathrm{a}}(\dot{Q}_{\mathrm{a}})^2 - v_{\mathrm{a}}Q_{\mathrm{a}} - e_{\mathrm{a}}Q_{\mathrm{a}}$$

and it can be easily shown (Fitzgerald and Kingsley, 1971) that the induced torque τ and emf e_{a} are given by

$$\begin{aligned} \tau &= Ki_{\mathrm{e}}i_{\mathrm{a}} \\ e_{\mathrm{a}} &= Ki_{\mathrm{e}}\omega \end{aligned}$$

for some constant K. Hence the general forces are

$$\begin{aligned} F_{Q_{\mathrm{a}}} &= -R_{\mathrm{a}}\dot{Q}_{\mathrm{a}} = -R_{\mathrm{a}}i_{\mathrm{a}} \\ F_{\theta} &= Ki_{\mathrm{e}}i_{\mathrm{a}} \end{aligned}$$

and so the equations of the d.c. motor are

$$\boxed{\begin{aligned} \frac{\mathrm{d}i_{\mathrm{a}}}{\mathrm{d}t} &= -\frac{R_{\mathrm{a}}}{L_{\mathrm{a}}}i_{\mathrm{a}} + \frac{1}{L_{\mathrm{a}}}v_{\mathrm{a}} + Ki_{\mathrm{e}}\omega \\ \frac{\mathrm{d}\omega}{\mathrm{d}t} &= -\frac{F\omega}{J} + \frac{K}{J}i_{\mathrm{a}}i_{\mathrm{e}} \end{aligned}} \tag{1.4.6}$$

For a discussion of other types of electromechanical devices, we refer the reader to Shinners (1978).

1.5 FINITE-DIMENSIONAL SYSTEMS

1.5.1 General Systems

In the above discussion we have seen that the derivation of differential equations describing electrical and mechanical systems is a simple matter of writing down the 'energy' of the system and applying Lagrange's equations. All the equations derived in the preceding sections have depended on a finite number of variables, i.e. generalized coordinates $q_1, \ldots, q_s$ for the mechanical system and $Q_1, \ldots, Q_r$ for the electrical system. Moreover, the equations of motion in these generalized coordinates are generally of second order, i.e. depend on $\ddot{Q}_i, \ddot{q}_j$. Any second-order equation of the form

$$\ddot{y} = g(y, \dot{y}, t)$$

for some function g can be changed into two first-order equations by the simple trick of replacing y and $\dot{y}$ by new variables z_1 and z_2 where

$$z_1 = y, \quad z_2 = \dot{y}$$

Then,

$$\begin{aligned} \dot{z}_1 &= z_2 \\ \dot{z}_2 &= g(z_1, z_2, t) \end{aligned}$$

It follows that we may introduce a set of coordinates $x_1, \ldots, x_n$, for some n, such that each of the above problems may be written in the form

$$\dot{x}_i(t) = f_i(x_1(t), \ldots, x_n(t), t), \quad 1 \leqslant i \leqslant n \tag{1.5.1}$$

for some functions f_i. Writing

$$\boldsymbol{x}(t) = (x_1(t), \ldots, x_n(t))^{\mathrm{T}} \in \mathbb{R}^n$$

where $\mathbb{R}^n$ is the n-dimensional Euclidean space (appendix 1), and

$$f = (f_1, \ldots, f_n)^{\mathrm{T}}$$

we may write (1.5.1) in the form

$$\boxed{\dot{\boldsymbol{x}}(t) = f(\boldsymbol{x}(t), t)} \tag{1.5.2}$$

This is the general equation for a lumped system. In control theory the function f depends on various inputs to the system which may be changed by the controller. For example, the voltage sources in an electrical circuit may be variable and in this case the control inputs are written explicitly into f. Then (1.5.2) becomes

$$\boxed{\dot{\boldsymbol{x}}(t) = f(\boldsymbol{x}(t), \boldsymbol{u}(t), t)} \tag{1.5.3}$$

where $\boldsymbol{u} = (u_1, \ldots, u_m)^{\mathrm{T}}$ is an m-dimensional control vector. This equation is the basis of general control theory. For the general theory of ordinary differential equations, see Codington and Levinson (1955) and Davis (1962).

1.5.2 Linear Systems

The solution of general non-linear equations of the form (1.5.2) is often very difficult or even impossible if closed form solutions are sought. For this reason special types of equations are usually considered which are of the form

$$\dot{x}(t) = A(t)x(t) \tag{1.5.4}$$

Here, $A(t)$ is a matrix-valued function of time t. If $A(t)$ is independent of time we obtain the **autonomous system**

$$\boxed{\dot{x}(t) = Ax(t)} \tag{1.5.5}$$

In the same way we consider the special form

$$\boxed{\dot{x}(t) = Ax(t) + Bu(t)} \tag{1.5.6}$$

of (1.5.3). The theory of such systems will be considered later. Very few (if any) real systems are of the form (1.5.6), although such an equation may be a reasonable approximation. The process of going from (1.5.3) to the linear equation (1.5.6) is called **linearization** and will be dealt with in chapter 5. Note also that even the equation (1.5.3) is a mathematical idealization of the real physical system which is often actually distributed i.e. does not consist of a finite number of isolated elements; such systems cannot be modelled precisely on finite-dimensional spaces. Of course, a finite-dimensional model may be a very accurate representation of the system, but it is usually important to derive as good a model as possible before making simplifications. An introduction to certain distributed systems will be given in the next section.

1.6 DISTRIBUTED PARAMETER SYSTEMS

1.6.1 Introduction

In the preceding discussion we have considered systems which contain a finite number of individual components and which may be described by a finite-dimensional set of equations. Such components are said to be discrete or **lumped**. In the next section, we shall consider the equation of motion of a string stretched between two points. We may try to 'lump' the string into a finite set of small masses connected by weightless springs and evaluate the equations governing this system by using the Lagrangian method (for example). This will again give rise to a finite-dimensional set of equations, the solutions of which will only approximately describe the motion of the string, however, since in particular the springs between the masses are assumed to be linear and will therefore not follow the general motion of the string in this region.

We can obtain a better approximation by increasing the number of masses and springs, leading, in the limit, to an infinite-dimensional set of equations. We shall see that these equations represent the 'modes' of the string which are the eigenfunctions of a partial differential operator.

In many cases, therefore, it is found that systems are more accurately described by partial differential equations; a general discussion of such systems is given in Banks (1983), Curtain and Pritchard (1978) and Lions (1970).

1.6.2 The Wave Equation

Consider a string stretched between $x=0$ and $x=l$ so that the tension in the string is τ (fig. 1.20). Since the 'system' now has an infinite number of points (which can be modelled, with increasing accuracy, by an increasing finite number of particles) it seems likely that the equation of motion of the string is not finite dimensional as in (1.5.2). However, it turns out that it is not necessary to consider each point (these are uncountable† in number) of the string independently. We shall see presently that only a countable number of 'coordinates' are required.

The vertical component of the tension is

$$\tau_u = \tau \sin\theta \approx \tau \tan\theta = \tau \frac{\partial u}{\partial x}$$

if θ is small. Hence the change in the vertical force F between x and $x+\mathrm{d}x$ is

$$\mathrm{d}F = [\tau_u]_{x+\mathrm{d}x} - [\tau_u]_x \approx \frac{\partial}{\partial x}\left(\tau \frac{\partial u}{\partial x}\right) \mathrm{d}x$$

For small changes in the position from equilibrium, τ is constant and so by

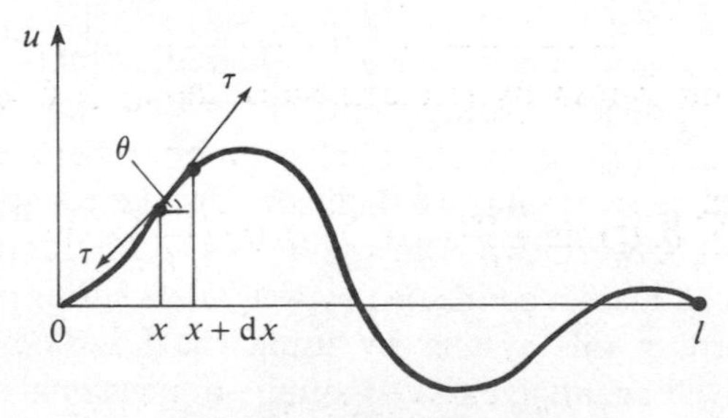

Fig. 1.20 Vibrating string.

†A set is **countable** if it can be put in a one-to-one correspondence with the natural numbers $\{0, 1, 2, \ldots\}$. Otherwise it is **uncountable**.

Newton's second law

$$\sigma \,\mathrm{d}x \frac{\partial^2 u}{\partial t^2} = \frac{\partial}{\partial x}\left(\tau \frac{\partial u}{\partial x}\right) \mathrm{d}x$$

where σ is the density of the string per unit length. Hence

$$\frac{\partial^2 u}{\partial t^2} = c\frac{\partial^2 u}{\partial x^2} \tag{1.6.1}$$

where $c = (\tau/\sigma)^{1/2}$.

This is the one-dimensional **wave equation**. Just as a finite-dimensional equation of second order requires two initial conditions, so (1.6.1) requires two initial conditions for each point on the string. Hence we must specify $u(x, 0)$ and $(\partial u/\partial t)(x, 0)$ for $x \varepsilon [0, l]$, say

$$\begin{aligned} u(x, 0) &= u_0(x) \\ \frac{\partial u}{\partial t}(x, 0) &= v_0(x) \end{aligned} \tag{1.6.2}$$

Moreover, since the string is fixed at the end-points we have the **boundary conditions**

$$u(0,t) = u(l, t) = 0 \tag{1.6.3}$$

If we assume that the solution of (1.6.1) is reasonably 'well behaved' it may be expressed in terms of a Fourier series (see appendix 1). Hence we shall write

$$u(x, t) = \sum_{n=1}^{\infty} q_n(t) \sin \frac{n\pi x}{l} \tag{1.6.4}$$

(The cosine terms vanish by (1.6.3).) Substituting (1.6.4) into (1.6.1) we have

$$\sum_{n=1}^{\infty} \ddot{q}_n(t) \sin \frac{n\pi x}{l} = - \sum_{n=1}^{\infty} q_n(t) \left(\frac{n\pi}{l}\right)^2 c \sin \frac{n\pi x}{l}$$

and so

$$\ddot{q}_n(t) = -q_n(t)\omega_n^2, \quad n \geqslant 1 \tag{1.6.5}$$

where $\omega_n^2 = (n\pi/l)^2 c$. Hence

$$q_n(t) = q_n(0) \cos \omega_n t + \frac{\dot{q}_n(0)}{\omega_n} \sin \omega_n t \tag{1.6.6}$$

where

$$\left.\begin{aligned} q_n(0) &= \frac{2}{l}\int_0^l u_0(x)\sin\frac{n\pi x}{l}\,\mathrm{d}x \\ \dot{q}_n(0) &= \frac{2}{l}\int_0^l v_0(x)\sin\frac{n\pi x}{l}\,\mathrm{d}x \end{aligned}\right\} \tag{1.6.7}$$

Notice that the problem posed by the boundary value problem (1.6.1)–(1.6.3) is equivalent to the solution of the infinite system of differential equations (1.6.5) together with the initial conditions (1.6.7). The equations (1.6.5) may be written in the simple form

$$\ddot{q}(t) = Aq(t)$$

where q is the infinite vector $(q_1, q_2, \ldots)^{\mathrm{T}}$ and A is the infinite diagonal matrix diag $(-\omega_1^2, -\omega_2^2, \ldots)$. Moreover, if we define

$$q_1(t) = q(t), \quad q_2(t) = \dot{q}(t)$$

then we have

$$\begin{pmatrix}\dot{q}_1(t)\\ \dot{q}_2(t)\end{pmatrix} = \begin{pmatrix}0 & I\\ A & 0\end{pmatrix}\begin{pmatrix}q_1(t)\\ q_2(t)\end{pmatrix} \tag{1.6.8}$$

The linear equation (1.5.5) is therefore valid even for some distributed systems, except that the 'state' vector may now be infinite dimensional.

It is also interesting to note that the string may be described by using the **normal modes** q_n as generalized coordinates and applying Lagrange's equations. The kinetic energy of the string is

$$\begin{aligned} T &= \int_0^l \frac{1}{2}\sigma\left(\frac{\partial u}{\partial t}\right)^2 \mathrm{d}x \\ &= \sum_{k=1}^{\infty}\frac{1}{4}\,l\sigma\dot{q}_k^2 \end{aligned}$$

by (1.6.4). The generalized force F_{q_k} may be found by noting that if the coordinate q_k is changed by δ_{q_k} while holding the others fixed then the point x of the string moves up a distance

$$\delta u = \delta_{q_k}\sin\frac{k\pi x}{l} \tag{1.6.9}$$

again by (1.6.4). The work done is then

$$W = F_{q_k}\,\delta_{q_k} = \int_0^l \frac{\partial}{\partial x}\left(\frac{\tau\partial u}{\partial x}\right)\delta u\,\mathrm{d}x$$

where the term $\partial/\partial x(\tau\partial u/\partial x)\,\mathrm{d}x$ is the force on the string element $\mathrm{d}x$ as above. Hence, substituting (1.6.9), we see that

$$F_{q_k} = -\frac{1}{2} l\tau \left(\frac{\pi k}{l}\right)^2 q_k$$

These forces are derivable from the potential

$$V = \sum_{k=1}^{\infty} \frac{1}{4} l\tau \left(\frac{\pi k}{l}\right)^2 q_k^2$$

Hence, by Lagrange's equations, applied to $L = T - V$, we have

$$\frac{1}{2} l\sigma \ddot{q}_k + \frac{1}{2} l\tau \left(\frac{\pi k}{l}\right)^2 q_k = 0, \quad k \geqslant 1$$

which is identical with (1.6.5).

Note that if the assumption of small changes in position from equilibrium does not hold, then the equation of the string is non-linear. Moreover, the general three-dimensional linear wave equation is

$$\boxed{\frac{\partial^2 u}{\partial t^2} - c\nabla^2 u} \qquad (1.6.10)$$

where

$$\nabla^2 = \frac{\partial^2}{\partial x^2} + \frac{\partial^2}{\partial y^2} + \frac{\partial^2}{\partial z^2}$$

is the **Laplace operator.**

1.6.3 The Heat Diffusion Equation

For simplicity we shall derive the one-dimensional heat equation.† Therefore, consider a slab of material of thickness $\mathrm{d}x$ and area A as in fig. 1.21. If the temperature on one side of the slab is T, then the rate of flow of heat across that boundary is

$$Q_x = -kA \frac{\partial T}{\partial x} \quad \text{(Fourier's law of heat conduction)}$$

where k is the coefficient of thermal conductivity, which may depend on x (and T). Then the rate of heat flow across the side at $x + \mathrm{d}x$ is

$$Q_{x+\mathrm{d}x} = -A(k + \mathrm{d}k) \frac{\partial}{\partial x}(T + \mathrm{d}T)$$

†This equation also governs other diffusion processes: the skin effect in electric currents $J(\nabla^2 J = k(\partial J/\partial t))$ and the diffusion of concentration $U(\nabla^2 U = k(\partial U/\partial t))$.

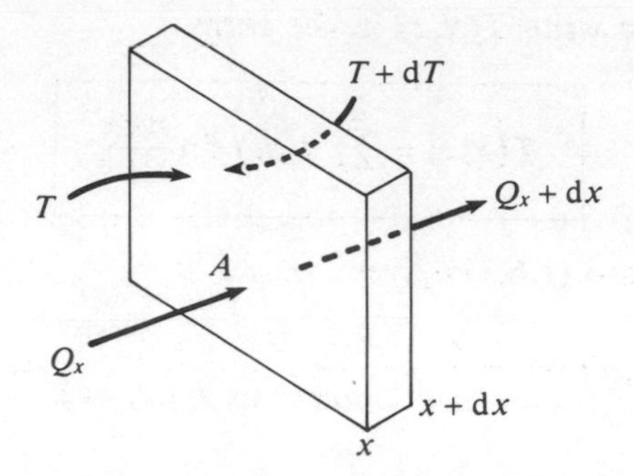

Fig. 1.21 Fourier's law of heat transfer.

Hence the net rate of heat flow into the slab is

$$Q_x - Q_{x+\mathrm{d}x} = kA\,\mathrm{d}x\,\frac{\partial^2 T}{\partial x^2} + A\,\mathrm{d}k\,\frac{\partial T}{\partial x}$$

If there are no heat sources in the slab, then this quantity must equal the rate of heat storage in the slab, which is

$$A\,\mathrm{d}x\rho c_\mathrm{p}\,\frac{\partial T}{\partial t}$$

where ρ is the density of the material and c_p is the specific heat at constant pressure. Hence,

$$\rho c_\mathrm{p}\,\frac{\partial T}{\partial t} = k\,\frac{\partial^2 T}{\partial x^2} + \frac{\partial k}{\partial x}\,\frac{\partial T}{\partial x}$$

$$= \frac{\partial}{\partial x}\left(k\,\frac{\partial T}{\partial x}\right)$$

If k is constant, then

$$\boxed{\frac{\partial T}{\partial t} = \frac{k}{\rho c_\mathrm{p}}\,\frac{\partial^2 T}{\partial x^2}} \tag{1.6.11}$$

As in the case of the wave equation this system requires one initial and two boundary conditions.

Suppose we consider a one-dimensional bar of length l held at zero temperature at the ends. Then

$$T(0, 1) = T(l, t) = 0 \quad \text{for all } t \geqslant 0 \tag{1.6.12}$$

The initial condition may be specified in the form

$$T(x, 0) = T_0(x) \tag{1.6.13}$$

for some given function T_0.

We may again write $T(x, t)$ in the form

$$T(x,t) = \sum_{n=1}^{\infty} T_n(t) \sin \frac{n\pi x}{l} \tag{1.6.14}$$

Substituting this into (1.6.11) gives

$$\sum_{n=1}^{\infty} \dot{T}_n(t) \sin \frac{n\pi x}{l} = -\sum_{n=1}^{\infty} T_n(t) \left(\frac{n\pi}{l}\right)^2 \frac{k}{\rho c_p} \sin \frac{n\pi x}{l}$$

and so

$$\dot{T}_n(t) = -\Omega_n T_n(t), \quad n \geqslant 1 \tag{1.6.15}$$

where

$$\Omega_n = \left(\frac{n\pi}{l}\right)^2 \frac{k}{\rho c_p}$$

Hence,

$$T_n(t) = T_n(0) \exp(-\Omega_n t)$$

However,

$$T_0(x) = \sum_{n=1}^{\infty} T_n(0) \sin \frac{n\pi x}{l}$$

and so

$$T_n(0) = \frac{2}{l} \int_0^l T_0(x) \sin \frac{n\pi x}{l} \tag{1.6.16}$$

Again the original system (1.6.11)–(1.6.13) is equivalent to the differential equation (1.6.15) together with the initial condition (1.6.16). Writing

$$T = (T_1, T_2, \ldots)^{\mathrm{T}}$$
$$A = -\operatorname{diag}(\Omega_1, \Omega_2, \Omega_2, \ldots)$$

we have 'reduced' the system to the form of (1.5.5):

$$\dot{T} = AT \tag{1.6.17}$$

Note, however, that the T_i are not generalized coordinates for this system.

1.6.4 Fluid Dynamics

Many physical systems contain fluids and it is important to be able to model such systems. The study of fluid mechanics also gives good insight into the

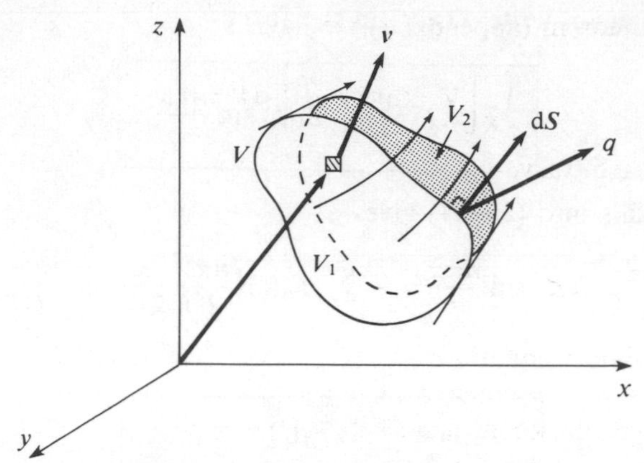

Fig. 1.22 A comoving volume of fluid.

modelling of general 'continuous' systems. By **continuous** we mean that the molecular structure of the fluid is ignored and the fluid is assumed to be infinitely divisible without any noticeable difference in the material composition of the fluid. The statistical description of turbulent flow is beyond the scope of the present work.

In contrast to the Lagrangian method studied above for lumped systems, where the motion of each individual particle (or solid body) is modelled completely, we now use the Euler method of fixing a volume in the fluid and observing the changing fluid as a whole moving through this volume. Consider therefore a volume V fixed in the fluid as in fig. 1.22. In a small time $\mathrm{d}t$ the particles in V move along streamlines and thus flow into another volume $\bar{V}$. If V and $\bar{V}$ are regarded as sets, then we can define

$$V_1 = V - \bar{V}, \quad \bar{V}_2 = \bar{V} - V$$

(see appendix 1 for set differences). Letting $m_V(t)$ denote the mass of fluid in the volume V at time t, it follows that

$$m_V(t) = m_V(t + \mathrm{d}t) - m_{V_1}(t + \mathrm{d}t) + m_{V_2}(t + \mathrm{d}t)$$

since the fluid that was in V at time t is in $(V - V_1) \cup V_2$ at time $t + \mathrm{d}t$. Now,

$$\lim_{t \to 0} \frac{m_V(t + dt) - m_V(t)}{\mathrm{d}t} = \frac{\partial}{\partial t} \int_V \rho \, \mathrm{d}V$$

where ρ is the fluid density. Moreover, we clearly have

$$\lim_{\mathrm{d}t \to 0} \left\{ \frac{m_{V_1}(t + \mathrm{d}t)}{\mathrm{d}t} - \frac{m_{V_2}(t + \mathrm{d}t)}{\mathrm{d}t} \right\} = - \int_S \rho \boldsymbol{v} \cdot \mathrm{d}\boldsymbol{S}$$

where S is the surface of V and $\boldsymbol{v}$ is the fluid velocity. Hence

$$\int_S \rho \boldsymbol{v} \cdot \mathrm{d}\boldsymbol{S} = - \frac{\partial}{\partial t} \int_V \rho \, \mathrm{d}V$$

Using Gauss' theorem (appendix 1) we have

$$\int_V \left[\nabla \cdot (\rho \boldsymbol{v}) + \frac{\partial \rho}{\partial t} \right] \mathrm{d}V = 0$$

and since V is arbitrary,

$$\boxed{\frac{\partial \rho}{\partial t} + \nabla \cdot (\rho \boldsymbol{v}) = 0} \tag{1.6.18}$$

This is the **continuity equation**, and states that the mass of the fluid is conserved, i.e. it is neither created nor destroyed.

Using Newton's second law,

$$\boldsymbol{f} = \frac{\mathrm{d}\boldsymbol{m}}{\mathrm{d}t}$$

where $\boldsymbol{m}$ is the linear momentum of a system, then denoting by $\boldsymbol{m}_V(t)$ the linear momentum of the fluid in V at time t and noting that the change in linear momentum of the fluid in V in time $\mathrm{d}t$ is

$$\mathrm{d}\boldsymbol{m} = \boldsymbol{m}_V(t + \mathrm{d}t) - \boldsymbol{m}_{V_1}(t + \mathrm{d}t) + \boldsymbol{m}_{V_2}(t + \mathrm{d}t) - \boldsymbol{m}_V(t)$$

it follows that

$$\boldsymbol{f} = \frac{\partial}{\partial t} \int_V \boldsymbol{v}\rho \, \mathrm{d}V + \int_S \boldsymbol{v}\rho \boldsymbol{v} \cdot \mathrm{d}\boldsymbol{S}$$

Generally, $\boldsymbol{f}$ is made up of the total surface force $\boldsymbol{f}_S$ (due to pressure and shear) and a body force $\boldsymbol{B}$ per unit volume. Hence we obtain the momentum equation

$$\boxed{\boldsymbol{f}_S + \int_V \boldsymbol{B} \mathrm{d}V = \frac{\partial}{\partial t} \int_V \boldsymbol{v}\rho \, \mathrm{d}V + \int_S \boldsymbol{v}\rho \boldsymbol{v} \cdot \mathrm{d}\boldsymbol{S}} \tag{1.6.19}$$

In just the same way we can derive the angular momentum equation

$$\boxed{\int_S \boldsymbol{r} \wedge \mathrm{d}\boldsymbol{f}_S + \int_V \boldsymbol{r} \wedge \boldsymbol{B} \, \mathrm{d}V = \frac{\partial}{\partial t} \int_V \boldsymbol{r} \wedge \boldsymbol{v}\rho \, \mathrm{d}V + \int_S \boldsymbol{r} \wedge \boldsymbol{v}\rho \boldsymbol{v} \cdot \mathrm{d}\boldsymbol{S}} \tag{1.6.20}$$

where $\boldsymbol{r}$ is the position vector of a fluid element relative to some inertial axes (fig. 1.22).

To express $\boldsymbol{f}_S$ in a more useful form we must recall the definition of the stress tensor. Consider an elemental cubic volume of fluid shown in fig. 1.23. The fluid surrounding this element exerts on each of its faces a net normal force and two net tangential forces. These forces, measured per unit

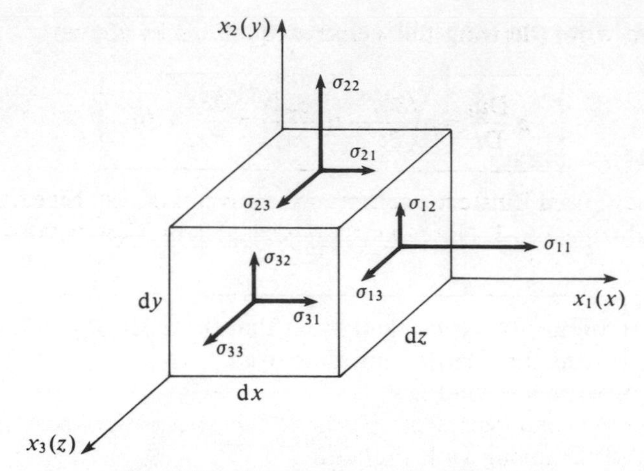

Fig. 1.23 Stresses on a cube of material.

area, are called the stresses and form a three by three matrix

$$\sigma_{ij} = \begin{pmatrix} \sigma_{11} & \sigma_{12} & \sigma_{13} \\ \sigma_{21} & \sigma_{22} & \sigma_{23} \\ \sigma_{31} & \sigma_{32} & \sigma_{33} \end{pmatrix}$$

called the **stress tensor.** Note that in fig. 1.23 the stresses on the hidden faces are equal and opposite to those on the corresponding opposing face. Hence σ is symmetric; i.e. $\sigma_{ij} = \sigma_{ji}$ (for otherwise an infinitesimal volume would have infinite angular velocity).

Now applying the momentum equation (1.6.19) to the elemental cube we have, in the x direction,

$$[\sigma_{11}(x+\mathrm{d}x) - \sigma_{11}(x)]\,\mathrm{d}y\,\mathrm{d}z + [\sigma_{21}(y+\mathrm{d}y) - \sigma_{21}(y)]\,\mathrm{d}x\,\mathrm{d}z + [\sigma_{31}(z+\mathrm{d}z) - \sigma_{31}(z)]\,\mathrm{d}x\,\mathrm{d}y + B_x\,\mathrm{d}x\,\mathrm{d}y\,\mathrm{d}z$$

$$= \frac{\partial}{\partial t}(u\rho)\,\mathrm{d}x\,\mathrm{d}y\,\mathrm{d}z + \mathrm{d}y\,\mathrm{d}z[(\rho u^2)(x+\mathrm{d}x) - (\rho u^2)(x)]$$

$$+ \mathrm{d}x\,\mathrm{d}z[(\rho uv)(y+\mathrm{d}y) - (\rho uv)(y)] + \mathrm{d}x\,\mathrm{d}y[(\rho uw)(z+\mathrm{d}z) - (\rho uw)(z)]$$

where $v = (u, v, w)$. Dividing by $\mathrm{d}x\,\mathrm{d}y\,\mathrm{d}z$ and using (1.6.18), we have

$$\rho\frac{\mathrm{D}u}{\mathrm{D}t} \triangleq \rho\left(\frac{\partial u}{\partial t} + u\frac{\partial u}{\partial x} + v\frac{\partial u}{\partial y} + w\frac{\partial u}{\partial z}\right)$$

$$= \frac{\partial \sigma_{11}}{\partial x} + \frac{\partial \sigma_{12}}{\partial y} + \frac{\partial \sigma_{13}}{\partial z} + B_x$$

If we write $v = (u_1, u_2, u_3)$, $(x, y, z) = (x_1, x_2, x_3)$ and $B = (B_1, B_2, B_3)$, then we can write (deriving the y and z equations as above)

$$\rho \frac{\mathrm{D}u_i}{\mathrm{D}t} = \rho\left(\frac{\partial u_i}{\partial t} + u_j \frac{\partial u_i}{\partial x_j}\right) = \frac{\partial \sigma_{ij}}{\partial x_j} + B_i \tag{1.6.21}$$

where we have used Einstein's summation convention, i.e. repeated indices are summed from 1 to 3. Note that the so-called **comoving** derivative $\mathrm{D}v/\mathrm{D}t$ is given by

$$\frac{\mathrm{D}v}{\mathrm{D}t} = \frac{\partial v}{\partial t} + (v \cdot \nabla)v = \frac{\partial v}{\partial t} + \nabla\left(\frac{v^2}{2}\right) - v \wedge \nabla \wedge v \tag{1.6.22}$$

where $v^2 = v \cdot v$, and represents physically the derivative of v at a point (x, y, z) as seen moving with the fluid.

If the fluid has no viscosity, then $\sigma_{ij} = 0$ for $i \neq j$ and

$$\sigma_{11} = \sigma_{22} = \sigma_{33} = -p$$

where p is the pressure and the equations of motion (in vector form) become

$$\rho\left[\frac{\partial v}{\partial t} + \nabla\left(\frac{v^2}{2}\right) - v \wedge \nabla \wedge v\right] = -\nabla p + B \tag{1.6.23}$$

If we integrate (1.6.23) along a streamline γ (i.e. v is parallel to the tangent to γ), then

$$\int_1^2 \frac{\partial v}{\partial t} \cdot \mathrm{d}s + \frac{v_2^2 - v_1^2}{2} + \int_1^2 \frac{\mathrm{d}p}{\rho} + \Psi_2 - \Psi_1 = 0$$

where $\Psi = gz$. Hence

$$\frac{v_2^2 - v_1^2}{2} + \int_1^2 \frac{\mathrm{d}p}{\rho} + g(z_2 - z_1) = 0$$

and for incompressible flow,

$$\frac{v^2}{2} + p + gz$$

is constant along a streamline. This is **Bernoulli's theorem.**

The term $\nabla \wedge v$ in (1.6.23) is called the rotation or **vorticity** of the fluid and if $\nabla \wedge v = 0$ we say that the fluid is irrotational. Recall that for any Cartesian tensor A_{ij} we can define the symmetric and antisymmetric parts of A_{ij} by

$$A_{ij}^{\mathrm{s}} = \frac{a_{ij} + a_{ji}}{2}$$

$$A_{ij}^{\mathrm{a}} = \frac{a_{ij} - a_{ji}}{2}$$

The tensor $\partial u_i/\partial x_j$ is called the **rate of deformation tensor** and we define

$$e_{ij} = \left(\frac{\partial u_i}{\partial x_j}\right)^{\mathrm{s}} = \text{strain rate tensor}$$

$$\omega_{ij} = \left(\frac{\partial u_i}{\partial x_j}\right)^{\mathrm{a}} = \text{rotation tensor}$$

Clearly,

$$\Omega = \tfrac{1}{2}\nabla \wedge \boldsymbol{v} \tag{1.6.24}$$

where $\Omega = (\omega_{32}, \omega_{13}, \omega_{21}) = -(\omega_{23}, \omega_{31}, \omega_{12})$.

We now define a **Newtonian fluid** as one in which the stress tensor is related to the strain tensor in a linear way. It can be shown that such a relation must be of the form ($\phi = \nabla \cdot \boldsymbol{v}$)

$$\begin{aligned}\sigma_{ij} &= -p\delta_{ij} + 2\mu e_{ij} + \delta_{ij}\lambda\phi \\ &= -p_{ij} + \sigma'_{ij}\end{aligned} \tag{1.6.25}$$

where μ and λ are coefficients of viscosity and δ_{ij} is the Kronecker delta (appendix 1). Hence, if $\zeta = \lambda + \frac{2}{3}\mu$ we have

$$\boxed{\sigma_{ij} = -p\delta_{ij} + \mu\left(\frac{\partial u_i}{\partial x_j} + \frac{\partial u_j}{\partial x_i} - \frac{2}{3}\delta_{ij}\frac{\partial u_k}{\partial x_k}\right) + \zeta\delta_{ij}\frac{\partial u_k}{\partial x_k}} \tag{1.6.26}$$

Now, using (1.6.21) we obtain the **Navier–Stokes equations**

$$\rho\left(\frac{\partial u_i}{\partial t} + u_j\frac{\partial u_i}{\partial x_j}\right) = -\frac{\partial p}{\partial x_i} + B_i + \frac{\partial}{\partial x_j}\left[\mu\left(\frac{\partial u_i}{\partial x_j} + \frac{\partial u_j}{\partial x_i} - \frac{2}{3}\delta_{ij}\frac{\partial u_k}{\partial x_k}\right)\right] + \frac{\partial}{\partial x_i}\left(\zeta\frac{\partial u_k}{\partial x_k}\right)$$

or, in vector form,

$$\boxed{\rho\frac{\mathrm{D}\boldsymbol{v}}{\mathrm{D}t} = -\nabla p + \boldsymbol{B} + \nabla \wedge [\mu(\nabla \wedge \boldsymbol{v})] + \nabla[(\zeta + \tfrac{4}{3}\mu)\nabla \cdot \boldsymbol{v}]} \tag{1.6.27}$$

If μ is constant and $\nabla \cdot \boldsymbol{v} = 0$ (i.e. the fluid is incompressible) then we obtain the important relation

$$\rho\frac{\mathrm{D}\boldsymbol{v}}{\mathrm{D}t} = -\nabla p + \boldsymbol{B} + \mu\nabla^2\boldsymbol{v} \tag{1.6.28}$$

Just as in finite-dimensional (lumped) systems, energy equations are important for distributed systems. Consider, therefore, a control volume V fixed in space as in fig. 1.22 with heat flux vector field $\boldsymbol{q}$ defined on the surface. The rate of heat flow into the volume V is

$$-\int_S \boldsymbol{q}\cdot \mathrm{d}\boldsymbol{S} = -\int_S q_i\,\mathrm{d}S_i$$

(summation convention). Moreover the rate of work done by the fluid in the volume V is

$$\frac{\mathrm{d}W}{\mathrm{d}t} = -\int_S u_i \sigma_{ji}\, \mathrm{d}S_j$$

(Note that the stresses do work only on S.) Let

$$e = U + \tfrac{1}{2}v^2 + \Psi$$

be the total energy per unit mass, where Ψ is the gravitational potential, and U is the internal energy per unit mass. Then, if Q denotes the internal heat generation rate per unit volume, we have the energy equation

$$\frac{\partial}{\partial t}\int_V \rho e\, \mathrm{d}V + \int_S \rho e u_i\, \mathrm{d}S_i = -\int_S q_i\, \mathrm{d}S_i + \int_S u_i \sigma_{ji}\, \mathrm{d}S_j + \int_V Q\, \mathrm{d}V \qquad (1.6.29)$$

i.e. the rate of increase of energy in V equals the rate of heat flow into V minus the rate of working of the fluid on the surrounding medium plus the rate of internal heat generation. By Gauss' theorem and the continuity equation it follows that

$$\rho\frac{\mathrm{D}e}{\mathrm{D}t} = -\frac{\partial}{\partial x_i} q_i - p\frac{\partial u_i}{\partial x_i} - u_i\frac{\partial p}{\partial x_i} + u_i\frac{\partial \sigma'_{ji}}{\partial x_j} + \Phi + Q \qquad (1.6.30)$$

where $\Phi = \sigma'_{ij}\partial u_j/\partial x_i$ is the **dissipation function,** representing the rate of working of the shear stresses on the fluid (σ'_{ij} is defined in (1.6.25)). However, from the equation of motion (1.6.21) we have

$$\rho u_i \frac{\mathrm{D}u_i}{\mathrm{D}t} = \rho\frac{\mathrm{D}}{Dt}\left(\frac{v^2}{2}\right)$$

$$= -u_i\frac{\partial p}{\partial x_i} + u_i\frac{\partial \sigma'_{ij}}{\partial x_j} - \rho u_i\frac{\partial \Psi}{\partial x_i}$$

assuming the body force B_i is due solely to the gravitational term $-\rho\partial\Psi/\partial x_i$, and so

$$\rho\frac{\mathrm{D}}{\mathrm{D}t}\left(\frac{v^2}{2} + \Psi\right) = -u_i\frac{\partial p}{\partial x_i} + u_i\frac{\partial \sigma'_{ij}}{\partial x_j} \qquad (1.6.31)$$

since Ψ is constant in time. Recalling the definition of e and subtracting (1.6.31) from (1.6.30), we obtain

$$\rho\frac{\mathrm{D}U}{\mathrm{D}t} = -p\frac{\partial u_i}{\partial x_i} - \frac{\partial q_i}{\partial x_i} + \Phi + Q \qquad (1.6.32)$$

This is the equation of evolution of the internal energy, and it should be noted that it was obtained by removing from the energy equation (1.6.30) the mechanical energy equation (1.6.31). Using Fourier's law as in section

1.6.2, we have

$$q = -k\nabla T$$

and if we assume a perfect gas with k constant, then $\mathrm{d}U = c_v\,\mathrm{d}T$ and so

$$\rho c_v \frac{\mathrm{D}T}{\mathrm{D}t} = -p\nabla\cdot v + k\nabla^2 T - \nabla\cdot q_r + \Phi + Q \tag{1.6.33}$$

where q_r is the heat flux vector due to radiation. Defining the **enthalpy** by $h = U + p/\rho$, we may write (1.6.32) in the form

$$\rho \frac{\mathrm{D}h}{\mathrm{D}t} = \frac{\mathrm{D}p}{\mathrm{D}t} + k\nabla^2 T - \nabla\cdot q_r + \Phi + Q$$

and so, again for a perfect gas, where $\mathrm{d}h = c_p\,\mathrm{d}T$, we have

$$\rho c_p \frac{\mathrm{D}T}{\mathrm{D}t} = \frac{\mathrm{D}p}{\mathrm{D}t} + k\nabla^2 T - \nabla\cdot q_r + \Phi + Q \tag{1.6.34}$$

For an incompressible fluid, the perfect gas law does not hold, but using $\nabla\cdot v = 0$ we may still write $\mathrm{d}U = c_v\,\mathrm{d}T$ and so we again obtain (1.6.33) with the term $-p\nabla\cdot v$ omitted.

We shall now study incompressible flow in more detail. Recall that

$$\omega = \nabla \wedge v$$

is called the vorticity of the fluid. The **circulation** Γ about any closed curve γ is defined by

$$\Gamma = \oint_\gamma v\cdot \mathrm{d}l = \int_S \omega\cdot \mathrm{d}\mathbf{S}$$

where S is a surface with boundary γ (using Stoke's theorem). A flow is irrotational if $\nabla\wedge v = 0$. In this case we can write v as the gradient of a potential, i.e.

$$v = -\nabla\phi$$

Note that, for an incompressible fluid,

$$0 = \nabla\cdot v = -\nabla\cdot(\nabla\phi) = \nabla^2\phi$$

and so ϕ is **harmonic**, i.e. ϕ satisfies Laplace's equation

$$\nabla^2\phi = 0 \tag{1.6.35}$$

If we consider a two-dimensional flow, which is independent of the z coordinate, we may define the **stream function** ψ by

$$u = -\frac{\partial\psi}{\partial y}, \quad v = \frac{\partial\psi}{\partial x}$$

Since the fluid is assumed to be irrotational $\nabla \wedge v = 0$. Hence,

$$\frac{\partial v}{\partial x} - \frac{\partial u}{\partial y} = \frac{\partial^2\psi}{\partial x^2} + \frac{\partial^2\psi}{\partial y^2} = 0$$

and so Ψ is also harmonic

$$\boxed{\nabla^2\psi = 0} \tag{1.6.36}$$

On a line of constant ϕ we have

$$\begin{aligned} 0 = \mathrm{d}\phi &= \frac{\partial\phi}{\partial x}\,\mathrm{d}x + \frac{\partial\phi}{\partial y}\,\mathrm{d}y \\ &= -u\,\mathrm{d}x - v\,\mathrm{d}y \end{aligned}$$

and so

$$\left.\frac{\mathrm{d}y}{\mathrm{d}x}\right|_{\text{const }\phi} = -\frac{u}{v}$$

Similarly,

$$\left.\frac{\mathrm{d}y}{\mathrm{d}x}\right|_{\text{const }\psi} = \frac{v}{u}$$

Hence lines of constant ϕ are orthogonal to lines of constant ψ. Note also that the **Cauchy–Riemann** equations

$$(u =) -\frac{\partial\phi}{\partial x} = -\frac{\partial\psi}{\partial y}, \quad (v =) -\frac{\partial\phi}{\partial y} = \frac{\partial\psi}{\partial x}$$

hold. This suggests that we should consider the **complex potential**

$$\boxed{F(z) = \phi(x, y) + j\psi(x, y)} \tag{1.6.37}$$

where $z = x + jy$. The **complex velocity** $-\mathrm{d}F/\mathrm{d}z$ is given by

$$\begin{aligned} -\frac{\mathrm{d}F}{\mathrm{d}z} &= -\left(\frac{\partial\phi}{\partial x} + j\frac{\partial\psi}{\partial x}\right) \\ &= u - jv \end{aligned}$$

Similarly,

$$-\frac{\mathrm{d}\bar{F}}{\mathrm{d}z} = u + jv$$

and so

$$\frac{\mathrm{d}F}{\mathrm{d}z}\frac{\mathrm{d}\bar{F}}{\mathrm{d}\bar{z}} = v^2$$

An important potential is

$$\boxed{F = -U_0\left(z + \frac{a^2}{z}\right)} \tag{1.6.38}$$

Clearly, from (1.6.37)

$$\psi = -U_0 y + \frac{a^2 U_0 y}{r^2}$$

$$\phi = -U_0 x\left(1 + \frac{a^2}{r^2}\right)$$

($r^2 = x^2 + y^2$). At ∞, $\phi = -U_0 x$ which represents a uniform flow with velocity U_0 in the x-direction. Moreover, when $r = a$, $\psi = 0$. Hence $r = a$ is a streamline and so (1.6.38) is the complex potential for the flow around an infinite cylinder (fig. 1.24).

Evaluating $\mathrm{d}F/\mathrm{d}z = -u + iv$ it is easy to see that

$$u = -U_0\left(\frac{a^2}{r^2}\cos 2\theta - 1\right)$$

$$v = -U_0\frac{a^2}{r^2}\sin 2\theta$$

and so

$$v^2 = U_0^2\left(1 + \frac{a^4}{r^4} + \frac{2a^2}{r^2}(\sin^2\theta - \cos^2\theta)\right)$$

Neglecting gravity, the pressure around the cylinder is given by Bernoulli's equation; in fact,

$$\frac{p}{\rho} + \frac{v^2}{2} = \frac{p_0}{\rho} = \text{constant}$$

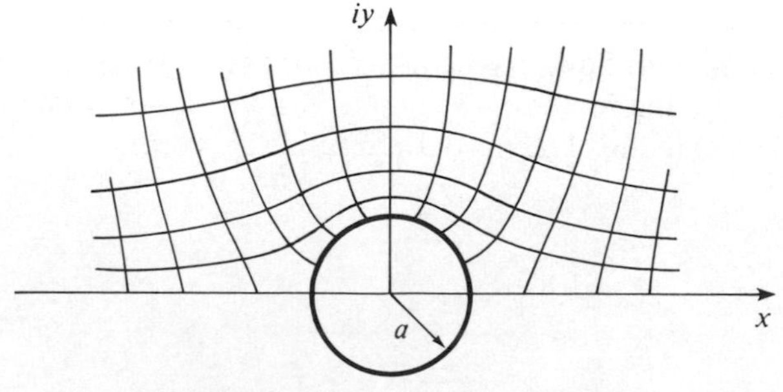

Fig. 1.24 Flow around a cylinder.

where p_0 is the pressure where $v^2 = 0$ (stagnation point). Hence

$$p\,|_{r=a} = p_0 - 2\rho U_0^2(1 + \sin^2\theta - \cos^2\theta)$$

The pressure is symmetric about the x- and y-axes and so there is no net force on the cylinder. This idealization may seem strange; however, in the real situation, separation would occur at the back of the cylinder producing drag. The idea of the aerofoil theory is to 'pull out' the rear end of the cylinder to reduce this drag. To illustrate this we must first consider the circulation around a cylinder.

Firstly, note that for the complex potential

$$\frac{j\Gamma}{2\pi}\ln\frac{z}{a}$$

we have

$$\phi = -\frac{\Gamma}{2\pi}\theta, \quad \psi = \frac{\Gamma}{2\pi}\ln\frac{r}{a}, \quad z = r\exp(j\theta)$$

and $\psi = 0$ when $r = a$. Lines of constant ψ (streamlines) are circles and so this complex potential represents a simple vortex. The circulation about any closed curve containing the origin is

$$\int_0^{2\pi} v_\theta\,|_{r=a}\, a\, \mathrm{d}\theta = \Gamma$$

Note that the circulation is independent of the path containing the origin since the flow is irrotational, i.e.

$$\nabla \wedge v = -\frac{1}{r}\frac{\partial}{\partial r}\left(r\frac{\Gamma}{2\pi r}\right) = 0$$

It follows that the complete complex potential

$$\boxed{F = -U_0\frac{z + a^2}{z} + j\frac{\Gamma}{2\pi}\ln\frac{z}{a}} \qquad (1.6.39)$$

represents the uniform flow around a cylinder with a vortex superimposed.

Consider now the forces on a cylinder with cross-sectional area in the form of an aerofoil (fig. 1.25). By Bernoulli's theorem,

$$p = p_0 - \tfrac{1}{2}\rho v^2$$

$$= p_0 - \frac{1}{2}\rho\frac{\mathrm{d}F}{\mathrm{d}z}\frac{\mathrm{d}\bar{F}}{\mathrm{d}\bar{z}}$$

where F is the complex potential of the flow. Hence, since the force $X + jY$

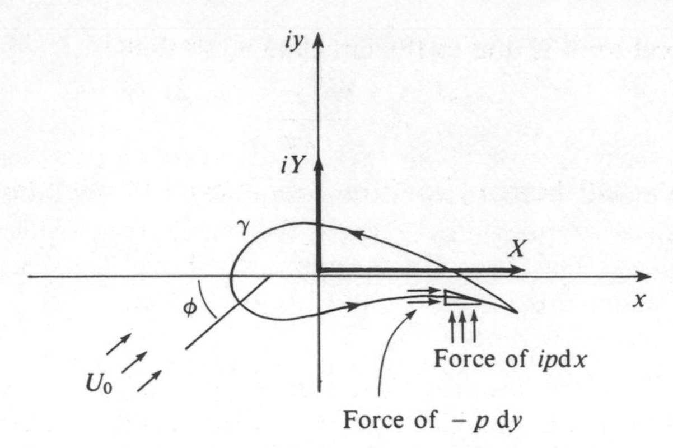

Fig. 1.25 Forces on an aerofoil.

on the aerofoil is determined by the pressure at the surface, we have

$$X - jY = -\oint_\gamma (p\,\mathrm{d}y + jp\,\mathrm{d}x)$$

$$= \oint_\gamma jp\,\mathrm{d}\bar{z}$$

$$= \oint_\gamma \frac{j}{2}\rho \frac{\mathrm{d}F\,\mathrm{d}\bar{F}}{\mathrm{d}z\,\mathrm{d}\bar{z}}\mathrm{d}\bar{z}$$

$$= \frac{j}{2}\rho \oint_\gamma \left(\frac{\mathrm{d}F}{\mathrm{d}z}\right)^2 \mathrm{d}z \quad \text{(Blasius' theorem)}$$

(Note that on the surface ψ = constant, so $\mathrm{d}\bar{F} = \mathrm{d}F$.) Similarly, the moment is

$$M = \oint_\gamma p(x\,\mathrm{d}x + y\,\mathrm{d}y)$$

$$= \mathrm{Re}\oint_\gamma pz\,\mathrm{d}\bar{z}$$

$$= -\mathrm{Re}\frac{1}{2}\rho\oint_\gamma z\left(\frac{\mathrm{d}F}{\mathrm{d}z}\right)^2 \mathrm{d}z$$

If we now expand $\mathrm{d}F/\mathrm{d}z$, for large $|z|$, in powers of z^{-1}, we have

$$-\frac{\mathrm{d}F}{\mathrm{d}z} = U_0 \exp(-j\phi) + \frac{A}{z} + \frac{B}{z^2} + \dots$$

and as $z \to \infty$, the complex velocity becomes $U_0 \exp(-j\phi)$, which is a free stream velocity at an angle ϕ to the x-axis. Integrating, we have

$$F = -U_0 \exp(-j\phi)_z - A \ln z + \frac{B}{z} + \dots$$

and the second term is due to the circulation, so that

$$A = -\frac{j\Gamma}{2\pi}$$

Hence, by Blasius' theorem, and Cauchy's integral theorem (appendix 1),

$$X - jY = \frac{1}{2}\, j\rho \left\{ 2\pi j \left[\frac{-j\Gamma U_0 \exp(j\phi)}{\pi} \right] \right\}$$

$$= j\rho\Gamma U_0 \exp(-j\phi)$$

Hence we see that the resultant force on the aerofoil is $-\rho U_0 \Gamma$, perpendicular to the free stream, and is therefore called the lift. The lift is positive if $\Gamma < 0$. Note that there is no drag force. This will only occur in the boundary layer or by separation at the trailing edge, which reduces the pressure there. However, two-dimensional potential theory as above will not predict this effect.

We can calculate the circulation Γ on simple types of aerofoil section by using conformal transformation (for more details of which see appendix 1). It can be shown that the conformal map

$$z = \zeta + \frac{l^2}{\zeta}$$

maps the circular cylinder shown in fig. 1.26 into the Joukowski aerofoil in the z plane. The flow about the cylinder in the ζ-plane (with circulation) is

$$F = -U_0 \left\{ [\zeta - b\exp(j\theta)] \exp(-j\phi) + \frac{a^2}{[\zeta - \mathrm{b}\exp(j\phi)] \exp(-j\theta)} + j\frac{\Gamma}{2\pi} \ln\left[\frac{\zeta - b\exp(j\theta)}{a}\right] \right\}$$

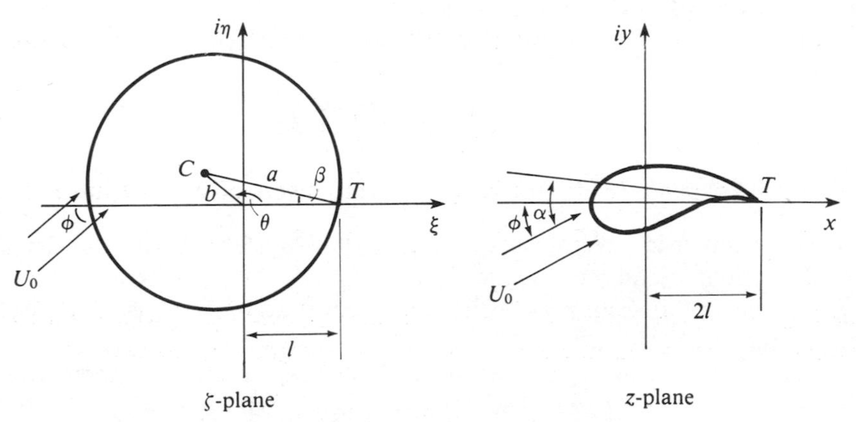

Fig. 1.26 The Joukowski transformation.

and since

$$\frac{dF}{dz} = \frac{dF}{d\zeta}\frac{d\zeta}{dz}$$

and $dz/d\zeta = 1 - l^2/\zeta^2 = 0$ at the trailing edge, $dF/d\zeta$ must be zero when $z = -2l$ so that dF/dz is finite at T. Thus

$$\frac{dF}{d\zeta} = -U_0\left[\exp(-j\phi) - \frac{a^2\exp(j\phi)}{(\zeta - b\exp j\theta)^2}\right] + j\frac{\Gamma}{2\pi}\,\frac{1}{[\zeta - b\exp(j\theta)]} = 0, \quad \zeta = l$$

Since $[\zeta - b\exp(j\theta)] = \{\zeta - l + a\exp[j(\pi - \beta)]\}$ it follows that

$$-U_0\{1 - \exp[-2j(\pi - \beta) + 2j\phi]\} + j\frac{\Gamma}{2\pi a}\exp[-j(\pi - \beta) + j\phi] = 0$$

and so

$$\Gamma = -4\pi a U_0 \sin(\beta + \phi)$$

The angle $\alpha = \beta + \phi$ is called the **angle of attack**. When $\alpha = 0$, $\Gamma = 0$ and since $L = -\rho U_0 \Gamma$, the lift is also zero. The coefficient

$$C_L = \frac{L}{c\rho U^2/2}$$

where L is the lift per unit length of the aerofoil and c is the chord length (width) of the aerofoil is called the **lift coefficient**. For the Joukowski aerofoil, $c \approx 4a$ and so

$$C_L \approx \frac{\rho U_0(4\pi a U_0)\sin\alpha}{4a\rho U_0^2/2} = 2\pi\sin\alpha$$

For small angles of attack the lift increases linearly with α. However, as α increases C_L falls off until boundary separation occurs, at which point the circulation is lost and the aerofoil stalls.

For a further discussion of fluid dynamics see Milne-Thomson (1968).

1.6.5 Systems Involving Delay

There is another important class of systems which are properly considered to be distributed and have applications in control theory. These are delay systems and in this section we shall give two illustrative examples. For more details see Hale (1977).

The first system comes from a biological model of population dynamics. We suppose that there are two species x and y of which x is the prey population and y is the predator population. Then if b is the rate of increase of the prey and d is the death rate of the predators, we have

$$\dot{x} = bx - K_1 xy$$
$$\dot{y} = K_2 xy - dy$$

where K_1 is the coefficient of predation effect on x and K_2 is the coefficient of predation effect on y. These equations can be solved numerically or by phase space arguments. However, it has been argued (Wangersky and Cunningham, 1957) that the two species cannot react immediately to a change in the other species and so the product term should be replaced by delayed versions of these terms. Hence we may write down the improved model

$$\dot{x} = bx(t) - K_1 x(t-\tau_1) y(t-\tau_1)$$
$$\dot{y} = K_2 x(t-\tau_2) y(t-\tau_2) - \mathrm{d}y(t)$$

For simplicity, Wangersky and Cunningham take $\tau_1 = 0$ and assume a single delay $\tau = \tau_2$. Moreover, if there is an external limiting factor on the population growth of the prey we may consider the equations

$$\boxed{\begin{aligned} \dot{x} &= bx(t)\left(\frac{\varkappa - x(t)}{\varkappa}\right) - K_1 x(t) y(t) \\ \dot{y} &= K_2 x(t-\tau) y(t-\tau) - \mathrm{d}y(t) \end{aligned}} \qquad (1.6.40)$$

As a second example consider an ideal delay line which may be regarded as a distributed sequence of reactances, a portion of which is shown in fig. 1.27. The loop equations for this section are

$$v(x,t) - \left[\frac{L\,\partial i}{\partial t}(x,t) + Ri(x,t)\right]\delta x - v(x,t) - \frac{\partial v(x,t)}{\partial x}\delta x = 0$$

and

$$i(x,t) - \left[\frac{C\,\partial v}{\partial t}(x,t) + Gv(x,t)\right]\delta x - i(x,t) - \frac{\partial i(x,t)\,\delta x}{\partial x} = 0$$

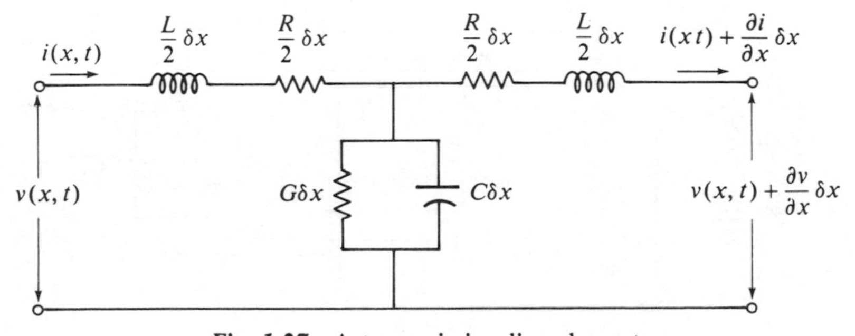

Fig. 1.27 A transmission line element.

Hence

$$\boxed{\begin{aligned} L\frac{\partial i}{\partial t} + Ri &= -\frac{\partial v}{\partial x} \\ C\frac{\partial v}{\partial t} + Gv &= -\frac{\partial i}{\partial x} \end{aligned}} \tag{1.6.41}$$

Suppose that a transmission line is terminated as in fig. 1.28 where $g(v)$ is a voltage dependent current source. We shall assume for simplicity that the line resistance R and admittance G are zero so that the partial differential equations (1.6.41) becomes

$$L\frac{\partial i}{\partial t} = -\frac{\partial v}{\partial x}, \quad C\frac{\partial v}{\partial t} = -\frac{\partial i}{\partial x}, \quad 0 < x < 1, t > 0 \tag{1.6.42}$$

with the boundary conditions

$$E = v(0, t) + R_1 i(0, t) \tag{1.6.43a}$$

$$C_1 \frac{\mathrm{d}y(1, t)}{\mathrm{d}t} = i(1, t) - g(v(1, t)) \tag{1.6.43b}$$

We shall show how to transform this partial differential equation with non-linear boundary conditions into a delay equation. In fact, the general solution of (1.6.42) is clearly

$$v(x, t) = \phi(x - st) + \psi(x + st)$$

$$i(x, t) = \frac{1}{z}[\phi(x - st) - \psi(x + st)]$$

for any functions ϕ, ψ where $s = (LC)^{-1/2}$, $z = (L/C)^{1/2}$. It follows that

$$\left.\begin{aligned} 2\phi(x - st) &= v(x, t) + zi(x, t) \\ 2\psi(x + st) &= v(x, t) - zi(x, t) \end{aligned}\right\} \tag{1.6.44}$$

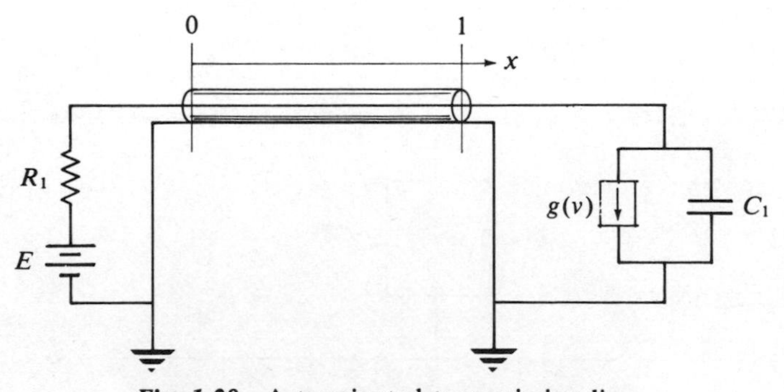

Fig. 1.28 A terminated transmission line.

and so

$$\left.\begin{aligned} 2\phi(-st) &= v\left(1, t+\frac{1}{s}\right) + zi\left(1, t+\frac{1}{s}\right) \\ 2\psi(st) &= v\left(1, t-\frac{1}{s}\right) - zi\left(1, t-\frac{1}{s}\right) \end{aligned}\right\} \tag{1.6.45}$$

by putting $x = 1$ and $t = t - 1/s$ in (1.6.44). Now setting $x = 0$ in (1.6.44) and $t = t - 1/s$ in both (1.6.44) and (1.6.45), the boundary condition (1.6.43a) gives

$$i(1,t) - Ki\left(1, t - \frac{2}{s}\right) = \alpha - \frac{1}{z}v(1,t) - \frac{K}{z}v\left(1, t-\frac{2}{s}\right)$$

where

$$K = \frac{z-R}{z+R}, \quad \alpha = \frac{2E}{z+R}$$

Now applying (1.6.43b), we obtain the equation

$$\boxed{\dot{u}(t) - K\dot{u}\left(t - \frac{2}{s}\right) = f\left(u(t), u\left(t - \frac{2}{s}\right)\right)} \tag{1.6.46}$$

where

$$u(t) = v(1,t)$$

and

$$f(u(t), u(t-r)) = \frac{\alpha}{C_1} - \frac{1}{zC_1}u(t) - \frac{K}{C_1 z}u(t-r) - \frac{1}{C_1}g(u(t)) + \frac{K}{C_1}g(u(t-r))$$

Hence we can obtain a non-linear differential delay equation for the voltage at the end of the delay line.

The two examples above are special cases of the general equation

$$\boxed{\dot{x}(t) = f(x(t), \dot{x}(t-\delta), x(t-\delta), t)} \tag{1.6.47}$$

where $x(t) \in \mathbb{R}^n$ for $t \geqslant 0$, and f is some function defined on $\mathbb{R}^{4n+1}$. It is natural to ask what kind of initial data is required to solve (1.6.47). This can be seen easily by considering the simple scalar delay equation

$$\dot{x}(t) = x(t-\delta), \quad t \geqslant 0$$

Integrating this we have

$$x(t) = x(0) + \int_0^t x(s-\delta)\, ds$$

In order to determine $x(t)$ for $t > 0$ it is clear that we require values for $x(t)$ on the interval $[-\delta, 0)$. However, this data is clearly sufficient (together with $x(0)$) to determine $x(t)$. In fact, we can evaluate x on the interval $[0, \delta)$ and then on the interval $[\delta, 2\delta)$ etc. Note that it is not necessary that

$$x(0) = \lim_{t \to 0^-} x(t)$$

Consider again the general equation (1.6.47). As before, we can suppose that the initial data which is specified is $\boldsymbol{x}(0)$ and $\boldsymbol{x}(t) = \boldsymbol{\phi}(t+\delta)$, $t \in [-\delta, 0)$ for some vector function $\boldsymbol{\phi}(t)$ defined on $[0, \delta)$. Hence, on the interval $[0, \delta)$, (1.6.47) may be written

$$\begin{aligned} \dot{\boldsymbol{x}}(t) &= f(\boldsymbol{x}(t), \dot{\boldsymbol{\phi}}(t), \boldsymbol{\phi}(t), t) \\ &= f_1(\boldsymbol{x}(t), t) \quad t \in [0, \delta) \end{aligned} \tag{1.6.48}$$

for some new function f_1. This ordinary differential equation can be solved (numerically) using the initial condition $\boldsymbol{x}(0) = \boldsymbol{x}_0$. Similarly, on the interval $[\delta, 2\delta)$ we have

$$\begin{aligned} \dot{\boldsymbol{x}}(t) &= f(\boldsymbol{x}(t), \dot{\phi}_1(t), \phi_1(t), t) \\ &= f_2(\boldsymbol{x}(t), t) \quad t \in [\delta, 2\delta) \end{aligned}$$

where $\boldsymbol{\phi}_1(t)$ is the solution of (1.6.48) which was just determined. This can again be solved using the initial condition

$$\boldsymbol{x}(\delta) = \boldsymbol{x}(0) + \int_0^\delta f_1(\boldsymbol{x}(s), s)\, \mathrm{d}s$$

The solution now proceeds in this way over the whole semi-axis $[-\delta, \infty)$.

1.7 STOCHASTIC SYSTEMS

1.7.1 Noise

All the systems which we have considered so far have been completely deterministic; that is, they are defined by various types of differential equations which, when supplied with the appropriate initial or boundary conditions, determine (at least theoretically) the future values of the system variables. However, as we said at the beginning of this chapter, any given real system will not agree exactly (in terms of the input–output structure) with any model, no matter how good the model is designed to be. Such inaccuracies can be regarded as coming from basically two sources; firstly, we may be forced to accept modelling error simply because we are not clever enough or do not have enough time to derive a model. For example, in deriving the gyroscope equations in this chapter it was implicitly assumed that the rotating disk of the gyroscope is perfectly circular and of uniform thickness. Any real disk, of course, is subject to small irregularities, no matter how precisely they are machined. We could spend more time measuring the disk with a microscope and deriving more accurate values for the moments of inertia (the principal axes of which will now not be along the obvious axes of the gyroscope). However, the improvement to be gained in the model will be so small as to make such a lengthy exercise almost futile. From these considerations we see that this type of uncertainty in the model is introduced at the point when we decide that our modelling effort is sufficient and any further attempts at a better model are subject to the law of diminishing returns.

The second type of uncertainty in our system models is due almost entirely to the random thermal motion of particles making up the system. This uncertainty is inevitable and cannot be removed, no matter how much time we are prepared to spend in the modelling exercise. Instead, we must use some kind of statistical model for the system. In most types of mechanical system this 'noise' is negligible, but in electronic systems it can be a serious problem and many ways of reducing the noise have been developed. Some of these will be discussed later.

Although the two types of modelling error discussed above appear to be somewhat different in character, they are normally regarded as being some kind of random variable (appendix 2), impressed on the system. Since the random variables used to model the noise may vary with time we are led to consider stochastic processes and differential equations containing noise terms.

1.7.2 Stochastic Processes

In order to define a stochastic process, recall that a random variable x is a real-valued function defined on a probability space Ω (see appendix 2). In any experiment involving x, the value of x cannot be determined a priori no matter how carefully the experiment is set up. Only the statistical moments can be known in advance, although even these may not be known precisely and may have to be modelled in some way. These statistical moments are, of course, the mean, variance, etc. defined by

$$\mu_x = \int_{-\infty}^{\infty} x f(x)\,\mathrm{d}x$$

$$v_x = \int_{-\infty}^{\infty} (x-\mu_x)^2 f(x)\,\mathrm{d}x$$

where $f(x)$ is the probability distribution of x.

A **stochastic process** is a vector-valued map $\boldsymbol{x}$: $\mathbb{R}^+ x\Omega \rightarrow \mathbb{R}^n$ each component of which is a random variable for each fixed t; i.e. $x_i(t, \cdot): \Omega \rightarrow R$ is a random variable. We have seen that any deterministic system which is finite-dimensional may be represented in the form (1.5.3), i.e.

$$\dot{\boldsymbol{x}} = \boldsymbol{f}(\boldsymbol{x}(t), \boldsymbol{u}(t), t)$$

If noise processes are present in the system then we may extend this model to include a stochastic process $\boldsymbol{\nu}(t)$; then we have the equation

$$\boxed{\dot{\boldsymbol{x}} = \boldsymbol{f}(\boldsymbol{x}(t), \boldsymbol{u}(t), \boldsymbol{\nu}(t, \omega), t)} \tag{1.7.1}$$

Any solution of this equation is a stochastic process $\boldsymbol{x}(t, \omega)$. However, we normally consider specific 'realizations' or **sample paths** of the noise process $\boldsymbol{\nu}(t, \omega)$, i.e. if ω: $\mathbb{R}^+ \rightarrow \Omega$ is a given map, then $\tilde{\boldsymbol{\nu}}(t, \omega(t))$ is a given value of the random variable $\boldsymbol{\nu}(t, \cdot)$ or each t. In this case the equation

$$\dot{\boldsymbol{x}} = \boldsymbol{f}(\boldsymbol{x}(t), \boldsymbol{u}(t), \tilde{\boldsymbol{\nu}}(t), t) \tag{1.7.2}$$

is deterministic and we can solve this equation for $\boldsymbol{x}(t) = \boldsymbol{x}(t, \tilde{\boldsymbol{\nu}}(t))$. By solving (1.7.2) for many different sample paths we can evaluate the moments of $\boldsymbol{x}(t, \omega)$. This procedure is called the Monte Carlo method.

In linear systems of the form (1.5.6) noise may appear additively in the form

$$\dot{\boldsymbol{x}} = A\boldsymbol{x}(t) + B\boldsymbol{u}(t) + \boldsymbol{\nu}(t)$$

or the elements of the matrices A and B may be uncertain and can be modelled as stochastic processes. The problem of obtaining values for these elements is called parameter estimation and will be studied later.

1.8 SYSTEM ELEMENTS AND SIGNAL FLOW GRAPHS

1.8.1 Trans- and Per-variables

From the preceding discussion, it can be seen that many types of system are composed of a number of **lumped system elements** which are interconnected in a variety of ways to produce the **system network**. Examples of system elements are resistors, capacitors, inductances, springs, dashpots and hydraulic reservoirs. The variables which characterize the behaviour of these elements may be of one of two types – **trans**- (or through) and **per**- (or across) variables. The names arise from the fact that, in order to measure a trans- (or per-) variable of an element, we must place a measuring device (meter) in series (respectively, in parallel) with the element. For example, a resistor has current as a trans-variable and voltage as a per-variable. The relation between trans- and per-variables is shown in table

Table 1.1

Element type	*Example*	*Trans-*		*Per-*
Inductive	Inductor	$\frac{di(t)}{dt}$	related to	$v(t)$
Dissipative	Resistor	$i(t)$	related to	$v(t)$
Capacitive	Capacitor	$i(t)$	related to	$\frac{dv(t)}{dt}$

1.1; note that it is standard to use the electrical analogy for the element types. (See also, MacFarlane, 1980; Wellstead, 1979; Shearer *et al.*, 1967; Seely, 1964.)

1.8.2 System Block Diagrams

In many cases, the equations describing a lumped parameter system may be represented in the form of a block diagram. Consider, for example, the simple RC circuit shown in fig. 1.29(a). Using the the familiar phasor notation, we have

$$V_o = V_i - V = iR = VCj\omega R = (V_i - V_o)Cj\omega R$$

or

$$\frac{V_o}{V_i} = \frac{j\omega RC}{1 + j\omega RC} \tag{1.8.1}$$

The ratio V_o/V_i is called the **transfer function** $T(j\omega)$ of the system,† and hence we write

$$T(j\omega) = \frac{j\omega RC}{1 + j\omega RC}$$

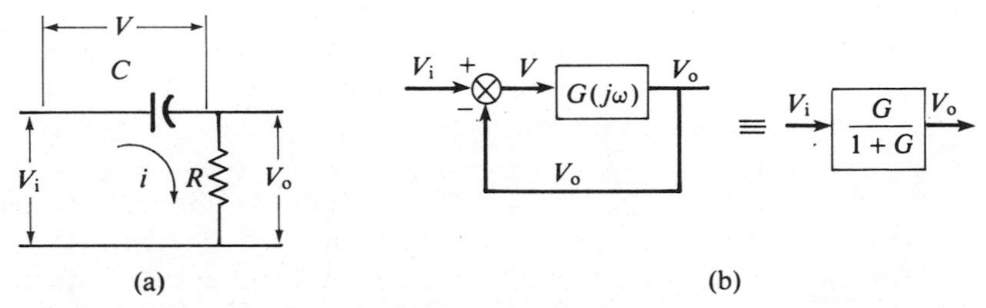

Fig. 1.29 A simple RC circuit (a) and its block diagram representation (b).

†We usually replace $j\omega$ by s – see Part 2 (chapter 4).

Now consider the block diagram in fig. 1.29(b). The circle containing a cross is a summing node and relates the output V to the inputs V_i and V_o, by

$$V = V_i - V_o \tag{1.8.2}$$

Also, the box containing $G(j\omega)$ is a 'generalized gain' and merely represents the operation of multiplication by $G(j\omega)$, i.e.

$$V_0 = G(j\omega)V \tag{1.8.3}$$

and so, by (1.8.2),

$$V_o = G(j\omega)V = G(j\omega)(V_i - V_o)$$

Hence

$$\frac{V_o}{V_i} = \frac{G(j\omega)}{1 + G(j\omega)}$$

and so the systems in figs. 1.29(a) and (b) are equivalent if

$$G(j\omega) = j\omega \mathrm{RC}$$

A more complex system is shown in fig. 1.30 (a). By splitting the system into two loops containing R_1, R_2 and C, R_3 respectively, it is easy to see that the system has the equivalent block diagram representation shown in fig. 1.30(b). Obtaining $V_o,/V_i$ is now more difficult, but this can be achieved by

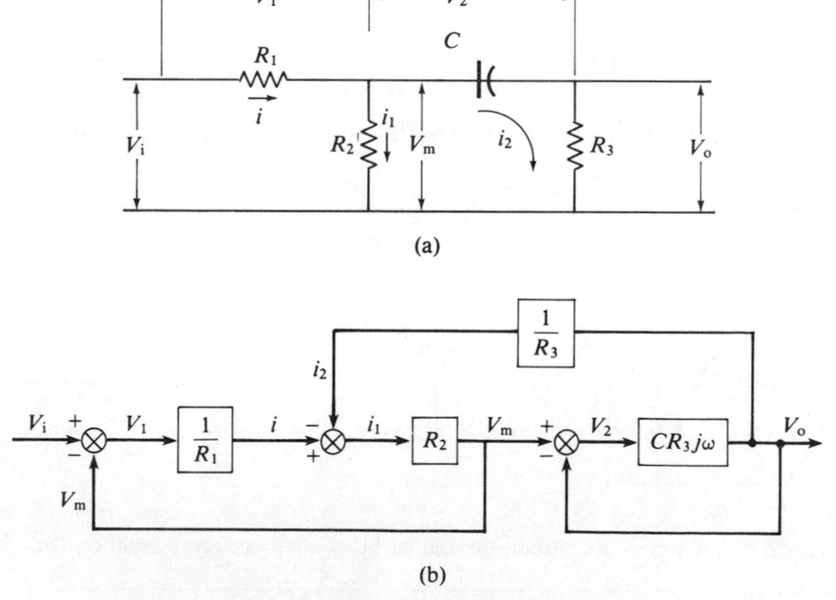

Fig. 1.30 A two stage RC circuit (a) and block diagram (b).

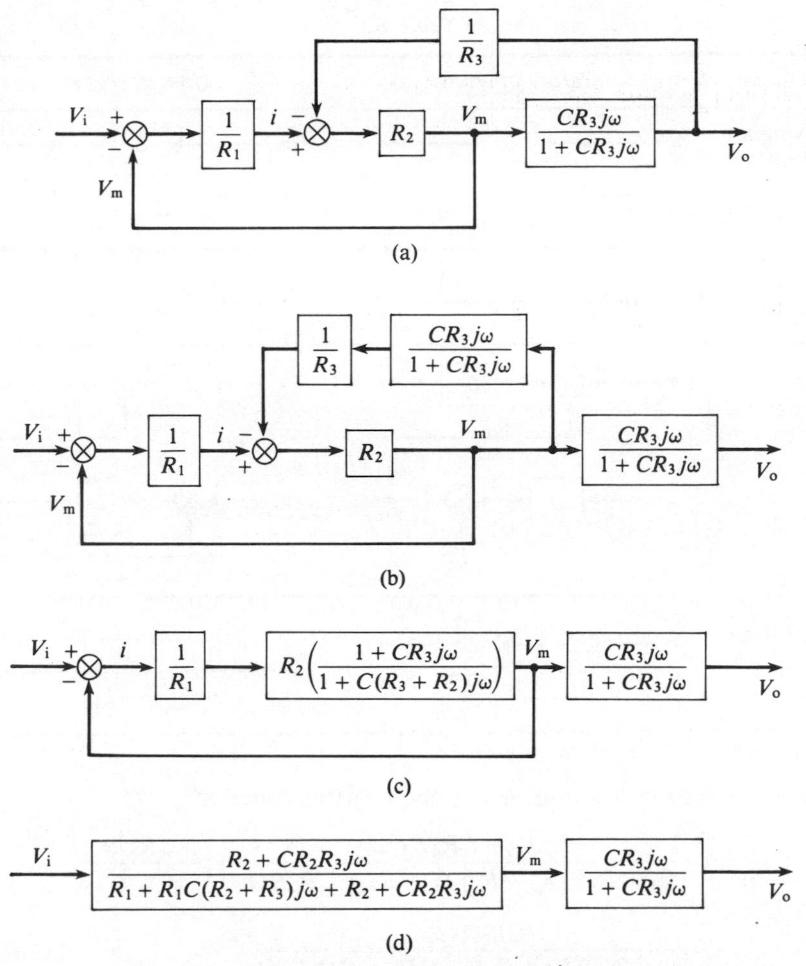

Fig. 1.31 Block diagram reduction of fig. 1.30(b).

performing a sequence of simple operations on the block diagram as shown in fig. 1.31. First remove the feedback of V_o to V_m, just as in fig. 1.30(b), and then move the upper feedback loop from V_o to V_m. Consider next the feedback loop from i to V_m. We have

$$R_2\left(i - \frac{Cj\omega V_m}{1 + CR_3j\omega}\right) = V_m$$

and so

$$V_m = R_2\left(\frac{1 + CR_3j\omega}{1 + (CR_3 + CR_2)j\omega}\right) i$$

This loop can now be removed as in fig. 1.31(c), and then simplifying the

Table 1.2

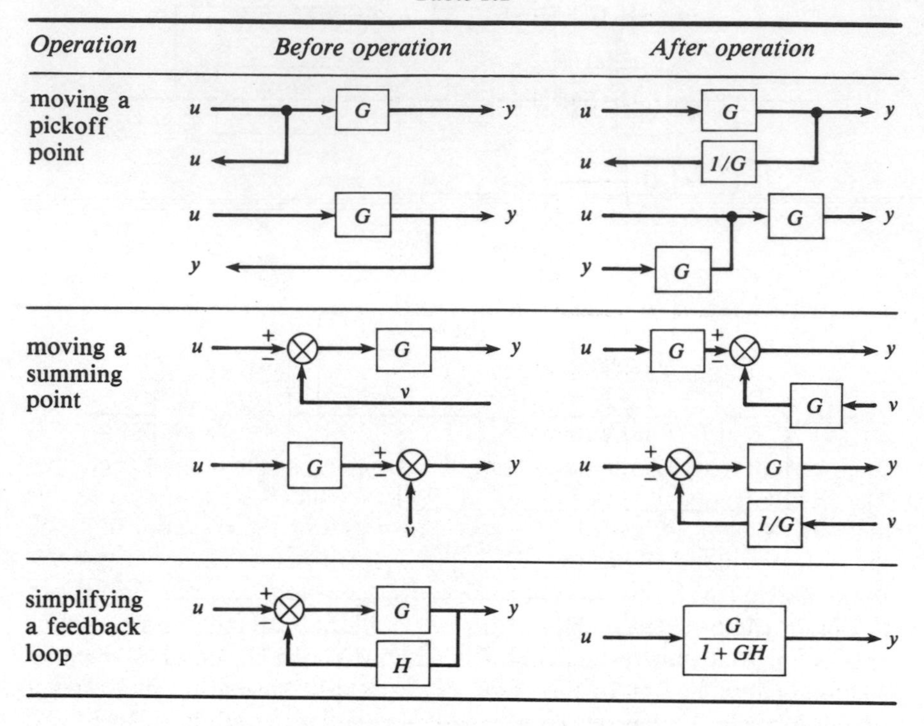

remaining feedback loop leads to the transfer function

$$\frac{V_o}{V_i} = \left(\frac{R_2 + CR_2R_3j\omega}{R_1 + R_1C(R_2 + R_3)j\omega + R_2 + CR_2R_3j\omega}\right)\left(\frac{CR_3j\omega}{1 + CR_3j\omega}\right)$$

$$= \frac{CR_2R_3j\omega}{R_1 + R_1C(R_2 + R_3)j\omega + R_2 + CR_2R_3j\omega} \tag{1.8.4}$$

as shown in fig. 1.31(d).

The simplifications of the block diagram used above are summarized in table 1.2, and are sufficient to obtain the transfer function of any block-structured system.

1.8.3 Signal-flow Graphs

An alterntive representation of the block diagram of a system can be found by the use of signal-flow graphs. For a simple block of the form shown in fig. 1.32(a), we draw a directed line with **nodes** at each end, which represent the system input and output, and the system 'gain' G is written along the

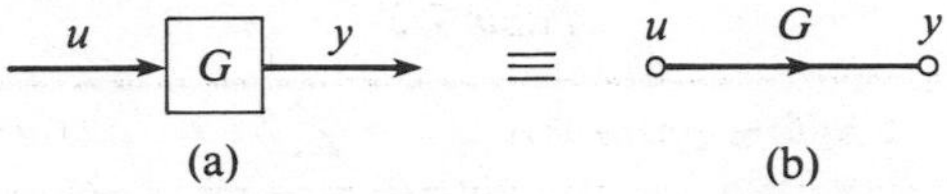

Fig. 1.32 Signal-flow graph of a simple block.

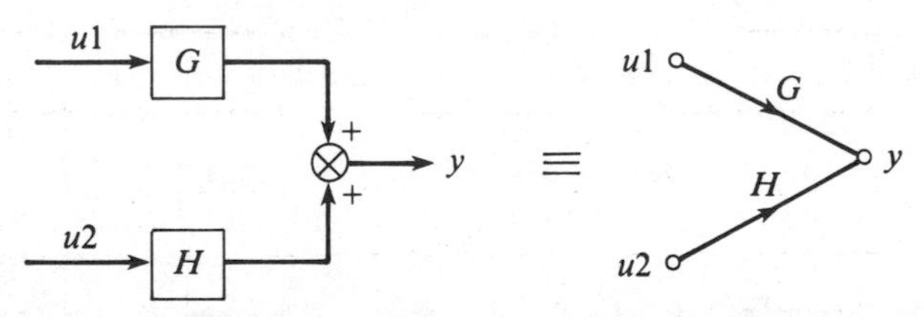

Fig. 1.33 Signal-flow graph of a summer.

line (fig. 1.32(b)). If more than one line is incident on a node, as in fig. 1.33, then we sum together the outputs of the blocks corresponding to each line. The simple feedback system in fig. 1.29(b) can thus be represented by the diagram in fig. 1.34, called the **signal-flow graph** of the system. Similarly, the system in fig. 1.30 has the signal-flow graph representation of fig. 1.35. In the previous section, we applied a sequence of block diagram operations to obtain the transfer function (or 'gain') of the system. For signal-flow graphs, the following result (**Mason's theorem**, Mason (1953, 1956)) enables a similar simplification of the graph. This theorem states that the transfer function of the system represented by a signal-flow graph is given by

$$G = \sum_i G_i \Delta_i / \Delta_i \quad \text{(sum over all forward paths)} \tag{1.8.5}$$

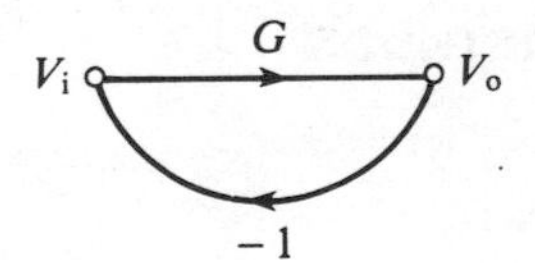

Fig. 1.34 Signal-flow graph of a unity-gain feedback system.

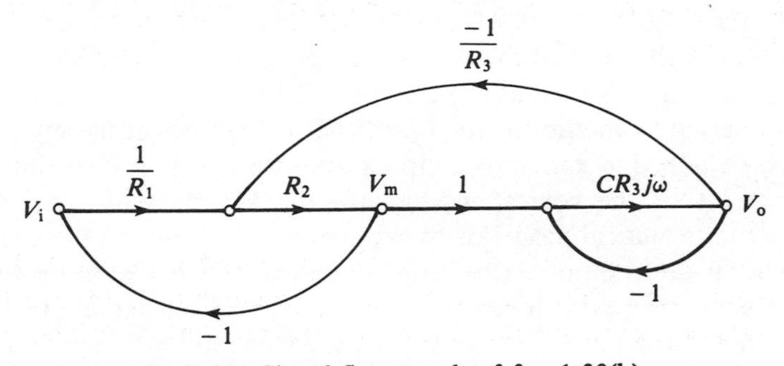

Fig. 1.35 Signal-flow graph of fig. 1.30(b).

where

$$\Delta = 1 - \sum L_1 + \sum L_2 - \sum L_3 + \ldots$$

and L_1 is the gain of each closed loop in the graph, L_2 is the product of the loop gains of two non-touching closed loops (i.e. loops which have no nodes in common), L_m is the product of the loop gains of any m non-touching loops, G_i is the gain of the forward path i and Δ_i is the value of Δ with the forward path i removed.

Note that a **forward path** is any consecutive sequence of directed lines from V_i to V_o which contain any vertex at most once. For the system in fig. 1.35, the application of this result is particularly simple, since there is only one forward path (with the gains $1/R_1$, R_2, 1, $CR_3j\omega$). Hence there is only one sum in (1.8.5) with $i = 1$ and $G_1 = CR_2R_3j\omega/R_1$. As for Δ_1, if we remove the forward path, then there are no loops left; i.e. $\Delta_1 = 1$. Finally, to evaluate Δ, note that there are three loops with gains $-R_2/R_1$, $-R_2Cj\omega$ and $-CR_3j\omega$, and two non-touching loops with gains $-R_2/R_1$ and $-CR_3j\omega$. Hence

$$\Delta = 1 + R_2Cj\omega + \frac{R_2}{R_1} + CR_3j\omega + \frac{R_2}{R_1}CR_3j\omega$$

and so the gain G of the graph is

$$G = \frac{R_2CR_3j\omega}{R_1 + R_1R_2Cj\omega + R_2 + R_1CR_3j\omega + R_2CR_3j\omega}$$

which is the same as the transfer function obtained in (1.8.4).

1.9 IDENTIFICATION METHODS

1.9.1 Introduction

All the methods of system modelling discussed so far involve the application of classical methods of analysis to the component parts of the system, resulting (usually) in a set of differential equations which describe the input–output behaviour of the system. Hence, given an input we can determine (computationally) what the output will be.

An alternative method is to regard the real system as merely a 'black box' into which one can inject inputs and from which one can obtain measurements of the corresponding outputs. We then use this data to specify a mathematical model of the system which produces the same outputs from the given inputs. The main advantage of this method is that one does not have to become involved in the physics of the actual system, but it does mean that a particular system structure has to be assumed, a priori. In most currently available methods of identification, the model has to be linear, although various methods of non-linear identification also exist (Billings, 1980). We shall discuss some elementary methods for linear models here – for a more complete description of the methods, see Graupe (1976) and also chapter 7, where stochastic and least squares estimation are discussed.

1.9.2 Impulse and Step Response Methods

The basic idea of identification of a linear (single-input, single-output) system S (which is defined by a differential equation) is that any such system is specified (as we shall see in chapter 4) by a single function of time $g(t)$ and that its output $y(t)$ for any given input $u(t)$ is given by

$$y(t) = \int_0^t g(\tau)u(t-\tau)\,\mathrm{d}\tau$$

(where we assume $u(t) = 0, t < 0$). The three most common types of input u which are used to test the system are:

(a) The unit impulse (δ-function).
(b) The unit step.
(c) Sinusoids.

In the case of an impulse, we have $u(t) = \delta(t)$ (the Dirac δ-function) where δ is 'defined' (somewhat loosely) as the limit of the functions

$$\delta_\varepsilon(t) = \begin{cases} \dfrac{1}{\varepsilon}, & t \in [0, \varepsilon] \\ 0, & t \notin [0, \varepsilon] \end{cases}$$

as $\varepsilon \to 0$. Note that

$$\int_0^\infty \delta_\varepsilon(t)\,\mathrm{d}t = 1, \quad \text{for any } \varepsilon > 0$$

and, using the inputs δ_ε, we have

$$\begin{aligned} y_\varepsilon(t) &= \int_0^t g(\tau)\,\delta_\varepsilon(t-\tau)\,\mathrm{d}\tau \\ &\approx g(t), \quad t \geqslant \varepsilon \end{aligned}$$

In the limit (assuming this is justified)

$$\begin{aligned} y(t) &= \int_0^t g(\tau)\,\delta(t-\tau)\,\mathrm{d}\tau \\ &= g(t), \quad t \geqslant 0 \end{aligned}$$

Hence, the system function g is the output from a δ-function input. Of course, we cannot produce a δ-function in reality and so we test the sytem with a rectangular pulse δ_ε for small ε. As we have seen, this will give a good approximation to g.

Using a (unit) step function input, defined by

$$u_s(t) = \begin{cases} 1, & t \geqslant 0 \\ 0, & t < 0 \end{cases}$$

we have

$$\begin{aligned} y(t) &= \int_0^t g(\tau) u_s(t-\tau)\, d\tau \\ &= \int_0^t g(\tau)\, d\tau \end{aligned}$$

and so

$$\frac{dy}{dt}(t) = g(t)$$

Hence the time derivative of the output resulting from a step input equals the system function g. Note, however, that for practical systems which are 'noisy', the process of differentiation tends to increase the noise, and so this method is not always feasible.

1.9.3 The Frequency Response Method

The final type of input with which we test linear systems is the sinusoid $\sin \omega t$ of all frequencies from 0 to ∞, although, of course, we can only consider a range (say 0.1 Hz to 20 kHz) in practice. It can be shown (chapter 4) that the output of a linear system to a sinusoidal input is a sinusoid of the same frequency, with an amplitude change and a phase shift. Hence, if $u_s(t) = \sin \omega t$, we can write

$$y_s(t) = A(\omega) \sin(\omega t + \theta(\omega))$$

for some functions A and θ of ω. Similarly, if $u_c(t) = \cos \omega t$,

$$y_c(t) = A(\omega) \cos(\omega t + \theta(\omega))$$

for the same functions A, θ. It is often convenient to combine these two inputs in the form

$$u_e(t) = \cos \omega t + j \sin \omega t$$

Then we clearly have, for the 'output' y_e† from the input u_e

$$\begin{aligned} y_e(t) &= A(\omega)\{\cos(\omega t + \theta(\omega)) + j\sin(\omega t + \theta(\omega))\} \\ &= A(\omega)\exp(j\theta(\omega))\exp(j\omega t) \end{aligned}$$

Since we also have

$$\begin{aligned} y_e(t) &= \int_0^\infty g(\tau)\exp(j\omega(t-\tau))\,\mathrm{d}\tau \\ &= \exp(j\omega t)\int_0^\infty g(\tau)\exp(-j\omega\tau)\,\mathrm{d}\tau \end{aligned}$$

it follows that

$$A(\omega)\exp(j\theta(\omega)) = \int_0^\infty g(\tau)\exp(-j\omega\tau)\,\mathrm{d}\tau$$

However, the right-hand side is the Fourier transform of $2\pi g$ and so (appendix 1)

$$g(t) = \frac{1}{2\pi}\int_0^\infty A(\omega)\exp(j\theta(\omega))\exp(j\omega t)\,\mathrm{d}\omega$$

Hence, to determine the system function g, we can test the system with sinusoidal inputs (over some frequency range) and take the inverse Fourier tansform of $A(\omega)\exp(j\theta(\omega))$. Frequency response analysis, as it is called, can be done practically using a sweep oscillator and a fast Fourier transform algorithm (Graupe, 1976).

1.10 SENSITIVITY THEORY

1.10.1 Introduction

When modelling a physical system we often obtain a differential equation of the form

$$\begin{aligned} \dot{x} &= f(x, t; p), \quad x \in \mathbb{R}^n, p \in \mathbb{R}^m \\ x(t_0) &= x_0 \end{aligned} \tag{1.10.1}$$

where p is a set of parameters, which may change with time, or for which

†This is a theoretical output, since we clearly cannot input a complex signal to a real system.

we have only approximate values. For example, in the linear model

$$\dot{x} = Ax \tag{1.10.2}$$

we may regard the elements a_{ij} of the A matrix as parameters, which may be slowly varying in the real system, even though our model assumes constant values of the a_{ij}.

It is important, therefore, to have some means of evaluating the effects of parameter variations on the system model (and hence on any controller which may be designed on the basis of the model). This is the purpose of sensitivity theory, and we shall give a brief outline below. A readable account of the general theory is given by Tomović and Vukobratović (1972).

1.10.2 Sensitivity Functions

The sensitivity of the solution $x(t;\, p)$ of (1.10.1), to variations in the parameters p, can be calculated from the matrix

$$S = \left[\frac{\partial x}{\partial p}\right] \triangleq \left[\frac{\partial x_i}{\partial p_j}\right], \quad 1 \leqslant i \leqslant n,\ 1 \leqslant j \leqslant m$$

which satisfies the equation

$$\begin{aligned} \dot{\delta} &= \frac{\partial f}{\partial x}\frac{\partial x}{\partial p} + \frac{\partial f}{\partial p} \\ &= \frac{\partial f}{\partial x} S + \frac{\partial f}{\partial p} \end{aligned} \tag{1.10.3}$$

with $S(t_0) = 0$, provided the parameter vector p does not contain the initial condition x_0. Note that the matrices

$$\phi = \frac{\partial f}{\partial x}, \quad \psi = \frac{\partial f}{\partial p}$$

are evaluated at the solution $x(t;\, p)$ for a nominal parameter p. Hence, for a given parameter p, knowing the solution $x(t;\, p)$ of (1.10.1), we can solve (1.10.3) (numerically) for the sensitivity matrix $S = \partial x/\partial p$, giving a measure of the variation of the solution with parameter variations.

1.10.3 Eigenvalue Sensitivity

As an example, we can consider the sensitivity of the eigenvalues of a matrix A, which will indicate the variation of the solutions of the linear system (1.10.2) to parameter changes. If the eigenvalues of A are distinct (which we assume for simplicity), then we have

$$Ax_i = \lambda_i x_i, \quad 1 \leqslant i \leqslant n$$

for n independent eigenvectors x_i. Also,

$$A^{\mathrm{T}} v_j = \lambda_j v_j$$

for the eigenvectors v_j of the transposed matrix. Hence, if the elements a_{ij} of A depend on a parameter vector p, then

$$\frac{\partial A}{\partial p} x_i + A \frac{\partial x_i}{\partial p} = \lambda_i \frac{\partial x_i}{\partial p} + \frac{\partial \lambda}{\partial p} x_i$$

and so

$$\left[\frac{\partial A}{\partial p} x_i\right]^{\mathrm{T}} V_i + \left[A \frac{\partial x_i}{\partial p}\right]^{\mathrm{T}} V_i = \lambda_i \left(\frac{\partial x_i}{\partial p}\right)^{\mathrm{T}} v_i + \frac{\partial \lambda_i}{\partial p} x_i^{\mathrm{T}} V_i$$

$$= \left(\frac{\partial x_i}{\partial p}\right)^{\mathrm{T}} A^{\mathrm{T}} v_i + \frac{\partial \lambda_i}{\partial p} x_i^{\mathrm{T}} V_i$$

Thus, say,

$$\frac{\partial \lambda_i}{\partial p} = \frac{\left[\dfrac{\partial A}{\partial p} x_i\right]^{\mathrm{T}}}{x_i^{\mathrm{T}} v_i} V_i = S_i$$

and so, knowing a nominal value of p, given a change $\mathrm{d}p$, we can find the corresponding change in λ_i from

$$\mathrm{d}\lambda_i = S_i \, \mathrm{d}p$$

1.11 EXERCISES

1. Prove equations (1.2.2).

2. Using (1.2.7), show that the kinetic energy

$$T = \frac{1}{2} \sum_{i=1}^{p} m_i v_i^2$$

in generalized coordinates is

$$T = \sum_{k=1}^{n} \sum_{l=1}^{n} \left[\frac{1}{2} \sum_{i=1}^{p} m_i (a_{ik} a_{il} + b_{ik} b_{il} + c_{ik} c_{il})\right] \dot{q}_k \dot{q}_l$$

$$+ \sum_{k=1}^{n} \left[\sum_{i=1}^{p} m_i (\alpha_i a_{ik} + \beta_i b_{ik} + \gamma_i c_{ik})\right] \dot{q}_k$$

$$+ \frac{1}{2} \sum_{i=1}^{p} m_i (\alpha_i^2 + \beta_i^2 + \gamma_i^2)$$

where

$$\dot{x}_1 = \sum_{k=1}^{n} a_{ik}\dot{q}_k + \alpha_i$$

$$\dot{y}_i = \sum_{k=1}^{n} b_{ik}\dot{q}_k + \beta_i$$

$$\dot{z}_i = \sum_{k=1}^{n} c_{ik}\dot{q}_k + \gamma_i$$

3. Prove the relation (1.2.20).

4. Prove the direction cosine relations in (1.2.29). *Hint:* consider a general rotation matrix as a product of three matrices of the form

$$\begin{pmatrix} 1 & 0 & 0 \\ 0 & \cos\psi & \sin\psi \\ 0 & -\sin\psi & \cos\psi \end{pmatrix}$$

5. Prove (1.2.34).

6. Write down the kinetic and potential energies for the double pendulum in fig. 1.36 and apply Lagrange's equations of motion to obtain the equation of motion of the system. If the strings are replaced by springs with constants k_1 and k_2, write down the kinetic energy, determine the generalized forces on the system and find the equations of motion. Show that the same result is obtained if the potential energy is found instead of the generalized forces.

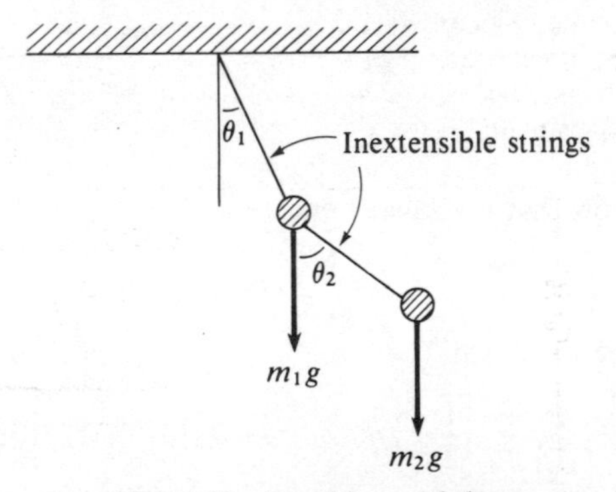

Fig. 1.36 Double pendulum.

7. Determine the equations of motion for the system in fig. 1.37 and discuss the motion of the system by solving these equations. (M is the mass of the disk and I is its moment of inertia.)

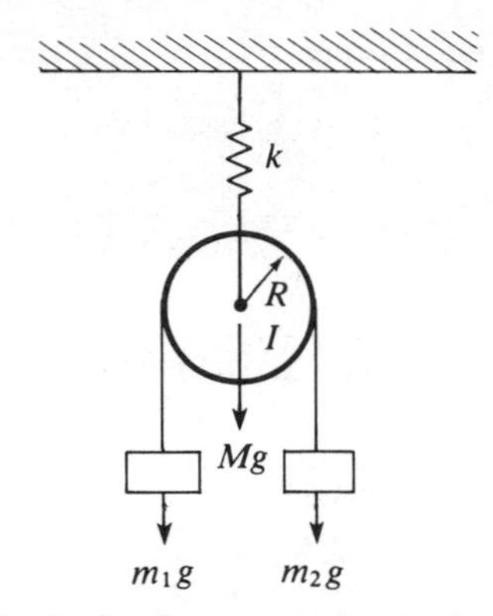

Fig. 1.37 A simple mass-spring-pulley system.

8. Write down the loop equations for the circuit in fig. 1.38 and apply Thévenin's theorem to reduce the system to an equivalent voltage source and impedance.

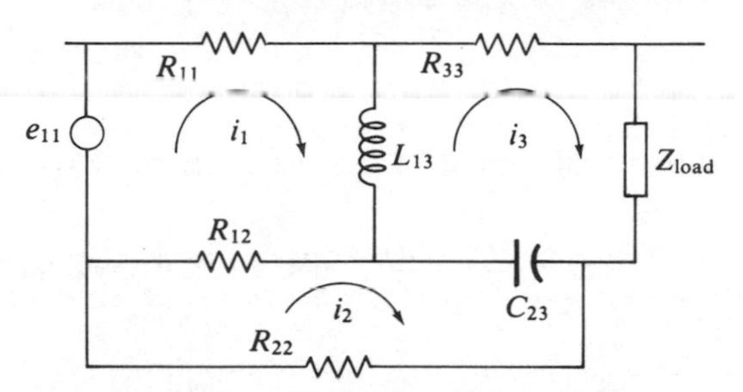

Fig. 1.38 A three loop RLC circuit.

9. A capacitor is formed by connecting two plates of area A to two springs of constants k_1,k_2 fixed at one end, and is placed in the circuit of fig. 1.39. Write out the equations of 'motion' of the system. (The capacitance is the ratio of A and the distance between the plates.)

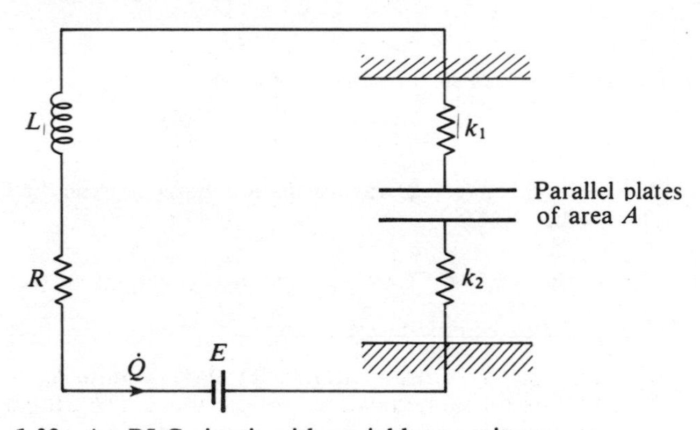

Fig. 1.39 An RLC circuit with variable capacitance.

10. Define the **generalized momentum**

$$p_r = \frac{\partial L}{\partial \dot{q}_r}$$

where L is the Lagrangian of the system. If $F_{q_r}^{\mathrm{d}}$ is the dissipative generalized force, show that

$$d\left[\sum_{r=1}^{n} p_r \dot{q}_r - L\right] = \sum_{r=1}^{n} [(F_{q_r}^{\mathrm{d}} - \dot{p}_r)\,\mathrm{d}q_r + \dot{q}_r\,\mathrm{d}p_r] - \frac{\partial L}{\partial t}\,\mathrm{d}t$$

and hence if

$$H \triangleq \sum_{r=1}^{n} p_r \dot{q}_r - L$$

derive Hamilton's canonical equations

$$\frac{\partial H}{\partial p_r} = \dot{q}_r$$

$$\frac{\partial H}{\partial q_r} = F_{q_r}^{\mathrm{d}} - \dot{p}_r$$

11. Generalize the results of section 1.6.2 to deal with the wave equation in two dimensions, i.e.

$$\frac{\partial^2 u}{\partial t^2} = c\left(\frac{\partial^2 u}{\partial x^2} + \frac{\partial^2 u}{\partial y^2}\right), \quad 0 \leqslant x \leqslant 1, 0 \leqslant y \leqslant 1$$

12. Consider the heat conduction equation

$$\frac{\partial T}{\partial t} = \frac{k}{\pi c_{\mathrm{p}}} \frac{\partial^2 T}{\partial x^2}$$

on the interval $[0,1]$ with boundary conditions

$$\frac{\partial T}{\partial x}(0) = \frac{\partial T}{\partial x}(1) = 0$$

Discuss this system in the spirit of section 1.6.3.

13. Show that the Joukowski transformation maps between the appropriate regions drawn in fig. 1.26.

14. Solve the delay equation

$$\dot{x}(t) = x(t) - x(t - \delta)$$

with initial data

$$x(0) = 1, \quad x(t) = 0, \quad t \in [-\delta, 0)$$

15. Simplify the system in fig. 1.40 by block diagram operations and by using the signal flow graph. Hence find the transfer function of the system.

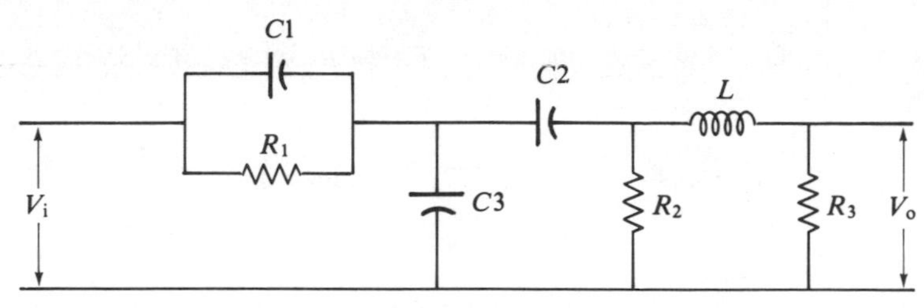

Fig. 1.40 A three stage RLC circuit.

REFERENCES

Banks, S. P. (1983) *State-Space and Frequency Domain Methods in the Control of Distributed Parameter Systems,* Peter Peregrinus.

Billings, S. A. (1980) Identification of nonlinear systems – a survey, *IEE Proc.*, 122, pt. D, 272–85.

Coddington, E. A. and Levinson, N. (1955) *Theory of Ordinary Differential Equations,* McGraw-Hill.

Curtain, R. F. and Pritchard, A. J. (1978) *Infinite Dimensional Linear Systems Theory,* Springer-Verlag.

Davis, H. T. (1962) *Introduction to Nonlinear Differential and Integral Equations,* Dover.

Fitzgerald, A. E. and Kingsley, C. (1971) *Electrical Machinery*, 3rd edn, McGraw-Hill.

Graupe, D. (1976) *Identification of Systems*, 2nd edn, Krieger.

Guillemin, E. A. (1957) *Synthesis of Passive Networks,* Wiley.

Hale, J. (1977) *The Theory of Functional Differential Equations, Applied Mathematical Sciences*, vol. 3, Springer-Verlag.

Kraus, J. D. and Carver, K. R. (1973) *Electromagnetics*, 2nd edn, McGraw-Hill.

Lions, J-L. (1970) *Optimal Control of Systems Governed by Partial Differential Equations,* Springer-Verlag.

MacFarlane, A. G. J. (1980) *Dynamical System Models,* Harrap.

Mason, S. J. (1953) Feedback theory: some properties of signal flow graphs, *Proc. IRE*, 41, 1144.

Mason, S. J. (1956) Feedback theory: further properties of signal flow graphs, *Proc. IRE,* 44, 920.

McRuer, D., Ashkenas, I. and Graham, D. (1973) *Aircraft Dynamics and Automatic Control*, Princeton University Press.

Milne-Thomson, L. M. (1968) *Theoretical Hydrodynamics,* 5th edn, Macmillan.

Seely, S. (1964) *Dynamic Systems Analysis,* Chapman and Hall.

Shearer, J. L., Murphy, A. T. and Richardson, H. H. (1967) *Introduction to System Dynamics,* Addision-Wesley.

Shinners, S. M. (1978) *Modern Control System Theory and Application,* 2nd edn, Addison-Wesley.

Stephenson, R. J. (1960) *Mechanics and Properties of Matter,* Wiley.
Tomović, R. and Vukobratović, M. (1972) *General Sensitivity Theory*, Elsevier.
Wangersky, P. J. and Cunningham, W. J. (1957) Delays in predator prey equations, *Ecology, 22*, 136–8.
Wellstead, P. E. (1979) *Introduction to Physical System Modelling*, Academic Press.

2 ANALOGUE SIMULATION

2.1 INTRODUCTION

In the last chapter we introduced some of the important methods for modelling physical systems and showed that, in many cases, the system may be represented by some form of differential equation. On the assumption that the model is accurate in the sense that it produces good approximations to the system outputs for a wide variety of inputs, it is clear that we may use the model to test the effect of various inputs to the real system by observing the outputs of the model obtained by integrating the model equations. This is desirable for many reasons, but mainly because tests on the real system may be costly or impractical and, although we may have used good modelling techniques, any model should be validated against the real system by performing input–output checks.

The method of integrating the model equations which we shall consider in this chapter is that of analogue simulation. This is based on the use of high-gain amplifiers with feedback which can be made to operate as summers or integrators. The differential equations of the model may therefore be integrated, with analogue voltages replacing the real system variables. This is in direct contrast to digital simulation, to be discussed in the next chapter, where the real system variables are represented at discrete times by binary numbers in a digital computer. One of the advantages of analogue simulation is that the results can be displayed directly on an oscilloscope or a chart recorder. However, as we shall see, non-linear systems require special circuitry for their analogue implementation. Also, because of the saturating properties of real amplifiers, scaling considerations must be carefully considered. For a modern treatment of electronics, see Horowitz and Hill (1980).

2.2 LINEAR ANALOGUE SYSTEMS

2.2.1 Basic Linear Computer Elements

The basis of most analogue computing devices is the operational amplifier

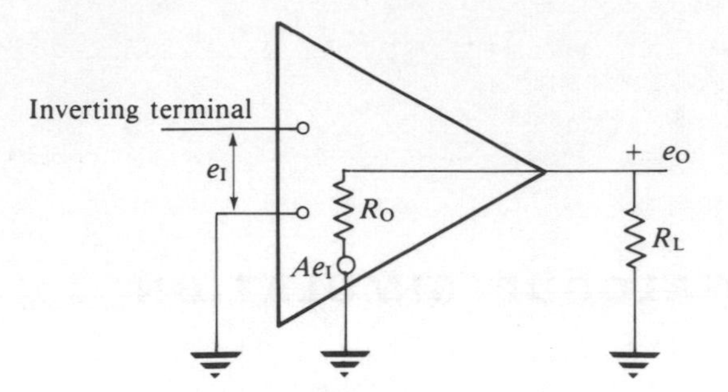

Fig. 2.1 An open-loop op-amp circuit.

(OP-amp), which is a high gain amplifier having very large input impedance and very small output impedance. A reasonable model of an operational amplifier is shown in fig. 2.1, where the input impedance is assumed to be infinity. Note that the input signal is usually applied to the inverting terminal, so that the amplifier gain $A = e_O/e_I$ is negative. R_O represents (for the present) the finite output resistance of the amplifier.

Suppose that we now input to the inverting terminal the total current flowing through a number of resistors as shown in fig. 2.2(a). Then, since the input impedance of the amplifier is infinite, all the input current flows through the feedback resistor R_F,† and so

$$\sum_{k=1}^{n} \left(\frac{e_k - e_I}{R_k} \right) = \frac{e_I - e_O}{R_F}$$

Moreover, we have

$$\frac{Ae_I - e_O}{R_O} + \frac{e_I - e_O}{R_F} = \frac{e_O}{R_L}$$

Hence

$$e_O \left[\left(\frac{1}{R_O} + \frac{1}{R_F} + \frac{1}{R_L} \right) \Big/ \left(\frac{A}{R_O} + \frac{1}{R_F} \right) - \frac{1}{R_F} \Big/ \left(\frac{1}{R_F} + \frac{1}{R_{1,n}} \right) \right]$$
$$= \left(\frac{e_1}{R_1} + \ldots + \frac{e_n}{R_n} \right) \Big/ \left(\frac{1}{R_F} + \frac{1}{R_{1,n}} \right)$$

where $1/R_{1,n} = 1/R_1 + \ldots + 1/R_n$. Thus, if $R_O \to 0$, then

$$e_O = -\left(\frac{e_1}{R_1} + \ldots + \frac{e_n}{R_n} \right) \Big/ \left[\frac{1}{R_F} - \frac{1}{A} \left(\frac{1}{R_F} + \frac{1}{R_{1,n}} \right) \right]$$

† The feedback through R_F effectively holds the inverting terminal at zero volts which is then called a **virtual ground**.

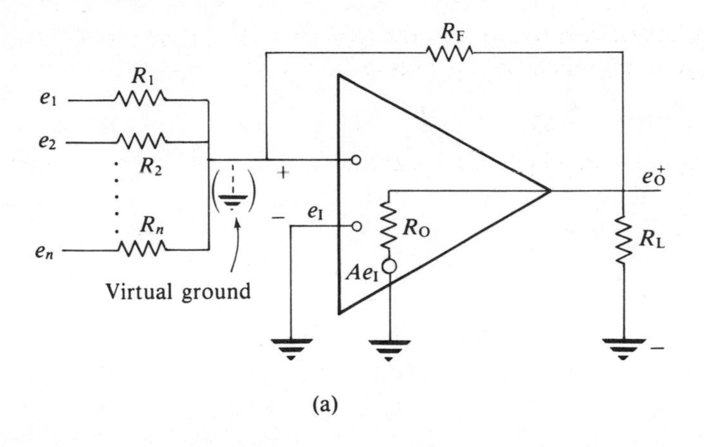

(a)

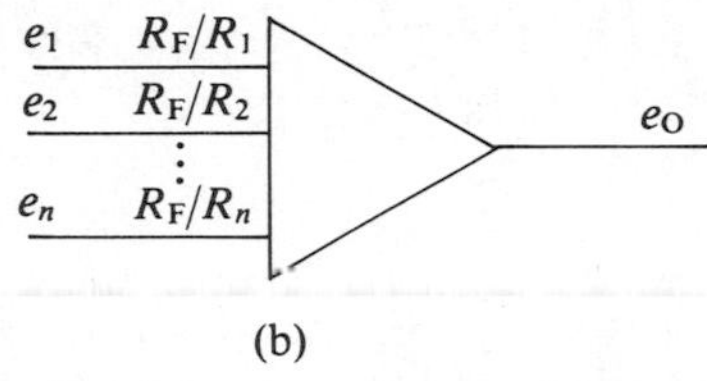

(b)

Fig. 2.2 An op-amp adder.

Moreover, as $A \to -\infty$, we have

$$e_O = -\left(\frac{R_F e_1}{R_1} + \ldots + \frac{R_F e_n}{R_n}\right) \tag{2.2.1}$$

i.e. the output voltage is (minus) a weighted sum of the input voltages $e_1, \ldots, e_n$. The circuit in fig. 2.2(a) is usually symbolized as in fig. 2.2(b).

Consider now the circuit in fig. 2.3(a) where we now assume that the output resistance is zero. Then

$$\sum_{k=1}^{n}\left(\frac{e_k - e_I}{R_k}\right) = C\frac{d}{dt}(e_I - e_O)$$

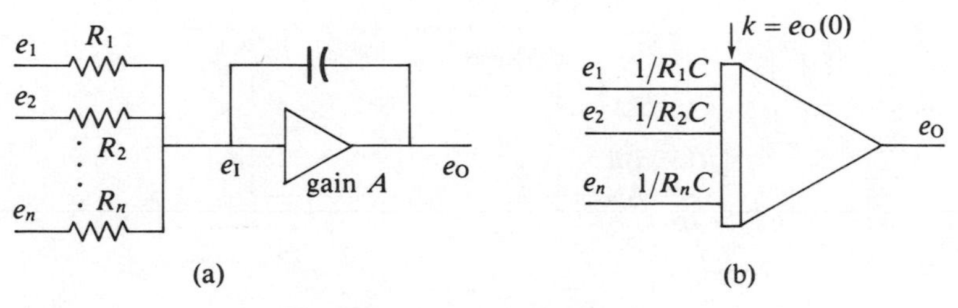

Fig. 2.3 An op-amp integrator.

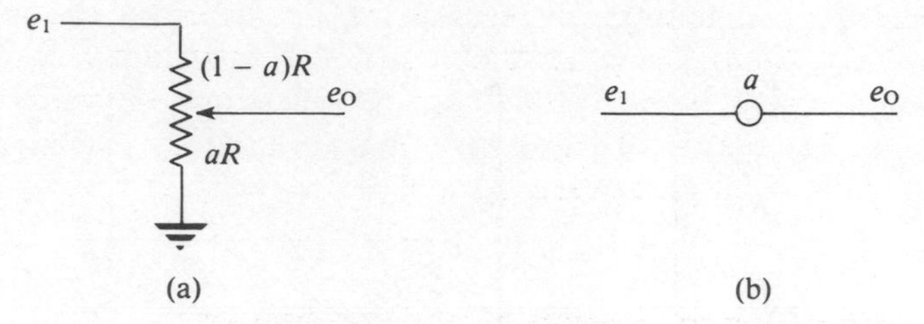

Fig. 2.4 A voltage divider.

and

$$e_O = Ae_I$$

Hence,

$$\sum_{k=1}^{n} \left\{\frac{e_k}{R_k}\right\} = -\left[C - \frac{1}{A}\left(C + \sum_{k=1}^{n} \frac{1}{R_k}\right)\right] \frac{de_O}{dt}$$

If $|A| \to \infty$, then

$$\boxed{e_O = -\frac{1}{R_1 C}\int e_1\,dt - \frac{1}{R_2 C}\int e_2\,dt - \ldots - \frac{1}{R_n C}\int e_n\,dt} \qquad (2.2.2)$$

and so capacitive feedback around the amplifier results in an output which is a weighted sum of the integrals of the inputs. Note that $e_O(0)$ is the initial value of the voltage across the capacitor. The symbolic form of the circuit is shown in fig. 2.3(b).

It is important to be able to multiply analogue voltages by constants, and this is made possible by the simple potentiometer shown in fig. 2.4. In fact, it is clear that

$$\boxed{e_O = ae_I} \qquad a \leqslant 1 \qquad (2.2.3)$$

(The effects of resistive loads are considered in exercises 2.5.1.) Using this

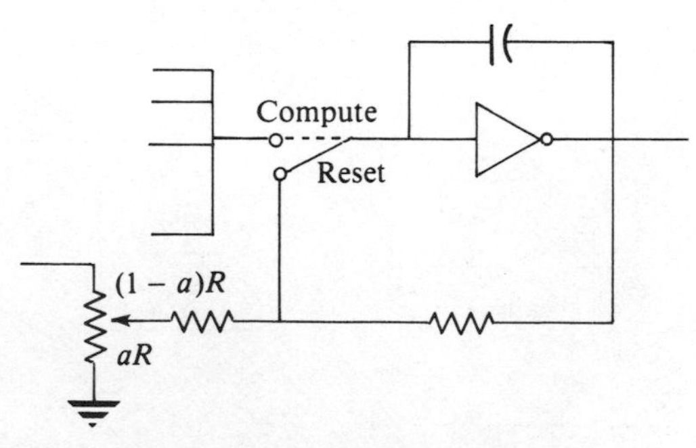

Fig. 2.5 Integrator reset circuit.

multiplier, it is possible to make a circuit for initializing the charge on the capacitor in fig. 2.3(a). Using a two-pole switch as in fig. 2.5 we can switch in the charging circuit before computation and then turn the switch to compute at $t = 0$. The output of the integrator is then set to $-aE$ at the start of computation. Multiplication is symbolized as in fig. 2.4(b).

2.2.2 Computation of Linear Differential Equations

Consider the second-order damped oscillator

$$\frac{d^2x}{dt^2} + 2\zeta\omega_n \frac{dx}{dt} + \omega_n^2 x = f(t) \tag{2.2.4}$$

A possible analogue implementation of this equation is shown in fig. 2.6. On the other hand, if x is not required during the simulation, we can dispense with one summer as in fig. 2.7. Although both of these circuits are theoretically correct, it should be noted that real amplifiers have a limited operational range. Voltages which are too high will drive the amplifier into saturation and outside the linear range. Moreover, the significant voltages in the system should not be so small (for all t) that they are comparable with the amplifier noise. This puts a minimum voltage limitation on the system voltages. Not only are we limited by the real physical parameters in the amplifiers with regard to the voltages, but also in the time scale with respect

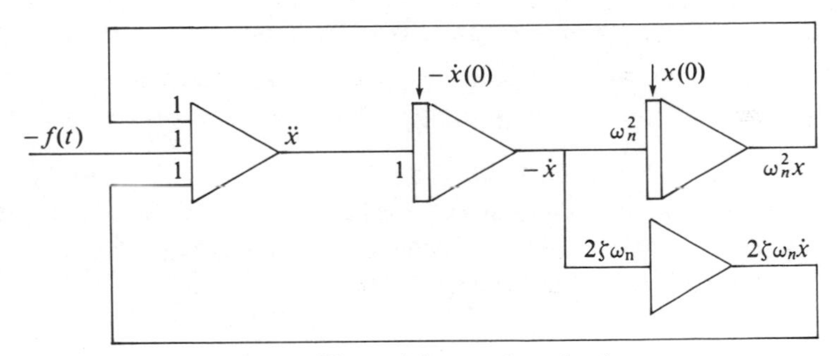

Fig. 2.6 Differential equation simulator.

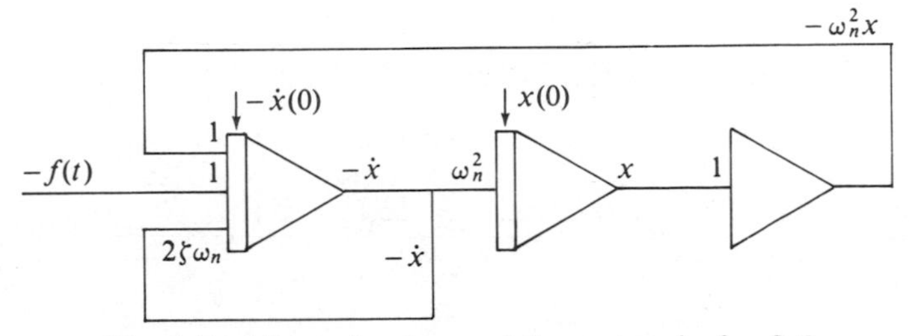

Fig. 2.7 Alternative form of the system in fig. 2.6.

to which we are considering the system. If the characteristic frequencies in the system are very small, then a large time will be required to integrate the equation over a significant number of periods. However, over long periods of time the errors can build up to intolerable levels. On the other hand, for very high frequency solutions, phase shifts in the analogue system components can become significant, thus placing an upper bound on the system frequencies.

The above restrictions can be accommodated by correct choice of the scaling factors in the system (2.2.4). In order to demonstrate the appropriate selection of scale factors consider the linear differential equation

$$35\frac{d^2x}{dt^2} + 15\frac{dx}{dt} + 750x = 600 \tag{2.2.5}$$

and suppose that the maximum output voltage of the computer amplifiers is V_m volts. For the system (2.2.5) the natural frequency is

$$\omega_n = \left(\frac{750}{35}\right)^{1/2}$$

which is well within the frequency range of most amplifiers. Hence no time scaling is necessary for this system. Now the system with output voltages e_1, e_2, e_3 and scale factors α_1, α_2, α_3, α_4 can be represented in the form of fig. 2.8. Suppose that the initial conditions are

$$x(0) = 4, \quad \dot{x}(0) = 0$$

Then $x(t)$ is initially 4 and eventually 600/750 from (2.2.5), since

$$\frac{d^2x}{dt^2} = \frac{dx}{dt} = 0$$

in the steady state. Since $\dot{x}(0) = 0$, the maximum value of x should be 4; however, for safety we often scale for x a little higher, say 4.5. Then, since the solution of (2.2.5) is a damped sinusoid of frequency $(750/35)^{1/2}$,

$$\dot{x}_{max} \approx \left(\frac{750}{35}\right)^{1/2} \times 4.5 = 20.83$$

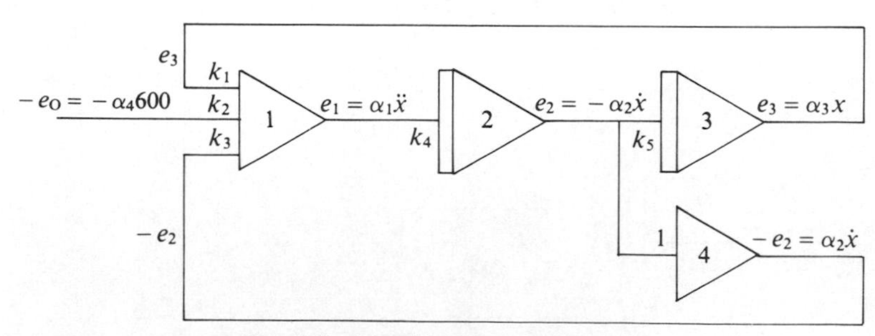

Fig. 2.8 Identifying system gains.

which can be rounded to 21. Similarly

$$\ddot{x}_{\max} \approx \omega_n^2 x_{\max} \leqslant 97$$

Now, recalling that the maximum output voltage of an amplifier is V_m, it follows that

$$\alpha_1 = \frac{V_{\mathrm{m}}}{\ddot{x}_{\max}} = \frac{V_{\mathrm{m}}}{97}$$

$$\alpha_2 = \frac{V_{\mathrm{m}}}{\dot{x}_{\max}} = \frac{V_{\mathrm{m}}}{21}$$

$$\alpha_3 = \frac{V_{\mathrm{m}}}{x_{\max}} = \frac{V_{\mathrm{m}}}{4.5}$$

(see fig. 2.8). However,

$$e_3 = \alpha_3 x = \frac{V_{\mathrm{m}} x}{4.5}$$

and so the input to amplifier 3 is $-V_{\mathrm{m}}\dot{x}/4.5k_5$, which must equal the output of amplifier 2. Hence,

$$\frac{-V_{\mathrm{m}}\dot{x}}{4.5k_5} = -\alpha_2 \dot{x}$$

or

$$k_5 = \frac{21}{4.5} \approx 4.7$$

which is the gain of the third amplifier. Similarly,

$$k_4 = \frac{\alpha_2}{\alpha_1} \approx 4.6$$

Since the output of amplifier 1 is $V_{\mathrm{m}}\ddot{x}/97$, its input must add up to $-V_{\mathrm{m}}\ddot{x}/97$, which is

$$\left(\frac{15}{35}\dot{x} + \frac{750}{35}x - \frac{600}{35}\right)\frac{V_{\mathrm{m}}}{97}$$

from (2.2.5). The term $(750/35)(xV_{\mathrm{m}}/97)$ is available as the output of amplifier 3. To obtain $(15/35)(\dot{x}V_{\mathrm{m}}/97)$ we must multiply the output of amplifier 2 by $-(21 \times 15)/(35 \times 97) \approx 0.093$. Hence we obtain the overall scaled system shown in fig. 2.9. Note that the initial conditions must also be scaled properly.

Returning to fig. 2.8 we see that

$$e_1 = -k_1 e_3 + k_2 e_0 + k_3 e_2$$

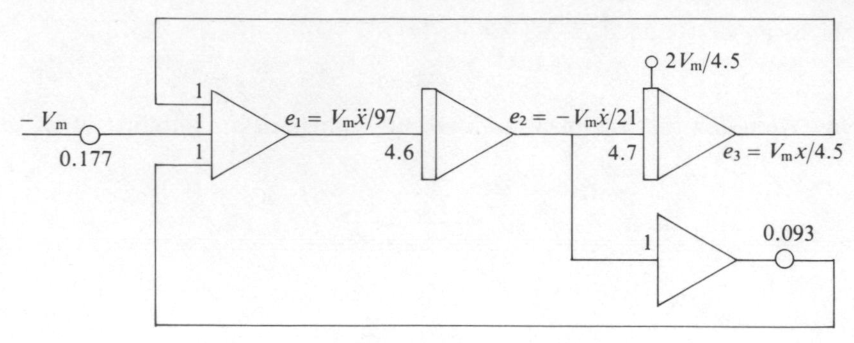

Fig. 2.9 Amplitude scaling of a simulator.

and so

$$\alpha_1\ddot{x} = -k_1\alpha_3 x - k_3\alpha_2\dot{x} + k_2\alpha_4 f(t) \tag{2.2.6}$$

However, equation (2.2.5) is of the form

$$\ddot{x} = -\frac{a_2 x}{a_1} - \frac{a_3 x}{a_1} + \frac{f(t)}{a_1}, \quad (a_1 = 35,\ a_2 = 15,\ a_3 = 750, f = 600)$$

whence, by (2.2.6),

$$\frac{a_3}{a_1} = \frac{k_1\alpha_3}{\alpha_1}, \quad \frac{a_2}{a_1} = \frac{k_3\alpha_2}{\alpha_1}, \quad \frac{1}{a_1} = \frac{k_2\alpha_4}{\alpha_1}$$

Moreover,

$$e_2 = -k_4 \int e_1 \, \mathrm{d}t$$

or

$$-\alpha_2\dot{x} = -k_4 \int \alpha_1\ddot{x} \, \mathrm{d}t$$

i.e. $\alpha_2 = k_4\alpha_1$.

Similarly, $\alpha_3 = k_5\alpha_2$ and so the 'loop gain'

$$k_1k_4k_5 = \left(\frac{a_3\alpha_1}{a_1\alpha_3}\right)\left(\frac{\alpha_2}{\alpha_1}\right)\left(\frac{\alpha_3}{\alpha_2}\right) = \frac{a_3}{a_1}$$

is independent of the scaling, which means that the stability of the system (to be defined precisely later) is unaffected by the scaling.

Returning to (2.2.4) we now consider time scaling. Suppose that the forcing function f is a sinusoid with high frequency, say $600 \sin \omega t$, with ω outside the frequency range of the amplifiers. Then we can define a new time scale by writing

$$\boxed{t\alpha_t = T} \tag{2.2.7}$$

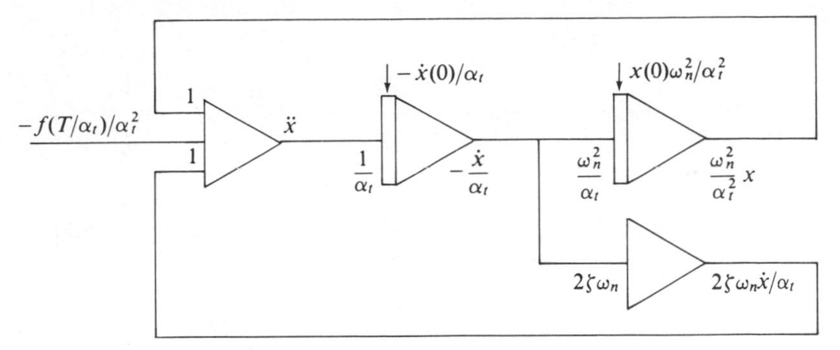

Fig. 2.10 Time scaling of a simulator.

The equation (2.2.4) then becomes

$$\alpha_t^2\frac{d^2x}{dT^2} + 2\zeta\omega_n\alpha_t\frac{dx}{dT} + \omega_n^2x = f\left(\frac{T}{\alpha_t}\right)$$

If we reduce the gain coefficient of each integrating amplifier by a factor of α_t it is clear that this is equivalent to the time scaling (2.2.7), since the integrating capacitor will charge $1/\alpha_t$ times slower than before. Hence, assuming that the circuit in fig. 2.6 is properly amplitude scaled, then the time scaled system is shown in fig. 2.10.

Note finally in this section that forcing functions $f(t)$ which can be determined as the solutions of simple linear differential equations may themselves be obtained as the outputs of analogue amplifiers. For example, we can produce

$$f(t) = M\exp(\alpha t)$$

as the solution of the equation

$$\frac{df(t)}{dt} = \alpha f(t), \quad f(0) = M$$

More general functions will be considered later.

2.2.3 General Linear Systems

Consider the general linear equation (with $a_n \neq 0$)

$$a_n\frac{d^nt}{dt^n} - a_{n-1}\frac{d^{n-1}x}{dt^{n-1}} + a_{n-2}\frac{d^{n-2}x}{dt^{n-2}} - \ldots \pm a_1\frac{dx}{dt} \mp a_0x = f(t) \quad (2.2.8)$$

where we assume, for simplicity, that each a_i is positive and $a_j/a_n \leqslant 1$.† The

† If some $a_i < 0$ we simply add an inverter in the appropriate feedback loop, and if $a_j/a_m > 1$ we alter the adder input multiplying factor.

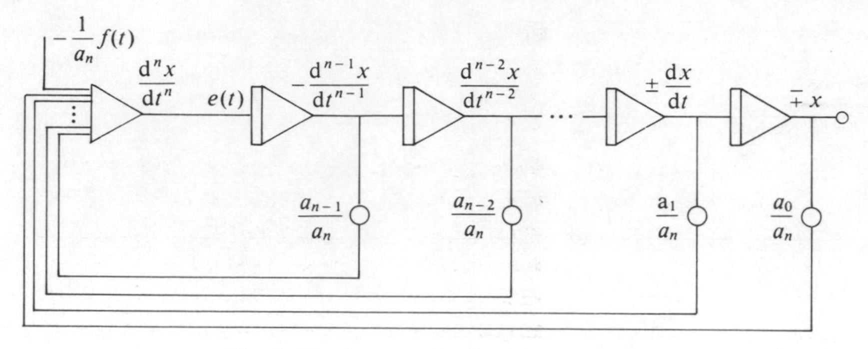

Fig. 2.11 A general nth order simulator.

computer implementation of this system is shown in fig. 2.11 which should be compared with fig. 2.6. From this it is clear that we can implement any linear system and it is easy to generalize the scaling methods introduced above.

If the function $f(t)$ is given by

$$f(t) = -\left(b_m \frac{d^m u}{dt^m} + b_{m-1} \frac{d^{m-1} u}{dt^{m-1}} + \ldots + b_0 u\right)$$

for some function u, the system becomes

$$\sum_{i=0}^{n} a_i \frac{d^i x}{dt^i} = -\sum_{j=0}^{m} b_j \frac{d^j u}{dt^j} \tag{2.2.9}$$

(Note that the signs in (2.2.8) have been changed.) If we write $s = d/dt$, then we can put

$$\frac{x}{u} = -\left(\frac{b_m s^m + b_{m-1} s^{m-1} + \ldots + b_0}{a_n s^n + a_{n-1} s^{n-1} + \ldots + a_0}\right) \tag{2.2.10}$$

(This has only formal significance, of course. More precise reasoning will be given in chapter 4.) The function of s on the right-hand side of (2.2.10)

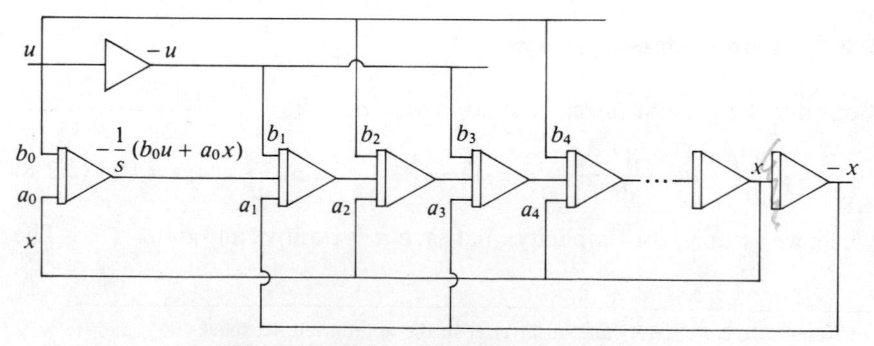

Fig. 2.12 Transfer function simulation.

is called the **transfer function** of (2.2.9). Since many systems in control theory are given in this form, it is convenient to be able to realize an analogue computer from (2.2.10). To do this we write

$$a_n x = -\frac{1}{s^n}(b_0 u + a_0 x) - \frac{1}{s^{n-1}}(b_1 u + a_1 x) - \ldots - \frac{1}{s}(b_{n-1} u + a_{n-1} x)$$

(assuming $m < n$, and writing $b_j = 0$ if $j > m$). Since $s = \mathrm{d}/\mathrm{d}t$ it is reasonable to replace $1/s$ by an integrator. We then obtain the implementation shown in fig. 2.12.

2.2.4 Linear Partial Differential Equations

The wave equation

$$\frac{\partial^2 \phi}{\partial t^2} = \frac{\partial^2 \phi}{\partial x^2} \tag{2.2.11}$$

was derived in chapter 2; we have taken the wave velocity to be unity for simplicity. Suppose that the equation is defined on the unit spatial interval [0, 1] and let the boundary and initial conditions be

$$\phi(0, t) = v(t), \quad \phi(1, t) = w(t)$$

and

$$\phi(x, 0) = V(x), \quad \frac{\partial \phi}{\partial t}(x, 0) = W(x)$$

In order to apply the above programming techniques we must replace the original equation by an approximating set of ordinary differential equations. This can be done by dividing the interval [0, 1] into N equal parts of length $\Delta x = 1/N$. Then, if $x_n = n\Delta x$

$$\left.\frac{\partial^2 \phi}{\partial x^2}\right|_{x = x_n} \approx \frac{v_{n+1} - 2v_n + v_{n-1}}{(\Delta x)^2}, \quad 1 \leqslant n \leqslant N - 1$$

where $v_n = \phi(x_n, t)$ and $v_0 = v(t)$, $v_N = w(t)$. Hence from (2.2.11), we obtain the following set of ordinary differential equations

$$\frac{\mathrm{d}^2 v_n}{\mathrm{d}t^2} = \frac{1}{(\Delta x)^2}(v_{n+1} - 2v_n + v_{n-1}), \quad 1 \leqslant n \leqslant N - 1$$

with the initial conditions $v_n(0) = V_n \triangleq V(x_n)$, $\dot{v}_n(0) = W_n \triangleq W(x_n)$. Hence we obtain the system shown in fig. 2.13.

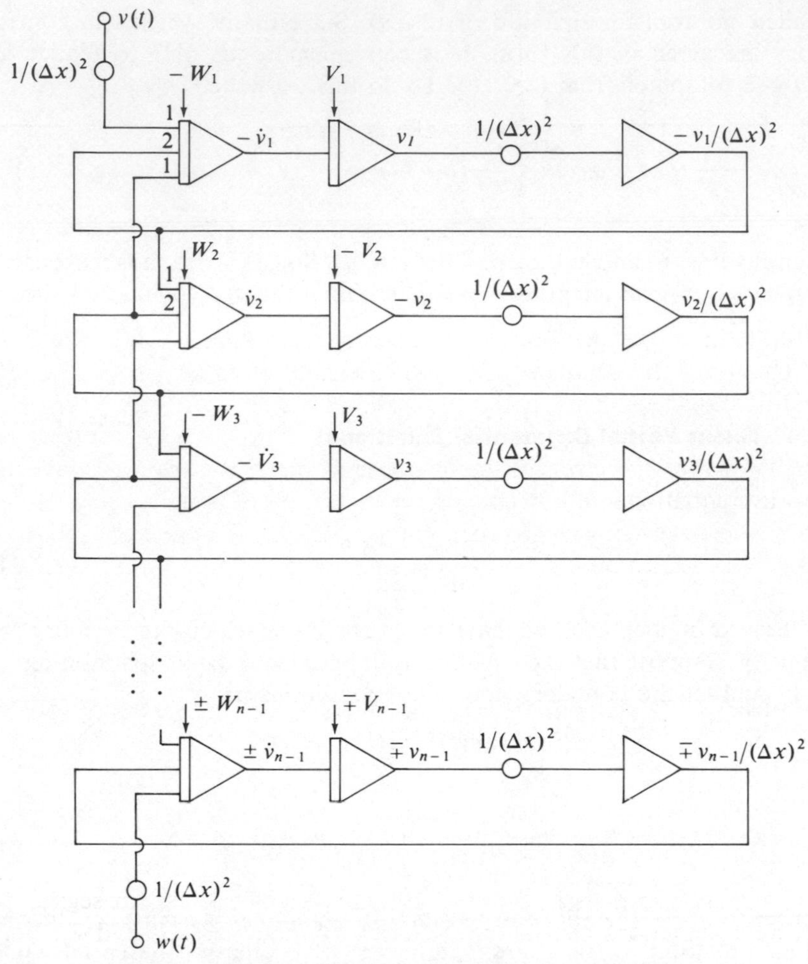

Fig. 2.13 Simulation of a partial differential equation.

2.3 NON-LINEAR SYSTEMS

2.3.1 The Servo Multiplier

Up to now we have considered only the analogue simulation of linear systems. If these were the only systems which we could simulate in this way, then analogue methods would not be particularly useful. The technique of analogue simulation is made much more powerful by the introduction of special circuitry for non-linear systems. Perhaps the most important type of non-linear operation is the multiplication of two system variables. A simple multiplier can be made with a potentiometer as shown in fig. 2.14. It effec-

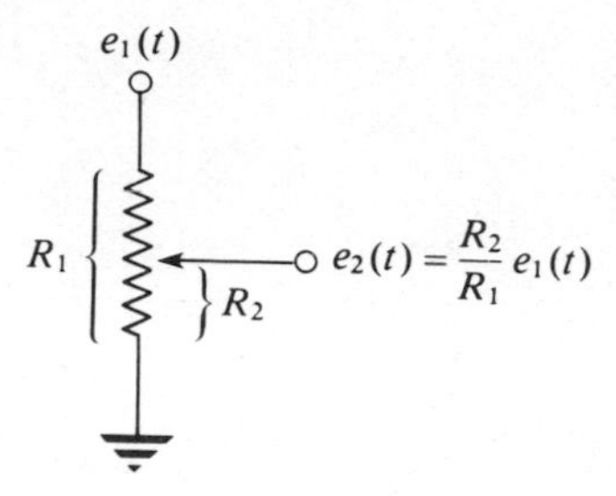

Fig. 2.14 A voltage divider circuit.

tively multiplies the quantities R_2/R_1 and the input voltage $e_1(t)$. The main drawback with this simple system is the difficulty of accurately positioning the potentiometer. This can be obviated by using a servomotor to automatically adjust the potentiometer slide (fig. 2.15). A differential amplifier senses the error between a desired input voltage $e_1(t)$ and the voltage tapped in a follow-up potentiometer. By linking a sequence of multiplying potentiometers to the same positioning servo, we can produce a polynomial function

$$P(x) = \sum_{i=1}^{n} a_i x^i$$

of an input x as in fig. 2.16. Note that integrated multipliers are also available – see Radiospares Catalogue.

2.3.2 Generating Non-linearities with Diode Switches

Many types of non-linearity may be approximated by straight line segments. Such approximations can be implemented by diode circuits and we shall now examine some basic types of these systems. Recall first that an ideal

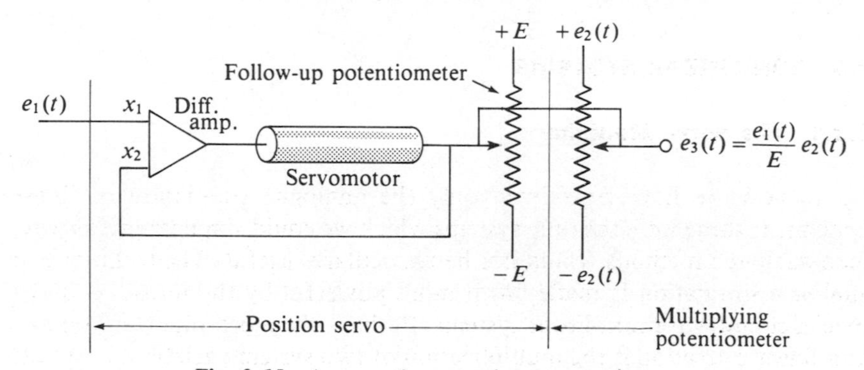

Fig. 2.15 Automatic potentiometer setting.

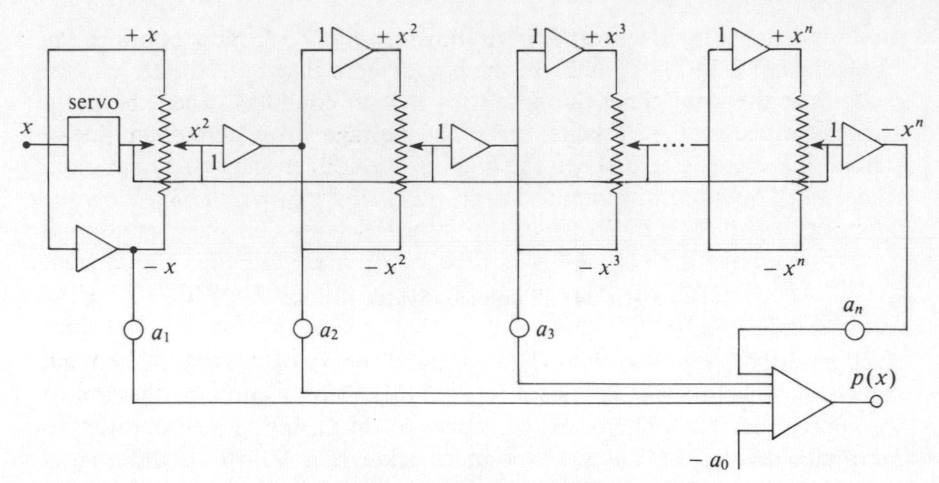

Fig. 2.16 Polynomial evaluation.

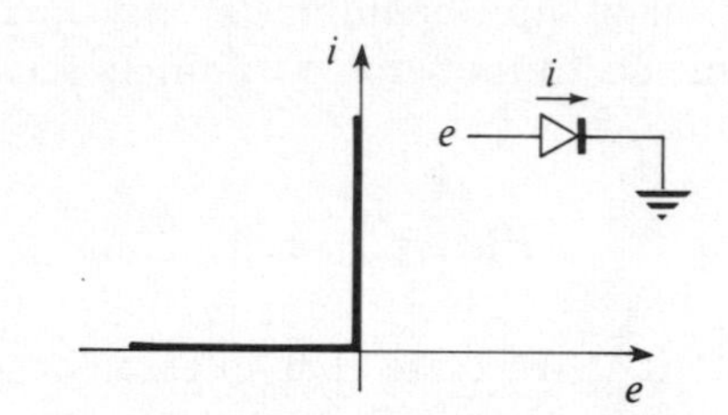

Fig. 2.17 Ideal diode characteristic.

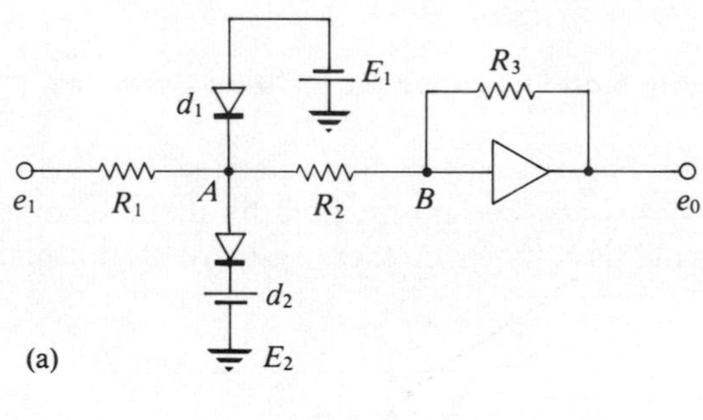

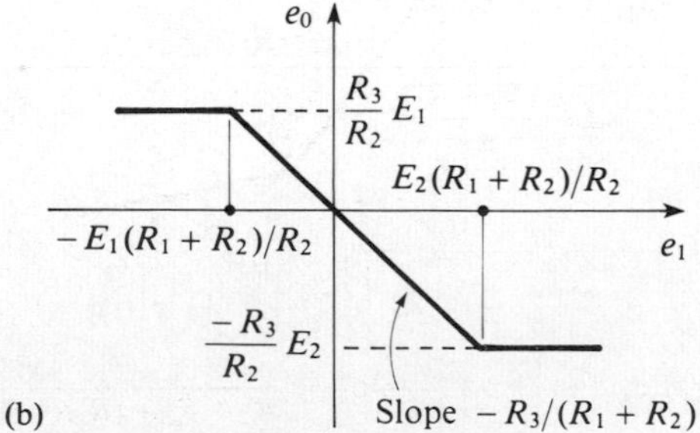

Fig. 2.18 Diode limiting circuit.

diode has the voltage–current curve shown in fig. 2.17. Now consider the circuit in fig. 2.18(a). To analyse the circuit note that both diodes cannot conduct at the same time. Suppose then that d_1 conducts. Then the point A is 'clamped' to $-E_1$ volts, since the voltage drop across an (ideal) conducting diode is zero. Then the high gain amplifier and resistors R_2 and R_3 act like a summer with input voltage $-E_1$. Hence $e_0 = E_1R_3/R_2$. Now the current through R_2 is E_1/R_2 (since the point B acts as a virtual ground) and that through R_1 is $-(E_1+e_1)/R_1$. Hence d_1 must conduct the current $E_1/R_2 + (E_1+e_1)/R_1$. This remains positive until $e_1 = -R_1E_1/R_2 - E_1 = -E_1(R_1+R_2)/R_2$. At this value of e_1, d_1 becomes reverse biased, and since d_2 is also reverse biased, neither diode conducts and the relation between e_1 and e_0 is linear until $e_1 = E_2(R_1+R_2)/R_2$ after which e_0 is clamped to

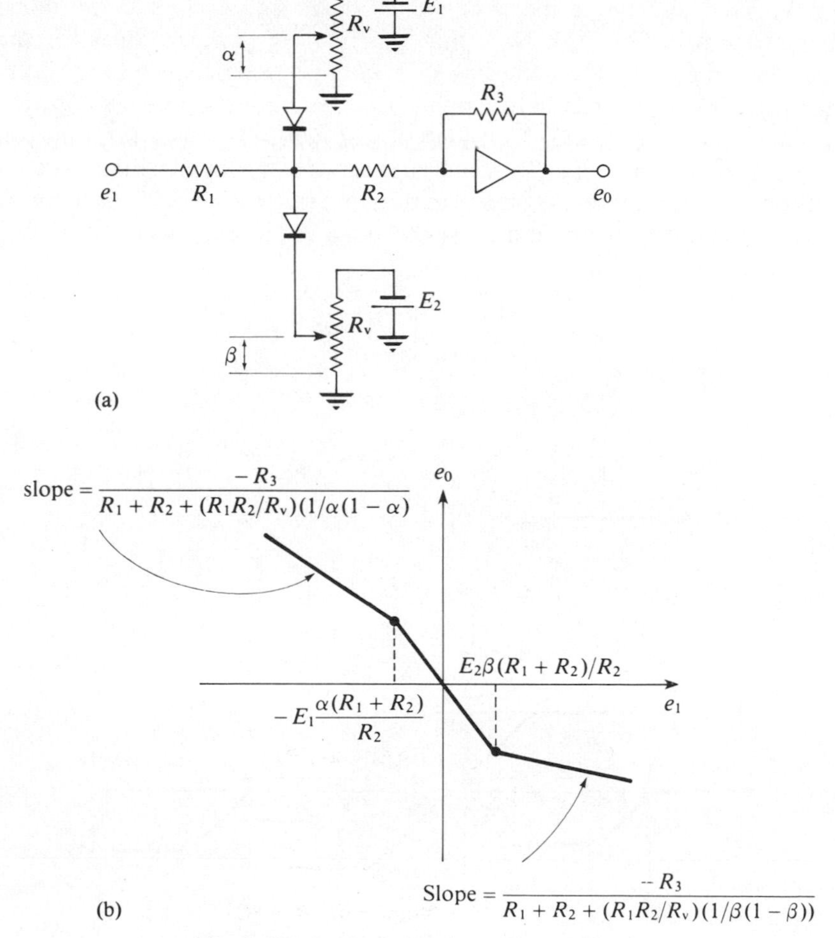

Fig. 2.19 A piecewise linear diode circuit.

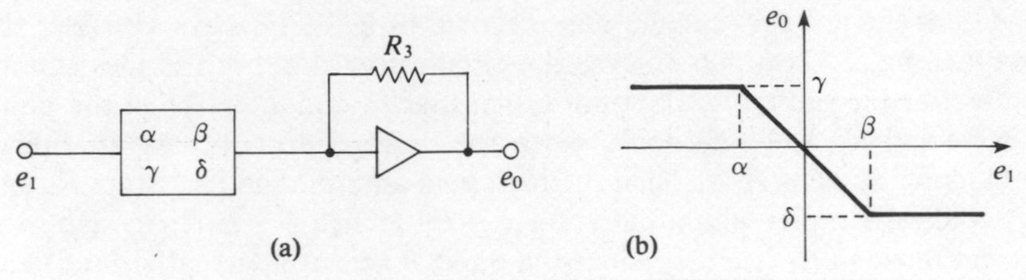

Fig. 2.20 A general limiter block.

$-E_1R_3/R_2$. Hence the graph of e_0 against e_1 takes the form in fig. 2.18(b). A more general circuit is shown in fig. 2.19(a) where variable resistors allow different slopes at the break-points. The analysis is similar to that given above and the reader is encouraged to verify the $e_0 - e_1$ characteristics in fig. 2.19(b).

We shall abbreviate the circuit diagram in fig. 2.18(a) to the block diagram form in fig. 2.20. Note that by choosing various values for the biases E_1 and E_2 we can obtain the $e_0 - e_1$ characteristics shown in fig. 2.21, and by placing an inverter before point B we can obtain a positive slope for the linear part (fig. 2.22). Consider now a general non-linear function $e_0 = f(e_1)$ as shown in fig. 2.23 and suppose that we approximate such a function by n straight line segments in a certain range $e_1^1 - e_1^{n+1}$. Each linear section i in the approximation passes through the e_0 axis (when extended, if necessary) at a point

$$\delta_i = \begin{vmatrix} e_1^i & e_1^{i+1} \\ e_0^i & e_0^{i+1} \end{vmatrix} \Big/ (e_1^i - e_1^{i+1})$$

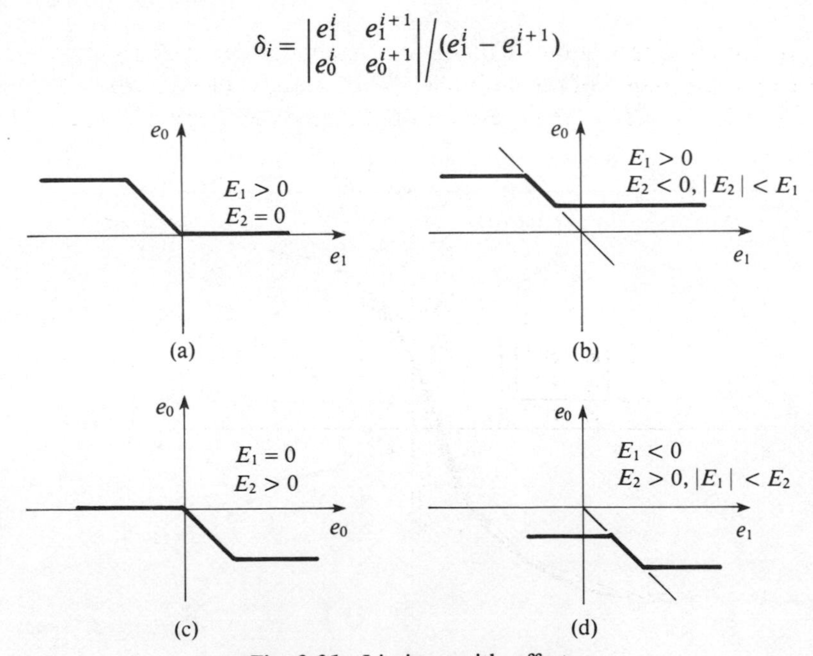

Fig. 2.21 Limiters with offsets.

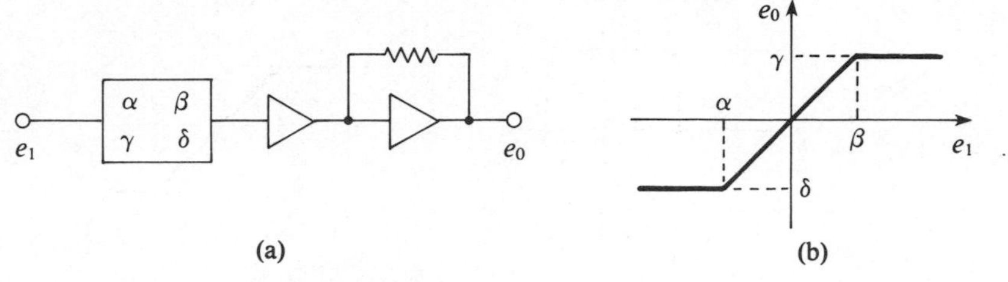

Fig. 2.22 Limiter with positive slope.

Hence the function $f - \delta_i$ can be realized in the interval $[e_1^i, e_1^{i+1}]$ by the system in fig. 2.22(a), with appropriate α, β, γ and δ. Doing the same for each line segment and summing results in a realization of

$$f - \left(e_0^1 - \sum_{i=1}^{n} (e_0^i - \delta_i)\right)$$

Hence, using the summing effect of the high gain amplifier in fig. 2.22, we can obtain the realization of the linear approximation to f shown in fig. 2.24. For other methods of non-linear function generation see Clayton (1979).

2.3.3 Multiplier Circuits

An important application of diode approximations to a function f is the electronic realization of the multiplier. This can be achieved with the circuit of fig. 2.25, where we require both polarities of the inputs x and y. The

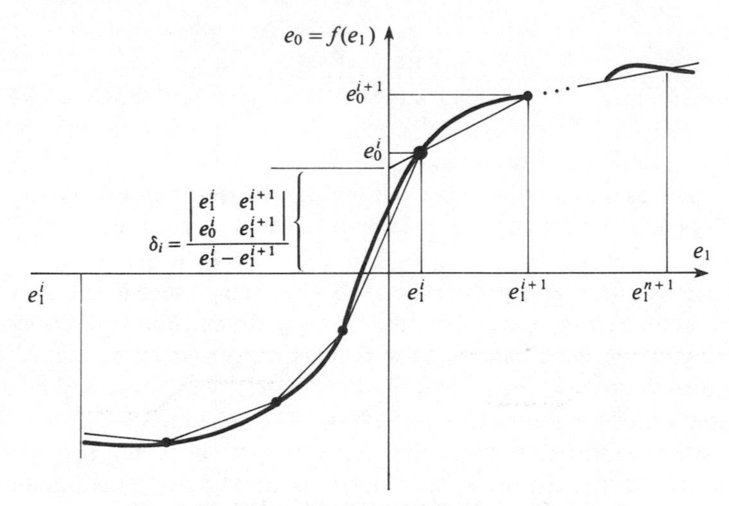

Fig. 2.23 General piecewise linear representation.

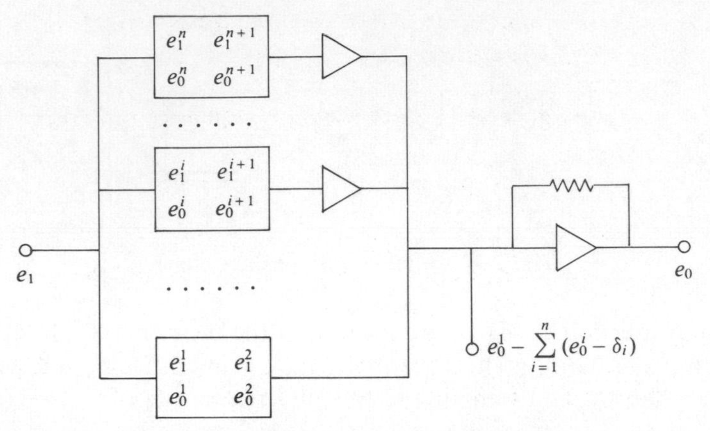

Fig. 2.24 Circuit for system in fig. 2.23.

voltage input v_1 to diode $d1$ when it is conducting, provided i is small, is given by

$$\frac{x-v}{R} = \frac{x-y}{2R}$$

i.e. $v_1 = (x+y)/2$. Similarly, the voltage input to diode d2 when conducting is

$$v_2 = -\frac{(x+y)}{2}$$

The biased diode circuits are designed to draw small currents relative to the current through R and then the input to the OP-amp is

$$i_1 + i_2 = \left(\frac{x+y}{2}\right)^2 - \left(\frac{x-y}{2}\right)^2 = xy \ \mu\text{A}$$

A large resistance R_1 then converts the current xy to a voltage $-xyR_1$.

A completely different type of circuit multiplier is based on a pulse-modulation technique. Consider first the circuit in fig. 2.26(a), which we shall show acts as a two-pole switch, and which we shall represent as in fig. 2.26(b). Suppose that the input y is known to satisfy $|y| \leqslant y_{max}$, then it is easy to see that if $g > y_{max}$, the output z is proportional to y, and if $g < -y_{max}$, the output z is proportional to $-y$. Hence the circuit operates as a switch (connecting y or $-y$ to the output) which is controlled by the voltages $\pm g$. The controlling voltages $\pm g$ can be produced by a fixed frequency multivibrator (Millman and Halkias, 1972) which we modify with a differential amplifier shown in fig. 2.27(a). When the inputs x and $-x$ to the differential amplifier are zero, the circuit is symmetric and the outputs are symmetric in time. However, as x increases above zero, a net imbalance in the collector currents results in a change in the time each transistor is

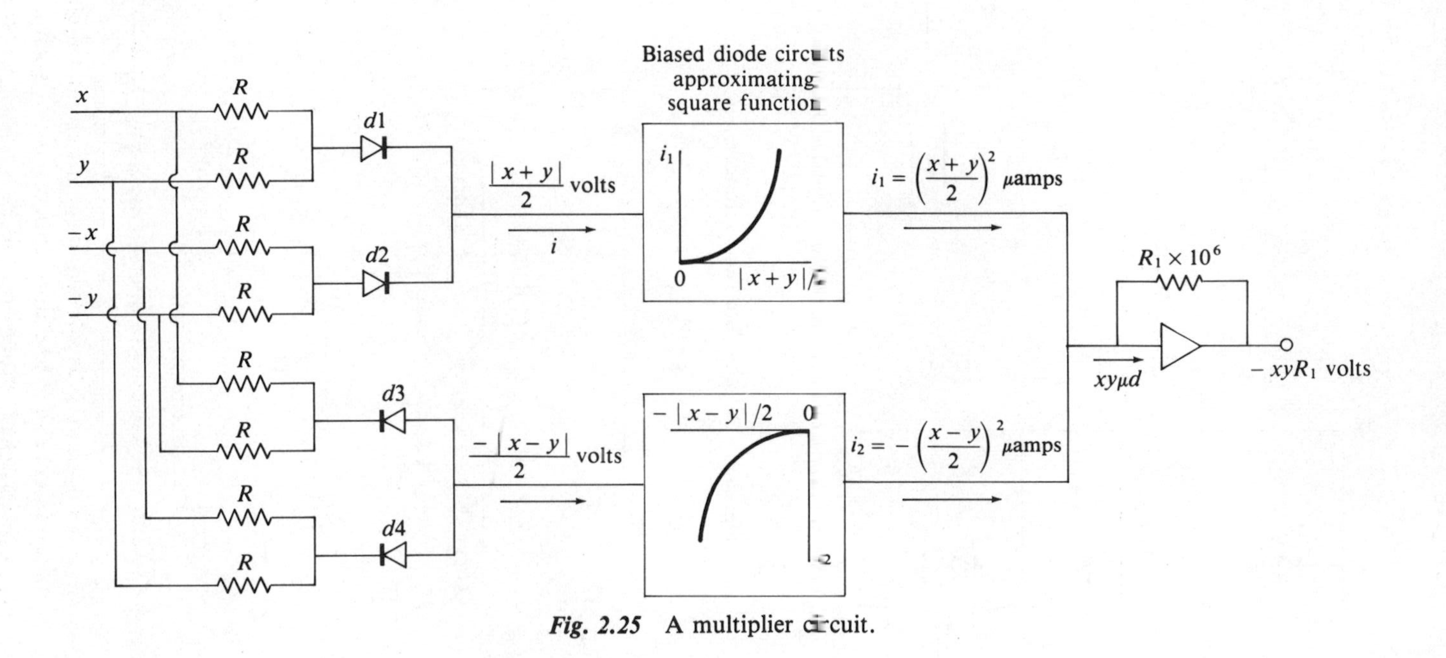

Fig. 2.25 A multiplier circuit.

CSE-D*

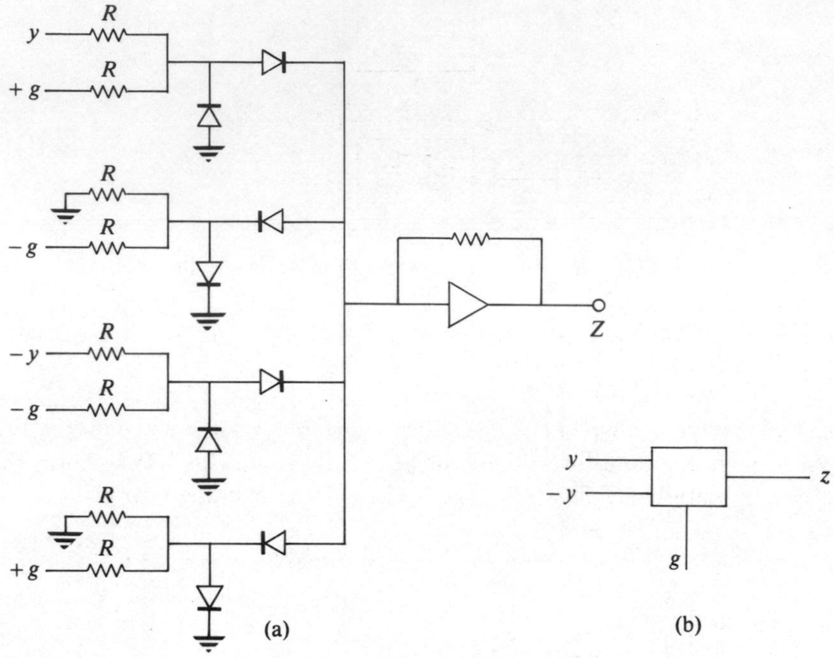

Fig. 2.26 An absolute value circuit.

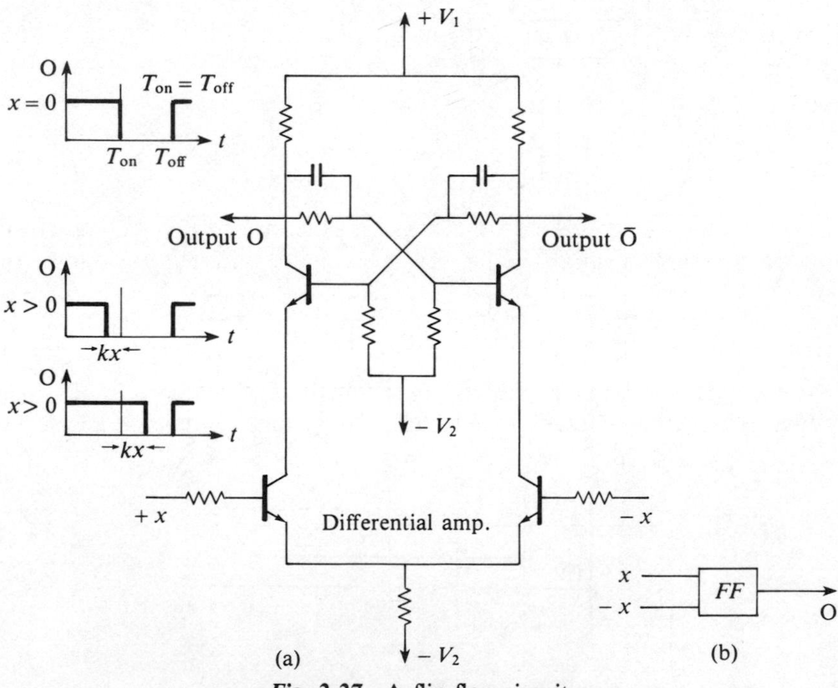

Fig. 2.27 A flip-flop circuit.

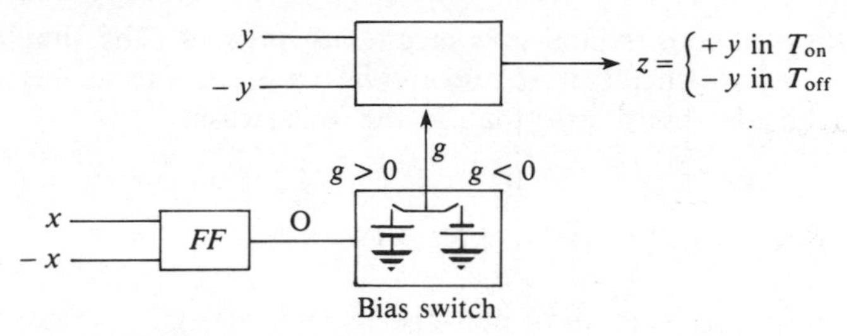

Fig. 2.28 Switching an absolute value circuit.

turned on, so the length of the time the output O is logical 1 is modulated by x. A symbolic representation of the (variable mark/space ratio) flip-flop is shown in fig. 2.27(b). If the output O drives a switch whose output is a positive voltage g when O is 'ON' and a negative voltage g when O is 'OFF' then we can combine the systems in figs. 2.26(b) and 2.27(b) to form the pulse modulation system in fig. 2.28. Clearly the average output

$$z = y\frac{T_{on} - T_{off}}{T_{on} + T_{off}} = kxy$$

for some constant k.

Consider as an example the equation

$$\dot{x} = -x^2, \quad x(0) = x_0 > 0 \tag{2.3.1}$$

Then using a multiplier circuit we can implement this equation simply as can be seen from fig. 2.29, where we have assumed that the multiplier with inputs x and y produces the inverted product $-xy$. Since the solution of (2.3.1) is

$$x(t) = (x_0^{-1} + t)^{-1}$$

it is clear that x decreases with increasing time and so the scaling of this non-linear system is simple. Note, however, that if we tried to simulate the system

$$\dot{x} = x^2, \quad x(0) = x_0 > 0 \tag{2.3.2}$$

then the solution is $x(t) = (x_0^{-1} - t)^{-1}$ which diverges as $t \to x_0^{-1}$ from below. Hence the voltages in a simulator become unbounded and will

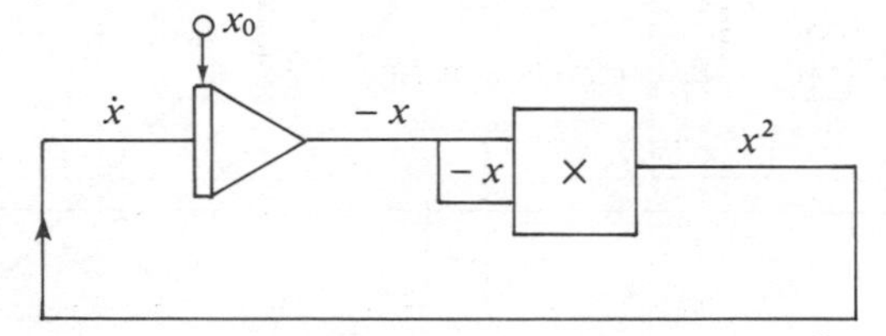

Fig. 2.29 Simulator for the system $\dot{x} = -x^2$

saturate the components and give erroneous answers. The simulation of non-linear systems therefore requires great care and the scaling is much more difficult and less precise than in the linear case.

2.3.4 Deadzone and Hysteresis Non-linearities

Two important special kinds of non-linearity which occur in many types of electromechanical systems are the deadzone and hysteresis non-linearities, which are due to finite frictional resistance and backlash respectively. Simple analogue circuits which simulate such systems can be made using Zener diodes. An ideal Zener diode has an input–output characteristic of the form shown in fig. 2.30. The forward-bias characteristic is like that of an ordinary diode (fig. 2.17), but under sufficient reverse bias the diode 'breaks down' and a maximum bias of e_z volts is possible producing arbitrary current. If we consider the circuit in fig. 2.31, then the output is clearly limited between the breakdown voltages $\pm e_z$ of the Zener pair, since if $|e_1|$ is sufficiently large then one of the diodes must be reverse biased with bias equal to $\pm e_z$ and since the other diode must be forward biased the diode pair will short out R_2. Letting $R_2 \to \infty$ produces a switch which gives an output of $+e_z$ if $e_1 < 0$ and $-e_z$ if $e_1 > 0$ (fig. 2.32).

Consider now the circuit in fig. 2.33. If $|e_1| < e_z$ then one diode is reverse biased, but not enough to produce a reverse current. Hence $i = 0$ and

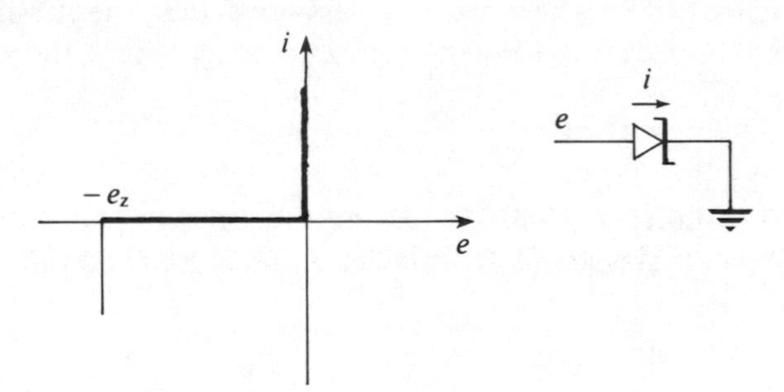

Fig. 2.30 Ideal zener diode characteristic

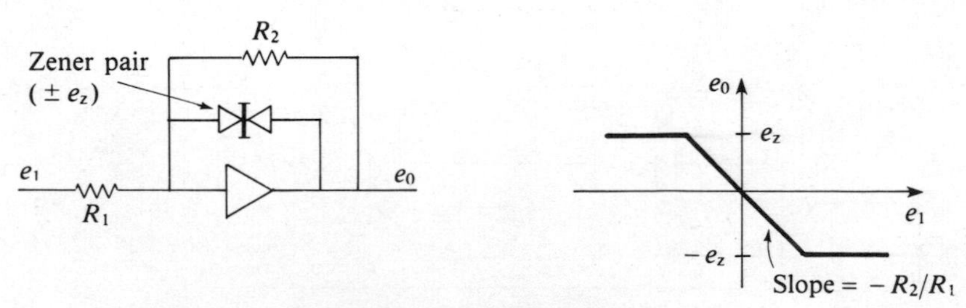

Fig. 2.31 Zener diode limiter.

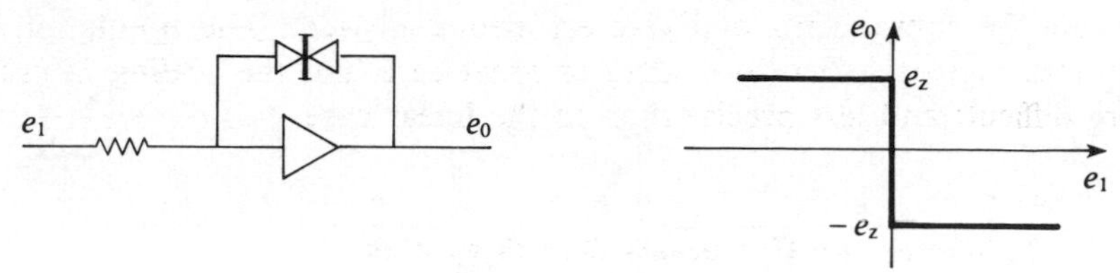

Fig. 2.32 Switching circuit.

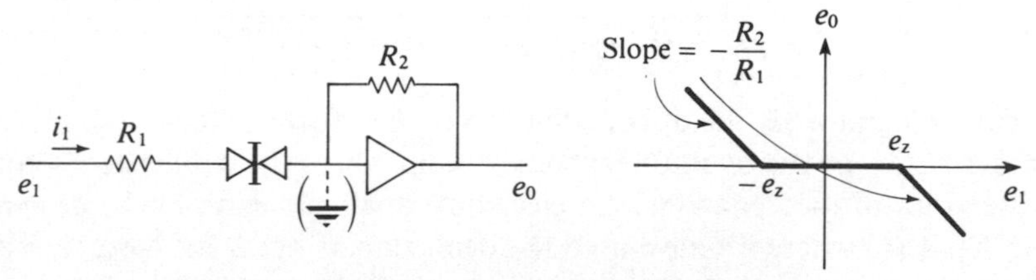

Fig. 2.33 Dead zone circuit.

so $e_0 = 0$. However, if $|e_1| > e_z$ then the Zener pair will conduct an arbitrary current and so we obtain a linear $e_1 - e_0$ graph with slope $-R_2/R_1$. This circuit therefore has the characteristic of a simple deadzone. A deadzone may also be simulated by using ordinary diodes as in fig. 2.34, where the circuit operation should now be clear.

The problem of designing a backlash simulator can be approached most easily by considering a deadzone and an integrator in a feedback system as shown in fig. 2.35. (More will be said about feedback systems in Part 2.) Suppose that we input a sinusoid to the system as shown. Then it can be shown that this system will oscillate (although in a non-linear fashion, cf. chapter 5). Hence the error signal $\varepsilon = -e_1 - e_0$ will also oscillate, driving the output y up and down the deadzone characteristic. On the linear portions of the deadzone, the $e_1 - e_0$ characteristic must also be linear. However, as ε crosses the deadzone we have $y = 0$ and so the output

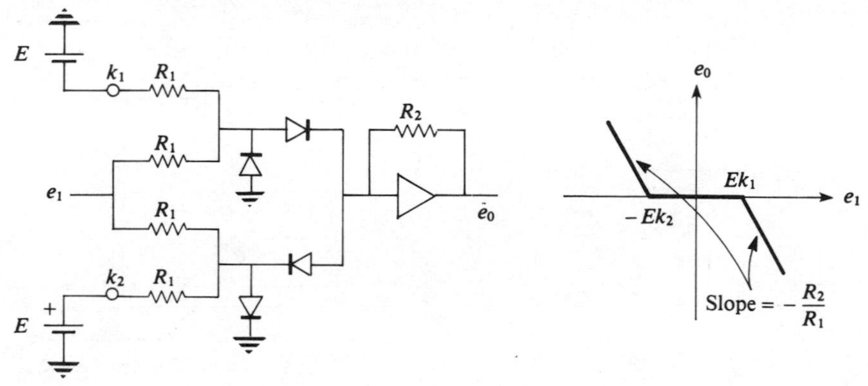

Fig. 2.34 Dead zone with shifted origin.

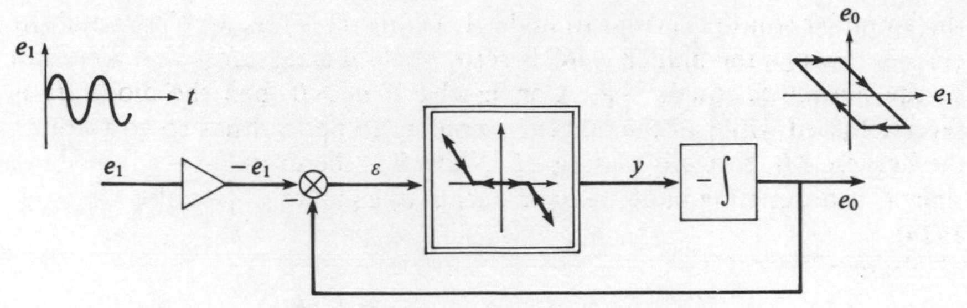

Fig. 2.35 Hysteresis block diagram.

of the integrator is held constant. We therefore obtain a hysteresis characteristic for the overall feedback system where the deadspace is equal to the width of the deadzone. An electronic analogue can now be designed as in fig. 2.36, where we have used the deadzone of fig. 2.34. Note that this configuration produces a frequency-dependent characteristic.

Another useful non-linear circuit appears in fig. 2.37 and has the absolute value characteristic. If $e_1 < 0$ then diode $d1$ is forward biased and

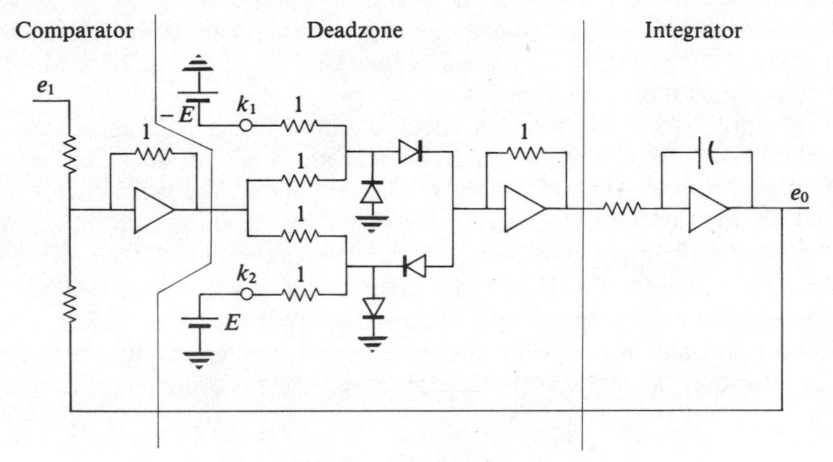

Fig. 2.36 Hysteresis circuit.

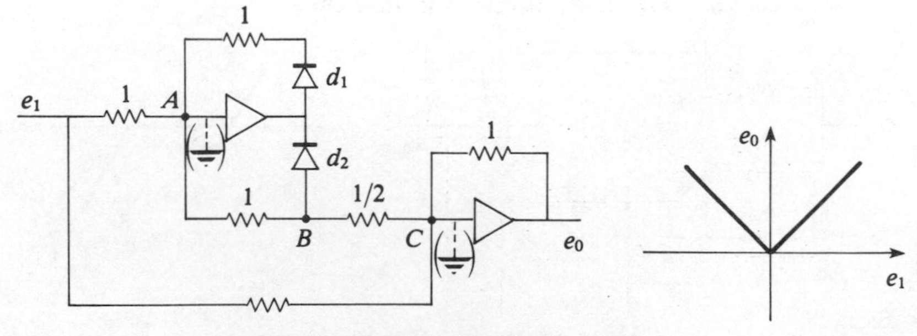

Fig. 2.37 Absolute value circuit.

the amplifier sources current to node A. Diode $d2$ is reverse biased and the current through the branch ABC is zero. Node B is therefore at 0 volts and so the output e_0 equals $-e_1$. Conversely, if $e_1 > 0$ then the diode $d1$ is reverse biased while all the current incoming to node A has to go through the branch AB, forward biasing $d2$. Node B is therefore at $-e_1$ volts and since C is a summing node we have output e_0 equal to e_1 (see also Clayton, 1979).

2.3.5 Inverse Function Simulation

Suppose that we have designed a system to produce a (non-linear) function f and consider the feedback system shown in fig. 2.38. Then

$$e_0 = -|A|(e_1 + f(e_0)) \tag{2.3.3}$$

If $|e_0| \ll |A||f(e_0)|$ then we can neglect e_0 and write

$$e_1 \approx -f(e_0) \tag{2.3.4}$$

or

$$e_0 \cong f^{-1}(-e_1) \tag{2.3.5}$$

Hence the feedback signal $e_2 \approx -e_1$ and $\varepsilon = e_1 + e_2 \approx 0$. This system is an example of non-unity feedback. In particular, consider the square root circuit in fig. 2.39. The squarer function in the feedback path is the system of fig. 2.25, where the inputs are arranged so that diodes $d1$ and $d4$ never conduct. When $z > 0$, diode 3 conducts and the output current is $-z^2$ μA and when $z < 0$, diode 2 conducts giving an output current of z^2 μA. The

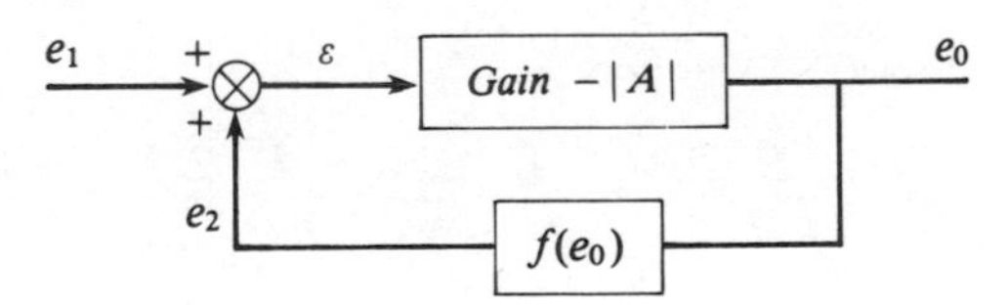

Fig. 2.38 Inverse function circuit.

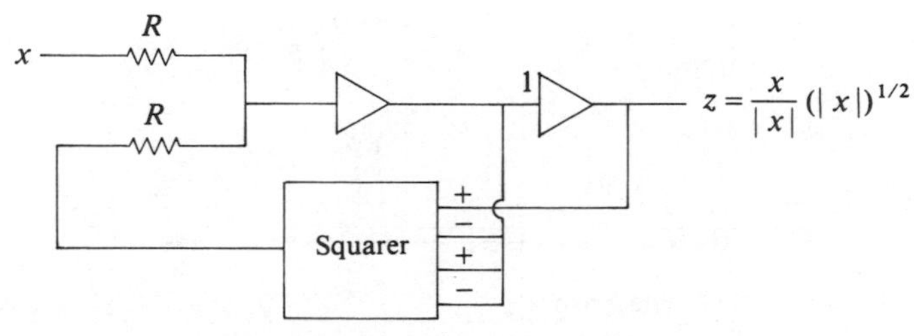

Fig. 2.39 Square-root circuit.

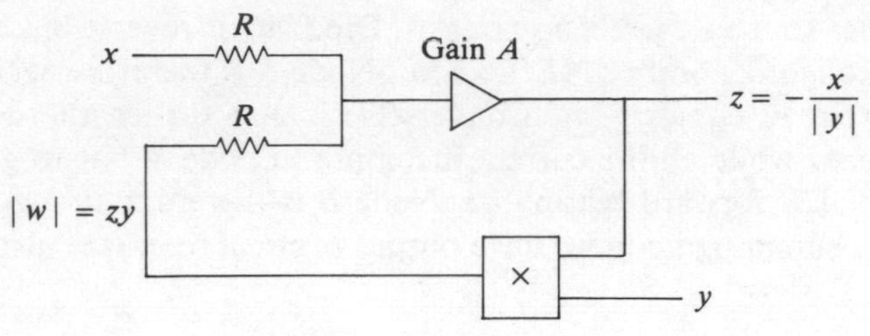

Fig. 2.40 Divider circuit.

current feedback to the summer is therefore always opposite in sign to that from the input x.

Finally note that by putting a multiplier in the feedback path we can make a divider circuit as in fig. 2.40.

2.3.6 Shaped Potentiometer Function Generators

Non-linear functions may be generated very simply by forming a potentiometer on a shaped card as in fig. 2.41. If $R(x)$ is the resistance of the wire between 0 and x, then

$$\frac{\Delta R(x)}{\Delta x} = \frac{2ry}{\Delta x}$$

where r is the wire resistance per unit length and Δx is the distance between successive turns. Applying E volts across the potentiometer, we must have

$$f(x) = \frac{E}{R(l)} R(x)$$

for a desired function $f(x)$. Hence

$$\frac{\mathrm{d}f}{\mathrm{d}x} = \frac{E}{R(l)} \frac{\mathrm{d}R}{\mathrm{d}x} \approx \frac{E}{R(l)} \frac{\Delta R}{\Delta x} = \frac{E}{R(l)} \frac{2ry}{\Delta x}$$

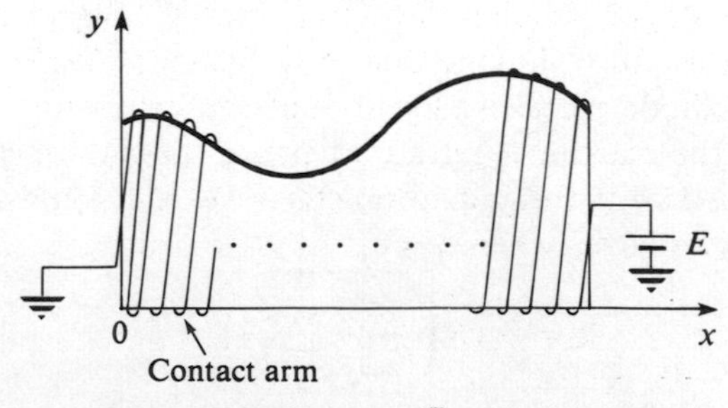

Fig. 2.41 Potentiometer function realisation.

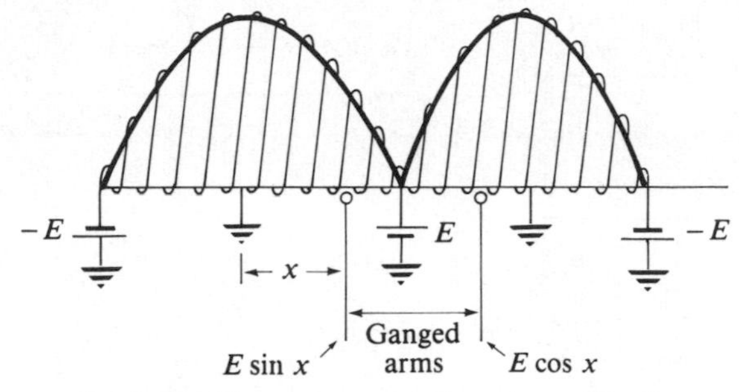

Fig. 2.42 Sine wave potentiometer realisation.

and so

$$y = \frac{R(l)}{2rE} \frac{\mathrm{d}f}{\mathrm{d}x} \Delta x$$

It follows that the card shape must be proportional to the derivative of the desired function. The main drawbacks with this method are the unwanted non-linearities introduced by errors in the card shape, non-uniform resistance wire, end resistances, temperature dependence, etc. Note also that $\mathrm{d}f/\mathrm{d}x > 0$ and so we can only make monotonic non-linearities by this method. For non-monotonic functions we can extend the method by tapping the potentiometer at the turning points with appropriate voltages. For example, the sine and cosine functions can be formed as in fig. 2.42.

2.3.7 Functions of Two Variables

Up to now we have considered only the generation of non-linear functions of a single variable. It is often necessary to be able to simulate functions of several variables. Many methods have been developed for the analogue simulation of such functions, but apart from the case of functions of two variables these methods are very impractical, especially in view of the availability of cheap microprocessors which can be used for digital simulation. It does seem appropriate here, however, to consider at least one method for the analogue simulation of functions of two variables and so we shall discuss a useful technique due to Meissinger (1955).

Consider the diode piecewise-linear function generator shown in fig. 2.43, where P_i is the fractional setting of the ith potentiometer measured from the x end. Then, if the potentiometer resistances are much smaller than R_1, R_2, . . . , we have, when $x = 0$,

$$z = -yR\left(\frac{1}{R_{0y}} + \sum \frac{P_i}{R_i}\right) \tag{2.3.6}$$

where the sum is taken over all conducting input branches. Moreover, for

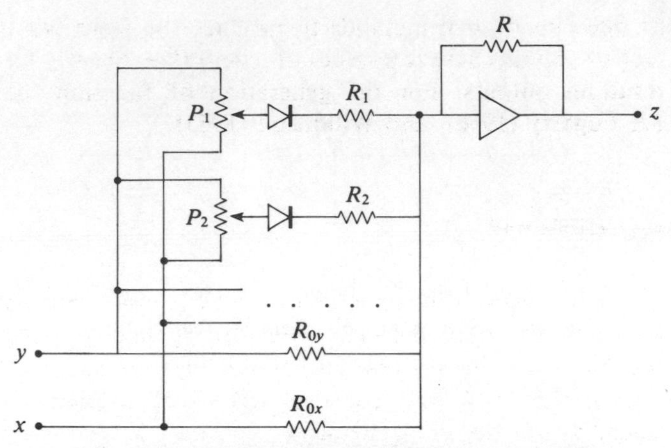

Fig. 2.43 Simulating a function of two variables.

a fixed input y, the x coordinate of the ith breakpoint is given by

$$(y - x_i)P_i + x_i = 0$$

or

$$x_i = -\frac{P_i}{1 - P_i}\,y \tag{2.3.7}$$

It follows from (2.3.6) and (2.3.7) that the curves of z against x move up the z-axis linearly with changes in y and that the breakpoints lie on straight lines through the origin as in fig. 2.44(a).

In this system the approximated function, say $z = f(x, y)$, is such that the curves $f(x, y_1)$, $f(x, y_2)$ for fixed values of y are parallel. By using functions $g_1(y)$, $g_2(y)$ instead of just y at the y end of each potentiometer we can put the breakpoints on general curves (rather than straight lines) and by putting offset voltages on y and x we can translate the function arbitrarily (fig. 2.44(b)). Alternatively, to generate a general function $f(x, y)$ we

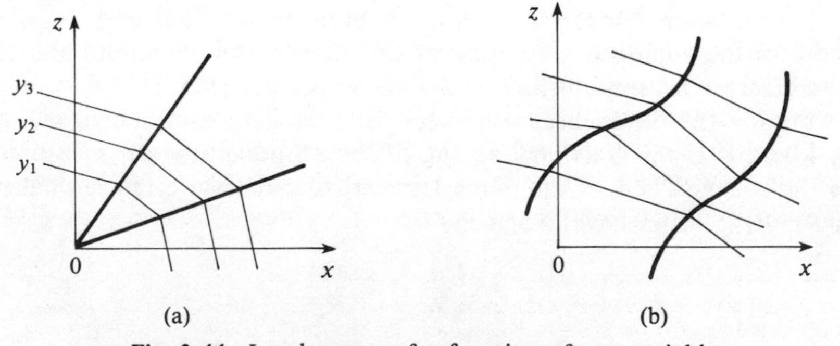

Fig. 2.44 Level curves of a function of two variables.

can use the one-dimensional methods to produce the functions $f(x, y_1)$, $f(x, y_2), \ldots$ for suitably selected values of y and then linearly interpolate between resulting outputs. For the generation of functions of several variables see Fogarty (1966), and Wilkinson (1963).

2.3.8 Delay Equations

As we have seen in chapter 1, many types of systems contain pure delays and it is desirable to be able to model such systems using analogue circuitry. In chapter 4 we shall show that the response of any linear system to a complex sinusoid† $\exp(j\omega t)$ is given by $G(j\omega)\exp(j\omega t)$ where $G(j\omega)$ is a complex number depending on the input frequency ω; i.e. the response is also a complex sinusoid of the same frequency multiplied by a complex function of this frequency. We can write

$$G(j\omega) = |G(j\omega)| \exp(j \angle G(j\omega))$$

G is called the **frequency response** of the system, while $|G(j\omega)|$ and $\angle G(j\omega)$ are called the **amplitude** and **phase responses** respectively. Suppose that the phase response $\angle G(j\omega)$ is linear, i.e.

$$\angle G(j\omega) = -\alpha\omega \tag{2.3.8}$$

for some $\alpha \geqslant 0$. Then the output produced by the input $\exp(j\omega t)$ is

$$\begin{aligned} G(j\omega)\exp(j\omega t) &= |G(j\omega)| \exp(j(\omega t - \alpha\omega)) \\ &= |G(j\omega)| \exp(j\omega(t-\alpha)) \end{aligned}$$

Hence, for a system with linear phase shift, sinusoids of all frequencies are delayed by the same amount α and therefore an arbitrary input signal will have each frequency component delayed by α. Thus, if

$$|G(j\omega)| = 1 \tag{2.3.9}$$

then the output of the system will be the same as the input delayed by α. Neither of the conditions (2.3.8) and (2.3.9) can be achieved in practice and so we approximate the pure delay system with finite dimensional systems. However, such approximations should be handled with care since the stability of the system defined by the approximations does not imply that of the pure delay (see Banks and Abbasi-Ghelmansarai, 1983). A possible sequence of approximations to the delay are given by Cunningham (1954)

† Using complex exponentials is just a convenience – for the response of a linear system to $\cos \omega t + j \sin \omega t$ is the same as its response to $\cos \omega t$ plus j times its response to $\sin \omega t$.

as follows:

$$\left.\begin{aligned}
H_1 &= \frac{s-a}{s+a} \\
H_2 &= \frac{s^2 - 2as + (a^2+b^2)}{s^2 + 2as + (a^2+b^2)} \\
H_3 &= \frac{[s^2 - 2as + (a^2+b^2)]\,(s-c)}{[s^2 + 2as + (a^2+b^2)]\,(s+c)} \\
H_4 &= \frac{[s^2 - 2as + (a^2+b^2)]\,[s^2 - 2cs + (c^2+d^2)]}{[s^2 + 2as + (a^2+b^2)]\,[s^2 + 2cs + (c^2+d^2)]} \\
&\vdots
\end{aligned}\right\} \qquad (2.3.10)$$

For each improved approximation we add a pole-zero pair symmetric with respect to the $j\omega$-axis, and so these are non-minimum phase systems, i.e. they have zeros (or poles) in the right half plane (see also Hausner and Furlani, 1966; Saucedo and Schining, 1968).

2.4 DESIGN OF COMPUTER AMPLIFIERS

2.4.1 Introduction

We have been concerned so far in this chapter with ideal computing elements from which we have seen how to construct analogue equivalents of many types of systems. All real elements (amplifiers, diodes, etc.) will depart to some extent from the ideal and so it is important to consider the effects of these inaccuracies on the computed solutions. We shall concentrate here on the design of amplifiers since this represents the greatest difficulty from the point of view of error estimation and reduction. Moreover, it will also provide an insight into the origins of control since most of the classical control of single-input, single-output systems was developed precisely for feedback amplifier design, and will therefore motivate much of what will be studied in chapter 4 on linear systems theory.

2.4.2 Ideal and Approximate Computing Elements

Consider the general circuit in fig. 2.45, where we have indicated the virtual ground at the amplifier input. If the amplifier is perfect (i.e. has infinite gain and infinite input impedance) then we obtain an ideal adder if $C_1 = C_2 = 0$, an ideal integrator if $C_1 = 0$, $R_2 = 0$ and an ideal differentiator if $R_1 = 0$, $C_2 = 0$. However, in any real circuits there will be leakage resistances and

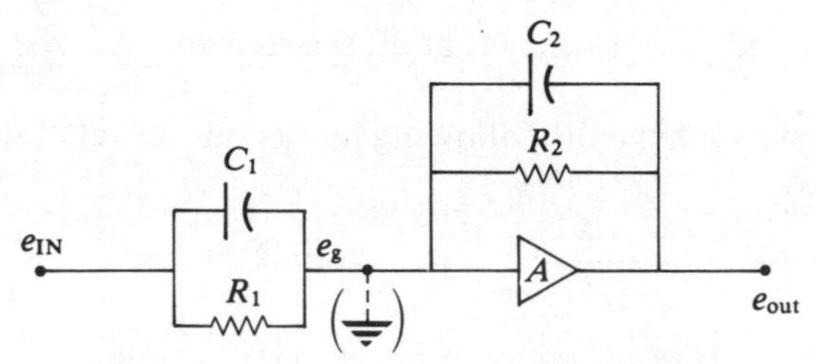

Fig. 2.45 A general op-amp feedback circuit.

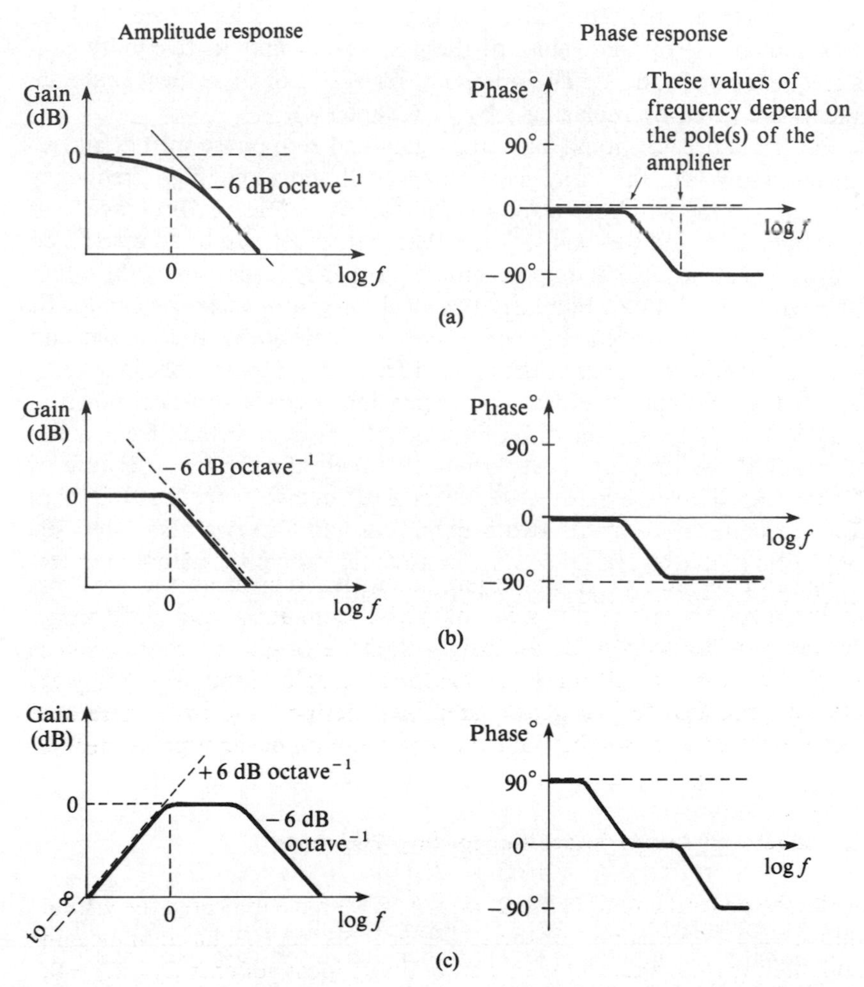

Fig. 2.46 Amplitude and phase responses; – – –, ideal characteristics; ———, real characteristics. (a) Adder; (b) integrator; (c) differentiator.

stray capacitances and so it is likely that all the elements R_1, R_2, C_1 and C_2 will be non-zero.

In the ideal case we have the following input–output relationships (i.e. transfer functions):

(a) *Adder* $e_{OUT} = e_{IN}$ or $E_{OUT}(s) = E_{IN}(s)$

(b) *Integrator* $e_{OUT} = \int e_{IN}$ or $E_{OUT}(s) = \frac{1}{s} E_{IN}(s)$

(c) *Differentiator* $e_{OUT} = \frac{de_{IN}}{dt}$ or $E_{OUT}(s) = sE_{IN}(s)$

where we have written, formally, $\int = 1/s$, and $d/dt = s$ as before, and we have assumed appropriate values of the passive elements to give unity gain at a frequency of 1 rad s^{-1}. The frequency responses of these ideal analogue elements are given by replacing s by $j\omega$ (chapter 4).

For the adder we should have unity gain and zero phase shift at all frequencies. However, the stray capacitance will short out high frequency signals and so the amplifier response will fall by at least 6 dB octave^{-1} at high frequencies, and we shall see later that there must also be an associated 90° phase lag. The ideal and approximate frequency responses of the adder are shown in fig. 2.46(a). Similarly the ideal integrator response cannot be achieved in practice since infinite gain is not possible at low frequencies and we cannot obtain zero phase shift at zero frequency (apart from in a delay line). Hence we obtain the ideal and approximate characteristics shown in fig. 2.46(b). Finally, for the differentiator infinite gain at high frequencies is impossible and the gain must flatten off eventually. In fact, because of the stray capacitance it must also fall at 6 dB octave^{-1} (at least) at high frequencies and so we obtain two breakpoints and also two associated 90° phase shifts (fig. 2.46(c)). Of course, in any real system the actual high frequency roll-off may be much higher than 6 db octave^{-1}.

2.4.3 Transfer Functions of Real Elements

The input–output relationship of the general system in fig. 2.45 is

$$\frac{E_{IN}(s) - E_g(s)}{R_1(1/sC_1)/(R_1 + 1/sC_1)} = \frac{E_g(s) - E_{OUT}(s)}{R_2(1/sC_2)/(R_2 + 1/sC_2)}$$

where we have again used the s-variable notation. (In this notation we conventionally use upper case letters rather than lower case ones for $e_{IN}(t)$, $e_{OUT}(t)$, etc.) We also have

$$E_{OUT}(s) = A(s)E_g(s)$$

and so

$$\frac{E_{OUT}(s)}{E_{IN}(s)} = -\frac{Z_2/Z_1}{1-(1/A)(1+Z_2/Z_1)} \tag{2.4.1}$$

where

$$Z_1 = \frac{R_1(1/sC_1)}{R_1+1/sC_1}, \quad Z_2 = \frac{R_2(1/sC_2)}{R_2+1/sC_2} \tag{2.4.2}$$

We can now obtain more accurate expressions for the transfer functions of the basic analogue computing elements by evaluating expressions for the computing amplifier $A(s)$. In the simplest case, if $A(s) = -\infty$ for each s, then we have

$$\left(\frac{E_{OUT}(s)}{E_{IN}(s)}\right)_{\text{adder}} = -\frac{R_2}{R_1}\frac{1+sT_1}{1+sT_2} \tag{2.4.3}$$

$$\left(\frac{E_{OUT}(s)}{E_{IN}(s)}\right)_{\text{integrator}} = -\frac{1}{R_1C_2}\frac{(1+sT_1)T_2}{1+sT_2} \tag{2.4.4}$$

and

$$\left(\frac{E_{OUT}(s)}{E_{IN}(s)}\right)_{\text{differentiator}} = -R_2C_1\frac{1+sT_1}{(1+sT_2)T_1} \tag{2.4.5}$$

where $T_1 = R_1C_1$ and $T_2 = R_2C_2$ are the time constants of the input and feedback RC networks.

More realistically, we can assume that $A(s) = -A_0/(1+sT_0)$ where A_0 is a large constant, since the term $1+sT_0$ gives rise to an amplitude asymptote of -6 dB octave^{-1}. (Further terms of this form can be added to produce further asymptotes.) In this case the adder transfer function becomes

$$\left(\frac{E_{OUT}(s)}{E_{IN}(s)}\right)_{\text{adder}} = -\frac{R_2}{R_1}\frac{(1+sT_1)/(1+sT_2)}{1+[(1+sT_0)/A_0]\{1+(R_2/R_1)[(1+sT_1)/(1+sT_2)]\}}$$

$$\approx -\frac{R_2}{R_1}\frac{1}{1+s[(T_0/A_0)(1+R_2/R_1)-(T_1-T_2)]}, \quad |s| \ll 1/T_1$$

where the latter expression is easily obtained by using the facts that $A_0 \gg 1$ and T_1, T_2 are small. In this approximation, the time constant of the adder is therefore

$$T_{\text{adder}} = \frac{T_0}{A_0}\left(1+\frac{R_2}{R_1}\right) - (T_1 - T_2)$$

for low frequency signals. To make T_{adder} small we require T_0, T_1, T_2 to be small, A_0 large and $1 + R_2/R_1 \ll A_0/T_0$. Note also that T_{adder} must be positive to maintain stability, since the poles must be in the left half plane (see chapter 4).

To analyse the integrator we write (2.4.1) in the form

$$\left(\frac{E_{\mathrm{OUT}}(s)}{E_{\mathrm{IN}}(s)}\right)_{\mathrm{integrator}} = \frac{1}{Z_1(s)}\,\frac{A}{1-A}\,\frac{1}{[1/Z_2(s)] + [1/Z_1(s)]\,[1/(1-A)]}$$

and we use (2.4.2) to obtain

$$\left(\frac{E_{\mathrm{OUT}}(s)}{E_{\mathrm{IN}}(s)}\right)_{\mathrm{integrator}} = -\frac{1}{R_1/(1+sR_1C_1)}\,\frac{A_0}{1+A_0+sT_0}\times$$

$$\frac{1}{1/R_2 + sC_2 + (1/R_1 + sC_1)[(1+sT_0)/(1+A_0+sT_0)]}$$

$$\approx -\frac{1}{R_1(1-sR_1C_1)[1+(T_0/A_0)s]}\times$$

$$\frac{1}{1/R_2 + sC_2 + (1/R_1 + sC_1)(1+sT_0)/(A_0+sT_0)}$$

for low frequencies, since R_1C_1 is small and A_0 is large. Also,

$$\frac{1+sT_0}{A_0+sT_0} \approx \frac{1}{A_0}\left[1 + sT_0\left(1-\frac{1}{A_0}\right) - s^2\,\frac{T_0^2}{A_0}\right]$$

$$\approx \frac{1}{A_0}(1+sT_0)$$

again for low frequencies. It is easy to check that we may approximate the integrator response by

$$\boxed{\left(\frac{E_{\mathrm{OUT}}(s)}{E_{\mathrm{IN}}(s)}\right)_{\mathrm{integrator}} = -\frac{1}{K}\,\frac{1}{1+sT_{\mathrm{H}}}\,\frac{\mathrm{T_L}}{1+sT_{\mathrm{L}}}} \tag{2.4.6}$$

where

$$K = R_1C_2$$

$$T_{\mathrm{H}} = \frac{T_0}{A_0} - R_1C_1$$

$$T_{\mathrm{L}} = [(A_0R_1C_2)^{-1} + (R_2C_2)^{-1}]^{-1}$$

and the subscripts H and L refer to high- and low-frequency time constants respectively.

Finally, a similar analysis may be applied to the differentiator element, but this will not be done here since the differentiator is not such a useful computing element and suffers from noise problems.

2.4.4 Error Analysis of Analogue Computer Solutions of Differential Equations

Consider again the general linear differential equation (2.2.8), i.e.

$$a_n \frac{\mathrm{d}^n x}{\mathrm{d}t^n} - a_{n-1} \frac{\mathrm{d}^{n-1} x}{\mathrm{d}t^{n-1}} + \ldots \pm a_1 \frac{\mathrm{d}x}{\mathrm{d}t} \mp a_0 x = f(t) \tag{2.4.7}$$

where, as stated before, we choose the a_i to be positive for simplicity. Then the analogue circuit for this system appears in fig. 2.11 and the adder output $e(t)$ is given, in the s-domain, by

$$E(s) = A(s)\left\{E(s)\left[I(s)\frac{a_{n-1}}{a_n} + I^2(s)\frac{a_{n-2}}{a_n} + \ldots + I^{n-1}(s)\frac{a_1}{a_n} + I^n(s)\frac{a_0}{a_n}\right] - \frac{F(s)}{a_n}\right\} \tag{2.4.8}$$

where $I(s)$ and $A(s)$ are the integrator and adder transfer functions respectively. In the ideal case we have

$$A(s) = -1, \quad I(s) = -\frac{1}{s}$$

and then it follows from (2.4.8) that

$$s^n - s^{n-1}\frac{a_{n-1}}{a_n} + s^{n-2}\frac{a_{n-2}}{a_n} - \ldots \pm s\frac{a_1}{a_n} \mp \frac{a_0}{a_n} = 0 \tag{2.4.9}$$

when $f = 0$. This is the familiar characteristic equation of the homogeneous form of the equation (2.4.7) and its roots specify the system behaviour.

However, if we substitute the approximate expressions for $I(s)$ and $A(s)$ derived in the last section, namely

$$I(s) = -\frac{T_{\mathrm{L}}}{1 + sT_{\mathrm{L}}}\frac{1}{1 + sT_{\mathrm{H}}}$$

and

$$A(s) = -\frac{1}{1 + sT_{\text{adder}}}$$

(ignoring the gain $1/K$ in the integrator), we have

$$[P(s)]^n - [P(s)]^{n-1}\frac{a_{n-1}}{a_n} + [P(s)]^{n-2}\frac{a_{n-2}}{a_n} - \ldots$$
$$\pm P(s)\frac{a_1}{a_n} \mp \frac{a_0}{a_n} = -sT_{\text{adder}}[P(s)]^n \tag{2.4.10}$$

where

$$P(s) = s^2 T_{\mathrm{H}} + s\left(1 + \frac{T_{\mathrm{H}}}{T_{\mathrm{L}}}\right) + \frac{1}{T_{\mathrm{L}}} \tag{2.4.11}$$

Hence, the characteristic equation of the real system even with the simple approximations to I and A above has $2n + 1$ poles instead of n. If n of the roots of (2.4.10) are close to the n roots of (2.5.9) while the other $n + 1$ have very large negative real parts, then the analogue circuit will be a good approximation to the ideal system. This will happen, of course, if T_L is large and T_H, T_{adder} are small.

We can write (2.4.10) in the form

$$C[P(s)] = -sT_{adder}[P(s)]^n \tag{2.4.12}$$

where $C(s)$ is the characteristic polynomial on the left of (2.4.9). Let the roots of $C(s) = 0$ be s_i, $1 \leqslant i \leqslant n$. Then, if for each i we have

$$\frac{1}{T_L} \ll |s_i| \ll \min\left\{\frac{1}{T_H}, \frac{1}{T_{adder}}\right\} \tag{2.4.13}$$

it follows that

$$\begin{aligned} |P(s_i) - s_i| &= \left| s_i^2 T_H + s_i\left(1 + \frac{T_H}{T_L}\right) + \frac{1}{T_L} - s_i \right| \\ &\leqslant |s_i|^2 T_H + |s_i| \frac{T_H}{T_L} + \frac{1}{T_L} \end{aligned}$$

which is small (relative to s_i). Hence we can expand $C[P(s)]$ in the form

$$C[P(s)] = [P(s) - s_i]C'(s_i) + \ldots$$

by Taylor's theorem (in a neighbourhood of each s such that $P(s) = s_i$). Thus

$$[P(s) - s_i]C'(s_i) = -sT_{adder}[P(s)]^n$$

by (2.4.12), and if we put $s = s_i + e_i$ where e_i is the error in the ith root, it is easy to see that

$$\left(s_i^2 T_H + 2s_i e_i T_H + e_i^2 T_H + e_i + s_i \frac{T_H}{T_L} + e_i \frac{T_H}{T_L} + \frac{1}{T_L}\right) C'(s_i)$$

$$= -T_{adder}(s_i + e_i)\left(s_i^2 T_H + 2s_i e_i T_H + e_i^2 T_H + s_i + e_i + s_i \frac{T_H}{T_L} + e_i \frac{T_H}{T_L} + \frac{1}{T_L}\right)^n$$

Using (2.4.13) this simplifies to

$$\left(s_i^2 T_H + e_i + \frac{1}{T_L}\right) C'(s_i) = -T_{adder}\, s_i^{n+1}$$

or

$$e_i = -\frac{T_{adder}\, s_i^{n+1}}{C'(s_i)} - T_H s_i^2 - \frac{1}{T_L}$$

This expression gives the errors in the n roots of (2.4.10) which are close to those of (2.4.9). Assuming that the remaining $n+1$ roots of (2.4.10) are far into the left half plane, then for s in a neighbourhood of any of these roots we have

$$|s| \gg 1, \quad |s|^n \gg |s|^{n-1} \gg |s|^{n-2} \gg \ldots$$

and so (2.4.10) simplifies to

$$(s^2 T_{\mathrm{H}} + s)^n \approx -sT_{\mathrm{adder}}(s^2 T_{\mathrm{H}} + s)^n$$

Hence

$$s^n(1 + sT_{\mathrm{H}})^n(1 + sT_{\mathrm{adder}}) \approx 0$$

and the remaining $n+1$ roots are close to $-1/T_{\mathrm{adder}}$ and $-1/T_{\mathrm{H}}$.

The above analysis is due to Macnee (1952). A more extensive analysis is given by Dow (1958) and Nelson (1963). We have seen in the analysis that when we solve an equation of the form (2.4.7) on an analogue computer, we introduce at least $n+1$ spurious roots into the system. These extra roots will introduce damping into the true solutions and may be tested on a computer by solving the simple harmonic equation

$$\ddot{x} = -\omega^2 x$$

which ideally, of course, has no damping.

2.4.5 Automatic Drift Stabilization

All real amplifiers are subject to drifting due to changes in the biasing conditions. Consider the simple feedback circuit in fig. 2.47, where the drift in the amplifier is modelled by an equivalent input ε and the feedback element β is some passive impedance. Then we have

$$e_0 = \frac{\varepsilon A_2}{1 - A_2\beta}$$

Suppose we add a drift-free amplifier A_1 in the input to A_2 as shown in fig. 2.48. Then

$$e_0 = \frac{\varepsilon A_2}{1 - A_2\beta - A_1 A_2 \beta}$$

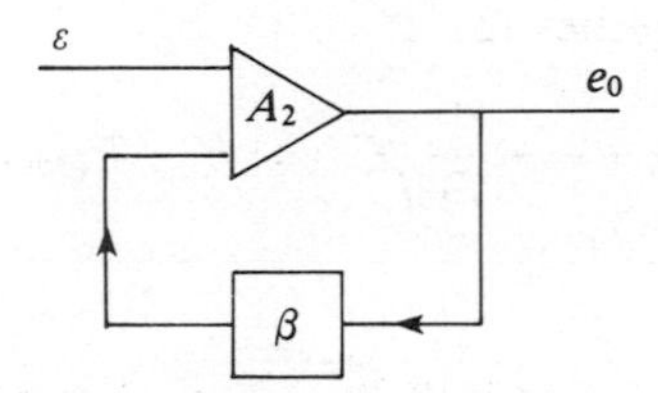

Fig. 2.47 Feedback amplifier with drift.

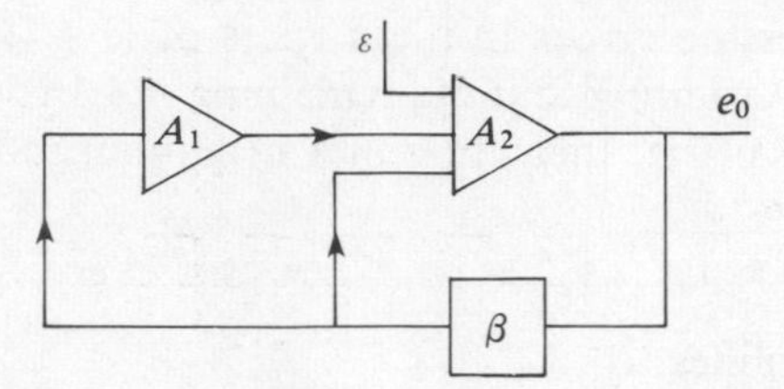

Fig. 2.48 Block diagram of automatic drift stabilisation.

If $\beta A_2 \gg 1$ and $A_1 \gg 1$, then in the first case the drift output is approximately ε/β and in the second $\varepsilon/A_1\beta$ and so the drift-free amplifier has the effect of reducing the drift output of A_2 by the factor $1/A_1$. Since drift is due to shifting d.c. biases in the amplifier, we can produce an effectively drift-free amplifier A_1 by capacitively connecting a high gain amplifier to A_2. However, the amplifier A_1 will not then respond to d.c. inputs and so the input to A_1 can be amplitude modulated on a square wave input as in Goldberg's arrangement shown in fig. 2.49. The amplitude modulation and demodulation is produced by a switch which grounds the inputs and outputs of the capacitively coupled A_1 amplifier, and the R_2C circuit is a low pass filter removing the high frequencies due to the chopper, which is usually a photoresistor and a flashing neon light. If the input e_a to A_2 is slowly varying with respect to the chopper frequency then we obtain the voltage waveforms at A, B and C shown in fig. 2.50. Note that the input and output at A and C are grounded in alternate time intervals. Hence if A_1 has no sign reversal then e_a and e_b have opposite signs as they should. Using this arrangement the drift output is reduced by a factor of $1/A_1$ as shown above.

Note that circuits of this kind are now available on precision ICs and instrumentation amplifiers; see, for example, Radiospares Catalogue.

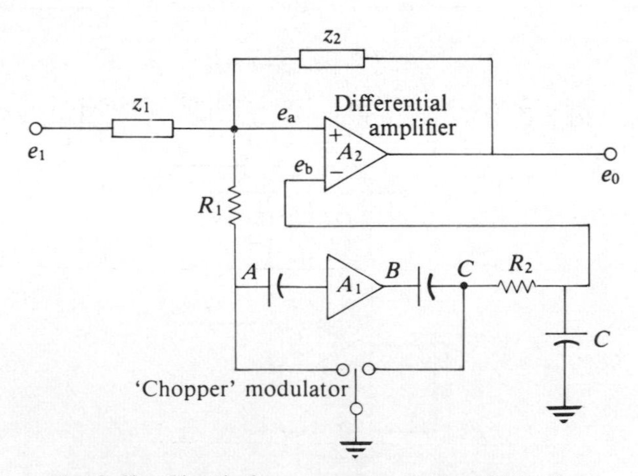

Fig. 2.49 Circuit for automatic drift stabilisation.

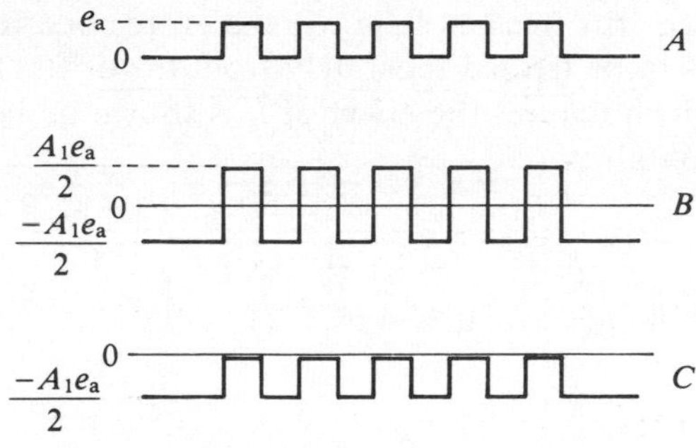

Fig. 2.50 Pulse waveforms for fig. 2.49.

2.4.6 Effect of Finite Input Impedance

Consider again the simple integrator circuit with a drift-free chopper modulated amplifier as shown in fig. 2.51 where we allow for the fact that the integrator amplifier has finite input impedance. A small input current i_g will then flow from the summing junction to the integrator amplifier and an elementary calculation shows that

$$E_0 \approx \frac{-E_1}{sCR} + \frac{I_g}{sC} - \frac{\varepsilon}{1 + A_1}\,\frac{1 + sCR}{sCR} \tag{2.4.14}$$

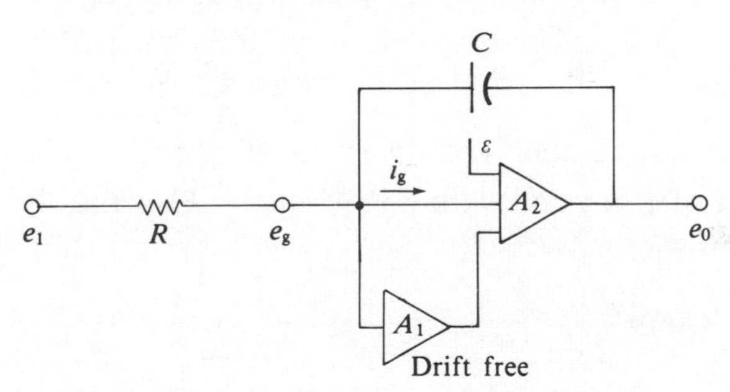

Fig. 2.51 Integrator with modulated amplifier.

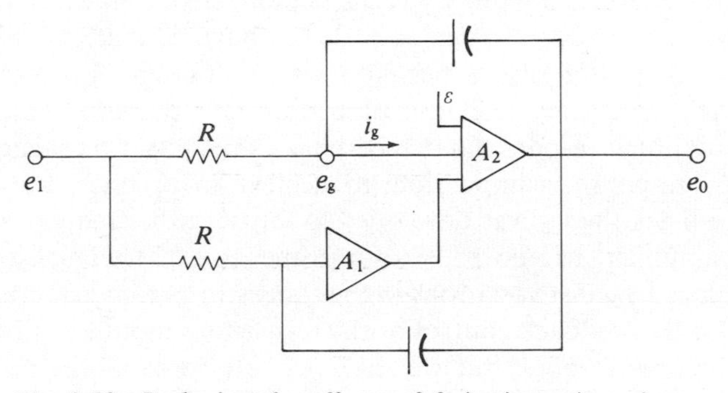

Fig. 2.52 Reducing the effects of finite input impedance.

for $A_1A_2 \gg 1$. The drift free amplifier will therefore not affect the input current i_g which has to be treated separately from the drifts. A system due to Hamer (1954) which reduces the effect of i_g is shown in fig. 2.52. Again, a simple calculation gives

$$E_0 \approx \frac{-E_1}{sCR} + \frac{I_g}{sC}\frac{1}{1+A_1} - \frac{\varepsilon}{1+A_1}\frac{1+sCR}{sCR} \tag{2.4.15}$$

and the effective input current is now $i_g/(1+A_1)$.

2.4.7 Amplifier Design

The satisfactory design of amplifiers depends on the use of compensation techniques which we shall discuss in chapter 4. However, to gain some insight into the application of these methods, we shall give an introduction to the basic ideas here. The general theory will then be motivated by the real amplifier examples.

In equation (1.3.12) we have seen that a general passive network may be described by a system of equations of the form

$$ZI = E$$

where $Z = (Z_{ij})$ is an impedance matrix and $I = (I_1, \ldots, I_N)^T$ and $E = (E_1, \ldots, E_N)^T$ are the vectors of loop currents and voltages respectively. (This is just a generalized Ohm's law.) If E_1 is regarded as the system input and $E_2 = \ldots = E_N = 0$, then the output current, say I_N, is given by

$$I_N = [Z^{-1}(E_1, 0, \ldots, 0)^T]_N$$

Hence the transfer function is the output voltage E_N across the output load (say Z_{NN}) divided by E_1, i.e.

$$G(s) = \frac{E_N}{E_1} = \frac{Z_{NN}I_N}{E_1} = \frac{a_0s^m + a_1s^{m-1} + \ldots + a_m}{b_0s^n + b_1s^{n-1} + \ldots + b_n}$$

for some constants a_i, b_j. If we put $s = j\omega$ then we obtain the frequency response $G(j\omega)$ of the system. By replacing an active element (a transistor) by a passive linear model we can extend the above derivation and obtain a transfer function $G(j\omega)$ for a passive network containing transistors.

The ideal frequency response of a transistor amplifier consists of a constant amplitude response and zero phase shift at all frequencies. Such a frequency response is impossible to achieve in practice, but much of classical feedback theory was developed to improve the frequency response of a basic amplifier. Moreover, with a proper use of feedback the system can be stabilized against variations in the transistor parameters. A complete discussion of the feedback control of the transistor amplifiers is beyond the scope of the present book but we shall indicate some of the main points involved (see also Millman and Halkias, 1972).

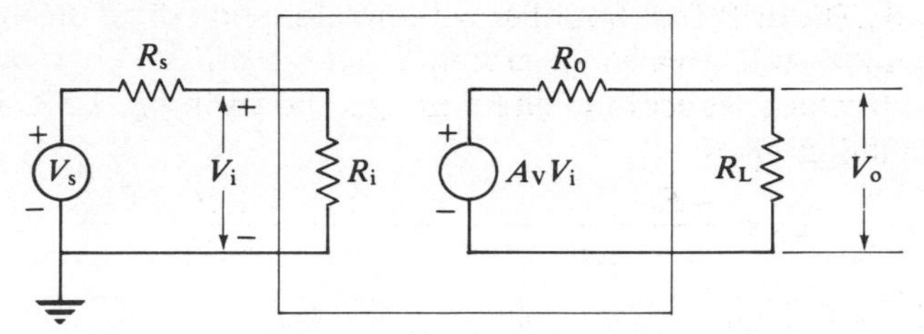

Fig. 2.53 An op-amp model.

Consider the voltage amplifier in fig. 2.53 where the input and output circuits have been replaced by their Thévenin equivalents (see section 1.3.3). Ideally, the input resistance R_i should be infinite and the output resistance R_o should be zero. Then

$$V_i \approx V_s \quad \text{and} \quad V_o \approx A_V V_i$$

and therefore

$$V_o \approx A_V V_s$$

so that A_V is the voltage amplification. A basic amplifier is usually incorporated in a feedback system of the form shown in fig. 2.54. Note that it is assumed that the input signal passes only through the amplifier and that the feedback network passes a signal only from output to input. Since both the amplifier and feedback networks are, in reality, bidirectional this will only be approximately true. The summing device at the input is implemented by using a differential amplifier. Now

$$X_d = X_s - X_f = X_s - \beta X_o$$

and $X_o = AX_d$. Hence

$$A_f \triangleq \frac{X_o}{X_s} = \frac{A}{1 + \beta A}$$

$1/(1 + \beta A)$ is called the sensitivity of the feedback amplifier while $D \triangleq 1 + \beta A$ is the **desensitivity**. Note that

$$\left| \frac{\delta A_f}{A_f} \right| = \frac{1}{|1 + \beta A|} \left| \frac{\delta A}{A} \right|$$

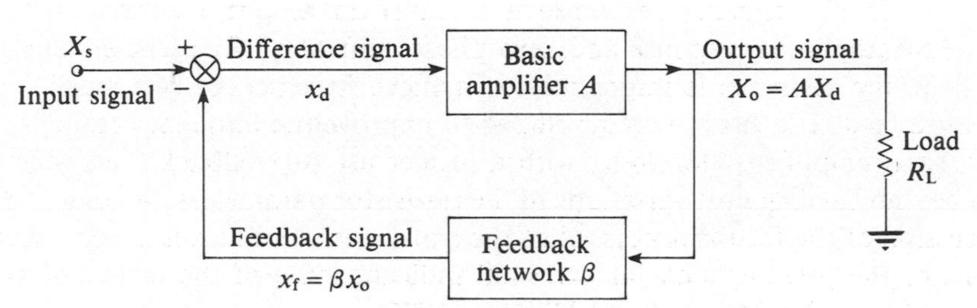

Fig. 2.54 General feedback amplifier configuration.

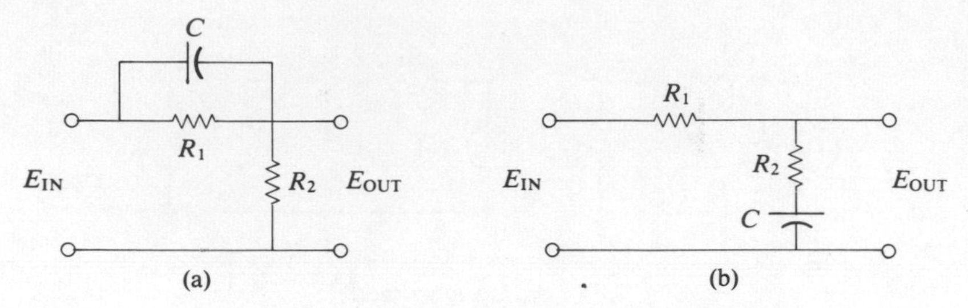

Fig. 2.55 Simple RC networks: (a) Lead; (b) lag.

for changes dA_f and dA in the feedback and open-loop amplifications, respectively. If $|\beta A|$ is large the sensitivity of the feedback amplifier to changes in the amplification is much smaller than that of the open-loop system. Also

$$A_f \cong \frac{A}{\beta A} = \frac{1}{\beta}$$

and so the amplification can be made (approximately) independent of the open-loop value A. It can also be shown that feedback has the effect of

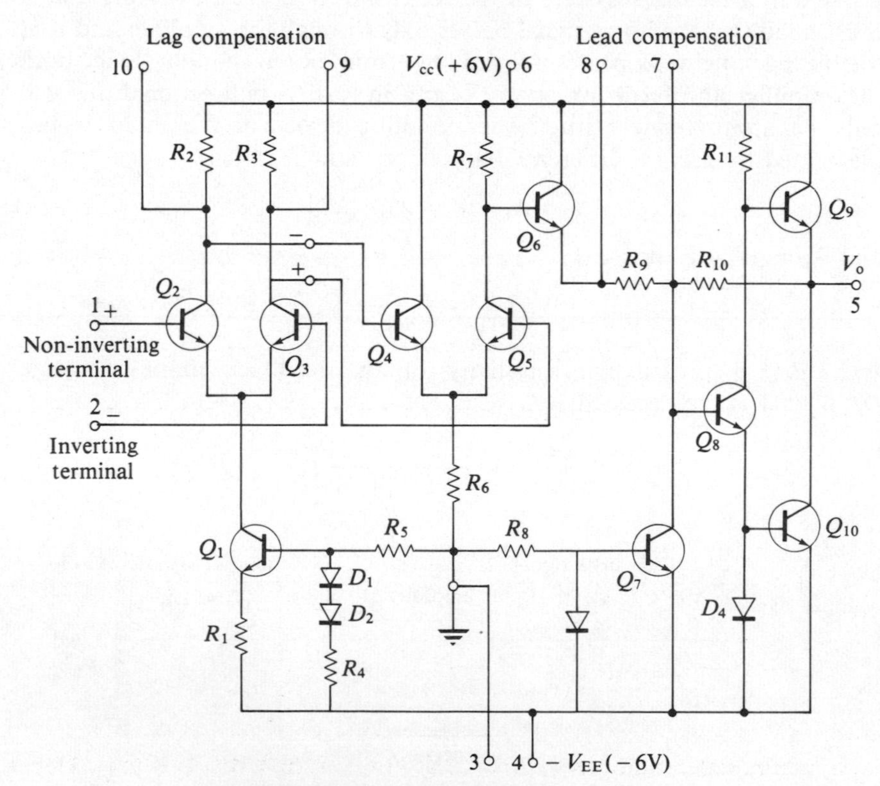

Fig. 2.56 The Motorola MC1530 op-amp.

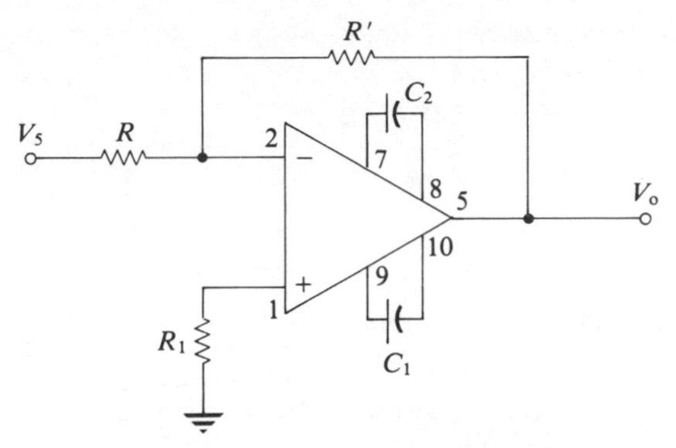

Fig. 2.57 Connecting the MC1530.

reducing non-linear distortion and noise by the same factor $1/|1+\beta A|$. Note that, for negative feedback, $|A_f| < |A|$ and so there is a drop in amplification.

Feedback is therefore very important in stabilizing an amplifier against changes in the system parameters. However, care must be taken against the introduction of unwanted oscillations at high frequencies. This involves an application of the Nyquist criterion which will be discussed in detail later.

Having achieved the correct basic feedback design, it is then necessary to examine the frequency response $G(j\omega)$ of the overall system. This is most conveniently shown by plotting $20 \log_{10} G(j\omega)$ and $\arg G(j\omega)$ against $\log_{10} \omega$. These two graphs constitute the **Bode diagrams** of the system, which again will be discussed in detail later. However, we can note now that the basic idea of amplifier 'compensation' is to add passive networks in the forward path (or by altering the β network) in order to change the frequency response to a more desirable form. The two basic types of compensators are shown in fig. 2.55, and are called the lead and lag networks. The lead network (as we shall see in chapter 4) affects the high frequency response, while the lag network improves low frequency response.

A typical operational amplifier is the Motorola MC1530 shown in fig. 2.56 and consists of a differential amplifier at the input feeding into another differential amplifier with single-ended output. Then an emitter follower ($Q6$) feeds the output stage which provides level translation and symmetric output swings. Inputs are provided for lead and lag compensators and a typical connection is shown in fig. 2.57.

2.5 EXERCISES

1. Consider the potentiometer in fig. 2.58 with resistive loading. Find the value of a so that we obtain $e_0 = ke_1$.

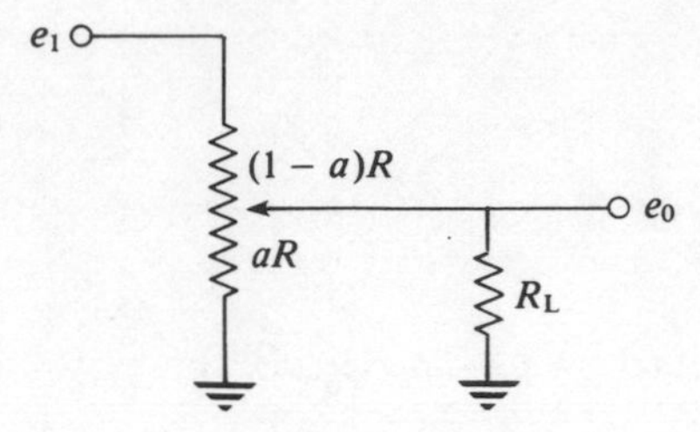

Fig. 2.58 A simple voltage divider with resistive load.

2. Design a computer circuit to integrate the equation

$$\frac{d^2x}{dt^2} + 35\frac{dx}{dt} + 210x = 500$$

and scale the system for maximum amplifier output voltages of 100 volts.

3. Design a computer system to simulate the equation

$$\frac{d^2x}{dt^2} + a\frac{dx}{dt} + bx = \sin t$$

4. Discuss an analogue network for the solution of the heat conduction equation

$$\frac{\partial\phi}{\partial t} = \frac{\partial^2\phi}{\partial x^2}$$

5. Design a circuit to model the transfer function

$$\frac{x}{u} = \frac{1}{s^2 + s + 1}$$

6. Using multipliers, design an analogue system to solve Van der Pol's equation

$$\frac{d^2x}{dt^2} + \left(\frac{dx}{dt}\right)^3 - \frac{dx}{dt} + x = 0$$

7. Discuss the flip-flop circuit in fig. 2.59 where A is a control input signal with magnitude $> ke_Z$. (*Hint*: use fig. 2.32).

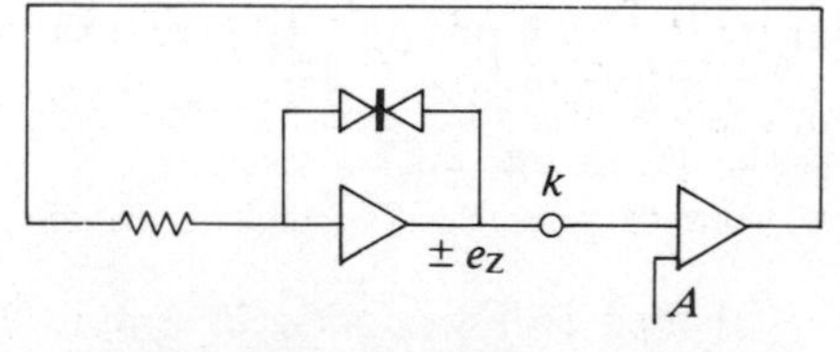

Fig. 2.59 A flip-flop circuit.

8. Repeat exercise 7 for the modified circuit in fig. 2.60 and show that we obtain the characteristics in fig. 2.61.

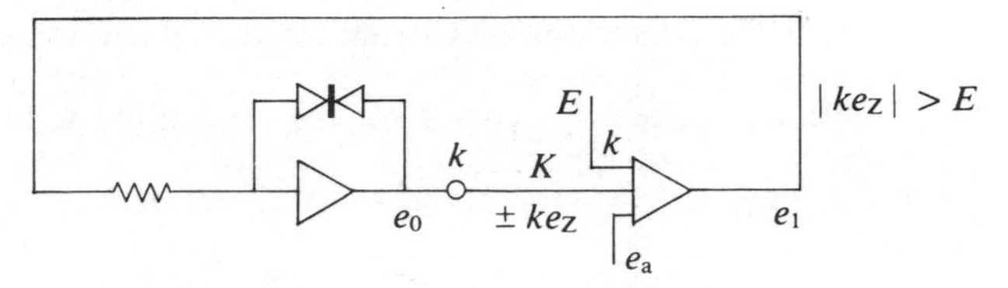

Fig. 2.60 A flip-flop circuit with hysteresis.

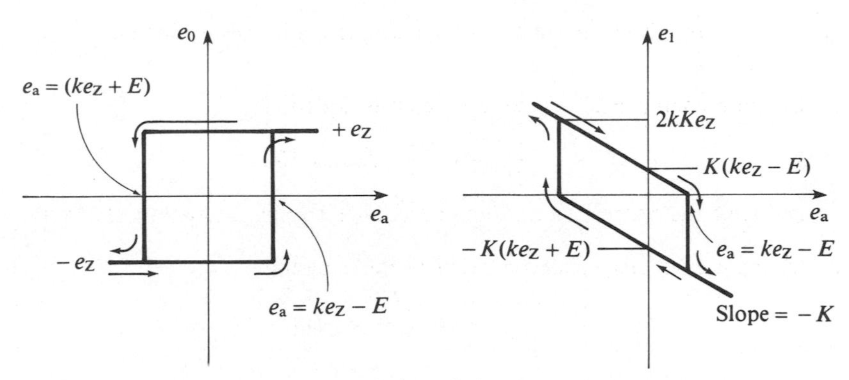

Fig. 2.61 Characteristics of the circuit in fig. 2.60.

9. Derive equation (2.4.6) in detail.

10. Discuss the effect of feedback (as in fig. 2.54) on the bandwidth of an amplifier with a single pole

$$A = \frac{A_0}{1 + j(f/f_H)}$$

where f_H is the 3-dB point.

REFERENCES

Banks, S. P. and Abbasi-Ghelmansarai, F. (1983) Delay equations, the left-shift operator and the infinite-dimensional root locus, *Int. J. Control*, 37, 235–49.

Clayton, G. B. (1979) *Operational Amplifiers*, 2nd edn, Butterworths.

Cunningham, W. J. (1954) Time-delay networks for an analog computer, *IRE Trans. Electron. Computers*, 3 (4), 16–18.

Dow, P. C. (1958) An analysis of certain errors in electronic differential analyzers, *IRE Trans. Electron. Computers*, 7 (1), 17–22.

Fogarty, L. E. (1966) Computer generation of arbitrary functions, *Simulation*, 7 (2), 80–9.

Hamer, H. (1954) A stabilised driftless analog integrator, *IRE Trans. Electron. Computers*, 3 (4), 19–20.

Haubner, A and Furlani, C. M. (1966) Chebyshev All-pass Approximants for Time-Delay Simulation, *IEEE Trans. Electron. Computers*, EC-*15*, 314–21.

Horowitz, P and Hill, W. (1980) *The Art of Electronics*, Cambridge University Press.

Macnee, A. B. (1952) Some limitations on the accuracy of electronic differential analysers, *Proc. IRE*, 40 (3), 303–8.
Meissinger, H. F. (1955) An electronic circuit for the generation of functions of several variables, *IRE Conv. Record*, part 4, 150–61.
Millman, J. and Halkias, C. C. (1972) *Integrated Electronics*, McGraw-Hill Kogukusha.
Nelson D. J. (1963) A fundamental error theory for analog computers, *IEEE Trans. Electron. Computers*, 12 (5), 541–50.
Radiospares Catalogue (1985) R. S. Components Ltd., Northants.
Saucedo, R. and Schining, E. E. (1968) *Introduction to Digital and Continuous Control Systems*, Collier–Macmillan.
Wilkinson, R. H. (1963) A method of generating functions of several variables using diode logic, *IEEE Trans. Electron. Computers*, 12 (2) 112–29, 550.

3 DIGITAL SIMULATION

3.1 INTRODUCTION

One of the most important aspects of control engineering is the digital simulation of the mathematical model of the system under consideration. In the last chapter we have seen that analogue computers are important for simulating differential equations in real time, the solutions being analogue signals. One of the main drawbacks with analogue computers is the need for special circuitry for each type of non-linearity in the system. In digital simulation we convert a differential equation, by some form of approximation, into a difference equation whose solutions are supposed to be sufficiently close to those of the original equation at certain discrete times. The problem of non-linearities in the equation is much simpler in digital than in analogue simulation, since the calculation of the solutions of the difference equation representing the physical system consists merely of a numerical evaluation of the right-hand side of the equation, which may be carried out on the general purpose hardware of a digital computer. The main disadvantage of digital simulation is the error which is introduced in the simulation procedure. This error originates from two sources: firstly, the fact that we have replaced a differential equation by a difference equation means that we have a **discretization error**, so that the solution of the true equation and that of the discrete model will not be equal even at the sampling times. Secondly, the numbers which we obtain for the solution of the discrete equation at the sampling times may be represented exactly by real numbers with many terms in the fractional part. Since a computer memory cannot store an arbitrarily large number of decimal digits at each sampling time we must **truncate** these numbers so that they will fit on a set of computer memories of fixed length. This second type of error is called quantization error.

These error processes are extremely important in digital simulation and a student must learn to examine the results of a simulation critically and not fall into the all too frequent trap of assuming that what is produced by a computer must be correct. In the absence of a complete error analysis of a particular difference equation (in terms of the two types of errors discussed above), the validity of any simulation results may only be judged by intuition and experience obtained by a large number of practical examples. Since a complete error analysis is usually very difficult in a complex system, these two qualities form the basis of any good systems analyst.

In order to understand more fully the process of digital simulation and the errors involved, it is necessary to study the basic operation of a digital computer. Therefore, before proceeding to the mathematical theory of simulation we shall continue this chapter with an introduction to digital computer structure and software. A more detailed study of microcomputer systems will be presented later.

3.2 DIGITAL COMPUTERS

3.2.1 Basic Computer Structure

Digital computers (whether mainframe, mini or micro†) usually consist of the main units shown in fig. 3.1. The central processor consists of the control and arithmetic/logic units and is the place where the overall control and arithmetic computations are performed. If the central processor is put on a single chip then we refer to it as a microprocessor. More will be said about the special structure of microprocessors later in the book.

The memory is just a collection of storage locations where the program and data are held until they are needed by the central processor. The main types of memory are magnetic disks (floppy disks in microcomputers), magnetic tapes, bubble memories and LSI memory chips (used in random access memory (RAM) and read only memory (ROM), see chapter 9).

We must be able to communicate with a computer, of course, and so various types of input and output devices are connected to the central computer and consist mainly of video screens and typewriter keyboards, magnetic tape units, punched cards and tapes, and printers. The method by which the central processor 'recognizes' these units and acquires data from them will be discussed in chapter 9 and is called interrupt handling.

It is important to recognize that a computer which consists of the above units has no intrinsic 'intelligence' and can only perform simple operations such as adding two numbers together and comparing a number with zero. The sophistication in computer applications is the **software** (i.e. the stored

† The distinction between mainframe (very large computers with many users), minicomputers (with typically about 20 users) and microcomputers (single-user systems) is somewhat arbitrary.

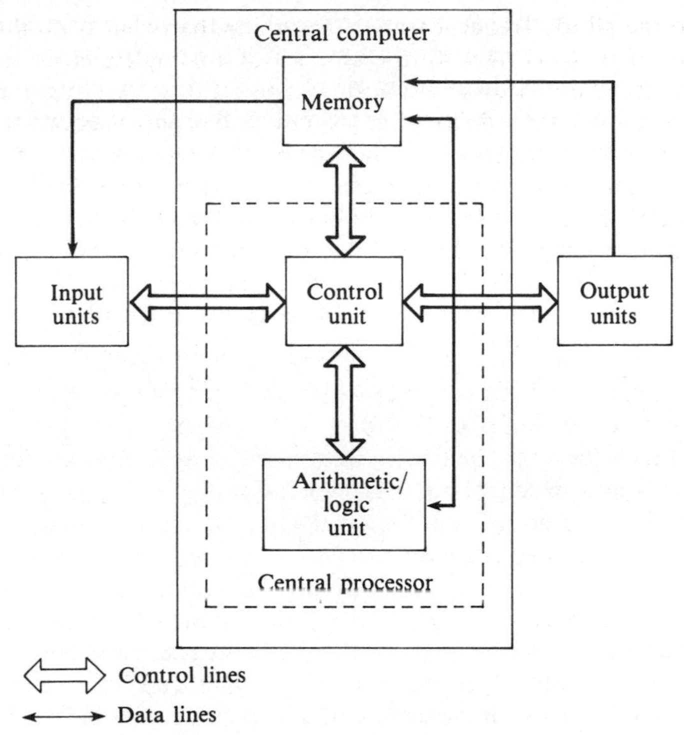

Fig. 3.1 The structure of a general digital computer.

program written by the user) which tells the computer exactly what to do. The computer system itself is called the **hardware**.

3.2.2 Number Systems and Codes

The set of natural (i.e. non-negative integral) numbers is defined by

$$N = \{\emptyset, \{\emptyset\}, \{\emptyset, \{\emptyset\}\}, \{\emptyset, \{\emptyset, \{\emptyset\}\}\}, \ldots\}$$

where $\emptyset$ is the empty set. If n is a natural number, then $n^+ = n \cup \{\emptyset\}$ denotes the **immediate successor** of n and so we may regard N as the set containing $\emptyset$, $\emptyset^+$, $(\emptyset^+)^+$, $((\emptyset)^+)^+$, ... This notation is clearly very cumbersome for large natural numbers and so in any counting system we single out a particular natural number, say r, and count in groups of r. r is called the **base**, or **radix**, of the number system and it may be shown that any natural number n may be written in the form

$$n = a_m r^m + a_{m-1} r^{m-1} + \ldots + a_1 r^1 + a_0 r^0 \tag{3.2.1}$$

(or $a_m a_{m-1} \ldots a_0$, if the base r is understood), where m is a natural number and a_i, $0 \leqslant i \leqslant m$ are natural numbers which are strictly less than r. By varying r, we may count in different bases, but it is important to understand that the same natural number n expressed in two different bases has an intrinsic existence independent of the base. If we wish to emphasize that the natural number n is expressed in the base r, we shall write n_r. For example, in the familiar base 10, the number $(4129)_{10}$ has the obvious representation (cf. (3.2.1))

$$(4129)_{10} = 4 \times 10^3 + 1 \times 10^2 + 2 \times 10^1 + 9 \times 10^0$$

In base ten, the numbers a_i $(0 \leqslant i \leqslant m)$ in (3.2.1) can take any of the values 0, 1, 2, . . . , 9. It should be noted, however, that the symbol 10 is not a single symbol in base ten, but is of the form (3.2.1) where $a_1 = 1$ and $a_0 = 0$.

The other important number systems used in computers are the **binary systems** with just two digits 0 and 1, the **octal system** with digits 0, 1, . . . , 7 (the symbol 8 does not exist in the octal system), and the **hexadecimal** (or **hex**) **system**. The latter has base sixteen and so we require sixteen distinct symbols which the a_i in (3.2.1) will take as values. We choose these symbols† to be 0, 1, 2, . . . , 9, A, B, C, D, E, F. Since binary numbers contain only 0s and 1s they are used in computer circuitry where transistors are (almost exclusively) operated in saturation (ON or 1) or OFF (0) states. Hence binary numbers can be represented by a group of transistor switches in appropriate states. Reading binary numbers is, however, difficult for human operators so the contents of memories are often translated (by the computer) to hexadecimal (or less commonly octal). For example, in an 8-bit‡ microprocessor (i.e. one whose data lines (see chapter 9) have 8 parallel paths), values are stored as 8-bit binary numbers which may be converted to 2-bit hexadecimal numbers. This is possible since four binary digits can correspond exactly to a single hexadecimal digit as can be seen from table 3.1.

Suppose now that we wish to convert the natural number n expressed in base r (i.e. n_r) to the corresponding representation n_ρ in base ρ. Then we have

$$\begin{aligned} n = n_r &= a_m r^m + a_{m-1} r^{m-1} + \ldots + a_0 r^0 \\ = n_\rho &= \alpha_\mu \rho^\mu + \alpha_{\mu-1} \rho^{\mu-1} + \ldots + \alpha_0 \rho^0 \end{aligned}$$

The numbers a_i are known and we wish to find the numbers α_i. Clearly

$$\frac{n_r}{\rho} = \alpha_\mu \rho^{\mu-1} + \alpha_{\mu-1} \rho^{\mu-2} + \ldots + \alpha_1 \rho^0 \quad \text{remainder } \alpha_0$$

† Strictly speaking, one should introduce entirely new symbols.
‡ A **bit** is a single **binary digit**.

Table 3.1

Hex	*Binary*	*Hex*	*Binary*
0H	0000	8H	1000
1H	0001	9H	1001
2H	0010	AH	1010
3H	0011	BH	1011
4H	0100	CH	1100
5H	0101	DH	1101
6H	0110	EH	1110
7H	0111	FH	1111

and

$$\left(\frac{n_r}{\rho} - \frac{\alpha_0}{\rho}\right)/\rho = \alpha_\mu \rho^{\mu-2} + \ldots + \alpha_2 \rho^0 \quad \text{remainder } \alpha_1$$

It is easy to see that we may continue this process, and so to obtain n_ρ from n_r we successively divide n_r by ρ and write down the remainders, which as we have just seen are $\alpha_0, \alpha_1, \ldots \alpha_\mu$ (in that order).

Up to now we have considered only integral numbers which have no fractional part. The expression (3.2.1) may be extended to include fractional numbers in base r in the obvious way, i.e. any real number R with terminating decimal in base r may be written

$$R = a_m r^m + a_{m-1} r^{m-1} + \ldots + a_0 r^0 + a_{-1} r^{-1} + a_{-2} r^{-2} + \ldots + a_{-l} r^{-l}$$

$$(= a_m a_{m-1} \ldots a_0 \cdot a_{-1} a_{-2} \ldots a_{-l})$$

The method of converting fractional numbers from one base to another is left to the reader. Note that a number may have a terminating fractional expansion in one base and not in another.

The arithmetic of numbers in a fixed base is very similar to that of numbers in the familiar base of ten and so we shall not discuss this in any detail. However, when performing subtraction in base 2, the standard method is often replaced by the method of complements which will now be described. (This method simplifies the computer hardware). Since we shall consider an 8-bit microprocessor later, the method of complements will be related to 8-bit binary numbers. The same ideas apply, of course, to binary numbers of arbitrary (but fixed) length. First we define the two's complement of an 8-bit binary number $n_2 = b_7 b_6 \ldots b_0$ as $2^8 - n_2 = 100000000 - b_7 b_6 \ldots b_0$ and it will be denoted by $\bar{\bar{n}}_2$. The two's complements of the (8-bit) binary equivalents of the numbers 1 to 128 are shown in table 3.2.

Alternatively, we may regard the numbers in the two columns of table 3.2 as the binary sign-magnitude representations of the decimal numbers

Table 3.2

Decimal	*Binary, b*	*Two's complement,* $\bar{\bar{b}} = 2^8 - b$
1	00000001	11111111
2	00000010	11111110
3	00000011	11111101
⋮	⋮	⋮
127	01111111	100000001
128	10000000	100000000

$-128, -127, \ldots, -1, 0, 1, \ldots, 127$ where the bit in the 2^7 place is the sign bit (0 for a positive number and 1 for a negative number). This gives a representation of positive and negative numbers where the sign is replaced by a code, as shown in table 3.3.

It is easy to verify now that the arithmetic (i.e. addition and subtraction) of signed numbers between -128 and 127 is equivalent to addition of the corresponding sign-magnitude representations, the answer again being the sign-magnitude equivalent of the correct decimal result, provided that the answer also lies between -128 and 127 and we neglect any overflow bit into the 2^8 place. Of course, if the answer is outside the range -128 to 127 then the result of adding the sign-magnitude equivalents will not be the correct sign-magnitude answer (which can only happen when adding numbers of the same sign) and can be detected by a change in the sign bit. The advantage of two's complement arithmetic, of course, is that computers do not require special subtraction circuitry, but require only circuitry for addition and complementation.

Binary digits may be used not only for direct representation of natural numbers as above, but also for indirect representation of numerical and alphabetic data in the form of **codes**. There are many types of computer codes, but we shall discuss only the most important ones. Since we are all

Table 3.3

Decimal		*Binary sign-magnitude representation*
−128		10000000
−127		10000001
⋮		⋮
−1	sign→	11111111
0	bit	00000000
1		00000001
2		00000010
⋮		⋮
127		011111111

Table 3.4

Decimal	*Binary*
0	0000
1	0001
2	0010
3	0011
4	0100
5	0101
6	0110
7	0111
8	1000
9	1001

familiar with the decimal number system it is important to be able to do decimal arithmetic on a computer. This is achieved by using **binary-coded decimal** (BCD) which consists of coding each decimal digit in a number by a four-bit binary equivalent as shown in table 3.4. Note that we require only equivalents for the digits $0, 1, 2, \ldots, 9$ and so the binary combinations $1010, \ldots, 1111$ are not used.

For example the BCD equivalent of $(385)_{10}$ is

$$(0011\ 1000\ 0101)_{BCD}$$

Addition is not entirely straightfoward, however, since for example $(5+8)_{10} = 13_{10}$ but if we convert 5 and 8 to BCD and add in binary, we obtain

$$(0101)_{BCD} + (1000)_{BCD} = (1101)_2$$
$$= (0001\ 0011)_{BCD}$$

Hence, we must have special circuitry which converts 1101 binary to the correct BCD answer $(0001\ 0011)_{BCD}$.

Another important code, used mainly in shaft position coding, is the **Gray code**. Its value lies in the fact that when counting in binary many bits may change on each count step, whereas in the Gray code only one bit changes each time. If

$$B = b_{n-1}b_{n-2} \ldots b_0$$

is a binary number, then the corresponding Gray code number is defined as follows:

(a) Add a 0 on the left-hand side, i.e. $0b_{n-1} \ldots b_0 = b_n b_{n-1} \ldots b_0$ where $b_n = 0$.
(b) Now define

$$g_i = \begin{cases} 0 & \text{if } b_{i+1} = b_i \\ 1 & \text{if } b_{i+1} \neq b_i \end{cases}$$

Table 3.5

Decimal	*Binary*	*Gray code*
0	0000	0000
1	0001	0001
2	0010	0011
3	0011	0010
4	0100	0110
5	0101	0111
6	0110	0101
7	0111	0100
8	1000	1100
9	1001	1101

It can easily be shown that the Gray code so defined has the required property that only one bit changes when we count in this code. For example, the Gray code equivalents of the first ten decimal numbers are shown in table 3.5.

Although computer circuits are now quite reliable and the probability of hardware faults is small, since we are frequently dealing with very large amounts of data some means of checking the validity of the data is important. We can provide a certain degree of checking by adding a **parity bit** onto each 'word' (i.e. a group of binary digits which can be stored in a single memory location). Suppose that we are using 7-bit words and we add an extra bit on the right-hand side as follows:

(a) For **odd parity** we add 1 or 0 so that the total number of 1s in the complete 8-bit word (including the parity bit) is odd.
(b) For **even parity** we add 1 or 0 so that the number of 1s is even.

For example, if we have a 7-bit word 0011100 then the completed word would be

(a) 00111000 for odd parity,
(b) 00111001 for even parity.

The computer can be made to check the parity of each word and (once the type of parity has been agreed on, i.e. even or odd) then a single bit error may be detected. Clearly, if two bits are in error, then no parity error will be detected and so the method relies on errors being very rare. Of course, even if only a single bit is in error this method will not be able to pinpoint the error in the word concerned, and so we may also add a **parity word** in every group of (say) seven data words and this will allow the computer not only to detect an error but also to determine which bit is in error (again assuming only a single error in that particular group of words). For example, if we have the data words (including an odd parity bit)

```
0111100|1  ← parity bit
1000011|0
1000111|1
0110111|0
0011000|1
0101100|0
1110011|0
-------|-
0110111|0  ← parity word
```

and suppose that the 0 in the box is changed to 1, then a parity error will be detected in the fourth row and the fourth column thus pinpointing the error.

The final type of coding which we shall mention is that which is required for the coding of input information from an input device, for example, a keyboard. Each keyboard character is changed to an 8-bit code which is then stored in (part of) a memory word. The main coding system used is the ASCII code (American National Standard Code for Information Exchange). A representative sample of ASCII codes is given in table 3.6.

Up to now we have discussed only memory words which store a binary representation without decimal point. In any real application, however, numbers with many decimal digits and a fractional part will be produced and have to be stored. In a computer a fixed number of memory words is chosen to store each real number and so a fixed finite number of digits (say n) may be stored in one location. Thus, if any real number has more than n digits the computer will 'round' the least significant digits from $n+1$ onwards. This introduces the **rounding error** (we say that the computer has **finite word length**) which is one of the main causes of error in a computer simulation.

For a further discussion of general computers and number systems the reader may consult the books of Nashelsky (1972) and Peatman (1972).

3.2.3 The Stored Program Concept and High Level Languages

We have mentioned above that the memory of a computer is organized into **words** which are groups of binary digits which can be processed in parallel by the central computer. The length of each word usually corresponds to the number of parallel lines in the data connections (or **buses**) between the memory and central computer. Each memory location can store a word of data which can be accessed by the central processor, operated on and replaced in the same or another memory location. However, the enormous power of the digital computer is due to the fact that we can use the memory

Table 3.6 ASCII (Hex) Codes of selected symbols

Character	*ASCII code*	*Character*	*ASCII code*
A	C1H	Blank	OOH
B	C2H	⊙	AEH
C	C3H	<	BCH
D	C4H	(	A8H
E	C5H	+	ABH
F	C6H	&	A6H
G	C7H	$	A4H
H	C8H	*	AAH
I	C9H	)	A9H
J	CAH	;	BBH
K	CBH	–	ADH
L	CCH	/	AFH
M	CDH	,	ACH
N	CEH	%	A5H
O	CFH	>	BEH
P	DOH	?	BFH
Q	D1H	:	BAH
R	D2H	#	A3H
S	D3H	@	COH
T	D4H	'	A7H
U	D5H	=	BDH
V	D6H	"	A2H
W	D7H		
X	D8H		
Y	D9H		
Z	DAH		
0	BOH		
1	B1H		
2	B2H		
3	B3H		
4	B4H		
5	B5H		
6	B6H		
7	B7H		
8	B8H		
9	B9H		

not only to store data but also to store the very operations which the computer is to apply to the data. Of course, each word of the memory consists of a group of binary digits and the computer cannot tell whether any given word in the memory is an operation or a piece of data. It is entirely the ordering of the data words, the operations and the place in the memory where we start the program which ensures that the operations and data are correctly interpreted.

The nature and number of operations available on a digital computer depend on the particular computer, but there are a great many similarities in the operations on different computers. At the lowest levels of programming (**machine language** or **assembly programming**) we must place the codes for these individual operations in the memory locations together with the data on which these operations are to act. Normally the program will proceed from one memory location to the next in serial order. However, operations exist for changing the order of the program. These are jump instructions which contain an address in memory to which the program must proceed. This memory location must, of course, contain an operation rather than data, for if not, the computer will interpret this data as if it were an operation, often with disastrous results. This need to be very careful with the exact contents of every memory address, although giving the programmer a great deal of flexibility, is also very tedious in large-scale programs. For this reason high level languages have been developed which are to a large extent independent of the particular computer and which automatically assign data to the correct memory locations. A program in such a high level language is **translated** to machine language by a special purpose program called a **compiler**. Much more will be said about assembly programming in chapter 8, but for now we shall use a particular high level language called FORTRAN 77 (**FOR**mula **Tran**slation) which is useful in simulation, and which is described by Ellis (1982).

3.3 NUMERICAL TECHNIQUES FOR ORDINARY DIFFERENTIAL EQUATIONS

3.3.1 Ordinary Differential Equations

In the following discussion on the numerical solution of differential equations, we shall use the standard notations of set theory. In particular, $\mathbb{R}$ will denote the real line or the set of real numbers, i.e. any number x of the form

$$x = (\pm\, a_n a_{n-1} \ldots a_1 a_0 \cdot a_{-1} a_{-2} \ldots)_{10}$$

written, for example, in base 10, where the fractional part may or may not terminate (i.e. $a_{-i} = 0$ for $i >$ some number j). Moreover, we shall denote, as usual, the set of n-dimensional vectors of real numbers by $\mathbb{R}^n$, i.e.

$$\mathbb{R}^n = \{x = (x_1, \ldots, x_n)^{\mathrm{T}} : x_i \in \mathbb{R}, 1 \leqslant i \leqslant n\}$$

(Note that we shall think of vectors in $\mathbb{R}^n$ as being columns, unless otherwise stated.)

A **function** f from $\mathbb{R}^n$ to $\mathbb{R}^m$ is a subset of $\mathbb{R}^n \times \mathbb{R}^m$ (the Cartesian product) such that if (x, y_1) and (x, y_2) are in f then $y_1 = y_2$.† If $(x, y) \in f$

† That is a function assigns a single vector in $\mathbb{R}^m$ to each vector in $\mathbb{R}^n$.

we normally write $y = f(x)$ and to indicate the sets of which f is defined we shall write

$$f: \mathbb{R}^n \to \mathbb{R}^m$$

If $f: \mathbb{R}^n \to \mathbb{R}^m$ is a function, then since $f(x) \in \mathbb{R}^m$ for any $x \in \mathbb{R}^n$, we can write

$$f(x) = [f_1(x), \ldots, f_m(x)]^{\mathrm{T}}$$

and hence we can associate with $f: \mathbb{R}^n \to \mathbb{R}^m$ m functions $f_i: \mathbb{R}^n \to \mathbb{R}$. We may now define a differential equation of the first order and dimension n as an expression of the form

$$\boxed{\frac{\mathrm{d}x}{\mathrm{d}t} = f(x, t)} \tag{3.3.1}$$

where $f: \mathbb{R}^{n+1} \to \mathbb{R}^n$ and

$$\frac{\mathrm{d}x}{\mathrm{d}t} = \left(\frac{\mathrm{d}x_1}{\mathrm{d}t}, \ldots, \frac{\mathrm{d}x_n}{\mathrm{d}t}\right)^{\mathrm{T}}$$

We shall think of t as **time** although it can be any independent variable. A **solution** of (3.3.1) is a function $x: \mathbb{R} \to \mathbb{R}^n$ which satisfies (3.3.1). We shall assume that all the differential equations to be considered here have unique solutions on the whole real line $(-\infty, \infty)$. Precise conditions for this to hold are given by Coddington and Levinson (1955). In order to specify the solution of (3.3.1) uniquely we must fix the solution at some time t_0, i.e. we specify $x(t_0)$ to be some given vector $x_0 \in \mathbb{R}^n$. This specification is called an **initial condition.**

An example of a differential equation of the form (3.3.1) for $n = 2$ is

$$\frac{\mathrm{d}x_1}{\mathrm{d}t} = x_1 x_2 + t^2$$

$$\frac{\mathrm{d}x_2}{\mathrm{d}t} = x_2 t$$

Here,

$$f_1(x_1, x_2) = x_1 x_2 + t^2$$

$$f_2(x_1, x_2) = x_2 t$$

If the function f in (3.3.1) is independent of explicit mention of t then the equation is said to be **autonomous** and we have

$$\boxed{\frac{\mathrm{d}x}{\mathrm{d}t} = f(x), \quad x(0) = x_0} \tag{3.3.2}$$

However, the simplest differential equations are the linear ones where f is given by

$$f(x) = Ax$$

and A is an $n \times n$ matrix. Then

$$\boxed{\frac{dx}{dt} = Ax, \quad x(0) = x_0} \tag{3.3.3}$$

and whereas the solution of (3.3.2) is difficult or impossible to write down, a solution of the linear equation (3.3.3) is easy to find. In fact, it is given by

$$\boxed{x(t) = [\exp(At)]\, x_0} \tag{3.3.4}$$

where

$$\boxed{\exp(At) = I + At + \frac{(At)^2}{2} + \frac{(At)^3}{3} + \dots} \tag{3.3.5}$$

It is easy to see that this series converges in the norm,† for any matrix A, just as in the scalar case. Here the **norm** of a matrix A is defined as

$$\| A \| = \sup_{\|x\| = 1} \| Ax \|$$

(See Exercise 7.)

As a final type of equation, we mention the linear nth order equations in the scalar variable ξ. These are of the form

$$\frac{d^n\xi}{dt^n} + a_{n-1}\frac{d^{n-1}\xi}{dt^{n-1}} + \dots + a_1\frac{d\xi}{dt} + a_0\xi = 0 \tag{3.3.6}$$

These equations are, in fact, special cases of the general linear equation (3.3.3) as can be seen by introducing the new variables

$$\boxed{x_1 = \xi,\ x_2 = \frac{d\xi}{dt},\ x_3 = \frac{d^2\xi}{dt^2}, \dots, x_n = \frac{d^{n-1}\xi}{dt^{n-1}}}$$

† That is, $\left\| \sum_{k=0}^{r} \frac{(At)^k}{k!} - \exp(At) \right\| \to 0$ as $k \to \infty$

Then we have

$$\left.\begin{aligned}\frac{dx_1}{dt} &= x_2 \\ \frac{dx_2}{dt} &= x_3 \\ &\vdots \\ \frac{dx_{n-1}}{dt} &= x_n \\ \frac{dx_n}{dt} &= -a_{n-1}x_n - \dots - a_1x_2 - a_0x_1\end{aligned}\right\} \qquad (3.3.7)$$

or

$$\frac{dx}{dt} = Ax$$

where $x = (x_1, \dots, x_n)^T$ and

$$A = \begin{bmatrix} 0 & 1 & & & & \\ & 0 & 1 & & & \\ & & 0 & 1 & & \\ & & & \ddots & \ddots & \\ & & & & 0 & 1 \\ -a_0 & -a_1 & \dots & & -a_{n-2} & -a_{n-1} \end{bmatrix}$$

(The elements not shown are understood to be zero.)

Even though we can solve linear ordinary differential equations exactly, evaluation of a matrix exponential is time consuming and it is usually better to use a general purpose numerical technique of the type which we shall discuss in the next section. This completes our introductory discussion on differential equations; for more details on the existence, uniqueness and qualitative properties of differential equations, see Coddington and Levinson (1955) and Lefschetz (1977).

3.3.2 Numerical Evaluation of Solutions

Since the analytic solution of general differential equations is not possible we must find other methods for the approximate evaluation of such solutions. These methods are based, essentially, on a Taylor series expansion of the right-hand side of the equation and the removal of the higher order terms. The resulting equation is then discretized and the difference equation obtained is easily solved by iteration. The accuracy of the method depends largely on the number of terms of the Taylor series retained in the difference equation and so we may consider a large number of different algorithms. However, we shall consider here only the ones which seem to be widely applied.

Consider therefore the differential equation (3.3.1) i.e.

$$\frac{dx}{dt} = f(x, t), \quad x(a) = x_0, \quad t \in [a, b] = I \tag{3.3.1}$$

where the initial value x_0 is prescribed. Suppose that we know a solution $x(t)$ of this equation exists, and assume that as a function of t it is real analytic. Then we can expand $x(t)$ in a Taylor series about some time t_1 (see appendix 1):

$$x(t) = x(t_1) + (t - t_1)\frac{dx(t_1)}{dt} + \frac{(t-t_1)^2}{2!}\frac{d^2x(t_1)}{dt^2} + \dots + \frac{(t-t_1)^n}{n!}\frac{d^nx(t_1)}{dt^n} + \dots \tag{3.3.8}$$

Although we do not know $x(t)$ a priori we can determine the derivatives in this expression from equation (3.3.1) assuming that f is sufficiently differentiable. In fact,

$$\frac{dx(t_1)}{dt} = f(x(t_1), t_1)$$

$$\frac{d^2x(t_1)}{dt^2} = \frac{df}{dt}(x(t_1), t_1) = \frac{\partial f}{\partial t}(x(t_1), t_1) + \frac{\partial f}{\partial x}(x(t_1), t_1)f(x(t_1), t_1)$$

$$\frac{d^3x(t_1)}{dt^3} = \frac{\partial^2 f}{\partial t^2} + \frac{2\partial^2 ff}{\partial x \partial t} + \frac{\partial^2 ff^2}{\partial x^2} + \frac{\partial f \partial f}{\partial t\, \partial x} + \left(\frac{\partial f}{\partial x}\right)^2 f$$

etc., where we have assumed for simplicity that the equation (3.3.1) is scalar, and we have omitted the arguments $(x(t_1), t_1)$ in the last expression. Hence, we can determine the derivatives of the solution in (3.3.8) at the time t_1 in terms of the right-hand side of (3.3.1) and its derivatives evaluated at $(x(t_1), t_1)$.

Now we define

$$T_k(x, t; h) = f(x, t) + \frac{h}{2!}f'(x, t) + \dots + \frac{h^{k-1}}{k!}f^{(k-1)}(x, t)$$

where $f^{(j)}(x, t)$ denotes the jth total derivative of $f(x(t), t)$ with respect to t, and we divide the time interval $I = [a, b]$ over which we wish to solve the equation into N equal subintervals of width $h = (b - a)/N$, i.e. put

$$t_0 = a, \quad t_n = a + nh$$

$$t_N = b$$

Then we write

$$x_n = x(t_n)$$

and we apply (3.3.8) with $t = t_{n+1}$ and $t_1 = t_n$ and we obtain

$$\boxed{x_{n+1} = x_n + hT_k(x_n, t_n; h) \quad n = 0, 1, \ldots, N-1} \tag{3.3.9}$$

Although (3.3.9) is only approximate, since we have omitted the terms in the Taylor series of order greater than $k-1$, we apply this as if it were an equality and we obtain **Taylor's algorithm** (of order k). From Taylor's theorem with remainder, it is easy to see that the local error is

$$E = \frac{h^{k+1} f^{(k)}(\xi, y(\xi))}{(k+1)!} = \frac{h^{k+1}}{(k+1)!} x^{(k+1)}(\xi)$$

for some $\xi \in [t_n, t_n + h]$.

This method can give very accurate approximations to the solution of equation (3.3.1) but suffers from the disadvantage of requiring (for $k > 1$) the numerical evaluation of the derivatives of f. Hence the method is used mainly when $k = 1$, in which case we obtain **Euler's method**

$$\boxed{x_{n+1} = x_n + hf(x_n, t_n)} \quad E = \frac{h^2}{2} x''(\xi) \tag{3.3.10}$$

Euler's method, however, has very poor accuracy and can be used with only very simple systems.

Note that, although we derived the above algorithms for scalar equations, it can be seen that they are also true for vector equations, but it is important to remember that the subscript n in (3.3.9) and (3.3.10) is an iteration parameter and does not represent the nth element of the vector x.

Example 3.3.1

As a simple example of Taylor's algorithm of order 2 consider the equation

$$\frac{dx}{dt} = \frac{1}{t^2} - \frac{x}{t} - x^2$$

Then

$$\frac{dx}{dt} = f(x, t) = \frac{1}{t^2} - \frac{x}{t} - x^2$$

$$\frac{df}{dt}(x, t) = -\frac{2}{t^3} - \frac{1}{t}\frac{dx}{dt} + \frac{x}{t^2} - \frac{2x\,dx}{dt}$$

$$= -\frac{3}{t^3} + \frac{3x^2}{t} + 2x^3$$

and so Taylor's algorithm of order 2 is

$$x_{n+1} = x_n + h\left[\frac{1}{t_n^2} - \frac{x_n}{t_n} - x_n^2 + \frac{h}{2}\left(-\frac{3}{t_n^3} + \frac{3x_n^2}{t_n} + 2x_n^3\right)\right]$$

As we have pointed out above, Taylor's method is either very inaccurate (as in the case $k = 1$) or requires the derivatives of f to be determined. In the **Runge–Kutta** methods we try to obtain accuracy without the need to evaluate the derivatives of f. This can be achieved by using information about f not only at the endpoints of the intervals $[t_n, t_{n+1}]$ but also at certain interior points of these intervals. We shall illustrate the procedure for the Runge–Kutta method of order 2. We wish to find a formula

$$x_{n+1} = x_n + ak_1 + bk_2 \tag{3.3.11}$$

where

$$k_1 = hf(x_n, t_n)$$

$$k_2 = hf(x_n + \beta k_1, t_n + \alpha h)$$

and we choose the constants a, b, α, β so that (3.3.11) agrees with the Taylor algorithm to as high a degree as possible. Now

$$x(t_{n+1}) = x(t_n) + hx'(t_n) + \frac{h^2}{2}x''(t_n) + \frac{h^3}{6}x'''(t_n) + \ldots$$

$$= x(t_n) + hf + \frac{h^2}{2}(f_t + ff_x)$$

$$+ \frac{h^3}{6}(f_{tt} + 2ff_{tx} + f_{xx}f^2 + f_tf_x + f_x^2f) + 0(h^4) \tag{3.3.12}$$

where all the functions in the last expression are evaluated at (x_n, t_n) and f_t, f_x etc. denote the appropriate partial derivatives.

On the other hand,

$$\frac{k_2}{h} = f(x_n + \beta k_1, t_n + \alpha h) = f + \alpha hf_t + \beta k_1 f_x$$

$$+ \frac{\alpha^2h^2}{2}f_{tt} + \alpha h\beta k_1 f_{tx} + \frac{\beta^2k_1^2}{2}f_{xx} + 0(h^3)$$

again evaluated at (x_n, t_n), by Taylor's theorem.

Hence, from (3.3.11),

$$x_{n+1} = x_n + (a+b)hf + bh^2(\alpha f_t + \beta ff_x)$$

$$+ bh^3\left(\frac{\alpha^2}{2}f_{tt} + \alpha\beta ff_{tx} + \frac{\beta^2}{2}f^2f_{xx}\right) + 0(h^4)$$

and this agrees with (3.3.12) to order h^2 if

$$a + b = 1, \quad b\alpha = b\beta = \tfrac{1}{2} \tag{3.3.13}$$

(Note that we cannot have agreement with (3.3.11) to order h^3.) One solution of (3.3.13) is $a = b = \frac{1}{2}$, $\alpha = \beta = 1$, in which case we obtain

$$\boxed{x_{n+1} = x_n + \frac{1}{2}(k_1 + k_2) = x_n + \frac{h}{2}\,[f(x_n, t_n) + f(x_n + hf(x_n, t_n), t_n + h)]}$$

which is the Runge–Kutta method of order 2.

In the same way, but with much more tedious computation we may obtain the **Runge–Kutta method of order 4**

$$\boxed{x_{n+1} = x_n + \frac{1}{6}(k_1 + 2k_2 + 2k_3 + k_4)} \tag{3.3.14}$$

where

$$\boxed{\begin{aligned} k_1 &= hf(x_n, t_n) \\ k_2 &= hf(x_n + \tfrac{1}{2}k_1, t_n + \tfrac{1}{2}h) \\ k_3 &= hf(x_n + \tfrac{1}{2}k_2, t_n + \tfrac{1}{2}h) \\ k_4 &= hf(x_n + k_3, t_n + h) \end{aligned}} \tag{3.3.15}$$

which has local error $0(h^5)$. Note that although the above derivation was again given only for scalar equations, the expressions (3.3.14) and (3.3.15) are also valid for vector equations. The Runge–Kutta method is very accurate and is widely used, although it requires four function evaluations per time step. It is therefore fairly time consuming.

The Runge–Kutta method of order 4 is easily implemented on a computer as the following FORTRAN program shows. It is written to solve the special system of equations called the Van der Pol oscillator

$$\begin{aligned} \dot{x} &= y - x^3 + x \\ \dot{y} &= -x \end{aligned} \tag{3.3.16}$$

FORTRAN PROGRAM TO IMPLEMENT THE RUNGE–KUTTA METHOD

```
      DIMENSION X(2),Y(2),RHS(2),SUM(2)
      READ (1,*) H,X(1),X(2),T,I
      J = 1
C     RHSS DEFINES THE EQUATIONS TO BE SOLVED
C     T IS THE INITIAL TIME
```

```
C  X(1),X(2) ARE THE INITIAL CONDITIONS,I IS NO. OF ITERATIONS
C  H IS STEP LENGTH. INPUTS ARE IN FREE FORMAT
10 CALL RUNGE(X,2,T,H,J,Y,RHS,SUM)
   WRITE(2,20) X(1),X(2)
20 FORMAT(2(E14.7,5X))
   J = J + 1
   IF( J.LE.I) GO TO 10
   END
   SUBROUTINE RUNGE(X,N,T,H,J,Y,RHS,SUM)
   DIMENSION X(N),Y(N),RHS(N),SUM(N)
C  SUM IS RUNNING TOTAL OF K
   CALL RHSS(X,N,T,RHS)
   DO 1 I = 1,N
   SUM(I) = RHS(I)*H
   Y(I) = X(I) + H*RHS(I)/2.0
 1 CONTINUE
   T = T + H/2
   CALL RHSS(Y,N,T,RHS)
   DO 2 I = 1,N
   SUM(I) = SUM(I) + H*RHS(I)*2.0
   Y(I) = X(I) + H*RHS(I)/2.0
 2 CONTINUE
   CALL RHSS(Y,N,T,RHS)
   DO 3 I = 1,N
   SUM(I) = SUM(I) + H*RHS(I)*2.0
   Y(I) = X(I) + H*RHS(I)
 3 CONTINUE
   T = T + H/2
   CALL RHSS(Y,N,T,RHS)
   DO 4 I + 1,N
   X(I) = (SUM(I) + H*RHS(I))/6.0 + X(I)
 4 CONTINUE
   RETURN
   END
   SUBROUTINE RHSS(Y,N,T,RHS)
   DIMENSION Y(N),RHS(N)
   RHS(1) = Y(2) - Y(1)*Y(1) + Y(1)
   RHS(2) = - Y(1)
   RETURN
   END
```

Note that the program has been written so that it includes the possibility of non-autonomous systems, although the particular system (3.3.16) does not require T in SUBROUTINE RHSS. Also, the subroutine RUNGE has been written for systems of general dimension N. In order to use the program for any other system we merely replace the main segment and the subroutine RHSS by the appropriate segments for the desired system.

The Runge–Kutta method has the main disadvantage of requiring four function evaluations for each time step and also gives no idea of the local error and so the only way we can check whether the solutions are correct is by choosing a smaller step length h, performing another simulation and comparing the two sets of results. (Of course, even this is not a proof of the validity of the simulation.) These difficulties may be overcome to some extent by the **predictor–corrector** methods. Consider again an equation of the form

$$\frac{dx}{dt} = f(x, t)$$

and suppose we have evaluated approximations to the solution $x(t)$ at the points $t_0, t_1, \ldots, t_n$. Then, integrating the equation from t_n to t_{n+1}, we obtain

$$x_{n+1} = x_n + \int_{t_n}^{t_{n+1}} f(x(t), t)\, dt$$

$$\approx x_n + \frac{h}{2}\,[f(x_n, t_n) + f(x_{n+1}, t_{n+1})] \tag{3.3.17}$$

by the trapezoidal rule for integration.

We cannot use this equation directly to find x_{n+1} since it is an implicit equation in this value. The technique of solution is therefore to iterate from some starting value $x_{n+1}^{(0)}$. However, we know that Euler's algorithm

$$\boxed{x_{n+1}^{(0)} = x_n + hf(x_n, t_n)} \tag{3.3.18}$$

will give an approximation to x_{n+1} using the previous value x_n. We may now write (3.3.17) as an iterative algorithm

$$\boxed{x_{n+1}^{(m)} = x_n + \frac{h}{2}\,[f(x_n, t_n) + f(x_{n+1}^{(m-1)}, t_{n+1})]} \tag{3.3.19}$$

and it can be shown that this will converge if

$$\left|\frac{\partial f}{\partial x}\right| h < 2$$

for all x, t. Equation (3.3.18) is called the **prediction formula** and (3.3.19) is the **correction formula**. The method then consists of predicting $x_{n+1}^{(0)}$ from (3.3.18) and iterating (3.3.19) to obtain

$$x_{n+1} = \lim_{m\to\infty} x_{n+1}^{(m)}$$

and then returning to (3.3.18) with n replaced by $n+1$ to obtain the new prediction $x_{n+2}^{(0)}$.

A much more accurate prediction–correction scheme is due to Adams–Moulton and it will be stated without proof as follows (see Conte and de Boor, 1972). Define

$$f_n = f(x(t_n), t_n)$$

where again $t_n = t_0 + nh$. Then the prediction algorithm is

$$x_{n+1}^{(0)} = x_n + \frac{h}{24}(55f_n - 59f_{n-1} + 37f_{n-2} - 9f_{n-3}) \tag{3.3.20}$$

and the corretion formula is

$$x_{n+1}^{(k)} = x_n + \frac{h}{24}(9f(x_{n+1}^{(k-1)}, t_{n+1}) + 19f_n - 5f_{n-1} + f_{n-2}) \tag{3.3.21}$$

for $k = 1, 2, \ldots$.

Note, however, that we require four starting values x_0, x_1, x_2, x_3 in this method. These are usually computed by using the fourth order Runge–Kutta algorithm. Since we require a method which has less than four function evaluations per time step, it is customary to apply the correction formula (3.3.21) only once, i.e. to obtain $x_{n+1}^{(1)}$. Now, by considering local errors it can be shown that the true error $x(t_{n+1}) - x_{n+1}^{(1)}$ is approximately $E = (x_{n+1}^{(0)} - x_{n+1}^{(1)})/14$ and so the correction formula also gives an estimate of the local error. If $|E|$ is not within some prescribed bounds on the first iteration of (3.3.21) then we usually change the step length h to $h/2$ if $|E|$ is too large or to $2h$ if $|E|$ is smaller than required. Each time this is done, of course, we must recompute four new starting values from the Runge–Kutta formula. The reader should have no difficulty now in writing a FORTRAN program to implement the Adams–Moulton algorithm.

3.4 ALGEBRAIC EQUATIONS

Although on most occasions a system is defined by a closed set of differential equations as in (3.3.1), we are sometimes faced with the situation where some of the variables are specified by a coupled set of differential and algebraic equations. For example, the two-dimensional system

$$\frac{dx_1}{dt} = f_1(x_1, x_2)$$

$$\frac{dx_2}{dt} = f_2(y, x_2)$$

$$g(x_1, y) = 0$$

is of this type. In the simplest case where we use Euler's algorithm we have

$$x_{1,n+1} = x_{1,n} + hf_1(x_{1,n}, x_{2,n})$$

$$x_{2,n+1} = x_{2,n} + hf_2(y_n, x_{2,n})$$

In order to solve for y_n we must invert the implicit relation

$$g(x_{1,n}, y_n) = 0 \quad (x_{1,n} \text{ known})$$

at each iteration n. Hence in this section we shall examine Newton's method for solving algebraic equations of the form

$$g(y) = 0, \quad g: \mathbb{R}^n \to \mathbb{R}^n \tag{3.4.1}$$

where g is any (reasonably well-behaved) function, since this method seems to be widely used in simulation languages. If g is differentiable then we can write, for any point y_0 close to a solution of (3.4.1),

$$g(y_1) = g(y_0) + (Jg)(y_0)(y_1 - y_0) + o(\| y_1 - y_0 \|), \quad y_1 \in \mathbb{R}^n$$

by Taylor's theorem, where Jg is the Jacobian matrix of g. In particular, if y_1 is (approximately) a solution of (3.4.1) then $g(y_1) = 0$ and we may write

$$y_1 = y_0 - (Jg)^{-1}(y_0)g(y_0)$$

This now leads to the iterative algorithm

$$\boxed{y_{n+1} = y_n - (Jg)^{-1}(y_n)g(y_n)} \tag{3.4.2}$$

in the obvious way. This generalizes the familiar one-dimensional method

$$y_{n+1} = y_n - \frac{g(y_n)}{g'(y_n)}$$

for $g: \mathbb{R} \to \mathbb{R}$.

Since the two-dimensional case occurs quite frequently we shall write out the vector equation (3.4.2) explicitly in this case. For simplicity of notation we shall write the two-dimensional y vector as (λ, μ). Hence, if we have the equations

$$g_1(\lambda, \mu) = 0$$

$$g_2(\lambda, \mu) = 0$$

then Newton's method for iterative solution becomes

$$\lambda_{n+1} = \lambda_n + \begin{vmatrix} -g_1 & g_{1\mu} \\ -g_2 & g_{2\mu} \end{vmatrix} \Bigg/ \begin{vmatrix} g_{1\lambda} & g_{1\mu} \\ g_{2\lambda} & g_{2\mu} \end{vmatrix}$$

$$\mu_{n+1} = \mu_n + \begin{vmatrix} g_{1\lambda} & -g_1 \\ g_{2\lambda} & -g_2 \end{vmatrix} \Bigg/ \begin{vmatrix} g_{1\lambda} & g_{1\mu} \\ g_{2\lambda} & g_{2\mu} \end{vmatrix}$$

where the functions g_1, g_2 and their partial derivatives are evaluated at (λ_n, μ_n).

It should be emphasized, however, that this method may not converge, although it does converge quadratically in some neighbourhood of the solution. Moreover, the equation (3.4.1) may have many solutions and for these reasons most simulation 'packages' will allow a certain amount of interactive capability between the computer and the user. Before discussing such simulation languages we must finally consider the situation when the right-hand side of our original differential equation $\dot{x} = f(x, t)$ is not defined entirely in terms of elementary functions but may contain terms which are only defined at certain discrete values. In this case we need to interpolate between these values and this will be considered in the next section.

3.5 INTERPOLATION

As was mentioned above, in most real simulations it is likely that $f(x, t)$ will not be given entirely in terms of closed form elementary functions. It is usually the case that f will contain at least some part which is not known at all values of x but which is determined experimentally at certain discrete values $x_1, x_2, \ldots, x_m$. For example, in aircraft flight simulation the aerodynamic derivatives depend on air pressure, Mach number, etc., and are usually determined at certain fixed values of these parameters in a wind tunnel or in flight testing. Thus in most simulation exercises we must be able to interpolate this data to find the intermediate values. Although there are many interpolation methods we shall mention only two which are widely used in practice. These are the Lagrange interpolation formula and the method of splines.

Suppose, then, that a scalar function of a single variable $f(x)$ is specified at $n+1$ distinct points $x_0, x_1, \ldots, x_n$. Then it is wellknown that we can fit a unique polynomial of degree less or equal to n through the points $f(x_i)$, $0 \leqslant i \leqslant n$. In order to derive this polynomial, note that if f has the particular form

$$f_k(x_i) = \begin{cases} 1 & i = k \\ 0 & i \neq k \end{cases}$$

then the polynomial

$$\boxed{g_k(x) = \prod_{\substack{i=0 \\ i \neq k}}^{n} \left\{ \frac{(x - x_i)}{(x_k - x_i)} \right\}}$$

is the desired interpolation polynomial for f_k. Hence for a general function f, the polynomial

$$\boxed{p(x) = \sum_{k=0}^{n} f(x_k) g_k(x)} \tag{3.5.1}$$

satisfies the desired relations

$$p(x_k) = f(x_k)$$

The polynomial (3.5.1) is called the **Lagrange form** of the interpolating polynomial.†

In many cases we will be presented with scalar functions of n variables $f(x_1, \ldots, x_n)$ which are again only known at certain fixed values of the variables. Although one could work out an n-dimensional version of the Lagrange form, there is a simpler algorithmic approach which may be taken and which we shall illustrate in the case $n = 2$. Suppose that we are given the values of the function $f(x, y)$ at the points (x_i, y_j), $0 \leqslant i \leqslant n$, $0 \leqslant j \leqslant m$. Then the graph of f is a surface as shown in fig. 3.2 on which we know the values $f(x_i, y_i)$. Consider a point (x^*, y^*) at which we want to find the value $f(x^*, y^*)$. For each x_i $(0 \leqslant i \leqslant n)$ fixed we have a function $f(x_i, y)$ which is known at the points y_j $(0 \leqslant j \leqslant m)$. Using the one-dimensional Lagrange form we can derive $n + 1$ polynomials $p_i(y)$ such that

$$p_i(y_j) = f(x_i, y_j)$$

and we write

$$p_i^* = p_i(y^*)$$

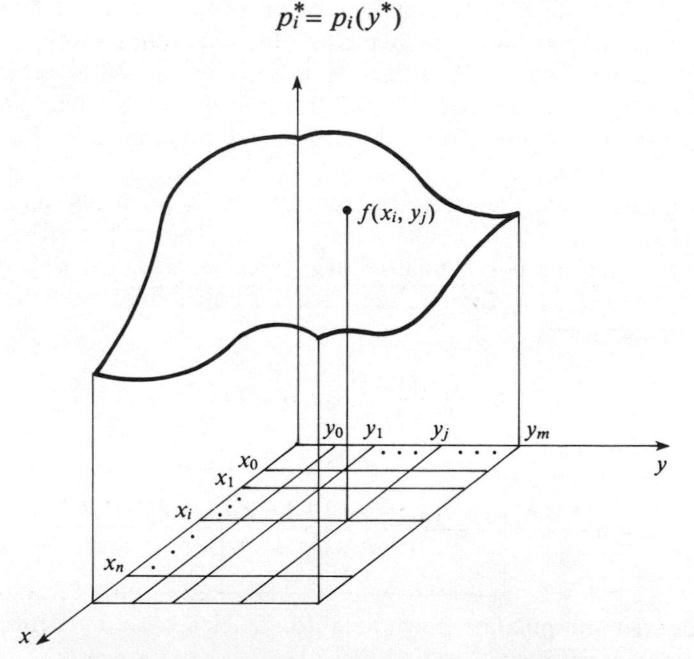

Fig. 3.2 Representing a function of two variables.

† Note that we cannot extrapolate using this method, and for high order polynomials there may be 'oscillations' between the data points, as well as numerical difficulties.

If we now use the data (x_i, p_i^*) and again interpolate in the x-direction we obtain a polynomial $q(x)$ such that

$$q(x_i) = p_i^*$$

The value $q(x^*)$ is now the desired approximation to $f(x^*, y^*)$.

Rather than interpolate through any number of points to obtain an approximating polynomial of degree n it is sometimes preferable to find an approximating function $S(x)$ which passes through the given values but which coincides on each interval $[x_{i-1}, x_i]$ with a polynomial of degree less than or equal to 3 only. If S is also twice continuously differentiable on the whole interval $[a, b]$, then S is called a **cubic spline**. In other words, a cubic spline consists of a set of cubic curves whose derivatives up to order 2 match at the data points, thus giving a smooth join.

In order to derive a cubic spline S interpolating the function values $(x_k, f(x_k))$, $0 \leqslant k \leqslant n$, we must find a vector $\alpha = (\alpha_0, \alpha_1, \ldots, \alpha_n)$ such that

$$S''(x) = \alpha_{i-1}\frac{(x_i - x)}{h_i} + \alpha_i \frac{(x - x_{i-1})}{h_i}, \quad 1 \leqslant i \leqslant n$$

$x \in [x_{i-1}, x_i]$, where $h_i = x_i - x_{i-1}$. Hence, integrating twice and imposing the conditions $S(x_i) = f(x_i)$, we have

$$S(x) = \alpha_{i-1}\frac{(x_i - x)^3}{6h_i} + \alpha_i \frac{(x - x_{i-1})^3}{6h_i} + \left[f(x_i) - \frac{\alpha_i h_i^2}{6}\right]\frac{x - x_{i-1}}{h_i}$$

$$+ \left[f(x_{i-1}) - \frac{\alpha_{i-1} h_i^2}{6}\right]\frac{x - x}{h_i}$$

Imposing the continuity of $S'(x)$ at the points x_i for $1 \leqslant i \leqslant n-1$ we obtain the equations

$$b_i\alpha_{i-1} + 2\alpha_i + c_i\alpha_{i+1} = d_i, \quad 1 \leqslant i \leqslant n-1$$

where

$$b_i = \frac{h_i}{h_i + h_{i+1}}, \quad c_i = 1 - b_i$$

and

$$d_i = 6\frac{\{[f(x_{i+1}) - f(x_i)]/h_{i+1} - [f(x_i) - f(x_{i-1})]/h_i\}}{h_i + h_{i+1}}$$

There are two further conditions required for α_0 and α_n and we can choose these arbitrarily to be

$$2\alpha_0 + \alpha_1 = 1, \quad \alpha_{n-1} + 2\alpha_n = 1$$

Then we have the system of equations

$$A\alpha = d$$

where

$$A = \begin{bmatrix} 2 & 1 & 0 & 0 & \dots & & & 0 \\ b_1 & 2 & c_1 & 0 & \dots & & & 0 \\ 0 & b_2 & 2 & c_2 & \dots & & & 0 \\ \vdots & & & & & & & \vdots \\ & & & & b_{n-1} & & 2 & c_{n-1} \\ 0 & & \dots & & & 0 & 1 & 2 \end{bmatrix}$$

and $d = (1, d_1, d_2, \dots, d_{n-1}, 1)^{\mathrm{T}}$. It can be seen that A is invertible and so $\alpha = A^{-1}d$.

We note finally that the cubic spline may be generalized to two or more dimensions as we did for the Lagrange polynomial.

3.6 SIMULATION LANGUAGES

The basic theory of simulation for ordinary differential equations has now been presented and the resulting algorithms may be programmed in FORTRAN for computer implementation. If the individual algorithms are programmed as subroutines and these are called by a main segment which allows the user to input the specific system equations, usually in a block structured form, we have a **simulation language**. In order to illustrate what is meant by being block structured, let us consider the two-dimensional system

$$\frac{\mathrm{d}x}{\mathrm{d}t} = x \sin y + x^2$$

$$\frac{\mathrm{d}y}{\mathrm{d}t} = \frac{y^3 x}{z}$$

$$z = \sin(zy)$$

Then

$$x(t) = x(0) + \int_0^t (x(s) \sin y(s) + x^2(s)) \, \mathrm{d}s$$

$$y(t) = y(0) + \int_0^t \frac{y^3(s)x(s)}{z(s)} \, \mathrm{d}s$$

and so that states x and y are the outputs of two integrators as shown in fig. 3.3. The right-hand sides of the differential equations and also the structure of the algebraic equation must be provided by special functions, summers, multipliers, etc. Usually, in any simulation language, the program sorts the variables and calculates the inputs to each integrator, solving any algebraic equations if necessary. An integration method (such as the fourth order Runge–Kutta or predictor–corrector techniques) is then

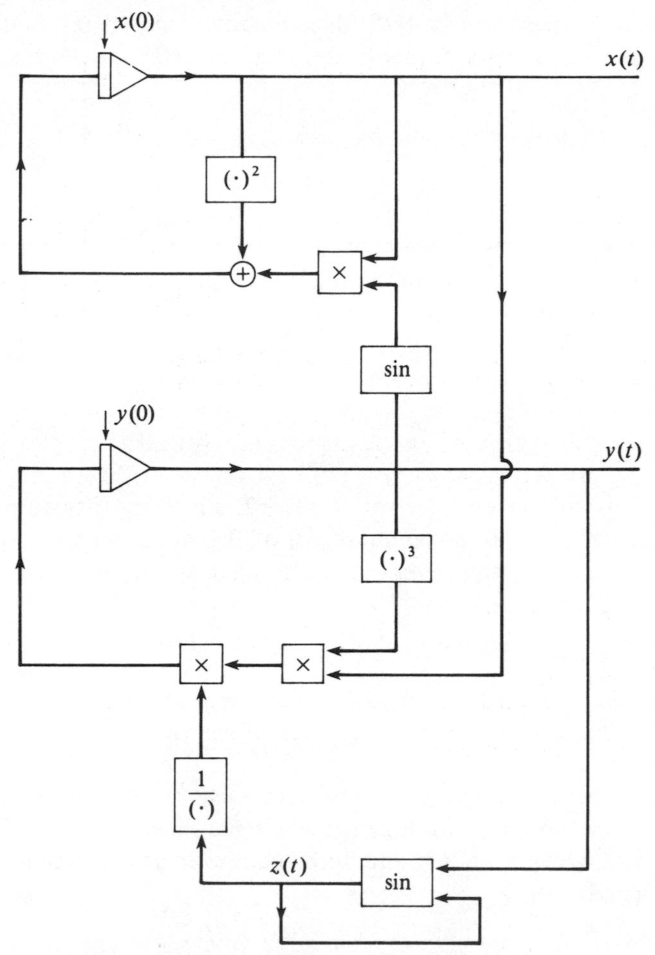

Fig. 3.3 Flow diagram for simulation.

applied to evaluate the integrator outputs at the next time step. An example of a widely used simulation language is ACSL (**A**dvanced **C**ontinuous **S**imulation **L**anguage) produced by Mitchell and Gauthier Associates.

The main reason for the block structured approach is to enable the user to simulate feedback systems which are usually defined as an interconnected system of input–output blocks. Simulation languages are therefore important in control engineering and play an important role in this field.

3.7 SIMULATION OF DISTRIBUTED PARAMETER SYSTEMS

3.7.1 Partial Differential Equations

Although many physical systems considered in control engineering are described by ordinary differential equations to a high degree of approximation, all real systems are in fact 'distributed'. This means that they are modelled more closely by partial differential (or delay) equations. We shall consider equations of the form

$$f(\phi, \phi_t, \phi_x, \phi_y, \phi_{tx}, \phi_{ty}, \phi_{tt}, \phi_{xx}, \phi_{xy}, \phi_{yy}, \ldots) = 0 \qquad (3.7.1)$$

where $\phi(x, y, t)$ is some function defined on a region $\Omega \subseteq \mathbb{R}^2$, and f is a function of the variables $\phi, \phi_t, \phi_x, \ldots$. The equation (3.7.1) can be generalized to more than two special variables x, y in the obvious way. Since we are dealing with simulation of systems in this chapter (which implies some form of time evolution) we are again labelling one variable explicitly as time t. In addition to equation (3.7.1) we must specify boundary conditions of the form

$$g\left(\phi, \phi_t, \frac{\partial \phi}{\partial n}, \ldots\right)\bigg|_{\partial\Omega} = 0, \quad \text{for all } t \qquad (3.7.2)$$

where $\partial/\partial n$ is the normal derivative on $\partial\Omega$, i.e. some function of $\phi, \phi_t, \partial\phi/\partial n, \ldots$ evaluated on the boundary $\partial\Omega$ of Ω is set to zero, and just as for ordinary differential equations we must also specify initial conditions, usually of the form

$$\phi(x, y, 0) = h_1(x, y), \phi_t(x, y, 0) = h_2(x, y), \ldots \qquad (3.7.3)$$

where each h_i is prescribed on Ω. The number of functions h_i which must be given is the same as the highest order of the time derivative appearing in (3.7.1).

If the functions f and g are linear in each variable then we say that (3.7.1), (3.7.2) and (3.7.3) together define a linear partial differential equation with linear boundary and initial conditions. (Recall that a function $f(\phi)$ is linear if

$$f(\alpha\phi + \beta\psi) = \alpha f(\phi) + \beta f(\psi)$$

for any scalars α and β.) An equation is called **quasilinear** if it is linear in the highest order derivatives of ϕ. Consider the pair of quasilinear equations (in the variables ϕ, ψ)

$$\left.\begin{aligned} a_1\phi_x + b_1\phi_t + c_1\psi_x + d_1\psi_t = f_1(x, t, \phi, \psi) \\ a_2\phi_x + b_2\phi_t + c_2\psi_x + d_2\psi_t = f_2(x, t, \phi, \psi) \end{aligned}\right\} \qquad (3.7.4)$$

and suppose that the solution of these equations is specified in a region of (x, t) space as shown in fig. 3.4. If Γ is a curve in (x, t) space joining the boundary points, then we may ask under what conditions the differential equations and the values of ϕ and ψ on Γ determine the derivatives ϕ_x, ϕ_t, ψ_x, ψ_t. Along Γ, we have

$$d\phi = \phi_x \, dx + \phi_t \, dt$$

$$d\psi = \psi_x \, dx + \psi_t \, dt$$

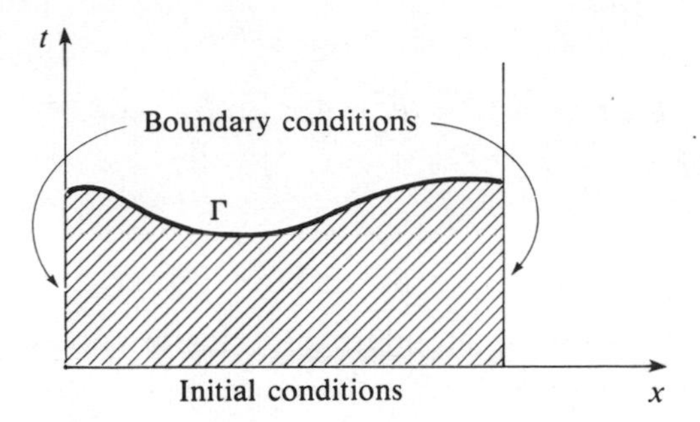

Fig. 3.4 Solution domain of a partial differential equation.

and so from (3.7.4)

$$\begin{bmatrix} a_1 & b_1 & c_1 & d_1 \\ a_2 & b_2 & c_2 & d_2 \\ \mathrm{d}x & \mathrm{d}t & 0 & 0 \\ 0 & 0 & \mathrm{d}x & \mathrm{d}t \end{bmatrix} \begin{bmatrix} \phi_x \\ \phi_t \\ \psi_x \\ \psi_t \end{bmatrix} = \begin{bmatrix} f_1 \\ f_2 \\ \mathrm{d}\phi \\ \mathrm{d}\psi \end{bmatrix} \tag{3.7.5}$$

where the matrix of coefficients and the RHS vector are known along Γ. Hence the derivatives of ϕ and ψ are uniquely determined if the matrix of coefficients is non-singular. In the contrary case we have

$$(a_1c_2 - a_2c_1)\left(\frac{\mathrm{d}t}{\mathrm{d}x}\right)^2 - (a_1d_2 - a_2d_1 + b_1c_2 - b_2c_1)\frac{\mathrm{d}t}{\mathrm{d}x} + (b_1d_2 - b_2d_1) = 0 \tag{3.7.6}$$

and so if the curve Γ has slope $\mathrm{d}t/\mathrm{d}x$ satisfying (3.7.6) at some point $P \in \Gamma$ then the derivatives of ϕ and ψ are not uniquely determined at P. The directions given by (3.7.6) are called the **characteristic directions** for the equations (3.7.4).

Let us now consider the second-order partial differential equation

$$a\phi_{xx} + b\phi_{xy} + c\phi_{yy} + d\phi_x + e\phi_y + f\phi = g \tag{3.7.7.}$$

where we emphasize that the y variable represents either a spatial co-ordinate or time. Introducing the transformations

$$\alpha = \alpha(x, y), \quad \beta = \beta(x, y)$$

it is easy to check that, in terms of the variables α, β, equation (3.7.7) is transformed to

$$A\phi_{\alpha\alpha} + B\phi_{\alpha\beta} + C\phi_{\beta\beta} + \ldots = 0 \tag{3.7.8}$$

where the dots denote terms in the first derivatives of ϕ and

$$\left.\begin{aligned} A &= a\alpha_x^2 + b\alpha_x\alpha_y + c\alpha_y^2 \\ B &= 2a\alpha_x\beta_x + b\alpha_x\beta_y + b\alpha_y\beta_x + 2c\alpha_y\beta_y \\ C &= a\beta_x^2 + b\beta_x\beta_y + c\beta_y^2 \end{aligned}\right\} \tag{3.7.9}$$

Moreover,

$$D \triangleq B^2 - 4AC = (b^2 - 4ac)(\alpha_x\beta_y - \alpha_y\beta_x)^2 \tag{3.7.10}$$

and so if

$$J = \begin{vmatrix} \alpha_x & \alpha_y \\ \beta_x & \beta_y \end{vmatrix} \neq 0$$

the sign of $B^2 - 4AC$ is invariant under this transformation. We classify the equation (3.7.7) as follows:†

(*a*) if $D > 0$ then (3.7.7) is called **hyperbolic**,
(*b*) if $D = 0$ then (3.7.7) is called **parabolic**,
(*c*) if $D < 0$ then (3.7.7) is called **elliptic**.

Consider the partial differential equations $A = 0$, $C = 0$, i.e.

$$a\gamma_x^2 + b\gamma_x\gamma_y + c\gamma_y^2 = 0 \quad \text{for } \gamma = \alpha \text{ or } \beta$$

Then, along the level curves $\gamma(x, y) = \text{constant}$, we have

$$\mathrm{d}\gamma = \gamma_x\,\mathrm{d}x + \gamma_y\,\mathrm{d}y = 0$$

i.e.

$$\frac{\mathrm{d}y}{\mathrm{d}x} = -\frac{\gamma_x}{\gamma_y}, \quad \gamma_y \neq 0$$

Thus,

$$a\frac{\mathrm{d}y^2}{\mathrm{d}x} - b\frac{\mathrm{d}y}{\mathrm{d}x} + c = 0$$

and so

$$\boxed{\frac{\mathrm{d}y}{\mathrm{d}x} = \frac{b \pm (b^2 - 4ac)^{1/2}}{2a}} \tag{3.7.11}$$

If $D > 0$ or $D < 0$ then this expression defines a pair of independent ordinary differential equations whose integrals

$$\alpha(x, y) = \text{constant}, \ \beta(x, y) = \text{constant}$$

define the **characteristic curves** of (3.7.7).

Choosing these functions as the transformations in (3.7.9) we have $A = C = 0$ and so (3.7.7) may be written in the form

$$\phi_{\alpha\beta} + \ldots = 0 \tag{3.7.12}$$

† This nomenclature arises from analytical geometry, where analogous equations give rise to the hyperboloids, paraboloids and ellipsoids.

In the hyperbolic case the characteristics are real and it is easy to see that the equation (3.7.12) may be written in the equivalent form

$$\phi_{\alpha\alpha} - \phi_{\beta\beta} + \ldots = 0 \tag{3.7.13}$$

for some new variables α', β' (given by $\alpha = \alpha' + \beta'$, $\beta = \alpha' - \beta'$). In the elliptic case the characteristics are complex and we can write (3.7.12) in the form

$$\phi_{\alpha'\alpha'} + \phi_{\beta'\beta'} + \ldots = 0 \tag{3.7.14}$$

with $\alpha' = \frac{1}{2}(\alpha + \beta)$, $\beta' = \frac{1}{2}i(\alpha - \beta)$. Finally, if $D = 0$ (parabolic case) then the equations (3.7.11) reduce to only the single equation

$$\frac{\mathrm{d}y}{\mathrm{d}x} = \frac{b}{2a}$$

and we take $\beta(x, y)$ to be given by the integral of this equation. Then $C = 0$ and so $B^2 = 4AC = 0$ and, if $\alpha(x, y)$ is independent of β, (3.7.8) reduces to

$$\phi_{\alpha\alpha} + \ldots = 0 \tag{3.7.15}$$

(since $A \neq 0$).

We have therefore shown that equation (3.7.7) may always be reduced to one of the standard forms

(a) $\phi_{xx} - \phi_{yy} + \ldots = 0$
(b) $\phi_{xx} + \ldots = 0$
(c) $\phi_{xx} + \phi_{yy} + \ldots = 0$

which have been extensively studied.

For a more detailed discussion of partial differential equations see Garabedian (1964), Ames (1977) and John (1971).

3.7.2 Finite Difference Methods for Parabolic Equations

The basic ideas of simulating partial differential equations by discretization methods similar to those for ordinary differential equations will now be discussed. In order to make the presentation reasonably straightforward the methods will be related to some fairly simple equations, but it should be noted that these ideas can be generalized to other linear and non-linear systems. To begin, let us consider the simplest parabolic equation

$$\frac{\partial \phi}{\partial t} = \frac{\partial^2 \phi}{\partial x^2}, \quad x \in [0, 1], t \geqslant 0 \tag{3.7.16}$$

where $\phi(0, t)$, $\phi(1, t)$ and $\phi(x, 0)$ are given. Divide the interval [0, 1] into subintervals of length Δx and let Δt denote a fixed time step. Then, if we define

$$\phi_{j,k} = \phi(j\Delta x, k\Delta t)$$

it is clear that the equation (3.7.16) may be discretized approximately as

$$\boxed{\frac{\phi_{j,k+1} - \phi_{j,k}}{\Delta t} = \frac{\phi_{j+1,k} - 2\phi_{j,k} + \phi_{j-1,k}}{\Delta x^2}} \tag{3.7.17}$$

Hence the value of ϕ at (j, k) depends on its values at $(j, k+1)$, $(j+1, k)$ and $(j-1, k)$ and we indicate this diagramatically by the **molecule** shown in fig. 3.5. If we define

$$r = \frac{\Delta t}{\Delta x^2}$$

then we obtain

$$\phi_{j,k+1} = r\phi_{j-1,k} + (1-2r)\phi_{j,k} + r\phi_{j+1,k}$$

and by a simple application of Taylor's theorem it can be seen that the solution of the finite difference equation converges to the solution of the partial differential equation (3.7.16) as $\Delta x \to 0$ and $\Delta t \to 0$ if

$$r \leqslant \tfrac{1}{2} \tag{3.7.18}$$

Moreover, if we impose the boundary conditions $\phi(0, t) = \phi(1, t) = 0$, then applying the molecule to the kth row of the grid in fig. 3.6 gives rise to the matrix equation

$$\begin{bmatrix} \phi_{1,k+1} \\ \phi_{2,k+1} \\ \phi_{3,k+1} \\ \phi_{4,k+1} \end{bmatrix} = \begin{bmatrix} (1-2r) & r & & \\ r & (1-2r) & r & \\ & r & (1-2r) & r \\ & & r & (1-2r) \end{bmatrix} \begin{bmatrix} \phi_{1,k} \\ \phi_{2,k} \\ \phi_{3,k} \\ \phi_{4,k} \end{bmatrix} \tag{3.7.19}$$

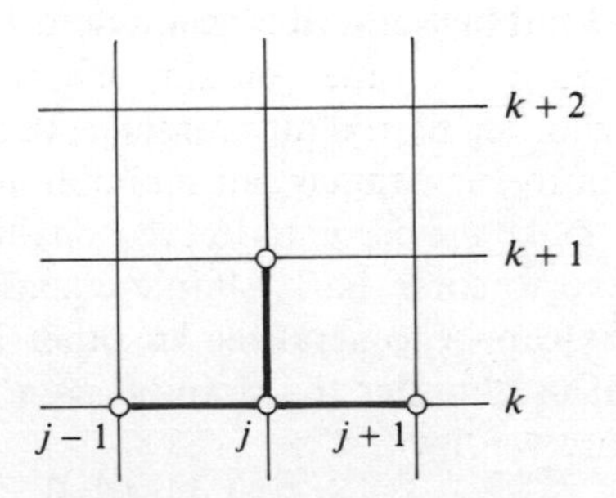

Fig. 3.5 A rectangular grid.

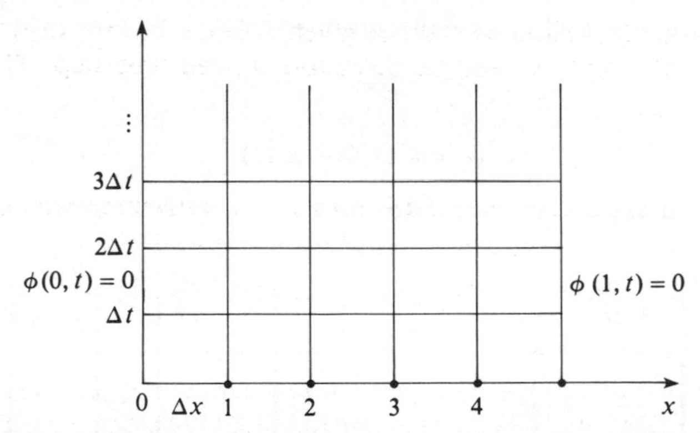

Fig. 3.6 Grid points for a partial differential equation simulation.

(In this and succeeding matrices, any missing elements will be assumed to be zero.) This equation may be written

$$\phi_{k+1} = A\phi_k \tag{3.7.20}$$

where A is the matrix in (3.7.19) and ϕ_k is the obvious four-dimensional vector. It can be shown that this discretization is **stable** (i.e. the quantization error introduced by solving (3.7.20) on a computer remains bounded) if the eigenvalues of A are less than or equal to 1 in modulus. Now $A = I + rT$, where

$$T = \begin{bmatrix} -2 & 1 & & \\ 1 & -2 & 1 & \\ & 1 & -2 & 1 \\ & & 1 & -2 \end{bmatrix}$$

and T has eigenvalues $-4\sin^2(s\pi/10)$, $s = 1, \ldots, 4$, and so A has the eigenvalues $1 - r4\sin^2(s\pi/10)$ which are bounded by 1 in magnitude if $r \leqslant \frac{1}{2}$. The formula (3.7.17) is said to be explicit since $\phi_{j,k+1}$ is written as an explicit function of ϕ at the other nodal mesh points, and as we have seen this leads to the convergence and stability criterion that $r \leqslant \frac{1}{2}$. This means that small step lengths are imposed in the time domain, and therefore implies large computation times. In order to try to overcome this restriction we may approximate $\partial^2\phi/\partial x^2$ at the new time $k+1$ and obtain, instead of (3.7.17), the **implicit** formula

$$\frac{\phi_{j,k+1} - \phi_{j,k}}{\Delta t} = \frac{1}{\Delta x^2}(\phi_{j-1,k+1} - 2\phi_{j,k+1} + \phi_{j+1,k+1}) \tag{3.7.21}$$

and defining r as before we have

$$-r\phi_{j-1,k+1} + (1+2r)\phi_{j,k+1} - r\phi_{j+1,k+1} = \phi_{j,k}$$

Note that this method gives the molecule

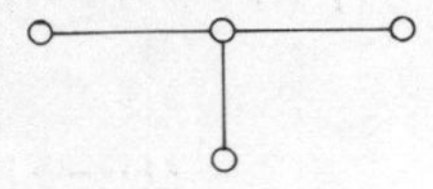

and if this is applied at the nodal points of a grid we obtain the matrix system

$$\begin{bmatrix} 1+2r & -r & & & \\ -r & 1+2r & -r & & \\ & \cdot & \cdot & \cdot & \\ & & & & -r \\ & & \cdot & -r & 1+2r \end{bmatrix} \begin{bmatrix} \phi_{1,k+1} \\ \cdot \\ \cdot \\ \cdot \\ \phi_{n,k+1} \end{bmatrix} = \begin{bmatrix} \phi_{1,k} \\ \cdot \\ \cdot \\ \cdot \\ \phi_{n,k} \end{bmatrix}$$

(where n is the number of nodes in the mesh), and so to find the vector $\phi_{k+1} = (\phi_{1,k+1}, \ldots, \phi_{n,k+1})^{\mathrm{T}}$ inductively we must invert a 'tridiagonal' matrix. This matrix may be reduced to an upper triangular matrix by Gaussian elimination† and so the system can be solved quite easily.

A compromise between the explicit and implicit methods gives rises to the formula

$$\frac{\phi_{j,k+1} - \phi_{j,k}}{\Delta t} = \frac{1}{\Delta x^2}\left[\theta(\phi_{j+1,k+1} - 2\phi_{j,k+1} + \phi_{j-1,k+1}) + (1-\theta)(\phi_{j+1,k} - 2\phi_{j,k} + \phi_{j-1,k})\right] \tag{3.7.22}$$

which has the molecule

θx ○———○———○ $k+1$

$(1-\theta)x$ ○———○———○ k

and reduces to the two previous algorithms for $\theta = 0, 1$. If $\theta = \frac{1}{2}$ we obtain the **Crank–Nicolson** formula. In matrix form it becomes

$$\begin{bmatrix} 1+2r\theta & -r\theta & & & \\ -r\theta & 1+2r\theta & -r\theta & \cdot & \\ & \cdot & \cdot & \cdot & -r\theta \\ & & \cdot & -r\theta & 1+2r\theta \end{bmatrix} \begin{bmatrix} \phi_{1,k+1} \\ \cdot \\ \cdot \\ \cdot \\ \phi_{n,k+1} \end{bmatrix}$$

† See also Roebuck and Barnett (1978) for a more efficient method.

$$= \begin{bmatrix} 1-2r(1-\theta) & r(1-\theta) & & & \\ r(1-\theta) & 1-2r(1-\theta) & r(1-\theta) & & \\ & \ddots & \ddots & \ddots & \\ & & \ddots & \ddots & r(1-\theta) \\ & & & r(1-\theta) & 1-2r(1-\theta) \end{bmatrix} \begin{bmatrix} \phi_{1,k} \\ \cdot \\ \cdot \\ \cdot \\ \phi_{n,k} \end{bmatrix}$$

$$+ \begin{bmatrix} r\theta\phi_{0,k+1} + r(1-\theta)\phi_{0,k} \\ 0 \\ \vdots \\ r\theta\phi_{n+1,k+1} + r(1-\theta)\phi_{n+1,k} \end{bmatrix}$$

where non-zero boundary conditions have been included. It can be shown that this method is stable under the conditions:

(*a*) for $0 \leqslant \theta < \frac{1}{2}$, we require $r \leqslant \frac{1}{2}(1-2\theta)$
(*b*) For $\frac{1}{2} \leqslant \theta \leqslant 1$, r can have any positive value (although we usually take $r < 8$),

and is the most accurate implicit method when $\theta = \frac{1}{2}$.

We may extend equation (3.7.16) to two or more dimensions; consider, for example, the equation

$$\frac{\partial \phi}{\partial t} = \frac{\partial^2 \phi}{\partial x^2} + \frac{\partial^2 \phi}{\partial y^2} \tag{3.7.23}$$

Then the explicit algorithm extends to

$$\boxed{\frac{\phi_{ijk+1} - \phi_{ijk}}{\Delta t} = \frac{\phi_{i-1jk} - 2\phi_{ijk} + \phi_{i+1jk}}{\Delta x^2} + \frac{\phi_{ij-1k} - 2\phi_{ijk} + \phi_{ij+1k}}{\Delta y^2}}$$

and has the molecule

For stability we require (assuming $\Delta x = \Delta y$)

$$r = \frac{\Delta t}{\Delta x^2} = \frac{\Delta t}{\Delta y^2} \leqslant \frac{1}{4}$$

The fully implicit method gives the relation

$$\boxed{\frac{\phi_{ijk+1} - \phi_{ijk}}{\Delta t} = \frac{\phi_{i-1jk+1} - 2\phi_{ijk+1} + \phi_{i+1jk+1}}{\Delta x^2} + \frac{\phi_{ij-1k+1} - 2\phi_{ijk+1} + \phi_{ij+1k+1}}{\Delta y^2}}$$

with the molecule

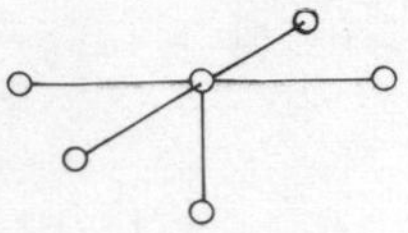

This leads to a matrix equation of the form

$$\left[\diagdown\!\!\diagdown\!\!\diagdown\right]\left[\phi\right] = f$$

However, by using the fully implicit method in the x-direction only followed by the same algorithm in the y-direction, we obtain the **alternating direction implicit (ADI) method**. It is defined by the formula

$$\frac{\phi_{ijk+1} - \phi_{ijk}}{\Delta t} = \frac{\phi_{i+1jk+1} - 2\phi_{ijk+1} + \phi_{i-1jk+1}}{\Delta x^2} + \frac{\phi_{ij+1k} - 2\phi_{ijk} + \phi_{ij-1k}}{\Delta y^2}$$

with the molecule

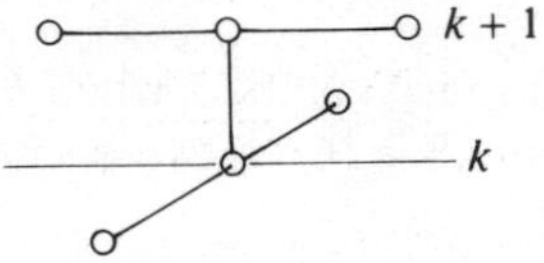

k + 1

k

followed by

$$\frac{\phi_{ijk+2} - \phi_{ijk+1}}{\Delta t} = \frac{\phi_{i+1jk+1} - 2\phi_{ijk+1} + \phi_{i-1jk+1}}{\Delta x^2} + \frac{\phi_{ij+1k+2} - 2\phi_{ijk+2} + \phi_{ij-1k+2}}{\Delta y^2}$$

with molecule

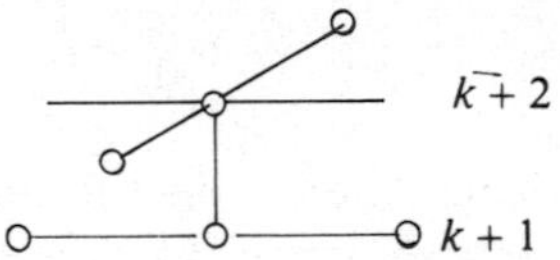

k + 2

k + 1

The ADI method is now regarded as the best of the above techniques.

3.7.3 Numerical methods for hyperbolic equations

In this section we shall consider hyperbolic equations of the form

$$A\frac{\partial^2\phi}{\partial x^2} + B\frac{\partial^2\phi}{\partial t^2} + 2H\frac{\partial^2\phi}{\partial x\partial t} + G = 0 \qquad (3.7.24)$$

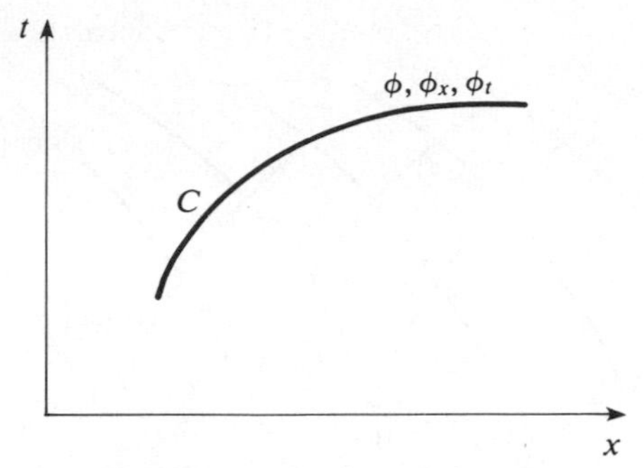

Fig. 3.7 A characteristic curve.

where $H^2 > AB$, and we first return to the idea of characteristics introduced above. Suppose that on some curve C in the xt-plane we know ϕ, ϕ_x and ϕ_t as in fig. 3.7. Then, for example,

$$d\phi = \frac{\partial \phi}{\partial x}\,dx + \frac{\partial \phi}{\partial t}\,dt = p\,dx + q\,dt$$

and we wish to determine ϕ_{xx}, ϕ_{tt}, ϕ_{xt} on C from equation (3.7.24) and the values of ϕ, ϕ_x and ϕ_t on C. Now

$$d\phi_x = \phi_{xx}\,dx + \phi_{xt}\,dt = r\,dx + s\,dt$$

for example, or

$$dp = r\,dx + s\,dt$$

and similarly

$$d\phi_t = \phi_{xt}\,dx + \phi_{tt}\,dt$$

or

$$dq = s\,dx + u\,dt$$

The equation (3.7.24) then gives

$$A\left(\frac{dp - s\,dt}{dx}\right) + B\left(\frac{dq - s\,dx}{dt}\right) + 2Hs + G = 0$$

or

$$A\,dt\,dp + B\,dx\,dq + s(2H\,dx\,dt - A\,dt^2 - B\,dx^2) + G\,dx\,dt = 0$$

Hence, if we choose curves for which

$$A\left(\frac{dt}{dx}\right)^2 - 2H\left(\frac{dt}{dx}\right) + B = 0 \qquad (3.7.25)$$

then (3.7.24) becomes an ordinary differential equation, namely

$$A\frac{dp}{dx} + B\frac{dq}{dt} + G = 0 \qquad (3.7.26)$$

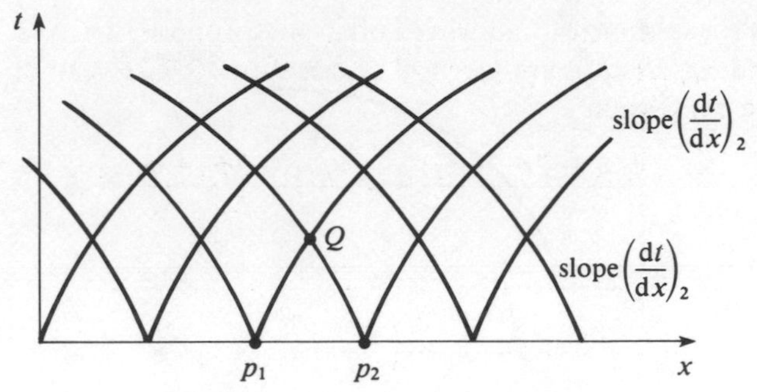

Fig. 3.8 A set of characteristics.

The solutions $u_1 = (dt/dx)_1$, $u_2 = (dt/dx)_2$ of (3.7.25) define two sets of curves called the **characteristics** which are straight lines if A, B and H are constants. The characteristics starting on the x-axis are shown in fig. 3.8. We can use the intersection of a finite set of characteristics as the grid points for a numerical solution of the original equation. To see how this proceeds note that (3.7.26) implies that

$$A\frac{dt}{dx}\frac{d}{dx}\left(\frac{\partial\phi}{\partial x}\right) + B\frac{d}{dx}\left(\frac{\partial\phi}{\partial t}\right) + G\frac{dt}{dx} = 0$$

and since u_1, u_2 are the roots of (3.7.25) we have

$$u_1 + u_2 = \frac{2H}{A}, \quad u_1u_2 = \frac{B}{A}$$

and so on characteristic 1 (2)

$$\frac{d}{dx}\left(\frac{\partial\phi}{\partial x}\right) + \underset{(1)}{u_2}\frac{d}{dx}\left(\frac{\partial\phi}{\partial t}\right) = -\frac{G}{A} \tag{3.7.27}$$

Now let $\partial\phi/\partial x = p$, $\partial\phi/\partial t = q$ be known on the x-axis and in particular at the points P_1, P_2 in fig. 3.8. Then by (3.7.27) we have

$$(p(Q) - p(P_1)) + u_2(q(Q) - q(P_1)) = -\frac{G}{A}(x(Q) - x(P_1))$$

$$(p(Q) - p(P_2)) + u_1(q(Q) - q(P_2)) = -\frac{G}{A}(x(Q) - x(P_2))$$

and so p and q can be determined at Q. Once $\partial\phi/\partial x$, $\partial\phi/\partial t$ are calculated at the grid points, we find ϕ quite easily.

The above method is referred to as the method of characteristics and can be applied to many types of system. We shall now examine the finite difference method for the particular system

$$\frac{\partial^2\phi}{\partial x^2} - \frac{\partial^2\phi}{\partial t^2} = 0, \quad x \in [0, 1] \tag{3.7.28}$$

which is the one-dimensional wave equation. Suppose that $\phi(0, t)$, $\phi(1, t)$, $\phi(x, 0)$ and $\partial\phi/\partial t(x, 0)$ are given. The equation (3.7.28) can be discretized to give the expression

$$\frac{\phi_{+1j} - 2\phi_{ij} + \phi_{i-1j}}{\Delta x^2} = \frac{\phi_{ij+1} - 2\phi_{ij} + \phi_{ij-1}}{\Delta t^2}$$

that is

$$\boxed{\phi_{ij+1} = 2\phi_{ij} - \phi_{ij-1} + \left(\frac{\Delta t}{\Delta x}\right)^2 (\phi_{i+1j} - 2\phi_{ij} + \phi_{i-1j})}$$

The process can be started from the initial conditions, since

$$\phi_{i1} = \phi_{i0} + \Delta t \left(\frac{\partial\phi}{\partial t}\right)_{i0}$$

and it can be shown that, for stability, we require $\Delta t \leqslant \Delta x$. We may also use the implicit formula

$$\boxed{\begin{aligned}\frac{\phi_{ij+1} - 2\phi_{ij} + \phi_{ij-1}}{\Delta t^2} = {} & [\theta(\phi_{i+1j+1} - 2\phi_{ij+1} + \phi_{i-1j+1}) \\ & + (1 - 2\theta)(\phi_{i+1j} - 2\phi_{ij} + \phi_{i-1j}) \\ & + \theta(\phi_{i+1j-1} - 2\phi_{ij-1} + \phi_{i-1j-1})]/\Delta x^2\end{aligned}}$$

with the molecule

$i-1$ $\quad i \quad$ $i+1$

$(j+1)x\theta$

$jx(1-2\theta)$

$(j-1)x\theta$

and if this is applied to the grid we obtain the equation

$$A\Phi_{j+1} = B\Phi_j - A\Phi_{j-1} + f \tag{3.7.29}$$

where

$$A = \begin{bmatrix} 1 + 2r\theta & -r\theta & & & \\ -r\theta & 1 + 2r\theta & -r\theta & & \\ & \ddots & \ddots & \ddots & \\ & & & & -r\theta \\ & & & -r\theta & 1 + 2r\theta \end{bmatrix}$$

$$B = \begin{bmatrix} 2-2(1-2\theta)r & (1-2\theta)r & & \\ (1-2\theta)r & \cdot & \cdot & \\ & \cdot & \cdot & \cdot \\ & & \cdot & (1-2\theta)r \\ & & (1-2\theta)r & 2-2(1-2\theta)r \end{bmatrix}$$

($r = \Delta t^2/\Delta x^2$) and

$$f = \begin{bmatrix} r\theta\phi_{0j+1} + (1-2\theta)r\theta\phi_{0j} + r\theta\phi_{0j-1} \\ 0 \\ \vdots \\ 0 \\ r\theta\phi_{n+1j+1} + (1-2\theta)r\theta_{n+1j} + r\theta\phi_{n+1j-1} \end{bmatrix}$$

$$\phi_{j+1} = \begin{bmatrix} \phi_{ij+1} \\ \vdots \\ \phi_{nj+1} \end{bmatrix}$$

(n is again the number of internal grid points on the x-axis).

3.7.4 Elliptic equations

In our consideration of ordinary differential equations, we also had to consider algebraic equations which are independent of time, and which can be thought of as some form of equilibrium condition. Similarly, in the theory of distributed systems, we have so far considered parabolic and hyperbolic equations which define the time evolution of certain systems. The counterpart of algebraic equations in distributed systems is taken by elliptic partial differential equations. It is therefore necessary to discuss the finite difference approximation of such systems, which we shall now proceed to do.

We shall illustrate the basic approach by considering Laplace's equation

$$\frac{\partial^2\phi}{\partial x^2} + \frac{\partial^2\phi}{\partial y^2} = 0, \quad 0 \leqslant x \leqslant 1, 0 \leqslant y \leqslant 1 \qquad (3.7.30)$$

subject to Dirichlet boundary conditions

$$\phi(x,0) = \phi(x,1) = 0$$

$$\phi(0,y) = y, \quad \phi(1,y) = 1 - y$$

(see fig. 3.9). It is easy to see that if we discretize equation (3.7.30) in the

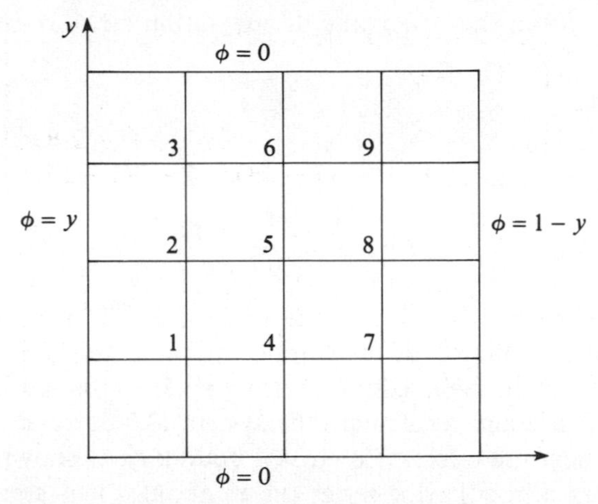

Fig. 3.9 A rectangular grid for the Laplace equation.

usual way, with equal step lengths in the x and y directions, then we obtain

$$\phi_{i-1j} - 2\phi_{ij} + \phi_{i+1j} + \phi_{ij-1} - 2\phi_{ij} + \phi_{ij+1} = 0$$

or

$$\boxed{\phi_{i-1j} + \phi_{i+1j} + \phi_{ij-1} + \phi_{ij+1} - 4\phi_{ij} = 0}$$

which leads to the molecule

$$\begin{array}{ccc} & 1 & \\ 1 & -4 & 1 \\ & 1 & \end{array}$$

If we apply this molecule to the grid in fig. 3.9, we obtain the equation

$$\begin{bmatrix} -4 & 1 & & 1 & & & & & \\ 1 & -4 & 1 & & 1 & & & & \\ & 1 & -4 & & & 1 & & & \\ 1 & & & -4 & 1 & & 1 & & \\ & 1 & & 1 & -4 & 1 & & 1 & \\ & & 1 & & 1 & -4 & & & 1 \\ & & & 1 & & & -4 & 1 & \\ & & & & 1 & & 1 & -4 & 1 \\ & & & & & 1 & & 1 & -4 \end{bmatrix} \begin{bmatrix} \phi_1 \\ \cdot \\ \cdot \\ \cdot \\ \\ \\ \\ \\ \phi_9 \end{bmatrix} = \begin{bmatrix} -\frac{1}{4} \\ -\frac{1}{2} \\ -\frac{3}{4} \\ 0 \\ 0 \\ 0 \\ -\frac{3}{4} \\ -\frac{1}{2} \\ -\frac{1}{4} \end{bmatrix} \tag{3.7.31}$$

Now it can be shown that if τ is the discretization error at any grid point, we have

$$|\tau| \leqslant \frac{Mh^4}{6}$$

where h is the grid size and

$$M = \max\left(\frac{\partial^4\phi}{\partial x^4}, \frac{\partial^4\phi}{\partial y^4}\right)$$

where the maximum is taken over the grid points. Hence we must solve (3.7.31) to find the solution of the elliptic equation and then determine M, from which we obtain the maximum error $|\tau|$. If $|\tau|$ is too large we must reduce the grid size and recalculate the system (3.7.31) and its solution.

Consider next the effect of a curved boundary as shown in fig. 3.10. Then expanding ϕ in a Taylor series (in x) about 0 it is seen that

$$\left(\frac{\partial^2\phi}{\partial x^2}\right)_0 = \frac{2[\phi_1 + \xi\phi_3 - (1+\xi)\phi_0]}{h^2\xi(1+\xi)} + 0(h) \tag{3.7.32}$$

and so the finite difference formula near the boundary is

$$\frac{\phi_2 + \phi_4 - 2\phi_0}{h^2} + \frac{2[\phi_1 + \xi\phi_3 - (1+\xi)\phi_0]}{h^2\xi(1+\xi)} + 0(h)$$

(The accuracy is lower at a curved boundary.)

Suppose now that, instead of having Dirichlet boundary conditions for equation (3.7.30), we know $\partial\phi/\partial x$ on $x = 1$ (fig. 3.11). Then instead of applying the molecule to just the nodes $1, \ldots, 9$ as in fig. 3.9 we consider nodes 10, 11, 12 on the boundary $x = 1$ and also add three 'fictitious' nodes 13, 14, 15 outside the original domain. Then, for example, applying the molecule at node 11 we obtain

$$\phi_8 + \phi_{12} + \phi_{10} + \phi_{14} - 4\phi_{11} = 0$$

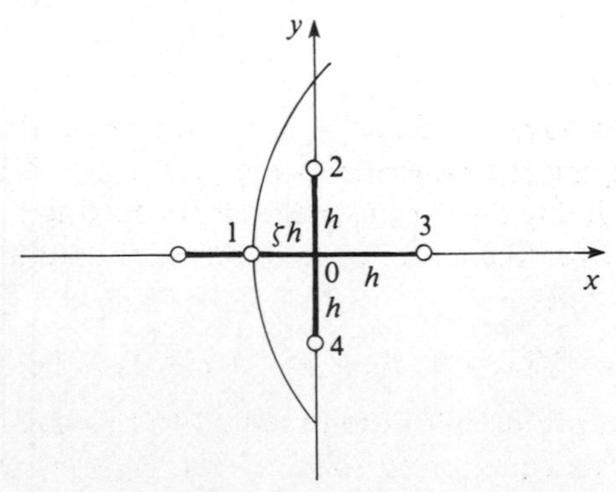

Fig. 3.10 Fitting a grid to the boundary.

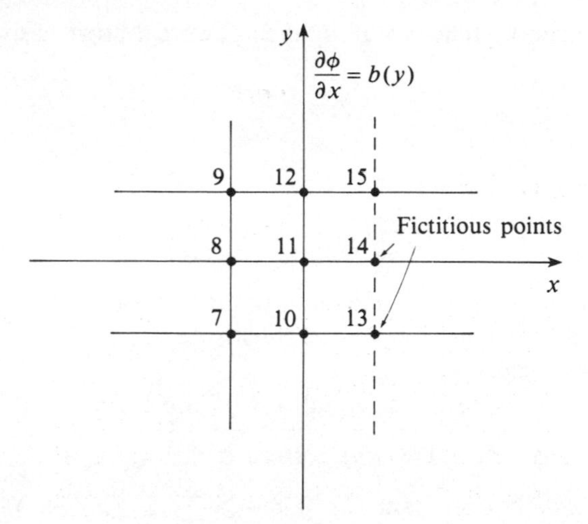

Fig. 3.11 Using fictitious points.

Since we do not know ϕ_{14} we use the derivative boundary condition to obtain

$$\frac{\partial \phi}{\partial x} = \frac{\phi_{14} - \phi_8}{2h} = b_{11}$$

and so

$$\phi_{14} = \phi_8 + 2hb_{11}$$

Hence, at node 11,

$$2\phi_8 + \phi_{12} + \phi_{10} - 4\phi_{11} + 2hb_{11} = 0$$

and the molecule on this boundary becomes

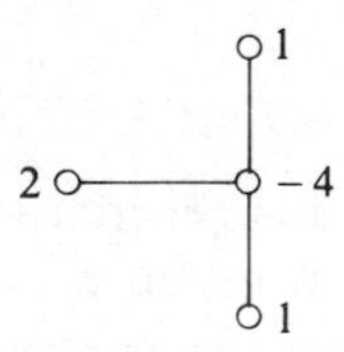

The reader should have no difficulty in writing down the matrix equation obtained by applying the molecule to the grid in fig. 3.11.

Before considering the various methods for solving the matrix equation which is derived from the above procedure, we should mention non-linear equations of the type

$$\phi_{xx} + f(x, y, \phi, \phi_x, \phi_y)\phi_{yy} = g(x, y, \phi, \phi_x, \phi_y) \tag{3.7.33}$$

If we apply the above discretization method to this equation, we obtain a non-linear algebraic equation to solve. However, if we guess a solution ϕ at each mesh point then we can calculate f and g and apply the standard methods for linear systems as if f and g were known functions. The molecule becomes

f_4

1 — $-2-2f_0$ — 1

f_2

and the nodal values vary from one mesh point to another. Once ϕ has been found from the matrix equation we can substitute this back into f and g and iterate until convergence is obtained.

Let us return now to linear equations of the form

$$A\phi = b, \quad A = (a_{ij}), \quad 1 \leqslant i \leqslant n, 1 \leqslant j \leqslant n$$

where A is diagonally dominant (i.e. $|a_{ii}| \geqslant \sum_{j \neq i} |a_{ij}|, \forall i$). If

$$D = \text{diag}(a_{11}, \ldots, a_{nn})$$

then we can write

$$D\phi = E\phi + b$$

where $E = D - A$, and so inductively we write

$$\boxed{D\phi^{(r+1)} = E\phi^{(r)} + b} \quad \text{(Jacobi iteration)} \qquad (3.7.34)$$

Then it can be shown that this process converges if the largest eigenvalue of $D^{-1}E$ is less than one. The method is called the **method of simultaneous displacement**.

Alternatively, if we assume that A is normalized so that $a_{ii} = 1$ for each i, then we can write

$$A = I - L - U$$

where $-L$ and $-U$ are lower and upper triangular parts of A respectively. Then we have the Gauss–Seidel iteration algorithm

$$\boxed{(I - L)\phi^{(r+1)} = U\phi^{(r)} + b} \qquad (3.7.35)$$

Applying this to the grid in fig. 3.9 we have

$$\begin{aligned}
\phi_1^{(r+1)} &= \tfrac{1}{4}\phi_2^{(r)} + \tfrac{1}{4}\phi_4^{(r)} + b_1 \\
-\tfrac{1}{4}\phi_1^{(r+1)} + \phi_2^{(r+1)} &= \tfrac{1}{4}\phi_3^{(r)} + \tfrac{1}{4}\phi_5^{(r)} + b_2 \\
-\tfrac{1}{4}\phi_2^{(r+1)} + \phi_3^{(r+1)} &= \tfrac{1}{4}\phi_6^{(r)} + b_3 \\
&\vdots \\
-\tfrac{1}{4}\phi_6^{(r+1)} - \tfrac{1}{4}\phi_8^{(r+1)} + \phi_9^{(r+1)} &= b_9
\end{aligned}$$

It is clear that we need only one set of values ϕ stored at any time since, for example, $\phi_1^{(r+1)}$ can be written on top of $\phi_1^{(r)}$ as soon as it is calculated. Equation (3.7.35) is called the **method of successive displacements**.

The two methods outlined above can be improved by adding an acceleration parameter ω as follows. If we assume $D = I$ in (3.7.34) for simplicity, then we obtain

$$\begin{aligned}\phi^{(r+1)} &= E\phi^{(r)} + b \\ &= (I - A)\phi^{(r)} + b \\ &= I\phi^{(r)} + (b - A\phi^{(r)})\end{aligned}$$

We accelerate the process by using

$$\boxed{\phi^{(r+1)} = I\phi^{(r)} + \omega(b - A\phi^{(r)})} \tag{3.7.36}$$

If μ_1 and μ_n are the largest and smallest eigenvalues of A, then we choose

$$\omega = \frac{2}{\mu_1 + \mu_n}$$

Similarly, from (3.7.35) we have

$$\phi^{(r+1)} = \phi^{(r)} + L\phi^{(r+1)} + U\phi^{(r)} - I\phi^{(r)} + b$$

and we replace this with the algorithm

$$\boxed{\phi^{(r+1)} = \phi^{(r)} + \omega(L\phi^{(r+1)} + U\phi^{(r)} - I\phi^{(r)} + b)} \tag{3.7.37}$$

If $\omega > 1$, this is called the **method of successive over-relaxation** (SOR). It can be shown that if μ is the largest eigenvalue of $D^{-1}E$ associated with the Jacobi method, then the best value of ω for (3.7.37) is given by

$$\omega_{\text{opt}} = \frac{2}{1 + (1 - \mu^2)^{1/2}} \tag{3.7.38}$$

The over-relaxation technique can be generalized as follows. Consider the grid in fig. 3.12. Then we obtain an equation of the form

$$\begin{bmatrix} B_1 & C_1 & & \\ A_2 & B_2 & C_2 & \\ & A_3 & B_3 & C_3 \\ & & A_4 & B_4 \end{bmatrix} \begin{bmatrix} \phi_{1-4} \\ \phi_{5-8} \\ \phi_{9-12} \\ \phi_{13-16} \end{bmatrix} = \begin{bmatrix} b_{1-4} \\ b_{5-8} \\ b_{9-12} \\ b_{13-16} \end{bmatrix} \tag{3.7.39}$$

for Laplace's equation. This has a block tridiagonal structure. We then extend the SOR method by writing

$$\boxed{B\Phi^{(r+1)} = B\Phi^{(r)} + \omega(L\Phi^{(r+1)} + U\Phi^{(r)} - B\Phi^{(r)} + b)}$$

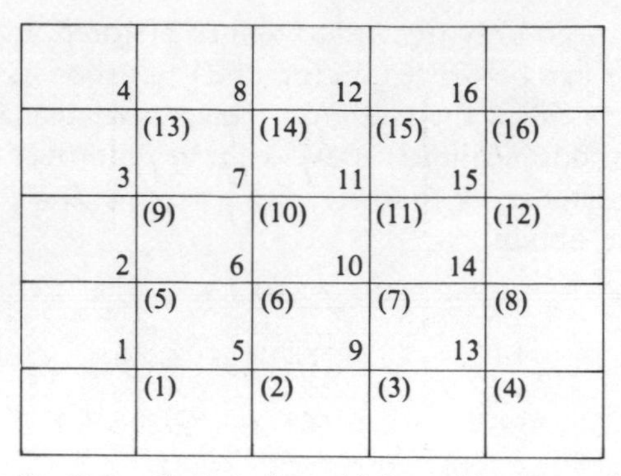

Fig. 3.12 Renumbering the grid for relaxation methods.

where

$$\boldsymbol{B} = \operatorname{diag}(B_1, B_2, B_3, B_4)$$

$$\boldsymbol{\Phi} = (\phi_{1-4}, \phi_{5-8}, \phi_{9-12}, \phi_{13-16})^{\mathrm{T}}$$

and $\boldsymbol{L}$, $\boldsymbol{U}$ are the block lower and upper triangular parts of the matrix of coefficients in (3.7.39).

We shall finally discuss the alternating direction method for elliptic equations. Consider again Laplace's equation on the grid of fig. 3.12. We then obtain the system

$$A\phi = b$$

where

$$A = \begin{bmatrix} A_1 & I & & \\ I & A_1 & I & \\ & I & A_1 & I \\ & & I & A_1 \end{bmatrix}$$

$$A_1 = \begin{bmatrix} -4 & 1 & & \\ 1 & -4 & 1 & \\ & 1 & -4 & 1 \\ & & 1 & -4 \end{bmatrix}$$

We split A into two matrices H and V associated with the molecules

1   −2   1
H

1
−2 V
1

Then

$$H = \begin{bmatrix} H_1 & I & & \\ I & H_1 & I & \\ & I & H_1 & I \\ & & I & H_1 \end{bmatrix}$$

$$V = \begin{bmatrix} V_1 & & & \\ & V_1 & & \\ & & V_1 & \\ & & & V_1 \end{bmatrix}$$

where

$$H_1 = \begin{bmatrix} -2 & & & \\ & -2 & & \\ & & -2 & \\ & & & -2 \end{bmatrix}$$

$$V_1 = \begin{bmatrix} -2 & 1 & & \\ 1 & -2 & 1 & \\ & 1 & -2 & 1 \\ & & 1 & -2 \end{bmatrix}$$

Thus A has the structurre

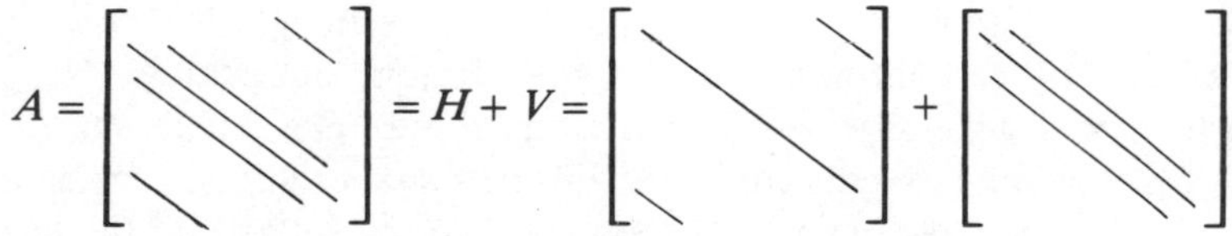

$$A = \begin{bmatrix} \ddots & \ddots & \ddots & \ddots \end{bmatrix} = H + V = \begin{bmatrix} \ddots & \ddots \end{bmatrix} + \begin{bmatrix} \ddots & \ddots & \ddots \end{bmatrix}$$

Suppose that we now order the grid row-wise as shown in fig. 3.12. Then we obtain the structure

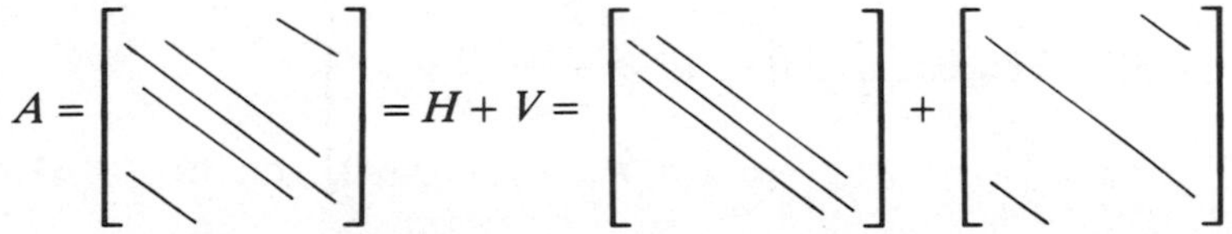

$$A = \begin{bmatrix} \ddots & \ddots & \ddots & \ddots \end{bmatrix} = H + V = \begin{bmatrix} \ddots & \ddots & \ddots \end{bmatrix} + \begin{bmatrix} \ddots & \ddots \end{bmatrix}$$

Note that H is easily inverted for row-wise ordering, while V is easily inverted for column-wise ordering. Hence we set

$$H\phi^{(r+1)} = b - V\phi^{(r)} \qquad \text{for row-wise ordering}$$

$$V\phi^{(r+2)} = b - H\phi^{(r+1)} \qquad \text{for column-wise ordering}$$

This is the **alternating direction implicit method**.

We can accelerate this method by adding a parameter r so that

$$(I + rH)\phi^{(r+1/2)} = (I - rV)\phi^{(r)} + rb \quad \text{for row-wise ordering}$$
$$(I + rV)\phi^{(r+1)} = (I - rH)\phi^{(r+1/2)} + rb \quad \text{for column-wise ordering}$$

It can be shown that if the eigenvalues of H and V are bounded by $a_1 \geqslant 0$ and $a_2 \geqslant 0$, i.e. $a_1 \leqslant \lambda_H, \eta_V \leqslant a_2$, then we should choose

$$r = \frac{1}{(a_1 a_2)^{1/2}} \tag{3.7.40}$$

A more detailed exposition of numerical methods for partial differential equations is given by Ames (1977).

3.7.5 The Finite Element Technique

The finite difference methods discussed above are based on the evaluation of approximations to the solutions of partial differential equations at certain discrete mesh points in the region of interest. The value of the solution at points not on the mesh is never considered. In the finite element technique we try to overcome this drawback by dividing the region of interest into 'small' areas (or volumes) called **elements** and approximating the solution to the equation by a polynomial in each element. We force these polynomials to be continuous across the element boundaries, but their derivatives will generally be discontinuous there. The advantage of this method is that the elements can have different shapes and so can be made to fit into irregular regions very accurately. We shall illustrate the method by considering the two dimensional static heat transfer equation

$$K_{xx}\frac{\partial^2 T}{\partial x^2} + K_{yy}\frac{\partial^2 T}{\partial y^2} + Q = 0, \quad \text{in } \Omega \subseteq \mathbb{R}^2 \tag{3.7.41}$$

with the boundary conditions

$$K_{xx}\frac{\partial T}{\partial x} lx + K_{yy}\frac{\partial T}{\partial y} ly + h(T - T_\infty) + q = 0, \quad \text{on } \partial\Omega \tag{3.7.42}$$

where K_{xx}, K_{yy} are the thermal conductivities in the x and y directions, Q is the heat generated in the body, l_x and l_y are the direction cosines of a point on $\partial\Omega$, h is the convection coefficient and T_∞ is the ambient temperature. This equation will bring out all the important aspects of the method and we shall show how to include the dynamic term $\partial T/\partial t$ later.

The finite element technique is based on variational principles which say that solving a partial differential equation on some region Ω together with boundary conditions on $\partial\Omega$ is equivalent to minimizing a certain integral (called the **action integral** by physicists) over Ω. To derive the action integral for (3.7.41) and (3.7.42) we shall first review the general theory of the calculus of variations. Suppose then that we have a functional

$$I = \int_{\Omega} F(x, y, \phi, \phi_x, \phi_y)\, \mathrm{d}\Omega + \int_{\partial\Omega} G(x, y, \phi)\, \mathrm{d}S \qquad (3.7.43)$$

where dS is an element of surface area on $\partial\Omega$, and consider the variation δI of I with respect to changes in ϕ, ϕ_x and ϕ_y. Then we have

$$\delta I = \int_{\Omega} \left(\frac{\partial F}{\partial \phi}\, \delta\phi + \frac{\partial F}{\partial \phi_x}\, \delta\phi_x + \frac{\partial F}{\partial \phi_y}\, \delta\phi_y \right) \mathrm{d}\Omega + \int_{\partial\Omega} \left(\frac{\partial G}{\partial \phi} \right) \delta\phi\, \mathrm{d}S$$

Hence, by Gauss' divergence theorem, we have

$$\delta I = \int_{\Omega} \left[\frac{\partial F}{\partial \phi} - \frac{\partial}{\partial x}\left(\frac{\partial F}{\partial \phi_x} \right) - \frac{\partial}{\partial y}\left(\frac{\partial F}{\partial \phi_y} \right) \right] \delta\phi\, \mathrm{d}\Omega$$

$$+ \int_{S} \left(l_x \frac{\partial F}{\partial \phi_x} + l_y \frac{\partial F}{\partial \phi_y} + \frac{\partial G}{\partial \phi} \right) \delta\phi\, \mathrm{d}S \qquad (3.7.44)$$

If this variation is set to zero, then, since, the variation is arbitrary, we must have

$$\frac{\partial F}{\partial \phi} = \frac{\partial}{\partial x} \frac{\partial F}{\partial \phi_x} + \frac{\partial}{\partial y} \frac{\partial F}{\partial \phi_y} \qquad (3.7.45)$$

$$l_x \frac{\partial F}{\partial \phi_x} + l_y \frac{\partial F}{\partial \phi_y} + \frac{\partial G}{\partial \phi} = 0 \qquad (3.7.46)$$

These equations determine the partial differential equation and the boundary conditions and so we must choose F and G so that (3.7.45) and (3.7.46) are equivalent to (3.7.41) and (3.7.42). The appropriate choices are easily seen to be

$$F = \frac{K_{xx}}{2} \left(\frac{\partial T}{\partial x} \right)^2 + \frac{K_{yy}}{2} \left(\frac{\partial T}{\partial y} \right)^2 - QT$$

$$G = qT + \tfrac{1}{2} h (T - T_\infty)^2$$

The action integral therefore becomes

$$I = \int_{\Omega} \left[\frac{K_{xx}}{2} \left(\frac{\partial T}{\partial x} \right)^2 + \frac{K_{yy}}{2} \left(\frac{\partial T}{\partial y} \right)^2 - QT \right] \mathrm{d}\Omega + \int_{\partial\Omega} \left[qT + \tfrac{1}{2} h (T - T_\infty)^2 \right] \mathrm{d}S \qquad (3.7.47)$$

We now forget the original equation and work exclusively with (3.7.47).

Having derived an action integral for our problem we must now con-

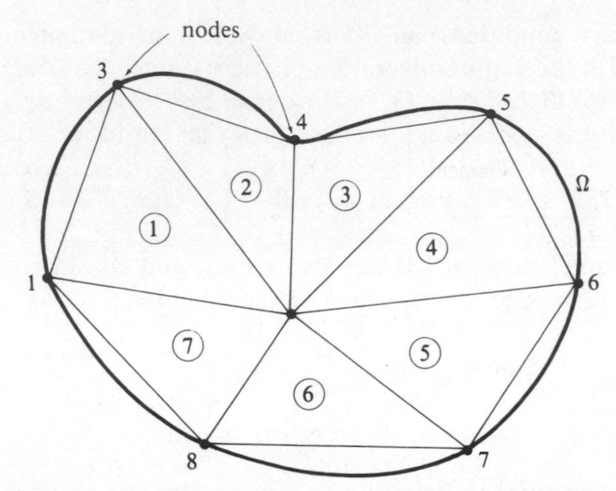

Fig. 3.13 Dividing a region into simplex approximations.

sider the approximation of T by polynomials in each element. For simplicity we shall use only triangular elements† although elements of almost any shape may be used. Consider a two-dimensional region Ω as in fig. 3.13 and divide it into a number of triangles (generally non-congruent) such that the boundary $\partial\Omega$ is modelled as accurately as desired. The vertices of the triangles are called **nodes** of the finite element approximation but it is important to note that we may also place nodes on the sides of the triangles, as in fig. 3.14(b). In this introduction to the finite element technique we shall consider only elements with nodes at the vertices (fig. 3.14(a)). The number of unknowns at each node is called the **degree of freedom** at the node. Since in our example we have only the unknown temperature T, the degree of freedom in this case is 1. Suppose that element α has nodal coordinates (x_β, y_β), $\beta = i, j$ or k, and that the unknown temperatures at these nodes are T_i, T_j, T_k. Since we have only three unknowns we can only fit a linear interpolating polynomial through these values. Hence if we choose the polynomial

$$T^\alpha = a_1 + a_2 x + a_3 y \tag{3.7.48}$$

then $T^\alpha(x_\beta, y_\beta) = T_\beta$, $\beta = i, j$ or k and so

$$(a_1, a_2, a_3)^{\mathrm{T}} = C^{-1}(T_i, T_j, T_k)^{\mathrm{T}}$$

where

$$C = \begin{bmatrix} 1 & x_i & y_i \\ 1 & x_j & y_j \\ 1 & x_k & y_k \end{bmatrix}$$

Note that det $C = 2 \times$ area of the triangle (i, j, k).

† Triangles, tetrahedra and higher dimensional objects of this kind are called **simplexes**.

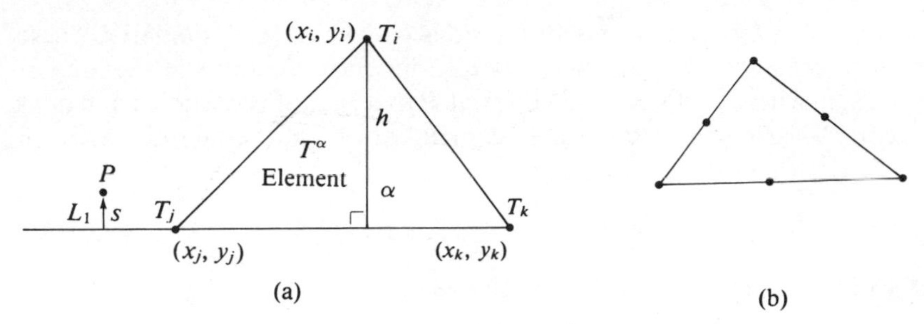

Fig. 3.14 Simplex coordinates.

Since a_1, a_2, a_3 depend linearly on T_i, T_j, T_k we may write (3.7.48) in the form

$$T^\alpha = N_i^\alpha T_i + N_j^\alpha T_j + N_k^\alpha T_k \tag{3.7.49}$$

where

$$N_i^\alpha = \frac{1}{|C|}[x_j y_k - x_k y_j + (y_j - y_k)x + (x_k - x_j)y]$$

$$N_j^\alpha = \frac{1}{|C|}[x_k y_i - y_k x_i + (y_k - y_i)x + (x_i - x_k)y]$$

$$N_k^\alpha = \frac{1}{|C|}[x_i y_j - x_k y_i + (y_i - y_j)x + (x_j - x_i)y]$$

and we have explicitly shown the dependence of each function on the element α. The functions N_β^α, $\beta = i, j, k$ are called the **shape functions** for the element α. Note that

$$\sum_{\beta = i, j, k} N_\beta^\alpha = 1$$

and

$$N_\beta^\alpha = 1 \quad \text{at node } \beta$$

$$N_\beta^\alpha = 0 \quad \text{at other nodes}$$

It is sometimes useful to consider local coordinates defined independently in each element, rather than the global coordinates x, y used above. A convenient local coordinate system is defined by the ratio of the distance s from a side of the triangle to the corresponding perpendicular height h of the triangle from that side (fig. 3.14(a)). Then we write, for the coordinate of a point P,

$$L_1 = \frac{s}{h}$$

and define L_2, L_3 associated with the sides (i, k) and (i, j) similarly. These are called area coordinates since they also measure the ratio of the area of the triangle with base (j, k) and vertex at P and that of the whole triangular element. A basic property of these coordinates is that, when expressed in global (x, y) coordinates, we have

$$L_1 = N_i^\alpha, \quad L_2 = N_j^\alpha, \quad L_3 = N_k^\alpha \tag{3.7.50}$$

and so (3.7.49) may be written locally as

$$T^\alpha = L_1 T_i + L_2 T_j + L_3 T_k$$

The main advantage of these coordinates is that we have the formulae

$$\int_l L_1^a L_2^b \, \mathrm{d}l = \frac{a!b!}{(a+b+1)!} l \tag{3.7.51}$$

$$\int_A L_1^a L_2^b L_3^c \, \mathrm{d}A = \frac{a!b!c!}{(a+b+c+2)!} 2A \tag{3.7.52}$$

for the integral of products of the L_i over a length l or an area A. These are of importance in evaluating the action integral in local coordinates as we shall see presently. Note that the area coordinates can be generalized to higher dimensions in an obvious way.

Having considered a single element we can now return to the region in fig. 3.13 and number the nodes as shown. (A careful numbering of nodes is important since this will reduce computer storage requirements.) Applying (3.7.49) to each element (i.e. for $\alpha = 1, 2, \ldots, 7$) we obtain the equation

$$\begin{bmatrix} T^1 \\ T^2 \\ T^3 \\ T^4 \\ T^5 \\ T^6 \\ T^7 \end{bmatrix} = \begin{bmatrix} N_1^1 & N_2^1 & N_3^1 & 0 & 0 & 0 & 0 & 0 \\ 0 & N_2^2 & N_3^2 & N_4^2 & 0 & 0 & 0 & 0 \\ 0 & N_2^3 & 0 & N_4^3 & N_5^3 & 0 & 0 & 0 \\ 0 & N_2^4 & 0 & 0 & N_5^4 & N_6^4 & 0 & 0 \\ 0 & N_2^5 & 0 & 0 & 0 & N_6^5 & N_7^5 & 0 \\ 0 & N_2^6 & 0 & 0 & 0 & 0 & N_7^6 & N_8^6 \\ N_1^7 & N_2^7 & 0 & 0 & 0 & 0 & 0 & N_8^7 \end{bmatrix} \begin{bmatrix} T_1 \\ T_2 \\ T_3 \\ T_4 \\ T_5 \\ T_6 \\ T_7 \\ T_8 \end{bmatrix}$$

Hence, in general, if there are n elements and m nodes we obtain

$$T^\alpha = \sum_{j=1}^{m} N_j^\alpha T_j, \quad 1 \leqslant \alpha \leqslant n \tag{3.7.53}$$

where N_j^α is zero for nodes j not in element α. Note that T^α is a function of x and y representing the approximation to the true solution in element α, whereas T_j is the value of this approximation at node j.

We must now substitute the equations (3.7.53) into the action integral (3.7.47) which can be split into integrals over each element α; then,

$$I = \sum_{\alpha=1}^{n} \int_{\Omega^\alpha} \left[\frac{K_{xx}}{2} \left(\frac{\partial T^\alpha}{\partial x} \right)^2 + \frac{K_{yy}}{2} \left(\frac{\partial T^\alpha}{\partial y} \right)^2 - QT^\alpha \right] \mathrm{d}\Omega$$

$$+ \sum_{\alpha=1}^{n} \int_{\partial\Omega^\alpha} [qT^\alpha + \tfrac{1}{2} h (T^\alpha - T_\infty)^2] \, \mathrm{d}S \qquad (3.7.54)$$

Note that the line integrals are evaluated only over those simplex boundaries which approximate $\partial\Omega$. If we write

$$D = \begin{pmatrix} K_{xx} & 0 \\ 0 & K_{yy} \end{pmatrix}$$

$$B = \begin{pmatrix} \dfrac{\partial N_1^\alpha}{\partial x} & \dfrac{\partial N_2^\alpha}{\partial x} & \cdots & \dfrac{\partial N_m^\alpha}{\partial x} \\ \dfrac{\partial N_1^\alpha}{\partial y} & \dfrac{\partial N_2^\alpha}{\partial y} & \cdots & \dfrac{\partial N_m^\alpha}{\partial y} \end{pmatrix}$$

$$T = (T_1 \; T_2 \ldots T_m)^{\mathrm{T}}$$

$$N^\alpha = (N_1^\alpha \; N_2^\alpha \ldots N_m^\alpha)$$

then (3.7.54) becomes

$$I = \sum_{\alpha=1}^{n} \Bigg(\int_{\Omega^\alpha} \tfrac{1}{2} T^{\mathrm{T}} (B^\alpha)^{\mathrm{T}} D B^\alpha T \, \mathrm{d}\Omega - \int_{\Omega^\alpha} Q N^\alpha T \, \mathrm{d}\Omega$$

$$+ \int_{\partial\Omega^\alpha} q N^\alpha T \, \mathrm{d}S + \int_{\partial\Omega^\alpha} \frac{h}{2} T^{\mathrm{T}} (N^\alpha)^{\mathrm{T}} N^\alpha T \, \mathrm{d}S$$

$$- \int_{\partial\Omega^\alpha} h T_\infty N^\alpha T \, \mathrm{d}S + \int_{\partial\Omega^\alpha} \frac{h}{2} T_\infty^2 \, \mathrm{d}S \Bigg)$$

The action integral now depends only on the unknown nodal values T_j, $1 \leqslant j \leqslant m$ and so we may now minimize I by differentiating with respect to T and setting the result to zero. Then we obtain

$$\sum_{\alpha=1}^{n} K^\alpha T = \sum_{\alpha=1}^{n} F^\alpha$$

where

$$\boxed{K^\alpha = \int_{\Omega^\alpha} (B^\alpha)^{\mathrm{T}} D B^\alpha \, \mathrm{d}\Omega + \int_{\partial\Omega^\alpha} h (N^\alpha)^{\mathrm{T}} N^\alpha \, \mathrm{d}S} \qquad (3.7.55)$$

and

$$F^\alpha = \int_{\Omega^\alpha} Q(N^\alpha)^{\mathrm{T}}\,\mathrm{d}\Omega - \int_{\partial\Omega^\alpha} q(N^\alpha)^{\mathrm{T}}\,\mathrm{d}S + \int_{\partial\Omega^\alpha} hT_\infty(N^\alpha)^{\mathrm{T}}\,\mathrm{d}S \tag{3.7.56}$$

or

$$\boxed{KT = F} \tag{3.7.57}$$

where

$$K = \sum_{\alpha=1}^{n} K^\alpha, \quad F = \sum_{\alpha=1}^{n} F^\alpha$$

We have therefore derived the matrix equation (3.7.57) which must be solved numerically for the unknown vector T. In order to evaluate the expressions (3.7.55) and (3.7.56) explicitly for each element α assume that this element has nodes i, j, k and write

$$N_\beta^\alpha = \frac{1}{2A^\alpha}(a_\beta^\alpha + b_\beta^\alpha x + c_\beta^\alpha y), \quad \beta = i, j, k$$

where A^α is the area of element α. Then,

$$\int_{\Omega^\alpha} (B^\alpha)^{\mathrm{T}} D B^\alpha\,\mathrm{d}\Omega = \int_{\Omega^\alpha} \frac{1}{4(A^\alpha)^2} \begin{bmatrix} b_i^\alpha & c_i^\alpha \\ b_j^\alpha & c_j^\alpha \\ b_k^\alpha & c_k^\alpha \end{bmatrix} \begin{bmatrix} K_{xx} & 0 \\ 0 & K_{yy} \end{bmatrix} \begin{bmatrix} b_i^\alpha & b_j^\alpha & b_k^\alpha \\ c_i^\alpha & c_j^\alpha & c_k^\alpha \end{bmatrix} \mathrm{d}\Omega^\alpha$$

$$= \frac{K_{xx}}{4A^\alpha} \begin{bmatrix} b_i^\alpha b_i^\alpha & b_i^\alpha b_j^\alpha & b_i^\alpha b_k^\alpha \\ b_j^\alpha b_i^\alpha & b_j^\alpha b_j^\alpha & b_j^\alpha b_k^\alpha \\ b_k^\alpha b_i^\alpha & b_k^\alpha b_j^\alpha & b_k^\alpha b_k^\alpha \end{bmatrix} + \frac{K_{yy}}{4A^\alpha} \begin{bmatrix} c_i^\alpha c_i^\alpha & c_i^\alpha c_j^\alpha & c_i^\alpha c_k^\alpha \\ c_j^\alpha c_i^\alpha & c_j^\alpha c_j^\alpha & c_j^\alpha c_k^\alpha \\ c_k^\alpha c_i^\alpha & c_k^\alpha c_j^\alpha & c_k^\alpha c_k^\alpha \end{bmatrix}$$

where we have omitted the zero elements of the B^α matrix.

Consider the integral

$$\int_{\partial\Omega^\alpha} h(N^\alpha)^{\mathrm{T}} N^\alpha\,\mathrm{d}S$$

and suppose that the boundary $\partial\Omega^\alpha$ over which we are integrating is the side ij of the element α. Since $N_k^\alpha = 0$ at i and j and is linear between these nodes, we must have $N_k^\alpha = 0$ along the whole side. Hence the integral becomes

$$\int_{(ij)} h \begin{bmatrix} N_i^\alpha N_i^\alpha & N_i^\alpha N_j^\alpha & 0 \\ N_j^\alpha N_i^\alpha & N_j^\alpha N_j^\alpha & 0 \\ 0 & 0 & 0 \end{bmatrix} \mathrm{d}l = \int_{(ij)} h \begin{bmatrix} L_1 L_1 & L_1 L_2 & 0 \\ L_2 L_1 & L_2 L_2 & 0 \\ 0 & 0 & 0 \end{bmatrix} \mathrm{d}l$$

when expressed in area coordinates. Using (3.7.51) we obtain

$$\frac{hl_{ij}}{6}\begin{bmatrix} 2 & 1 & 0 \\ 1 & 2 & 0 \\ 0 & 0 & 0 \end{bmatrix}$$

for the value of this integral, where l_{ij} is the length of the side (ij).

The integral

$$Q\int_{\Omega^\alpha} (N^\alpha)^{\mathrm{T}}\,\mathrm{d}\Omega = Q\int_{\Omega^\alpha} [L_1 L_2 L_3]^{\mathrm{T}}\,d\Omega$$

can be evaluated by putting $a=1$, $b=0$, $c=0$ in (3.7.52) and we obtain

$$\frac{QA^\alpha}{3}(1,1,1)^{\mathrm{T}}$$

The two remaining integrals in (3.7.56) contain the term

$$\int_{(ij)} [N_i^\alpha \quad N_j^\alpha \quad 0]^{\mathrm{T}}\,\mathrm{d}l = \frac{l_{ij}}{2}[1 \quad 1 \quad 0]^{\mathrm{T}}$$

and this completes the determination of K^α and F^α for element α. When the corresponding values for the other elements have been found it only remains to solve the linear system (3.7.57) and this can be done easily by Gaussian elimination, or by any of the standard acceleration methods.

We shall conclude this introductory discussion of the finite element method by noting that we may extend the method to dynamic equations quite easily. Hence, if we have, instead of (3.7.41), the equation

$$K_{xx}\frac{\partial^2 T}{\partial x^2} + k_{yy}\frac{\partial^2 T}{\partial y^2} + Q = \lambda\frac{\partial T}{\partial t}$$

then we merely replace F in the action integral by

$$\frac{K_{xx}}{2}\left(\frac{\partial T}{\partial x}\right)^2 + \frac{K_{yy}}{2}\left(\frac{\partial T}{\partial y}\right)^2 - QT + \lambda\frac{\partial T}{\partial t}T$$

and proceed as before. Then (3.7.57) is replaced by

$$\boxed{C\frac{\partial T}{\partial t} = -KT + F} \tag{3.7.58}$$

where

$$C = \sum_{i=1}^{n} C^\alpha$$

and

$$C^\alpha = \lambda \int_{\Omega^\alpha} (N^\alpha)^T N^\alpha \, d\Omega$$

$$= \lambda \int_{\Omega^\alpha} [L_1 \quad L_2 \quad L_3]^T [L_1 \quad L_2 \quad L_3] \, d\Omega$$

$$= \frac{\lambda A^\alpha}{12} \begin{bmatrix} 2 & 1 & 1 \\ 1 & 2 & 1 \\ 1 & 1 & 2 \end{bmatrix}$$

again using (3.7.52).

The finite element method introduced above can be generalized in many ways; in particular, we can consider higher order elements (with nodes not necessarily at the vertices) or different shaped elements. See, for example, Zienkiewicz (1977), Segerlind (1976), Tong and Rossettos (1977) and for the modern developments of the boundary element technique see Brebbia (1980).

3.8 SIMULATION OF STOCHASTIC SYSTEMS

3.8.1 Noisy Systems

In the previous sections we have discussed the simulation of systems which are completely determined by their initial conditions and are therefore known as **deterministic systems**. In this last section of the present chapter we shall briefly consider the simple system

$$\boxed{\dot{x} = f(x, t) + \nu(t), \quad x(0) = x_0} \tag{3.8.1}$$

where $x \in \mathbb{R}$ and $\nu(t)$ is a Gaussian random variable (see appendix 2) for each t with mean $\mu(t)$ and variance $\sigma(t)$. The case of general n-dimensional systems and coloured noise can be considered similarly. (Coloured noise is generated by passing Gaussian white noise through a dynamical system such as the linear system

$$\dot{\xi} = A\xi + b\nu \tag{3.8.2}$$

where ν is Gaussian white noise.)

If the function f in (3.8.1) is linear and time invariant, then it is easy to show that the state x is also Gaussian and so is determined by its mean μ_x and variance σ_x. Dynamic equations for μ_x, σ_x can be obtained from (3.8.1) and we then have a pair of deterministic systems to solve for these parameters. This is the approach which is applied in linear filtering theory and will be described later.

In the general non-linear case we do not expect x to be characterized by the first statistical moments and so we may resort to simulation of (3.8.1). However, in contrast to the deterministic case, the initial value x_0 does not yield a unique solution of (3.8.1), since if the system is operated at different times the noise term $\nu(t)$ will vary from one run to another. Each such possible solution of (3.8.1) is called a **sample path** and we can obtain many sample paths by numerically solving the equation with given values of $\nu(t)$. To determine these values of $\nu(t)$ we must generate sequences of Gaussian random variables at each time. The generation of such random variables is considered in the next section.

3.8.2 Random Number Generation

The most widely applied method of **(pseudo-) random** number† generation is via the multiplicative congruential procedure

$$\boxed{R_{n+1} \equiv \alpha R_n \pmod{m}} \tag{3.8.3}$$

for suitable parameters α and m. The choice of α and m will now be considered.

First note that if α and m are relatively prime (integers) then there is a smallest positive integer x such that

$$\alpha^x \equiv 1 \pmod{m}$$

It follows that $\alpha^{x+k} = \alpha^k \pmod{m}$ for any k and so the sequence of positive remainders modulo m of successive powers of α is periodic with period precisely equal to x. Moreover, it is easy to see that if α is of the form $4t + 3$ (non-negative t) and if 2^r is the largest power of 2 dividing $4t + 4$, then, with $m = 2p$, we have

$$x = 2^{p-r}$$

provided $p > r$ (see Moshman, 1954).

Suppose that we have a computer with a 35-bit binary word length. Then if we choose $\alpha = 2^{18} + 3$ and $m = 2^{35}$ we have $r = 2$ and so $x = 2^{33}$. The mod 2^{35} arithmetic is easy to perform if the computer ignores any overflow and so with these parameters we can generate a sequence of pseudo-random numbers which are uniformly distributed between 0 and 1 if we put

$$R_n^* = \frac{R_n}{2^{35}}$$

† Pseudo-random numbers satisfying (3.8.3) are, in fact, deterministic, but are periodic with a very long period and are regarded as being (almost) random (see chapter 7).

The sequence is, of course, deterministic, but repeats itself only after the 2^{35}th term. The correlation between two sequences of random numbers generated in this way can be shown to be of order 2^{-18} (see Greenberger, 1961).

In order to generate Gaussian random variables with mean μ and variance σ, Box and Muller (1958) have shown that if we define

$$S_n = \sigma(-2 \ln R_n^*)^{1/2} \cos 2\pi R_{n+1}^{*}{}^* + \mu \tag{3.8.4}$$

$$S_{n+1} = \sigma(-2 \ln R_n^*)^{1/2} \sin 2\pi R_{n+1}^* + \mu \tag{3.8.5}$$

for successive values of R_n^*, R_{n+1}^*, then (S_n, S_{n+1}) is a pair of normally distributed random variables.

3.8.3 Simulation

Returning now to equation (3.8.1), we may replace this equation with a discrete version (by any of the methods considered above) and obtain, for example,

$$x_{n+1} = F(x_n, t_n) + \nu(t_n) \tag{3.8.6}$$

To find the statistical moments of x we may use the Monte Carlo type technique of evaluating the recursive equation (3.8.6) over m sample paths and then average the solutions. At each time step t_n we use either of the equations (3.8.4) or (3.8.5) to generate m random numbers with Gaussian parameters $\mu(t_n)$ and $\sigma(t_n)$. In this way we calculate m sequences $(x_n^i)_{n>0}$, $1 \leqslant i \leqslant m$ and so the mean of x is given (at the sample times t_n) by

$$\mu_x(t_n) = \frac{1}{m} \sum_{i=1}^{m} x_n^i$$

The other moments of $x(t)$ may be found similarly.

3.9 EXERCISES

1. Explain the difference between discretization error and quantization error.

2. Using the division theorem of arithmetic (which states that, for any natural numbers n, m we may write $n = qm + r$, i.e. dividing n by m gives remainder r) prove that the expression (3.2.1) is valid for any numbers n, r.

3. (a) Using table 3.1, convert the following binary numbers to hex:

   (i) 1110011011011 (ii) 111100001111 (iii) 1111111

(b) Convert the following hex numbers to binary:

(i) AF3 (ii) FF1 (iii) ABC

4. Determine a method to convert the fractional base r number

$$n_r = a_{-1}r^{-1} + a_{-2}r^{-2} + \ldots + a_{-m}r^{-m}$$

to base ρ.

5. Prove that only a single bit changes at each stage of a Gray code count.

6. If P reduces the matrix A to Jordan form by a similarity transformation (so that $\Lambda = P^{-1}AP$, where Λ is in Jordan form), show that

$$\exp(At) = P\exp(\Lambda t)P^{-1}$$

Evaluate $\exp(\Lambda t)$ for any Jordan matrix Λ.

7. (For mathematicians) Define a **norm** on the vector space $\mathbb{R}^n$ by

$$\|x\| = \sum_{i=1}^{n} (x_i^2)^{1/2}, \quad x \in \mathbb{R}^n$$

and for any matrix put

$$\|A\| = \sup_{x \neq 0} \frac{\|Ax\|}{\|x\|}$$

Show that $\|\cdot\|$ defines a norm on the space of all $n \times n$ matrices; in particular, show that $\|A\| < \infty$ for any $n \times n$ matrix A. Hence show that $\exp(At)$ exists when defined by the series (3.3.5).

8. Evaluate

$$\exp\left\{\begin{pmatrix} a & -b \\ b & a \end{pmatrix} t\right\}$$

9. Write out, explicitly, Euler's algorithm for the system

$$\dot{x}_1 = x_1^2 + x_2^2$$

$$\dot{x}_2 = \sin x_1 \cos x_2$$

10. Using the Runge–Kutta subroutine given in the text, solve the equations

$$\dot{x}_1 = -10x_1 + 10x_2$$

$$\dot{x}_2 = -x_1x_3 + 28x_1 - x_2$$

$$\dot{x}_3 = x_1x_2 - \frac{8}{3}x_3$$

for various initial conditions on a digital computer and plot the phase plane trajectories on the screen. (These equations are called Lorentz' equations, or the equations of the strange attractor, see Hirsch and Smale (1971).

11. Write a FORTRAN program subroutine to apply the Adams–Moulton predictor corrector method.

12. Write down Newton's algorithm to solve the equations

$$x_1^2 + x_2^2 - 1 = 0$$

$$x_1 + \sin x_2 = 0$$

Discuss the results when this is applied numerically.

13. Prove (3.7.8) and (3.7.10).

14. Prove that the eigenvalues of the $n \times n$ tridiagonal matrix

$$T = \begin{bmatrix} b & c & & & \\ a & b & c & & \\ & \cdot & \cdot & \cdot & \\ & & \cdot & \cdot & c \\ & & & a & b \end{bmatrix}$$

are

$$\lambda_k = b + 2(ac)^{1/2} \cos\left(\frac{k\pi}{n+1}\right), \quad k = 1, \ldots, n$$

15. Prove (3.7.50) by considering the simplex with vertices (0, 0), (1, 0), (0, 1).

16. Prove the relations (3.7.51) and (3.7.52).

17. Simulate the transistor oscillator circuit in fig. 3.15. (Hint: First write down the differential equations for i_B, V_2, V_1, V_5, $\dot{V}_2$ using input and output characteristics of the transistor. This gives a five-dimensional system of non-linear equations. Write an interpolation subroutine for the transistor characteristics which should be input as finite look-up tables.)

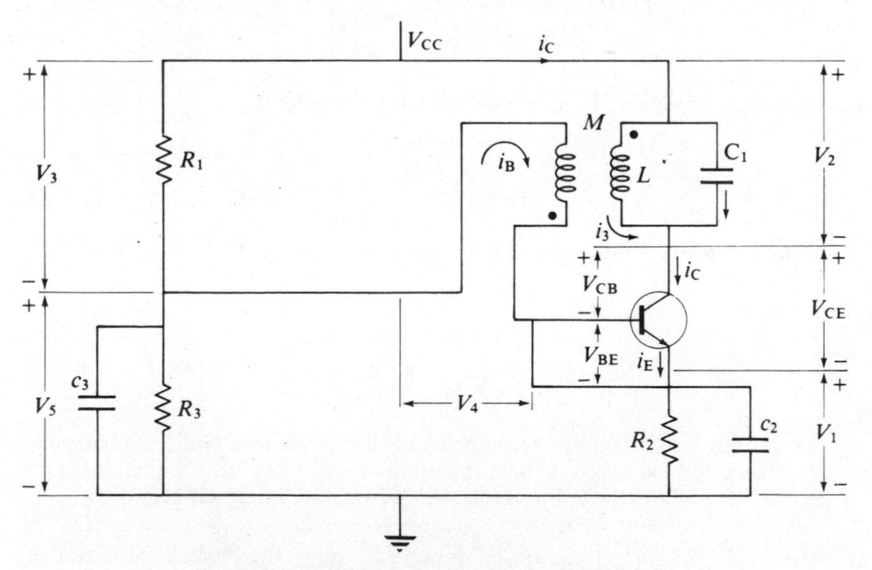

Fig. 3.15 An oscillator circuit for simulation.

REFERENCES

Ames, W. F. (1977) *Numerical Methods for Partial Differential Equations*, 2nd edn, Nelson.

Brebbia, C. A. (ed.) (1980) *New Developments in Boundary Element Methods*, CML Publications.

Box, G. and Muller, M. E. (1958) A note on the Generation of Random Normal Deviates, The Annals of Mathematical Statistics, 29, 610–611.

Coddington, E. A. and Levinson, N. (1955) *Theory of Ordinary Differential Equations*, McGraw-Hill.

Conte, S. D. and de Boor, C. (1972) *Elementary Numerical Analysis*, 2nd edn, McGraw-Hill, Kogakusha.

Ellis, T.M.R. (1982) *A Structured Approach to FORTRAN 77 Programming*, Addison-Wesley.

Garabedian, P. R. (1964) *Partial Differential Equations*, Wiley.

Greenberger, M. (1961) Notes on a new pseudo-random number generator, *J. Assoc. Comp. Mach.*, 8, 163–7.

Hirsch, M. W. and Smale, S. (1971) *Differential Equations and Linear Algebra*, Academic Press.

John, F. (1971) *Partial Differential Equations*, Springer-Verlag.

Lefschetz, S. (1977) *Differential Equations; Geometric Theory*, Dover Publications.

Moshman, J. (1954) The generation of pseudo random numbers on a decimal calculator, *J. Assoc. Comp. Mach.* 1, 88–91.

Nashelsky, L. (1972) *Introduction to Digital Computer Technology*, Wiley.

Peatman, J. B. (1972) *The Design of Digital Systems*, McGraw-Hill.

Roebuck, D and Barnett, S. (1978). A survey of Toeplitz and Related Matrices, Int. J. Systems Sci., *9*, 921–34.

Segerlind, L. J. (1976) *Applied Finite Element Analysis*, Wiley.

Tong, P. and Rossettos, J. N. (1977) *Finite Element Method*, MIT Press.

Zienkiewicz, O. C. (1977) *The Finite Element Method*, 3rd edn, McGraw-Hill.

PART TWO

Control Theory

4 LINEAR SYSTEMS THEORY

4.1 INTRODUCTION

In chapter 1 we discussed in some detail the modelling of several types of physical systems and it was shown that many systems may be represented by an equation of the form

$$\dot{x}(t) = f(x(t), u(t), t), \quad x \in \mathbb{R}^n, u \in \mathbb{R}^l \tag{4.1.1}$$

The state vector x may not be directly measurable and so we may also have a measurement equation of the form

$$y(t) = g(x(t), u(t)), \quad y \in \mathbb{R}^m \tag{4.1.2}$$

for some function g. As we have seen earlier, general non-linear systems of this kind are extremely difficult to deal with and so we usually consider simpler linear systems, which can be used to model many kinds of processes, at least in a neighbourhood of an operating point. These linear systems are of the form

$$\left.\begin{aligned} \dot{x}(t) &= Ax(t) + Bu(t) \\ y(t) &= Cx(t) + Du(t) \end{aligned}\right\} \tag{4.1.3}$$

and constitute the basis of this chapter. These equations are in the so-called state-space or time domain form. However, following the classical development of control, we shall first consider systems from the 's-domain' viewpoint and after defining the Laplace transform we shall study the classical compensation techniques for single-input single-output systems. We shall then discuss the more modern developments for the state space formulation and finally we shall return to the s-domain to give a brief introduction to multivariable control (see Takahashi *et al.*, 1972).

4.2 LAPLACE TRANSFORMATION

4.2.1 Definition and Elementary Properties

The basic mathematical tool in classical control theory is the Laplace transform, which essentially converts a differential equation (in the time domain) into an algebraic equation (in the s- or complex frequency domain). Suppose that f is a function of time which satisfies $|f(t)| \leqslant K\exp(\bar{\sigma}t)$, for $t \geqslant 0$ and some real $\bar{\sigma}$. Then we define the (one-sided) **Laplace transform** $\mathcal{L}f$ of f by

$$\boxed{\mathcal{L}f = \int_0^{\infty} f(t)\exp(-st)\,\mathrm{d}t} \tag{4.2.1}$$

where $s = \sigma + j\omega$ and $\sigma > \bar{\sigma}$. The smallest value of $\bar{\sigma}$ for f is called the **abscissa of convergence**. The Laplace transform of $f(t)$ is often denoted with the corresponding capital letter $F(s)$. The **inverse transform** of $F(s)$ can be shown (Widder, 1972) to be

$$\boxed{f(t) = \frac{1}{2\pi j}\int_{\sigma_1 - j\infty}^{\sigma_1 + j\infty} F(s)\exp(st)\,\mathrm{d}s} \tag{4.2.2}$$

where F is analytic (appendix 1) to the right of the line $\mathrm{Re}(s) = \bar{\sigma} < \sigma_1$. Some of the important properties of the Laplace transform are given in table 4.1, the proofs of which will be left as exercises. In this table, $1(t)$ denotes the unit step function defined by†

$$\boxed{1(t) = \begin{cases} 0 & \text{if } t < 0 \\ 1 & \text{if } t \geqslant 0 \end{cases}} \tag{4.2.3}$$

Apart from the unit step function, one of the most important 'functions' in classical control is the δ-function. It is not, in fact, an ordinary function but is defined loosely as a function which is zero everywhere except at $t = 0$, and is such that

$$\int_{-\varepsilon}^{\varepsilon} \delta(t) f(t)\,\mathrm{d}t = f(0), \quad \text{for all } \varepsilon > 0 \tag{4.2.4}$$

for any function f which is continuous. (For a precise definition of $\delta(t)$, see Schwartz (1950) or Banks (1983).) From (4.2.4) it is clear that

$$\boxed{\mathrm{L}(\delta) = 1} \tag{4.2.5}$$

Although we have the inversion integral (4.2.2), if $F(s)$ is a **rational function** of s, i.e. has the form

$$F(s) = \frac{a_m s^m + a_{m-1}s^{m-1} + \ldots + a_1 s + a_0}{s^n + b_{n-1}s^{n-1} + \ldots 2 + b_0} = \frac{A(s)}{B(s)}$$

† Occasionally, $1(t)$ is defined as $\frac{1}{2}$ when $t = 0$.

Table 4.1 Properties of the Laplace Transform

	time(t)-domain	*s-domain*
1. Delay	$1(t-a)f(t-a)$	$\exp(-as)F(s)$
2. Multiplier		
(a) exp	$\exp(-\alpha t)f(t)$	$F(s+\alpha)$
(b) t^n n integer	$t^n f(t)$	$(-1)^n \frac{\mathrm{d}^n}{\mathrm{d}s^n}(F(s)), n \geqslant 0$
(c) $1/t$	$f(t)/t$	$\int_s^{\infty} F(\eta)\,\mathrm{d}\eta$
3. Time scaling	$f(at)$	$\frac{1}{a}F\left(\frac{s}{a}\right)$
4. Differentiation	$\mathrm{d}^n f/\mathrm{d}t^n$	$s^n F(s) - s^{n-1}f(0^+) - \ldots - f^{(n-1)}(0^+)$
5. Integration	$\int_{-\infty}^{t} f(\tau)\,\mathrm{d}\tau$	$\frac{F(s)}{s} + \frac{1}{s}\int_{-\infty}^{0^+} f(\tau)\,\mathrm{d}\tau$
6. Convolution	$\int_0^t f_1(\tau)f_2(t-\tau)\,\mathrm{d}\tau$ $f_1(t) = f_2(t) = 0, t < 0$	$F_1(s)F_2(s)$
7. Periodic function	$f(t) = f(t+T)$	$\left\{\frac{\int_0^T f(t)\exp(-st)\,\mathrm{d}t}{[1-\exp(-sT)]}\right\}$
8. Initial value theorem	$\lim_{t\to 0} f(t)$	$\lim_{s\to\infty} sF(s)$
9. Final value theorem	$\lim_{t\to\infty} f(t)$	$\lim_{s\to 0} sF(s)$, if $sF(s)$ is analytic in right half plane
10. Product	$f_1(t)f_2(t)$	$\frac{1}{2\pi j}\int_{\sigma_1 - j\infty}^{\sigma_1 + j\infty} F_1(p)F_2(s-p)\,\mathrm{d}p$ $\bar{\sigma}_1 < \sigma_1 < \sigma - \bar{\sigma}_2$ $\sigma = \mathrm{Re}\, s$

then $f(t)$ may be found by the method of **partial fractions**. We can write F in the form

$$F(s) = \sum_{i=1}^{N} \sum_{j=1}^{m_i} \frac{K_{ij}}{(s - p_i)^j} \tag{4.2.6}$$

where the p_i are the roots of $B(s)$ (with multiplicities m_i), and

$$K_{ij} = \left[\frac{1}{(m_i - j)!} \frac{\mathrm{d}^{m_i - j}}{\mathrm{d}s^{m_i - j}} (s - p_i)^{m_i} F(s)\right]\Bigg|_{s=p_i} \tag{4.2.7}$$

$(1 \leqslant i \leqslant N,\ 1 \leqslant j \leqslant m_i)$. Hence,

$$\boxed{f(t) = \sum_{i=1}^{N} \sum_{j=1}^{m_i} K_{ij} \frac{t^{j-1}}{(j-1)} \exp(p_i t)} \tag{4.2.8}$$

Examples

(a) If

$$F(s) = \frac{(s+2)}{(s+1)(s+3)}$$

then

$$F(s) = \frac{A}{s+1} + \frac{B}{s+3}$$

where

$$A = F(s)(s+1)\Bigg|_{s=-1} = \frac{(s+2)}{(s+3)}\Bigg|_{s=-1} = \frac{1}{2}$$

and

$$B = F(s)(s+3)\Bigg|_{s=-3} = \frac{(s+2)}{(s+1)}\Bigg|_{s=-3} = \frac{1}{2}$$

(b) If

$$F(s) = \frac{1}{(s+1)s^2}$$

then

$$F(s) = \frac{A}{s+1} + \frac{B}{s} + \frac{C}{s^2}$$

where

$$A = F(s)(s+1)\Bigg|_{s=-1} = \frac{1}{s^2}\Bigg|_{s=-1} = 1$$

$$C = F(s)s^2\Bigg|_{s=0} = \frac{1}{s+1}\Bigg|_{s=0} = 1$$

$$B = \frac{d}{ds}(F(s)s^2)\bigg|_{s=0} = \frac{d}{ds}\left(\frac{1}{s+1}\right)\bigg|_{s=0} = -1$$

4.2.2 Transfer Function of a System

Consider a system S with input $u(t)$ and output $c(t)$ as in fig. 4.1. S is said to be

(a) **linear** if $S(\alpha u_1 + \beta u_2) = \alpha S(u_1) + \beta S(u_2)$ for any inputs u_1, u_2 and scalars α, β,
(b) **time-invariant** if $c(t) = S(u)(t)$ implies $c_\Delta(t) = S(u_\Delta)(t)$ where $f_\Delta(t) = f(t-\Delta)$,
(c) **causal** if $S(u)(t)$ depends only on $u(\tau)$ for $\tau \leqslant t$.

In particular, let $g(t)$ be the impulse response of the system; i.e. $g(t) = S(\delta)(t)$. Clearly, if S is causal then $g(t) = 0$ for $t < 0$, for if not then the output would be anticipating the impulse at $t = 0$. Hence, if S is also linear and time invariant we have

$$\begin{aligned} c(t) &= S(u)(t) \\ &= \int_{-\infty}^{\infty} u(\tau)S(\delta(t-\tau))\,d\tau \\ &= \int_{-\infty}^{\infty} u(\tau)g(t-\tau)\,d\tau \end{aligned}$$

since

$$u(t) = \int_{-\infty}^{\infty} u(\tau)\,\delta(t-\tau)\,d\tau$$

Hence, if $u(t) = 0$ for $t < 0$ we have

$$\begin{aligned} c(t) &= \int_0^t u(\tau)g(t-\tau)\,d\tau \\ &= \int_0^t g(\tau)u(t-\tau)\,d\tau \end{aligned} \tag{4.2.9}$$

or

$$\boxed{c(t) = (u * g)(t)} \tag{4.2.10}$$

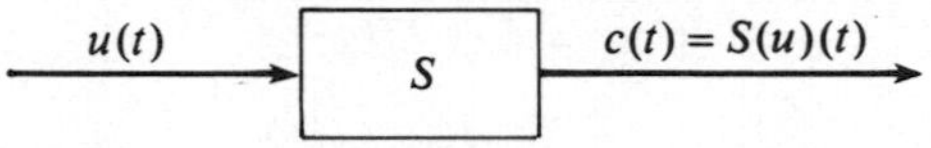

Fig. 4.1 A single input-output block.

where $u*g$ is the **convolution** of u and g, defined by (4.2.9). From table 4.1, we have

$$C(s) = U(s)G(s)$$

or

$$\boxed{G(s) = \frac{C(s)}{U(s)}} \qquad (4.2.11)$$

G is called the **transfer function of** S.

For example, consider the differential equation

$$\frac{d^n}{dt^n}c(t) + b_{n-1}\frac{d^{n-1}}{dt^{n-1}}c(t) + \ldots + b_0c(t)$$
$$= a_m\frac{d^m}{dt^m}u(t) + \ldots + a_0u(t), \quad m \leqslant n \qquad (4.2.12)$$

If the initial conditions are zero, then

$$G(s) = \frac{a_ms^m + a_{m-1}s^{m-1} + \ldots + a_0}{s^n + b_{n-1}s^{n-1} + \ldots + b_0}, \quad m \leqslant n$$
$$= a_m\left[\prod_{i=1}^{m}(s - z_i)/\prod_{i=1}^{n}(s - p_i)\right]$$

for example. The complex numbers z_i are called the **zeros** of the system and p_i are the **poles** of the system S.

It should be noted at this point that we obtain the transfer function of (4.2.12) under the assumption that

$$\frac{d^{m-1}u}{dt^{m-1}}(0) = \ldots = u(0) = 0$$

and then apply $G(s)$ to a step function $u(t)$ for which this clearly does not hold. This apparent contradiction can be explained rigorously using 'distribution theory', but is beyond the scope of this book. However, in section 4.8 we shall write the equation (4.2.12) as a first-order system where the derivatives of u do not appear. The application of the Laplace transform is then completely justified.

When considering feedback systems we usually require the overall closed-loop transfer function. If $G(s)$ and $H(s)$ are given transfer functions, then for the system in fig. 4.2, we have

$$E(s) = R(s) - H(s)C(s) = R(s) - H(s)G(s)E(s)$$

and so

$$C(s) = E(s)G(s)$$
$$= \frac{R(s)G(s)}{1 + G(s)H(s)}$$

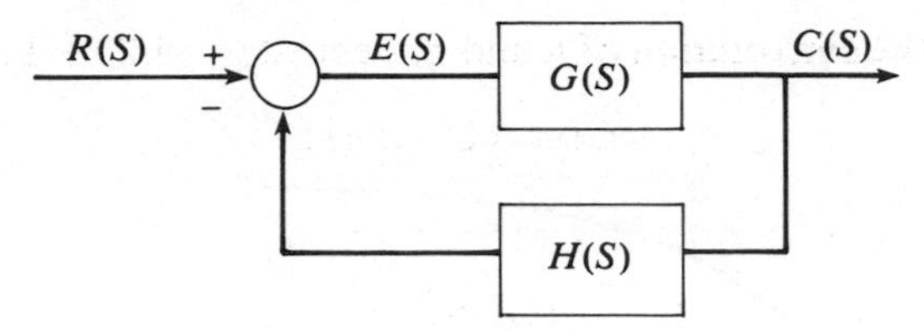

Fig. 4.2 A simple feedback system.

that is

$$\boxed{\frac{C(s)}{R(s)} = \frac{G(s)}{1 + G(s)H(s)}} \tag{4.2.13}$$

4.2.3 Response of Simple Systems

In classical control theory the quality of a feedback system is usually judged by its response to a variety of simple inputs. These inputs are the unit step $1(t)$, the unit ramp $t1(t)$ and possibly the 'unit acceleration' input $(t^2/2)1(t)$. Higher order inputs are rarely considered. The idea is that if the system responds well to these inputs, then it should respond well to any input. We shall now consider some simple systems and their responses to certain of these inputs.

Consider first the first-order system given by the transfer function $G(s) = a/(s + a)$, $a > 0$. For a step input $u(t) = 1(t)$, we have $U(s) = 1/s$ and so

$$\begin{aligned} C(s) &= G(s)U(s) \\ &= \frac{a}{s(s+a)} \\ &= \frac{1}{s} - \frac{1}{s+a} \end{aligned}$$

from which we obtain

$$c(t) = (1 - \exp(-at))1(t)$$

The **time constant** of the system is defined as $1/a$. The response $c(t)$ is shown in fig. 4.3. Note that $c(\infty) = u(\infty) = 1$, and so the system has zero steady-state error. For the ramp input $U(s) = 1/s^2$ and so

$$C(s) = G(s)U(s) = \frac{a}{s^2(s+a)}$$

The error is $E(s) = U(s) - C(s) = 1/(s(s + a))$. By the final value theorem (table 4.1) we have $e(\infty) = \lim_{s\to 0} sE(s) = 1/a$. Hence the system does not

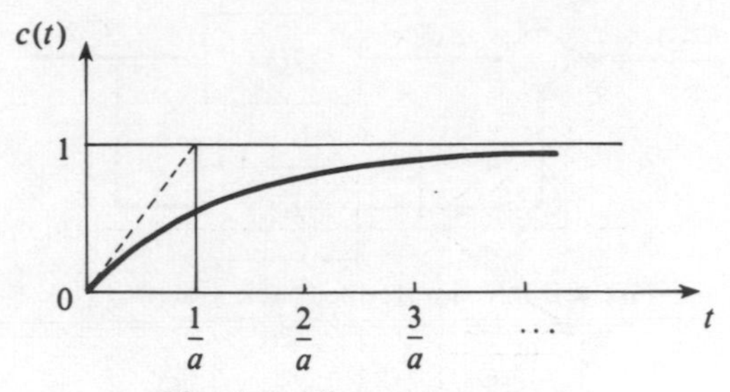

Fig. 4.3 First order response.

have zero steady-state error to a ramp. The **steady state** (or d.c.) **gain** of the system $G(s)$ is, by definition, $G(0)$.

The most important system for classical control is defined by the second-order transfer function

$$\boxed{G(s) = \frac{\omega_n^2}{s^2 + 2\zeta\omega_n s + \omega_n^2}} \qquad \omega_n > 0, \zeta \geqslant 0$$

ω_n is the **natural frequency** and ζ is the **damping ratio**. As shown in fig. 4.4. the poles of the system are at

$$s = -\zeta\omega_n \pm j\omega_n(1-\zeta^2)^{1/2}$$

It is easy to show that the response to a step input is

$$\boxed{c(t) = \left[1 - \exp(-\zeta\omega_n t)\,\frac{\sin(\omega_n(1-\zeta^2)^{1/2}t + \cos^{-1}\zeta}{(1-\zeta^2)^{1/2}}\right]1(t)} \qquad (4.2.14)$$

for $\zeta < 1$. Similar expressions follow for general ζ. The response is shown in fig. 4.5 for various values of ζ. By solving $\mathrm{d}c(t)/\mathrm{d}t = 0$ for t it follows that the time to the first peak is $t_1 = \pi/(\omega_n(1-\zeta^2)^{1/2})$ and so the size of the 'overshoot' is

$$c(t_1) = 1 + \exp\left[-\frac{\zeta\pi}{(1-\zeta^2)^{1/2}}\right]$$

$c(t) - 1$ is the percentage overshoot and for $0.3 \leqslant \zeta \leqslant 0.7$ we obtain a percentage overshoot approximately between 7 per cent and 35 per cent. We usually try to produce a control system with a ζ in this range, although this does depend, to a large extent, on the particular application. Note that the steady-state error of the system with step input is zero and the error to a ramp is

$$\lim_{s\to 0} \frac{(s + 2\zeta\omega_n)}{(s^2 + 2\zeta\omega_n s + \omega_n^2)} = \frac{2\zeta}{\omega_n}$$

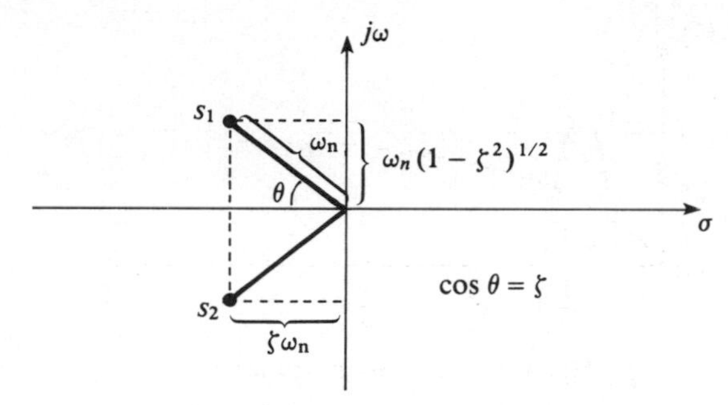

Fig. 4.4 Dominant pole configuration.

Consider finally a second-order system with an added zero

$$G(s) = \frac{(s+a)(\omega_n^2/a)}{s^2 + \zeta\omega_n s + \omega_n^2}$$

Note that $G(s)$ has been chosen so that $G(0) = 1$. The response of this system to a step is

$$C_1(s) = \frac{(s+a)(\omega_n^2/a)}{s(s^2 + 2\zeta\omega_n s + \omega_n^2)} = \frac{\omega_n^2}{s(s^2 + 2\zeta\omega_n s + \omega_n^2)} + \frac{s}{a}\,\frac{\omega_n^2}{s(s^2 + 2\zeta\omega_n s + \omega_n^2)}$$

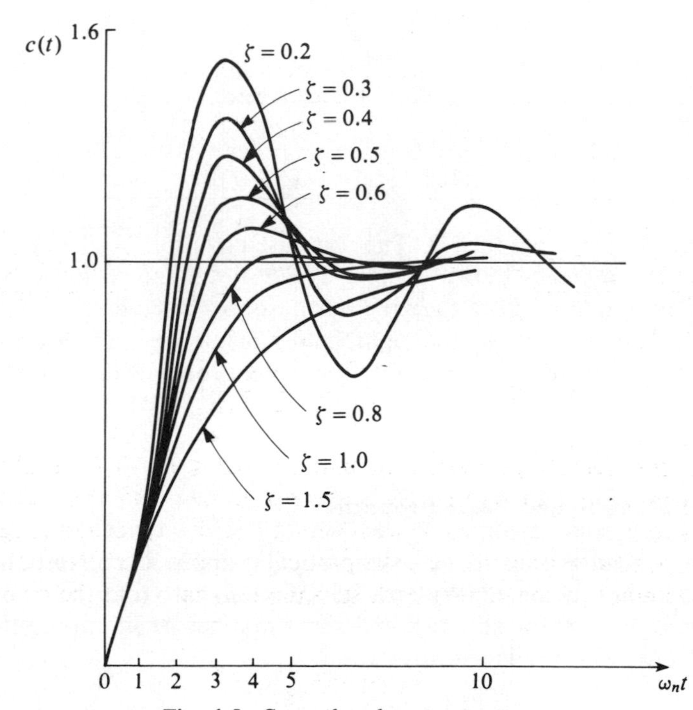

Fig. 4.5 Second order responses.

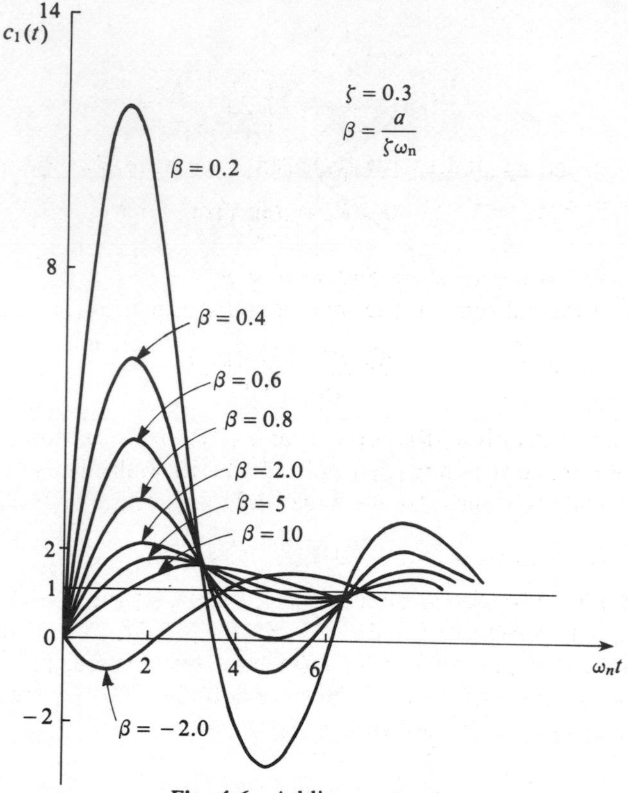

Fig. 4.6 Adding a zero.

Hence,

$$c_1(t) = c(t) + \frac{1}{a}\frac{\mathrm{d}}{\mathrm{d}t}(c(t))$$

where $c(t)$ is given by (4.2.14). The response $c_1(t)$ for $\zeta = 0.3$ and various values of $\beta = a/\zeta\omega_n$ is shown in fig 4.6. For $a < 0$ the response 'goes the wrong way' initially before showing second-order behaviour. Systems with zeros (or indeed poles) in the right half s plane are said to be of **non-minimum phase** and are responsible for many problems in control theory.

4.3 LINEAR SYSTEM STABILITY

4.3.1 Definition and Basic Properties

A linear system is said to be asymptotically stable if the response to a bounded input is bounded (Willems, 1970). (A signal $x(t)$, $t \geqslant 0$ is bounded if $|x(t)| \leqslant M < \infty$ for all $t \geqslant 0$ and some number M.) Suppose that $x(t)$ is a function with Laplace transform

$$X(s) = \frac{a_m s^m + a_{m-1}s^{m-1} + \ldots + a_0}{s^n + b_{n-1}s^{n-1} + \ldots + b_0}, \quad m \leqslant n$$

Then

$$X(s) = K_\infty + \sum_{i=1}^{l} \sum_{j=1}^{m_i} \frac{K_{ij}}{(s - p_i)^j} \tag{4.3.1}$$

by partial fractions and so for bounded x we have $K_\infty = 0$ and

$$0 = K_\infty = \lim_{s \to \infty} X(s)$$

that is $X(s)$ has a zero at ∞ and so $m < n$.

Now a typical term in the inverse transform of (4.3.1) is

$$\frac{K_{im_i} t^{m_i - 1} \exp(p_i t)}{(m_i - 1)!}$$

This is bounded only if $\mathrm{Re}(p_i) \leqslant 0$, and if $\mathrm{Re}(p_i) < 0$ then x is bounded while if $\mathrm{Re}\, p_i = 0$ it is bounded only if m_i (the multiplicity of p_i) is 1.

Now for a system $G(s)$ we have the input–output relation

$$C(s) = G(s)U(s)$$

and if $c(t)$ is bounded for bounded $u(t)$ then $G(s)$ must have poles in the open left half plane i.e. for Re $(s) < 0$. For, if $G(s)$ has a pole p in the closed right half plane, then let $U(s)$ have an identical pole if Re $p = 0$ or $U(s) = 1/s$ otherwise; $c(t)$ is then unbounded. Hence the system G is asympotically stable if the poles p_i of G satisfy

$$\boxed{\mathrm{Re}\, p_i < 0 \quad \text{for all } i} \tag{4.3.2}$$

4.3.2 Routh's Criterion

A simple method of calculating stability conditions for a linear system is given by Routh's criterion. If the system is defined by the transfer function

$$G(s) = \frac{B(s)}{a_n s^n + a_{n-1} s^{n-1} + \ldots + a_0}$$

then we have shown that the stability is determined entirely by the roots of $A(s) = a_n s^n + \ldots + a_0$. If any coefficient a_i is not strictly positive then there certainly exist roots in the closed right half plane. Hence $a_i > 0$ for $0 \leqslant i \leqslant n$ is a necessary condition for stability. In this case we form the table

$n+1$ rows					
	1	a_n	a_{n-2}	a_{n-4}	$\ldots$
	2	a_{n-1}	a_{n-3}	a_{n-5}	$\ldots$
	3	b_{n-1}	b_{n-3}	b_{n-5}	$\ldots$
	4	c_{n-1}	c_{n-3}	c_{n-5}	$\ldots$
	$\vdots$	$\vdots$			
	$n+1$	r_{n-1}			

where the b are defined in terms of rows 1 and 2 by

$$b_{n-i} = \begin{vmatrix} a_{n-1} & a_{n-i-2} \\ a_n & a_{n-i-1} \end{vmatrix} \Bigg/ a_{n-1}, \quad i = 1, 3, 5, \ldots$$

and the remaining rows are defined in terms of the preceding two rows in the same way. The Routh criterion states that

(a) the system is stable if the elements of the first column are strictly positive,
(b) the number of sign changes in the first column is equal to the number of roots of $A(s)$ in the right half plane.

Moreover, if a zero appears in the first column and non-zero elements in the same row then we can replace the zero by a small positive number ε. For example, if $A(s) = s^4 + s^3 + s^2 + s + 2$ we obtain the array

$$\begin{array}{c|ccl} 1 & 1 & 1 & 2 \\ 2 & 1 & 1 & (0) \leftarrow \text{'pad out' the row with a zero if necessary} \\ 3 & 0 \rightarrow \varepsilon & 2 & \\ 4 & (\varepsilon - 2)/\varepsilon & (0) & \\ 5 & 2 & & \end{array}$$

The first column is $(1, 1, 0, -\infty, 2)^{\mathrm{T}}$ as $\varepsilon \rightarrow 0$ and so there are two sign changes, i.e. the system has two right half plane poles.

If all the elements in a row are zero then it can be shown that the polynomial has an even factor. Thus the roots are symmetric with respect to the $j\omega$ and σ axes and so the system is unstable. Also the row above the zero row contains the coefficients of the even factor. The array may be continued by replacing the zero row by the coefficients of the derivative of the even factor. For example, consider

$$A(s) = s^6 + 2s^5 + 5s^4 + 8s^3 + 8s^2 + 8s + 4$$

The array is

$$\begin{array}{llll} s^6 & 1 & 5 & 8 & 4 \\ s^5 & 2 & 8 & 8 \\ s^4 & 1 & 4 & 4 \\ \boxed{s^3 \;\; 0 \;\; 0} & & & \text{hence even factor } s^4 + 4s^2 + 4 \\ s^3 & 4 & 8 \\ s^2 & 2 & 4 \\ \boxed{s^1 \;\; 0} & & & \text{even factor } 2s^2 + 4 \\ s^1 & 4 \\ s^0 & 4 \end{array}$$

Since there are no sign changes there are no poles in the (open) right half plane. This method can be used to determine the values of variable coefficients of a system polynomial for stability.

4.3.3 Nyquist Criterion

Suppose that a stable linear system with impulse response $g(t)$ has an input† of the form $\exp(j\omega t)$. Then the steady-state output is

$$\int_0^\infty g(\tau)\exp[j\omega(t-\tau)]\,d\tau = \exp(j\omega t)G(j\omega)$$

Hence the steady-state response to a complex sinusoid is also a complex sinusoid of amplitude $|G(j\omega)|$ and phase $\angle G(j\omega)$. The transfer function $G(s)$ evaluated on the imaginary axis (i.e. $G(j\omega)$) is therefore called the **frequency response** of the system. Note that, for real systems, $G(-j\omega)=\overline{G(j\omega)}$ and so we need consider positive frequencies only. The image of the function $\omega \to G(j\omega)$ $(0 \leqslant \omega < \infty)$ is called the **Nyquist plot** of the system. The plots of two important systems are shown in fig. 4.7. Both are first order, but the output of the first always lags the input while that of the second leads the input. Hence we call these the first-order **lag** and **lead** networks. The Nyquist plot of the second-order system $G(s)=\omega_n^2/(s^2+2\zeta\omega_n s+\omega_n^2)$ is shown in fig. 4.8(a), and it is interesting to consider the effect of a pure delay on this system, in which case $G(s)=\omega_n^2\exp(-as)/$

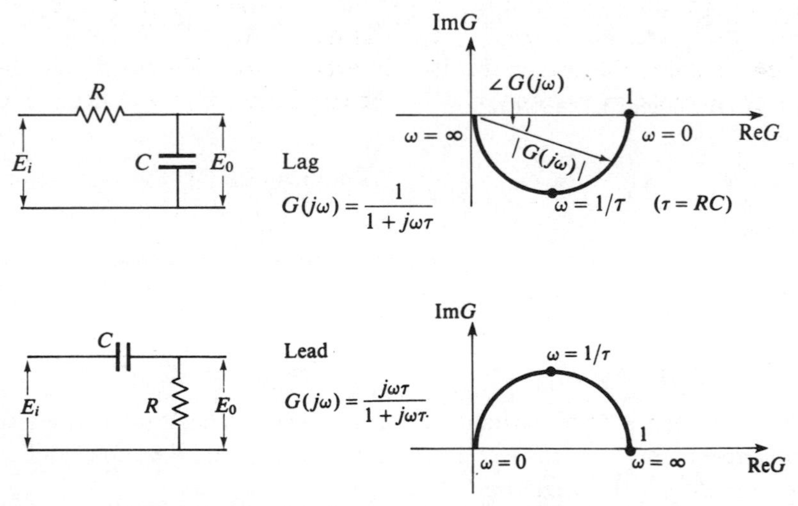

Fig. 4.7 Nyquist diagrams for lag and lead networks.

† It is convenient to use the complex exponential. Of course, the real signals are cos ωt and sin ωt.

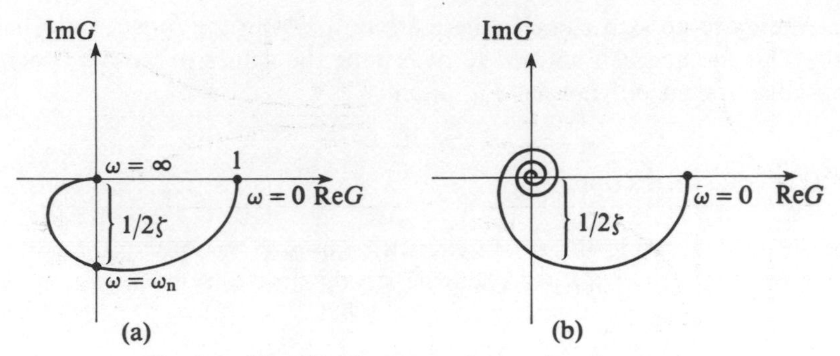

Fig. 4.8 Nyquist diagrams of second order systems.

$(s^2 + 2\zeta\omega_n s + \omega_n^2)$ and the Nyquist plot appears in fig. 4.8(b). In the latter case, the plot 'wraps around' the origin an infinite number of times. (For simplicity we have assumed that $a = 2\pi/\omega_n$.)

A problem occurs if $G(s)$ has a pole of order n on the $j\omega$-axis, say at $s = j\omega_0$. Then, close to ω_0,

$$G(s) \cong \frac{K\omega_0}{(s - j\omega_0)^n} + \{\text{term analytic at } j\omega_0\}$$

with

$$K\omega_0 = \lim_{s \to j\omega_0} [(s - j\omega_0)^n G(s)]$$

Hence if ω is close to ω_0, $G(j\omega) \cong K\omega_0/j^n(\omega - \omega_0)^n$, and so as $\omega \to \omega_0$, $|G(j\omega)| \to \infty$ and

(a) $\angle G(j\omega) \to -n\pi/2$ for n even

(b) $\angle G(j\omega) \to \begin{cases} -(n+2)\pi/2 & \text{for } \omega < \omega_0 \\ -n\pi/2 & \text{for } \omega > \omega_0 \end{cases}$ and n odd

(see fig. 4.9.) The Nyquist plot can be made continuous at $s = j\omega_0$ by replacing the part of the $j\omega$-axis near $j\omega_0$ by a small semicircle in the right half plane† as in fig. 4.10. Then

$$G(s) = K\omega_0\varepsilon^{-n}\exp(-jn\theta) + (\text{small term near } j\omega_0)$$

and $|G(s)| = |K\omega_0\varepsilon^{-n}| = \text{constant}$, $\angle G(s) = -n\theta$. Hence, as we traverse the semicircle from $-\pi/2$ to $+\pi/2$, then G moves on a circular arc of radius $|K\omega_0\varepsilon^{-n}|$, through the angle $n\pi$.

We can now demonstrate the Nyquist stability criterion for the feed-

† or the left half plane, see Willems (1970).

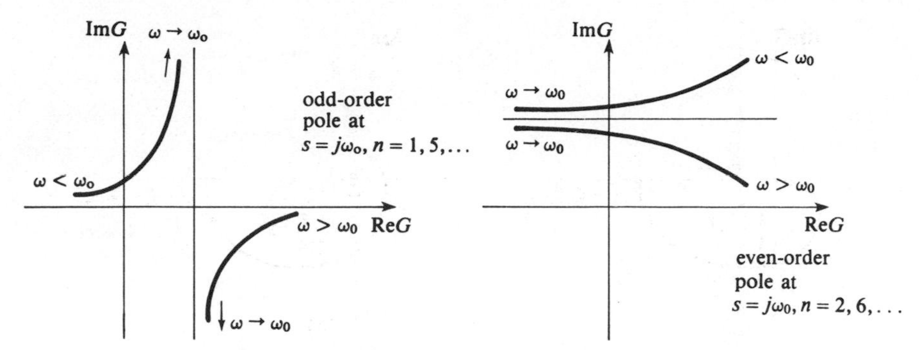

Fig. 4.9 Asymptotic properties of Nyquist diagrams.

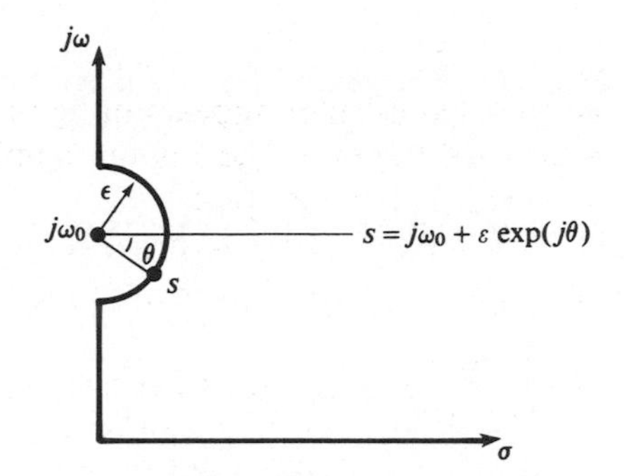

Fig. 4.10 Avoiding a pole on the imaginary axis.

back system in (4.2.12). First, we define the **Nyquist contour** in the s-plane, related to to (4.2.12), to be the contour shown in fig. 4.11(a), which consists of part of the $j\omega$-axis (indented at poles of GH if necessary) together with a semicircle of large radius, so that the resulting closed curve contains all the right half plane poles and zeros of GH. We denote by γ the curve oriented as shown. Now consider the complex function $s \rightarrow G(s)H(s)$. Then γ maps into a closed curve γ' in the GH-plane (fig. 4.11(b)). We recall the **principle of the argument** (appendix 1) which states that if f is a function which is analytic on a region D in the complex s-plane, except for poles and zeros, and γ is a closed (clockwise oriented) curve in D, then $f_o\gamma$ (where $(f_o\gamma)(z) = f[\gamma(z)]$) encircles the origin $P_f - Z_f$ times, where Z_f and P_f are the numbers of zeros and poles of f inside γ (counting multiplicities). Hence the number of times N that γ' encircles the point -1, or equivalently, the number of times the image of γ under the map $s \rightarrow 1 + G(s)H(s)$ encircles the origin, equals the number of poles minus the number of zeros of $1 + G(s)H(s)$ in the right half plane. Therefore, if

$$N_p\dagger = \text{number of right half plane poles of } 1 + GH$$

† N_p also equals the number of right half plane poles of GH.

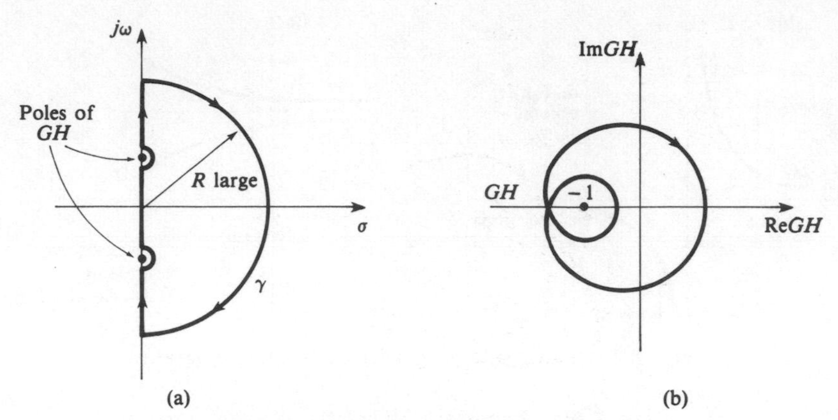

Fig. 4.11 The general Nyquist diagram.

and

$$N_z = \text{number of right half plane zeros of } 1 + GH$$
$$= \text{number of right half plane closed-loop poles}$$

then

$$\boxed{N_z = N_p - N} \tag{4.3.3}$$

and if $N_p - N \geqslant 1$ the system must be unstable. Note that if the closed-loop system has a pole on the $j\omega$-axis then $1 + G(j\omega)H(j\omega) = 0$ and so GH passes through the -1 point, and conversely. Hence the closed-loop system is stable if and only if the Nyquist plot of G does not pass through the -1 point and

$$\boxed{N_p = N} \tag{4.3.4}$$

In particular, if GH is stable, i.e. the open-loop system is stable, then $N_p = 0$ and so the closed-loop system is stable if and only if the $G(j\omega)H(j\omega)$ contour does not pass through or encircle the point -1.

4.3.4 Gain and Phase Margins and Error Constants

Two qualitative measures of the margin of stability are provided by the gain and phase margins. Suppose that $1 + G(s)H(s)$ has no right half plane poles and so, for stability of the closed-loop system, $G(j\omega)H(j\omega)$ must not encircle or pass through the point -1. Moreover, if GH has more poles than zeros then

$$\lim_{s \to \infty} G(s)H(s) = 0$$

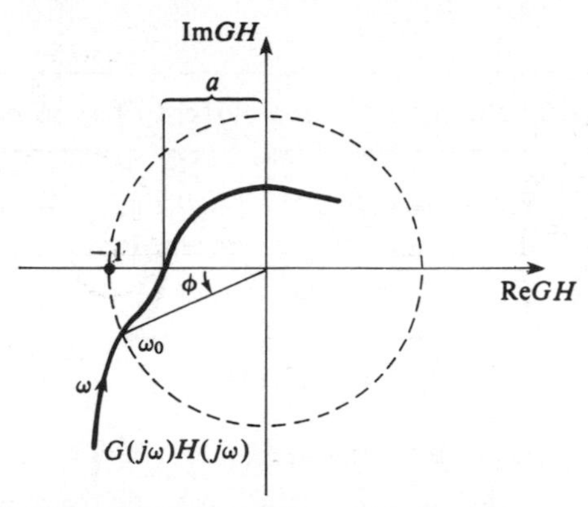

Fig. 4.12 Graphical stability criterion.

and so the Nyquist plot tends to 0 as $\omega \to \infty$. Suppose that the plot of a stable closed-loop system appears as in fig. 4.12. Then, if the gain of the open-loop system $G(s)H(s)$ is multiplied by $1/a$, the Nyquist plot passes through -1 and the closed-loop system becomes unstable. Hence $1/a$ is called the system **gain margin**. Similarly, if we add a phase shift of ϕ to GH the closed-loop system becomes unstable, and so ϕ is called the **phase margin**. Typical acceptable values of gain and phase margin are 4–12 dB and 30°–60° respectively.

Another important measure of system performance is the steady-state error to various inputs. Consider again the system defined by (4.2.12), where the error E is given by $E(s) = R(s)/(1 + G(s)H(s))$. By the final value theorem (table 4.1), the steady-state error to an input R is

$$e_f = \lim_{s\to 0} \frac{sR(s)}{1 + G(s)H(s)}$$

Hence, for an input $R(s)$ of the form $1/s^n$ we have

$$e_f = \begin{cases} 1/(1 + \lim_{s\to 0} G(s)H(s)) & \text{if } n = 1 \\ 1/(\lim_{s\to 0} s^{n-1}G(s)H(s)) & \text{if } n > 1 \end{cases} \tag{4.3.5}$$

The **type** N of the system $G(s)H(s)$ is the number of poles of the system at the origin, i.e. GH has a factor s^{-N}. Hence for an input $1/s^n$ we have

$$e_f = 0 \quad \text{for a system of type } n \tag{4.3.6}$$

Moreover, if $n > 1$, then

$$e_f = \infty \quad \text{for a system of type } n - 2 \tag{4.3.7}$$

and finally

$$e_f \text{ is finite and non-zero for a system of type } n-1 \tag{4.3.8}$$

For $n = 1, 2, 3$ these errors are called the **position, velocity and acceleration** errors respectively. The corresponding **error constants** K_p, K_v, K_a are defined by

$$\begin{aligned} K_p &= \lim_{s\to 0} G(s)H(s) \\ K_v &= \lim_{s\to 0} sG(s)H(s) \\ K_a &= \lim_{s\to 0} s^2G(s)H(s) \end{aligned} \tag{4.3.9}$$

Hence, by (4.3.5),

$$\begin{aligned} e_f(\text{step}) &= \frac{1}{1+K_p} \\ e_f(\text{ramp}) &= \frac{1}{K_v} \\ e_f(\text{acceleration}) &= \frac{1}{K_a} \end{aligned}$$

and so large values of the error constants imply small steady-state errors.

4.4 FREQUENCY RESPONSE ANALYSIS

4.4.1 Bode Diagrams

We have already shown that the frequency response of a system $G(s)$ is just $G(j\omega)$, i.e. the Laplace transform of the impulse response evaluated on the imaginary axis, and we have considered a graphical representation of $G(j\omega)$ in the form of the Nyquist plot. We could equally well plot the magnitude $|G|$ and phase $\angle G$ of $G(j\omega)$ separately against frequency. However, it is usually convenient to plot $20 \log_{10}|G(j\omega)|$ and $\angle G(j\omega)$ against $\log_{10}\omega$; the two resulting graphs are called the **Bode diagrams** of the system (fig. 4.13). If we factorize $G(s)$ in the form

$$G(s) = K_N \prod_{i=1}^{\alpha} (sa_i + 1) \Bigg/ \left[s^N \prod_{j=1}^{\beta} (sb_j + 1) \prod_{k=1}^{\gamma} \left(1 + \frac{2\zeta_k}{\omega_k} s + \frac{s^2}{\omega_k^2}\right)\right]$$

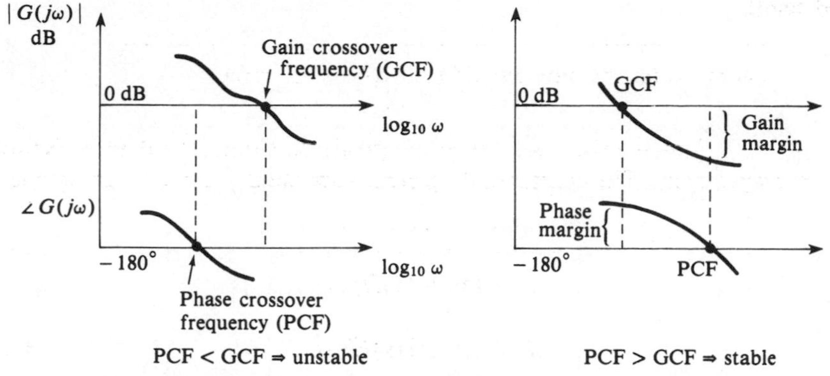

Fig. 4.13 Gain and phase response—Bode diagrams.

then we have

$$20 \log_{10} |G(j\omega)| = 20\left[\log_{10} K_N + \sum_{i=1}^{\alpha} \log_{10} |j\omega a_i + 1| - N \log_{10} |\omega| - \sum_{j=1}^{\beta} \log_{10} |j\omega b_j + 1| - \sum_{k=1}^{\gamma} \log_{10} \left| 1 + j\frac{2\zeta\omega}{\omega_k} - \left(\frac{\omega}{\omega_k}\right)^2 \right|\right] \tag{4.4.1}$$

and

$$\angle G(j\omega) = \sum_{i=1}^{\alpha} \tan^{-1}(\omega a_i) - N\frac{\pi}{2} - \sum_{j=1}^{\beta} \tan^{-1}(\omega b_j) - \sum_{k=1}^{\gamma} \tan^{-1}\left(\frac{2\zeta\omega\omega_k}{\omega_k^2 - \omega^2}\right) \tag{4.4.2}$$

Hence the contribution from each pole or zero can be drawn separately and the results added together. For example, when $\omega \to 0$, $20 \log_{10} |G(j\omega)| \to 20 \log_{10} K_N - 20N \log_{10} |\omega|$ and so the low frequency asymptote has slope $-20N$ dB/decade. (A **decade** is a change in $\log_{10}\omega$ of 1.) Moreover, for a second-order term of the form $(1 + j(2\zeta/\omega_n)\omega - \omega^2/\omega_n^2)^{-1}$, we have the magnitude response

$$-20 \log_{10} \left| 1 + \frac{2\zeta j\omega}{\omega_n} - \frac{\omega^2}{\omega_n^2} \right|$$

which has a maximum M_m when

$$\omega_m = \omega_n(1 - 2\zeta^2)^{1/2} \tag{4.4.3}$$

Then

$$M_m = -20 \log_{10} 2\zeta(1-\zeta^2)^{1/2} \tag{4.4.4}$$

These relations are important in frequency response compensation.

We can use the Bode plots to determine the stability of a closed-loop system. Suppose that $G(s)H(s)$ has no right half plane poles. Then, for stability, the frequency response should not encircle -1. If $G(j\omega)$ passes to the left of -1, i.e. $\angle G(j\omega) = -180^\circ$, and $|G(j\omega)| > 1$, then the system is unstable, while if $\angle G(j\omega) = -180^\circ$ when $|G(j\omega)| < 1$, the system is stable.

4.4.2 Nichol's Chart

Another method of representing a frequency response $G(j\omega)$ graphically is by plotting $x = 20 \log_{10} |G(j\omega)|$ against $y = \angle G(j\omega)$. We then obtain a **Nichol's chart** of the system (fig. 4.14). The advantage of this plot is that we can superimpose, on the rectangular (x, y) axes, curvilinear (ξ, η) axes given by

$$\xi = 20 \log_{10} \left| \frac{G(j\omega)}{1+G(j\omega)} \right|, \quad \eta = \angle \left[\frac{G(j\omega)}{1+G(j\omega)} \right] \tag{4.4.5}$$

We will then be able to read off directly the unity feedback closed-loop response. A typical Nichol's chart for the system

$$G(j\omega) = \frac{1+j\omega}{j\omega(1+5j\omega)(1+j\omega/4-\omega^2/10)}$$

is shown in fig. 4.15. Note that the maximum closed-loop gain is at a point where the open-loop plot is tangent to a ξ-curve.

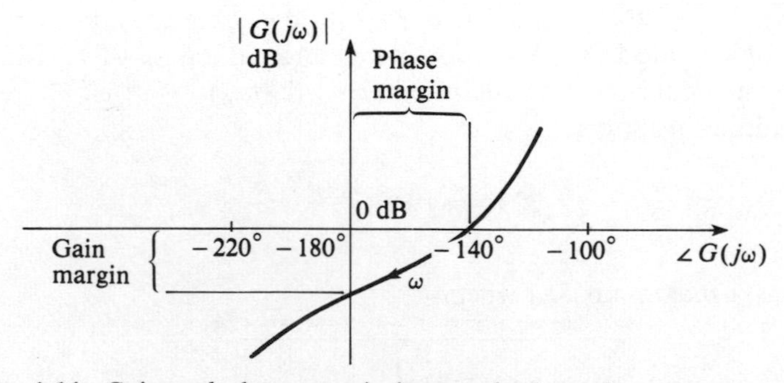

Fig. 4.14 Gain and phase margin in magnitude-angle representation.

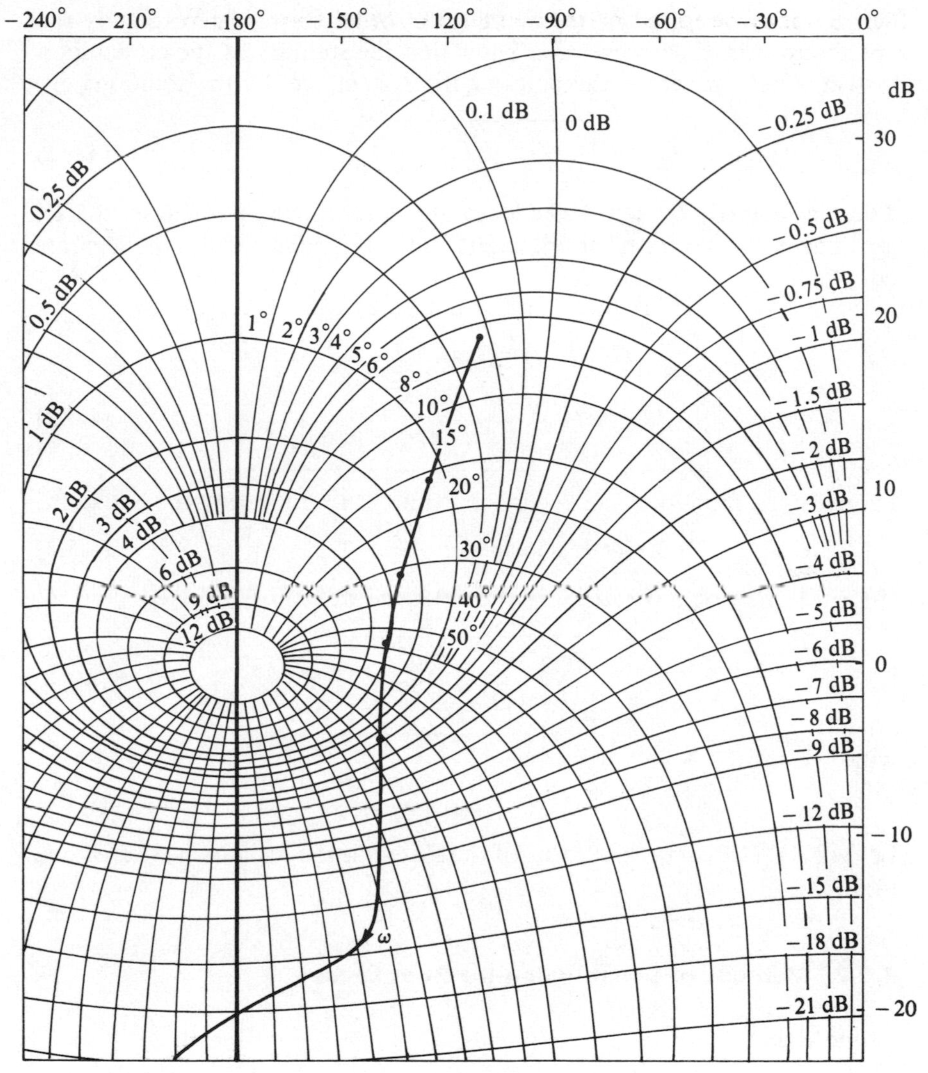

Fig. 4.15 Nichol's chart.

4.5 ROOT LOCUS THEORY

4.5.1 Definition

Consider again the feedback system shown in fig. 4.2. We have shown before that the closed-loop transfer function is

$$\frac{kG(s)}{1 + kG(s)H(s)}$$

where we have replaced $G(s)$ by $kG(s)$ in order to indicate explicitly the open loop gain k. However, we know that the stability of the closed-loop system is determined by the poles of the system, i.e. by the solutions of

$$\boxed{kG(s)H(s) = -1} \tag{4.5.1}$$

The **root locus** is defined as the locus of the solutions of equation (4.5.1), as the gain k varies from 0 to ∞. Clearly, (4.5.1) is equivalent to the separate equations

$$\boxed{\begin{aligned} &\angle G(s)H(s) = \pm \pi i, \quad i = 1, 3, 5, \ldots \\ &|G(s)H(s)| = \frac{1}{k} \end{aligned}} \tag{4.5.2}$$

However, to draw the root locus we require only to satisfy the equation

$$\angle G(s)H(s) = \pm \pi i, \quad i = 1, 3, 5, \ldots \tag{4.5.3}$$

since for any point s_1 satisfying this relation the gain

$$\boxed{k = \frac{1}{|G(s_1)H(s_1)|}} \tag{4.5.4}$$

will then satisfy

$$kG(s_1)H(s_1) = -1$$

i.e. (4.5.1). Hence the root locus of the system is the set of points satisfying (4.5.3).

4.5.2 Methods of Constructing the Root Locus

Suppose that

$$G(s)H(s) = \prod_{i=1}^{m} (s - z_i) \Big/ \prod_{j=1}^{n} (s - p_j), \quad n > m$$

Note that, as $s \to \infty$, $G(s)H(s) \sim 1/s^{n-m}$ and so we say that GH has $\boldsymbol{n - m}$ **zeros at infinity**. Then

$$\angle G(s)H(s) = \sum_{i=1}^{m} \angle (s - z_i) - \sum_{j=1}^{n} \angle (s - p_j) \tag{4.5.5}$$

Hence, once the poles and zeros of the open-loop transfer function have been obtained (perhaps numerically) then for any point s we can perform the sums in (4.5.5) and then check (4.5.3) to see if s lies on the root locus.

Having plotted the root locus, we can use (4.5.4) to determine the value of k at any point s_1 on the locus. However, at s_1

$$k = \prod_{j=1}^{n} |s_1 - p_j| \Big/ \prod_{i=1}^{m} |s_1 - z_i| \tag{4.5.6}$$

which can again be determined graphically (or numerically). The importance of the root locus is that we obtain closed-loop information from the open-loop poles and zeros.

Although the preceding ideas are sufficient to plot the root locus, the following rules are useful in determining the main structural form of the locus of a given system $G(s)H(s) = A(s)/B(s)$, say, for some polynomials A, B of orders m, n $(n > m)$.

RULE 1: The root locus has n branches which are symmetric about the real axis, where n is the number of open-loop poles.
Proof: $1 + kGH = (B + kA)/B$ and $(B + kA)$ is of degree n for finite k.

RULE 2: The root locus branches begin on open-loop poles (for $k = 0$) and end on open-loop zeros (for $k = \infty$).
Proof: $1 + kGH = k(1/k + A/B) = 0$ when $A/B = -1/k$.
If k is very small, then the solutions of this equation must be close to the zeros of B. As $k \to \infty$ we can only make A/B small near the zeros of A.

RULE 3: The root locus contains a point s_r on the real axis if there are an odd number of (open-loop) poles and zeros on the real axis to the right of this point.
Proof: For each pole and zero on the real axis to the right of s_r, we have $\angle(s_r - z_i) = \angle(s_r - p_i) = 180°$ and to the left of s_r, $\angle(s_r - z_i) = \angle(s_r - p_i) = 0$. Since poles and zeros occur in conjugate pairs the angle contribution to $\angle G(s_r)H(s_r)$ of complex poles and zeros is 0. Hence,

$$\angle G(s_r)H(s_r) = 180° \times (\text{no. of open-loop zeros} - \text{no. of open-loop poles to right of } s_r) = -180°$$

if the difference of the numbers in parentheses is odd, which is equivalent to their sum being odd.

RULE 4: Branches of the root locus which diverge to ∞ (i.e. to open-loop zeros at ∞) are asymptotic to the lines with angle $\theta = l(180°)/(n - m)$ for $l = \pm 1, \pm 3, \ldots$ which intersect the real axis at the common point, called the 'pivot' or 'hub',

$$\boxed{\sigma_0 = \left(\sum_{j=1}^{n} p_j - \sum_{i=1}^{m} z_i\right) \Big/ (n - m)} \tag{4.5.7}$$

Proof:

$$G(s)H(s) = \frac{s^m - \left(\sum_{i=1}^{m} z_i\right)s^{m-1} + \ldots + \prod_{i=1}^{m}(-z_i)}{s^n - \left(\sum_{j=1}^{n} p_j\right)s^{n-1} + \ldots + \prod_{j=1}^{n}(-p_j)}$$

$$= \frac{1}{s^{n-m} - (\sum p_j - \sum z_i)s^{n-m-1} + \ldots}$$

and so, for large s, GH behaves like

$$\frac{1}{(s-\sigma_0)^{n-m}} = \frac{1}{s^{n-m} - (n-m)\sigma_0 s^{n-m-1} + \ldots}$$

with σ_0 as above. However, if $\angle(s-\sigma_0) = \theta$, then $\angle 1/(s-\sigma_0)^{n-m} = -(n-m)\theta$, and so the root locus of $1/(s-\sigma_0)^{n-m}$ is given by $\theta = l(180^\circ)/(n-m)$.

RULE 5: The root locus branches intersect on the real axis when k is an extremum for real s, i.e. write $k = -1/G(s)H(s)$ for real s and evaluate the points s on the real axis where $dk/ds = 0$.
Proof: Consider a segment of the locus on the real axis between two poles. Since the branches start on poles this segment must consist of two different branches. Now branches end on zeros, so the branches must break away from the real axis somewhere between the poles. This point is clearly a maximum of k. Similarly a breakaway point between two zeros on the real axis is where k is a minimum.

RULE 6: The angle of departure from a complex pole (or entry to a complex zero) equals 180°–(angle contribution at this pole or zero of all other finite poles and zeros).
Proof: Consider the complex pole p_1. Then for any point s on the root locus,

$$\sum_{i=1}^{m}(s-z_i) - \sum_{j=1}^{n}(s-p_j) = \pm 180^\circ$$

If s is close to p_1, then $\angle(s-z_i) \approx \angle(p_1-p_j)$ and $\angle(s-p_j) \approx \angle(p_1-p_j)$ for $j \neq 1$. Hence

$$-\angle(s-p_1) \approx \pm 180^\circ - \left[\sum_{i=1}^{n}(p_1-z_i) - \sum_{j=2}^{n}(p_1-p_j)\right]$$

RULE 7: The $j\omega$-axis crossings may be found from the Routh table of $B(s) + kA(s)$ (where $A/B = GH$).
Proof: This is just an application of the Routh stability criterion.

Note finally that the root locus can be generalized to multivariable systems (Owens, 1985) and even to distributed systems (Banks, 1983; Banks and Abbasi Ghelmansarai, 1983).

Example
Consider the open-loop system

$$G(s)H(s) = \frac{1}{s(s+1)(s+2)}$$

By rules 1 and 2, the root locus has three branches beginning at $s = 0$, -1 and -2 and ending at ∞ (no open-loop zeros). By rule 3, the intervals $(-\infty, -2]$ and $[-1, 0]$ of the real axis are part of the locus. By rule 4, the branches are asymptotic to the lines with angles $\theta = 60°, 180°, 300°$ intersecting at $s = -1$. By rule 5, a breakpoint on the real axis occurs when $k = -s(s+1)(s+2)$ is an extremum for real s. Now,

$$\frac{dk}{ds} = -3s^2 - 6s - 2 = 0 \quad \text{when } s = -1 \pm (1/3)^{1/2}$$

However, $-1 - (1/3)^{1/2}$ is not on the root locus and so $s = -1 + (1/3)^{1/2}$ is the real axis breakpoint. Rule 6 is not particularly useful here, and by rule 7 we have the Routh table for $s^3 + 3s^2 + 2s + k$:

$$\begin{array}{lccl}
s^3 & 1 & 2 & \\
s^2 & 3 & k & \\
s^1 & (6-k)/3 & \leftarrow & \text{zero row if } k = 6 \\
s^0 & k & &
\end{array}$$

The row above the zero row (when $k = 6$) gives the factor $3s^2 + 6$ and so the $j\omega$ crossings are at $s = \pm j\sqrt{2}$. The complete root locus can now be drawn as in fig. 4.16. The gain at any point on the locus can be found by using (4.5.6).

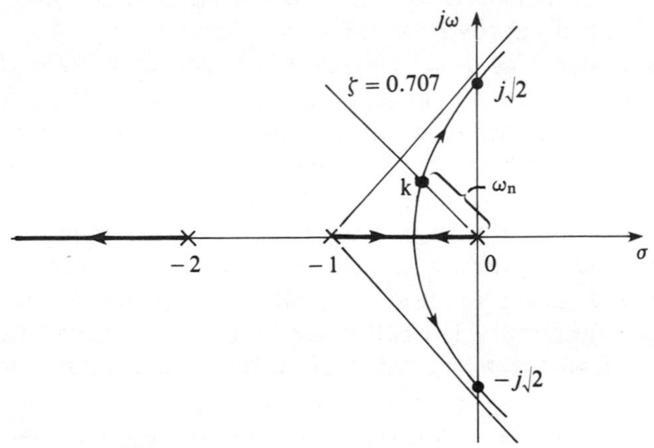

Fig. 4.16 Root locus of 1/s(s + 1)(s + 2).

4.6 CLASSICAL COMPENSATION

4.6.1 Lead-lag Compensation

A given feedback system may have poor transient response or large steady-state errors. The transient response is the immediate behaviour of the system following changes in the input. For a step change in the input, the output will not change instantaneously because of the system dynamics. However, a second-order system with damping ratio in the range 0.3 to 0.8 is regarded as having good transient response, since the overshoot is not too high (between 5 per cent and 35 per cent)† and the settling time is small (i.e. the system is not too oscillatory). Hence for good transient response we try to design the feedback system to have two dominant poles giving the desired response, while the other poles are pushed as far into the left half plane as possible and thereby contribute little to the response. The steady-state response is the response at large times when the system variables have become time independent or periodic, and is measured by the error constants K_p, K_v and K_a defined earlier. Large values of the parameters imply small steady-state errors.

Consider now a transfer function $G_p(s)$ which is a model of a physical system called the **plant**. The basic idea of classical control is to add a 'compensating system' $G_c(s)$, as in fig. 4.17, of the form

$$G_c(s) = A\frac{(s - z_1)(s - z_2)}{(s - p_1)(s - p_2)} \tag{4.6.1}$$

where z_1, z_2 are real zeros and p_1, p_2 are real poles. We assume that $z_1 < p_1 < 0$ and $p_2 < z_2 < 0$, in which case $(s - z_1)/(s - p_1)$ is called a **lag compensator** and $(s - z_2)/(s - p_2)$ a **lead compensator** (A is an amplifier gain). Lag compensation is used to improve steady-state response, while lead compensation improves the transient response.

As an example of lead compensation, consider again the system $1/s(s+1)(s+2)$ with root locus as in fig. 4.16. Suppose that the uncompensated system has gain k and natural frequency ω_n when the damping ratio is $\zeta = 0.707$. By using lead compensation, we try to increase ω_n while maintaining ζ at 0.707. This will increase the speed of response, but the overshoot will remain the same. By placing a zero in the region of the open-loop poles we try to alter the two 'unstable branches' of the root locus so that they are pulled further into the left half plane. The pole of the compensator is then placed far into the left half plane so that it will have little effect on the response. Since the pole is to the left of the zero, it follows from (4.5.7)

---

† 35 per cent overshoot is too high, of course, for certain applications.

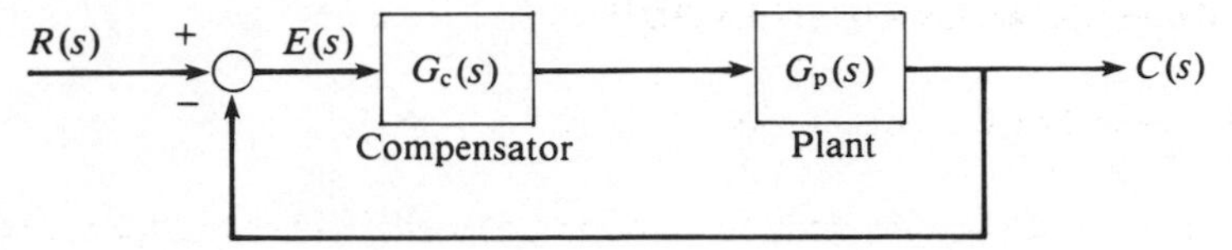

Fig. 4.17 A unity gain feedback system with compensator.

that the intersection σ_0 of the asymptotes moves to the left, leaving the asymptote directions fixed. Hence, in fig. 4.16 we place a zero between the poles -2 and -1 and a pole far into the left half plane and we obtain the new root locus in fig. 4.18. As can be seen, the new natural frequency ω_n' is much larger than the original ω_n.

Consider now the effect of lag compensation. Referring to (4.3.9) we recall that the error constants K_n for the plant $G_p(s)$ are generally given by

$$K_n = \lim_{s \to 0} s^n G_p(s) \quad (K_1 = K_p, K_2 = K_v, K_3 = K_a)$$

Suppose that, for a particular system $K_1 \ldots, K_{n-1}$ tend to ∞ while K_n is finite and too small (i.e. the steady-state error to the corresponding input t^n is too large). If we could add a pole to the system then the error constant k_n would tend to ∞ and so we would have reduced the error to 0. However, a single pole is likely to destabilize the system and so we add a zero close to the origin. In fact, the pole and zero of the compensator are usually shifted slightly away from the origin with the zero to the left of the pole. Adding such a compensator to the system in fig. 4.16 leads to the root locus shown in fig. 4.19. Note that σ_0 is shifted to the right slightly and so the natural frequency ω_n'' may be reduced from the original ω_n. However, using a combination of lag–lead compensators we can improve both the steady-state and transient responses.

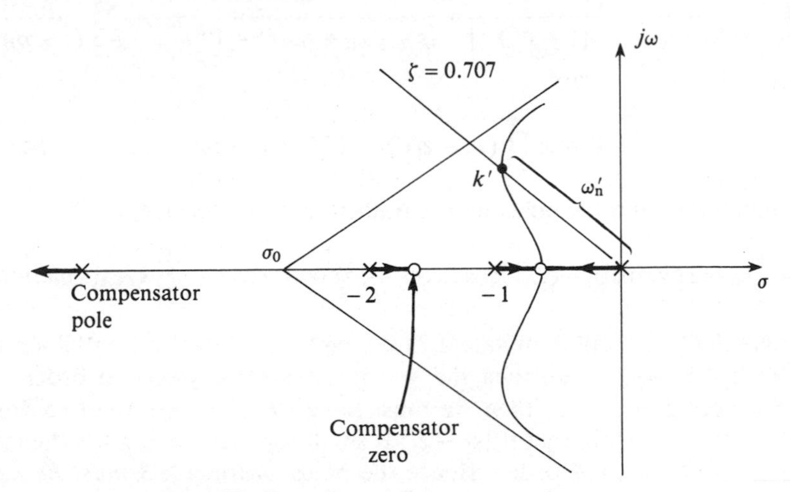

Fig. 4.18 Root locus of compensated system.

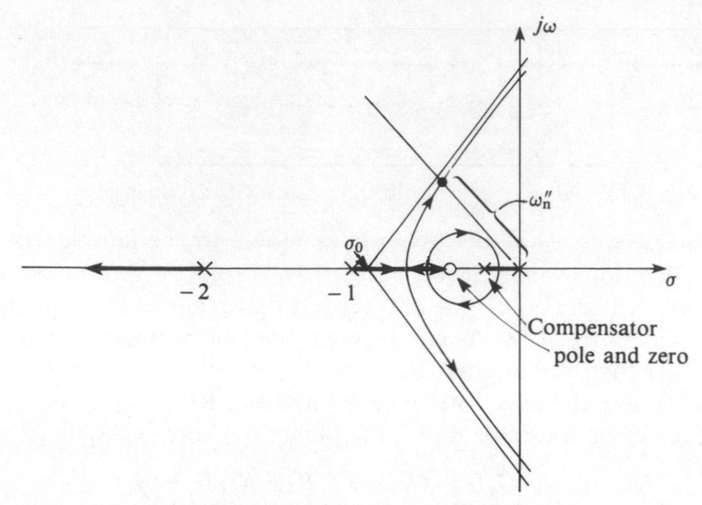

Fig. 4.19 Example of lag compensation.

4.6.2 Root Locus Compensation

Suppose that the closed-loop system in fig. 4.17 is given by

$$T(s) = k \prod_{i=1}^{m} (s - z_i) \Big/ \prod_{j=1}^{n} (s - p_j)$$

Then the response to a step input is

$$C(s) = \frac{1}{s} T(s) = \frac{k}{s} \prod_{i=1}^{m} (s - z_i) \Big/ \prod_{j=1}^{n} (s - p_j)$$

$$= \frac{k_0}{s} + \frac{k_1}{[s + \zeta\omega_n - j\omega_n(1 - \zeta^2)^{1/2}]} + \frac{\bar{k}_1}{[s + \zeta\omega_n + j\omega_n(1 - \zeta^2)^{1/2}]} + \sum_{j=3}^{n} \frac{k_j}{(s - p_j)}$$

where

$$k_j = k \prod_{i=1}^{m} (p_j - z_i) / p_j \prod_{i=1,\, i \neq j}^{n} (p_j - p_i) \tag{4.6.2}$$

assuming two complex poles with parameters ζ, ω_n. Hence,

$$c(t) = \left\{ k_0 + 2|k_1| \exp(-\zeta\omega_n t) \cos[\omega_n(1 - \zeta^2)^{1/2} t + \angle k_1] + \sum_{j=3}^{n} k_j \exp(p_j t) \right\} 1(t)$$

The complex poles will dominate if $2|k_1|\exp(-\zeta\omega_n t) \gg |k_j|\exp[\mathrm{Re}(p_j)t]$ and so if Re $p_j \ll -\zeta\omega_n$ then the system is essentially second order. If, however, Re $p_j \approx -\zeta\omega_n$ then we must have $|k_1| \gg |k_j|$, and so from (4.6.2) it follows again that if $|p_j - z_i|$ is small for each i and $j \geq 3$ then the system is again second order. Hence the poles p_j for $j \geq 3$ *must be near zeros*.

Consider first the design of a lead compensator $G_c(s)$ using the root locus, and suppose that $G_c(s)$ is of the form

$$G_c(s) = A\frac{s + 1/\tau}{s + \alpha/\tau} \quad A, \tau > 0, \alpha > 1 \tag{4.6.3}$$

Then the total open-loop transfer function $G_T(s)$ equals $G_c(s)G_p(s)$. We now fix a pair of complex poles p_d, $\bar{p}_d$ with desired values of ω_n and ζ, so that a system with these as dominant poles will have a particular second-order response. If we find $\angle G_p(p_d)$, then we have

$$\pm 180^\circ = \angle G_T(p_d) = \angle G_p(p_d) + \angle G_c(p_d)$$

if p_d is on the root locus of $G_T(s)$, and so

$$\angle G_c(p_d) = \pm 180^\circ - \angle G_p(p_d) \tag{4.6.4}$$

If we fix the compensator zero position $-1/\tau$ by placing it on or near a plant pole near the $j\omega$ axis, then the compensator pole $-\alpha/\tau$ is determined by (4.6.4), as in fig. 4.20.

Next suppose that

$$G_c(s) = A\frac{s + 1/\tau}{s + 1/(\alpha\tau)} \quad A, \tau > 0, \alpha > 1 \tag{4.6.5}$$

is a lag compensator, i.e. the pole is to the right of the zero. Usually we take $A \approx 1$ and we choose the pole and zero to be close to the origin relative to the plant poles and zeros, so that they do not alter the general form of the root locus. For, if we have already designed a lead compensator then the lag compensator will not affect the former design too much. It has been found that if the plant pole or zero nearest the $j\omega$-axis is at s then we can take $1/\tau \leqslant |\mathrm{Re}\, s|/10$. Note that a non-zero finite error constant is multiplied by α and so if we wish to increase a given error constant by a certain factor then the compensator pole is fixed by the position of the zero.

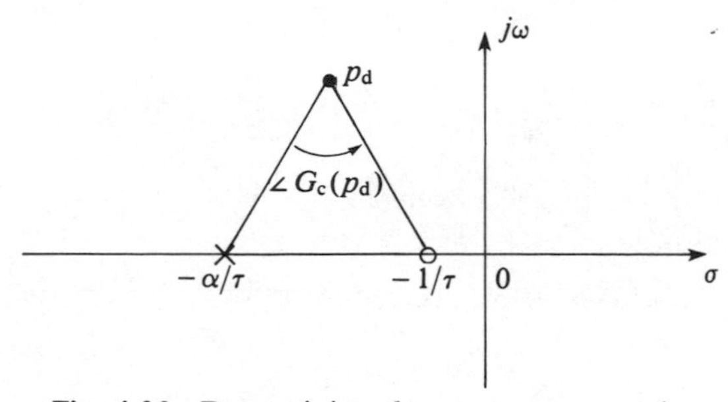

Fig. 4.20 Determining the compensator pole.

Once a lead–lag compensator had been obtained for a particular plant, the closed-loop response should be calculated and if it is not as required the design should be repeated with different parameters until a satisfactory response is obtained.

4.6.3 Frequency Response Compensation

Frequency response compensation techniques are important because of the ease of measuring the frequency response of a system. The basic idea is the same as before; we add a lead compensator to improve transient response and a lag compensator to improve the steady-state error constants. In the latter case, if

$$K_m = \lim_{\omega \to 0} (j\omega)^m G(j\omega)$$

is a finite error constant, then

$$G_T(0) = G_c(0) G_p(0)$$

will determine the compensator low frequency response to give the desired value of K_m. For high frequencies, we require a lag compensator to satisfy

$$|G_c(j\omega)| \ll |G_p(j\omega)|, \quad |\angle G_c(j\omega)| \ll |\angle G_p(j\omega)|$$

which is equivalent to choosing the pole and zero near the origin.

Suppose we use the Nichol's chart for the compensation of a plant $G_p(s)$ (we could also use the Bode diagrams). In the Nichol's chart an open-loop frequency response is tangent to some closed-loop magnitude curve M at some frequency ω. If the frequency response of the open-loop system is

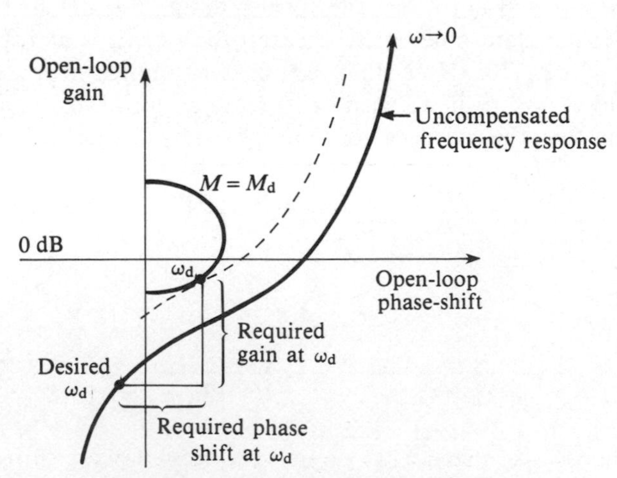

Fig. 4.21 Nichol's chart compensation.

reasonably smooth and is tangent only to the curve M, then the closed-loop system will have roughly second-order behaviour with parameters ω_n and ζ given by (4.4.3) and (4.4.4). (Note that this method is not as precise as the root locus method.) If the desired closed-loop response is tangent to the curve M_d at frequency ω_d then we must choose a compensator to shift the uncompensated open-loop response to the desired response (fig. 4.21). However, we have

$$G_T(j\omega) = G_c(j\omega)G_p(j\omega)$$

and so

$$20\log_{10}|G_T(j\omega_d)| = 20\log_{10}|G_c(j\omega_d)| + 20\log_{10}|G_p(j\omega_d)|$$

$$\angle G_T(j\omega_d) = \angle G_c(j\omega_d) + \angle G_p(j\omega_d)$$

Hence

$$\boxed{\begin{aligned}&\text{required gain at } \omega_d = 20\log_{10}|G_c(j\omega_d)|\\ &\text{required phase shift at } \omega_d = \angle G_c(j\omega_d)\end{aligned}} \tag{4.6.6}$$

Consider the lead compensator

$$G_c(s) = \frac{A}{\alpha}\,\frac{1+s\tau}{1+s\tau/\alpha} = \frac{A}{\alpha}\,G_c^{\text{lead}}(s), \quad \text{say } (\alpha > 1) \tag{4.6.7}$$

Then

$$20\log_{10}|G_c^{\text{lead}}(j\omega)| = 20\log_{10}|1+j\omega\tau| - 20\log_{10}\left|1+\frac{j\omega\tau}{\alpha}\right|$$

$$\angle G_c^{\text{lead}}(j\omega) = \tan^{-1}\omega\tau - \tan^{-1}\frac{\omega\tau}{\alpha}$$

Using these relations we can draw the frequency response of $G_c^{\text{lead}}(j\omega)$ as $20\log_{10}$ magnitude versus phase plot as in fig. 4.22(a) for various values of α. It is easy to show that the maximum phase shift is given by

$$\boxed{\phi_{\max} = \tan^{-1}\left(\frac{\alpha-1}{2\sqrt{\alpha}}\right)}$$

and occurs at

$$\boxed{\omega_{\max} = \frac{\sqrt{\alpha}}{\tau}}$$

Now we choose α so that $\phi_{\max}(\alpha)$ equals the desired phase shift and choose τ so that this occurs at ω_d, i.e. $\omega_{\max} = \omega_d = \sqrt{\alpha}/\tau$. To finish the design we

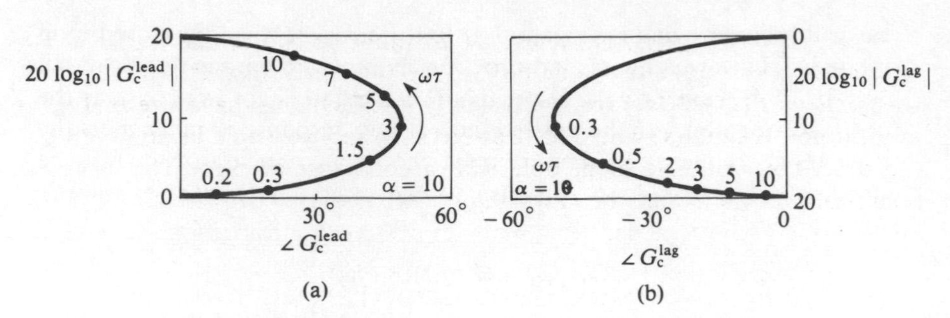

Fig. 4.22 Lead and lag magnitude-angle characteristics.

choose A from the relation

$$20 \log_{10} \frac{A}{\alpha} = 20 \log_{10} | G_c(j\omega_d) | - 20 \log_{10} | G_c^{\text{lead}}(j\omega_d) |$$

(from (4.6.7)), noting that $20 \log_{10} | G_c(j\omega_d) |$ is the required gain at ω_d and $20 \log_{10} | G_c^{\text{lead}}(j\omega_d) |$ is fixed when α and τ are chosen.

For a lag compensator

$$G_c(s) = A\alpha \frac{1 + s\alpha}{1 + s\alpha\tau} = A\alpha G_c^{\text{lag}}(s) \quad (\alpha > 1)$$

The frequency response for $G_c^{\text{lag}}(s)$ is just the graph in fig. 4.22(a) rotated through 180°, as in fig. 4.22(b) (the $\omega\tau$ scale is changed, of course). From the open-loop frequency response of $G_p(j\omega)$ on the Nichol's chart we find $\omega_{\max}$ and then choose τ so that the phase shift of $G_c^{\text{lag}}(j\omega_{\max})$ is just 5°. It can be shown that this is when $\tau \approx 10/\omega_{\max}$ for all α (α is chosen as before to give the desired error constant). Finally, the gain A is chosen so that the value of M_m is the same for the compensated system as it was for the uncompensated system.

Again, with some adjustment we can combine the lead and lag compensators to provide a reasonable closed-loop response.

4.6.4 Design by *s*-plane Synthesis

A different method of compensator design is to specify the closed-loop transfer function $T(s)$ and then find the compensator $G_c(s)$ which achieves this function. If $G_p(s)$ is the plant, then

$$T(s) = \frac{G_c(s)G_p(s)}{1 + G_c(s)G_p(s)}$$

and so

$$\boxed{G_c(s) = \frac{T(s)}{[1 - T(s)]\,G_p(s)} = \frac{N(s)}{[F(s) - N(s)]\,G_p(s)}} \qquad (4.6.8)$$

where $T(s) = N(s)/F(s)$. However, $T(s)$ cannot be chosen arbitrarily and must satisfy certain restrictions. Firstly, if $G_p(s)$ has a right half plane pole at s_p, then $[F(s) - N(s)]$ (or, equivalently $1 - T(s)$) must have a zero at s_p. For, if not, then $G_c(s)$ would have a zero at s_p and we would be attempting to cancel the unstable plant pole with a compensator zero. This is not physically possible and the closed-loop system would be unstable. Often this restriction is removed by designing an inner loop to stabilize $G_p(s)$. Secondly, if the plant has a right half plane zero s_z (i.e. a non-minimum phase system) then $N(s)$ (and therefore also $T(s)$) must have a zero at the same point. For, if not, then $G_c(s)$ has a right half plane pole at s_z and so is not stable. Finally, for $G_c(s)$ to be realizable, it must have at least as many poles as zeros and so the number of excess poles of $N(s)/[F(s) - N(s)]$ must be at least as many as those of $G_p(s)$. But $N/(F - N)$ has the same number of excess poles as $T = N/F$ and so we must require

$$\boxed{\text{number of excess poles of } T(s) \geqslant \text{number of excess poles of } G_p(s)} \tag{4.6.9}$$

For a minimum phase system we can choose $T(s)$ to be a product of factors containing two dominant poles, a pole-zero pair near the origin (for good steady-state response) and further LHP poles to satisfy (4.6.9).

This completes our outline of the classical compensation of continuous systems. For a more detailed discussion see Shinners (1978) and Gupta and Hasdorff (1970).

4.7 DISCRETE SYSTEMS

4.7.1 The *Z*-transform

In most modern control systems, the control algorithm or compensator is implemented on a computer and so it is important to consider the behaviour of discrete or sampled-data systems. Very often, the error $e(t)$ in a feedback loop is sampled and digitized and fed into the computer which processes this data to produce the control input to the plant (after digital to analogue conversion). We shall consider only the process of sampling in this chapter – digitization will be considered later. Suppose then that $f(t)$ is a continuous signal which is sampled by closing a switch every T seconds at the times nT, for integer n. Let $f^*(t)$ be the sampled signal. Then we obtain a sequence of pulses of finite width h and of height $f(nT)$ (fig. 4.23). If the switch con-

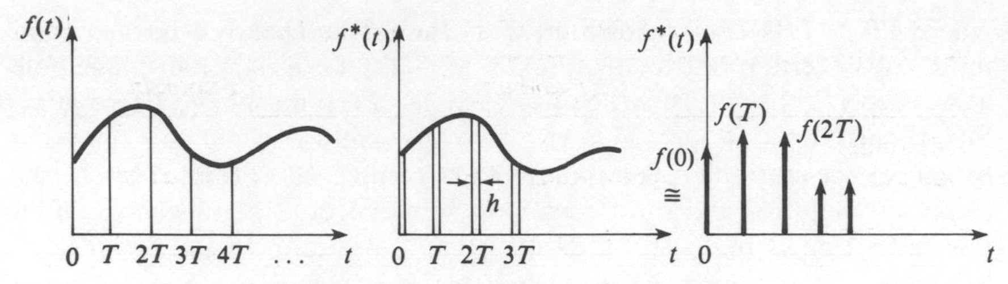

Fig. 4.23 Sampling a function.

tact time is small relative to T, then we have approximately,

$$\mathcal{L}(f^*) = \int_0^\infty f^*(t) \exp(-st)\, \mathrm{d}t$$

$$\approx \sum_{n=0}^{\infty} hf(nT) \exp(-snT)$$

We normally regard h as being incorporated in the following plant and model $f^*(t)$ as a sequence of impulses of strength $f(nT)$. Then

$$\mathcal{L}(f^*) = \mathcal{L}\left\{\sum_{n=0}^{\infty} f(nT)\delta(t - nT)\right\}$$

$$= \sum_{n=0}^{\infty} f(nT) \exp(-snT) \tag{4.7.1}$$

which is the same as before (modulo the factor h).

Once the sampling has been performed we may regard f^* as just a sequence of numbers $\{f(nT)\}_{-\infty<n<\infty} = \{\ldots, f(-2T), f(-T), f(0), f(1), \ldots\}$. In just the same way as a continuous time function, whose value at time t is $f(t)$, is denoted by f, we denote the sequence whose value at n is $h(nT)$ by h or $\{h(nT)\}_{-\infty<n<\infty}$. In other words, when we write $h(nT)$ we mean the value of the sequence at n and when we write just h we are thinking of the sequence as a whole (h may be regarded as a function $h\colon \mathbb{Z} \to \mathbb{R}$ from the integers to the real numbers). Given a sequence h we define the (one-sided) **Z-transform** of h by

$$\boxed{Z\{h\} = \sum_{n=0}^{\infty} h(nT) z^{-n}} \tag{4.7.2}$$

From (4.7.1) we see that if f_s is the sequence $\{f(nT)\}_{-\infty<n<\infty}$ associated with the function f, then

$$\boxed{\mathrm{L}(f^*) = Z\{f_s\}\,|_{z=\exp(sT)}} \tag{4.7.3}$$

In table 4.2 we list some basic properties of the Z-transform.

Table 4.2 Properties of the Z-Transform

	nT-domain	*z-domain*
1. Multiplier		
(a) $\exp(-anT)$	$[\exp(-anT)f(nT)]$	$F[\exp(aT)z]$
(b) $(nT)^n$	$[(nT)^n f(nT)]$	$(-1)^n T^n z^n \frac{d}{dz} F(z)$
2. Delay	$\{f[(n-m)T]\}$	$z^{-m}F(z) + z^{-(m-1)}f(-T) + \dots + f(-mT)$
3. Convolution	$\left\{\sum_{m=0}^{n} f_1(mT)f_2[(n-m)T]\right\}$ $f_1(nT) = f_2(nT) = 0, n < 0$	$F_1(z)F_2(z)$
4. Initial value theorem	$f(0)$	$\lim_{z\to\infty} F(z)$
5. Final value theorem	$\lim_{n\to\infty} f(nT)$	$\lim_{z\to 1} (z-1)F(z)$
6. Inversion integral	$f(nT)$	$(=)\frac{1}{2\pi j}\oint_C F(z)z^{n-1}\,dz$ where C is a circle, centre 0, enclosing poles of $F(z)z^{n-1}$

4.7.2 Transfer Functions

Let u_0 denote the **unit impulse** defined by

$$u_0(nT) = \begin{cases} 1 & n = 0 \\ 0 & n \neq 0 \end{cases}$$

and for any sequence v let v^{n_0} denote the sequence delayed by n_0 sampling periods, i.e.

$$v^{n_0}(nT) = v((n-n_0)T)$$

We can regard a discrete system ϕ as an algorithm or a function which converts an input sequence into an output sequence y, i.e.

$$y = \phi(x)$$

Note that we cannot write $y(nT) = \phi(x(nT))$ unless ϕ is **memoryless**. The system ϕ is **linear** if

$$\phi(\alpha x_1 + \beta x_2) = \alpha\phi(x_1) + \beta\phi(x_2)$$

for sequences x_1, x_2 and scalars α, β; ϕ is **time-invariant** if

$$y=\phi(x) \Rightarrow y^{n_0}=\phi(x^{n_0}), \quad \forall n_0$$

and ϕ is **causal** if $y(nT)$ depends only on $x(mT)$ for $-\infty < m \leqslant n$. Now, if ϕ is linear and time-invariant we define

$$h=\phi(u_0) \quad (h \text{ is the } \textbf{impulse response})$$

and since for any input x we have

$$x(nT)=\sum_{m=-\infty}^{\infty} x(mT)u_0((n-m)T)$$

or

$$x=\sum_{m=-\infty}^{\infty} x(mT)u_0^m$$

it follows that

$$\begin{aligned} y=\phi(x) &= \phi\left(\sum_{m=-\infty}^{\infty} x(mT)u_0^m\right) \\ &= \sum_{m=-\infty}^{\infty} x(mT)\phi(u_0^m) \quad \text{(linearity)} \\ &= \sum_{m=-\infty}^{\infty} x(mT)h^m \quad \text{(time-invariance)} \end{aligned}$$

Hence

$$\boxed{\begin{aligned} y(nT) &= \sum_{m=-\infty}^{\infty} x(mT)h((n-m)T) \\ &= \sum_{m=-\infty}^{\infty} x((n-m)T)h(mT) \end{aligned}} \tag{4.7.4}$$

It follows from (4.7.4) that ϕ is causal if and only if $h(nT)=0$ for $n<0$. In contrast to continuous systems, discrete system models are often given in the form of a difference equation such as

$$\begin{aligned} y(k)+b_1y(k-1)+\ldots+b_my(k-m) &= a_0x(k) \\ &\quad +a_1x(k-1)+\ldots+a_nx(k-n) \end{aligned} \tag{4.7.5}$$

Using the properties of the Z-transform in table 4.2, and assuming zero initial conditions, we have

$$Y(z)\left(1+\sum_{i=1}^{m} b_iz^{-i}\right)=X(z)\sum_{j=0}^{n} a_jz^{-j}$$

Then we define the **transfer function** $H(z)$ by

$$H(z) \triangleq \frac{Y(z)}{X(z)} = \sum_{j=0}^{n} a_j z^{-j} \bigg/ \left(1 + \sum_{i=1}^{m} b_i z^{-i}\right) \tag{4.7.6}$$

4.7.3 Frequency Response and Aliasing

Let ϕ be linear, time-invariant and causal and suppose we input a complex sinusoid $x(nT) = \exp(j\omega nT)$ to the system. Then

$$\begin{aligned} y(nT) &= (\phi(x))(nT) \\ &= \sum_{m=0}^{\infty} h(mT) \exp[j\omega(n-m)T] \\ &= \exp(j\omega nT) \sum_{m=0}^{\infty} h(mT) \exp(-j\omega mT) \\ &= x(nT) H(\exp(j\omega T)) \end{aligned}$$

where

$$H(\exp(j\omega T)) = \sum_{m=0}^{\infty} h(mT) \exp(-j\omega mT)$$

i.e. the transfer function evaluated on the unit circle of the z-plane.

It is easy to show that $H(\exp(j\omega T))$ is periodic with frequency $2\pi/T$. Note, however, that $\omega_s \triangleq 2\pi/T$ is the sampling frequency and this periodicity corresponds to the fact that, for a sampled system, an input of frequency ω is indistinguishable from one of frequency $\omega + m\omega_s$, for all integers m. It is also interesting to note that $H(\exp(j\omega T))$ is symmetric about $\omega_s/2$ while $\angle H(\exp(j\omega T))$ is antisymmetric about $\omega_s/2$.

Now let the sequence $\{x(nT)\}$ be derived from the continuous signal $x(t)$ by sampling. By the Fourier transform theorem (appendix 1), we have

$$X_{\mathcal{C}}(j\Omega) = \int_{-\infty}^{\infty} x(t) \exp(-j\Omega t)\, \mathrm{d}t$$

$$x(t) = \frac{1}{2\pi} \int_{-\infty}^{\infty} X_{\mathcal{C}}(j\Omega) \exp(j\Omega t)\, \mathrm{d}\Omega$$

where $X_{\mathcal{C}}(j\Omega)$ is the frequency spectrum of the continuous signal. (In the appendix we have identified $\hat{f}(\omega) = \frac{1}{2\pi} X_{\mathcal{C}}(j\Omega)$, $f(t) = x(t)$ and $\omega = \Omega$.) Then

$$\begin{aligned} x(nT) &= \frac{1}{2\pi} \sum_{m=-\infty}^{\infty} \int_{(2m-1)\pi/T}^{(2m+1)\pi/T} X_{\mathcal{C}}(j\Omega) \exp(j\Omega nT)\, \mathrm{d}\Omega \\ &= \frac{T}{2\pi} \int_{-\pi/T}^{\pi/T} \left\{ \frac{1}{T} \sum_{m=-\infty}^{\infty} X_{\mathcal{C}}\left[j\left(\omega + \frac{2\pi m}{T}\right)\right]\right\} \exp(j\omega nT)\, \mathrm{d}\omega \end{aligned} \tag{4.7.7}$$

(putting $\Omega = \omega + 2\pi m/T$). However, denoting by $X(\exp(j\omega T))$ the frequency spectrum of the sequence $\{x(nT)\}$, as above, we have

$$x(nT) = \frac{T}{2\pi}\int_{-\pi/T}^{\pi/T} X(\exp(j\omega T))\exp(j\omega nT)\,d\omega \tag{4.7.8}$$

By the uniqueness of Fourier series, we have, from (4.7.7) and (4.7.8),

$$\boxed{X(\exp(j\omega T)) = \frac{1}{T}\sum_{m=-\infty}^{\infty} X_{\mathrm{c}}\left[j\left(\omega + \frac{2\pi m}{T}\right)\right]} \tag{4.7.9}$$

Hence the frequency spectrum of the discrete signal $X(\exp(j\omega T))$ equals that of the continuous signal plus the 'sidebands' $X_{\mathrm{c}}(j(\omega + \omega_s m))$, in the interval $[-\omega_s/2, \omega_s/2]$. Suppose that $X_{\mathrm{c}}(j\omega)$ is band limited i.e. $|X_{\mathrm{c}}(j\omega)| \approx 0$ for $|\omega| > \omega_c$, where $\omega_c \leqslant \omega_s/2$. Then, in the interval $[-\omega_s/2, \omega_s/2]$, the sidebands are zero and so we have

$$X(\exp(j\omega T)) \approx \frac{1}{T}X_{\mathrm{c}}(j\omega), \quad \omega \in \left[-\frac{\omega_s}{2}, \frac{\omega_s}{2}\right] \tag{4.7.10}$$

Recalling that $X(\exp(j\omega T))$ is periodic we therefore obtain the frequency spectrum plots shown in fig. 4.24. The sidebands are inevitably introduced by the sampling process. If $X_{\mathrm{c}}(j\omega)$ is not band limited then (4.7.10) does not hold and the frequencies in $X_{\mathrm{c}}(j\omega)$ above $\omega_s/2$ 'fold back' during sampling to produce **aliasing** of the frequency response $X(\exp(j\omega T))$, as in fig. 4.25.

If the continuous signal $x(t)$ is band limited and (4.7.10) holds, then it is clear from fig. 4.24 that we could recover $x(t)$ from $x(nT)$ exactly with

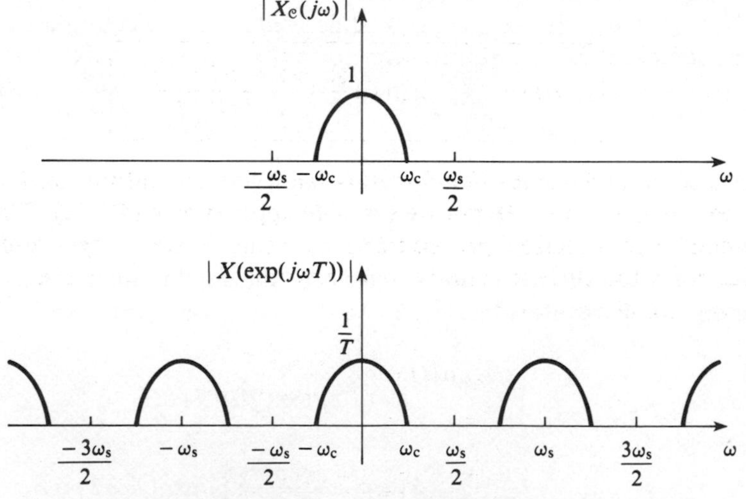

Fig. 4.24 Frequency response of a sampled band-limited signal.

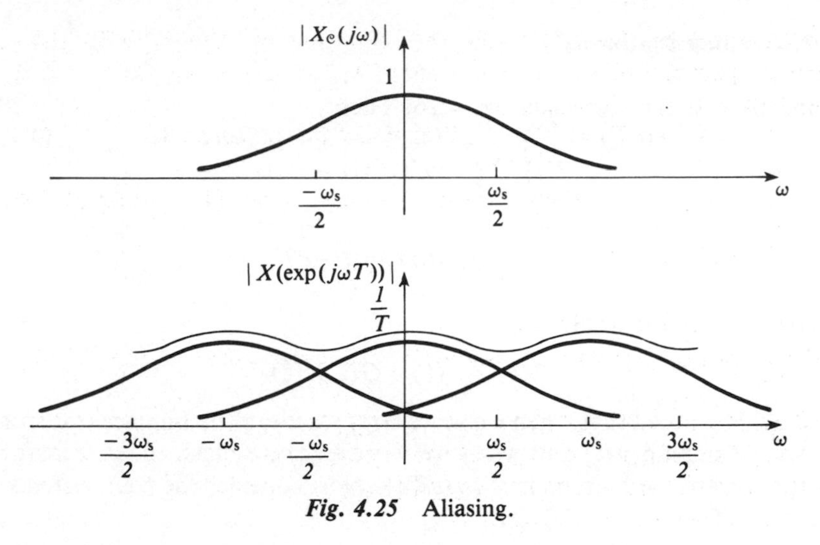

Fig. 4.25 Aliasing.

an ideal filter $F(s)$ with frequency response $F(j\omega) = 1$, for $\omega \in [-\omega_s/2, \omega_s/2]$. Such a filter has impulse response

$$f(t) = \frac{\sin(\pi t/T)}{\pi t/T}$$

Then

$$x(t) = (f * x^*)(t)$$

where

$$x^*(t) = \sum_{n=-\infty}^{\infty} x(nT)\,\delta(t - nT)$$

Hence

$$\boxed{x(t) = \sum_{n=-\infty}^{\infty} x(nT)\,\frac{\sin(\pi(t - nT)/T)}{\pi(t - nT)/T}} \tag{4.7.11}$$

Of course, in practice no signal is truly band limited and we cannot make an ideal low-pass filter. Hence we can only approximate (4.7.11). The most common approximation procedure is by using a **zero-order hold**. This device holds the current digital output constant for the sampling period T and has impulse response

$$g_{ZOH}(t) = \begin{cases} 1 & 0 \leqslant t < T \\ 0 & \text{otherwise} \end{cases}$$

and so

$$\mathcal{L}(g_{ZOH}) = \frac{1 - \exp(-sT)}{s}$$

4.7.4 Block Diagrams

Consider first the system in fig. 4.26. Then

$$c(t) = \int_{-\infty}^{\infty} u^*(\tau)g(t-\tau)\,d\tau$$

$$= \sum_{n=0}^{\infty} u(nT)g(t-nT)$$

if $u(t)=0$ for $t<0$. Hence,

$$Z(c) = C(z) = G(z)U(z)$$

It should be noted that when we write $G(s)$ we mean the Laplace transform of some function $g(t)$ and when we write $G(z)$ we mean the Z-transform of the corresponding sequence $\{g(nT)\}$. Now consider the four systems in fig. 4.27. Then in fig. 4.27(a),

$$C(s) = G_1(s)G_2(s)U^*(s)$$

and†

$$C(z) = Z\{G_1(s)G_2(s)\}U(z)$$

$$C(z) = (G_1G_2)(z)U(z)$$

Note that $(G_1G_2)(z) \neq G_1(z)G_2(z)$
In fig. 4.27(b),

$$B(z) = U(z)G_1(z)$$

and

$$C(z) = G_1(z)G_2(z)u(z)$$

and so we can 'separate' $(G_1G_2)(z)$ if there is a sampler between them. In fig. 4.27(c),

$$C(s) = G(s)E^*(s) = G(s)E(z) \quad \text{with } z = \exp(sT)$$

and

$$E(s) = R(s) - H(s)C(s) = R(s) - H(s)G(s)E(z)$$

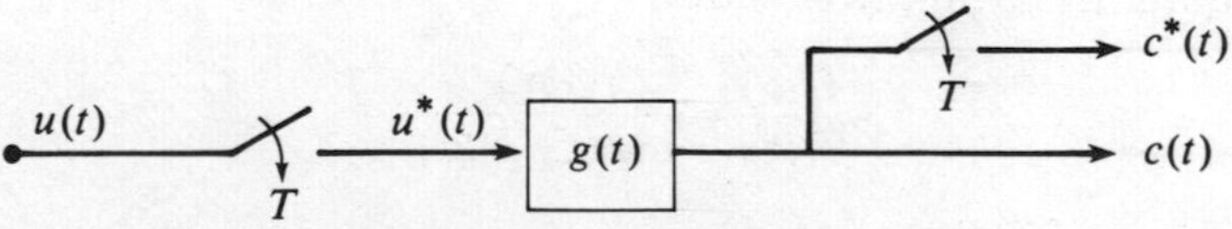

Fig. 4.26 A discrete open-loop system.

---

†$Z[G_1(s)G_2(s)]$ is shorthand for $Z[(g_1 * g_2)(nT)]$

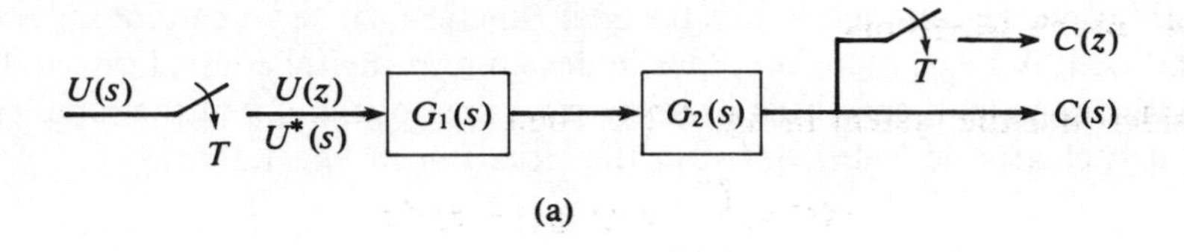

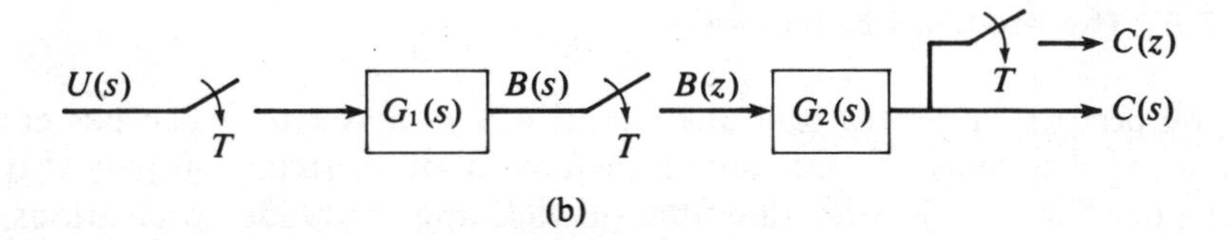

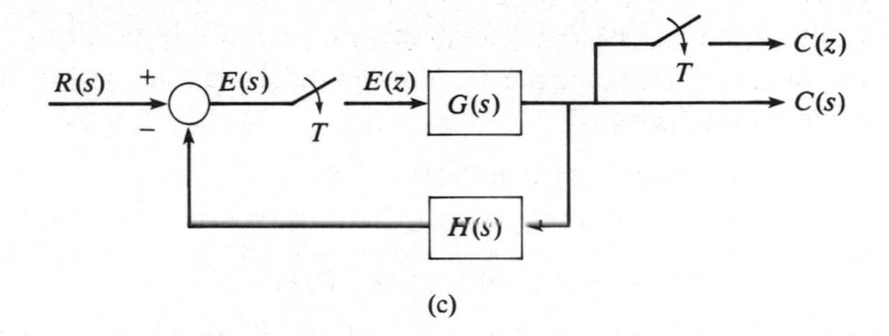

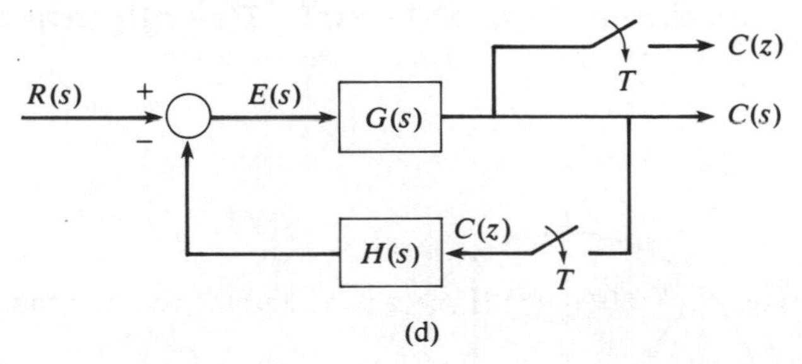

Fig. 4.27 Various discrete system configurations.

Hence

$$\frac{C(z)}{G(z)} = E(z) = \frac{R(z)}{1 + (GH)(z)}$$

and we obtain the transfer function

$$\frac{C(z)}{R(z)} = \frac{G(z)}{1 + (GH)(z)}$$

Finally, in fig. 4.25(d),

$$C(z) = \frac{(RG)(z)}{1 + (GH)(z)}$$

and this time we cannot obtain a transfer function, since we cannot separate R and G. It is clear, therefore, that in designing a digital control system for a continuous plant, care must be taken over the positioning of the samplers, and it is clearly desirable to avoid the situation in fig. 4.27(d).

4.7.5 The Modified *Z*-transform

If we perform a *z*-transform analysis of a digital system which has continuous subsystems, we only obtain the continuous systems' outputs at the sampling times. It is often desirable (in checking for hidden oscillations,† for example) to obtain these outputs for all times between the sampling instants. To show how this may be done consider first a sampled signal $u^*(t)$. Then

$$\begin{aligned}\mathcal{L}(u^*) &= U^*(s)\\ &= \mathcal{L}[u(t)\textstyle\sum \delta(t-nT)]\\ &= \frac{1}{2\pi j}\int_{\sigma_1-j\infty}^{\sigma_1+j\infty} \frac{U(p)}{1-\exp[-T(s-p)]}\,\mathrm{d}p \end{aligned} \tag{4.7.12}$$

from table 4.1(10), where $\sigma^{\mathrm{I}} < \sigma_1 < \sigma - \sigma^{\mathrm{II}}$, and σ^{I}, σ^{II} are the abscissae of convergence in the p-plane of $U(p)$, $1/\{1-\exp[-T(s-p)]\}$, respectively (fig. 4.28).

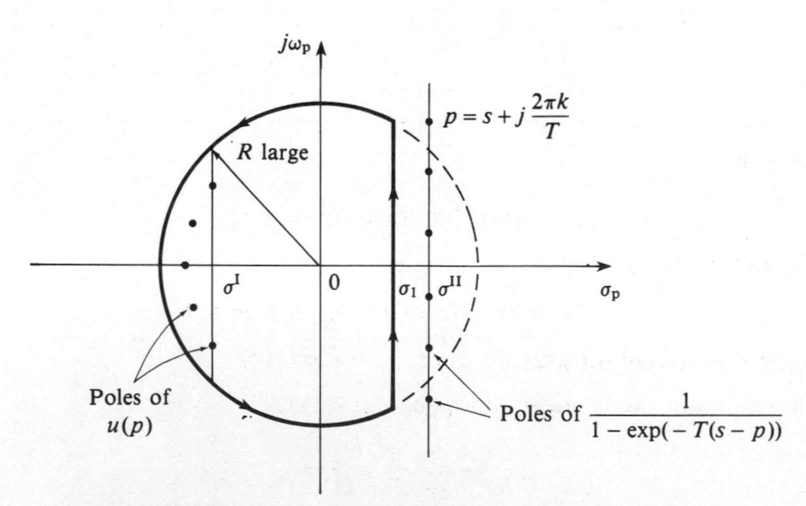

Fig. 4.28 Region of convergence for modified Z-transform.

† Hidden oscillations are oscillations which may occur in a system which has frequencies commensurate with the sampling rate.

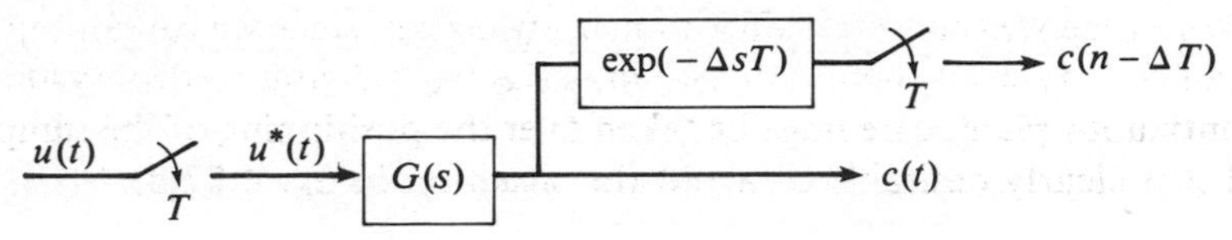

Fig. 4.29 Finding inter-sample values.

Now suppose that we have a system of the form shown in fig. 4.29 with a continuous output $c(t)$ and we add a fictitious delay before sampling this signal. We define $G(s, \Delta) = G(s)\exp(-\Delta sT)$ for $0 \leqslant \Delta \leqslant 1$. However, the integral on the left semicircle (from $\sigma_1 + jR$ to $\sigma_1 - jR$ for large R) does not converge and so we put $\Delta = 1 - m$, with $0 \leqslant m \leqslant 1$. Then

$$G(s, m) \triangleq G(s, \Delta)\,|_{\Delta = 1-m}$$

$$= G(s)\exp(-sT)\exp(msT)$$

or

$$G(s, m)\exp(sT) = G(s)\exp(msT)$$

and so

$$zG(z, m) = \frac{1}{2\pi j}\int_{\sigma_1 - j\infty}^{\sigma_1 + j\infty} \frac{G(p)\exp(mpT)}{1 - \exp[-T(s-p)]}\,\mathrm{d}p, \quad 0 \leqslant m \leqslant 1$$

Clearly, convergence on the left semicircle is assured if $G(p)$ has at least one more pole than zero. We define the **modified *Z*-transform** $Z_m(g(nT))$ as

$$Z_m(g) \triangleq G(z, m) = \frac{z^{-1}}{2\pi j}\int_{\sigma_1 - j\infty}^{\sigma_1 + j\infty} \frac{G(p)\exp(mpT)}{1 - \exp[-T(s-p)]}\,\mathrm{d}p \quad 0 < m < 1$$

Note that

$$G(z) = \lim_{m \to 0} zG(z, m)$$

The output in fig. 4.29 is

$$C(z, m) = G(z, m)U(z), \quad 0 \leqslant m \leqslant 1$$

Using the inverse integral we have

$$c(nT, m) = c((n + m - 1)T)$$

$$= \frac{1}{2\pi j}\oint C(z, m)z^{n-1}\,\mathrm{d}z$$

and by varying m between 0 and 1 we can find the output between sampling periods. For the feedback system in fig. 4.30 we have

$$C(s) = E(z)G(s) \quad \text{or} \quad C(z, m) = E(z)G(z, m)$$

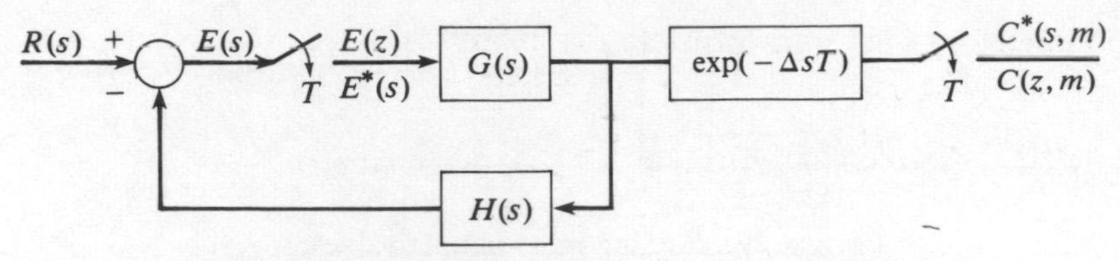

Fig. 4.30 Inter-sample values in a feedback system.

and since

$$E(z) = \frac{1}{1 + GH(z)} R(z)$$

we have

$$C(s) = \frac{G(s)}{1 + GH(z)} R(z)$$

whence

$$\frac{C(z, m)}{R(z)} = \frac{G(z, m)}{1 + GH(z)}$$

4.7.6 Complex Transformations

We have seen that if we associate with a continuous function $f(t)$ the 'sampled function'

$$f^*(t) = \sum_{i=-\infty}^{\infty} \delta(t - nT) f(nT)$$

then the Laplace transform of f^* equals the Z-transform of the sequence $\{f(nT)\}_{-\infty < n < \infty}$ evaluated at $z = \exp(sT)$. It is instructuve to consider the function $z = \exp(sT)$ as a transformation from the s to the z-plane. Since

$$\exp[T(s + jk\omega_s)] = \exp(sT), \quad -\infty < k < \infty$$

where ω_s is the sampling frequency, we see that strips of width ω_s parallel to the real s-axis map onto the same set in the z-plane. This set is, in fact, the whole z-plane as is easy to see. Moreover, each such strip in the left half s-plane is mapped into the interior of the unit circle in the z-plane. The strip $-\omega_s/2 \leqslant \operatorname{Im} s < \omega_s/2$ is called the **primary strip**. Note that the imaginary s axis wraps around the unit circle in the z-plane, under the mapping $z = \exp(sT)$, an infinite number of times. It is precisely this multiple covering which causes aliasing.

In order to remove the aliasing effect of the above mapping we often consider the transformation of the z-plane to a different 'continuous' complex domain w. This transformation is defined by

$$\boxed{w = \frac{z-1}{z+1}} \qquad (4.7.13)$$

and is called the **bilinear transform**. It is easy to show that this maps the z-plane in a one-to-one manner onto the w-plane and the interior of the unit z circle maps onto the left half of the w-plane.

4.7.7 *Z*-domain Analysis and Compensation Techniques

All the methods for analysis and compensation of continuous systems described above can be modified to deal with discrete systems. We shall show later that a discrete system is stable if the poles are inside the unit circle. Moreover, Routh's criterion can be modified into Jury's criterion for the stability of the system

$$F(z) = \sum_{i=0}^{n} a_i z^i$$

(see Gupta and Hasdorf, 1970).

Since the stability boundary in the z-plane is the unit circle, the Nyquist contour is a pair of circles, one of radius 1 which avoids poles on $|z| = 1$ and the other of large radius, connected together along the positive real z-axis for $\operatorname{Re} z \geqslant 1$ (fig. 4.31). Then, for the open-loop system $GH(z)$, the closed-loop system is stable if

$$\boxed{N = N_z - N_p} \qquad (4.7.14)$$

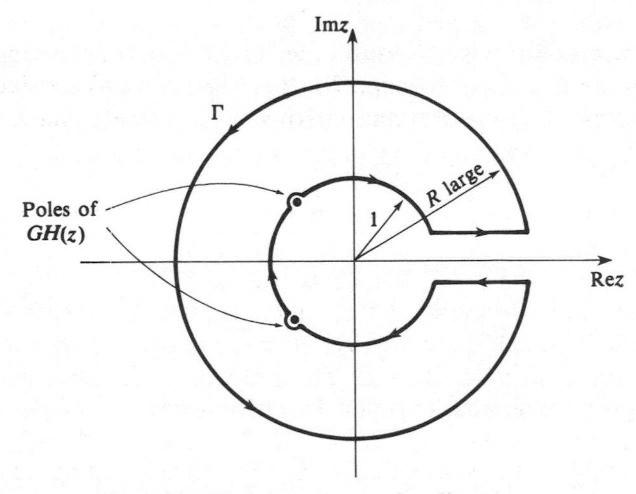

Fig. 4.31 Nyquist contour for discrete systems.

(cf.(4.3.3)) where N is the number of counter-clockwise rotations of the Nyquist plot of $GH(z)$ about -1, N_z is the number of zeros of $1+GH(z)$ outside the unit circle, and N_p is the number of poles of $GH(z)$ outside the unit circle. If $N_z = 0$, then we require, as before

$$\boxed{N = -N_p} \tag{4.7.15}$$

Note also that the Bode plots and root locus methods are straightforward for discrete systems, the rules for the latter being similar to the continuous case. The bilinear transform (4.7.13) is also useful, since we can map the discrete transfer function $GH(z)$ into the w-plane and obtain $GH(w)$ which has frequency response on the jv-axis (where $w = u + jv$), just as for continuous systems. We can therefore use the Nichol's chart just as before and then return to the z-plane using the inverse transform

$$z = \frac{1+w}{1-w}$$

All the design methods discussed earlier can be used for discrete systems in the z-domain, bearing in mind the difference from the s-domain approach in the interpretation of the pole positions. However, by using the w-transformation we can use exactly the same ideas for the design as in the s-domain. For example, consider the system in fig. 4.32 where the compensator $G_c(z)$ is to be implemented on a microprocessor, the output from which will be passed through a digital to analogue (D/A) converter which we model with a zero order hold. To design the compensator $G_c(z)$ we first determine the (impulse invariant†) open-loop transfer function $(G_hG)(z)$, noting that the z-domain function corresponding to $\{[1-\exp(-sT)]/s\}F(s)$ is $[(z-1)/z]Z[\mathcal{L}^{-1}(F(s)/s)(nT)]$. We can then transfer G_hG to the w-plane and then design $G_c(w)$ for $G_hG(w)$ as in the continuous case (using Bode or Nichol's charts, or root locus design) and then return to the z-domain to obtain $G_c(z)$. The only restriction which must be placed on $G_c(z)$ is that it is realizable and stable. A complete discussion of discrete system compensation is given in Gupta and Hasdorf (1970).

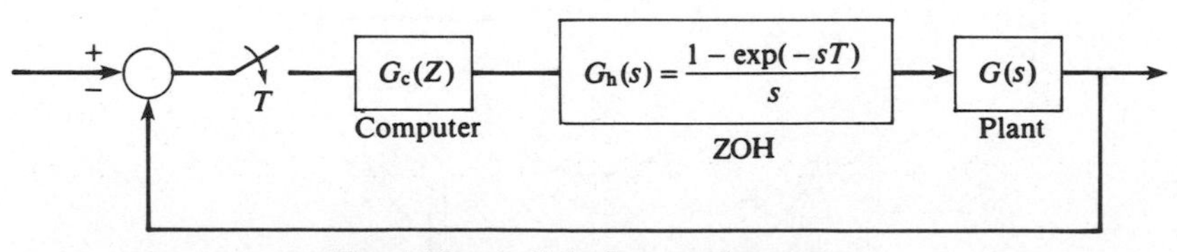

Fig. 4.32 A digital control system.

†By this we mean that $Z^{-1}(G_hG(z))$ and $Z^{-1}(G_h(s)G(s))$ are equal at sampling instants.

4.8 STATE VARIABLE APPROACH

4.8.1 State Variable Models

In chapter 1 we have shown that many physical systems can be modelled by a linear time-invariant equation of the form

$$\boxed{\dot{x} = Ax + Bu} \tag{4.8.1}$$

where $x = (x_1, \ldots, x_n)^T \in \mathbb{R}^n$ is the state vector and $u = (u_1, \ldots, u_m)^T \in \mathbb{R}^m$ is the control or input vector. Frequently, we cannot observe the states directly but we have some measurements $(y_1, \ldots, y_p)^T \in \mathbb{R}^p$ which we assume depend on x and u; then

$$\boxed{y = Cx + Du} \tag{4.8.2}$$

(See fig. 4.33). For the present, we shall assume that the system is single-input single-output, so that $m = p = 1$. Suppose that the system is specified in the form

$$y^{(n)} + b_{n-1}y^{(n-1)} + \ldots + b_1 y^{(1)} + b_0 y = a_m u^{(m)} + a_{m-1}u^{(m-1)} + \ldots + a_0 u \tag{4.8.3}$$

where $m < n$. We define the states $(x_1, \ldots, x_n)^T$ so that x_1 satisfies

$$x_1^{(n)} + b_{n-1}x_1^{(n-1)} + \ldots + b_1 x_1^{(1)} + b_0 x_1 = u \tag{4.8.4}$$

and

$$x_2 = \dot{x}_1,\ x_3 = \dot{x}_2,\ \ldots,\ x_n = x_1^{(n-1)}$$

Then we clearly have

$$\dot{x} = Ax + bu \tag{4.8.5}$$

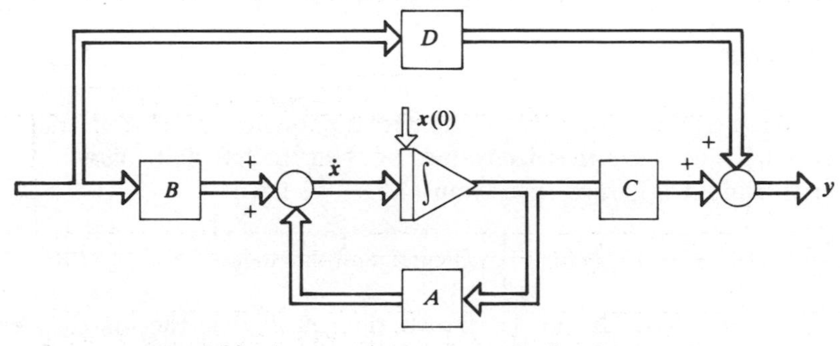

Fig. 4.33 A general state-space feedback system.

where

$$A = \begin{bmatrix} 0 & 1 & \cdot & \cdot & \cdot & 0 \\ 0 & 0 & 1 & & & \vdots \\ \cdot & \cdot & 0 & 1 & & \vdots \\ \cdot & \cdot & \vdots & & & 1 \\ \cdot & \cdot & \vdots & & & \\ -b_0 & -b_1 & \cdot & \cdot & \cdot & -b_{n-1} \end{bmatrix}$$

$$\boldsymbol{b} = (0, 0, \ldots, 0, 1)^{\mathrm{T}}$$

Differentiating (4.8.4) m times and substituting $u^{(i)}$ into (4.8.3) it is easy to show that

$$\boxed{y = \boldsymbol{c}^{\mathrm{T}}\boldsymbol{x}} \tag{4.8.6}$$

where

$$\boldsymbol{c}^{\mathrm{T}} = (a_0, a_1, \ldots, a_m, 0, \ldots, 0)$$

Hence we can obtain a model of the form (4.8.1) and (4.8.2) with $D = 0$ for equation (4.8.3). From (4.8.5) and (4.8.6) we obtain, by Laplace transformation,

$$sX(s) - x(0) = AX(s) + \boldsymbol{b}U(s)$$
$$Y(s) = \boldsymbol{c}^{T}X(s)$$

Hence

$$X(s) = (sI - A)^{-1}\boldsymbol{x}(0) + (sI - A)^{-1}\boldsymbol{b}U(s) \tag{4.8.7}$$

if s is not an eigenvalue of A. By definition, the transfer function is $Y(s)/U(s)$ with $\boldsymbol{x}(0)$ set to zero, and so

$$T(s) \triangleq \frac{Y(s)}{U(s)}$$
$$= \boldsymbol{c}^{\mathrm{T}}(sI - A)^{-1}\boldsymbol{b}$$

and T is an analytic function apart from at the eigenvalues of A. Note that $(sI - A)^{-1} = \mathrm{adj}(sI - A)/\det(sI - A)$ and so the denominator of the transfer function is $\det(sI - A)$, i.e. the characteristic polynomial of A. Hence for stability we must have the eigenvalues of A in the left half plane.

If the input u is zero then from (4.8.7) we have

$$X(s) = (sI - A)^{-1}\boldsymbol{x}(0) = \frac{1}{s}\left[I + \frac{1}{s}A + \frac{1}{s^2}A^2 + \frac{1}{s^3}A^3 + \ldots\right]\boldsymbol{x}(0)$$

for large enough s. (It can be shown rigorously that the matrix series converges.) Hence

$$x(t) = 1(t)\left[I + At + \frac{(At)^2}{2!} + \dots\right]x(0)$$
$$= \exp(At)x(0), \quad t \geqslant 0$$

where we have defined

$$\exp(At) \triangleq \sum_{i=0}^{\infty} \frac{(At)^i}{i!} \tag{4.8.8}$$

Now, again from (4.8.7), if u is non-zero, we have

$$\boxed{x(t) = \exp(At)x(0) + \int_0^t \exp[A(t-\tau)]\,bu(\tau)\,d\tau} \tag{4.8.9}$$

$(= \exp(At)x(0) + \exp(At)*bu(t))$. We can now see clearly what is lost in the input–output representation. By setting $x(0) = 0$ the step response is

$$y(t) = c^T \int_0^t \exp[A(t-\tau)]\,bu(\tau)\,d\tau$$

and we have neglected the term $c^T \exp(At)x(0)$. This does not affect classical compensation theory since we are only interested in changes in the output for changes in the input. However, there are cases where the lack of knowledge of the initial states is important. For example, suppose we consider a simple delay system shown in fig. 4.34. Now delays tend to destabilize systems and degrade the feedback system performance and so it would be useful to be able to feedback y_τ instead of y, so that the delay just appears in the output. However y_τ is unknown, since we only observe y and y_τ would require knowledge of the output T seconds in advance. We therefore feed ke into a model of G_p, say G_m, and use the output z of G_m as the feedback (fig. 4.35). Now, even if we can make G_m exactly equal to G_p, $y_\tau \neq z$ generally since we do not know the internal state initial conditions of G_p. However, if k is chosen to stabilize G_m (and G_p) then the error will tend to zero. The system in fig 4.35 is called a **Smith predictor**.

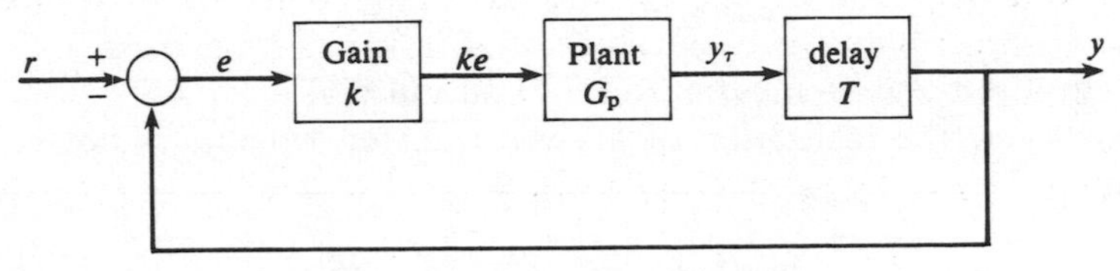

Fig. 4.34 A system with delay.

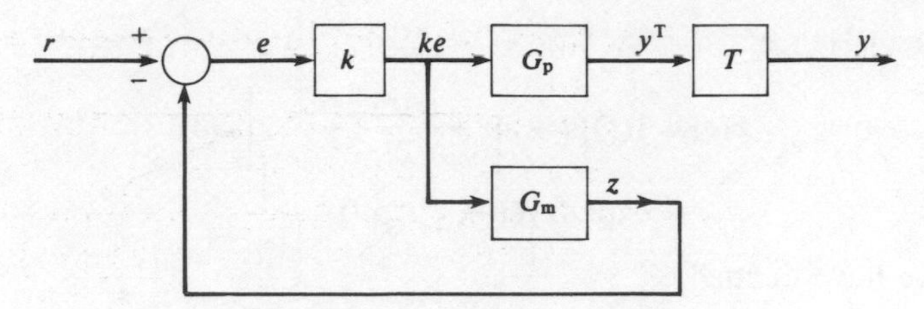

Fig. 4.35 The Smith predictor.

4.8.2 Realization Theory

We have seen how to translate a state-space model into an input–output (s-domain) description. The inverse problem of deriving a state-space model of the form (4.8.5) and (4.8.6) given a transfer function is called the **realization problem**. Let the input–output relation of a linear system be

$$Y(s) = T(s)U(s)$$

where T is the transfer function

$$T(s) = \frac{a_m s^m + \ldots + a_0}{s^n + b_{n-1}s^{n-1} + \ldots + b_0}, \quad m < n$$

Let

$$E(s) = \frac{U(s)}{1 + b_{n-1}s^{-1} + \ldots + b_0 s^{-n}}$$

Then

$$Y(s) = a_m s^{m-n}E(s) + a_{m-1}s^{m-n-1}E(s) + \ldots + a_0 s^{-n}E(s)$$

Hence we obtain the system shown in fig. 4.36. Clearly, if we define

$$\dot{x}_i = x_{i+1}, \quad 1 \leqslant i \leqslant n-1$$

then

$$\dot{x}_n = -b_0x_1 - b_1x_2 - \ldots - b_{n-1}x_n + u$$

and

$$y = a_m x_{m+1} + \ldots + a_0 x_1$$

These are the same as equations (4.8.5) and (4.8.6).

However, the realization of a given transfer function is not unique. For, if

$$\dot{\boldsymbol{x}} = A\boldsymbol{x} + \boldsymbol{b}u, \quad y = \boldsymbol{c}^{\mathrm{T}}\boldsymbol{x} + du \tag{4.8.10}$$

is one realization, then for a non-singular matrix P we have, by transform-

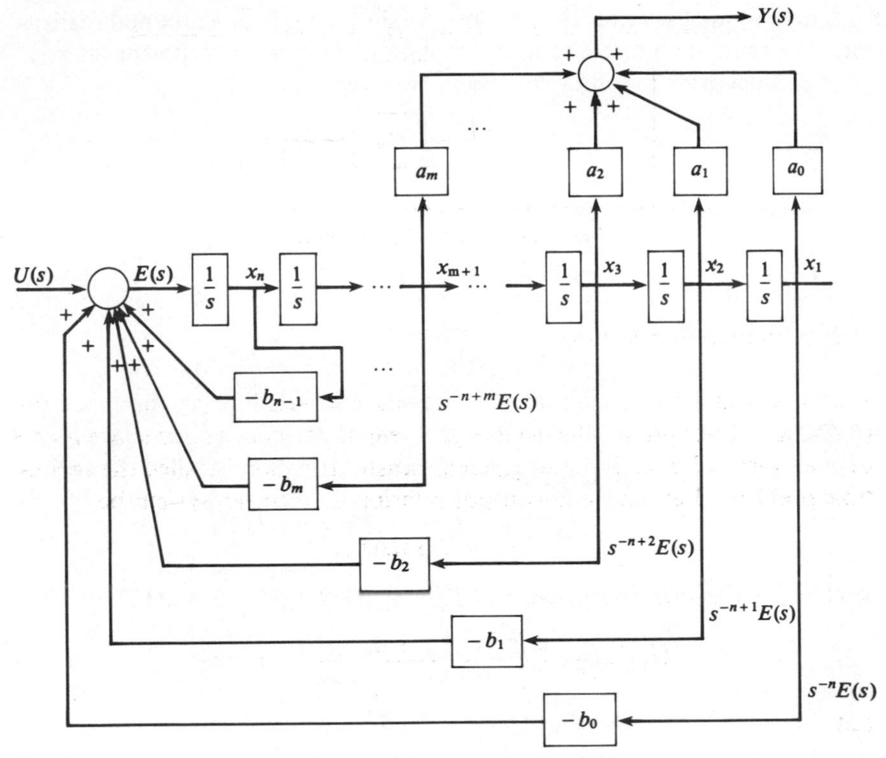

Fig. 4.36 State assignment.

ing the state $\boldsymbol{x}$ into $\boldsymbol{z} = P\boldsymbol{x}$,

$$\dot{\boldsymbol{z}} = P\dot{\boldsymbol{x}} = PAP^{-1}\boldsymbol{z} + P\boldsymbol{b}u = A_1\boldsymbol{z} + \boldsymbol{b}_1 u$$

$$y = \boldsymbol{c}^{\mathrm{T}}\boldsymbol{x} + du = \boldsymbol{c}^{\mathrm{T}}P^{-1}\boldsymbol{z} + du = \boldsymbol{c}_1^{\mathrm{T}}\boldsymbol{z} + du$$

which is 'equivalent' in the sense that it generates the same input–output representation. Moreover, if we have, in the case of (4.8.10) (with $x \in \mathbb{R}^p$, $p > n$)

$$Y(s) = \boldsymbol{c}^{\mathrm{T}}(sI - A)^{-1}\boldsymbol{b}U(s) + dU(s)$$

where

$$A = \begin{pmatrix} A_1 & 0 \\ 0 & A_2 \end{pmatrix}$$

with $\boldsymbol{c}_2^{\mathrm{T}}(sI - A)^{-1}\boldsymbol{b}_2 = 0$, it follows that

$$Y(s) = \boldsymbol{c}_1^{\mathrm{T}}(sI - A_1)^{-1}\boldsymbol{b}_1 U(s) + dU(s)$$

when $\boldsymbol{b} = (\boldsymbol{b}_1\ \boldsymbol{b}_2)^{\mathrm{T}}$, $\boldsymbol{c} = (\boldsymbol{c}_1\ \boldsymbol{c}_2)^{\mathrm{T}}$. Hence the matrix A may be reduced to A_1 giving the same transfer function and so the submatrix A_2 is redundant.

Roughly, we may say that the smallest possible n leads to a **minimal** realization. We shall say more about the 'minimal' realization problem later.

Consider now the discrete system transfer function

$$G(z) = k\frac{(z - z_1)(z - z_2)\dots(z - z_m)}{(z - p_1)(z - p_2)\dots(z - p_n)} \tag{4.8.11}$$

Define the states $x_2, \dots, x_n$ by

$$X_1(z) = \frac{k}{z - p_1}U(z), X_2(z) = \frac{z - z_1}{z - p_2}X_1(z), \dots, X_{m+1}(z) = \frac{z - z_m}{z - p_{m+1}}X_m(z)$$

$$X_{m+2}(z) = \frac{1}{z - p_{m+2}}X_{m+1}(z), \dots, X_n(z) = \frac{1}{z - p_n}X_{n-1}(z)$$

assuming $m < n$. Then

$$\begin{aligned}
&x_1((k+1)T) = p_1x_1(kT) + ku(kT)\\
&x_2((k+1)T) = p_2x_2(kT) + x_1((k+1)T) - z_1x_1(kT)\\
&\vdots\\
&x_{m+1}((k+1)t) = p_{m+1}x_{m+1}(kT) + x_m((k+1)T) - z_mx_m(kT)\\
&x_{m+2}((k+1)T) = p_{m+2}x_{m+2}(kT) + x_{m+1}(kT)\\
&\vdots\\
&x_n((k+1)T) = p_nx_n(kT) + x_{n-1}(kT)
\end{aligned}$$

and $y(kT) = x_n(kT)$. Hence

$$\boxed{\begin{aligned}\boldsymbol{x}((k+1)T) &= A\boldsymbol{x}(kT) + \boldsymbol{b}u(kT)\\ y(kT) &= \boldsymbol{c}^{\mathrm{T}}\boldsymbol{x}(kT)\end{aligned}} \tag{4.8.12}$$

where $\boldsymbol{x} = (x_1, \dots, x_n)^{\mathrm{T}}$, $\boldsymbol{c} = (0, 0, \dots, 0, 1)^{\mathrm{T}}$, $\boldsymbol{b} = (\underbrace{k, k, \dots, k}_{m+1}, 0, \dots, 0)^{\mathrm{T}}$

and

$$A = \begin{bmatrix}
p_1 & 0 & 0 & \dots & & & & & 0\\
p_1 - z_1 & p_2 & 0 & \dots & & & & & 0\\
p_1 - z_1 & p_2 - z_2 & p_3 & 0 & \dots & & & & 0\\
\vdots & \vdots & & & & & & & \\
p_1 - z_1 & p_2 - z_2 & \dots & & p_{m+1} & & 0 & \dots & 0\\
0 & 0 & \dots & & 1 & p_{m+2} & 0 & \dots & 0\\
\vdots & \vdots & & & & \vdots & & & \vdots\\
0 & 0 & \dots & & & 0 & \dots & 1 & p_n
\end{bmatrix}$$

We can also obtain an equation of the form (4.8.12) for a transfer function

$$G(z) = \frac{a_mz^m + a_{m-1}z^{m-1} + \dots + a_0}{z^n + b_{n-1}z^{n-1} + \dots + b_0}$$

by replacing s^{-1} by z^{-1} in fig. 4.36 and defining

$$x_1((k+1)T) = x_2(kT)$$

$$x_2((k+1)T) = x_3(kT)$$

$$\vdots$$

$$x_{n-1}((k+1)T) = x_n(kT)$$

$$x_n((k+1)T) = -b_0x_1(kT) - b_1x_2(kT) - \ldots - b_{n-1}x_n(kT) + u(kT)$$

and

$$y(nT) = a_0x_1(kT) + \ldots + a_mx_{m+1}(kT)$$

In this case the matrices and vectors in (4.8.12) are exactly the same as in the continuous case.

The equation (4.8.12) of a discrete system is a difference equation and can be solved easily by Z-transformation, just as we solved the continuous equation by Laplace transformation. Without control (i.e. $u = 0$) we obtain

$$\boxed{x(kT) = A^k x(0)} \tag{4.8.13}$$

and for the general equation

$$\boxed{x(kT) = A^k x(0) + \sum_{l=0}^{k-1} A^{k-1-l} bu(lT)} \tag{4.8.14}$$

From (4.8.13) we see that the stability of a discrete system is determined by the behaviour of A^k as $k \to \infty$. If we assume, for simplicity, that the eigenvalues of A are distinct, then we can find a non-singular matrix P such that

$$PAP^{-1} = \Lambda$$

where Λ is the diagonal matrix containing the eigenvalues of A (appendix 1). Hence,

$$PA^kP^{-1} = PAP^{-1} \cdot PAP^{-1} \ldots PAP^{-1} = \Lambda^k$$

and so $A^k \to 0$ if and only if $\Lambda^k \to 0$ as $k \to \infty$. (Here we say that a matrix function converges if and only if each element converges.) However, $\Lambda^k \to 0$ if and only if every eigenvalue of A is less than one in modulus.

4.8.3 State-variable Feedback

Suppose for the moment that all the states of the system

$$\left.\begin{aligned} \dot{x} &= Ax + bu \\ y &= c^T x \end{aligned}\right\} \tag{4.8.15}$$

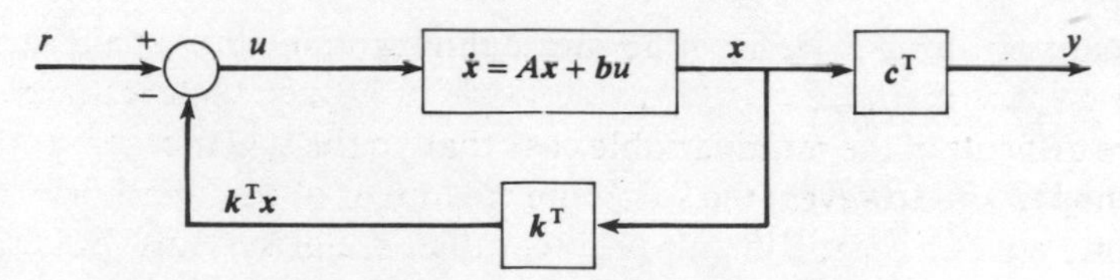

Fig. 4.37 General state feedback system.

are available to the controller. (In the next section we shall consider a method for estimating the states from the output y.) Then we may feed back a weighted sum of the states $\boldsymbol{x}$ as in fig. 4.37. From (4.8.15) we have

$$\dot{\boldsymbol{x}} = A\boldsymbol{x} + \boldsymbol{b}(r - \boldsymbol{k}^{\mathrm{T}}\boldsymbol{x}) = (A - \boldsymbol{b}\boldsymbol{k}^{\mathrm{T}})\boldsymbol{x} + \boldsymbol{b}r$$

and so

$$\frac{Y(s)}{R(s)} = T(s) = \boldsymbol{c}^{\mathrm{T}}(sI - A + \boldsymbol{b}\boldsymbol{k}^{\mathrm{T}})^{-1}\boldsymbol{b}$$

Suppose that the open-loop transfer function is

$$G(s) = \frac{Y(s)}{U(s)} = \frac{a_m s^m + \ldots + a_0}{s^n + b_{n-1}s^{n-1} + \ldots + b_0}$$

and we choose the states so that A, $\boldsymbol{b}$, $\boldsymbol{c}$ are in the 'canonical form' given in section (4.8.1), then

$$T(s) = (a_0 \; a_1 \ldots a_m \; 0 \ldots 0)$$

$$\begin{bmatrix} s & -1 & 0 & & \cdots & 0 \\ 0 & s & -1 & 0 & \cdots & 0 \\ \vdots & \vdots & & \vdots & & \vdots \\ 0 & 0 & \cdots & 0 & s & -1 \\ b_0 + k_1 & b_1 + k_2 & \cdots & & b_{n-2} + k_{n-1} & s + b_{n-1} + k_n \end{bmatrix}^{-1} \begin{bmatrix} 0 \\ 0 \\ \vdots \\ 0 \\ 1 \end{bmatrix}$$

$$= \frac{a_m s^m + \ldots + a_0}{s^n + (b_{n-1} + k_n)s^{n-1} + \ldots + (b_0 + k_1)} \qquad (4.8.16)$$

Hence, if we can feed back all the tates, then we can keep the open-loop zeros fixed and place the poles arbitrarily by a suitable choice of the k_i. This solves the so-called pole assignment problem for state feedback. The more general problem of pole assignment by output feedback is more difficult, but can be solved by an appropriate observer design, which we discuss later.

4.9 MULTIVARIABLE SYSTEMS

4.9.1 System Representations

We have been primarily concerned, up to now, with single-input single-output systems. This assumption is very restrictive and it is obviously important to be able to deal with systems having many inputs and outputs. The state-space methods of optimal control and other such techniques are no more difficult in the multivariable case than in the scalar case, as we shall see in chapter 6. However, the s-domain treatment of multivariable systems represents a much more difficult problem than scalar systems. Some of the many techniques which have been developed to generalize the preceding scalar theory will now be outlined.

The general state-space model for multivariable systems defined by differential equations has been given in (4.8.1) and (4.8.2). It is easy to see that the transfer function of this system is

$$\boxed{G(s) = C(sI - A)^{-1}B + D} \tag{4.9.1}$$

The poles of $G(s)$ are the roots of $|sI - A| = 0$ provided no cancellations occur in the product of C, $(sI - A)^{-1}$ and B. If $D = 0$ we say that $G(s)$ is strictly proper if $|G(s)| \to 0$ as $|s| \to \infty$.

A slight generalization of the case of systems consisting of differential equations of the general form (4.2.12) is to include linear algebraic equations. These systems have been considered by Rosenbrock (1970) and Wolovich (1974) and by taking Laplace transforms in the usual way, with zero initial conditions, we arrive at a system of equations of the form

$$\left.\begin{aligned} T(s)\boldsymbol{\xi}(s) &= U(s)\boldsymbol{u}(s) \\ \boldsymbol{y}(s) &= V(s)\boldsymbol{\xi}(s) + W(s)\boldsymbol{u}(s) \end{aligned}\right\} \tag{4.9.2}$$

or

$$\begin{bmatrix} T(s) & U(s) \\ -V(s) & W(s) \end{bmatrix} \begin{bmatrix} \boldsymbol{\xi}(s) \\ -\boldsymbol{u}(s) \end{bmatrix} = \begin{bmatrix} 0 \\ -\boldsymbol{y}(s) \end{bmatrix} \tag{4.9.3}$$

where T, U, V and W are polynomial matrices. We call

$$P(s) = \begin{bmatrix} T(s) & U(s) \\ -V(s) & W(s) \end{bmatrix}$$

the system matrix or, somewhat loosely, just refer to P as 'the system'. In this case the transfer function is

$$\boxed{G(s) = V(s)T^{-1}(s)U(s) + W(s)} \tag{4.9.4}$$

which clearly reduces to (4.9.1) for systems of the form (4.8.1) and (4.8.2).

An important concept is that of system equivalence which we have met in the case of scalar systems in terms of a similarity transformation. A more general definition of equivalence is required in the case of transfer functions of the form (4.9.4). We first define two polynomial matrices $M(s)$ and $N(s)$ to be **relatively left prime** if the factorizations

$$M(s) = L(s)M_1(s), \quad N(s) = L(s)N_1(s)$$

where $L(s)$ is invertible, can only hold if $L(s)$ is **unimodular**, i.e. det $L(s)$ is a non-zero constant, independent of s. (Hence M and N cannot both be divided on the left by the same non-unimodular matrix.) Relative right primeness is defined similarly. Also the **order** of the system P is defined by

$$\text{ord } P = \deg\{\det T(s)\}$$

Now, two systems P_1 and P_2 are defined to be **strictly equivalent** if there exist polynomial matrices M, N, X, Y such that M and N are unimodular and we have

$$\boxed{\begin{bmatrix} M(s) & 0 \\ X(s) & I \end{bmatrix} \begin{bmatrix} T_1(s) & U_1(s) \\ -V_1(s) & W_1(s) \end{bmatrix} \begin{bmatrix} N(s) & Y(s) \\ 0 & I \end{bmatrix} = \begin{bmatrix} T_2(s) & U_2(s) \\ -V_2(s) & W_2(s) \end{bmatrix}} \tag{4.9.5}$$

(Here the matrices have appropriate dimensions so that this makes sense.)

It can be shown that any system $P(s)$ giving rise to a strictly proper $G(s)$ is strictly equivalent to

$$\bar{P}(s) = \left[\begin{array}{cc|c} I_{r-n} & 0 & 0 \\ 0 & sI - A & B \\ \hline 0 & -C & 0 \end{array}\right]$$

where $n = \dim A$, i.e. P is equivalent to a state space representation. In the case where ord $P = n > \dim T(s) = r$ we can form

$$\tilde{P} = \begin{bmatrix} I_{n-r} & 0 & 0 \\ 0 & T & U \\ 0 & -V & W \end{bmatrix}$$

with

$$\dim \begin{bmatrix} I_{n-r} & 0 \\ 0 & T \end{bmatrix} \geqslant \text{ord } \tilde{P}$$

If, in the system P, $T(s)$ and $U(s)$ are relatively left prime and $V(s)$ and $T(s)$ are relatively left prime, then P is said to be of **least order**. Note that if this condition does not hold, there exists non-unimodular invertible matrices $L(s)$ and $M(s)$ such that

$$T(s) = L(s)T_1(s), \quad U(s) = L(s)U_1(s) \tag{4.9.6}$$

or

$$V(s) = V_1(s)M(s), \quad T(s) = T_2(s)M(s)$$

for some matrices T_1, U_1, V_1. Then

$$\begin{aligned} G(s) &= V(s)T_1^{-1}(s)L^{-1}(s)L(s)U_1(s) + W(s) \\ &= V(s)T_1^{-1}(s)U_1(s) + W(s) \end{aligned}$$

or

$$G(s) = V_1(s)T_2^{-1}(s)U(s) + W(s)$$

and so some cancellation of the polynomials in $G(s)$ will occur. The roots of the non-trivial polynomials $p_1(s) = \det L(s)$ and $p_2(s) = \det M(s)$ are called the **input and output decoupling zeros** of P respectively. The roots of the highest common factor $q(s)$ of $p_1(s)$ and $p_2(s)$ are called the **input–output decoupling zeros**. The zeros of p_1/q and p_2 (or equivalently p_1 and p_2/q) are the **decoupling zeros**. Clearly, from (4.9.6), we have that the poles of the least order system are those of $P(s)$ minus the decoupling zeros, i.e. the zeros of $[\det T(s)]q(s)/p_1(s)p_2(s)$.

4.9.2 Controllability and Observability

Two important concepts in linear systems theory are those of controllability and observability. In the former case, we say that a system†

$$\left.\begin{aligned} \dot{x} = Ax(t) + Bu(t) \\ y(t) = Cx(t) \end{aligned}\right\} \tag{4.9.7}$$

is **controllable** if, given x_0, $x_1 \in \mathbb{R}^n$, there exists a (piecewise continuous) control $u(t)$ and a time t_1 such that if $x(t)$ is the solution of (4.9.7) starting at x_0, then $x(t_1) = x_1$, i.e. we can drive any initial point x_0 to any other point x_1. Likewise, we say that (4.9.7) is **observable** if we can find a time t_1 such that given the measurement $y(t)$ and the input $u(t)$ over the interval $[0, t_1]$ we can determine the initial state x_0. If we form the matrices

$$\boxed{\begin{aligned} R_c &= [B \ AB \ A^2B \dots A^{n-1}B] \\ R_o &= [C^T \ A^TC^T \ (A^2)^TC^T \dots (A^{n-1})^TC^T] \end{aligned}} \tag{4.9.8}$$

then it can be shown (Wonham, 1979; Lee and Markus, 1967) that (4.9.7) is controllable if and only if

$$\boxed{\operatorname{rank} R_c = n} \tag{4.9.9}$$

and that (4.9.7) is observable if and only if

$$\boxed{\operatorname{rank} R_o = n} \tag{4.9.10}$$

More generally, we can define the set

$$\langle A \mid \mathcal{B} \rangle \triangleq \mathcal{B} + A\mathcal{B} + \dots + A^{n-1}\mathcal{B}$$

† We often refer to (4.9.7) as the 'system' $[A, B, C]$.

where $\mathcal{B} = \text{Im } B = \{x \in \mathbb{R}^n : x = Bu \text{ for some } u \in \mathbb{R}^m\}$ and for two sets X, Y we write $X + Y = \{z : z = x + y,\ x \in X, y \in Y\}$. Then we can show, as for controllability, that the set of states which can be 'reached' by a control $u(t)$ in some time t_1 is $\langle A \mid \mathcal{B} \rangle$. For any feedback control $u(t) = Fx(t)$ in (4.9.7) we have the system

$$\dot{x} = (A + BF)x(t)$$

Suppose for simplicity that the eigenvalues of A are distinct and that $\text{Re } \lambda_1, \ldots, \text{Re } \lambda_i$ are positive and $\text{Re } \lambda_{i+1}, \ldots, \text{Re } \lambda_n$ are negative. If $x_1, \ldots, x_i, x_{i+1}, \ldots, x_n$ are the corresponding eigenvectors, then let $\mathbb{R}_u^i$ be the subspace of $\mathbb{R}^n$ spanned by the vectors $x_1, \ldots, x_i$ and $\mathbb{R}_s^{n-i}$ be the subspace spanned by $x_{i+1}, \ldots, x_n$ (see appendix 1). Then it can be shown (Wonham, 1979) that we can find a feedback matrix F such that the eigenvalues of $A + BF$ can be assigned to have negative real parts if and only if

$$\boxed{\mathbb{R}_u^i \subseteq \langle A \mid \mathcal{B} \rangle} \tag{4.9.11}$$

i.e. if and only if the 'unstable modes' of A are controllable. (Of course, the desired eigenvalues must be symmetric, i.e. must occur in conjugate pairs.) In this case the system is said to be **stabilizable**. The problem of specifying desired feedback poles is, naturally, called pole assignment and has been studied by many authors (see, for example, Patel and Munro, 1982).

If the system (4.9.7) is controllable then rank $R_c = n$ and so we can find a matrix Γ with linearly independent columns of the form

$$\Gamma = [b_1, Ab_1, \ldots, A^{\mu_1 - 1}b_1, b_2, Ab_2, \ldots, A^{\mu_2 - 1}b_2, \ldots, A^{\mu_m - 1}b_m] \tag{4.9.12}$$

where b_i is the ith column of B. Note that

$$\sum_{i=1}^{m} \mu_i = n$$

and we assume, without loss of generality, that the columns of B are linearly independent. Now define

$$k_1 = \mu_1, k_2 = \mu_1 + \mu_2, k_3 = \mu_1 + \mu_2 + \mu_3, \ldots, k_m = \sum_{j=1}^{m} \mu_j$$

and let r_i^{T}, $1 \leqslant i \leqslant m$ be the k_ith row of Γ^{-1}. Then we introduce the matrix

$$P = [r_1\ A^{\mathrm{T}}r_1 \ldots (A^{\mathrm{T}})^{\mu_1 - 1}r_1, r_2, A^{\mathrm{T}}r_2, \ldots, (A^{\mathrm{T}})^{\mu_2 - 1}r_2 \ldots (A^{\mathrm{T}})^{\mu_m - 1}r_m]^{\mathrm{T}} \tag{4.9.13}$$

It can be shown that P is invertible and that the system $[\tilde{A}, \tilde{B}, \tilde{C}]$ defined by

$$\boxed{\tilde{A} = PAP^{-1}, \quad \tilde{B} = PB, \quad \tilde{C} = CP^{-1}} \tag{4.9.14}$$

has the form

$$\widetilde{A} = [\widetilde{A}_{ij}]_{1 \leqslant i,j \leqslant m}$$

where the blocks $\widetilde{A}_{ij}$ have the form

$$\widetilde{A}_{ii} = \begin{pmatrix} 0 & 1 & 0 & \dots & & 0 \\ 0 & 0 & 1 & 0 & \dots & 0 \\ \vdots & \cdot & \cdot & & & \vdots \\ & \cdot & \cdot & & & \\ 0 & \cdot & \cdot & & & 1 \\ * & * & * & \dots & & * \end{pmatrix} \quad (\dim \mu_i \times \mu_i)$$

and

$$\widetilde{A}_{ij} = \begin{pmatrix} & & \mathbf{0} & & \\ * & * & * & \dots & * \end{pmatrix} \quad (\dim \mu_i \times \mu_j), i \neq j$$

Moreover we have $\widetilde{B} = (\widetilde{B}_1^{\mathrm{T}}, \dots, \widetilde{B}_m^{\mathrm{T}})^{\mathrm{T}}$, where

$$\widetilde{B}_i = \begin{pmatrix} & & & \mathbf{0} & & & \\ 0 & 0 & \dots & 1 & * & \dots & * \end{pmatrix} \quad (\dim \mu_i \times m)$$

$$\uparrow$$

ith column

The asterisks can be any real numbers. The parameters μ_i are called the **controllability indices**; for a proof of the above see Wonham (1979). The system (4.9.14) is equivalent to the system (4.9.7) and is called the **controllable canonical form**. A similar result gives the **observable canonical form** (see also Kailath, 1980).

We should also note that a different definition of controllability has been introduced by Rosenbrook (1970). We say that a system is **functionally controllable** if any vector of outputs $\boldsymbol{y}(t)$ can be generated for $t > 0$ by an appropriate (piecewise continuous) control $\boldsymbol{u}(t)$ starting with zero initial conditions $\boldsymbol{x}(0) = \mathbf{0}$. For a system of the form (4.9.7), with $\dim \boldsymbol{y} = \dim \boldsymbol{u}$ we have functional controllability if and only if $\det G(s) \not\equiv 0$. Note, however, that functional controllability does not imply and is not implied by controllability.

A discussion of controllability for distributed systems is given by Banks (1983) (but requires a knowledge of functional analysis).

We have already met the concept of state-space realization of a single-input single-output transfer function $G(s)$. We may obtain a similar realization for a multivariable system in the form (4.9.7). However such a realization may not be 'minimal' in the sense that the system may have uncontrollable or unobservable modes which do not appear in the transfer function $G(s)$. Hence we define the realization to be **minimal** if $[A, B, C]$ is controllable and observable (Kalman, 1963). A computational procedure for calculating a minimal realization of a transfer function has been developed by Munro (1971). Note, finally, that all minimal realizations of a given system are equivalent in the sense that if $[\tilde{A}, \tilde{B}, \tilde{C}]$ is another such realization, then $[A, B, C]$ and $[\tilde{A}, \tilde{B}, \tilde{C}]$ are related by (4.9.14) for some matrix P.

4.9.3 Observers

We have seen that it is often desirable to use state feedback in a system $[A, B, C]$. However, in the system (4.9.7) we do not measure the states $\boldsymbol{x}(t)$ but a linear combination $\boldsymbol{y}(t) = C\boldsymbol{x}(t)$ of these variables. If C is of dimension $l \times n$ with $l < n$, then we cannot determine the states $\boldsymbol{x}(t)$ directly, and so it is not possible to implement the feedback $F\boldsymbol{x}(t)$.

A method for estimating the states $\boldsymbol{x}(t)$ from a knowledge of $\boldsymbol{y}(t)$ and $\boldsymbol{u}(t)$ has been obtained by Luenberger (1966) and consists of specifying a system of the form

$$\boxed{\dot{\boldsymbol{z}}(t) = D\boldsymbol{z}(t) + T\boldsymbol{y}(t) + R\boldsymbol{u}(t)} \qquad (4.9.15)$$

where $\boldsymbol{z}(t) \in \mathbb{R}^{n-l}$ for each t. If we choose D to have eigenvalues to be distinct from those of A and further in the left half plane than any of A, then it can be shown (Barnett and Storey, 1970) that we can solve the equation

$$LA - DL = TC$$

uniquely for L. Now setting $R = LB$ in (4.9.15) we have

$$\begin{aligned}\dot{\boldsymbol{z}}(t) - L\dot{\boldsymbol{x}}(t) &= D\boldsymbol{z}(t) + T\boldsymbol{y}(t) - LA\boldsymbol{x}(t)\\ &= D(\boldsymbol{z}(t) - L\boldsymbol{x}(t))\end{aligned}$$

and so

$$\boldsymbol{z}(t) \to L\boldsymbol{x}(t) \quad \text{as } t \to \infty$$

Now let K_1 and K_2 be matrices satisfying

$$K_1C + K_2L = I_l$$

It can be shown that if the system $[A, B, C]$ is observable, then we can choose L so that

$$\begin{pmatrix} C \\ \hline L \end{pmatrix}$$

has full rank. Then K_1 and K_2 are specified by

$$(K_1K_2) = \begin{pmatrix} C \\ \hline L \end{pmatrix}^{-1}$$

The estimated state $\hat{x}(t)$ can now be obtained as

$$\boxed{\hat{x}(t) = K_1 y(t) + K_2 z(t)} \tag{4.9.16}$$

and we may consider the estimated state feedback $u(t) = F\hat{x}(t)$. Then we have

$$\begin{bmatrix} \dot{x}(t) \\ \dot{z}(t) \end{bmatrix} = \begin{bmatrix} A + BFK_1C & BFK_2 \\ TC + LBFK_1C & D + LBFK_2 \end{bmatrix} \begin{bmatrix} x(t) \\ z(t) \end{bmatrix}$$

or

$$\boxed{\begin{bmatrix} \dot{x}(t) \\ \dot{e}(t) \end{bmatrix} = \begin{bmatrix} A + BF & BFK_2 \\ 0 & D \end{bmatrix} \begin{bmatrix} x(t) \\ e(t) \end{bmatrix}} \tag{4.9.17}$$

where $e(t) = z(t) - Lx(t)$. Hence the poles of the closed-loop system are those of $A + BF$ and D and so we can assign the closed-loop poles as before by using the estimated feedback (see fig. 4.38).

Many other results on pole assignability and observers are given by Patel and Munro (1982) and O'Reilly (1984).

4.9.4 Multivariable Feedback Systems, Poles and Zeros

Consider the multivariable feedback system shown in fig. 4.39. Then if we cut the feedback loops at y and inject the signal α at this point the returned

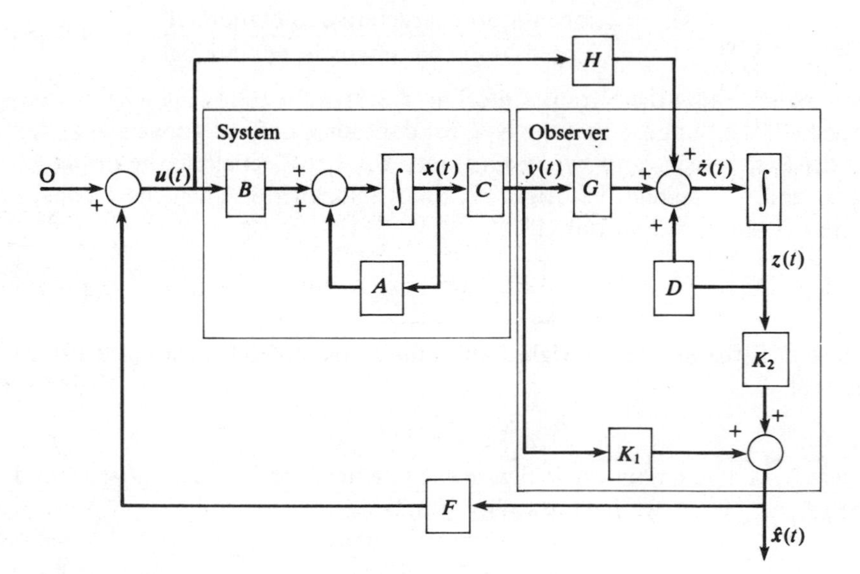

Fig. 4.38 State feedback system with dynamic observer.

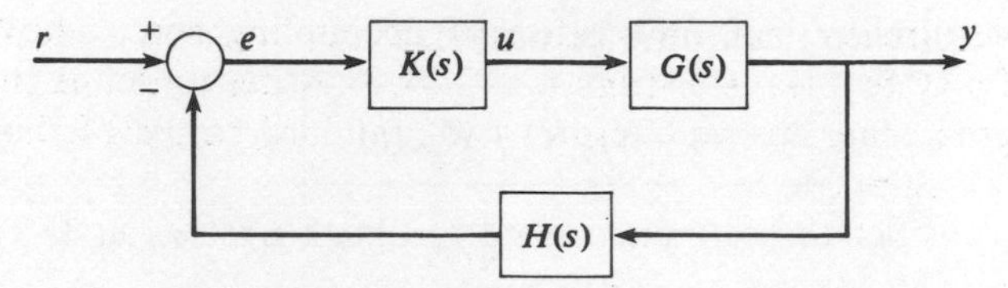

Fig. 4.39 Non-unity gain feedback system.

signal is $-G(s)K(s)H(s)\alpha$ and so

$$\boxed{F_y(s) = I + G(s)K(s)H(s)}$$

is called the return difference matrix. It is easy to see that the closed-loop transfer function $R(s)$ is given by

$$\begin{aligned} R(s) &= [I + G(s)K(s)H(s)]^{-1}G(s)K(s) \\ &= F_y^{-1}(s)G(s)K(s) \\ &= F_y^{-1}(s)Q(s) \end{aligned} \tag{4.9.18}$$

where

$$Q(s) = G(s)K(s)$$

is the open-loop transfer function. Hence we have

$$\begin{aligned} \det F_y(s) &= \frac{\det Q(s)}{\det R(s)} \\ &= \frac{\text{open-loop characteristic polynomial}}{\text{closed-loop characteristic polynomial}} \end{aligned} \tag{4.9.19}$$

We can generalize the Nyquist criterion (4.3.3) in the following way (Rosenbrock, 1974): Suppose that the Nyquist contour γ defined above is mapped by $\det R(s)$ into Γ_c and by $\det Q(s)$ into Γ_o. Let Γ_c encircle the origin N_c times and Γ_o encircle the origin N_o times (clockwise). Then the feedback system is stable if and only if

$$\boxed{N_c - N_o = p_o} \tag{4.9.20}$$

where p_o is the number of right half plane zeros of $Q(s)$. From (4.9.19) we see that

$$N_f = N_o - N_c$$

where N_f is the number of clockwise encirclements of $\Gamma_f = \det F_y(\gamma)$ around the origin. Hence we have stability if and only if

$$\boxed{N_f = -p_o} \tag{4.9.21}$$

We have already met the notion of decoupling zeros above in the definition of a system transfer function $G(s)$. In order to define the system poles and zeros generally we consider two standard forms of the transfer function $G(s)$. Suppose first that

$$G(s) = \frac{N(s)}{d(s)}$$

where $N(s)$ is an $l \times m$ polynomial matrix and $d(s)$ is the (monic) least common denominator of the elements of $G(s)$. Then it can be shown that, using unimodular matrices L_1 and L_2, we can write $N(s)$ in the form $L_1^{-1}(s)S(s)L_2^{-1}(s)$ where

$$S(s) = \begin{cases} (\operatorname{diag}(\psi_i(s)) \quad 0_{l,m-l}) & m > l \\ (\operatorname{diag}(\psi_i(s))) & m = l \\ \begin{pmatrix} \operatorname{diag}(\psi_i(s)) \\ 0_{l-m,m} \end{pmatrix} & m < l \end{cases} \tag{4.9.22}$$

(see Rosenbrock, 1970). This is called the **Smith form** of $N(s)$ and if $f_i(s)$ is the monic highest common factor of all non-zero $i \times i$ minors of $N(s)$, then

$$\boxed{\psi_i(s) = \frac{f_i(s)}{f_{i-1}(s)}} \tag{4.9.23}$$

The **McMillan form** of $G(s)$ is then

$$M(s) = \begin{cases} \left(\operatorname{diag} \dfrac{(\alpha_i(s))}{\beta_i(s)} \; 0_{l,m-l}\right) & m > l \\ \left(\operatorname{diag} \dfrac{(\alpha_i(s))}{\beta_i(s)}\right) & m = l \\ \begin{pmatrix} \operatorname{diag} \dfrac{(\alpha_i(s))}{\beta_i(s)} \\ 0_{l-m,m} \end{pmatrix} & m < l \end{cases} \tag{4.9.24}$$

where $M(s) = S(s)/d(s)$ and we have made any root cancellations possible in forming the ratios $\alpha_i(s)/\beta_i(s)$.

We can now define the **poles** of the system $G(s)$ to be the roots of all the polynomials $\beta_i(s)$, counted with appropriate multiplicity. The roots of the polynomials $\alpha_i(s)$ are called the **transmission zeros** and represent complex frequencies at which system transmission paths vanish. The **invariant zeros** of a system matrix $P(s)$ are the roots of the greatest common divisor of all maximal order minors of $P(s)$ (see Kouvaritakis and MacFarlane,

1976). Finally, we define the **system zeros** to the the transmission zeros together with the decoupling zeros introduced earlier (MacFarlane and Karcanias, 1976).

4.9.5 Inverse Nyquist Array and Compensator Design

The closed-loop transfer function $R(s)$ is related to that of the open loop (i.e. $Q(s)$) in a complicated way, via equation (4.9.18). If these matrices are not identically singular it is more convenient to consider the simpler relation

$$\boxed{R^{-1}(s) = Q^{-1}(s) + H(s)} \tag{4.9.25}$$

We usually write

$$\hat{R}(s) \triangleq R^{-1}(s) = (\hat{r}_{ij}(s))$$

$$\hat{Q}(s) \triangleq Q^{-1}(s) = (\hat{q}_{ij}(s))$$

For simplicity we shall assume that $H(s) = \text{diag}(k_1, \ldots, k_m)$ and then we have

$$\hat{r}_{ii}(s) = \hat{q}_{ii}(s) + k_i$$

$$\hat{r}_{ij}(s) = \hat{q}_{ij}(s), \qquad i \neq j$$

We say that an $m \times m$ matrix $A = (a_{ij})$ is **diagonally (row) dominant** if

$$| a_{ii} | > \sum_{j=1}^{m} | a_{ij} |, \quad \text{for each } i \in \{1, \ldots, m\}$$

The **inverse Nyquist array** is the set of m^2 loci of $\hat{q}_{ij}(j\omega)$, $-\infty < \omega < \infty$. It can then be shown (Rosenbrock, 1974) that the closed-loop system (4.9.18) is asymptotically stable if $\hat{R}(s)$ and $\hat{Q}(s)$ are diagonally dominant on the Nyquist locus γ and

$$\sum_{i=1}^{m} N^0_{q_{ii}} - \sum_{i=1}^{m} N^{-1}_{q_{ii}} = p_0, \quad \text{for each } i$$

where $N^0_{q_{ii}}$ is the number of encirclements of the origin of $\hat{q}_{ii}(s)$, $N^{-1}_{q_{ii}}$ is the number of encirclements of $(-k_i, 0)$ of $\hat{q}_{ii}(s)$ and p_0 is the number of open-loop zeros in the RHP (see fig. 4.40).

Now let

$$d_i(s) = \sum_{\substack{j=1 \\ j \neq i}}^{m} | \hat{q}_{ij}(s) | \tag{4.9.26}$$

It can then be shown (Ostrowski, 1952) that $r_{ii}^{-1}(s) - k_i$ ($\neq \hat{q}_{ii}(s)$, generally) is contained in the circles of radius $d_i(s)$ about $\hat{q}_{ii}(s)$. Hence, if the (**Gershgorin**) bands made up of circles of radius $d_i(s)$ on the inverse

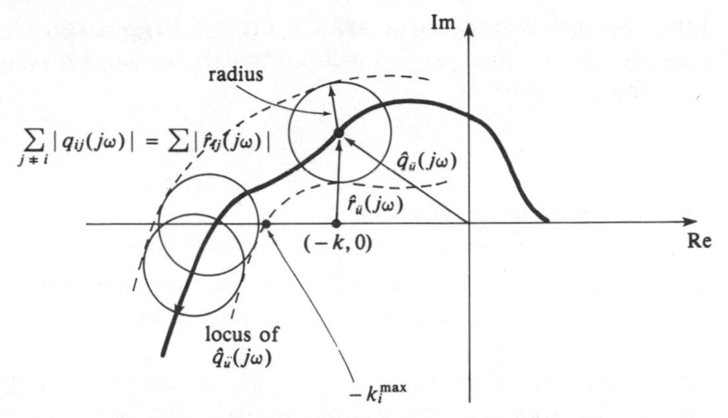

Fig. 4.40 The inverse Nyquist stability criterion.

Nyquist loci $\hat{q}_{ii}(s)$ do not cross the lines joining 0 to $(-k_i^{max}, 0)$ in the m complex planes, then the feedback system is stable (if $p_0 = 0$) for all gains k_i between 0 and k_i^{max}. We are therefore guaranteed system 'integrity' in the sense that if any feedback loop is broken (i.e. has gain 0) then the system is still stable.

The result of Ostrowski also implies that $r_{ii}^{-1}(s) - k_i$ is, in fact, contained in the band generated by circles with centres on $\hat{q}_i(s)$ and radii $d_i'(s)$, where

$$\boxed{d_i'(s) = \left[\max_{\substack{j \\ j \neq i}} \frac{d_j(s)}{k_j + \hat{q}_{jj}(s)}\right] d_i(s)} \tag{4.9.27}$$

Hence, once we have achieved stability using Gershgorin bands the 'Ostrowski' bands give a measure of gain margin for each loop, although it should be noted that the gains k_j are all assumed to be known.

The inverse Nyquist design technique consists of determining the compensator $K(s)$ as a product $K_1(s)K_2(s)$ where $K_1(s)$ is chosen to make $G(s)K_1(s)$ diagonally dominant and $K_2(s)$ is a diagonal matrix of compensators which is chosen as in the classical case to improve the performance of the systems seen from the ith input to the ith output (for $1 \leqslant i \leqslant m$). Usually we make $K_1(s)$ as simple as possible and it is found, in many cases, that a constant matrix is sufficient. Several methods have been proposed to find K_1 in any particular situation (see, for example, Patel and Munro, 1982). The most commonly used techniques are as follows:

(a) Use elementary row operations to reduce $G(s)$ to a diagonally dominant form and take K_1 to be the product of the associated elementary matrices.
(b) If $G(0)$ is non-singular choose $K_1 = G^{-1}(0)$, which is a matrix of real constants.

(c) Find the optimum real matrix K_1 which 'nearly' diagonalizes $G(j\omega_d)$ at some desired frequency ω_d, i.e. which minimizes the sums of squares of the off-diagonal elements.†

4.9.6 Characteristic Locus Method

A different approach to multivariable design is the **characteristic locus method** (MacFarlane and Belletrutti, 1973; Kouvaritakis, 1974). Consider the feedback system in fig. 4.39 with $H(s) = I$. The basic problem with multivariable control is that of interaction between the ith and jth loops $(1 \leqslant i \neq j \leqslant m)$. In the inverse Nyquist method the compensator was chosen to make the open-loop system diagonally dominant and thereby reduce the effects of interaction. In the characteristic locus method the idea is to try to use the eigenvalues of $Q(s) \triangleq G(s)K(s)$ to design an appropriate non-interacting controller. Let, therefore, $q_i(s)$ $(1 \leqslant i \leqslant m)$ be the m solutions of the equation

$$\det(q(s)I_m - Q(s)) = 0 \tag{4.9.28}$$

Note that the functions $q_i(s)$ are not rational, but may be considered as algebraic functions defined on a Riemann surface (see Knopp, 1945). The eigenvalues $q_i(s)$ of $Q(s)$ are called the **characteristic gain functions**. If we assume for simplicity that, for each s, the corresponding eigenvectors $w_i(s)$ are linearly independent, then we have

$$Q(s) = W(s) \operatorname{diag} \{q_i(s)\} V(s) \tag{4.9.29}$$

where

$$W(s) = [w_1(s) \quad w_2(s) \quad \dots \quad w_m(s)]$$

and

$$V(s) = W^{-1}(s) \triangleq [v_1^T(s) \quad v_2^T(s) \quad \dots \quad v_m^T(s)]^T$$

(See appendix 1). The vectors $w_i(s)$ are called the **characteristic directions** (of $Q(s)$). Denoting

$$\det(q(s)I_m - Q(s)) = \Delta_1(s) \dots \Delta_l(s)$$

where the factors $\Delta_i(s)$ are rational functions and are irreducible (i.e. cannot be written as a product of two non-constant rational functions), then we can prove (MacFarlane and Postlethwaite, 1977, 1979) the **generalized**

---

† Note, however, that none of the above methods guarantees diagonal dominance, which may, of course, be impossible (see Mees, 1981; Limebeer, 1982).

Nyquist stability criterion:

Each equation $\Delta_i(s) = 0$ generates an algebraic function $\bar{q}_i(s)$ on a Riemann surface, consisting of, say, l_i copies of the complex plane. If γ is the standard Nyquist contour, then we call the **characteristic loci** the images of l_i copies of γ (on the Riemann surface) in the q plane, for $1 \leqslant i \leqslant l$. Then the system in fig. 4.39 (with $H = I$) is stable if and only if each characteristic locus encircles the point $(-1, 0)$ p_{oi} times, where p_{oi} is the number of right half plane poles of $\Delta_i(s)$.

Now we have, for the closed-loop transfer function $R(s)$,

$$R(s) = (I_m + Q(s))^{-1}Q(s)$$

and so by (4.9.29) we have

$$\boxed{H(s) = \sum_{i=1}^{m} \left(\frac{q_i(s)}{1 + q_i(s)} \right) w_i(s) v_i^{\mathrm{T}}(s)} \tag{4.9.30}$$

Hence $q_i(s)/(1 + q_i(s))$ is the ith closed-loop characteristic function, and the characteristic directions are the same as for the open-loop system. This important observation implies that, for small interaction at any particular frequency ω, the characteristic directions $w_i(s)$ should be close to the standard basis vector e_i, and so if the angles θ_i are given by† (MacFarlane and Kouvaritakis, 1977)

$$\cos \theta_i(j\omega) = \frac{|(w_i(j\omega), e_i)|}{\|w_i(j\omega)\|}, \quad 1 \leqslant i \leqslant m \tag{4.9.31}$$

then for small interactions the numbers in (4.9.31) should all be close to 1. Note that, by (4.9.30) if $|q_i(j\omega)|$ is large for all i at a frequency ω, then

$$H(s) \approx \sum_{i=1}^{m} w_i(s) v_i^{\mathrm{T}}(s) = \mathrm{I}$$

and so we have low interaction. This is always possible for low frequencies and so we can reduce interaction arbitrarily at low frequencies by making $|q_i(j\omega)|$ large. However, at high frequencies we normally have $|q_i(j\omega)| \to 0$ and so we must use the criterion stated above for the angles θ_i in this frequency range.

The basic characteristic locus design method may be summarized as follows:

(a) A real controller K_{h} is chosen so that the so-called misalignment angles defined by (4.9.31) are small at some high frequency ω. A simple

† This definition assumes a suitable ordering imposed on the eigenvectors.

procedure for obtaining the controller K_h is given by Kouvaritakis (1974) and MacFarlane and Kouvaritakis (1977).

(b) The new system $G_1(s) = G(s)K_h$ is now considered and two real matrices $W_r(j\omega_m)$ and $V_r(j\omega_m)$ are found which approximate $W(j\omega_m)$ and $V(j\omega_m)$ at the intermediate frequency ω_m. An intermediate frequency controller $K_m(s)$ is then defined as

$$K_m(s) = W_r(j\omega_m)\,\text{diag}\{k_i(s)\}\,V_r(j\omega_m)$$

where the k_i are chosen to compensate the 'approximately decoupled' single loops.

(c) Finally $G_2(s) = G(s)K_hK_m(s)$ is compensated at low frequencies by considering two further real approximations $W_r(j\omega_l)$ and $V_r(j\omega_l)$ to $W(j\omega_l)$ and $V(j\omega_l)$ and forming the controller

$$K_l(s) = \frac{\alpha}{s}W_r(j\omega_l)\,\text{diag}\{k_i\}\,V_r(j\omega_l) + I$$

where the level α of the integral control is chosen so that the first term is small at ω_m, and the gains k_i alter the magnitudes of the characteristic functions at ω_l.

In this last section of the chapter we have given a very short introduction to multivariable control theory. Many of the details have been omitted, of course, and the reader should consult the literature for a full discussion of the subject, where many examples can be found, and also other design techniques. Because of the complexity of most of the multivariable design methods, they are usually most easily implemented by using a special purpose CAD (computer-aided design) package. Many such packages have been developed and are now widely available. For more on computer controlled systems, see Astrom and Wittenmark (1984).

4.10 EXERCISES

1. Prove the results of table 4.1.

2. Prove (4.2.6) and (4.2.7).

3. Determine the Laplace transform of $t^n\exp(\alpha t)$ and hence prove (4.2.8).

4. Prove (4.2.13) and determine expressions for $c(t)$ for $\zeta \geqslant 1$.

5. Determine the transfer function (from u to y) of the system in fig. 4.41.

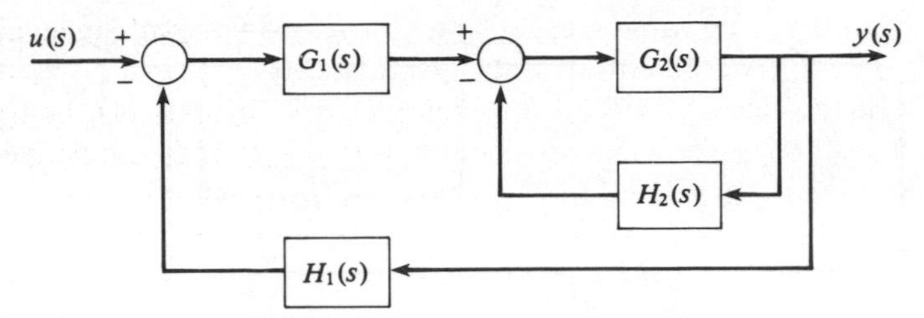

Fig. 4.41 Multiple-loop system.

6. Apply Routh's criterion to the polynomial equation

$$s^5 + s^4 + 2s^3 + 2s^2 + 4s + 1 = 0$$

and determine whether there are any right half plane zeros.

7. Plot Nyquist diagrams for the systems $G(s)H(s)$ given by

(a) $k_a/s(1+T_1s)(1+T_2s)$
(b) $K_a/(1+T_1s)(1+T_2s)(1+T_3s)$
(c) $K_a/(-1+T_1s)$

for various values of K_a, T_1, T_2 and T_3.

8. Plot the Bode diagrams for each of the individual terms in (4.4.1) and (4.4.2), i.e. $\log_{10}K_N$, $\log_{10}|1+j\omega a_i|$, $N\log_{10}|\omega|$, etc., and show that each zero contributes a slope of 20 db/decade while a pole contributes −20 db/decade. Plot the Bode diagrams for the system

$$\frac{(1+s)}{s(1+10s)(1+s/8+s^2/16)}$$

9. Plot the root locus of the system

$$\frac{1}{(s+4)(s+6)(s^2+2s+2)}$$

and find the gain where it crosses the $j\omega$ axis.

10. Consider the system

$$G(s) = \frac{1}{s(s+1)(s+4)}$$

Draw a root locus and find the gain where $\zeta = 0.5$ (i.e. 16 per cent overshoot). Also determine ω_n at this value of ζ and design a lead compensator to increase ω_n to 2. Determine the velocity error constant K_v for the uncompensated system and design a lag compensator to increase this by a factor of 3. Now combine the two compensators, draw a root locus and determine the response characteristics.

11. Prove the results of table 4.2.

12. Determine the z-domain transfer function of the system in fig. 4.42.

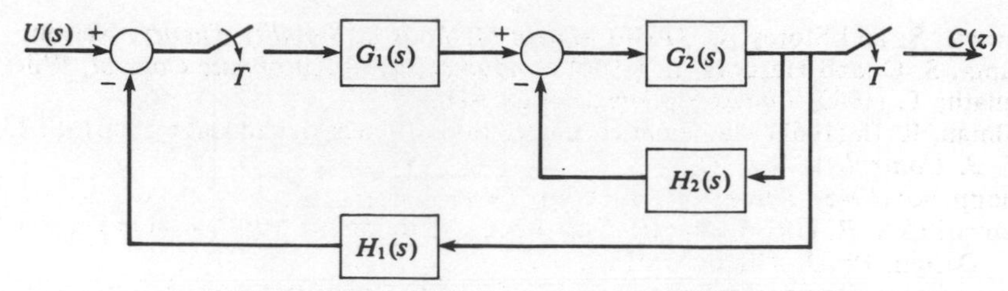

Fig. 4.42 Multiple-loop system with sampling.

13. Prove that the bilinear mapping (4.7.13) maps the interior of the unit disk in the z-plane onto the left half of the w-plane.

14. Using (a) w-plane method and (b) the root locus, design a digital controller for the system $G_h(s)G(s)$, where $G_h(s)$ is a zero-order hold and

$$G(s) = \frac{8}{(s+1)(s+2)}$$

to give a phase margin of about 50° and approximately 25 per cent overshoot.

15. Design a controller for the multivariable system

$$G(s) = \begin{bmatrix} \dfrac{1-s}{(1+s)^2} & \dfrac{2-s}{(1+s)^2} \\[2ex] \dfrac{1-3s}{3(1+s)^2} & \dfrac{1-s}{(1+s)^2} \end{bmatrix}$$

using the inverse Nyquist array (see Rosenbrock, 1969).

16. Use the characteristic locus method to design a controller for the system

$$G(s) = \begin{bmatrix} \dfrac{0.04}{s+0.4} & \dfrac{s}{s^2+0.4s+1\times10^{-4}} \\[2ex] \dfrac{1\times10^{-4}s+1\times10^{-5}}{s^2+0.4s+1\times10^{-4}} & \dfrac{-0.01}{s^2+0.4s+1\times10^{-4}} \end{bmatrix}$$

(See MacFarlane and Belletrutti (1973) for a similar example.)

17. Show how to evaluate $\exp(At)$ by writing A in Jordan canonical form (appendix 1).

18. Prove (4.8.16).

REFERENCES

Astrom, K. and Wittenmark, B. (1984) *Computer Controlled Systems*, Prentice-Hall.

Banks, S. P. (1983) *State Space and Frequency Domain Methods in the Control of Distributed Parameter Systems*, Peter Peregrinus.

Banks, S. P. and Abbasi-Ghelmansarai, F. (1983) Realisation theory and the infinite dimensional root locus, *Int. J. Control*, 38, 589–606.

Barnett, S. and Storey, C. (1970) *Matrix Methods in Stability Theory*, Nelson.
Gupta, S. C. and Hasdorf, L. (1970) *Fundamentals of Automatic Control*, Wiley.
Kailath, T. (1980) *Linear Systems*, Prentice-Hall.
Kalman, R. E. (1963) Mathematical description of linear dynamical systems, *SIAM J. Control*, 1, 152–92.
Knopp, K. (1945) *Theory of Functions*, Dover.
Kouvaritakis, B. (1974) Characteristic Locus Methods for Multivariable Feedback Design, Ph.D. Thesis, UMIST.
Kouvaritakis, B and MacFarlane, A. G. J. (1976) Geometric approach to analysis and synthesis of system zeros, *Int. J. Control*, 23, 149–81.
Lee, E. B. and Markus, L. (1967) *Foundations of Optimal Control Theory*, Wiley.
Limebeer, G. (1982). The application of a generalized diagonal dominance to linear system stability theory, *Int. J. Control*, *36*, 135.
Luenberger, D. G. (1966) Observers for multivariable systems, *IEEE Trans. Aut. Control*, 11, 190–7.
MacFarlane, A. G. J. and Belletrutti, J. J. (1973) The characteristic locus design method, *Automatica*, 9, 575–88.
MacFarlane, A. G. J. and Karcanias, N. (1976), Poles and zeros of linear multivariable systems: a survey of the algebraic, geometric and complex-variable theory, *Int. J. Control*, 24, 33.
MacFarlane, A. G. J. and Kouvaritakis, B. (1977) A design technique for linear multivariable feedback systems, *Int. J. Control*, 25, 837.
MacFarlane, A. G. J. and Postlethwaite, I. (1977) The generalised Nyquist stability criterion and multivariable root loci, *Int. J. Control*, 25, 81–127.
MacFarlane, A. G. J. and Postlethwaite, I. (1979) *Complex Variable Methods for Linear Multivariable Systems*, Springer-Verlag.
Mees, A. I. (1981) Optimal scaling of linear systems, *Systems and Control Letters*, *1*, 155.
Munro, N. (1971) Minimal realisation of transfer-function matrices using the system matrix, *Proc. IEE*, 118, 1298–301.
O'Reilly, J. (1984) *Observers for Linear Systems*, Academic Press.
Ostrowski, A. M. (1952) Note on bounds for determinants with dominant principal diagonal, *Proc. Am. Math. Soc.*, 3, 26–30.
Owens, D. H. (1985) *Multivariable Root Locus: Compensation and Computation Theory*, Research Studies Press, to be published.
Patel, R. V. and Munro, N. (1982) *Multivariable System Theory and Design*, Pergamon Press.
Rosenbrock, H. H. (1969) Design of multivariable control systems using the inverse Nyquist array, *Proc. IEE*, 116, 1929.
Rosenbrock, H. H. (1970) *State Space and Multivariable Theory*, Nelson.
Rosenbrock, H. H. (1974) *Computer Aided Control System Design*, Academic Press.
Schwartz, L. (1950) *Théorie des Distributions*, Hermann.
Shinners, S. M. (1978) *Modern Control Systems Theory and Application*, Addison-Wesley.
Takahashi, Y., Rabins, M. J. and Auslander, D. M. (1972), *Control and Dynamic Systems*, Addison-Wesley.
Widder, D. V. (1972) *The Laplace Transform*, Princeton University Press.
Willems, J. L. (1970) *Stability Theory of Dynamical Systems*, Nelson.
Wolovich, W. A. (1974) *Linear Multivariable Systems*, Vol. 11, *Applied Mathematical Sciences*, Springer-Verlag.
Wonham, W. M. (1979) *Linear Multivariable Control: A Geometric Approach*, 2nd edn., Vol. 10, *Applications of Maths. Series*, Springer-Verlag.

5 NON-LINEAR SYSTEMS

5.1 INTRODUCTION

The theory of linear systems which has been presented in chapter 4 applies equally well to any linear system of the form

$$\dot{x} = Ax + Bu$$
$$y = Cx$$

In direct contrast to this it is found that many types of non-linear systems do not fall into a general class and each may have to be treated individually. Only when one can say that the non-linearity is 'bounded' by a linear system, in some sense, can one consider whole classes of non-linearities together, and then the methods of analysis resemble the linear techniques. For this reason, there is no 'non-linear theory' which will apply to any non-linear system and up to now research has concentrated on discovering methods for studying particular aspects of as broad a class of non-linear systems as possible. There is always a trade-off between the generality of the theory and the preciseness of the results – in fact, the more types of non-linearities we try to include in our theory, the more conservative the results. Some non-linear systems are so strange that it appears that the only way to treat them is by methods specially developed for these systems. In section 3.9. exercise 10 we have encountered such an example which has been given the name of the 'strange attractor' and has been used in models of (deterministic) chaos.

In spite of the preceding remarks, significant advances have been made in the study of non-linear systems, particularly in the areas of stability theory, two-dimensional systems, local structures and the theory of oscillations. It is these aspects of such systems which we shall consider in this chapter. The importance of a thorough study of linear systems is brought out clearly by the fact that many non-linear systems may be studied locally, usually in a neighbourhood of an equilibrium point,† by linearizing the

† Precise definitions of terms used in this introduction will be given in the following sections.

system about such a point. This leads *inter alia* to the phase plane method which we shall study in detail below. Most of the synthesis methods for designing non-linear control systems are based on stability studies of the system and so we shall present the most useful results in this area, together with the approximation methods for the estimation of limit cycles.

A particular type of control system which arises in practice and which has received considerable attention recently is the bilinear system of the form

$$\dot{x} = Ax + Bxu$$

and we shall end this chapter with an introduction to systems of this type.

5.2 LINEARIZATION OF NON-LINEAR SYSTEMS

5.2.1 Ordinary Differential Equations

An important method for the study of non-linear systems, certainly for an initial attack on the problem, is the linearization of the system about some 'operating point'. We begin by giving a precise definition of the latter concept.

Definition 5.2.1

Consider the differential equation

$$\dot{x} = f(x, t), \quad x(t_0) = x_0 \tag{5.2.1}$$

where f is defined on some open subset† $\Omega \times I$ of $\mathbb{R}^n \times \mathbb{R}$. An **isolated equilibrium point** $x_e \in \mathbb{R}^n$ of (5.2.1) is a point such that

(a) $f(x_e, t) = 0$ for all $t \in I$

(b) $f(x, t) \neq 0$ for all $t \in I$ and all $x \neq x_e$ in some neighbourhood of x

Since x_e is independent of t, we may write

$$\begin{aligned}\frac{dx}{dt} &= \frac{d}{dt}(x - x_e)\\ &= f((x - x_e) + x_e, t)\\ &\triangleq g(x - x_e, t)\end{aligned}$$

---

† $\Omega \times I$ is assumed to be open to avoid difficulties with the existence theory of differential equations.

for some function g. Hence

$$\frac{dy}{dt} = g(y, t), \quad y \triangleq x - x_e \tag{5.2.2}$$

and $y = 0$ is an equilibrium point of (5.2.2). We can therefore always regard the origin as the equilibrium point by a simple change of coordinates, and we shall assume that this has been done in the following discussion. Moreover, we shall consider only the case of autonomous systems, where f is independent of explicit mention of t.

Let $x = 0$ be an isolated equilibrium point of the system

$$\dot{x} = f(x) \tag{5.2.3}$$

and assume that f is a (real) analytic function, i.e. f has a convergent Taylor series expansion about $x = 0$. Then, by Taylor's theorem, we may write

$$f(x) = f(0) + J_0 x + o(\|x\|^2)$$

where

$$J_a = \left.\frac{\partial f_i(x)}{\partial x_j}\right|_{x=a}$$

is the Jacobian matrix of f. Hence, for small x, we can approximate (5.2.3) in a neighbourhood of 0 by the linear equation

$$\boxed{\dot{x} = Ax = J_0 x} \tag{5.2.4}$$

Of course, we require $|J_0| \neq 0$ in order that the linear term in the expansion of f dominates and thus the system (5.2.4) is a reasonable local representation of the non-linear system (5.2.3) close to $x = 0$. The following examples illustrate the problems when $|J_0| = 0$.

Example 5.2.1
The scalar system

$$\dot{x} = x^2$$

has a unique equilibrium point at $x = 0$ and $f(x) = x^2$ is real analytic but the system has the linearization

$$\dot{x} = 0$$

which does not have an isolated equilibrium at 0.

Example 5.2.2
The system

$$\dot{x}_1 = x_1$$
$$\dot{x}_2 = x_2^2$$

has an isolated equilibrium at (0, 0), but the linearization

$$\dot{x}_1 = x_1$$
$$\dot{x}_2 = 0$$

now has the line of equilibria $\{(0, x_2): x_2 \in \mathbb{R}\}$.

5.2.2 Partial Differential Equations

The linearization of a non-linear system may be used in order to study the local structure of dynamical systems or to control non-linear systems in the neighbourhood of an equilibrium point. Before going on to study the former of these topics for two-dimensional systems we should mention that non-linear partial differential equations may be linearized in a similar way and so we may also investigate the behaviour of non-linear distributed systems in a 'neighbourhood' of an equilibrium point. The mathematical technicalities involved are, however, beyond the scope of this book and so we shall present a heuristic approach to this problem. It should be noted that the discussion to follow can be made precise but requires familiarity with infinite dimensional vector spaces. Consider the partial differential equation

$$\boxed{\frac{\partial \phi}{\partial t} = f(\phi, \phi_x, \phi_{xx}, \phi_y, \phi_{yy})} \qquad (5.2.5)$$

where $f: \mathbb{R}^5 \to \mathbb{R}$ is a given function. An **equilibrium point** or **equilibrium solution** of (5.2.5) is a solution ϕ of the equation

$$\boxed{f(\phi, \phi_x, \phi_{xx}, \phi_y, \phi_{yy}) = 0} \qquad (5.2.6)$$

with the same boundary conditions as those imposed on (5.2.5). Hence an equilibrium point of a partial differential equation is a solution of a non-linear stationary partial differential equation. If ϕ^e is an 'isolated' solution of (5.2.6) (this is where we need topological notions of closeness of functions, but we shall leave the term 'isolated' as an undefined intuitive concept) then it can be shown that we may write

$$f(\phi, \phi_x, \phi_{xx}, \phi_y, \phi_{yy}) = \left(\frac{\partial f}{\partial \phi}\right)_e \phi + \left(\frac{\partial f}{\partial \phi_x}\right)_e \phi_x + \left(\frac{\partial f}{\partial \phi_{xx}}\right)_e \phi_{xx}$$
$$+ \left(\frac{\partial f}{\partial \phi_y}\right)_e \phi_y + \left(\frac{\partial f}{\partial \phi_{yy}}\right)_e \phi_{yy} + \text{h.t.}$$

where h.t. denotes higher order terms and $(\cdot)_e$ indicates that the expression in parentheses is to be evaluated at the equilibrium value ϕ_e. (These terms are therefore just functions of x and y.) We therefore finally obtain the linearization of (5.2.5) specified by the equation

$$\boxed{\frac{\partial\phi}{\partial t}=\left(\frac{\partial f}{\partial\phi}\right)_{\mathrm{e}}\phi+\left(\frac{\partial f}{\partial\phi_x}\right)_{\mathrm{e}}\phi_x+\left(\frac{\partial f}{\partial\phi_{xx}}\right)_{\mathrm{e}}\phi_{xx}+\left(\frac{\partial f}{\partial\phi_y}\right)_{\mathrm{e}}\phi_y+\left(\frac{\partial f}{\partial\phi_{yy}}\right)_{\mathrm{e}}\phi_{yy}} \tag{5.2.7}$$

valid in a neighbourhood of ϕ^{e}. However, we must emphasize again that the notion of neighbourhoods of ϕ^{e} may be given several interpretations depending on which topological structure we place on the space of solutions of (5.2.5). For simplicity we may think of two functions ϕ_1, ϕ_2 defined on a region Ω as being 'close' if

$$\sup_{x\in\Omega}|\phi_1(x)-\phi_2(x)| \tag{5.2.8}$$

is small, although this is by no means necessary.

Example 5.2.3

As a simple example of linearization of distributed systems, consider the equation

$$\frac{\partial\phi}{\partial t}=\frac{\partial^2\phi}{\partial x^2}(\phi+1) \tag{5.2.9}$$

defined on the interval $[0, 1]$, with initial condition $\phi(x,0)=\phi_0(x)=0$. Then the equilibrium solutions are given by

$$\frac{\mathrm{d}^2\phi}{\mathrm{d}x^2}=0, \quad \text{or } \phi=-1$$

that is

$$\phi(x)=ax+b, \quad \phi(0)=\phi(1)=0$$

and so $\phi(x)=0$. However, the equilibrium solution must satisfy the boundary conditions and so we must rule out the possibilty $\phi=-1$ and the system has a unique equilibrium solution $\phi=0$. Now by (5.2.7) we have

$$\begin{aligned}\frac{\partial\phi}{\partial t}&=\left.\frac{\partial^2\phi}{\partial x^2}\right|_{\phi=0}\phi+(\phi+1)\bigg|_{\phi=0}\frac{\partial^2\phi}{\partial x^2}\\&=\frac{\partial^2\phi}{\partial x^2}\quad\text{at } \phi=0,\ \phi(x,0)=\phi_0(x)\end{aligned} \tag{5.2.10}$$

Hence the system (5.2.9) has a linearization at $\phi=0$ which is just the linear heat equation. Note, however, that $\phi_0(x)$ must be small in the sense of (5.2.8), for example.

5.3 PHASE PLANE ANALYSIS AND CLASSIFICATION OF LINEAR SYSTEMS

5.3.1 The Phase Plane

The study of second-order systems is important for two reasons; firstly, they occur frequently in practical systems and, secondly, their qualitative structures may be obtained from a two-dimensional graphical representation in the **phase plane.** To introduce the phase plane, consider the second-order system

$$\ddot{x} = f(x, \dot{x}) \tag{5.3.1}$$

and define

$$\left.\begin{aligned} x_1 &= x \\ x_2 &= \dot{x} \end{aligned}\right\} \tag{5.3.2}$$

where we may think of x_2 as the 'velocity'.

If we knew the solution of (5.3.1) with initial conditions $x(0) = x_0$, $\dot{x}(0) = v_0$ then we could plot graphs of x and $\dot{x}$ against time t. In general, however, the solutions of (5.3.1) will not be known and so we introduce the transformation (5.3.2) and then we have the equivalent system

$$\left.\begin{aligned} \dot{x}_1 &= x_2 \\ \dot{x}_2 &= f(x_1, x_2) \end{aligned}\right\} \tag{5.3.3}$$

The map

$$(x_1, x_2) \to (x_2, f(x_1, x_2)) \tag{5.3.4}$$

is called a **vector field** and assigns to each point $(x_1, x_2) \in \mathbb{R}^2$ the vector $(x_2, f(x_1, x_2))$. (Note that we are regarding elements of $\mathbb{R}^2$ in two ways, as points and also as vectors.) A solution of (5.3.3) through a given point (x_1^0, x_2^0) is just a function $t \to (x_1(t), x_2(t))$ and we may plot such a solution in (x_1, x_2) space as a curve parameterized by t. The two-dimensional plane (x_1, x_2) consisting of these parameterized solutions is called the phase plane. We shall denote a particular solution trajectory in the phase plane by the generic symbol λ, and regard λ either as the map $t \to (x_1(t), x_2(t))$ or as the set of points in the (x_1, x_2) plane which are the values of this map.

The fundamental property of the phase plane is that if we plot the vector field (5.3.4) in the plane by drawing the vector $(x_2, f(x_1, x_2))$ based at (x_1, x_2) (cf. fig. 5.1), then this vector is tangent to the solution trajectory λ through (x_1, x_2). This follows immediately from (5.3.3) by finding, dx_2/dx_1, and gives a practical method of constructing phase plane trajectories for the system (5.3.3.), as shown in fig. 5.2. We plot a large number of the vectors $(x_2, f(x_1, x_2))$ at various points and draw smooth curves tangent to these vectors.

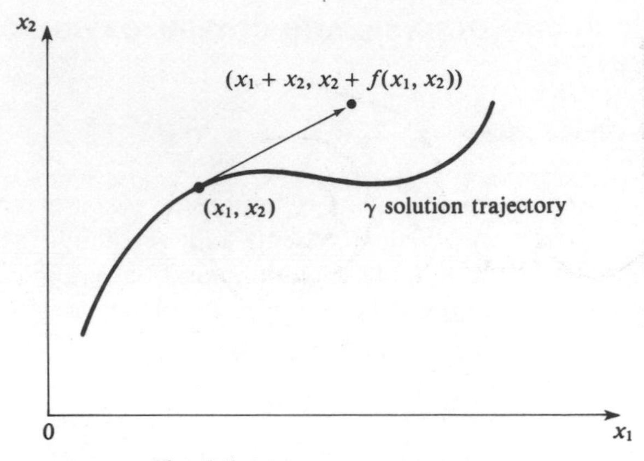

Fig. 5.1 A tangent vector.

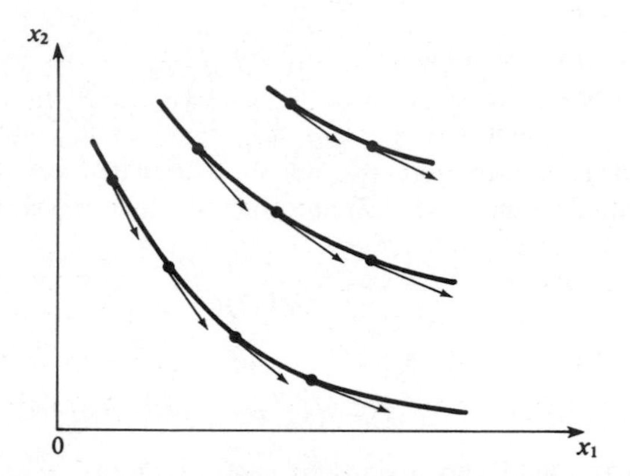

Fig. 5.2 Vector field defining solution trajectories.

Consider now the various forms which a trajectory γ may assume. An equilibrium point of (5.3.3) is a point where the vector $(x_2, f(x_1, x_2))$ equals the zero vector (0, 0) and is often called a **singularity** of the vector field. Since at such a point (5.3.3) reduces to the equation

$$\dot{x}_1 = 0, \quad \dot{x}_2 = 0$$

it follows that x_1 is constant, x_2 is constant and these constants are just the initial values assigned to x_1 and x_2. Hence, in the phase plane an equilibrium point gives rise to a stationary trajectory at a fixed point, as indicated in fig. 5.3, by trajectories of the form γ^0. A trajectory may start at an equilibrium point and continue unboundedly as exemplified by γ^1 or may join two equilibria γ^0 which is illustrated by γ^2. Finally, a trajectory γ^3 may not contain an equilibrium point, but there may be smallest positive

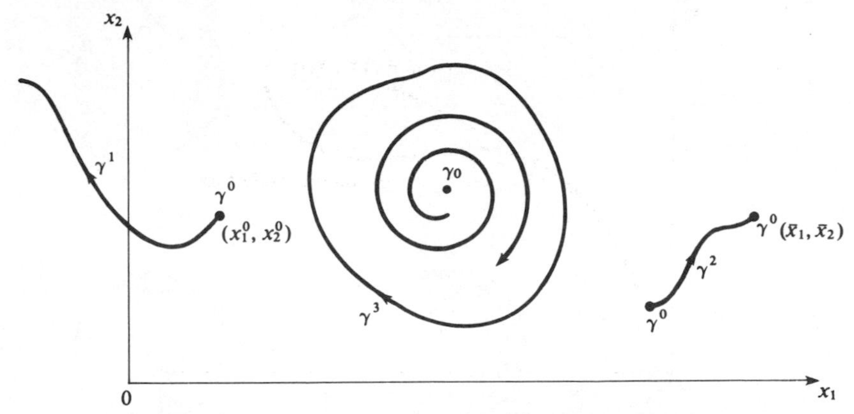

Fig. 5.3 Global structures of a two-dimensional dynamical system.

time p such that†

$$\boxed{\gamma^3(t+p) = \gamma^3(t) \quad \text{for all } t \geqslant 0}$$

and we obtain the important concept of **limit cycle** (see fig. 5.3). Limit cycles are very important in non-linear systems theory since they correspond to (non-linear) oscillations in the system and have been studied extensively. In order to discuss the nature of limit cycles in more detail, we must introduce the concept of the ω-limit set $\omega(x_0)$ of a point x_0. If $\gamma_{x_0}(t)$ is the trajectory of the equation (5.3.3) through x_0, then we define

$$\omega(x_0) = \{x \in \mathbb{R}^2 : \gamma_{x_0}(t_n) \to x \text{ for some sequence } t_n \to \infty\}$$

Intuitively, this states that x is in the ω-limit set of x_0 if the trajectory through x_0 passes close to x for some arbitrarily large times. The basic theorem concerning two-dimensional limit cycles can best be stated for equations defined on a sphere. Any system defined in the phase plane can be converted into a system defined on a sphere by the following topological trick. Consider a plane sheet of some material which we shall assume may be distorted in any way without tearing the sheet. Then we turn up the sides of the sheet and shorten their lengths as in fig. 5.4 (a) and (b), and continue as in fig. 5.4 (c) and (d) until we are left with a sphere with a small hole. Finally, if we shrink the diameter of this hole to 0 and add an extra point (fig. 5.4(e)) then we obtain an ordinary two-dimensional sphere. Note that in this process we have lost the precise notion of distance in the plane, but we still have a qualitative sense of nearness.

† i.e. the trajectory γ^3 is closed (or periodic) and returns to any starting point on it after p seconds.

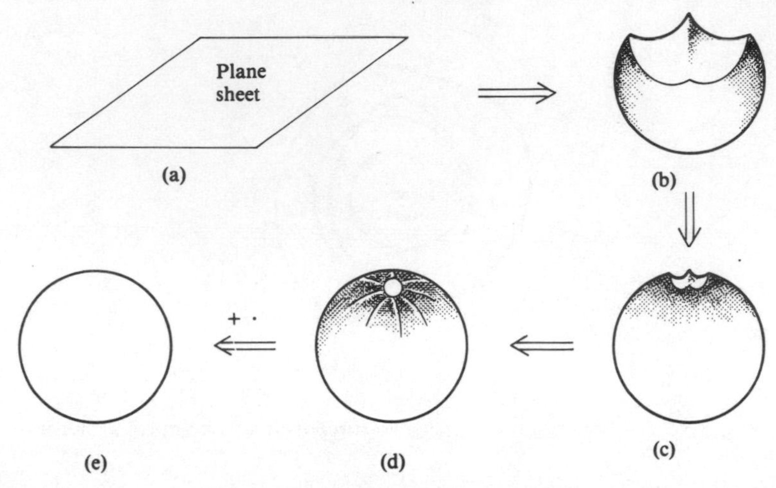

Fig. 5.4 'Compactifying' the plane.

The first point to note now is that it can be shown (from topological arguments) that any vector field defined on a two-sphere must have at least one singular point (i.e. equilibrium point of the corresponding differential equation). Hence any system in the phase plane which has no (finite) singularity must have one 'at infinity'. We may now characterize the ω-limit sets of points on a sphere S on which is defined a vector field (see Coddington and Levinson, 1955).

Theorem 5.3.1 (Poincaré–Bendixson)
The ω-limit set of a point x_0 on S may consist of any of the following sets:

(a) a singularity
(b) a limit cycle
(c) a closed curve made up of singularities and complete trajectories.

We have already seen examples of (a) and (b) in fig. 5.3; in particular, the trajectory γ^1 has the ω-limit set consisting of the point at infinity, γ^2 has the ω-limit set consisting of the point $(\bar{x}_1, \bar{x}_2)$ and γ^3 is a limit cycle. An example of the situation in (c) is shown in fig. 5.5, where x_1, x_2, x_3 are singularities and $\gamma_1, \gamma_2, \gamma_3$ are complete trajectories. A simple consequence of theorem 5.3.1 is the following.

Corollary 5.3.1
If a vector field defined on a sphere has only two singularities, both of which are unstable† (i.e. all trajectories leave the singularities), then there must exist at least one limit cycle.

---

† Precise definitions of stability will be given later.

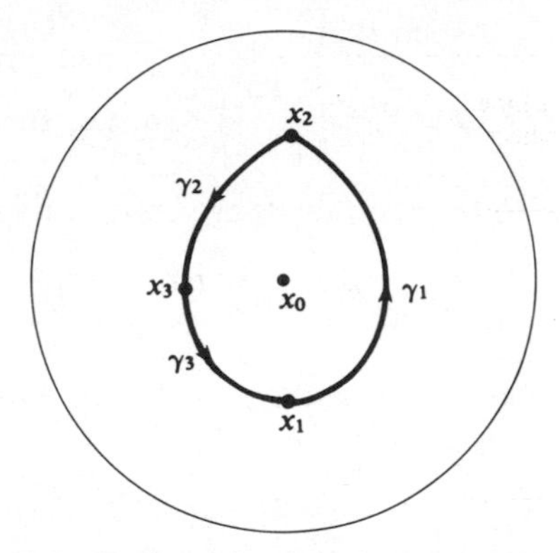

Fig. 5.5 Trajectories forming a closed curve.

Before giving an example of this type of system, we must first consider the local behaviour of singular points, which is the object of the next section.

5.3.2 Local Structure of Linear Systems

We have shown above that any non-linear system defined by an analytic function may be represented in a neighbourhood of an equilibrium point by a linear system. It follows that we may characterize the local behaviour of such a system entirely in terms of linear theory. In fact, if

$$\dot{x} = \begin{pmatrix} \dot{x}_1 \\ \dot{x}_2 \end{pmatrix} = Ax \tag{5.3.5}$$

is a two-dimensional linear system, we shall show that the local qualitative behaviour of (5.3.5) is determined entirely by the eigenvalues of A, which we denote by λ_1, λ_2. These numbers may, of course, be complex. First we recall the canonical structure theorem for two-dimensional matrices (cf. appendix 1).

Theorem 5.3.2
If A is a 2×2 real matrix, then we have the following possibilities:

(a) λ_1, λ_2 are real and A is diagonalizable. Then there exists a real non-singular matrix D such that

$$D^{-1}AD = \Lambda_D \triangleq \begin{pmatrix} \lambda_1 & 0 \\ 0 & \lambda_2 \end{pmatrix}$$

(Of course, λ_1 may equal λ_2).

(b) $\lambda_1 = \lambda_2 = \lambda$ (say) and A is not diagonalizable. Then there exists a real non-singular matrix J such that

$$J^{-1}AJ = \Lambda_J \triangleq \begin{pmatrix} \lambda & 0 \\ 1 & \lambda \end{pmatrix} \quad \text{(Jordan form)}$$

(c) $\lambda_1 = \bar{\lambda}_2$. If $\lambda_1 = a + ib$ $(b \neq 0)$ then there exists a real matrix R such that

$$R^{-1}AR = \Lambda_R \triangleq \begin{pmatrix} a & -b \\ b & a \end{pmatrix}$$

Using this theorem we can transform equation (5.3.5) into a simpler system. In fact if we write (in case (a))

$$y_D = D^{-1}x \tag{5.3.6}$$

then we have

$$\dot{y}_D = D^{-1}\dot{x} = D^{-1}Ax = D^{-1}ADy_D = \Lambda_D y_D \tag{5.3.7}$$

Similarly in cases (b) and (c), we obtain

$$\dot{y}_J = \Lambda_J y_J, \quad \dot{y}_R = \Lambda_R y_R \tag{5.3.8}$$

where $y_J = J^{-1}x$, $y_R = R^{-1}x$.

The solutions of the equations (5.3.7) and (5.3.8) are easy to find; in fact, we have

(a)
$$y_D(t) = \begin{pmatrix} \exp(\lambda_1 t) & 0 \\ 0 & \exp(\lambda_2 t) \end{pmatrix} y_D(0) \tag{5.3.9}$$

(b)
$$y_J(t) = \begin{pmatrix} \exp(\lambda t) & 0 \\ t\exp(\lambda t) & \exp(\lambda t) \end{pmatrix} y_J(0) \tag{5.3.10}$$

(c)
$$y_R(t) = \exp(at) \begin{pmatrix} \cos bt & -\sin bt \\ \sin bt & \cos bt \end{pmatrix} y_R(0) \tag{5.3.11}$$

We may now classify the local structure of two-dimensional systems by plotting the phase plane trajectories of the systems (5.3.7) and (5.3.8) for the different types of eigenvalues which may occur. These phase plane trajectories are shown in fig. 5.6 in the two-dimensional y_D, y_J or y_R space; however, for simplicity of notation we have omitted the subscripts D, J and R. We can describe these pictures as follows:

CASE 1: The eigenvalues of A are real and of opposite signs. Then by (5.3.9) we obtain the phase portrait of a **saddle** where $\lambda_1 < 0 < \lambda_2$.

CASE 2: The eigenvalues of A have negative real parts. These systems are **sinks**.

(a) If the eigenvalues of A are real and A is diagonalizable, then by (5.3.9) we obtain a **focus** if $\lambda_1 = \lambda_2 < 0$ or more generally a **node** if $\lambda_1 < \lambda_2 < 0$.

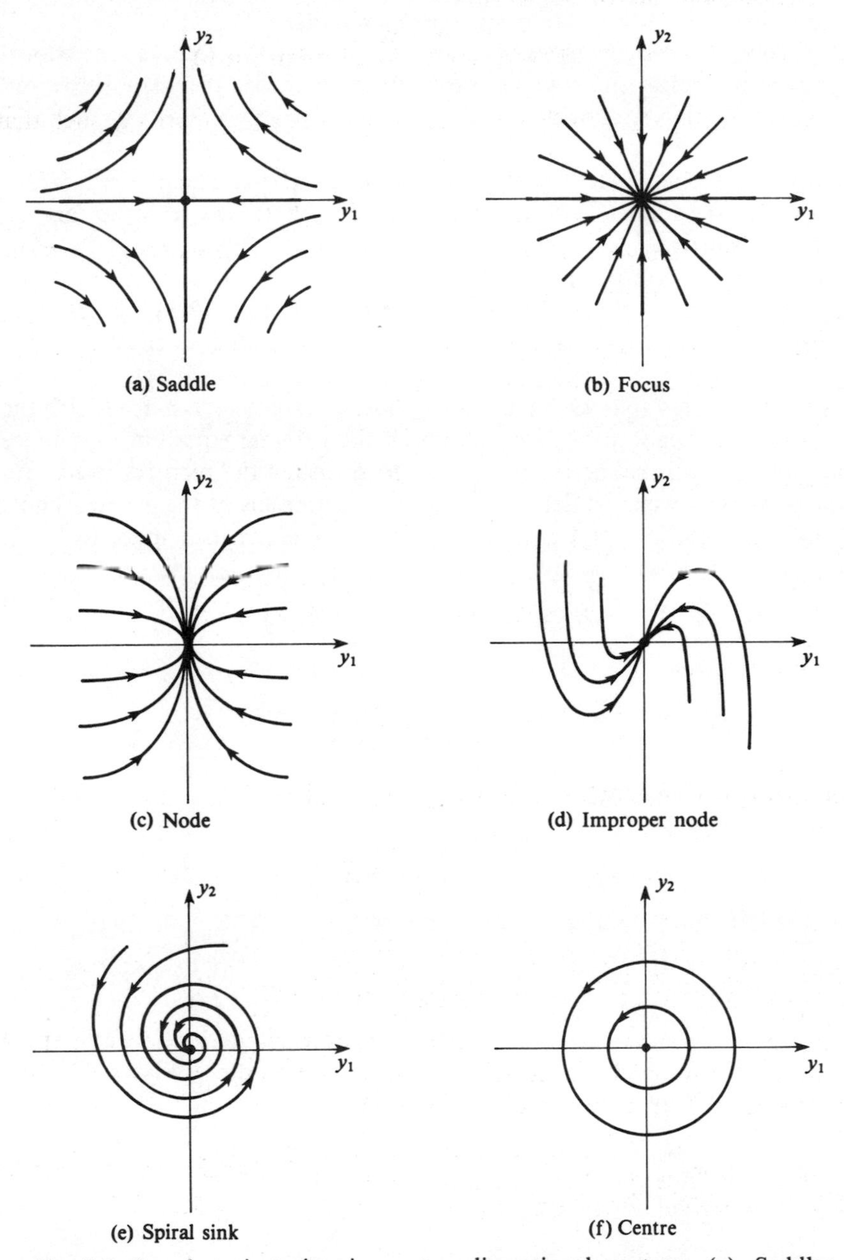

Fig. 5.6 Local trajectories in a two-dimensional system: (a) Saddle; (b) focus; (c) node; (d) improper node; (e) spiral sink; (f) centre.

(b) If A is not diagonalizable, then A must have a single real eigenvalue λ and by (5.3.10) we obtain an **improper node.**

(c) Finally, if the eigenvalues of A are complex then by (5.3.11) the trajectories are spiral sinks which are counter-clockwise if b is positive and clockwise if b is negative.

CASE 3: The eigenvalues of A have positive real parts. The phase portraits in this case are similar to those in case 2 with the arrows reversed, and so we obtain **sources.**

CASE 4: The eigenvalues of A are pure imaginary. By (5.3.11), with $a = 0$, we obtain concentric circular trajectories. This is called a **centre.**

It should be noted that we have drawn phase portraits in y-space where the A matrix has a canonical form, although the original equation in x-space may not have this property. If we wish to translate the pictures in fig. 5.6 back to x-space we must determine the y-axes in terms of the x-axes. Thus, for example, (5.3.6) implies that the y-axes are given in x-space by

$$\left.\begin{aligned} \bar{d}_{21}x_1 + \bar{d}_{22}x_2 = 0 \quad (y_1\text{-axis}) \\ \bar{d}_{11}x_1 + \bar{d}_{12}x_2 = 0 \quad (y_2\text{-axis}) \end{aligned}\right\} \tag{5.3.12}$$

where

$$D^{-1} = \begin{pmatrix} \bar{d}_{11} & \bar{d}_{12} \\ \bar{d}_{21} & \bar{d}_{22} \end{pmatrix}$$

For example, consider the system

$$\begin{aligned} \dot{x}_1 &= 5x_1 + 3x_2 \\ \dot{x}_2 &= -6x_1 - 4x_2 \end{aligned}$$

Then it is easy to see that we may choose

$$D = \begin{pmatrix} 1 & 1 \\ -1 & -2 \end{pmatrix}$$

and then

$$\dot{y}_D = \begin{pmatrix} 2 & 0 \\ 0 & -1 \end{pmatrix} y_D$$

which is a saddle (fig. 5.6(a)) and (5.3.12) becomes

$$\begin{aligned} -x_1 - x_2 &= 0 \quad (y_1\text{-axis}) \\ 2x_1 + x_2 &= 0 \quad (y_2\text{-axis}) \end{aligned}$$

Hence we obtain the phase portrait in x-space shown in fig. 5.7.

The nature of the eigenvalues of the matrix A may be visualized more clearly in the following way. First note that the characteristic polynomial

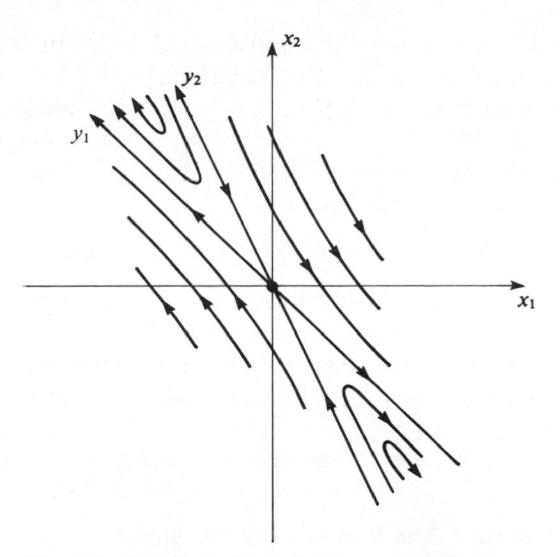

Fig. 5.7 Saddle trajectories in y-coordinates.

of A is

$$\lambda^2 - (\text{tr } A)\lambda + \det A$$

where tr A is the sum of the diagonal elements of A. The **discriminant** Δ is defined by

$$\Delta = (\text{tr } A)^2 - 4 \det A$$

and the eigenvalues of A are

$$\tfrac{1}{2}(\text{tr } A \pm \sqrt{\Delta})$$

Hence, in terms of tr A and det A the phase portraits in fig. 5.6 may be classified as follows:

(a) Saddle. Eigenvalues real and of opposite sign, so that $\Delta > 0$ and $\sqrt{\Delta} > \text{tr } A$, i.e. $\det A < 0$.
(b) Focus or improper node. Eigenvalues are equal, i.e. $\Delta = 0$.
(c) Node. Eigenvalues are real and of same sign, so that $\Delta > 0$ and $\det A > 0$.
(d) Same as for (b).
(e) Spirals. Eigenvalues are imaginary, i.e. $\Delta < 0$.
(f) Centres. Eigenvalues are pure imaginary, so that $\Delta < 0$ and $\text{tr } A = 0$.

If we regard det A and tr A as orthogonal coordinates (as A varies) then we obtain the diagram in fig. 5.8. Note that the equation $\Delta = 0$ is given by the parabola

$$(\text{tr A})^2 = 4 \det \text{A}$$

in these coordinates.

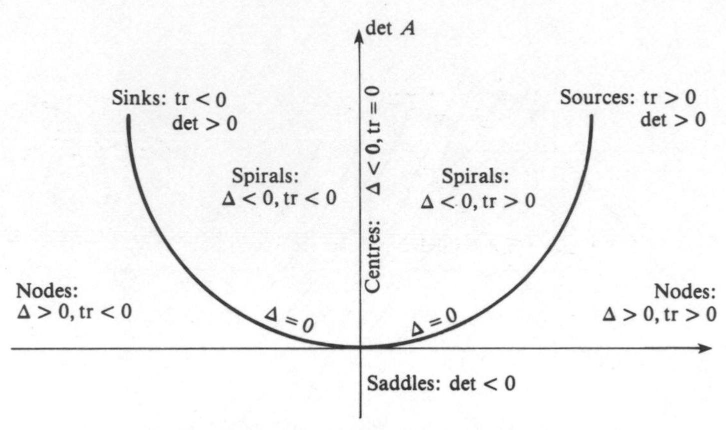

Fig. 5.8 Classifying local structures.

5.3.3 An Example – The Van der Pol Oscillator

We shall now show how the above theory may be used to obtain a qualitative picture of the solution trajectories of the Van der Pol oscillator defined by the equations

$$\left.\begin{aligned}\frac{dx_1}{dt} &= x_2 - x_1^3 + x_1\\ \frac{dx_2}{dt} &= -x_1\end{aligned}\right\} \tag{5.3.13}$$

We shall apply corollary 5.3.1 and so we must consider the equations (5.3.13) to be defined on a sphere, which as we have just stated above may be regarded as the finite plane with an extra point added. In order to study the structure of (5.3.13) 'at ∞' we consider the parallelogram $ABCD$ in fig. 5.9(a) where BC is part of the line $x_2 = x_1 - a$ for large a. We shall show that if a is large enough, all trajectories point into this region. On the line AB we have $\dot{x}_2 = -x_1 < 0$, and so the trajectory certainly points downwards on AB. However, by considering the graph of $-x_1^3 + x_1$ it is easy to see that, for large enough a, $\dot{x}_1$ is positive at A and negative at B with a single turning point between. Next, on the line BC we have

$$x_2 = x_1 - a$$

and so

$$\begin{aligned}\dot{x}_1 &= 2x_1 - x_1^3 - a, \quad x_1 \geqslant -1\\ \dot{x}_2 &= -x_1\end{aligned}$$

Again, considering the graph of $2x_1 - x_1^3 - a$ for large $a > 0$ we see that x_1 is always negative on BC. For $x_1 < 0$, $\dot{x}_2$ points upwards and so on

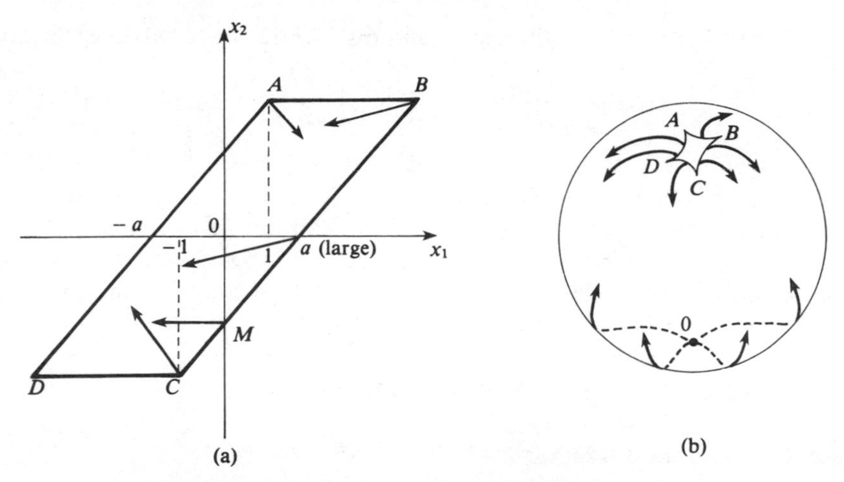

Fig. 5.9 The Van der Pol trajectories 'near ∞'.

CM the trajectories point into the parallelogram $ABCD$. On BM, $\dot{x}_2$ points downwards. However,

$$|2x_1 - x_1^3 - a| > x_1, \quad x_1 > 0$$

for large a and so again the trajectories point inwards on the whole of BC. A similar argument applies on CD and DA and so we have proved the above assertion.

If we now map the plane onto the sphere shown in fig. 5.9(b), the parallelogram $ABCD$ maps to a small hole about ∞ on the sphere from which all trajectories emanate if a is large. Now although the vector field defined by the equations (5.3.13) becomes infinitely large at ∞, it is clear that we can replace the original equation with one having the same qualitative (topological) behaviour at ∞ and a finite vector field, i.e. one with an unstable equilibrium at ∞ from which all trajectories radiate. (This statement is intuitively clear, but requires differential geometry to prove it rigorously.) By letting $a \to \infty$ it follows that we may regard ∞ as a source.

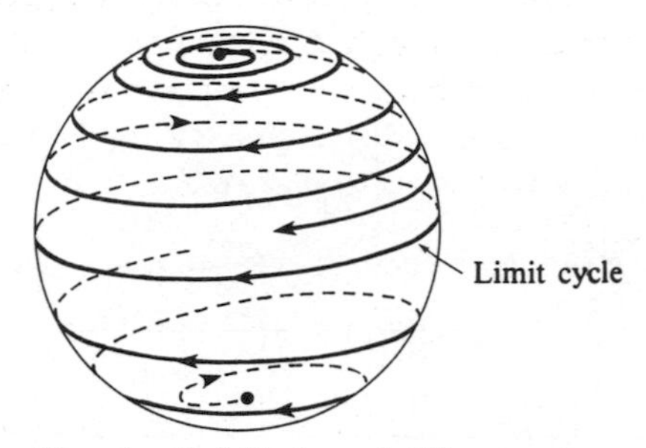

Fig. 5.10 The Van der Pol limit cycle illustrated on the sphere.

Moreover, at the origin, the equations (5.3.13) have the linearization

$$\frac{\mathrm{d}}{\mathrm{d}t}\begin{pmatrix} x_1 \\ x_2 \end{pmatrix} = \begin{pmatrix} 1 & 1 \\ -1 & 0 \end{pmatrix}\begin{pmatrix} x_1 \\ x_2 \end{pmatrix}$$

which again represents a source. Since (0, 0) and ∞ are the only equilibria of the (modified) system, corollary 5.3.1 now implies the existence of a limit cycle in the plane. The trajectories of the Van der Pol oscillator on the sphere can then be seen to be as shown in fig. 5.10. Note that a quantitative picture of the limit cycle may be obtained from the simulation in chapter 3.

5.4 NON-LINEAR SYSTEM STABILITY

5.4.1 Lyapunov Stability†

In the previous sections of this chapter and in the last chapter we have used the concept of stability in a fairly intuitive way. When studying the asymptotic behaviour of non-linear systems, one must be much more precise and so we shall discuss in detail the theory of stability of non-linear systems in this section. This theory is now very extensive and so only a small part of it can be given here; however, the ideas which we present should be representative of most aspects of the theory.

We shall begin by assuming that a non-linear system

$$\dot{x} = f(x, t) \tag{5.4.1}$$

is given and, for simplicity, we assume that this system has solutions defined for all time t and all initial states $x_0 \in \mathbb{R}^n$, specified at time t_0. Such a solution is denoted by $x(t)$ or $x(t;\, x_0,\, t_0)$ in order to show the explicit dependence on (x_0, t_0). Intuitively, for a stability study of the system, we wish to examine the effects of a small change in the initial condition x_0. Hence we consider a reference solution $x_r(t) = x(t; x_r, t_0)$, where $x_r \in \mathbb{R}^n$ is fixed, and examine the relative behaviour of the solution $x(t;\, x_0,\, t_0)$ when x_0 is close to x_r. However, if $c \in \mathbb{R}^n$ is any constant vector we have

$$\begin{aligned} \frac{\mathrm{d}}{\mathrm{d}t}(x(t) - x_r(t)) &= f(x, t) - f(x_r, t) \\ &= f((x - x_r + c) + (x_r - c), t) - f(x_r, t) \\ &= g(x - x_r + c, t) \end{aligned}$$

for some new function g. Hence, if $y(t) = x(t) - x_r(t) + c$, we have

$$\dot{y}(t) = g(y, t)$$

† See also Rouche *et al.* (1977).

and $g(c, t) = 0$. It follows that studying perturbations of solutions of non-linear systems about reference trajectories is equivalent to studying perturbations about equilibrium points. This is the view which we shall take and before giving a definition of stability we remind the reader of the concept of a norm in $\mathbb{R}^n$ (cf. appendix 1). If $x \in \mathbb{R}^n$ we define its length $\|x\|$ to be

$$\|x\| = \left(\sum_{i=1}^{n} x_i^2\right)^{1/2}$$

where x_i are the components of x. The notion of 'nearness' in $\mathbb{R}^n$ can be stated in terms of the **balls** $B_\varepsilon(\bar{x})$ defined by

$$\boxed{B_\varepsilon(\bar{x}) \triangleq \{x \in \mathbb{R}^n: \|x - \bar{x}\| \leqslant \varepsilon\}}$$

We can now introduce a precise definition of stability.

Definition 5.4.1

The (equilibrium) state x_e of the system (5.4.1) is **stable** if for any $\varepsilon > 0$, there exists $\delta > 0$ (which depends on ε and t_0) such that

$$x_0 \in B_\delta(x_e) \Rightarrow x(t; x_0, t_0) \in B_\varepsilon(x_e) \quad \text{for all } t \geqslant t_0$$

(see fig. 5.11). If δ is independent of t_0, the stability is said to be **uniform**.

Definition 5.4.2

The solutions of (5.4.1) are **bounded** if for any $\delta > 0$, there exists $M > 0$ (depending on δ and t_0) such that

$$x_0 \in B_\delta(x_e) \Rightarrow x(t; x_0, t_0) \in B_M(x_e), \quad \text{for all } t \geqslant t_0.$$

If M is independent of t_0, the solutions are **uniformly bounded.**

We frequently require that solutions are not only bounded but actually tend to the equilibrium vector as $t \to \infty$. Hence we introduce the following definition.

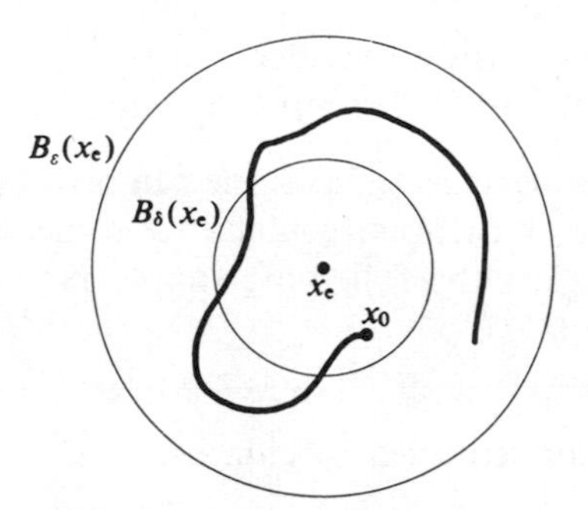

Fig. 5.11 Illustrating stability.

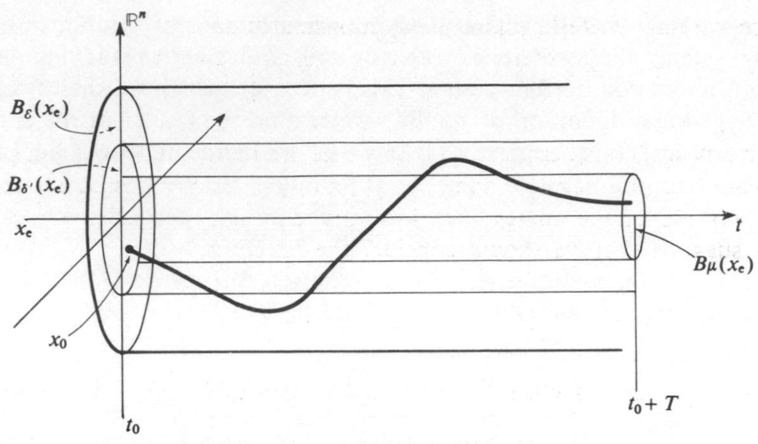

Fig. 5.12 Asymptotic stability.

Definition 5.4.3

The state x_e is **asymptotically stable** if it is stable and for a given $\mu > 0$, there exists $\delta > 0$ (independent of μ) and a time $T(=T(\mu, \delta, t_0))$ such that

$$x_0 \in B_\delta(x_e) \Rightarrow x(t;\ x_0,\ t_0) \in B_\mu(x_e), \quad \text{for all } t \geqslant t_0 + T \tag{5.4.2}$$

In other words, given any μ, no matter how small, we can find a sufficiently large T so that the solution is in the ball $B_\mu(x_e)$ of arbitrarily small radius μ. Since x_e is stable, given any $\varepsilon > 0$, there exists $\delta_1(\varepsilon)$ such that

$$x_0 \in B_{\delta_1}(x_e) \Rightarrow x(t;\ x_0,\ t_0) \in B_\varepsilon(x_e)$$

Hence, if $\delta' = \min\{\delta, \delta_1\}$ we can illustrate asymptotic stability as in fig. 5.12.

If the number δ in definition 5.4.3 can be taken arbitrarily large then x_e is said to be **asymptotically stable in the large**. Moreover, we say that the equilibrium solution x_e is **uniformly asymptotically stable in the large** if

(a) x_e is uniformly stable,
(b) every solution is uniformly bounded,
(c) The condition (5.4.2) holds for any $\delta > 0$ and for T independent of t_0.

The question now arises as to how one can prove stability of a system. Before giving a general theorem which, to some extent, answers this question we shall consider the following simple example.

Example 5.4.1

Let us consider the damped linear oscillator

$$\ddot{x} = -2\nu\omega_n\dot{x} - \omega_n^2 x \tag{5.4.3}$$

which we may write in phase-plane coordinates as

$$\begin{pmatrix} \dot{x}_1 \\ \dot{x}_2 \end{pmatrix} = \begin{pmatrix} 0 & 1 \\ -\omega_n^2 & -2\nu\omega_n \end{pmatrix} \begin{pmatrix} x_1 \\ x_2 \end{pmatrix} \tag{5.4.4}$$

From physical considerations we know that the energy of this system cannot increase (for non-negative damping ν) and must strictly decrease with time if $\nu > 0$. Hence the energy function must decrease along the trajectories. This suggests that we should consider the function

$$\boxed{V(x_1, x_2) = \omega_n^2 x_1^2 + x_2^2} \tag{5.4.5}$$

which, as is easily seen, is proportional to the total energy of the system. Now we have

$$V(x_1, x_2) > 0, \quad \text{if } \|x\|^2 = x_1^2 + x_2^2 > 0 \tag{5.4.6}$$

and

$$V(0, 0) = 0 \tag{5.4.7}$$

Moreover, the decay of energy with time may be demonstrated simply by differentiating V 'along the trajectories' of (5.4.4). By this we mean that if $x_1(t)$, $x_2(t)$ are the solutions of (5.4.4) then we have

$$\begin{aligned} \frac{\mathrm{d}}{\mathrm{d}t} V(x_1(t), x_2(t)) &= \frac{\partial V}{\partial x_1}\frac{\mathrm{d}x_1}{\mathrm{d}t} + \frac{\partial V}{\partial x_2}\frac{\mathrm{d}x_2}{\mathrm{d}t} \\ &= 2\omega_n^2 x_1(t)\dot{x}_1(t) + 2x_2(t)\dot{x}_2(t) \\ &= -4\nu\omega_n x_2^2(t) \leqslant 0 \end{aligned} \tag{5.4.8}$$

since $\dot{x}_1$, $\dot{x}_2$ (on a solution) are given by (5.4.4). Note that (5.4.8) only states that the energy along a solution is not increasing. However, we know physically that when $x_2(t) = 0$ for t in some time interval t_1, t_2, so that $\dot{x}_1 = 0$ and hence x_1 is constant, then the system must be at the equilibrium position. Hence, along any non-trivial (i.e. non-equilibrium) solution, (5.4.8) implies

$$\frac{\mathrm{d}}{\mathrm{d}t} V(x_1(t), x_2(t)) < 0 \tag{5.4.9}$$

for all t apart from some isolated values. Note that the function

$$\frac{\mathrm{d}V}{\mathrm{d}t}(x_1(t), x_2(t))$$

is independent of explicit mention of t and so may be regarded as just a function of the spatial variables x_1, x_2. We therefore write it as $\dot{V}(x_1, x_2)$.

The function V defined by (5.4.5) is a paraboloid as shown in fig. 5.13, together with the level curves where V is a constant, which have been projected into the plane (x_1, x_2). Since the time derivative of V along a

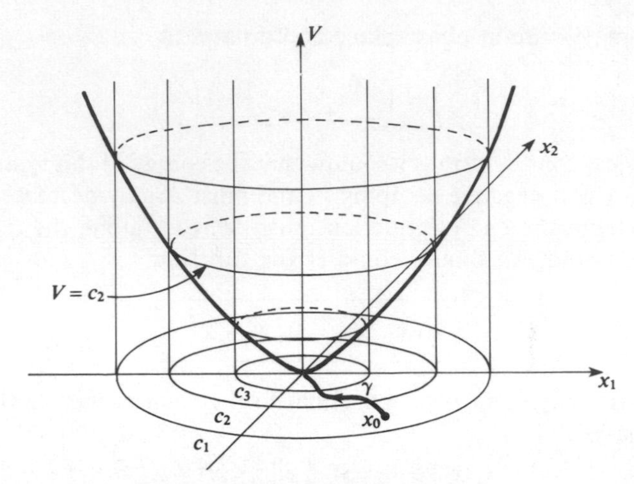

Fig. 5.13 An energy function.

trajectory is negative, V must decrease along such a trajectory. Hence, the phase plane curve γ must cross the level curves of V in decreasing order and so the system is forced to the equilibrium point (0, 0).

The function V obtained in this example leads us to the following generalization:

Definition 5.4.4
A function $V: \mathbb{R}^n \to \mathbb{R}^+$ associated with the system (5.4.1) which satisifies the properties

(a) $V(x, t) > 0 \quad \text{if } x \neq 0 \text{ and } V(0, t) = 0, \text{ for all } t \geqslant t_0$

(b) $\dot{V} \triangleq \sum_{i=1}^{n} \frac{\partial V(x, t)}{\partial x_i} f_i(x, t) + \frac{\partial V}{\partial t} < 0$

is called a **Lyapunov function.**

The following theorem essentially states that the converse of example 5.4.1 is true; namely, if we can find a Lyapunov function then the system is stable. Note, however, that we shall assume only that local solutions of the equation exist, i.e. for any (x_0, t_0) there exists a time τ such that the equation (5.4.1) has a unique solution on the interval $[t_0, t_0 + \tau)$. The fact that the solutions actually exist for all times $t \geqslant t_0$ will be a consequence of the other assumptions. Let C_0 denote the set of continuous non-decreasing scalar functions α such that $\alpha(0) = 0$.

Theorem 5.4.1 (Lyapunov's main stability theorem)
Consider the system defined by the differential equation

$$\dot{x} = f(x, t)$$

and assume that local solutions exist for all initial conditions (x_0, t_0). Suppose that the origin is an isolated equilibrium point, so that $f(0, t) = 0$ for all t, and assume that there exists a function $V: \mathbb{R}^n \times \mathbb{R}^+ \to \mathbb{R}^+$ such that V is differentiable and satisfies

(a) $V(x, t)$ is positive definite, i.e. $V(0, t) = 0$ and there exists $\alpha \in C_0$ such that $V(x, t) > \alpha(\|x\|) > 0$ for all $x \neq 0$ and all t,

(b) $$\dot{V}(x, t) = \sum_{i=1}^{n} \frac{\partial V(x, t)}{\partial x_i} f_i + \frac{\partial V}{\partial t}(x, t)$$

is negative definite, so that there exists $\gamma \in C_0$ such that $\dot{V}(x, t) \leqslant -\gamma(\|x\|) < 0$ for all $x \neq 0$ and all t,

(c) there exists $\beta \in C_0$ such that $V(x, t) \leqslant \beta(\|x\|)$,

(d) $\alpha(s) \to \infty$ as $s \to \infty$.

Then the origin $x = 0$ is uniformly asymptotically stable in the large.

Proof

Since β is continuous and $\beta(0) = 0$, if $\varepsilon > 0$ is given, there exists $\delta(-\delta(\varepsilon)) > 0$ such that $\beta(\delta) < \alpha(\varepsilon)$ (see fig. 5.14). Now

$$V(x(t; x_0, t_0), t) - V(x_0, t_0) = \int_{t_0}^{t} \dot{V}(x(\tau; x_0, t_0), \tau)\, d\tau < 0, \quad t > t_0$$

for arbitrary t_0. Thus, if $\|x_0\| < \delta$, we have

$$\begin{aligned} \alpha(\varepsilon) &> \beta(\delta) \\ &\geqslant V(x_0, t_0) \\ &\geqslant V(x(t; x_0, t_0), t) \\ &\geqslant \alpha(\|x(t; x_0, t_0)\|), \quad t \geqslant t_0 \end{aligned}$$

Since α is non-decreasing, we obtain

$$\|x(t; x_0, t_0)\| < \varepsilon \quad \text{for } t \geqslant t_0,\ \|x_0\| \leqslant \delta$$

This proves uniform stability, and that any solution starting in $B_\delta(0)$ exists for all $t \geqslant t_0$, because of the assumption of local existence of solutions.

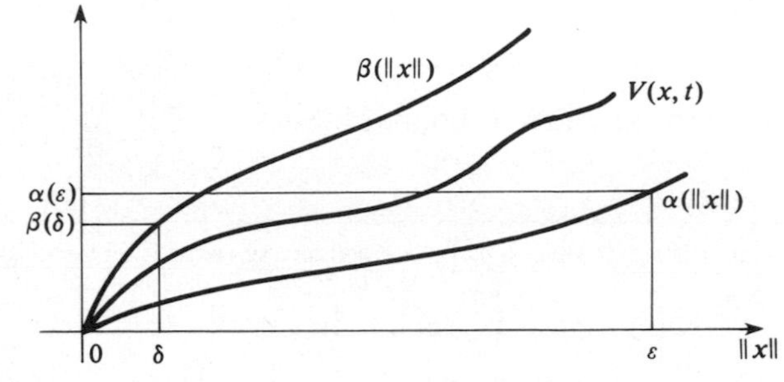

Fig. 5.14 Upper and lower bounds of a Lyapunov function.

We now prove† that $\|x(t; x_0, t_0)\| \to 0$ as $t \to \infty$ uniformly in t_0 for $\|x_0\| \leqslant \delta$, where δ is as in the first part. Let $0 < \mu < \|x_0\|$ and find $\nu(=\nu(\mu)) > 0$ such that $\beta(\nu) < \alpha(\mu)$. Define $\varepsilon' = \gamma(\nu)$ and set

$$T(=T(\mu, \delta)) = \frac{\beta}{\varepsilon'} > 0$$

Then, if $\|x(t, x_0, t_0)\| > \nu$ for $t_0 \leqslant t \leqslant t_1 \triangleq t_0 + T$, we have

$$\begin{aligned}
0 &< \alpha(\nu) \\
&\leqslant V(x(t_1; x_0, t_0), t_1) \\
&\leqslant V(x_0, t_0) - \int_{t_0}^{t_1} \gamma(\|x(\tau; x_0, t_0)\|)\, d\tau \\
&\leqslant V(x_0, t_0) - \int_{t_0}^{t_1} \gamma(\nu)\, d\tau \\
&= V(x_0, t_0) - (t_1 - t_0)\, \varepsilon' \\
&\leqslant \beta(\delta) - T\varepsilon' = 0
\end{aligned}$$

This is a contradication and so there must exist $t_2 \in [t_0, t_1]$ such that

$$\|x_2\| \triangleq \|x(t_2; x_0, t_0)\| \leqslant \nu$$

Thus,

$$\begin{aligned}
\alpha(\|x(t; x_2, t_2)\|) &\leqslant V(x(t; x_2, t_2), t) \\
&\leqslant V(x_2, t_2) \\
&\leqslant \beta(\nu) \\
&< \alpha(\mu)
\end{aligned}$$

for all $t \geqslant t_2$, and so

$$\|x(t; x_0, t_0)\| < \mu \quad \text{for all } t \geqslant t_0 + T \geqslant t_2$$

which proves uniform asymptotic stability.

Since $\alpha(s) \to \infty$ as $s \to \infty$ it follows that for any δ_1, no matter how large, there exists $\varepsilon(=\varepsilon(\delta_1))$ such that $\beta(\delta_1) < \alpha(\varepsilon)$. As above we can show that

$$\|x(t; x_0, t_0)\| \leqslant \varepsilon, \quad \text{if } \|x_0\| \leqslant \delta_1$$

and so we have proved uniform boundedness.

---

† The idea here is that if $x \to 0$, then $\gamma(\|x\|) \geqslant a > 0$ and so (writing $V(x) = V(x, t)$),

$$V(x) - V(0) = \int_0^t \dot{V}\, dt \leqslant -\int_0^t \gamma \leqslant -ta$$

thus $0 \leqslant V(x) \leqslant V(0) - ta$, giving a contradiction if t is large enough.

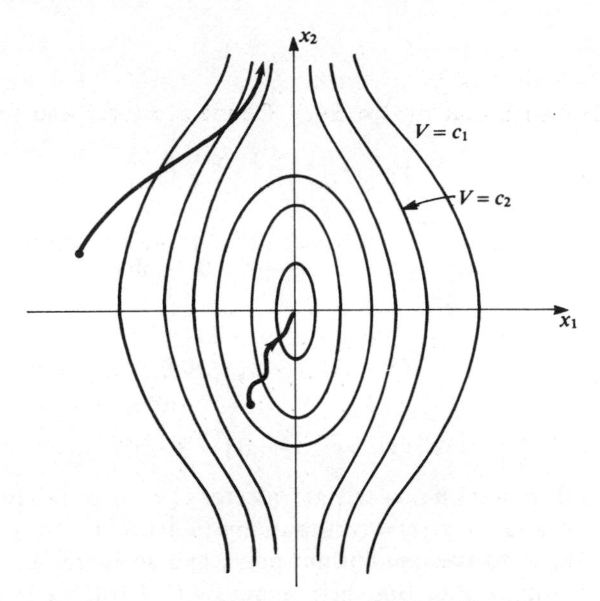

Fig. 5.15 A system with non closed trajectories.

Note that condition (d) in theorem 5.4.1 is necessary as the example

$$V(x_1, x_2) = x_1^2 + \frac{x_2^2}{1 + x_2^2}$$

shows, since the level curves $V = c$ of V are closed only if $c \leqslant 1$ and so some trajectories with $\|x_0\| > 1$ may escape to $x_2 = \infty$ (fig. 5.15). Condition (d) is not satisfied in this case since if $x_1 = \beta = \text{constant}$, $x_2 \to \infty$, then $V(x_1, x_2) \to \beta^2 + 1 < \infty$.

In example 5.4.1 we used the energy function V, but we could only show that $\dot{V} \leqslant 0$ (rather than the strict inequality). However, the set there $\dot{V} = 0$ was not a trajectory of the system and so we could still prove asymptotic stability. We can easily generalize theorem 5.4.1 in this way to obtain the following.

Theorem 5.4.2

Let the assumptions of theorem 5.4.1 hold except that we assume only $\dot{V}(x) \leqslant 0$ and suppose that the set

$$\Omega = \{x \in \mathbb{R}^n \colon \dot{V}(x) = 0\}$$

does not contain a non-trivial trajectory of the system. Then the origin is globally asymptotically stable.

Theorem 5.4.2 is a form of **La Salle's invariance principle** (see Banks, 1983).

Example 5.4.2
Consider the system

$$\left.\begin{aligned}\dot{x}_1 &= x_2\\ \dot{x}_2 &= -Kg(x_1) - \omega x_2\end{aligned}\right\} \tag{5.4.10}$$

where $x_1 g(x_1) > 0$ if $x_1 \neq 0$ and $g(0) = 0$. Consider the function

$$V(x_1, x_2) = K\int_0^{x_1} g(x_1)\,\mathrm{d}x_1 + \frac{x_2^2}{2} > 0 \quad \text{unless } x_1 = x_2 = 0$$

Then

$$\dot{V}(x_1, x_2) = -\omega_1 x_2^2 \leqslant 0$$

Now,

$$\dot{\Omega} = \{(x_1, x_2): x_2 = 0\} = x_1\text{-axis}$$

Suppose that Ω contains a non-trivial trajectory $(x_1(t), x_2(t))$ of the system. Then $x_2(t) = 0$ and so $x_1(t) = \text{constant}$, by (5.4.10). If $x_1(t) = 0$ then the trajectory reduces to the equilibrium point and so is trivial. Hence $x_1(t)$ must be a constant $c \neq 0$. But then, again by (5.4.10), we have

$$0 = -Kg(c)$$

which contradicts the definition of g. Hence, by theorem 5.4.2, the system (5.4.10) is globally asymptotically stable in the large.

Finally in this section we shall show that one can use Lyapunov theory to determine the existence of limit cycles in two-dimensional systems. In fact, we have the following result, which is a simple corollary of the Poincaré–Bendixson theorem.

Theorem 5.4.3
Consider again the two-dimensional system

$$\dot{x} = f(x, t), \quad x \in \mathbb{R}^2$$

and assume that $x = 0$ is the unique equilibrium point of this system. Suppose that there exists a function V with the properties

(a) $V(0) = 0$ and $V(x) > 0$ for all $x \neq 0$,
(b) $\Omega \triangleq \{x: \dot{V}(x) = 0\} = \Omega_1 \cup \Omega_2$ is the union of two one-dimensional manifolds (smooth curves in $\mathbb{R}^2$) such that Ω_1 is a closed curve and Ω_2 contains no trajectory of the system.
(c) inside† Ω_1, $\dot{V}(x) \geqslant 0$, and outside Ω_1, $\dot{V}(x) \leqslant 0$.

† This makes sense by the Jordan curve theorem, which states that any simple closed curve divides the plane into two distinct parts, called the 'inside' and 'outside' (see Saff and Snider, 1976).

If

$$W_1 = \{x: V(x) = c_1\}, \quad W_2 = \{x: V(x) = c_2\}$$

where the constants c_1 and c_2 are chosen so that W_1 (respectively W_2) is the largest (smallest) such set inside (outside) Ω_1, then there is at least one limit cycle between W_1 and W_2.

Example 5.4.3

Consider the system in fig. 5.16, and let $x_1 = \theta$, $x_2 = \dot{\theta}$. Then we have

$$\theta = \frac{\omega_n}{s^2 - \omega_n s + \omega_n^2} x = -\frac{\omega_n}{s^2 - \omega_n s + \omega_n^2}(\dot{\theta}^{2n-1} + \theta\dot{\theta}^{2n-2})$$

that is

$$\ddot{\theta} - \omega_n\dot{\theta} + \omega_n^2\theta = -\omega_n(\dot{\theta}^{2n-1} + \theta\dot{\theta}^{2n-2})$$

and so

$$\begin{aligned} \dot{x}_1 &= x_2 \\ \dot{x}_2 &= \omega_n x_2 - \omega_n^2 x_1 - \omega_n(x_2^{2n-1} + x_1^{2n-2}x_2) \end{aligned}$$

We now try the function

$$V(x) = \tfrac{1}{2}(\omega_n^2 x_1^2 + x_2^2)$$

Then

$$\dot{V} = \omega_n^2 x_1 \dot{x}_1 + x_2 \dot{x}_2 = \omega_n x_2^2(1 - x_2^{2n-2} - x_1^{2n-2})$$

Hence, $\dot{V} = 0$ when $x_2 = 0$ or $1 = x_2^{2n-2} + x_1^{2n-2}$. Clearly, $x_2 = 0$ is not a solution of the system, and so if we let

$$\begin{aligned} \Omega_1 &= \{(x_1, x_2): 1 = x_2^{2n-2} + x_1^{2n-2}\} \\ \Omega_2 &= \{(x_1, x_2): x_2 = 0\} \end{aligned}$$

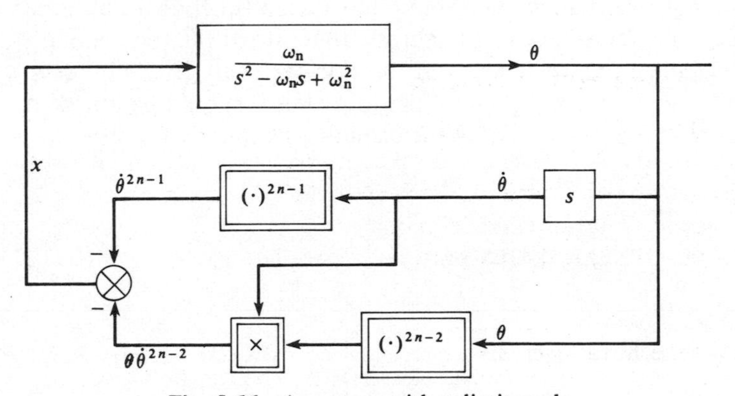

Fig. 5.16 A system with a limit cycle.

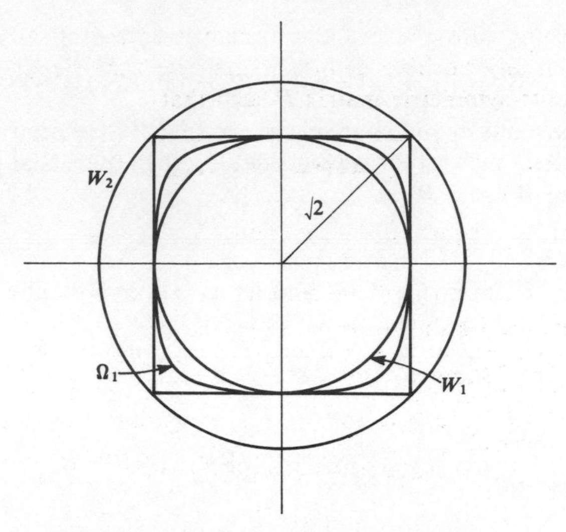

Fig. 5.17 Identifying the limit cycle for the system in fig. 5.16.

it follows from theorem 5.4.3 that a limit cycle must exist between the sets

$$W_1 = \{(\theta, \dot{\theta}): \theta^2 + \dot{\theta}^2 = 1\}$$
$$W_2 = \{(\theta, \dot{\theta}): \theta^2 + \dot{\theta}^2 = 2\}$$

Note that Ω_1 is a square with the corners 'rounded off' for large n, and W_1 and W_2 are two circles (see fig. 5.17). If $n = 2$ then W_1 is a limit cycle.

5.4.2 Generation of Lyapunov Functions

In the previous section we have shown how the existence of a certain function V, associated with a differential equation, implies stability of the equilibrium points or the existence of limit cycles of the system. However, we have not shown how one should find such a function V, although in the examples considered above a quadratic function of the state variables seems to be a good thing to try. In this section we shall give some less *ad hoc* methods for finding Lyapunov functions, but it should be noted that there is no general method which will determine a Lyapunov function for a stable system.

METHOD 1 – LINEAR SYSTEMS

Theorem 5.4.4
Consider the linear system

$$\dot{x} = Ax, \quad x \in \mathbb{R}^n$$

The equilibrium state $x=0$ of this system is asymptotically stable if and only if, given any positive definite symmetric† matrix Q, there exists a positive definite symmetric matrix P such that

$$\boxed{A^{\mathrm{T}}P+PA=-Q} \tag{5.4.11}$$

Also, $V=x^{\mathrm{T}}Px$ is a Lyapunov function.

Proof

If there exists such a matrix P for Q given, then

$$V(x)=x^{\mathrm{T}}Px>0 \quad \text{for } x\neq 0,\ V(0)=0$$

and

$$\begin{aligned}\dot{V}(x)&=\dot{x}^{\mathrm{T}}Px+x^{\mathrm{T}}P\dot{x}\\&=x^{\mathrm{T}}A^{\mathrm{T}}Px+x^{\mathrm{T}}PAx\\&=-x^{\mathrm{T}}Qx<0\end{aligned}$$

and so the system is asymptotically stable.

Conversely, if $x=0$ is asymptotically stable, then consider the matrix differential equation

$$\dot{X}=A^{\mathrm{T}}X+XA, \quad X(0)=Q$$

The solution is clearly $X=\exp(A^{\mathrm{T}}t)\,Q\exp(At)$. However, by integrating the equation we have

$$X(\infty)-X(0)=A^{\mathrm{T}}\int_0^\infty X\,\mathrm{d}t+\int_0^\infty X\,\mathrm{d}t\,A$$

Since the system is stable, $X(\infty)=0$ and

$$\int_0^\infty X\,\mathrm{d}t$$

exists and so

$$-Q=A^{\mathrm{T}}P+PA$$

where

$$\begin{aligned}P&=\int_0^\infty X\,\mathrm{d}t\\&=\int_0^\infty \exp(A^{\mathrm{T}}t)\,Q\exp(At)\,\mathrm{d}t\end{aligned}$$

† That is $x^{\mathrm{T}}Qx>0$ for all $x\neq 0$ and $Q^{\mathrm{T}}=Q$.

Clearly, $P^{\mathrm{T}} = P$ and

$$x^{\mathrm{T}}Px = \int_0^\infty (\exp(At)x)^{\mathrm{T}} Q(\exp(At)x)\, \mathrm{d}t > 0 \quad \text{for } x \neq 0$$

The result now follows.

Note that, since Q is any positive definite symmetric matrix, we usually choose $Q = I$. The positive definiteness of P can be checked by applying Sylvester's criterion (see Wilkinson, 1965):

A symmetric matrix $P = (p_{ij})$ is positive definite if and only if

$$p_{11} > 0, \quad \begin{vmatrix} p_{11} & p_{12} \\ p_{21} & p_{22} \end{vmatrix} > 0, \ldots, \quad \begin{vmatrix} p_{11} & \cdots & p_{1n} \\ \vdots & & \vdots \\ p_{n1} & \cdots & p_{nn} \end{vmatrix} > 0$$

that is all the leading principle minors of P are positive.

Example 5.4.4

Consider the system

$$\begin{aligned} \dot{x}_1 &= -x_1 - 2x_2 \\ \dot{x}_2 &= x_1 - 4x_2 \end{aligned}$$

The unique equilibrium state is $x = 0$. If $A^{\mathrm{T}}P + PA = -I$, then

$$\begin{aligned} -2p_{11} + 2p_{12} &= -1 \\ -2p_{11} - 5p_{12} + p_{22} &= 0 \\ -4p_{12} - 8p_{22} &= -1 \end{aligned}$$

and so

$$p_{11} = \frac{23}{60}, \quad p_{12} = \frac{-7}{60}, \quad p_{22} = \frac{11}{60}$$

Hence

$$P = \begin{pmatrix} \frac{23}{60} & \frac{-7}{60} \\ \frac{-7}{60} & \frac{11}{60} \end{pmatrix} > 0$$

and so the system is stable. Note that

$$V(x) = x^{\mathrm{T}}Px = \frac{1}{60}(23x_1^2 - 14x_1x_2 + 11x_2^2)$$

and

$$\dot{V}(x) = -\|x\|^2$$

METHOD 2 – KRASOVSKII'S METHOD

In an attempt to generalize the linear theory above, Krasovskii considers the autonomous system

$$\dot{x} = f(x), \quad x \in \mathbb{R}^n$$

where $f(0) = 0$ and f is differentiable. Define $F(x)$ to be the Jacobian matrix of the system, i.e.

$$F(x) = \left(\frac{\partial f_i}{\partial x_j}\right)_{1 \leqslant i \leqslant n,\, 1 \leqslant j \leqslant n} = (\nabla f_i)_{1 \leqslant i \leqslant n}$$

where ∇f_i is the row vector

$$\left(\frac{\partial f_i}{\partial x_1}, \ldots, \frac{\partial f_i}{\partial x_n}\right)$$

Then we have the following theorem.

Theorem 5.4.5

Suppose that $f(0) = 0$ (although we do not *assume* that 0 is the only equilibrium point), and set $\hat{F}(x) = F^{\mathrm{T}}(x) + F(x)$ (the symmetrized Jacobian). If $\hat{F}$ is negative definite, for each x, then $x = 0$ is asymptotically stable. A Lyapunov function for this system is

$$\boxed{V(x) = f^{\mathrm{T}}(x) f(x)}$$

If, in addition, $f^{\mathrm{T}}(x) f(x) \to \infty$ as $\| x \| \to \infty$, then the origin is asymptotically stable in the large.

Proof

We first show that 0 is, in fact, the only equilibrium point. Suppose the contrary, then $f(x_1) = 0$ for some $x_1 = 0$. Let γ denote the straight path parameterized by arc length s from 0 to x_1. Then

$$\begin{aligned} f_i(x_1) &= \int_\gamma \frac{\mathrm{d}f_i}{\mathrm{d}s} \\ &= \int_0^{\|x_1\|} \frac{\mathrm{d}}{\mathrm{d}s} f_i(x(s))\, \mathrm{d}s \\ &= \int_0^{\|x_1\|} \nabla f_i(x) \frac{\mathrm{d}x}{\mathrm{d}s}\, \mathrm{d}s \end{aligned}$$

for $1 \leqslant i \leqslant n$. Hence,

$$\begin{aligned} 0 &= f(x_1) \\ &= \int_0^{\|x_1\|} \frac{x_1^{\mathrm{T}}}{\|x_1\|} F \frac{x_1}{\|x_1\|}\, \mathrm{d}s \\ &= \int_0^{\|x_1\|} \frac{x_1^{\mathrm{T}}}{\|x_1\|} F^{\mathrm{T}} \frac{x_1}{\|x_1\|}\, \mathrm{d}s \end{aligned}$$

Therefore

$$0 = \int_0^{\|x_1\|} \frac{x_1^{\mathrm{T}}}{\|x_1\|} \hat{F}(x) \frac{x_1}{\|x_1\|} \, \mathrm{d}s < 0$$

since F is negative definite. This contradiction implies that $x = 0$ is the only equilibrium point, and so

$$\begin{aligned} V(x) &= f^{\mathrm{T}}(x) f(x) \\ &= \|f(x)\|^2 > 0 \quad \text{if } x \neq 0 \end{aligned}$$

Also

$$\begin{aligned} \dot{V}(x) &= \dot{f}^{\mathrm{T}}(x) f(x) + f^{\mathrm{T}}(x) \dot{f}(x) \\ &= f^{\mathrm{T}}(x) \hat{F}(x) f(x) < 0, \quad x \neq 0 \end{aligned}$$

The result now follows from theorem 5.4.1.

METHOD 3 – THE VARIABLE GRADIENT METHOD (Schultz and Gibson, 1962)
Consider the autonomous system

$$\dot{x} = f(x), \quad x \in \mathbb{R}^n$$

and suppose that V is a Lyapunov function for this system. Then, if x: $\mathbb{R}^+ \to \mathbb{R}^n$ is any differentiable function, with $x(0) = 0$, we have

$$\dot{V}(x) = (\nabla V)^{\mathrm{T}} \dot{x} \tag{5.4.12}$$

and so

$$\begin{aligned} V(x(t)) &= \int_0^t (\nabla V)^T \frac{\mathrm{d}x}{\mathrm{d}\tau} \, \mathrm{d}\tau \\ &\triangleq \int_0^{x(t)} \nabla^{\mathrm{T}} V \, \mathrm{d}x \end{aligned}$$

(where $\int_0^{x(t)}$ denotes a line integral†).

The line integral is independent of the path in $\mathbb{R}^n$ from 0 to $x(t)$ if

$$\frac{\partial^2 V}{\partial x_i \, \partial x_j} = \frac{\partial^2 V}{\partial x_j \, \partial x_i} \quad (i, j = 1, \ldots, n) \tag{5.4.13}$$

Under these conditions, we can define, unambiguously, the function

$$\boxed{V(x) = \int_L (\nabla^{\mathrm{T}} V)(x_1) \, \mathrm{d}x_1, \quad x \in \mathbb{R}^n} \tag{5.4.14}$$

for any line L joining 0 to x.

† Along the line $L : t \to x(t)$. This is also denoted $\int_L$.

Since we expect V to be similar to a quadratic form we assume that ∇V may be written

$$\nabla V = \begin{bmatrix} \alpha_{11}x_1 + \alpha_{12}x_2 + \ldots + \alpha_{1n}x_n \\ \alpha_{21}x_1 + \alpha_{22}x_2 + \ldots + \alpha_{2n}x_n \\ \vdots \quad\quad \vdots \quad\quad \vdots \\ \alpha_{n1}x_1 + \alpha_{n2}x_2 + \ldots + 2x_n \end{bmatrix}$$

However, to increase our freedom in the choice of V we allow α_{ij} to depend on $x_1, \ldots, x_{n-1}$. Now, $\dot{V} = (\nabla V)^{\mathrm{T}}\dot{x}$ and so $\dot{V} = (\nabla V)^{\mathrm{T}}f$ and we can find $\dot{V}$ in terms of the α_{ij} and f. We then choose some of the α_{ij} to make $\dot{V}$ negative (semi-definite) and choose the remaining α_{ij} to satisfy (5.4.13). Finally, V is found from (5.4.14) by choosing a suitable path of integration L. It is usually most convenient to choose a path which is parallel to each axis in turn (for general x). Then we have

$$V = \int_0^{x_1} (\nabla V)_1(x_1', 0, \ldots, 0)\,\mathrm{d}x_1' + \int_0^{x_2} (\nabla V)_2(x_1, x_2', 0, \ldots, 0)\,\mathrm{d}x_2'$$

$$+ \ldots + \int_0^{x_n} (\nabla V)_n(x_1, x_2, \ldots, x_n')\,\mathrm{d}x_n'$$

where $(\nabla V)_i = \partial V/\partial x_i$.

Example 5.4.5

Consider the system in fig. 5.18, where $n(e) = e^3$ and

$$G(s) = \frac{1}{s(s+1)}$$

If we set $x_1 = c(t)$, $x_2 = \dot{c}(t)$, then

$$\dot{x}_1 = x_2$$
$$\dot{x}_2 = -x_1^3 - x_2$$

Let

$$\nabla V = \begin{bmatrix} \alpha_{11}x_1 + \alpha_{12}x_2 \\ \alpha_{21}x_1 + 2x_2 \end{bmatrix}$$

Hence

$$\dot{V} = (\alpha_{11} - 2x_1^2 - \alpha_{21})x_1x_2 + (\alpha_{12} - 2)x_2^2 - \alpha_{21}x_1^4$$

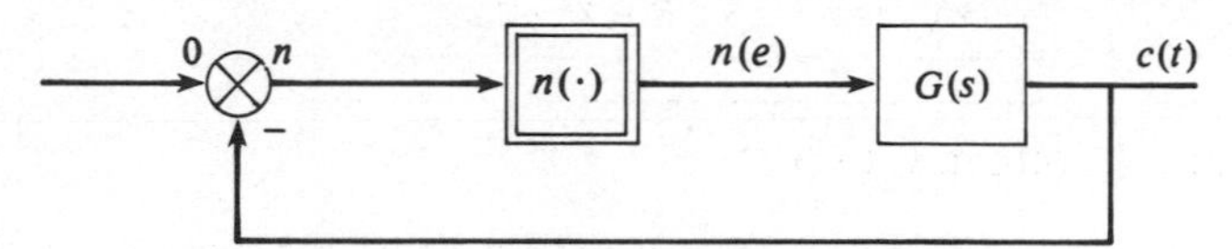

Fig. 5.18 A simple non-linear system.

We may choose $\alpha_{21} > 0, 0 < \alpha_{12} < 2$ and $\alpha_{11} = \alpha_{21} + 2x_1^2$ to make $\dot{V}$ negative semi-definite. Then, if we take $\alpha_{12} = 1$,

$$\dot{V} = -x_2^2 - \alpha_{21} x_1^4$$

Hence

$$\nabla V = \begin{bmatrix} \alpha_{21} x_1 + 2x_1^3 + x_2 \\ \alpha_{21} x_1 + 2x_2 \end{bmatrix}$$

and so

$$\frac{\partial^2 V}{\partial x_2\, \partial x_1} = 1$$

$$\frac{\partial^2 V}{\partial x_1\, \partial x_2} = \alpha_{21} + x_1 \frac{\partial \alpha_{21}}{\partial x_1}$$

Then (5.4.13) is satisfied if $\alpha_{21} = 1$, in which case

$$\nabla V = \begin{bmatrix} x_1 + 2x_1^3 + x_2 \\ x_1 + 2x_2 \end{bmatrix}$$

Hence, finally,

$$V = \int_0^{x_1} (x_1' + 2x_1'^3)\, dx_1' + \int_0^{x_2} (x_1 + 2x_2')\, dx_2'$$

$$= \frac{x_1^2}{2} + \frac{x_1^4}{2} + x_1 x_2 + x_2^2$$

and it may easily be seen that this is a Lyapunov function for the system.

Many other methods exist for the generation of Lyapunov functions, see, for example, Zubov (1961) and Lurie (1957).

5.4.3 Absolute Stability: Popov and Circle Criteria

Many control systems have the structure of a linear forward path transfer function and a non-linear element in the feedback path as shown in fig. 5.19(a). We are often interested in obtaining stability theorems for f in a whole class of non-linearities, and we then speak of **absolute stability** of the system. Suppose that, for example, the non-linearity f is assumed to lie in a sector (cf. fig. 5.20), so that

$$\boxed{k_1 < \frac{f(y)}{y} < k_2 \quad y \neq 0, f(0) = 0} \tag{5.4.15}$$

for some constants k_1, k_2. Then we may be tempted to replace the non-

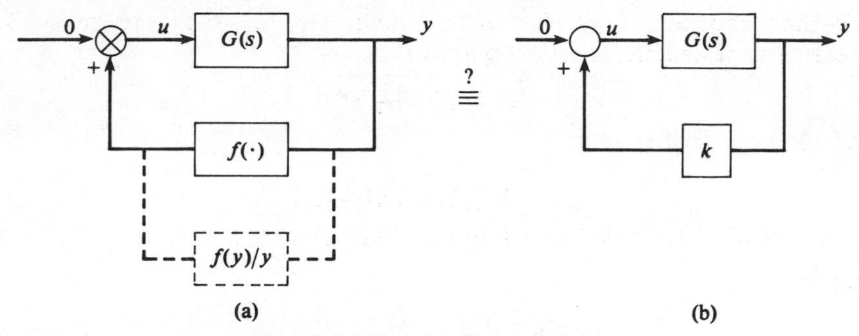

Fig. 5.19 Aizerman's conjecture.

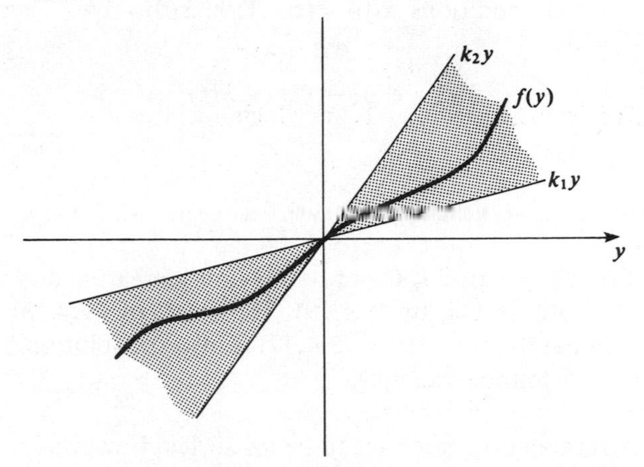

Fig. 5.20 Cone-bounded function f.

linearity by a linear gain k as in fig. 5.19(b) where $k_1 < k < k_2$. Aizerman conjectured that the stability of the linear system for all $k \in (k_1, k_2)$ implies that of the non-linear system for any f satisfying (5.4.15). However, this turns out to be false as the following counter-example shows.

Example 5.4.6

Consider the system

$$\left.\begin{aligned} &\ddot{x} + u = 0, \quad u = f(y) \\ &y = \dot{x} + x \end{aligned}\right\} \tag{5.4.16}$$

where

$$f(y) \begin{cases} = \dfrac{y}{e(1+e)}, & y \leqslant 1 \\[2ex] = \dfrac{e^{-y}}{1+e^{y}}, & y \geqslant 1 \end{cases}$$

Then

$$0 < \frac{f(y)}{y} \leqslant \frac{1}{e(1+e)}$$

and the linearized system

$$\ddot{x} + k(\dot{x} + x) = 0$$

is asymptotically stable for all $k > 0$. Now, if $\dot{x} + x \geqslant 1$, then (5.4.16) becomes

$$\ddot{x} = -\frac{\exp[-(\dot{x}+x)]}{1+\exp(\dot{x}+x)} \tag{5.4.17}$$

Consider the initial conditions $x(0) = (e-1)/e$, $\dot{x}(0) = 1/e$. Then, integrating (5.4.17) yields

$$\dot{x}(t) = \exp[-x(t) - \dot{x}(t)] \tag{5.4.18}$$

so that

$$\exp x(t) = \frac{1}{\dot{x}(t)\exp[\dot{x}(t)]} \tag{5.4.19}$$

since $\dot{x} \neq 0$. From (5.4.17), $\ddot{x} \leqslant 0$ and so $\dot{x}$ is non-increasing and so (5.4.17) is not asymptotically stable, by (5.4.19). However, by (5.4.18), $x + \dot{x} \geqslant 1$ for all $t \geqslant 0$, so that the solution of (5.4.17) is also the solution of (5.4.16), and this gives our counter-example.

The idea of Aizerman does not have to be discarded, however, since we can obtain stability conditions in a similar way by putting more restrictions on the linear element. Before stating the result, we remind the reader of the Fourier transform $F(j\omega)$ of a function $f(t)$ which vanishes for $t < 0$ and tends to zero exponentially as $t \to \infty$; namely,

$$F(j\omega) = \int_0^\infty f(t)\exp(-j\omega t)\,\mathrm{d}t$$

Then we shall require the following lemmas (see section 5.7, exercise 11).

Lemma 5.4.1
Let $H: \mathbb{R} \to \mathbb{R}$ be continuous with $H(y) > 0$ for all $y \neq 0$ and $H(0) = 0$. If $f(t), \dot{f}(t)$ are bounded for all $t \geqslant 0$, then if

$$\int_0^\infty H(f(t))\,\mathrm{d}t$$

is finite, we must have

$$\lim_{t \to \infty} f(t) = 0$$

Lemma 5.4.2
Let $f_1(t), f_2(t), f_3(t)$ be real functions which vanish for $t < 0$ and tend to zero exponentially as $t \to \infty$, and denote their Fourier transforms by $F_m(j\omega)$, $m = 1,2,3$. If

(a) $F_1(j\omega) = -H(j\omega)F_3(j\omega) + F_2(j\omega)$
(b) $\operatorname{Re} H(j\omega) \geqslant \delta > 0$ for all $\omega \in \mathbb{R}$

then

$$\int_0^\infty f_1(t) f_3(t)\,\mathrm{d}t \leqslant \frac{1}{8\pi\delta} \int_{-\infty}^{\infty} |F_2(j\omega)|^2\,\mathrm{d}\omega$$

$$= \frac{1}{4\delta} \int_0^\infty f_2^2(t)\,\mathrm{d}t$$

(The last equality is by Parseval's theorem; see appendix 1.)

We may now state the above-mentioned result which is due to Popov. In the result, which we shall prove, we shall assume that the non-linearity f is continuous and bounded in a sector as in fig. 5.20 where $k_1 = \varepsilon > 0$. The result may be proved under the weaker condition of $k_1 = 0$ and discontinuous f, but the proof given here is simple and instructive. It can be generalized to deal with the case $k_1 = 0$. For further details see La Salle and Lefschetz (1961), Siljak (1969) and Popov (1961). The Popov criterion can also be generalized to multivariable systems as shown by Anderson and Moore (1968), Banks (1982) and Banks and Collingwood (1979).

Theorem 5.4.6 (Popov's Criterion)
Consider the system

$$\begin{aligned} \dot{x} &= Ax + bu, \quad x(0) = x_0 \\ y &= c^{\mathrm{T}} x \end{aligned} \tag{5.4.20}$$

where $u = f(y)$ with f continuous. If the linear system is asymptotically stable and

(a) $f(0) = 0$, $0 < \varepsilon \leqslant f(y)/y < k$ for all $y \neq 0$
(b) there exists $\alpha > 0$ such that

$$\operatorname{Re}(1 + \alpha i\omega) G(i\omega) + \frac{1}{k} \geqslant \delta > 0$$

for all $\omega \in \mathbb{R}$ and some $\delta > 0$, then the non-linear system is asymptotically stable in the large.

Proof
From (5.4.20), we have

$$x(t) = \exp(At)x_0 + \int_0^t \exp[A(t-\theta)]\ bu(\theta)\ d\theta$$

$$u(t) = f(y(t))$$

Let

$$y = c^{T}x \triangleq a(t) + b(t)$$

where

$$a(t) = c^{T}\int_0^t \exp[A(t-\theta)]\ bu(\theta)\ d\theta$$

$$b(t) = c^{T}\exp(At)x_0$$

and define

$$u_\tau(t) = \begin{cases} u(t) & \text{for } t \leqslant \tau \\ 0 & \text{for } t > \tau \end{cases}$$

Let

$$x_\tau(t) = \exp(At)x_0 + \int_0^t \exp[A(t-\theta)]\ bu_\tau(\theta)\ d\theta$$

and

$$a_\tau(t) = c^{T}\int_0^t \exp[A(t-\theta)]\ bu_\tau(\theta)\ d\theta$$

Then

$$a_\tau(t) \leqslant \int_0^t \|b\|\ \|c\| M \exp[-\rho(t-\theta)]\ |u_\tau(\theta)|\ d\theta$$

$$\leqslant \exp(-\rho t)\|b\|\ \|c\| M \int_0^\tau \exp(\rho\theta)\ |u_\tau(\theta)|\ d\theta$$

where we have used the asymptotic stability of the linear system to write

$$\|\exp(At)\| \leqslant M\exp(-\rho t) \quad \text{for some } M, \rho$$

Hence, $|a_\tau(t)|$, $|y_\tau(t)| \to 0$ exponentially as $t \to \infty$ and so are Fourier transformable. Denote the transforms of such functions by $A_\tau(j\omega)$, $Y_\tau(j\omega)$, etc. Then, defining

$$f_1(t) = b(t) + \alpha\dot{b}(t)$$

$$f_2(t) = y_\tau(t) + \alpha\dot{y}_\tau(t) - \frac{u_\tau(t)}{k}$$

we have

$$A_\tau(j\omega) = -G(j\omega)U_\tau(j\omega)$$

where

$$G(j\omega) = -c^{\mathrm{T}}(j\omega I - A)^{-1}b$$

and

$$F_1(j\omega) = B_\tau(j\omega) + \alpha i\omega B_T(j\omega) - \alpha c^{\mathrm{T}}x_0$$

$$F_2(j\omega) = Y_\tau(j\omega) + ai\omega Y_\tau(j\omega) - \frac{U_\tau(j\omega)}{k} - \alpha c^{\mathrm{T}}x_0$$

$$Y_\tau(j\omega) = A_\tau(j\omega) + B_\tau(j\omega)$$

Thus

$$F_2(j\omega) = F_1(j\omega) - \left[(1 + \alpha j\omega)G(j\omega) + \frac{1}{k}\right]U_\tau(j\omega)$$

and since $\mathrm{Re}(1 + \alpha_j\omega)G(j\omega) + \dfrac{1}{k} \geqslant \delta > 0$, by assumption, lemma 5.4.2 implies that

$$\int_0^\infty u_\tau(t)\left[y_\tau(t) + \alpha\dot{y}_\tau(t) - \frac{u_\tau(t)}{k}\right]\mathrm{d}t \leqslant \frac{1}{4\delta}\int_0^\infty |f_1(t)|^2\,\mathrm{d}t$$

$$= C(x_0)$$

for some function C of x_0 such that $C(x_0) \to 0$ as $\|x_0\| \to 0$.

Hence

$$\int_0^T f(y(t))\left[y(t) - \frac{f(y(t))}{k}\right]\mathrm{d}t + \alpha\int_{y(0)}^{y(\tau)} f(y)\,\mathrm{d}y \leqslant C$$

and since each integral is positive we have

$$\int_0^T f(y(t))\left[y(t) - \frac{f(y(t))}{k}\right]\mathrm{d}t \leqslant C, \quad \text{for all } \tau > 0 \qquad (5.4.21)$$

and

$$\alpha\int_{y(0)}^{y(\tau)} f(y)\,\mathrm{d}y \leqslant C$$

From the last inequality, we have

$$\int_0^{y(\tau)} f(y)\,\mathrm{d}y \leqslant \frac{C}{\alpha} + \int_0^{y(0)} f(y)\,\mathrm{d}y$$

and since $f(y)/y \geqslant \varepsilon$ we have

$$y^2(\tau) \leqslant \frac{2C}{\alpha\varepsilon} + \frac{2}{\varepsilon}\int_0^{y(0)} f(y)\,\mathrm{d}y, \quad \text{for all } \tau > 0$$

Hence y is bounded and since $y(0) = c^{\mathrm{T}}x_0$, y and hence u and x must tend to zero (for all t) as $\|x_0\| \to 0$; the system is therefore stable.

Since x and u are bounded, it follows that $\dot{y} = c^{\mathrm{T}}\dot{x} = c^{\mathrm{T}}(Ax + bu)$ is also bounded and so by lemma 5.4.1 and (5.4.21) (with $\tau = \infty$) we obtain

$$y(t) \to 0 \quad \text{as } t \to \infty$$

Hence $u(t) \to 0$ as $t \to \infty$, and so, for any $\varepsilon_1 > 0$, there exists T_1 such that $|u(t)| < \varepsilon_1$ for $t > T_1$. If $t > 2T_1$, then

$$\begin{aligned}\|x(t)\| &\leqslant M \exp(-\rho t)\|x_0\| + M \int_0^t \exp[-\rho(t-s)]\,\|b\|\,\|u(s)\|\,\mathrm{d}s \\ &\leqslant M \exp(-\rho t)\|x_0\| + MC_1 \exp(-\rho t/2) + MC_2\varepsilon_1\end{aligned}$$

for some constants C_1, C_2 independent of t, and so $\|x(t)\| \to 0$ as $t \to \infty$.

Popov's criterion may be interpreted graphically by noting that the frequency condition is just that

$$[\operatorname{Re} G(j\omega)] - \alpha\omega[\operatorname{Im} G(j\omega)] + \frac{1}{k} \geqslant \delta > 0$$

and so if we write (for real ξ, η)

$$\begin{aligned}\zeta(\omega) &= \xi(\omega) + j\eta(\omega) \\ &\triangleq \operatorname{Re} G(j\omega) + j\omega \operatorname{Im} G(j\omega)\end{aligned}$$

and define the line

$$L = \left\{(\xi, \eta) \colon \xi - \alpha\eta + \frac{1}{k} = 0\right\}$$

then we require the polar plot of $\zeta(\omega)$ in the ζ-plane to lie strictly to the right of L as shown in fig. 5.21.

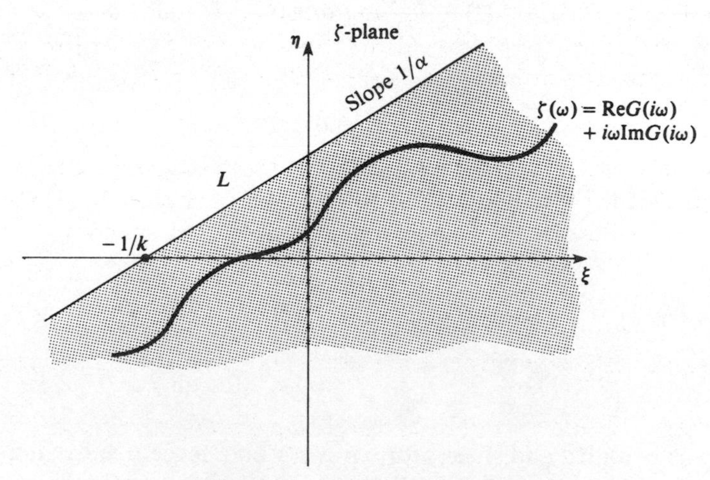

Fig. 5.21 Popov's criterion.

The Popov criterion is a kind of generalization of the Nyquist criterion where the linear feedback gain is replaced by the non-linearity. A more direct generalization is contained in the circle theorem which replaces the critical point $-1/k$ by a critical disk determined by the non-linearity. The circle theorem has a long history beginning with Sandberg (1964) and Zames (1966) with generalizations for multivariable systems by Rosenbrock (1972) and Cook (1972) and for infinite dimensional systems by Banks (1981) and Freedman *et al.* (1969). The simplest proofs require some functional analysis which is beyond the scope of the present book (see, for example, Banks (1983)) and so we shall content ourselves by just stating a slightly more general version of the single-input single-output result called the off-axis circle theorem, due to Cho and Narendra (1968).

Consider the system

$$\left.\begin{aligned} F(g*e)(t) + e(t) &= x(t) \\ (g*e)(t) &= y(t) \end{aligned}\right\} \tag{5.4.22}$$

where F satisfies

$$\left.\begin{aligned} &F(0) = 0 \\ &K_1 \leqslant \frac{F(y_1) - F(y_2)}{y_1 - y_2} \leqslant K_2 \quad \forall y_1, y_2 \text{ and for } K_1, K_2 \in (-\infty, \infty) \end{aligned}\right\} \tag{5.4.23}$$

Then the circle theorem may now be stated as follows.

Theorem 5.4.7

Consider the system (5.4.22) satisfying the conditions (5.4.23) and let $G(s)$ be the Laplace transform of g. Then the system is asymptotically stable under one of the following conditions (see fig. 5.22)

(a) If $K_1 = 0$, then the polar plot of $G(i\omega)$ lies to the right of the line through $(-(1/K_2) + \delta, 0)$, for any $\delta > 0$.

(b) If $K_1 \neq 0$, then $G(i\omega)$ lies outside (for $K_1 > 0$) or inside (for $K_1 < 0$) a circle passing through $(-(1/K_2) + \delta, 0)$ and $(-(1/K_1) - \delta, 0)$ for some $\delta > 0$.

Finally we mention a generalization of absolute stability, due to Popov (1961), called **hyperstability**. (This will be used in chapter 8 to study adaptive control.) Consider the single-input single-output system

$$\left.\begin{aligned} \dot{x} &= Ax + bu(t) \\ y(t) &= c^{\mathrm{T}}x + u(t) \end{aligned}\right\} \tag{5.4.24}$$

and assume that the controls u are restricted to satisfy the inequality

$$\int_0^T u(t)f(t)\,\mathrm{d}t \leqslant \delta \sup_{0 \leqslant t \leqslant T} \|x(t)\| \tag{5.4.25}$$

for any $T > 0$, for some function f and some $\delta > 0$.

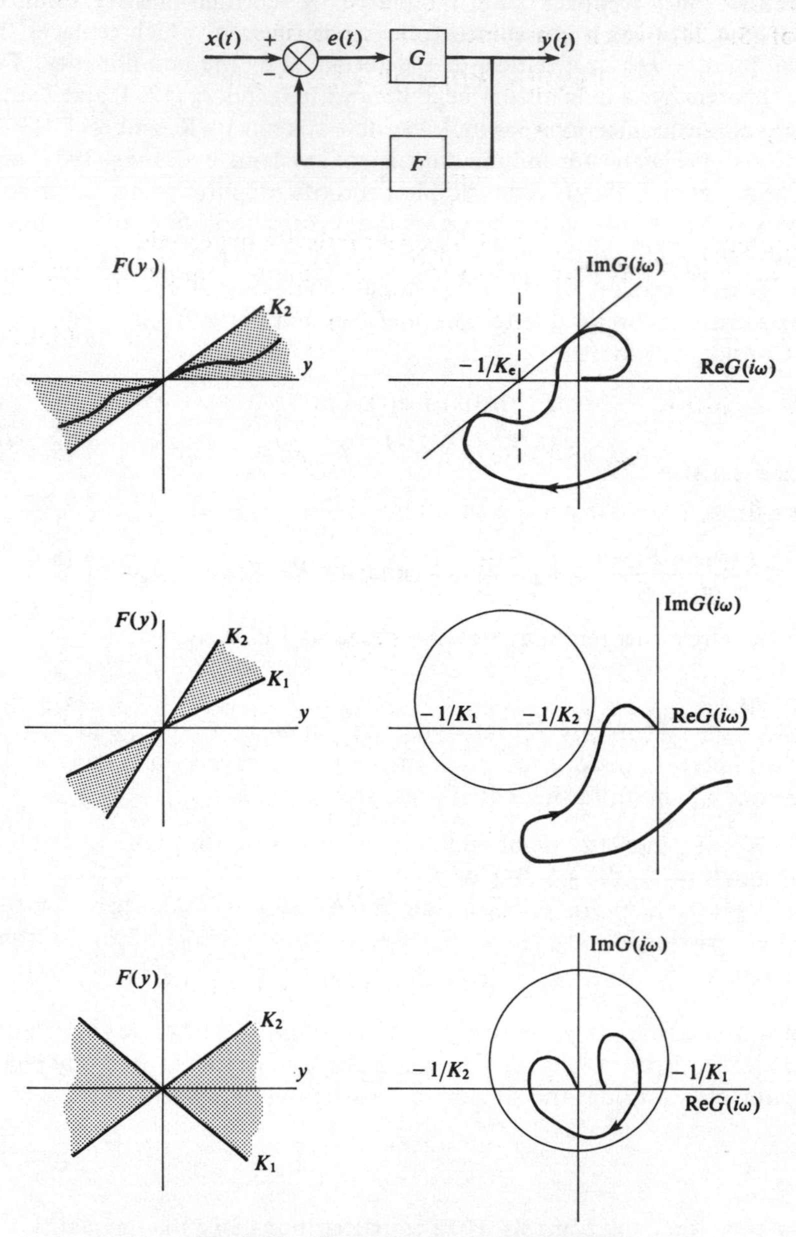

Fig. 5.22 Illustrating the circle theorem.

Then we say that (the solution $x = 0$, $u = 0$ of) the system (5.4.24) is **hyperstable** (with respect to f) if there exists $K > 0$ such that the solution $x(t)$ of (5.4.24) (which is assumed to exist) satisfies

$$\|x(t)\| \leqslant K(\|x(0)\| + \delta), \quad t \geqslant 0$$

for any control which satisfies (5.4.25). If

$$\lim_{t\to\infty} x(t) = 0$$

in addition, we say that (5.4.24) is **asymptotically hyperstable**.

To see how this is related to absolute stability, consider the system

$$\left.\begin{aligned} \dot{x} &= A_1 x + b_1 \phi(\sigma) \\ \sigma &= -c_1^{\mathrm{T}} x \end{aligned}\right\} \tag{5.4.26}$$

where ϕ satisfies

$$0 \leqslant \phi(\sigma)\sigma \leqslant h\sigma^2, \quad \text{for all } \sigma \tag{5.4.27}$$

Associate with this system the function

$$\begin{aligned} y(t) &= -\sigma - \nu\frac{\mathrm{d}\sigma}{\mathrm{d}t} + \frac{1}{h}\phi(\sigma(t)) \\ &= (c_1^{\mathrm{T}} + \nu c_1^{\mathrm{T}}\mathbf{A}_1)x + \left(\nu c_1^{\mathrm{T}} b_1 + \frac{1}{h}\right)\phi(\sigma(t)) \end{aligned} \tag{5.4.28}$$

Suppose that (5.4.26) is hyperstable with respect to y defined by (5.4.28). Then, for any ϕ satisfying (5.4.27), we have

$$\begin{aligned} \int_0^T \phi(\sigma(t))y(t)\,\mathrm{d}t &= -\int_0^T \phi(\sigma(t))\sigma(t)\,\mathrm{d}t - \nu\int_0^T \phi(\sigma(t))\frac{\mathrm{d}\sigma}{\mathrm{d}t}\,\mathrm{d}t \\ &\quad + \frac{1}{h}\int_0^T \phi^2(\sigma(t))\,\mathrm{d}t \\ &= \int_0^T \phi((\phi/h) - \sigma)\,\mathrm{d}t - \nu\int_0^T \phi(\sigma)\frac{\mathrm{d}\sigma}{\mathrm{d}t}\,\mathrm{d}t \\ &\leqslant \nu\left|\int_{\sigma(0)}^{\sigma(T)} \phi(\sigma)\,\mathrm{d}\sigma\right| \\ &\leqslant \delta \sup_{0\leqslant t\leqslant T}\|x(t)\| \end{aligned}$$

if

$$\delta = \nu\left|\int_{\sigma(0)}^{\sigma(T)} \phi(\sigma)\,\mathrm{d}\sigma\right| \Big/ \|x(0)\|$$

(Note that the integral is bounded by hyperstability.) Hence, hyperstability with respect to y implies absolute stability.

Now let $G(s) = c^{\mathrm{T}}(sI - A)^{-1}b + \gamma(\gamma = G(\infty))$. G is said to be **positive real** if

$$\text{Re } s \geqslant 0 \Rightarrow \text{Re } G(s) \geqslant 0$$

Then Popov's two main theorems on hyperstability may be stated as follows.

Theorem 5.4.8
Suppose that the system (5.4.24) is controllable and observable (see chapter 4), then it is hyperstable (with respect to y) if and only if $G(s)$ is positive real.

Theorem 5.4.9
If (5.4.24) is controllable and observable then it is asymptotically hyperstable (with respect to y) if and only if $G(s)$ has no poles in the closed right half plane and

$$\text{Re } G(j\omega) > 0 \quad \text{for all real } \omega$$

We note also that Anderson and Moore (1968) have shown that these results may be generalized to multivariable systems. To state the generalizations we must first define positive reality of a matrix transfer function $G(s)$. We say that $G(s)$ is **positive real** (respectively, **strictly** positive real) if

(a) $G(s)$ has elements which are analytic for $\text{Re } s > 0$ (respectively $\text{Re } s \geqslant 0$),
(b) $G^*(s) = G(s^*)$ for $\text{Re } s > 0$,
(c) $G^{\mathrm{T}}(s^*) + G(s)$ is non-negative (respectively, positive) definite for $\text{Re } s > 0$,
(*denotes complex conjugation).

Theorem 5.4.10
$G(s)$ defines a hyperstable system if and only if $G(s)$ is positive real. Moreover, $G(s)$ defines an asymptotically hyperstable system if and only if $G(s)$ is strictly positive real.

5.5 APPROXIMATE METHODS FOR NON-LINEAR OSCILLATIONS

5.5.1 The Describing Function

Since general non-linear systems have proved so difficult to study, many types of approximate methods have been proposed. The local linearization of systems discussed earlier is such a technique. A different form of linearization which one may apply is to consider the response of a non-

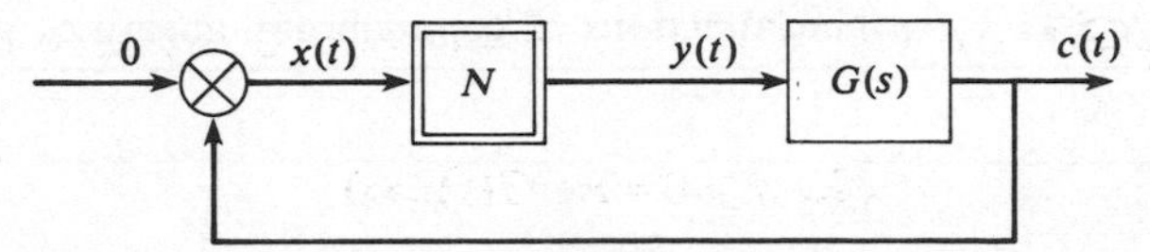

Fig. 5.23 A non-linear system for describing a function analysis.

linear system to a sinusoidal input, and retain only certain of the output harmonics in the analysis. In the simplest case, one assumes that the system is sufficiently low-pass to consider only the fundamental frequency. This leads to the **describing function**. More precisely, if we have a non-linear system of the form shown in fig. 5.23, then we assume that the input x of the non-linearity is given by

$$x(t) = M \sin \omega t$$

Then the output y is given by the Fourier series

$$y(t) = \frac{A_0}{2} + \sum_{k=1}^{\infty} A_k \cos k\omega t + \sum_{k=1}^{\infty} B_k \sin k\omega t$$

where

$$A_k = \frac{1}{\pi} \int_{-\pi}^{\pi} y(t) \cos k\omega t \, \mathrm{d}(\omega t), \quad k \geqslant 0$$

$$B_k = \frac{1}{\pi} \int_{-\pi}^{\pi} y(t) \sin k\omega t \, \mathrm{d}(\omega t), \quad k \geqslant 1$$

If N is an odd non-linearity, i.e. $N(x) = -N(-x)$ then $A_k = 0, k \geqslant 0$, because $\cos k\omega t$ is symmetric about $\omega = 0$.

The **describing function** (DF), $N(M, \omega)$, of the non-linearity N is defined as the ratio of the fundamental component of the output (in phasor form) to that of the input, i.e.

$$N(M, \omega) = \frac{B_1 + jA_1}{M} \tag{5.5.1}$$

or

$$N(M, \omega) = |N| \exp(j \angle N)$$

where

$$|N| = \frac{(B_1^2 + A_1^2)^{1/2}}{M}, \quad \angle N = \arctan \frac{A_1}{B_1}$$

The describing functions of some common non-linearities are given in table 5.1 (see Shinners, 1978).

(Note that the DF of a static non-linearity, i.e. one which depends on the input x and not its derivatives $\dot{x}, x, \ldots$, is independent of the input frequency ω.)

Table 5.1 Describing functions of common non-linearities

Non-linearity N	$\lvert N \rvert$	$\angle N$
(1) Dead zone	$\frac{2K}{\pi}\left(\frac{\pi}{2} - \frac{D}{M}\cos\sin^{-1}\frac{D}{M} - \sin^{-1}\frac{D}{M}\right)$	0
(2) Saturation	$\frac{2K}{\pi}\left(\frac{S}{M}\cos\sin^{-1}\frac{S}{M} + \sin^{-1}\frac{S}{M}\right)$	0
(3) Backlash	$A_1 = \frac{2D}{\pi M}\left(\frac{2D}{M} - 2\right)M$ $B_1 = \frac{1}{\pi}\left[\frac{\pi}{2} - \sin^{-1}\left(\frac{2D}{M} - 1\right) - \left(\frac{2D}{M} - 1\right) \times \cos\sin^{-1}\left(\frac{2D}{M} - 1\right)\right]M$ $\lvert N \rvert = \frac{1}{M}(A_1^2 + B_1^2)^{1/2}$	$\arctan\frac{A_1}{B_1}$
(4) On-off element with hysteresis	$A_1 = -\frac{2K}{\pi}\left(\frac{h}{M}\right)$ $B_1 = \frac{2K}{\pi}\left[\cos\sin^{-1}\frac{D+h}{M} - \cos\left(\pi - \sin^{-1}\frac{D}{M}\right)\right]$ $\lvert N \rvert = \frac{D}{M}(A_1^2 + B_1^2)^{1/2}$	$\arctan\frac{A_1}{B_1}$

Returning to the system in fig. 5.23, suppose again that

$$x(t) = M\sin\omega t = \operatorname{Im}[M\exp(j\omega t)]$$

Then the output of N is approximately

$$y(t) = \operatorname{Im}\{\lvert N(M,\omega)\rvert M\exp[j(\omega t + \phi_1)]\} + (\text{terms of higher frequencies})$$

where $\phi_1 = \angle N(M, \omega)$, and so the output c of G is given approximately by†

$$c(t) = \text{Im}\{\, |N(M, \omega)| \, M \, |G(j\omega)| \exp[j(\omega t + \phi_1 + \phi_2)]\,\}$$

where $\phi_2 = \angle G(j\omega)$. For a sustained oscillation, we require $x(t) = -c(t)$ and so

$$1 + N(M, \omega)\, G(j\omega) = 0$$

or

$$\boxed{G(j\omega) = -\frac{1}{N(M, \omega)}} \tag{5.5.2}$$

is the condition for a limit cycle. Hence any oscillation with amplitude M and frequency ω satisfying (5.5.2) 'probably' gives rise to a limit cycle. A great deal of work has been done justifying the describing function method rigorously, and precise conditions for the validity of the condition (5.5.2) can be obtained (see, for example, Bergen and Franks (1971) and Mees and Bergen (1975), and for multi-input describing functions – (where instead of a single sinusoidal input $M \sin \omega t$ we take a sum of the form

$$\sum_{k=1}^{K} a \sin(k\omega t + \theta_k)$$

– see Mees (1972) and Swern (1983)).

The stability of the limit cycles determined by the DF method may easily be determined from the amplitude–phase plot shown in fig. 5.24. For example, consider the limit cycle corresponding to the point 1 in this diagram. Any increase in the amplitude of the limit cycle M will decrease $|N(M)|$ and thus lead to a stable response which will tend to return to the limit cycle, and conversely any decrease in M will produce an unstable response which will grow and again return to the limit cycle. Thus point 1 corresponds to a stable limit cycle. Similar arguments apply to points 2 and 3.

From (5.5.2) we note that we can plot the loci of $G(j\omega)$ and $-1/N(M, \omega)$ in the Nichol's chart or a Nyquist diagram for various values of M and ω. If these loci intersect we 'probably' have a limit cycle in the system. This diagram can be used in a design technique, since we can introduce a compensator $KG_1(j\omega)$ into the system and plot the loci of $KGG_1(j\omega)$ and $-1/N(M, \omega)$ and search for intersections, for various values of the gain K and different poles and zeros of G_1.

Another important aspect of non-linear systems which we have not yet mentioned is the possibility of subharmonic oscillations, i.e. when a sinusoidal input of a certain frequency ω produces an output containing sinusoids with frequencies which are submultiples of ω. In order to study

† provided G is a low-pass filter.

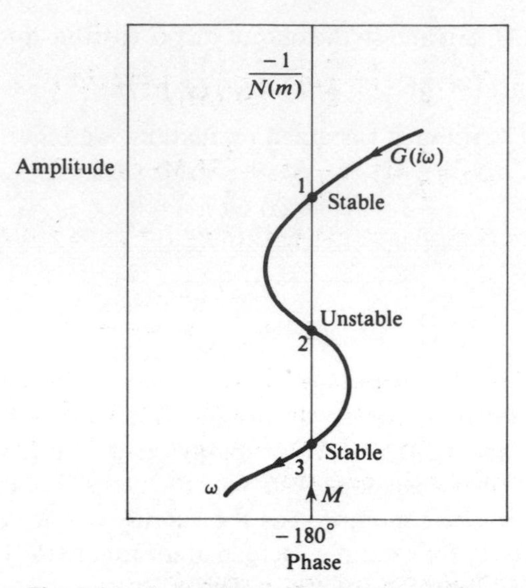

Fig. 5.24 Determining limit cycle stability.

this phenomenon in the simplest case we consider the **two input describing function,** where the input is now

$$x(t) = M_1 \sin(\omega_1 t + \theta_1) + M_2 \sin(\omega_2 t + \theta_2)$$

Generalizing (5.5.1), we define the describing functions

$$N_1 = \frac{\text{phasor representation of output component of frequency } \omega_1}{\text{phasor representation of input component of frequency } \omega_1}$$

$$N_2 = \frac{\text{phasor representation of output component of frequency } \omega_2}{\text{phasor representation of input component of frequency } \omega_2}$$

Note that N_1 and N_2 depend on M_1, M_2, ω_1, ω_2, θ_1 and θ_2. We shall illustrate subharmonic oscillations in the case of the cubic non-linearity

$$y = x^3$$

discussed by West *et al.* (1956) and Gelb and Vander Velde (1968). Suppose that the input is of the form

$$x = M_1 \cos \omega t + M_2 \cos\left(\frac{\omega t}{3} + \theta\right)$$

Then it is easy to show that the output is

$$y = \left(\frac{3M_1^3}{4} + \frac{3M_1M_2^2}{2}\right)\cos \omega t + \left(\frac{3M_1^2M_2}{2} + \frac{3M_2^3}{4}\right)\cos\left(\frac{\omega t}{3} + \theta\right)$$

$$+ \frac{3M_1M_2^2}{4}\cos\left(\frac{-\omega t}{3} + 2\theta\right) + \frac{M_2^3}{4}\cos(\omega t + 3\theta) + (\text{terms of other frequencies})$$

Hence

$$N_1 = \frac{3}{4}M_1^2 + \frac{3}{2}M_2^2 + \frac{1}{4}\left(\frac{M_2^3}{M_1}\right)\exp(3j\theta)$$

$$N_2 = \frac{3}{2}M_1^2 + \frac{3}{4}M_2^2 + \frac{3}{4}M_1M_2\exp(-3j\theta)$$

N_2 is the one-third subharmonic describing function and has phase shift β given by

$$\tan\beta = \frac{-M\sin 3\theta}{2 + M^2 + M\cos 3\theta}$$

where $M = M_2/M_1$. The maximum and minimum values of β are approximately $\pm 21^\circ$. Hence, if a system contains the cubic non-linearity and a linear element whose phase shift is less than 159° in magnitude, then the system has no one-third subharmonics.

5.5.2 The Dual-input Describing Function

It is important to be able to deal with systems of the form shown in fig. 5.23 which have a non-zero input $r(t)$ (we have assumed so far that $r = 0$). This allows us to consider forced oscillations and can be studied approximately with the **dual-input describing function** (DIDF). In this case we assume that the input r is slowly varying so that we may consider r to be constant and consider the input

$$x(t) = M_1 + M_2\sin\omega t$$

to the non-linear element where M_1 is 'small' relative to M_2 and the non-linear characteristic. Then approximating the output y of the non-linear element by

$$y = \frac{A_0}{2} + A_1\cos\omega t + B_1\sin\omega t$$

we define the describing functions

$$N_1 = \frac{A_0}{2M_1}$$

$$N_2 = \frac{B_1 + jA_1}{M_2}$$

where, as before,

$$A_k = \frac{1}{\pi}\int_{-\pi}^{\pi} y(M_1 + M_2\sin\omega t)\cos k\omega t\,\mathrm{d}(\omega t), \quad k = 0,1$$

$$B_1 = \frac{1}{\pi}\int_{-\pi}^{\pi} y(M_1 + M_2\sin\omega t)\sin\omega t\,\mathrm{d}(\omega t)$$

Table 5.2. Common dual input describing functions

Non-linearity N	N_1	N_2
(1) Ideal relay 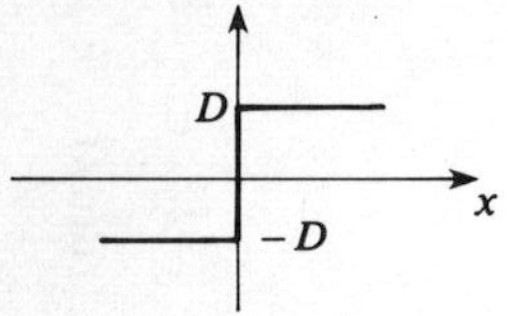	$\frac{2D}{\pi M_1}\sin^{-1}\frac{M_1}{M_2}$	$\frac{4D}{\pi M_2}\left(1-\left(\frac{M_1}{M_2}\right)^2\right)^{1/2}$
(2) Hysteresis 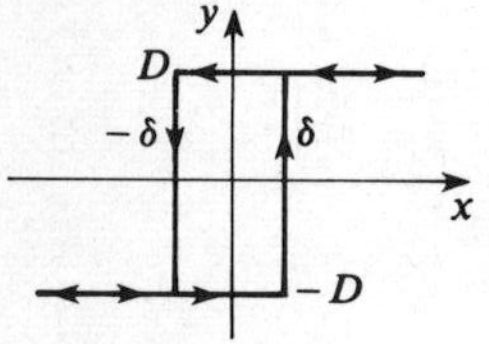	$\frac{D}{\pi M_1}\left(\sin^{-1}\left(\frac{\delta+M_1}{M_2}\right)-\sin^{-1}\left(\frac{\delta-M_1}{M_2}\right)\right]$	$\frac{2D}{\pi M_2}(\exp\{-j\sin^{-1}[(\delta-M_1)/M_2]\}$ $+\exp\{-j\sin^{-1}[(\delta+M_1)/M_2]\})$
(3) Piecewise-linear element 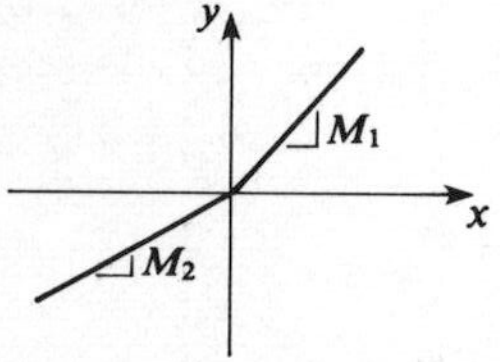	$\frac{M_1+M_2}{2}+\frac{M_1-M_2}{\pi}$ $\times\left[\sin^{-1}\left(\frac{M_1}{M_2}\right)+\frac{M_2}{M_1}\left(1-\left(\frac{M_1}{M_2}\right)^2\right)^{1/2}\right]$	$\frac{M_1+M_2}{2}+\frac{M_1-M_2}{\pi}$ $\times\left[\sin^{-1}\left(\frac{M_1}{M_2}\right)+\frac{M_1}{M_2}\left(1-\left(\frac{M_1}{M_2}\right)^2\right)^{1/2}\right]$

Table 5.2 gives some of the common DIDFs (see Gelb and Vander Velde, 1968).

The value N_1 essentially represents the open-loop d.c. gain of the non-linearity and N_2 represents the limit cycle gain. We may show that a system which is limit cycling is inherently adaptive in the following way. Suppose, for simplicity, that the non-linearity is an ideal relay, so that we have

$$N_1 \approx \frac{2D}{\pi M_2}, \quad N_2 \approx \frac{4D}{\pi M_2}$$

from table 5.2 and so $N_1/N_2 \approx \frac{1}{2}$. Now the open-loop signal transfer function of the system in fig. 5.23 is $N_1 G(j\omega) \approx \frac{1}{2} N_2 G(j\omega)$. However, for a sustained limit cycle of frequency ω_0, we must have

$$N_2 G(j\omega_0) = -1$$

and so the open-loop signal transfer function is constrained to pass through the point $\frac{1}{2} \angle -180^\circ$, independent of any parameter variations in G. This will be useful later in our study of adaptive control.

5.5.3 Artificial Dither

An important application of the dual-input describing function is to the idea of artificial dither, which can be introduced into a non-linear system to remove limit cycles (called signal stabilization). Consider an ideal relay with input equal to the signal x plus a sinusoidal dither waveform,† which we assume has frequency much higher than that of x (cf. fig. 5.25(a)). Then, if $M_1 = x_{\text{average}}$, we have

$$e(t) = M_1 + M_d \sin \omega_d t,$$

over a period of the dither signal. Using the dual-input describing function for this system we define the equivalent signal gain as

$$N_1 = \frac{y_{av}}{M_1}$$

(where av is average). From table 5.2, we have

$$N_1 = \begin{cases} \dfrac{2D}{\pi M_1} \sin^{-1} \dfrac{M_1}{M_2} & |M_1| < D \\ +D & M_1 \geqslant D \\ -D & M_1 \leqslant -D \end{cases}$$

---

† Note that dither is not necessarily sinusoidal; see Mossaheb (1983) for an exact analysis of dither systems.

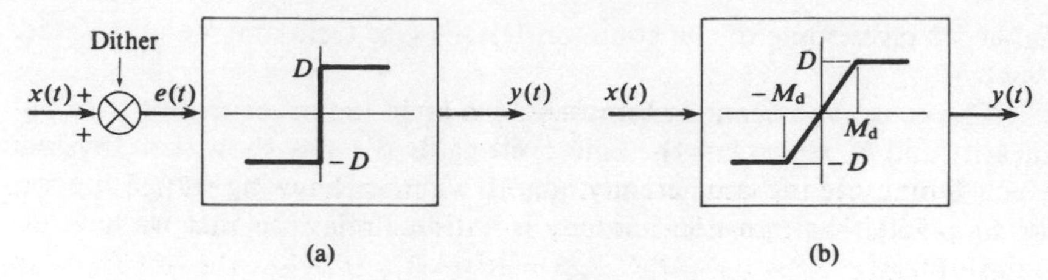

Fig. 5.25 Artificial dither.

and so we may regard the system in fig. 5.25(a) with dither as being equivalent to the system with the modified non-linear element and no dither shown in fig. 5.25(b). We may then consider the stability of any system containing the original ideal relay plus dither by using the equivalent non-linearity and the ordinary single-input describing function. To illustrate this method consider the system discussed by Oldenburger and Boyer (1962) (see Gelb and Vander Velde, 1968) and shown in fig. 5.26. Note that the equivalent non-linearity has also been included. The describing function of the equivalent circuit is determined from

$$N_2 = \frac{4}{\pi M}\int_0^{\pi/2} \frac{2D}{\pi} \sin^{-1}(k \sin \omega t) \sin \omega t \, \mathrm{d}(\omega t)$$

$$= \frac{8D}{\pi^2 M}\frac{1}{k}\left(\int_0^{\pi/2} (1-k^2\sin^2\psi)^{1/2}\,\mathrm{d}\psi + \frac{k^2-1}{k}\int_0^{\pi/2}\frac{\mathrm{d}\psi}{(1-k^2\sin^2\psi)^{1/2}}\right)$$

where $k = M/M_{\mathrm{d}} (< 1)$. Hence

$$N_2 = \frac{8D}{\pi^2 M_{\mathrm{d}} k^2}\,[E(k) + (k^2-1)K(k)] \quad (k < 1)$$

where E and K are elliptic integrals (see Abramowitz and Stegun, 1965). For

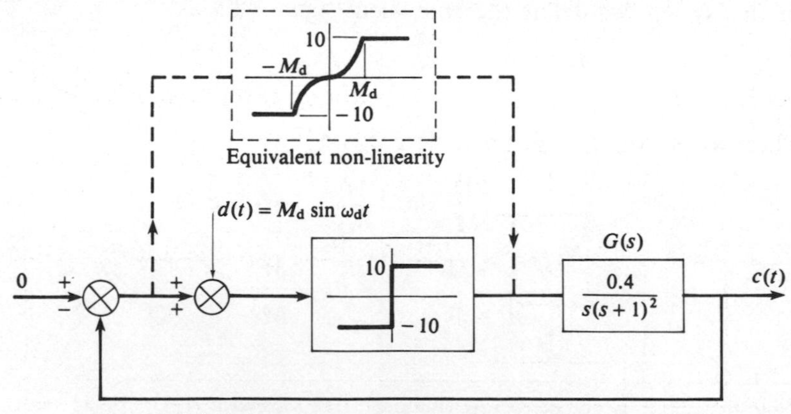

Fig. 5.26 Equivalent non-linearity for a dither system.

$k > 1$ we have

$$N_2 = \frac{8D}{\pi^2 M_d k} E\left(\frac{1}{k}\right)$$

For a limit cycle to occur at, say, $\omega_0 = 1$ we must have $N_2 \mid G(i) \mid = 1$ and so $N_2 = 5$. Using numerical values for the elliptic integrals we see that $(M_d N_2/D)_{max} = 0.85$, i.e. $(M_d N_2)_{max} = 8.5$ if D = 10. Now the $-1/N_2$ locus will not cut the $G(j\omega)$ locus if $N_{max} < 5$ and so no limit cycle will be present if

$$M_d \geqslant \frac{8.5}{5} = 1.7$$

5.5.4 Describing Functions for Sampled Systems

Systems which contain digital elements may also be analysed by the describing function method but the technique is much more delicate than for continuous systems. We shall illustrate the method with the system shown in fig. 5.27, which contains a two level relay, a sampler and zero order hold (ZOH) and a linear element $G(s)$ which may contain discrete factors. First consider the effect of sampling. If $x(t)$ is an input sinusoid of frequency ω then y has frequencies $\pm k\omega$, $k = 0,1,\ldots,$ due to the non-linearity. Now (referring to the sampling theorem in section 4.7.3) we have

$$Y^*(j\omega) = \frac{1}{T_s} \sum_{l=-\infty}^{\infty} Y[j(\omega + l\omega_s)] \tag{5.5.3}$$

where T_s and ω_s are the sampling time and frequency respectively. It follows that $y^*(t)$ contains frequencies $\pm k\omega \pm l\omega_s$, and so by (5.5.3), $y^*(t)$ may contain frequencies lower than ω unless $n\omega = \omega_s$ for some positive integer n. We shall assume that this condition holds in future, although it is as well

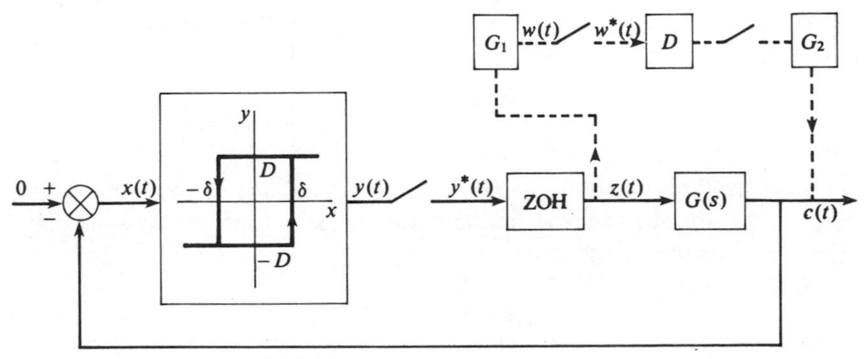

Fig. 5.27 A system with hysteresis.

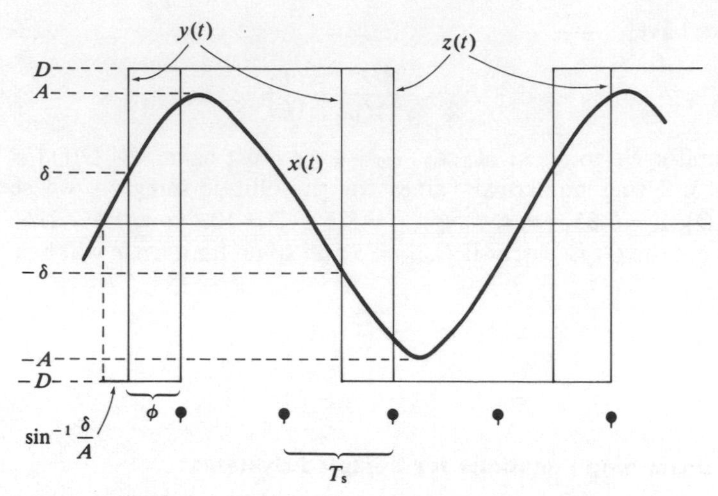

Fig. 5.28 Evaluation of the describing function.

to be aware of the possibility of subharmonics. In fact, we assume even more; n will be even since these modes are to be expected if $G(s)$ contains an integrator.

Since y^* is not $1/T_s$ times the fundamental component of y we must evaluate the describing function between x and y^* and not x and y. Consider therefore an input $x(t)$ with period $4T_s$. Then the outputs y and z are shown in fig. 5.28 and so the describing function between x and z is clearly

$$N(\phi) = \frac{4D}{\pi M} \angle \left(-\sin^{-1} \frac{\delta}{M} - \phi \right), \quad 0 < \phi < \frac{\pi}{n}, M > \delta$$

since the phase lag ϕ relative to x can be any value between 0 and π/n (i.e. 0 to T_s in time).

Suppose now that G takes the form shown connected by dotted lines between z and c in fig. 5.27, where D is a digital filter and G_2 contains a hold. Then

$$C(j\omega) = \frac{1}{T_s} G_2(j\omega)\, D(j\omega) \sum_{l=-\infty}^{\infty} G_1(j(\omega + l\omega_s))\, Z(j(\omega + l\omega_s))$$

Hence, if we ignore higher harmonics than ω in Z due to the sampler at y then $\omega < \frac{1}{2}\omega_s$ implies that the fundamental transfer function between z and c (i.e. of frequency ω) is just

$$\frac{1}{T_s} G_1(j\omega) D(j\omega)\, G_2(j\omega)$$

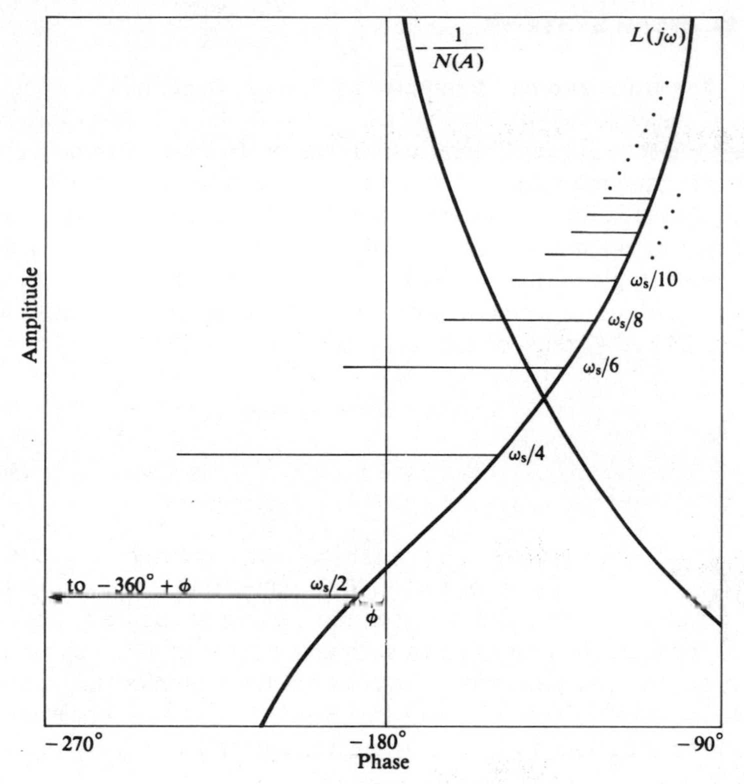

Fig. 5.29 Estimation of limit cycles.

The condition for a limit cycle is again

$$1 + N(\phi)\, G(j\omega) = 0$$

and if we include the phase lag ϕ in $G(j\omega)$, then we must have

$$\boxed{-\frac{1}{N} = \exp(j\phi)\, G(j\omega), \quad 0 < \phi < \frac{\pi}{n}} \tag{5.5.4}$$

where N is just the continuous system describing function for the two level relay.

If we plot (5.5.4) on an amplitude-phase diagram we obtain fig. 5.29 where the unknown phase shift ϕ becomes a series of lines parallel to the phase axis starting at the points $\omega = 1/2n(\omega_s)$ on $G(j\omega)$. This graph indicates two possible limit cycles at $\omega = \frac{1}{6}\,\omega_s$ and $\omega = \frac{1}{8}\,\omega_s$.

For a more detailed discussion of discrete describing functions see Gelb and Vander Velde (1968), Vidal (1969).

5.6 BILINEAR SYSTEMS

5.6.1 The Structure and Examples of Bilinear Systems

As we have seen above, general non-linear systems are extremely difficult to deal with and most results which can be obtained are of an approximate nature. This is in direct contrast to the theory of linear systems, which we considered in chapter 4, where a complete theory of control (at least for single-input single-output systems) can be worked out. In an attempt to recapture some of the interesting properties of linear systems, many authors have considered systems of the form

$$\begin{aligned} \dot{x} &= Ax + Bu + \sum_{i=1}^{p} N_i x u_i \\ &= Ax + Bu + Nxu \\ y &= Cx \end{aligned} \qquad (5.6.1)\dagger$$

where $x \in \mathbb{R}^n$, A, B, C, N_i are matrices of appropriate sizes and $u = (u_1, \ldots, u_p)^{\mathrm{T}}$ is a control vector. In a sense, this is the simplest non-linear system which one may consider and contains a quadratic term in the control and state. It turns out that such an equation models many types of real systems such as a d.c. motor, nuclear reactor dynamics and many types of chemical and biological systems (see Bruni *et al.*, 1974). For example, a well-known model of a d.c. motor is (see chapter 1):

$$\frac{\mathrm{d}i_{\mathrm{a}}}{\mathrm{d}t} = -\frac{R_{\mathrm{a}}}{L_{\mathrm{a}}} i_{\mathrm{a}} + \frac{1}{L_{\mathrm{a}}} v_{\mathrm{a}} + K i_{\mathrm{e}} \omega$$

$$\frac{\mathrm{d}\omega}{\mathrm{d}t} = -\frac{F\omega}{J} + \frac{K}{J} i_{\mathrm{a}} i_{\mathrm{e}}$$

or

$$\frac{\mathrm{d}}{\mathrm{d}t}\begin{pmatrix} i_{\mathrm{a}} \\ \omega \end{pmatrix} = A\begin{pmatrix} i_{\mathrm{a}} \\ \omega \end{pmatrix} + B\begin{pmatrix} i_{\mathrm{e}} \\ v_{\mathrm{a}} \end{pmatrix} + N_1\begin{pmatrix} i_{\mathrm{a}} \\ \omega \end{pmatrix} i_{\mathrm{e}}$$

where

$$A = \begin{pmatrix} -R_{\mathrm{a}}/L_{\mathrm{a}} & 0 \\ 0 & -F/J \end{pmatrix}$$

$$B = \begin{pmatrix} 0 & 1/L_{\mathrm{a}} \\ 0 & 0 \end{pmatrix}$$

$$N_1 = \begin{pmatrix} 0 & K \\ K/J & 0 \end{pmatrix}$$

† Here we interpret Nxu as $u_1 N_1 x + u_2 N_2 x + \ldots + u_p N_p x$.

$$\left.\begin{aligned} i_e &= \text{stator current} \\ v_a &= \text{rotor voltage} \end{aligned}\right\}\text{controls}$$

$$\left.\begin{aligned} i_a &= \text{rotor current} \\ \omega &= \text{axis speed} \end{aligned}\right\}\text{state variables}$$

and where R_a, L_a are the rotor electric parameters, F, J are mechanical parameters of the load and K is the torque constant.

Another example is derived from the neutron level control problem in a fission reactor (Mohler, 1973):

The neutron population is described by the equations

$$\left.\begin{aligned} \frac{dn}{dt} &= \frac{u-\beta}{l}n + \lambda c \\ \frac{dc}{dt} &= \frac{\beta}{l}n - \lambda c \end{aligned}\right\} \tag{5.6.2}$$

where $n, c > 0$ and

$$\left.\begin{aligned} n &= \text{neutron population} \\ c &= \text{precursor population} \end{aligned}\right\}\text{state variables}$$
$$u = \text{reactivity (control)}$$

Choosing non-dimensional state variables

$$x_1 = \frac{n - n_e}{n_e} \quad (> -1)$$

$$x_2 = \frac{c - c_e}{c_e} \quad (> -1)$$

we obtain the bilinear system

$$\dot{x} = Ax + bd(x)u \tag{5.6.3}$$

where

$$A = \begin{pmatrix} -\beta/l & \beta/l \\ \lambda & -\lambda \end{pmatrix}$$

$$d(x) = \frac{x_1 + 1}{l}$$

$$b = (1 \quad 0)^T$$

Another important description of a system is in terms of the input–output map as we have seen in the case of the linear system

$$\begin{aligned} \dot{x} &= Ax + Bu, \quad x(0) = x \\ y &= Cx \end{aligned}$$

where the variation of constants formula leads to the input–output relation

$$y(t) = C \exp(At)\, x + \int_0^t C \exp[A(t-\tau)]\; Bu(\tau)\, d\tau$$

We can obtain the analogue of this for the bilinear system (5.6.1) by noting that the sequence of solutions of the equations

$$\begin{aligned} \dot{x}_0 &= Ax_0 + Bu, \quad x(0) = \hat{x} \\ \dot{x}_i &= Ax_i + Nx_{i-1}u + Bu, \quad x_i(0) = \hat{x},\ i = 1,2,\ldots \end{aligned}$$

converges to the solution of (5.6.1), as shown by Bruni *et al.* (1974). Hence the identity

$$x = x_0 + \sum_{i=1}^{\infty} (x_i - x_{i-1})$$

leads to the input–output relation

$$\begin{aligned} y(t) = C \exp(At)_{\hat{x}} + \sum_{i=1}^{\infty} \int_0^t \int_0^{t_1} \cdots \int_0^{t_{i-1}} C \exp[A(t-t_1)] \\ N(\exp[A(t_1-t_2)]N(\ldots \exp[A(t_{i-1}-t_i)]N \exp(At_i)\hat{x})\, u(t_i))\ldots) \\ u(t_1)\, dt_1 \ldots dt_i \\ + \sum_{i=1}^{\infty} \int_0^t \int_0^{t_l} \cdots \int_0^{t_{i-1}} C \exp[A(t-t_1)]\; N(\exp[A(t_1-t_2)] \\ N(\ldots N(\exp[A(t_{i-1}-t_i)]\; Bu(t_i))\ldots)\, u(t_1)\, dt_i \ldots dt_1 \end{aligned} \tag{5.6.4}$$

The third term on the right is the zero initial state response, and can be written in the form

$$\sum_{i=1}^{\infty} \underbrace{\int_0^t \cdots \int_0^t}_{i} \sum_{j_1, j_2, \ldots, j_i = 1}^{p} v_i^{(j_1, j_2, \ldots, j_i)}(\tau_1, \tau_2, \ldots, \tau_i) \times u_{j_1}(t-\tau_1) u_{j_2}(t-\tau_2) \ldots u_{j_i}(t-\tau_i)\, d\tau_1\, d\tau_2 \ldots d\tau_i \tag{5.6.5}$$

where

$$v_i^{(j_1, j_2, \ldots, j_i)}(\tau_1, \tau_2, \ldots, \tau_i) = \begin{cases} C \exp(A\tau_1) N_{j_1} \exp[A(\tau_2-\tau_1)]\; N_{j_2} \ldots N_{j_i - 1} \exp[A(\tau_i - \tau_{i-1})]\, b_{j_i} \\ \tau_1 \leqslant \tau_2 \leqslant \ldots \leqslant \tau_i \\ 0 \quad \text{otherwise} \end{cases} \tag{5.6.6}$$

(b_i, $i = 1, \ldots, p$ are the columns of B). If we set

$$w_i^{(j_1, j_2, \ldots, j_i)}(\tau_1, \tau_2, \ldots, \tau_i) = \frac{1}{i!} \sum\nolimits^{*} v_i^{(j_1, j_2, \ldots, j_i)}(\tau_1, \tau_2, \ldots, \tau_i)$$

where $\sum^{*}$ is taken over all permutations of the indices $(1, \ldots, i)$, then we obtain the input–output relation (zero initial conditions)

$$y(t) = \sum_{i=1}^{\infty} \underbrace{\int_0^t \cdots \int_0^t}_{i} \sum_{j_1, j_2, \ldots, j_i = 1}^{p} w_i^{(j_1, j_2, \ldots, j_i)}(\tau_1, \tau_2, \ldots, \tau_i)$$

$$\times u_{j_1}(t-\tau_1)u_{j_2}(t-\tau_2)\ldots u_{j_i}(t-\tau_i)\,d\tau_1\,d\tau_2\ldots d\tau_i \qquad (5.6.7)$$

Such an expression is called a **Volterra series** and it can be shown (Brilliant, 1958) that any non-linear functional relation can be approximated arbitrarily closely by a Volterra series.

In the case of a scalar control, (5.6.7) becomes

$$y(t) = \sum_{i=1}^{\infty} \underbrace{\int_0^t \cdots \int_0^t}_{i} w_i(\tau_1, \ldots, \tau_i)u(t-\tau_1)u(t-\tau_2)\ldots u(t-\tau_i)\,d\tau_1\ldots d\tau_i \qquad (5.6.8)$$

5.6.2 Generalized Laplace Transform

One may generalize the notion of frequency response to a non-linear system by writing

$$y_i(t_1, \ldots, t_i) = \int_0^{t_1} \cdots \int_0^{t_i} w_i(\tau_1, \ldots, \tau_i)u(t_1-\tau_1)\ldots u(t_i-\tau_i)\,d\tau_1\ldots d\tau_i$$

Note that $y_i(t, \ldots, t)$ is just the ith term of the series (5.6.8). Then, taking the i-dimensional Laplace transform, we have

$$Y(s_1, \ldots, s_i) = \underbrace{\int_0^{\infty} \cdots \int_0^{\infty}}_{i} y_1(t_1, \ldots, t_i) \exp[-(s_1t_1 + \ldots + s_it_i)]\,dt_1\ldots dt_i$$

$$= W_i(s_1, \ldots, s_i)U(s_1)\ldots U(s_i)$$

where

$$W_i(s_1, \ldots, s_i) = \int_0^{\infty} \cdots \int_0^{\infty} w_i(t_1, \ldots, t_i) \exp[-(s_1t_1 + \ldots + s_it_i)]\,dt_1\ldots dt_i$$

$$U(s_j) = \int_0^{\infty} u(t) \exp(-s_jt)\,dt$$

The ith homogeneous term therefore has the **transfer function** $H_i(s_1, \ldots, s_i)$, defined by

$$H_i(s_1, \ldots, s_i) = W_i(s_1, \ldots, s_i) = \frac{Y(s_1, \ldots, s_i)}{U(s_1)\ldots U(s_i)}$$

If the kernel† is given by (5.6.6), then the ith transfer function is clearly

$$H_i(s_1, \ldots, s_i) = c(s_iI - A)^{-1}N(s_{i-1}I - A)^{-1}N \ldots N(s_1I - A)^{-1}b \qquad (5.6.9)$$

for a single-input single-output system (i.e. $p = 1$, $N = N_1$, $b = b_1$, $c = C$). The stability of the overall system is therefore determined by the eigenvalues of A.

Given a sequence of transfer functions of the form (5.6.9), one may also ask the converse question; namely, does there exist a bilinear state-space representation which has these transfer functions? The answer is in the affirmative, although it is beyond the scope of the present book to discuss this aspect of bilinear systems theory. This is part of **realization theory** and the reader is referred to Mitzel *et al.* (1979) and Isidori (1973) for a full discussion.

5.6.3 Stabilizing Controllers for Bilinear Systems

We shall illustrate the control of a simple bilinear system by considering the neutron level problem (5.6.2). Our discussion will follow Gutman (1981). Referring to (5.6.3), we see that A has a zero eigenvalue and so the stabilization is non-trivial. Note that $x_1 > -1$ and so

$$d(x) = \frac{x_1 + 1}{l} > 0$$

and we may define the control

$$u = \frac{k^{\mathrm{T}}x}{d(x)}$$

for suitable k. The controlled system (5.6.3) then becomes

$$\dot{x} = Ax + bk^{\mathrm{T}}x$$

and we must choose k so that $A + bk^{\mathrm{T}}$ is a stable matrix.

With the parameters

$$l = 10^{-5}, \quad \beta = 0.0065, \quad \lambda = 0.4$$

it is found that

$$k^{\mathrm{T}} = (648.4 \quad -650.9)$$

produces a reasonable stable controller.

---

† The **kernel** of an integral operator

$$f(t) \to \int_0^t k(t, \tau)f(\tau)\,\mathrm{d}\tau$$

is, by definition, the function $k(t, \tau)$.

Gutman (1981) also considers a more general control law for a bilinear system of the form

$$\dot{x} = Ax + b(c^{\mathrm{T}}x + 1)u$$

by defining

$$S_+ = \{x : d(x) > 0\}$$
$$S_0 = \{x : d(x) = 0\}$$
$$S_- = \{x : d(x) < 0\}$$

leading to the control

$$u_+ = \frac{k_+^T u}{d(x)} \quad \text{if } x \in S_+$$

$$u_0 = 0 \quad \text{if } x \in S_0$$

$$u_- = \frac{k_-^{\mathrm{T}} u}{d(x)} \quad \text{if } x \in S_-$$

called a **division controller.** (A more refined version is presented by Gutman (1981).) There is a strong connection between such a controller and variable structure systems as we shall see later. Also, for a discussion of the controllability of bilinear systems, see Murthy (1979).

5.7 EXERCISES

1. Find all the equilibrium points of the differential equation

$$\dot{x}_1 = \sin x_2 + x_1 \tag{1}$$

$$\dot{x}_2 = x_1 + 1 \tag{2}$$

and linearize the equations about each point.

2. Repeat exercise 1 when equation (1) is replaced by

$$\dot{x}_1 = \sin \frac{1}{x_2} + x_1 \tag{1$'$}$$

and (2) is as before.

3. Does the equation

$$\dot{x} = \exp\left(-\frac{1}{x^2}\right)$$

have a linearization? (Is $\exp(-1/x^2)$ real analytic at $x = 0$?)

4. Linearize the non-linear wave equation

$$\frac{\partial^2 \phi}{\partial t^2} = \frac{\partial^2 \phi}{\partial x^2}(1 + \phi), \quad x \in [0, 1]$$

with $\phi(0, t) = \phi(1, t) = 0$. (*Hint:* put $\psi_1 = \phi$, $\psi_2 = \partial\phi/\partial t$)

5. Draw the phase plane trajectories near each equilibrium point of the system

$$\begin{aligned}\dot{x}_1 &= x_1^2 + x_2^2 - 4 + x_3^3\\ \dot{x}_2 &= x_1x_2 + x_2x_3\\ \dot{x}_3 &= -x_3\end{aligned}$$

6. Repeat exercise 5 for the system

$$\frac{d^4x}{dt^4} = \left(\frac{d^2x}{dt^2}\right)^2 + \cos\left(\frac{dx}{dt}\right) + x - 1$$

7. (a) Draw the **global** phase plane trajectories of the non-linear oscillator

$$\begin{aligned}\dot{\theta} &= \omega\\ \dot{\omega} &= -a\sin\theta - b\omega\end{aligned}$$

(b) (For mathematicians) Can you interpret this as a flow on a compact manifold?

8. Prove that the system

$$\ddot{x} + \omega\dot{x} + Kx^3 = 0, \quad K, \omega > 0$$

is asymptotically stable at $x = 0$.

9. Show that the system

$$\ddot{\theta} - \omega_n\dot{\theta} + \omega_n^2\theta = -\omega_n(\alpha\dot{\theta}^3 + \beta\theta^2\dot{\theta})$$

has a limit cycle between the circles

$$\theta^2 + \dot{\theta}^2 = \alpha^2$$

and

$$\theta^2 + \dot{\theta}^2 = \beta^2$$

10. Apply Krasovskii's method to show that the system

$$\begin{aligned}\dot{x}_1 &= f_1(x_1) + f_2(x_2)\\ \dot{x}_2 &= x_1 + ax_2\end{aligned}$$

is globally asymptotically stable if we have

(a) $f_1'(x_1) < 0$, for all x_1,

(b) $4af_1'(x_1) - [1 + f_2'(x_2)]^2 > 0$ for all $x = (x_1, x_2) \neq 0$,

(c) $[f_1(x_1) + f_2(x_2)]^2 + [x_1 + ax_2]^2 \to \infty$ as $\|x\| \to \infty$.

11. Prove lemmas 5.4.1 and 5.4.2.

12. Apply the Popov criterion to the system

$$\dddot{y} + a\ddot{y} + b\dot{y} + f(y) = 0, \quad a, b > 0$$

where $0 < f(y)/y < k$ and obtain a condition on k for the system to be asymptotically stable.

13. Derive the describing functions in tables 5.1 and 5.2.

14. Consider the system in fig. 5.23 and suppose that N is a dead zone with slope K_1 (table 5.1 (1)) and

$$G(j\omega)=\frac{K_2}{j\omega(j\omega+0.05)(j\omega+0.1)}$$

Find conditions on K_1, K_2 and ω for a limit cycle by using the describing function and a gain-phase plot.

15. Derive (5.6.4) in detail.

REFERENCES

Abramowitz, M. and Stegun, I. A. (1965) *Handbook of Mathematical Functions*, Dover.

Anderson, B. D. O. and Moore, J. B. (1968) Algebraic structure of generalized positive real matrices. *SIAM J. Control*, 6, 615.

Banks, S. P. (1981) The circle theorem for nonlinear parabolic systems, *Int. J. Control*, 34, 843.

Banks, S. P. (1982) Expansion of linear and semilinear systems and the Popov criterion, *Int. J. Control*, 35, 1085.

Banks, S. P. (1983) *State-space and Frequency Domain Methods in the Control of Distributed Parameter Systems, Peter Perigrinus.*

Banks, S. P. and Collingwood, P. C. (1979) Stability of nonlinearly interconnected systems and the small gain theorem, *Int. J. Control*, 30, 901.

Bergen, A. R. and Franks, R. L. (1971) Justification of the describing function method, *SIAM J. Control*, 9, 568.

Brilliant, M. B. (1958) Theory of the analysis of nonlinear systems, *MIT Research Laboratory of Electronics Report 345.*

Bruni, C., Di Pillo, G. and Koch, G. (1974) Bilinear systems: an appealing class of 'nearly linear' systems in theory and applications, *IEEE Trans. Aut. Control*, 19, 334.

Cho, Y-S. and Narendra, K. S. (1968) An off-axis circle criterion for the stability of feedback systems with a monotonic nonlinearity, *IEEE Trans. Aut. Control*, 13, 413.

Coddington, E. A. and Levinson, N. (1955) *Theory of Ordinary Differential equations*, McGraw-Hill.

Cook, P. A. (1972) Modified multivariable circle theorem, *Proc. 2nd IMA Conf. on Recent Mathematical Developments in Control.*

Freedman, M. I., Falb, P. L. and Zames, G. (1969) A Hilbert space stability theorem, *SIAM J. Control*, 3, 479.

Gelb, A. and Vander Velde, W. E. (1968) *Multiple-input Describing Functions and Nonlinear System Design*, McGraw-Hill.

Gutman, P-O., (1981) Stabilising controllers for bilinear systems, *IEEE Trans. Aut. Control*, 26, 917.

Isidori, A. (1973) Direct construction of minimal bilinear realisations, *IEEE Trans. Aut. Control*, 18, 626.

La Salle, J. P. and Lefschetz, S. (1961) *Stability by Lyapunov's Direct Method*, Academic Press.

Lurie, A. I. (1957) *Some Nonlinear Problems in the Theory of Automatic Control*, Her Majesty's Stationary Office.

Mees, A. I. (1972), The describing function matrix, *J. Inst. Maths. Appl.*, 10, p. 49.
Mees, A. I. and Bergen, A. R. (1975) Describing functions revisited, *IEEE Trans. Aut. Control*, 20, 473.
Mitzel, G. E., Clancy, S. J. and Rugh, W. J. (1979) On transfer function representations for homogeneous nonlinear systems, *IEEE Trans. Aut. Control*, 24, 242.
Mohler, R. R. (1973) *Bilinear Control Processes*, Academic Press.
Mossaheb, A. (1983) Application of a method of averaging to the study of dither in non-linear systems, *Int. J. Control*, 38 (3), 557.
Murthy, D. N. P. (1979), Controllability of a discrete-time bilinear system, *IEEE Trans. Aut. Control*, 24, 974.
Oldenburger, R. and Boyer, R. C. (1962) Effects of extra sinusoidal inputs to nonlinear systems, *Trans. ASME, Ser. D, J. Basic Eng.*, 84, 559.
Popov, V. M. (1961) Absolute stability of nonlinear systems of automatic control. *Automatika i. Telemekhanika*, 22, 961.
Rosenbrock, H. H. (1972) Multivariable circle theorems, *Proc. 2nd IMA Conf. on Recent Mathematical Developments in Control.*
Rouche, N., Habets, A. and Laloy, G. (1977) *Stability Theory by Lyapunov's Direct Method*, Springer-Verlag. 2nd IMA Conf on Recent Mathematical Developments in Control.
Saff, E. B. and Snider, A. D. (1976) *Fundamentals of Complex Analysis*, Prentice-Hall.
Sandberg, I. W. (1964) A frequency-domain condition for the stability of feedback systems containing a single time-varying nonlinear element, *Bell Syst. Tech. J.*, 43, pt. 2, 1601.
Schultz, D. G. and Gibson, J. E. (1962) The variable gradient method for generating Lyapunov functions, *Trans. AIEE*, 81, pt. 11, 203.
Shinners, S. N. (1978) *Modern Control System Theory and Application*, Addison-Wesley.
Siljak, D. D. (1969) *Nonlinear Systems*, Wiley.
Swern, F. (1983) Analysis of oscillations in systems with polynomial type nonlinearities using describing functions, *IEEE Trans. Aut. Control*, 28, 31.
Vidal, P. (1969) *Non-linear Sampled-data Systems*, Gordon and Breach.
West, J. C., Douce, J. L. and Livesley, R. K. (1956) The dual input describing function and its use in the analysis of nonlinear feedback systems, *J. IEE*, B-103, 463.
Wilkinson, J. H. (1965) *The Algebraic Eigenvalue Problem*, Clarendon Press.
Zames, G. (1966) On the input–output stability of time-varying nonlinear feedback systems, *IEEE Trans. Aut. Control*, 11, 228, 465.
Zubov, V. I. (1961) *Mathematical Methods for the Study of Automatic Control Systems*, Academic Press.

6 OPTIMAL CONTROL THEORY

6.1 INTRODUCTION

In chapter 4 we have considered the theory of control of linear systems essentially from a complex frequency-domain approach. This led to the design of compensators which make the overall closed-loop system have certain desirable features, relative to a certain class of inputs. The compensators are usually simple to implement using passive or active analogue components or digital software. However, the control system obtained is not the 'best' in the sense that it is guaranteed to minimize some performance criterion, which measures how close the system variables come to some desired values. In the case of spacecraft, for example, we may also wish to minimize fuel consumption; indeed this may be critical for the success of the flight.

We shall discuss these problems in this chapter, and in particular Pontryagin's maximum principle will be considered in some detail. This gives necessary conditions for the existence of extremals of a given performance criterion subject to certain constraints. Although a complete mathematical proof of the criterion is beyond the scope of the present book (see, for example, Girsanov (1972) and Fleming and Rishel (1975)), we shall try to motivate the proof by considering first the basic finite-dimensional Lagrange multiplier theory and then see how this can be generalized to obtain the calculus of variations and constrained optimal control.

Pontryagin's principle can be used to obtain time-optimal controllers and optimal controls for linear systems with a quadratic cost criterion (the linear-quadratic regulator problem). However, it is also instructive to consider an alternative approach called dynamic programming and this will be introduced later. Moreover, we shall extend the linear-quadratic regulator problem to the case of so-called receding horizon control which produces a non-linear feedback law; this has the property of reacting quickly to large errors but more slowly than linear feedback to small errors which may be due to spurious noise signals.

A final comment should be made about the relative merits of the essentially classical approach in chapter 4 and the optimal theory developed here. Two major criticisms of optimal control theory are firstly that it tends to produce complex controllers which are difficult (if not impossible) to implement practically and secondly that optimality does not imply system integrity in the sense that failure of one feedback loop may lead to an unstable system (see Rosenbrock and McMorran, 1971). The inverse Nyquist array method was shown to cope with the latter problem in chapter 4. In the case of the former objection, many authors have considered simplified sub-optimal controllers, which often work with considerable success. With due regard to these criticisms, the optimal control approach to systems theory has become an important tool in many cases and should always be considered as a complement rather than a replacement of classical control design techniques.

6.2 CLASSICAL OPTIMIZATION THEORY

6.2.1 Unconstrained Optimization

In the classical theory of optimization (without constraints) we consider the minimization (or maximization) of a real-valued function $f: \mathbb{R}^n \to \mathbb{R}$ of n variables. For simplicity we shall write the function as $f(x_1, \ldots, x_n)$ (or $f(\boldsymbol{x})$) and we shall not always distinguish between the function f and its value $f(x_1, \ldots, x_n)$ at the point $\boldsymbol{x} = (x_1, \ldots, x_n)^{\mathrm{T}}$.

The basic topology of $\mathbb{R}^n$ will be assumed to be understood (see appendix 1) in the following discussion. Suppose that f is defined on a set $S \subseteq \mathbb{R}^n$. If $\boldsymbol{x}^* \in S$ is such that

$$f(\boldsymbol{x}^*) \geqslant f(\boldsymbol{x}), \quad \text{for all } \boldsymbol{x} \in S \tag{6.2.1}$$

then $f(\boldsymbol{x}^*)$ is called the **global or absolute maximum** of f on S. If (6.2.1) holds only on a neighbourhood of $\boldsymbol{x}^*$ then $f(\boldsymbol{x}^*)$ is called a **local maximum** of f. Similar definitions apply for minima. We have the following well-known necessary condition:

Theorem 6.2.1

If $f(\boldsymbol{x})$ has a local maximum at $\boldsymbol{x}_0$, which is an interior point of S, then

$$\boxed{\operatorname{grad} f(\boldsymbol{x}_0) = \mathbf{0}} \tag{6.2.2}$$

Proof

There exists an $\varepsilon > 0$ such that

$$f(\boldsymbol{x}) \leqslant f(\boldsymbol{x}_0) \quad \text{for } \boldsymbol{x} \in \{\boldsymbol{x} : ||\boldsymbol{x} - \boldsymbol{x}_0|| \leqslant \varepsilon\} \triangleq B_\varepsilon(\boldsymbol{x}_0)$$

and so if $\boldsymbol{x} = \boldsymbol{x}_0 + h\boldsymbol{e}_j$, where

$$\boldsymbol{e}_j = (0, \ldots, 0, 1, 0, \ldots, 0)^{\mathrm{T}}$$

and $0 < |h| < \varepsilon$, then

$$f(\boldsymbol{x}_0 + h\boldsymbol{e}_j) - f(\boldsymbol{x}_0) \leqslant 0, \quad 1 \leqslant j \leqslant n$$

Dividing by h and letting $h \to 0$ we obtain

$$\frac{\partial f(\boldsymbol{x}_0)}{\partial x_j} \leqslant 0, \quad \text{if } h > 0$$

and

$$\frac{\partial f(\boldsymbol{x}_0)}{\partial x_j} \geqslant 0 \quad \text{if } h < 0$$

Hence

$$\text{grad}\, f(\boldsymbol{x}_0) = \boldsymbol{0}$$

The condition in theorem 6.2.1 is not sufficient, however; the next result provides such a condition:

Theorem 6.2.2
If grad $f(\boldsymbol{x}_0) = \boldsymbol{0}$ and the Hessian matrix

$$\boxed{H(\boldsymbol{x}_0) \triangleq \left(\frac{\partial^2 f}{\partial x_i\, \partial x_j}\right)(\boldsymbol{x}_0)}$$

of f evaluated at $\boldsymbol{x}_0$, is negative definite, then $f(\boldsymbol{x}_0)$ is a local maximum of f.

Proof
By Taylor's theorem in n dimensions (appendix 1), we have

$$f(\boldsymbol{x}) - f(\boldsymbol{x}_0) = \frac{1}{2}\,\mathrm{d}^2 f(\boldsymbol{x}_0; \boldsymbol{h}) + \frac{1}{3!}\,\mathrm{d}^3 f(\boldsymbol{z}; \boldsymbol{h}) \quad \boldsymbol{h} = \boldsymbol{x} - \boldsymbol{x}_0$$

for some $\boldsymbol{z}$ on the line joining $\boldsymbol{x}$ and $\boldsymbol{x}_0$. If $\mathrm{d}^2 f(\boldsymbol{x}_0; \boldsymbol{h}) = \boldsymbol{h}^T H(\boldsymbol{x}_0)\boldsymbol{h}$ is negative definite, then $\mathrm{d}^3 f(\boldsymbol{z}; \boldsymbol{h})$ (which is $o(\|\boldsymbol{h}\|^3)$ as $\|\boldsymbol{h}\| \to 0$) is dominated by the first term. Hence

$$f(\boldsymbol{x}) - f(\boldsymbol{x}_0) \leqslant 0$$

in some neighbourhood of $\boldsymbol{x}_0$. (A similar result holds for minimization of f, which the reader should formulate.)

In order to determine whether $H(\boldsymbol{x}_0)$ is negative definite, we may use Sylvestor's criterion. For any square $n \times n$ matrix $A = (a_{ij})$, define the **leading principal minors** M_k of A as the determinants

$$M_k \triangleq \det A_k, \quad 1 \leqslant k \leqslant n$$

where A_k is the submatrix

$$A_k = (a_{ij})_{1 \leqslant i \leqslant k, 1 \leqslant j \leqslant k}$$

of A. Then Sylvestor's criterion states that A is negative definite if and only if we have

$$(-1)^k M_k > 0$$

(i.e. $M_1 < 0, M_2 > 0, M_3 < 0, \ldots$). If A is symmetric then there exists a real matrix P such that $P^T = P^{-1}$ and

$$\Lambda = PAP^{-1}$$

where Λ is the diagonal matrix of (real) eigenvalues of A. Now A is negative definite if and only if

$$\boldsymbol{x}^T A \boldsymbol{x} < 0, \quad \boldsymbol{x} \neq \boldsymbol{0}$$

and so A is negative definite if and only if

$$\boldsymbol{x}^T P^{-1} PAP^{-1} P\boldsymbol{x} = \boldsymbol{y}^T \Lambda \boldsymbol{y}$$

for all $\boldsymbol{y}$ $(= P\boldsymbol{x})$. The matrix A is therefore negative definite if and only if all its eigenvalues are negative.

Consider a function f such that grad $f(\boldsymbol{x}_0) = 0$ at some interior point $\boldsymbol{x}_0$ of its domain S. Then, as above, we can choose an orthogonal change of coordinates $\boldsymbol{y} = P\boldsymbol{x}$ and write $H(\boldsymbol{x}_0)$ in the diagonal form

$$\begin{pmatrix} \lambda_1 & & & & \\ & \lambda_2 & & \boldsymbol{0} & \\ & & \cdot & & \\ & & & \cdot & \\ & \boldsymbol{0} & & & \cdot \\ & & & & \lambda_n \end{pmatrix}$$

where the λ are the eigenvalues of $H(\boldsymbol{x}_0)$. By renumbering the coordinates $\boldsymbol{y}$ if necessary we can assume that $\lambda_1, \ldots, \lambda_i < 0$, $\lambda_{i+1}, \ldots, \lambda_j > 0$, $\lambda_{j+1}, \ldots, \lambda_n = 0$, where i or $j - i$ may be zero. Put

$$\boldsymbol{y}_1 = (y_1, \ldots, y_i), \quad \boldsymbol{y}_2 = (y_{i+1}, \ldots, y_j), \quad \boldsymbol{y}_3 = (y_{j+1}, \ldots, y_n)$$

Then, intuitively speaking, f (at $\boldsymbol{x}_0$) has maxima in the $\boldsymbol{y}_1$ variables, minima in the $\boldsymbol{y}_2$ variables and generalized 'inflection points' in the $\boldsymbol{y}_3$ variables. Also in $(\boldsymbol{y}_1, \boldsymbol{y}_2)$ space $f(\boldsymbol{x}_0)$ is a general saddle point.

6.2.2 Constrained Optimization

Consider next the problem of maximizing the function $f(\boldsymbol{x})$ subject to the $m(< n)$ constraints

$$g_i(\boldsymbol{x}) = b_i, \quad 1 \leqslant i \leqslant m \tag{6.2.3}$$

or more concisely, $\boldsymbol{g}(\boldsymbol{x}) = \boldsymbol{b}$. Define the **Jacobian matrix** of $\boldsymbol{g}$ to be

$$G = \left(\frac{\partial g_i}{\partial x_j}\right)_{1 \leqslant i \leqslant m,\, 1 \leqslant j \leqslant n}$$

$$= \frac{\partial \boldsymbol{g}}{\partial \boldsymbol{x}}$$

and suppose for now that if $\boldsymbol{x}_0$ is a local maximum of $f(\boldsymbol{x})$ subject to the constraints (6.2.3), then G has full rank at $\boldsymbol{x}_0$ (appendix 1). We shall obtain a necessary condition for $\boldsymbol{x}_0$ to be such a local maximum.

By renumbering the coordinate axes (if necessary) we can assume that the first m columns of $G(\boldsymbol{x}_0)$ are linearly independent. Write

$$\boldsymbol{x} = (\boldsymbol{\xi}, \boldsymbol{\eta}) = ((x_1, \ldots, x_m), (x_{m+1}, \ldots, x_n))$$

Then by the implicit function theorem, we may solve the equations $\boldsymbol{g}(\boldsymbol{x}) = \boldsymbol{b}$ for $\boldsymbol{\xi}$ in terms of $\boldsymbol{\eta}$ (in some neighbourhood of $\boldsymbol{\eta}_0$ where $\boldsymbol{x}_0 = (\boldsymbol{\xi}_0, \boldsymbol{\eta}_0)$), i.e.

$$\boldsymbol{\xi} = \boldsymbol{\phi}(\boldsymbol{\eta})$$

for some differentiable function $\boldsymbol{\phi}$. Clearly, the function

$$h(\boldsymbol{\eta}) = f(\boldsymbol{\phi}(\boldsymbol{\eta}), \boldsymbol{\eta})$$

has a local (unconstrained) maximum at $\boldsymbol{\eta}_0$ and so

$$\mathbf{0} = \frac{\partial h}{\partial \boldsymbol{\eta}} = \frac{\partial f}{\partial \boldsymbol{\xi}} \frac{\partial \boldsymbol{\phi}}{\partial \boldsymbol{\eta}} + \frac{\partial f}{\partial \boldsymbol{\eta}} \quad \text{at } \boldsymbol{\eta}_0 \tag{6.2.4}$$

(Note that $\partial f/\partial \boldsymbol{\xi}$ is regarded as a row vector). However, we also have

$$\boldsymbol{g}(\boldsymbol{\phi}(\boldsymbol{\eta}), \boldsymbol{\eta}) = \boldsymbol{b}$$

and so

$$\frac{\partial \boldsymbol{g}}{\partial \boldsymbol{\xi}} \frac{\partial \boldsymbol{\phi}}{\partial \boldsymbol{\eta}} + \frac{\partial \boldsymbol{g}}{\partial \boldsymbol{\eta}} = \mathbf{0}$$

By assumption, $\partial \boldsymbol{g}/\partial \boldsymbol{\xi}$ (at $\boldsymbol{x}_0$) exists and so

$$\frac{\partial \boldsymbol{\phi}}{\partial \boldsymbol{\eta}} = -\left(\frac{\partial \boldsymbol{g}}{\partial \boldsymbol{\xi}}\right)^{-1} \frac{\partial \boldsymbol{g}}{\partial \boldsymbol{\eta}} \quad \text{again at } \boldsymbol{x}_0 \tag{6.2.5}$$

Substituting (6.2.5) into (6.2.4) we have

$$\mathbf{0} = \frac{\partial f}{\partial \boldsymbol{\xi}} \left(\frac{\partial \boldsymbol{g}}{\partial \boldsymbol{\xi}}\right)^{-1} \frac{\partial \boldsymbol{g}}{\partial \boldsymbol{\eta}} + \frac{\partial f}{\partial \boldsymbol{\eta}} \quad \text{at } \boldsymbol{x}_0$$

Introducing the m-dimensional vector $\boldsymbol{\lambda}_0$ where

$$\boldsymbol{\lambda}_0 = \frac{\partial f}{\partial \boldsymbol{\xi}}(\boldsymbol{x}_0) \left(\frac{\partial \boldsymbol{g}}{\partial \boldsymbol{\xi}}(\boldsymbol{x}_0)\right)^{-1} \tag{6.2.6}$$

we have

$$\left.\begin{aligned}\frac{\partial f}{\partial \eta} - \lambda_0 \frac{\partial g}{\partial \eta} = 0\\ \frac{\partial f}{\partial \xi} - \lambda_0 \frac{\partial g}{\partial \xi} = 0\end{aligned}\right\} \text{at } x_0$$

or

$$\boxed{\frac{\partial f}{\partial x_i}(x_0) - \sum_{j=1}^{m} \lambda_{0j} \frac{\partial g_j}{\partial x_i}(x_0) = 0, \quad 1 \leqslant i \leqslant n} \tag{6.2.7}$$

Equations (6.2.7) (plus the constraints (6.2.3)) are the necessary conditions for x_0 to be a maximum (or minimum) of f subject to the constraints (6.2.3). The numbers $\lambda_{01}, \ldots, \lambda_{0m}$ are called **Lagrange multipliers**. Defining the **Lagrangian** $F: \mathbb{R}^{n+m} \to \mathbb{R}$ of the problem by

$$F(x, \lambda) = f(x) + \sum_{j=1}^{m} \lambda_j (b_j - g_j(x))$$

we can obtain the necessary conditions for f subject to constraints from the classical unconstrained necessary conditions for F, i.e.

$$\boxed{\begin{aligned}\frac{\partial F}{\partial x}(x_0, \lambda_0) = 0\\ \frac{\partial F}{\partial \lambda}(x_0, \lambda_0) = 0\end{aligned}} \tag{6.2.8}$$

Example 6.2.1

What is the minimum fencing required to surround a rectangular field of unit area?

In this case, the objective function is

$$f(l_1 l_2) = 2(l_1 + l_2)$$

where l_1, l_2 are the lengths of the sides of the field. The constraint is

$$g(l_1, l_2) = l_1 l_2 = 1$$

and so

$$F(l_1, l_2, \lambda) = 2(l_1 + l_2) + \lambda(1 - l_1 l_2)$$

whence

$$\left.\begin{aligned}\frac{\partial F}{\partial l_1} &= 2 - \lambda l_2 = 0\\ \frac{\partial F}{\partial l_2} &= 2 - \lambda l_1 = 0\end{aligned}\right\} \Rightarrow l_1 l_2$$

$$\frac{\partial F}{\partial \lambda} = 1 - l_1 l_2 = 0 \quad \text{i.e. } l_1 = l_2 = 1$$

It follows that the optimal shape is a square. (Note that if we remove the restriction on the shape of the field as being rectangular and allow an arbitrary (curved) boundary, then we require the calculus of variations, introduced later, to solve the problem.)

We assumed above that rank $(G(\boldsymbol{x}_0)) = m$ to obtain the necessary condition (6.2.7), which holds for some vector $\boldsymbol{\lambda}_0$. In general we can consider the matrix

$$G_f = \begin{bmatrix} G \\ \cdots \\ \nabla f \end{bmatrix}$$

which has rank at most $m + 1$. It can be shown (Hadley, 1964) that if rank $(G_f(\boldsymbol{x}_0)) = m + 1$ then $\boldsymbol{x}_0$ cannot be an optimum of f. Hence, if $\boldsymbol{x}_0$ is an optimum it is necessary that rank $(G_f(\boldsymbol{x}_0)) \leqslant m$. If rank $(G_f(\boldsymbol{x}_0)) =$ rank $G(\boldsymbol{x}_0)$, then ∇f can be written as a linear combination of the rows of G, i.e. (6.2.7) holds for some non-zero vector $\boldsymbol{\lambda}_0$. If, however, rank $G(\boldsymbol{x}_0) =$ rank $G(\boldsymbol{x}_0) + 1$, then the rows of $G(\boldsymbol{x}_0)$ must be linearly dependent; for if they were linearly independent then rank $G(\boldsymbol{x}_0) = m$ and so rank $G_f(\boldsymbol{x}_0) = m + 1$, contradicting the above statement. Hence, in this case there must exist a non-zero vector $\boldsymbol{\lambda}_0$ such that

$$\sum_{j=1}^{m} \lambda_{0j} \frac{\partial g_j}{\partial x_i}(\boldsymbol{x}_0) = 0, \quad 1 \leqslant i \leqslant n$$

However, this means that we obtain the necessary condition

$$\boxed{\lambda_{0,m+1} \frac{\partial f}{\partial x_i}(\boldsymbol{x}_0) - \sum_{j=1}^{m} \lambda_{0j} \frac{\partial g_j}{\partial x_i}(\boldsymbol{x}_0) = 0, \quad 1 \leqslant i \leqslant n} \tag{6.2.9}$$

for an extremum of f at $\boldsymbol{x}_0$, where $(\lambda_{01}, \lambda_{02}, \ldots, \lambda_{0m}, \lambda_{0m+1})$ is a non-zero vector and

$$\boxed{\lambda_{0,m+1} = \begin{cases} 1 & \text{if rank } G_f(\boldsymbol{x}_0) = \text{rank } G(\boldsymbol{x}_0) \\ 0 & \text{if rank } G_f(\boldsymbol{x}_0) = \text{rank } G(\boldsymbol{x}_0) + 1 \end{cases}} \tag{6.2.10}$$

Note that in the latter case (when $\lambda_{0,m+1} = 0$), the necessary condition (6.2.9) is independent of the **objective** function f. We illustrate this curious fact in the following example.

Example 6.2.2

Suppose we consider the scalar function f shown in fig. 6.1 as the objective function, and let

$$g(x) = x(\sin x - 1) = 0$$

be the constraint. Then the solutions of the constraint equation are $x = 0$ or $x = \pm\pi/2, \pm 5\pi/2, \ldots$ and we have

$$g'(x) = (\sin x - 1) + x\cos x$$

Consider the points $x = 0, \pi/2$. Then

(a) $f'(0) = 0,\ g'(0) = -1$

(b) $f'(\pi/2) = 0,\ g'(\pi/2) = 0$

In case (a) we have

$$\lambda_{02} f'(0) - 0(-1) = 0$$

so that (6.2.9) holds with $\lambda_{02} = 1$. However, for (b),

$$\lambda_{02} f'(\pi/2) - \lambda_{01}(0) = 0$$

and so (6.2.9) holds with $\lambda_{02} = 0$ and λ_{01} arbitrary. (Note that $n = m$ in this example, although we began by assuming that $m < n$. However, it is clear that the necessary conditions (6.2.9) still hold in this case, but G_g has maximum rank of only $n = m = 1$.) We therefore see that if we had not included the possibility of $\lambda_{0,m+1} = 0$ in (6.2.9) then we would not have obtained the two constrained maxima for this problem at $\pm\pi/2$. Only the local maximum at $x = 0$ would have been found which we might then have mistakenly believed to be the global maximum.

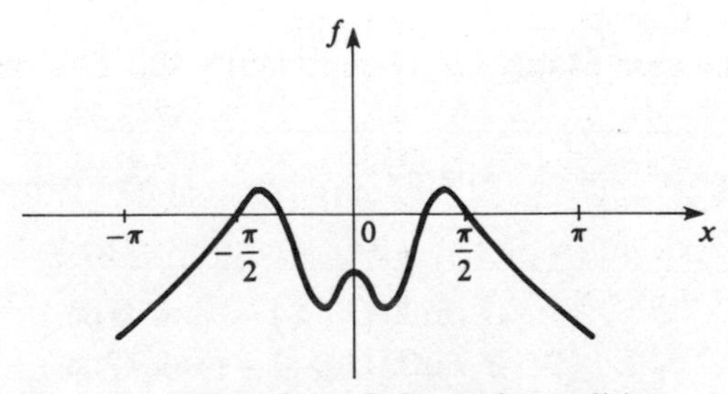

Fig. 6.1 Illustration of the rank conditions.

6.2.3 Kuhn–Tucker Theory

Returning to the necessary conditions (6.2.8) we have seen that it is necesary (for an extremum of f) that the Lagrangian F satisfies the classical (unconstrained) necessary conditions as a function of the $n+m$ variables $(\boldsymbol{x}, \boldsymbol{\lambda})$. In fact, it is often the case that at an extremum of f the point $(\boldsymbol{x}_0, \boldsymbol{\lambda}_0)$ is a saddle point of F. If we generalize the problem in section 6.2.2 to include inequality constraints then we obtain a problem of the form

extremise $f(\boldsymbol{x})$

subject to

$$\left.\begin{aligned} g_i(\boldsymbol{x}) &\leqslant b_i, \quad i = 1, \ldots, u \\ g_i(\boldsymbol{x}) &\geqslant b_i, \quad i = u+1, \ldots, v \\ g_i(\boldsymbol{x}) &= b_i, \quad i = v+1, \ldots, m \end{aligned}\right\} \tag{6.2.11}$$

It can be shown quite easily that there exists a Lagrange multiplier $\boldsymbol{\lambda}_0$ such that necessary conditions of the form (6.2.7) hold (assuming for simplicity that rank G = rank G_f at $\boldsymbol{x}_0$). However, if

$$\left.\begin{aligned} g_i(\boldsymbol{x}) &< b_i, \quad i = 1, \ldots, u \\ \text{or} \quad g_i(\boldsymbol{x}) &> b_i, \quad i = u+1, \ldots, v \end{aligned}\right\} \tag{6.2.12}$$

then

$$\boxed{\lambda_{0i} = 0} \tag{6.2.13}$$

This says that if the constraint i is 'inactive' (i.e. strict inequality) then we can ignore that constraint. Moreover, one can show that

$$\boxed{\begin{aligned} \lambda_{0i} &\geqslant 0, \quad i = 1, \ldots, u \\ \lambda_{0i} &\leqslant 0, \quad i + u + 1, \ldots, v \end{aligned}} \tag{6.2.14}$$

always holds at an extremum. Finally, one can add the further constraint of non-negative variables $\boldsymbol{x} \geqslant \boldsymbol{0}$ (i.e. $x_i \geqslant 0$, $1 \leqslant i \leqslant n$), but these can be added on as further inequality constraints.

Even in the case of inequality constraints and non-negative variables, therefore, we can form the Lagrangian F as before and obtain necessary conditions for an extremum. Consider now the assumption (which often holds, as pointed out above) that F has a saddle point at $(\boldsymbol{x}_0, \boldsymbol{\lambda}_0)$, and suppose that F is a maximum in $\boldsymbol{x}$ and a minimum in $\boldsymbol{\lambda}$ at this point. Then if $(\boldsymbol{x}_0, \boldsymbol{\lambda}_0)$ is an interior point of the constraint set we must have

$$\frac{\partial F}{\partial x_i}(\boldsymbol{x}_0, \boldsymbol{\lambda}_0) = \frac{\partial F}{\partial \lambda_j}(\boldsymbol{x}_0, \boldsymbol{\lambda}_0) = 0, \quad \text{each } i, j$$

However, if the ith coordinate x_{0i} of $\boldsymbol{x}_0$ is on the boundary of the constraint set, i.e. $x_{0i} = 0$, then we must have

$$\frac{\partial F}{\partial x_i}(\boldsymbol{x}_0, \boldsymbol{\lambda}_0) \leqslant 0$$

if x_i is constrained to be non-negative, and

$$\frac{\partial F}{\partial x_i}(\boldsymbol{x}_0, \boldsymbol{\lambda}_0) \geqslant 0$$

if x_i is constrained to be non-positive. Similarly, if $\lambda_i \geqslant 0$ (or $\lambda_i \leqslant 0$) then

$$\frac{\partial F}{\partial \lambda_i}(\boldsymbol{x}_0, \boldsymbol{\lambda}_0) \geqslant 0 \quad (\leqslant 0)$$

Hence, if F is defined in $(\boldsymbol{x}, \boldsymbol{\lambda})$-space and we constrain the vector $(\boldsymbol{x}, \boldsymbol{\lambda})$ so that

$$\boxed{\begin{array}{l} x_i \geqslant 0, \quad 1 \leqslant i \leqslant s \\ x_i \leqslant 0, \quad s+1 \leqslant i \leqslant t \\ x_i \text{ unconstrained} \quad \text{for } t+1 \leqslant i \leqslant n \\ \lambda_i \geqslant 0, \quad 1 \leqslant i \leqslant u \\ \lambda_i \leqslant 0, \quad u+1 \leqslant i \leqslant v \\ \lambda_i \text{ unconstrained} \quad \text{for } v+1 \leqslant i \leqslant m \end{array}}$$

then we obtain the following **Kuhn–Tucker necessary conditions** for a saddle point of F:

$$\boxed{\begin{array}{l} \dfrac{\partial F}{\partial x_i}(\boldsymbol{x}_0, \boldsymbol{\lambda}_0) \begin{cases} \leqslant 0 & 1 \leqslant i \leqslant s \\ \geqslant 0 & s+1 \leqslant i \leqslant t \\ = 0 & t+1 \leqslant i \leqslant n \end{cases} \\ \\ \dfrac{\partial F}{\partial \lambda_i}(\boldsymbol{x}_0, \boldsymbol{\lambda}_0) \begin{cases} \geqslant 0 & 1 \leqslant i \leqslant u \\ \leqslant 0 & u+1 \leqslant i \leqslant v \\ = 0 & v+1 \leqslant i \leqslant m \end{cases} \end{array}} \tag{6.2.15}$$

Note that we have strict inequality in these conditions only if the corresponding coordinate is zero. Hence we can also state the following additional necessary condition for a saddle point:

$$\boxed{\begin{array}{l} \boldsymbol{x}_0^{\mathrm{T}} \nabla_{\boldsymbol{x}} F(\boldsymbol{x}_0, \boldsymbol{\lambda}_0) = 0 \\ \boldsymbol{\lambda}_0^{\mathrm{T}} \nabla_{\boldsymbol{\lambda}} F(\boldsymbol{x}_0, \boldsymbol{\lambda}_0) = 0 \end{array}} \tag{6.2.16}$$

The importance of the conditions (6.2.15) and (6.2.16) is that if F is the Lagrangian function of the optimization problem (6.2.12) then under a certain condition, the relations (6.2.15) and (6.2.16) are not only necessary conditions for a saddle point of F but also sufficient for an extremum of f subject to the constraints. The condition referred to above was introduced by Kuhn and Tucker (see Hadley, 1964) and is called the **constraint qualification**. It may be stated as follows:

Consider the optimum point $\boldsymbol{x}_0$ and suppose that it is on the boundaries of each constraint set, i.e. each set

$$C_i \triangleq \{\boldsymbol{x} : g_i(\boldsymbol{x}) \leqslant (\text{or } \geqslant, \text{ or } =) b_i\}$$

(If $\boldsymbol{x}_0$ is not on the boundary of a particular constraint set then we just ignore that constraint and consider the same problem with one less constraint.) Then $\boldsymbol{x}_0$ is on the boundary of the intersection

$$\bigcap_{i=1}^{m} C_i$$

of all the constraint sets. Let T_i be the tangent hypersurface to C_i at $\boldsymbol{x}_0$. Then we also have

$$\boldsymbol{x}_0 \in \bigcap_{i=1}^{m} T_i$$

(fig. 6.2(a)). Finally, let H_i be the closed half-space which is bounded by t_i and such that the normal to T_i at $\boldsymbol{x}_0$ points into C_i. The constraint qualification then states that any vector in

$$\bigcap_{i=1}^{m} H_i$$

based on $\boldsymbol{x}_0$ must be the tangent vector of a (differentiable) curve starting at $\boldsymbol{x}_0$ and constrained in

$$\bigcap_{i=1}^{m} C_i.$$

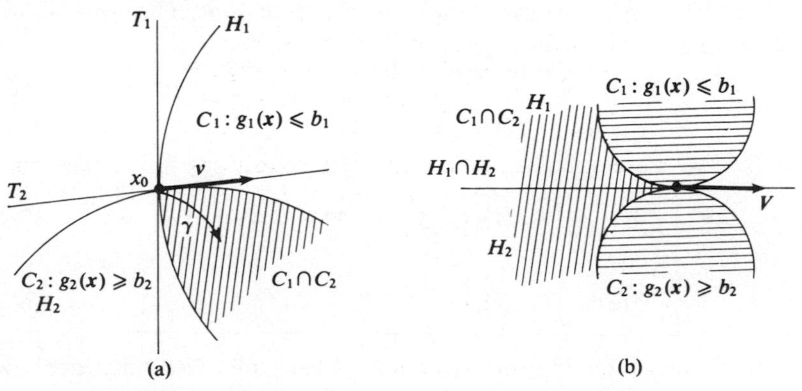

Fig. 6.2 The Kuhn-Tucker constraint qualification.

Such a vector v and a curve γ are shown in fig. 6.2(a). Figure 6.2(b) shows a case where the constraint qualification is violated.

6.3 CLASSICAL CALCULUS OF VARIATIONS

6.3.1 Three Equivalent Problems

We shall continue towards our goal (i.e. the maximum principle) by considering the most simple generalization of finite-dimensional optimization discussed in the last section. This is the classical calculus of variations in which one desires to optimize a functional $J(x)$ which does not just depend on a vector variable $x \in \mathbb{R}^n$, but on a whole vector-valued function $x(t)$. The distinction between the calculus of variations and modern optimal control is that in the former case we normally assume that the functions $x(t)$ involved in the optimization and the functional J itself (and possibly the constraints) are differentiable in an appropriate sense. In the case of 'hard' control constraints, this is of course not a valid assumption.

In order to be more specific about the functional J it is convenient to introduce a class of vector spaces which do not have finite dimension (see appendix 1 for a fuller discussion). Let $C_n^k[t_0, t_f] = \{x : [t_0, t_f] \to \mathbb{R}^n \mid x(t)$ is k times continuously differentiable$\}$ i.e. $C_n^k[t_0, t_f]$ (or C_n^k for brevity) is the set of functions x defined on the interval $[t_0, t_f]$ with values in $\mathbb{R}^n$ which have k continuous derivatives. Clearly C_n^k is a vector space which does not have a finite basis. Usually, we take $k = 2$ in the classical calculus of variations, and J is defined as a real-valued function on C_n^k, i.e.

$$J : C_n^k[t_0, t_f] \to \mathbb{R} \tag{6.3.1}$$

Before defining what we meant above by saying that J is differentiable, we consider three examples of such a function as in (6.3.1). Let $\theta : \mathbb{R}^{n+1} \to \mathbb{R}$ and $\phi : \mathbb{R}^{2n+1} \to \mathbb{R}$ be two twice differentiable functions. Then the following functions J are of the form (6.3.1):

$$J(x) = \int_{t_0}^{t_f} \phi(x(t), \dot{x}(t), t)\, dt \tag{6.3.2}$$

$$J(x) = [\theta(x(t), t)]_{t_0}^{t_f} + \int_{t_0}^{t_f} \phi(x(t), \dot{x}(t), t)\, dt \tag{6.3.3}$$

$$J(x) = [\theta(x(t), t]_{t_0}^{t_f} \tag{6.3.4}$$

The problems of finding an extremum of each of these functionals with respect to $x(t) \in C_n^2$ has been studied by many people, but they have become known as the Lagrange problem (6.3.2), the Bolza problem (6.3.3) and the Mayer problem 6.3.4, respectively. In fact, they are equivalent, since the transformation

$$\Omega = \phi + \frac{d\theta}{dt}$$

changes the Bolza problem into a Lagrange problem with integrand Ω. Similarly, the Mayer problem becomes a Bolza problem by writing

$$\theta(x(t), t) = \psi(x(t), t) + \int_{t_0}^{t_f} \phi(x(t), \dot{x}(t), t)\, dt$$

(Transforming a Lagrange problem into a Bolza problem or a Bolza problem into a Mayer problem is, of course, trivial.) Since the problems are equivalent we may study any one of them.

6.3.2 Unconstrained Optimization

We now have an extremum problem (i.e. extremizing J) on the vector space C_n^k. At first, we may be tempted to approach this problem as in the finite dimensional case, where we recall that a necessary condition for the optima of f is given by (6.2.2), i.e.

$$\frac{\partial f}{\partial x_i}(x_0) = 0$$

This would entail giving a meaning to

$$\left\{\frac{\partial J}{\partial x(t)}\right\}$$

if indeed such a thing existed. In fact, we can define the derivative of a functional J with respect to its argument function and this is called the Fréchet derivative. It is a direct generalization of the Jacobian matrix to infinite-dimensional spaces. However, in order to define this derivative we must put a norm (i.e. a length) on the vectors in C_n^k, and the conditions which we would have to impose on J are far too restrictive for most problems.

Fortunately, we can consider the problem of extremizing J from a slightly different point of view by defining a much weaker derivative. Again with reference to the finite dimensional case, the necessary condition derived above states that the partial derivatives of f vanish at x_0. However, a partial derivative is just the derivative of f in an axial direction, and so the necessary condition merely states that the derivative of f in any direction is zero. We therefore say that if J is defined on a vector space V, then J has the **directional derivative** $\delta J(u; v)$ at u in the direction v if $J(u + \varepsilon v)$ is differentiable at $\varepsilon = 0$. Then we write

$$\delta J(u;v) = \frac{\mathrm{d}}{\mathrm{d}\varepsilon} J(u + \varepsilon v)\,|_{\varepsilon=0} \tag{6.3.5}$$

$\delta J(u;v)$ is also called the **Gateau derivative** or **variational** of J with respect to the variation εv. Just as in finite-dimensional problems, we may consider adding constraints to the problem, in which case the constraint set may be any subset X of V. Then we should only define $\delta J(u;v)$ for those directions v such that $u + \varepsilon v \in X$ for all sufficiently small ε. In either case we also define

$$\delta^2 J(u;v) = \frac{\mathrm{d}^2}{\mathrm{d}\varepsilon^2} J(u + \varepsilon v)\,|_{\varepsilon=0} \tag{6.3.6}$$

Since the map $\varepsilon \to J(u + \varepsilon v)$ is just a real-valued function of a real variable, we see that if the directional derivative is defined at u^* then a necessary condition for a minimum of J at u^* is

$$\boxed{\delta J(u^*;\, v) = 0, \quad \delta^2 J(u^*;\, v) \geqslant 0} \tag{6.3.7}$$

and if the directional derivative is only defined in certain directions at u^*, then a necessary condition for a minimum is

$$\boxed{\delta J(u^*;\, v) \geqslant 0} \tag{6.3.8}$$

in any direction v for which the derivative is defined.

Before proceeding with the development of classical calculus of variations we mention two elementary results which, together with the concept of directional derivative, constitute the necessary apparatus for the method.

Theorem 6.3.1
If $G(t, \varepsilon)$ is any function which has a partial derivative $G_\varepsilon(t, \varepsilon)$ which is continuous, then

$$\frac{\mathrm{d}}{\mathrm{d}\varepsilon} \int_{t_1}^{t_2} G(t, \varepsilon)\,\mathrm{d}t = \int_{t_1}^{t_2} G_\varepsilon(t, \varepsilon)\,\mathrm{d}t, \quad \text{for all } t_1, t_2, \varepsilon$$

Theorem 6.3.2
If $\phi(t)$ is continuous on $[t_0, t_f]$ and

$$\int_{t_0}^{t_f} \phi(t)\psi(t)\,\mathrm{d}t = 0$$

for all continuously differentiable functions ψ, then $\phi(t)$ is identically zero on $[t_0, t_f]$.

Consider now the Lagrange problem (6.3.2). Then

$$J(x) = \int_{t_0}^{t_f} \phi(x(t), \dot{x}(t), t)\, dt$$

and so

$$J(x + \varepsilon y) = \int_{t_0}^{t_f} \phi(x + \varepsilon y(t), \dot{x}(t) + \varepsilon \dot{y}(t), t)\, dt$$

Hence, using theorem 6.3.1,

$$\delta J(x; y) = \int_{t_0}^{t_f} (\phi_x y + \phi_{\dot{x}} \dot{y})\, dt$$

$$= \int_{t_0}^{t_f} \left(\phi_x - \frac{d}{dt}\phi_{\dot{x}}\right) y\, dt + \phi_{\dot{x}} y \Big|_{t_0}^{t_f}$$

(integrating by parts). Now, by theorem 6.3.2, if y is any admissible variation (i.e. $y \in C_n^2$) we have the necessary condition

$$\phi_x - \frac{d}{dt}\phi_{\dot{x}} = 0 \tag{6.3.9}$$

$$\phi_{\dot{x}} y \Big|_{t_0}^{t_f} = 0 \tag{6.3.10}$$

(6.3.9) is called Euler's equation and is a (non-linear) differential or algebraic equation for $x(t)$. The condition (6.3.10) is called the transversality condition and can be satisfied in a variety of ways. If either of the endpoints $x_0 = x(t_0)$ and $x_f = x(t_f)$ are fixed then we must have y_0 or y_f equal to zero at the corresponding point which will satisfy (6.3.10). However, if, for example, x_f is free then it follows that

$$\phi_{\dot{x}} = 0, \quad \text{at } t = t_f$$

Example 6.3.1

In chapter 1 we considered the modelling of finite-dimensional dynamical systems and we derived Lagrange's equations (1.2.19). If the system has no dissipative forces operating on it then these equations become

$$\frac{d}{dt}\left(\frac{\partial L}{\partial \dot{q}_j}\right) - \frac{\partial L}{\partial q_j} = 0 \tag{6.3.11}$$

where L is the Lagrangian of the system.

The above results now demonstrate **Hamilton's principle of least action**, i.e. the motion of a dynamical system is such that it makes the functional

$$J = \int_{t_1}^{t_2} L(q_i, \dot{q}_i, t)\, \mathrm{d}t$$

a minimum. This is because equation (6.3.11) are just the same as Euler's equations (6.3.9) with $\phi = L$.

Suppose now that we allow the terminal time t_f to vary. Then J is a functional depending on $\boldsymbol{x}(t)$ and t_f and so we write

$$J(\boldsymbol{x}, t_f) = \int_{t_0}^{t_f} (\boldsymbol{x}, \dot{\boldsymbol{x}}, t)\, \mathrm{d}t$$

Thus,

$$J(\boldsymbol{x} + \varepsilon \boldsymbol{y}, t_f + \varepsilon\tau) = \int_{t_0}^{t_f + \varepsilon\tau} \phi(\boldsymbol{x} + \varepsilon\boldsymbol{y}, \dot{\boldsymbol{x}} + \varepsilon\dot{\boldsymbol{y}}, t)\, \mathrm{d}t$$

Now we recall the following simple result.

Lemma 6.3.1
If

$$G(a) = \int_{u_1(a)}^{u_2(a)} f(t, a)\, \mathrm{d}t$$

where u_1, u_2 and f are differentiable, then G is differentiable and

$$\frac{\mathrm{d}G}{\mathrm{d}a} = \int_{u_1(a)}^{u_2(a)} \frac{\partial f}{\partial a}(t, a)\, \mathrm{d}t + f(u_2(a), a)\frac{\partial u_2}{\partial a} - f(u_1(a), a)\frac{\partial u_1}{\partial a}$$

Using this result it is easy to show that

$$\begin{aligned}\delta J(\boldsymbol{x}, t_f; \boldsymbol{y}, \tau) &= \int_0^{t_f} (\phi_{\boldsymbol{x}}\boldsymbol{y} + \phi_{\dot{\boldsymbol{x}}}\dot{\boldsymbol{y}})\, \mathrm{d}t + \tau\phi(\boldsymbol{x}, \dot{\boldsymbol{x}}, t)\Big|_{t=t_f} \\ &= \int_0^{t_f} \left(\phi_{\boldsymbol{x}} - \frac{\mathrm{d}}{\mathrm{d}t}\phi_{\dot{\boldsymbol{x}}}\right)\boldsymbol{y}\, \mathrm{d}t + \phi_{\dot{\boldsymbol{x}}}\boldsymbol{y}\Big|_{t_0}^{t_f} + \tau\phi(\boldsymbol{x}, \dot{\boldsymbol{x}}, t)\Big|_{t=t_f}\end{aligned}$$

Assuming that $\boldsymbol{x}(t_f)$ is constrained onto a **terminal manifold** defined by the function† $\boldsymbol{c}(t_f) = (c_1(t_f), \ldots, c_n(t_f))^{\mathrm{T}}$ of t_f, then at t_f the trajectories and the terminal manifold must intersect, and so

$$\boldsymbol{x}(t_f + \varepsilon\tau) + \varepsilon\boldsymbol{y}(t_f + \varepsilon\tau) = \boldsymbol{c}(t_f + \varepsilon\tau)$$

Hence

$$\tau\dot{\boldsymbol{x}}(t_f) + \boldsymbol{y}(t_f) = \tau\dot{\boldsymbol{c}}(t_f)$$

† That is $\boldsymbol{x}(t_f)$ must lie on the curve $\boldsymbol{c}(t_f)$ parameterized by t_f.

and so we obtain the necessary conditions

$$\begin{aligned} \phi_x - \frac{d}{dt}\phi_{\dot{x}} &= 0 \\ \phi_{\dot{x}} y \Big|_{t=t_0} &= 0 \\ \phi + \phi_{\dot{x}} \Big|_{t=t_f} (\dot{c} - \dot{x}) &= 0 \end{aligned} \tag{6.3.12}$$

6.3.3 Constrained Optimization

In the last section we derived necessary conditions for the unconstrained extrema of a functional J defined on C_n^2. We shall continue in our search towards the maximum principle by considering the constrained Lagrange problem, i.e. the optimization of

$$J = \int_{t_0}^{t_f} \phi(\boldsymbol{x}, \dot{\boldsymbol{x}}, t)\, dt$$

subject to the constraint $\boldsymbol{G}(\boldsymbol{x}, \dot{\boldsymbol{x}}, t) = 0$, $\boldsymbol{G} = (G_1, \ldots, G_m)^T$ $(m \leqslant n)$. Under conditions similar to those introduced in section 6.2 on finite dimensional optimization, we can show that the above problem is equivalent to optimizing the functional

$$J_\lambda = \int_{t_0}^{t_f} [\phi(\boldsymbol{x}, \dot{\boldsymbol{x}}, t) + \boldsymbol{\lambda}^T(t)\boldsymbol{G}(\boldsymbol{x}, \dot{\boldsymbol{x}}, t)]\, dt$$

where $\boldsymbol{\lambda}^T = (\lambda_1, \ldots, \lambda_m)$ is a Lagrange multiplier vector. Hence, in the case of fixed initial and terminal conditions we obtain the **augmented Euler equations**

$$(\phi + \boldsymbol{\lambda}^T\boldsymbol{G})_{\boldsymbol{x}} - \frac{d}{dt}(\phi + \boldsymbol{\lambda}^T\boldsymbol{G})_{\dot{\boldsymbol{x}}} = 0 \tag{6.3.13}$$

These equations (together with the constraints) constitute the necessary conditions for the constrained optimization.

Note that we can also consider inequality constraints of the form

$$\boldsymbol{G}_{\min} \leqslant \boldsymbol{G}(\boldsymbol{x}, \dot{\boldsymbol{x}}, t) \leqslant \boldsymbol{G}_{\max}$$

by introducing 'slack' variables $\boldsymbol{\xi}(t)$ so that the constraints become

$$\boldsymbol{G}'(\boldsymbol{x}, \dot{\boldsymbol{x}}, \boldsymbol{\xi}, t) \triangleq (\boldsymbol{G}_{\max} - \boldsymbol{G})(\boldsymbol{G} - \boldsymbol{G}_{\min}) - \boldsymbol{\xi}^2 = 0$$

where $\boldsymbol{\xi}^2 \triangleq (\xi_1^2, \ldots, \xi_m^2)$ is a non-negative vector. $\boldsymbol{\xi}$ is then regarded as another state and we obtain the Euler equations (6.3.13) with $\boldsymbol{G}$ replaced by $\boldsymbol{G}'$, together with the extra equations

$$\boxed{\boldsymbol{\lambda}^{\mathrm{T}}\boldsymbol{G}\boldsymbol{\xi} = 0} \tag{6.3.14}$$

Example 6.3.2
As a simple example of the application of (6.3.13) we shall consider the control problem

$$\dot{x} = u \tag{6.3.15}$$

and find the control which minimizes the functional

$$J = \int_{t_0}^{t_f} \tfrac{1}{2}\{[d - x(t)]^2 + ru^2\}\, \mathrm{d}t \quad (r > 0) \tag{6.3.16}$$

The initial and terminal conditions are fixed as $x(t_0) = x_0$, $x(t_f) = d$. Then the Euler equations for the augmented functional

$$J_\lambda = \int_{t_0}^{t_f} (\tfrac{1}{2}\{[d - x(t)]^2 + ru^2\} + \lambda(\dot{x} - u))\, \mathrm{d}t$$

are

$$-[d - x(t)] - \dot{\lambda}(t) = 0$$
$$ru(t) - \lambda(t) = 0$$

These must be solved together with the constraint

$$\dot{x} = u$$

subject to the end conditions $x(t_0) = x_0$, $x(t_f) = d$, i.e. we must solve

$$\ddot{x}(t) - \frac{1}{r}x(t) = -\frac{d}{r}, \quad x(t_0) = x_0,\, x(t_f) = d$$

This type of differential equation, with conditions at two different end-points is called a **two point boundary value problem**, and is difficult to solve in general. Here the solution is easy, for if we denote $\dot{x}(t_0)$ by x_1, then we have the solution

$$x(t) = d + (x_0 - d)\cosh\left[\frac{t - t_0}{\sqrt{r}}\right] + \sqrt{r}\, x_1 \sinh\left[\frac{t - t_0}{\sqrt{r}}\right]$$

Since $x(t_f) = d$ we have

$$x(t) = d - (d - x_0)\left[1 - \coth\frac{(t_f - t_0)}{\sqrt{r}} \tanh\frac{1}{\sqrt{r}}(t - t_0)\right] \cosh\frac{(t - t_0)}{\sqrt{r}}$$

and so

$$u(t) = \dot{x}(t) = \frac{1}{\sqrt{r}}(d - x_0)\left[\coth \frac{(t_f - t_0)}{\sqrt{r}} - \tanh \frac{(t - t_0)}{\sqrt{r}}\right] \cosh \frac{(t - t_0)}{\sqrt{r}}$$

At t_0 we obtain the control

$$\begin{aligned} u(t_0) &= \frac{1}{\sqrt{r}}(d - x_0) \coth \frac{(t_f - t_0)}{\sqrt{r}} \\ &= K(t_0)(d - x_0) \end{aligned} \qquad (6.3.17)$$

where

$$K(t_0) = \frac{1}{\sqrt{r}} \coth \left(\frac{t_f - t_0}{\sqrt{r}}\right)$$

Now the control (6.3.17) is optimal for the cost (6.3.16) at time t_0. If we consider the cost

$$J = \int_{t_1}^{t_f} \tfrac{1}{2}\{[d - x(t)]^2 + ru^2\}\, dt$$

for any $t_1 \in [t_0, t_f)$ it is clear that

$$\begin{aligned} u(t_1) &= \frac{1}{\sqrt{r}}[d - x(t_1)] \coth \frac{(t_f - t_1)}{\sqrt{r}} \\ &= K(t_1)[d - x(t_1)] \end{aligned}$$

must be the optimal control at time t_1. (This is essentially the dynamic programming argument to be introduced precisely later.) Hence

$$u(t) = K(t)[d - x(t)]$$

is the optimal control for all $t \in [t_0, t_f]$, and we obtain the feedback system in fig. 6.3 with time-varying gain $K(t)$.

The above problem is a special case of the tracking problem which we shall consider in detail later.

6.3.4 Non-smooth Optimization – The Weierstrass–Erdmann Corner Conditions

So far we have considered the optimization of a functional J defined on C_n^2. In many cases it happens that J does not have an optimum on C_n^2 although

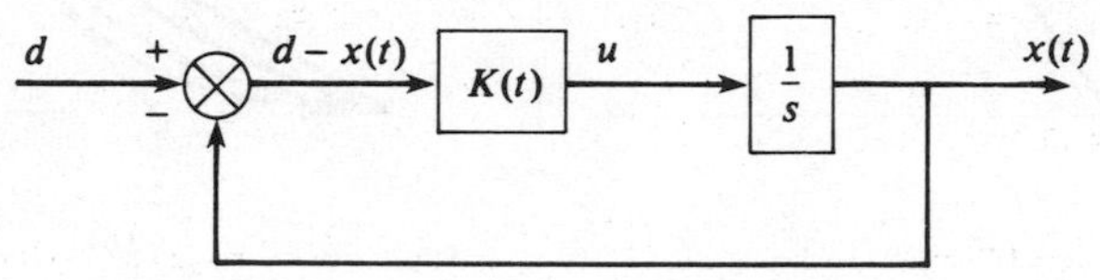

Fig. 6.3 A system with time varying gain.

an optimizing sequence does exist. The problem is that the limit of this sequence does not exist in C_n^2, but it may exist in a bigger space. For example, suppose we try to minimize

$$J = \int_0^1 x^2(2-\dot{x})^2\, dx, \quad \text{with } x(0)=0, x(1)=1 \tag{6.3.18}$$

Clearly, $J \geqslant 0$ for all continuous functions $x(t)$ and $J=0$ when

$$x(t) = \begin{cases} 0 & t \in [0, \frac{1}{2}] \\ 2t-1 & t \in [\frac{1}{2}, 1] \end{cases}$$

(fig. 6.4(a)). The latter function x is clearly the unique continuous function which minimizes (6.3.18), and although a sequence of functions exists (e.g. the sequence x_n in fig. 6.4.(b)) in C_1^2 which converges (pointwise) to $x(t)$, the limit is not in C_1^2. Note however that the Euler equation is satisfied by $x(t)$ on the intervals $[0, \frac{1}{2})$, $(\frac{1}{2}, 1]$.

For this reason we extend the set of functions over which J is optimized to the set of piecewise smooth functions $x(t)$, i.e. those which are in C_n^2 apart from at a finite number of points where $x(t)$ may be discontinuous. Consider the problem of optimizing

$$J = \int_a^b \phi(\boldsymbol{x}(t), \dot{x}(t), t)\, dt$$

where the external arcs $\boldsymbol{x}(t)$ are in C_n^2 except at a point $c \in (a,b)$, which is unknown. As before, $\boldsymbol{x}(t)$ must satisfy Euler's equation on (a,c) and (c,b). Write

$$J = J_1 + J_2 \triangleq \int_a^c \phi\, dt + \int_c^b \phi\, dt$$

Since c is unknown we must take the variational of each of J_1 and J_2 with variable final and initial time respectively. Then

$$\delta J_1(\boldsymbol{x}, c; y, \tau) = \int_a^c \left(\phi_x - \frac{d}{dt}\phi_{\dot{x}}\right) y\, dt + \phi_{\dot{x}} y \Big|_a^c + \tau\phi(\boldsymbol{x}, \dot{\boldsymbol{x}}, t)\Big|_{t=c}$$

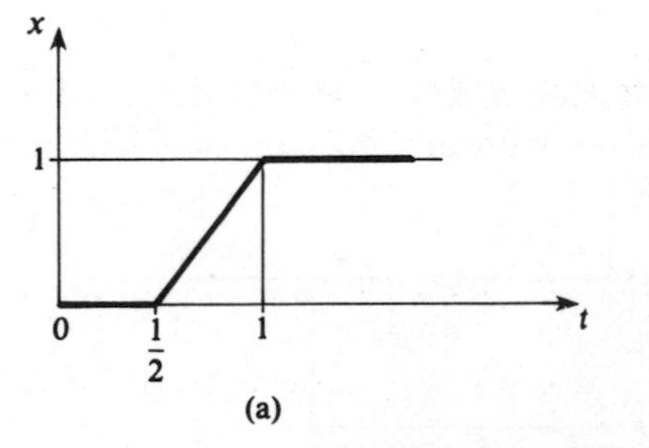

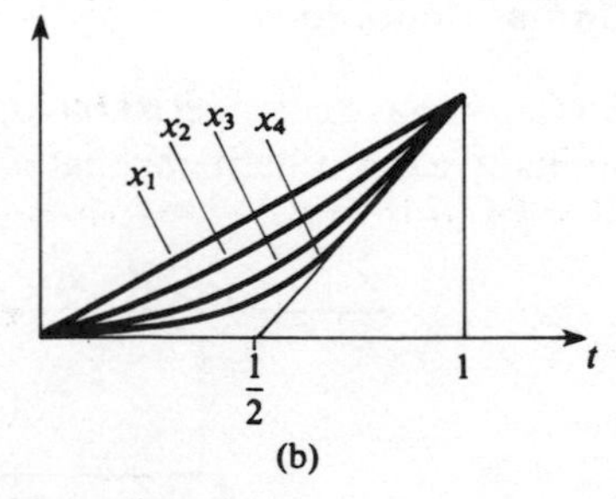

Fig. 6.4 Non-differentiable optimum.

Now we write

$$\delta x(c) \triangleq x(c+\tau) + y(c+\tau) - x(c)$$
$$\approx \dot{x}(c)\tau + y(c)$$

and so

$$\delta J_1(x, c; y, \tau) = \phi_{\dot{x}}\, \delta x\Big|_{c-} + \tau(\phi - \phi_{\dot{x}}\dot{x})\Big|_{c-}$$

(ignoring the first term which must satisfy Euler's equation at an extremum). Similarly

$$\delta J_2(x, c; y, \tau) = -\phi_{\dot{x}}\, \delta x\Big|_{c+} - \tau(\phi - \phi_{\dot{x}}x)\Big|_{c+}$$

and since $\delta J = \delta J_1 + \delta J_2 = 0$ at an extremum we have the Weierstrass–Erdmann corner conditions

$$\boxed{\begin{aligned} \phi_{\dot{x}}\Big|_{c-} &= \phi_{\dot{x}}\Big|_{c+} \\ (\phi - \phi_{\dot{x}}\dot{x})\Big|_{c-} &= (\phi - \phi_{\dot{x}}\dot{x})\Big|_{c+} \end{aligned}} \tag{6.3.19}$$

These conditions must be satisfied at every discontinuity of x. It can be shown (Fleming and Rishel, 1975) that there are no discontinuities in x if $\phi_{\dot{x}\dot{x}}$ is positive definite for all $(x, \dot{x}, t)$. Such a problem is called **regular**.

6.3.5 The Jacobi Condition and the Second Variation

In the finite dimensional optimization problem, we saw that a sufficient condition for a local minimum of a function $f(x)$ is $\partial^2 f/\partial x^2 > 0$. It is also necessary that $\partial^2 f/\partial x^2 \geqslant 0$. A similar necessary condition for a minimum of J at x_0 in the calculus of variations is clearly

$$\delta^2 J(x_0; y) = \frac{d^2}{d\varepsilon^2} J(x_0 + \varepsilon y)\Big|_{\varepsilon=0} \geqslant 0 \tag{6.3.20}$$

If $\phi \in C_n^4$, then it is easy to see that

$$\delta^2 J(x_0; y) = \int_{t_0}^{t_f} (y^T L_{xx} y + 2\dot{y}^T L_{xx} y + \dot{y}^T L_{\dot{x}\dot{x}} \dot{y})\, dt$$
$$\triangleq 2\int_{t_0}^{t_f} \Omega(y, \dot{y}, t)\, dt$$

say. Note that

$$\Omega_{\dot{y}\dot{y}} = L_{\dot{x}\dot{x}}$$

and so if the original problem is regular, so is the **secondary minimum problem** of minimizing $\delta^2 J(x_0; y)$ with respect to y. The Euler equations for the secondary problem are

$$\boxed{\Omega_y = \frac{\mathrm{d}}{\mathrm{d}t}\Omega_{\dot{y}}} \tag{6.3.21}$$

Consider the fixed endpoint problem of minimizing

$$J = \int_{t_0}^{t_f} \phi(x, \dot{x}, t)\,\mathrm{d}t$$

with $x_0 = x(t_0)$, $x_f = x(t_f)$ fixed. A point $(t', x'(t'))$, where $t_0 < t' < t_f$ and x' is an extremal of the problem, is said to be **conjugate** to (t_0, x_0) if there is a non-zero secondary extremal y (i.e. a solution of (6.3.21)) such that $y(t_0) = y(t') = \mathbf{0}$. Then we have the Jacobi necessary condition as follows.

Theorem 6.3.3
If J has a minimum at x_0, then there are no conjugate points to (t_0, x_0) with $t_0 < t' < t_f$.

This result is important in providing further information in the selection of a minimum from a set of extrema which satisfies the primary Euler equation.

6.4 THE PONTRYAGIN MAXIMUM PRINCIPLE

6.4.1 The Bolza Problem Without Inequality Constraints

Before considering the general maximum principle we shall motivate the optimal control problem by examining the situation where there are no 'hard' inequality constraints on the control. (The latter conditions are the cause of most of the problems as we shall see presently.) Suppose, therefore, that we have a control system

$$\dot{x} = f(x, u, t), \quad t \in [t_0, t_f]$$

where $x(t) \in \mathbb{R}^n$, $u(t) \in \mathbb{R}^m$ for each t and that we wish to minimize

$$J = \theta(x(t), t)\Big|_{t_0}^{t_f} + \int_{t_0}^{t_f} \phi(x(t), u(t), t)\,\mathrm{d}t$$

(Although this problem is equivalent to a Lagrange problem, as noted earlier, it is now convenient to consider a cost function of this form.) As before, we adjoin the constraints with n Lagrange multipliers $\boldsymbol{\lambda}(t)$ to obtain the new cost functional

$$J_{\lambda} = \theta(\boldsymbol{x}(t), t) \int_{t_0}^{t_f} + \int_{t_0}^{t_f} \{\phi(\boldsymbol{x}(t), \boldsymbol{u}(t), t) + \boldsymbol{\lambda}^{\mathrm{T}}(t)[f(\boldsymbol{x}, \boldsymbol{u}, t) - \dot{\boldsymbol{x}}]\}\,\mathrm{d}t$$

We define the **Hamiltonian function** H by

$$H(\boldsymbol{x}(t), \boldsymbol{u}(t), \boldsymbol{\lambda}(t), t) = \phi(\boldsymbol{x}(t), \boldsymbol{u}(t), t) + \boldsymbol{\lambda}^{\mathrm{T}}(t) f(\boldsymbol{x}(t), \boldsymbol{u}(t), t)$$

(Note that many authors define $H = -\phi + \boldsymbol{\lambda}^{\mathrm{T}} f$.) Then

$$J_{\lambda} = \theta(\boldsymbol{x}(t), t)\Big|_{t_0}^{t_f} + \int_{t_0}^{t_f} \{H(\boldsymbol{x}(t), \boldsymbol{u}(t), \boldsymbol{\lambda}(t), t) - \boldsymbol{\lambda}^{\mathrm{T}}(t)\dot{\boldsymbol{x}}(t)\}\,\mathrm{d}t$$

Taking the directional derivative at $(\boldsymbol{x}, \boldsymbol{u})$ in the direction $(\boldsymbol{y}, \boldsymbol{v})$ we have

$$\delta J_{\lambda}(\boldsymbol{x}, \boldsymbol{u}; \boldsymbol{y}, \boldsymbol{v}) = \left[\boldsymbol{y}^{\mathrm{T}}\left(\frac{\partial \theta}{\partial \boldsymbol{x}} - \boldsymbol{\lambda}\right)\right]\Bigg|_{t_0}^{t_f} + \int_{t_0}^{t_f}\left[\boldsymbol{y}^{\mathrm{T}}\left(\frac{\partial H}{\partial \boldsymbol{x}} + \dot{\boldsymbol{\lambda}}\right) + \boldsymbol{v}^{\mathrm{T}}\frac{\partial H}{\partial \boldsymbol{u}}\right]\mathrm{d}t$$

(assuming t_0 and t_f fixed). For arbitrary variations $\boldsymbol{y}$ and $\boldsymbol{v}$, it follows that we obtain the following necessary conditions for an extremum

$$\boldsymbol{y}^{\mathrm{T}}\left(\frac{\partial \theta}{\partial \boldsymbol{x}} - \boldsymbol{\lambda}\right) = 0 \quad \text{at } t = t_0, t = t_f \tag{6.4.1a}$$

$$\dot{\boldsymbol{\lambda}} = -\frac{\partial H}{\partial \boldsymbol{x}} \tag{6.4.1b}$$

$$\frac{\partial H}{\partial \boldsymbol{u}} = \boldsymbol{0} \tag{6.4.1c}$$

Of course, we also have the original constraint equation

$$\dot{\boldsymbol{x}} = f(\boldsymbol{x}, \boldsymbol{u}, t) = \frac{\partial H}{\partial \boldsymbol{\lambda}} \tag{6.4.1d}$$

Equation (6.4.1c) implies that we seek an extremum of H in $\boldsymbol{u}$ – this is a special case of Pontryagin's maximum principle.

Note that if the initial and final conditions are specified by manifolds $\boldsymbol{M}(\boldsymbol{x}(t_0), t_0) = \boldsymbol{0}$ and $\boldsymbol{N}(\boldsymbol{x}(t_f), t_f) = \boldsymbol{0}$ where $\boldsymbol{M}$ and $\boldsymbol{N}$ take values in $\mathbb{R}^p$ and $\mathbb{R}^q$ respectively, then we can adjoin these to the terminal value θ and obtain the cost

$$J_{\lambda, \xi, \nu} = \theta(\boldsymbol{x}(t), t)\Big|_{t_0}^{t_f} - \boldsymbol{\xi}^{\mathrm{T}}\boldsymbol{M}\Big|_{t_0} + \boldsymbol{\nu}^{\mathrm{T}}\boldsymbol{N}\Big|_{t_f} + \int_{t_0}^{t_f}(H - \boldsymbol{\lambda}^{\mathrm{T}}\dot{\boldsymbol{x}})\,\mathrm{d}t$$

for some Lagrange multipliers $\boldsymbol{\xi}$ and $\boldsymbol{\nu}$. Extremizing J in the usual way we obtain the general transversality conditions

$$\boxed{\begin{aligned} \boldsymbol{\lambda}(t_0) &= \frac{\partial \theta}{\partial \boldsymbol{x}} + \left(\frac{\partial \boldsymbol{M}^{\mathrm{T}}}{\partial \boldsymbol{x}}\right)\boldsymbol{\xi} \quad \text{at} \quad t = t_0 \\ \boldsymbol{\lambda}(t_f) &= \frac{\partial \theta}{\partial \boldsymbol{x}} + \left(\frac{\partial \boldsymbol{N}^{\mathrm{T}}}{\partial \boldsymbol{x}}\right)\boldsymbol{\nu} \quad \text{at} \quad t = t_f \end{aligned}} \tag{6.4.2}$$

(plus the constraints $\boldsymbol{M} = \boldsymbol{0}$, $\boldsymbol{N} = \boldsymbol{0}$). In a similar way, if t_f is free we obtain the necessary conditions (6.4.1b)–(6.4.1d) together with the transversality conditions

$$\boxed{\begin{aligned} \boldsymbol{\lambda}(t_f) &= \frac{\partial \theta}{\partial \boldsymbol{x}} + \left(\frac{\partial \boldsymbol{N}^{\mathrm{T}}}{\partial \boldsymbol{x}}\right)\boldsymbol{\nu} \quad \text{at } t = t_f \\ &\boldsymbol{N}(\boldsymbol{x}(t_f), t_f) = \boldsymbol{0} \\ H + \frac{\partial \theta}{\partial t_f} &+ \left(\frac{\partial \boldsymbol{N}^{\mathrm{T}}}{\partial t_f}\right)\boldsymbol{\nu} = 0 \quad \text{at } t = t_f \end{aligned}} \tag{6.4.3}$$

6.4.2 The Maximum Principle

We are now ready to state the general maximum principle. As we have said previously the problem comes with inequality constraints in the control variables. The above 'weak' variations of the form $\boldsymbol{x} + \varepsilon \boldsymbol{y}$ are not sufficient to prove the maximum principle which requires the notion of cones and separating hyperplanes. A general proof is given by Girsanov (1972), Pontryagin *et al.* (1964) and Fleming and Rishel (1975); we shall merely state the maximum principle in the following form.

Theorem 6.4.1
Using the above notation, consider the problem of minimizing the cost functional

$$J = \theta(\boldsymbol{x}(t_f), t_f) + \int_{t_0}^{t_f} \phi(\boldsymbol{x}(t), \boldsymbol{u}(t), t)\,\mathrm{d}t$$

subject to the constraint

$$\dot{\boldsymbol{x}}(t) = \boldsymbol{f}(\boldsymbol{x}(t), \boldsymbol{u}(t), t)$$

and assume that $(t_0, \boldsymbol{x}_0)$ is fixed. Suppose, moreover, that t_f is free and that $\boldsymbol{x}(t_f)$ is constrained to the terminal manifold defined by the equation

$$\boldsymbol{N}(\boldsymbol{x}(t_f), t_f) = \boldsymbol{0}$$

and finally that $\boldsymbol{u}$ is constrained to belong to some closed set $\Omega \subseteq \mathbb{R}^m$. Then we have the following necessary conditions for a minimum of J at $\hat{\boldsymbol{x}}$, $\hat{\boldsymbol{u}}$

subject to the constraints:

$$H(\hat{x}(t), \hat{u}(t), \lambda(t), t) \leqslant H(x(t), u(t), \lambda(t), t) \quad \text{all } t$$

$\hat{x}$ and $\hat{u}$ satisfy (6.4.4a)

$$\frac{\partial H}{\partial \lambda} = \dot{x} \tag{6.4.4b}$$

$$\frac{\partial H}{\partial x} = -\dot{\lambda} \tag{6.4.4c}$$

$$\left.\begin{aligned} \lambda &= \frac{\partial \theta}{\partial x} + \left(\frac{\partial N^{\mathrm{T}}}{\partial x}\right)\nu \quad &(6.4.4\text{d}) \\ -H &= \frac{\partial \theta}{\partial t} + \left(\frac{\partial N^{\mathrm{T}}}{\partial t}\right)\nu \quad &(6.4.4\text{e}) \end{aligned}\right\} \text{ at } t = t_f$$

As before, H denotes the Hamiltonian defined by

$$H = \phi + \lambda^{\mathrm{T}} f \tag{6.4.5}$$

and λ, ν are Lagrange multipliers. In Pontryagin's formulation, as stated earlier, $H = -\phi + \lambda^{\mathrm{T}} f$ and then the inequality in (6.4.4a) is reversed. In our case we are minimizing H among the admissible controls with values in Ω; strictly speaking we have therefore stated the 'minimum' principle. The Lagrange multipliers λ are often called the costate (adjoint or dual) variables and equation (6.4.4c) is called the costate or adjoint equation.

Example 6.4.1

Consider the system defined by the equation

$$\ddot{x} + \dot{x} = u(t)$$

and let $x_1 = x$, $x_2 = \dot{x}$. Then

$$\dot{x}_1 = x_2, \quad \dot{x}_2 = -x_2 + u \tag{6.4.6}$$

Suppose that we wish to minimize the cost functional

$$J = \int_0^1 \tfrac{1}{2}(x_1^2 + au^2)\, \mathrm{d}t$$

subject to (6.4.6) with $|u| \leqslant 1$. The Hamiltonian for this problem is

$$H = \phi + \lambda^{\mathrm{T}} f = \tfrac{1}{2}(x_1^2 + au^2) + \lambda_1 x_2 + \lambda_2(-x_2 + u)$$

which we must minimize in u. Note that $\partial H/\partial u = au + \lambda_2 = 0$ if $u = -\lambda_2/a$, and so if $|\lambda_2| \leqslant a$ we obtain the classical minimum

$$u = -\frac{\lambda_2}{a}$$

However, if $|\lambda_2| > a$ then by (6.4.4a) we must choose

$$u = \begin{cases} +1 & \text{if } \lambda_2 < 0 \\ -1 & \text{if } \lambda_2 > 0 \end{cases}$$

Having chosen the control we must determine λ_2 which may be found from the equations (6.4.4b) and (6.4.4c), i.e.

$$\left.\begin{aligned} \dot{x}_1 &= x_2 \\ \dot{x}_2 &= x_2 + u \\ \dot{\lambda}_1 &= -x_1 \\ \dot{\lambda}_2 &= -(\lambda_1 - \lambda_2) \end{aligned}\right\} \tag{6.4.7}$$

The transversality condition is given by (6.4.1a) to be $\boldsymbol{\lambda}(1) = \mathbf{0}$. The equations (6.4.7) together with the boundary conditions $\boldsymbol{\lambda}(1) = \mathbf{0}$ and $\boldsymbol{x}(0) = \boldsymbol{x}_0$ (given) form a two-point boundary value problem. Such a problem can be solved numerically by methods such as Bellman's invariant imbedding technique (see Bellman (1964) for example) or we may solve the equations backwards in time to obtain a picture of the initial $\boldsymbol{x}$ state which may be driven to some final $\boldsymbol{x}$ state. For example, suppose $\boldsymbol{x}(1) = \boldsymbol{x}_1$ is given and consider the equations (6.4.7) running backwards in time, i.e.

$$\frac{\mathrm{d}}{\mathrm{d}\tau}\begin{pmatrix} \boldsymbol{x}(1-\tau) \\ \boldsymbol{\lambda}(1-\tau) \end{pmatrix} = \begin{pmatrix} 0 & -1 & 0 & 0 \\ 0 & +1 & 0 & 0 \\ +1 & 0 & 0 & 0 \\ 0 & 0 & 1 & -1 \end{pmatrix} \begin{pmatrix} \boldsymbol{x}(1-\tau) \\ \boldsymbol{\lambda}(1-\tau) \end{pmatrix} - \begin{pmatrix} 0 \\ u(1-\tau) \\ 0 \\ 0 \end{pmatrix}$$

with $(\boldsymbol{x}(1), \boldsymbol{\lambda}(1))^{\mathrm{T}} = (\boldsymbol{x}_1, \mathbf{0})^{\mathrm{T}}$ or

$$\frac{\mathrm{d}}{\mathrm{d}\tau}\begin{pmatrix} \boldsymbol{x}' \\ \boldsymbol{\lambda}' \end{pmatrix} = A \begin{pmatrix} \boldsymbol{x}' \\ \boldsymbol{\lambda}' \end{pmatrix} - \begin{pmatrix} 0 \\ u' \\ 0 \\ 0 \end{pmatrix} \tag{6.4.8}$$

for an appropriate matrix A, where $(\boldsymbol{x}'(\tau), \boldsymbol{\lambda}'(\tau)) = (\boldsymbol{x}(1-\tau), \boldsymbol{\lambda}(1-\tau))$, and $\boldsymbol{u}'(\tau) = \boldsymbol{u}(1-\tau)$. Since $\lambda_2(1) = 0$ and λ_2 is continuous there must be some maximum interval $[0, \tau_1)$ in which $u'(\tau) = -\lambda_2/a$ and so (6.4.8) reduces to the equation

$$\frac{\mathrm{d}}{\mathrm{d}\tau}\begin{pmatrix} \boldsymbol{x}'(\tau) \\ \boldsymbol{\lambda}'(\tau) \end{pmatrix} = A_1 \begin{pmatrix} \boldsymbol{x}' \\ \boldsymbol{\lambda}' \end{pmatrix}$$

where

$$A_1 = A + \begin{pmatrix} 0 & 0 & 0 & 0 \\ 0 & 0 & 0 & 1/a \\ & & \mathbf{0} & \end{pmatrix}$$

Hence, for $\tau \in [0, \tau_1)$,

$$\begin{pmatrix} \boldsymbol{x}'(\tau) \\ \boldsymbol{\lambda}'(\tau) \end{pmatrix} = \exp(A_1 \tau) \begin{pmatrix} \boldsymbol{x}_1 \\ \mathbf{0} \end{pmatrix}$$

At τ_1 we must use (6.4.8) with $u' = \pm 1$ depending on $\lambda_2(\tau_1)$. This solution may be continued back to $t = 0$ ($\tau = 1$) to find $\boldsymbol{x}(0)$. The next example shows that, in some cases, we can evaluate the control switches explicitly.

Example 6.4.2

Suppose we now try to determine the admissible control which transfers the system

$$\ddot{x} = u, \quad |u| \leqslant 1$$

to the origin from any initial state (ξ_1, ξ_2) in minimum time. (It can be shown that such a control exists and is unique.) The cost functional is

$$J = \int_0^{t_f} 1 \mathrm{d}t$$

and the Hamiltonian is $H = 1 + \lambda_1 x_2 + \lambda_2 u$, since the system may be written in the form $(x = x_1, \ x = x_2)$

$$\dot{x}_1 = x_2, \quad \dot{x}_2 = u$$

Hence H is minimized by choosing the control

$$u = -\operatorname{sgn}\{\lambda_2\}$$

$$= \begin{cases} -1 & \text{if } \lambda_2 > 0 \\ +1 & \text{if } \lambda_2 < 0 \end{cases}$$

The adjoint equations are easily seen to be

$$\dot{\lambda}_1 = 0, \quad \dot{\lambda}_2 = -\lambda_1 \tag{6.4.9}$$

Rather than apply transversality conditions at $t = t_f$, it is simpler to proceed as follows. We can solve equations (6.4.9) to obtain

$$\lambda_1 = \lambda_{10}, \quad \lambda_2 = \lambda_{20} - \lambda_{10} t$$

where $(\lambda_{10}, \lambda_{20})$ is the (unknown) initial value of $\boldsymbol{\lambda}$. Since λ_2 is linear in time it is either a positive or negative gradient and is therefore of a constant sign or has a single sign change on the entire half axis $[0, \infty)$. Hence there are only four possibilities for the control u, namely

$$u(t) = +1, \quad t \in [0, \infty) \tag{6.4.10a}$$

$$u(t) = -1, \quad t \in [0, \infty) \tag{6.4.10b}$$

$$u(t) = \begin{cases} +1, & t \in [0, t_1) \\ -1, & t \in [t_1, \infty) \end{cases} \tag{6.4.10c}$$

$$u(t) = \begin{cases} -1, & t \in [0, t_2) \\ +1, & t \in [t_2, \infty) \end{cases} \tag{6.4.10d}$$

for some times t_1 or t_2. Now $\ddot{x} = u$ and so

$$x_2 = \xi_2 + \beta t$$

$$x_1 = \xi_1 + \xi_2 t + \frac{\beta t^2}{2}$$

where $x_1(0) = \xi_1$, $x_2(0) = \xi_2$ and $\beta = u = \pm 1$. Eliminating t gives

$$x_1 = \xi_1 + \frac{1}{2\beta}(x_2^2 - \xi_2^2)$$

These phase plane trajectories are parabolas passing through (ξ_1, ξ_2) as shown in fig. 6.5. Consider control (6.4.10a). The set of points in the plane which can be driven to the origin using the single non-switching control $u = +1$ consists of the lower portion of the parabola in fig. 6.5(a) which passes through $(0, 0)$. Call this curve γ_+ (see fig. 6.6). Similarly, let γ_- denote the curve of points which can be driven to $(0, 0)$ using the control $u = -1$. The combined curve $\gamma_- \cup \gamma_+$ separates the (x_1, x_2)-plane into two open regions H_+, H_-. Any point $\boldsymbol{\xi}^+$ in H_+ can be driven to $(0, 0)$ using a control of the form (6.4.10c), where t_1 depends on $\boldsymbol{\xi}^+$, and a point $\boldsymbol{\xi}^- \in H_-$ requires a control of the form (6.4.10d). The non-linear control

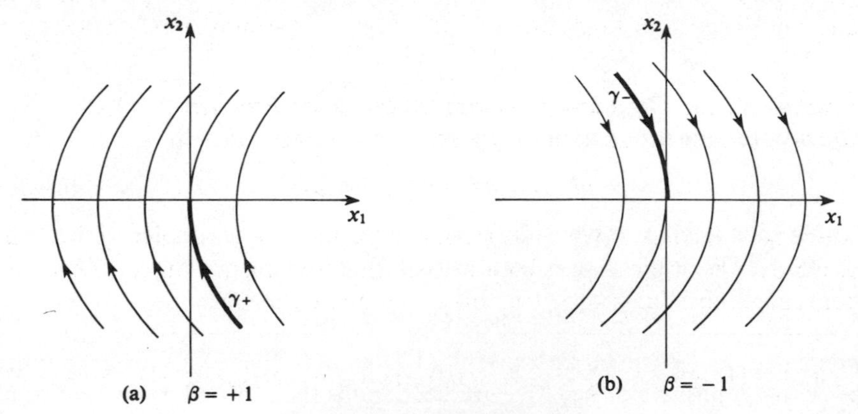

Fig. 6.5 Switching curves.

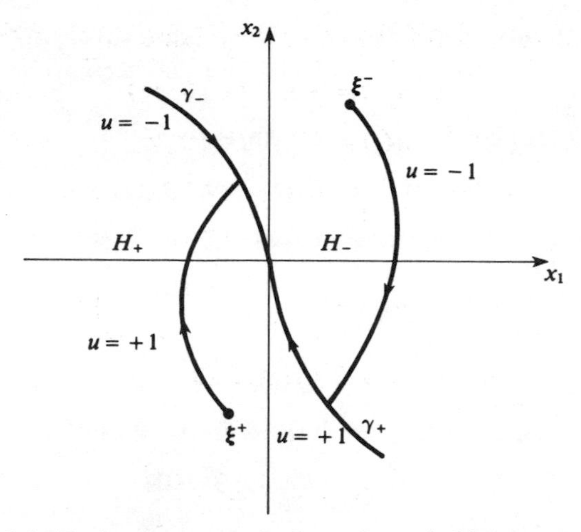

Fig. 6.6 Evaluating controls from the switching curves.

$u(x_1, x_2)$ can be synthesized as

$$u(x_1, x_2) = \begin{cases} +1 & \text{if } (x_1, x_2) \in \gamma_+ \cup H_+ \\ -1 & \text{if } (x_1, x_2) \in \gamma_- \cup H_- \end{cases}$$

This example can be generalized to the time optimal control of any linear system

$$\frac{\mathrm{d}x}{\mathrm{d}t} = Ax + Bu$$

when $\boldsymbol{u}$ is constrained to a polyhedron U provided

$$Bv, ABv, \ldots, A^{n-1}Bv$$

are linearly independent for $\boldsymbol{v}$ along any edge of U. Then it can be shown that there are only a finite number of switchings of u at the vertices of U (see, for example, Pontryagin *et al.*, 1964). Moreover if U is a parallelepiped and the eigenvalues of A are real, then there are at most $n-1$ switchings.

Example 6.4.3: The Linear–quadratic Regulator Problem

In this example we shall consider the general linear system

$$\dot{\boldsymbol{x}}(t) = A(t)\boldsymbol{x}(t) + B(t)\boldsymbol{u}(t) \tag{6.4.11}$$

where $\boldsymbol{u}$ is unconstrained. The regulator problem requires $\boldsymbol{x}(t)$ to be kept as close to the origin as possible without using 'too much' control. The most convenient cost functional for this is the quadratic form

$$\boxed{J = \tfrac{1}{2}\boldsymbol{x}^{\mathrm{T}}(t_f)F\boldsymbol{x}(t_f) + \tfrac{1}{2}\int_{t_0}^{t_f} [\boldsymbol{x}^{\mathrm{T}}(t)Q(t)\boldsymbol{x}(t) + \boldsymbol{u}^{\mathrm{T}}(t)R(t)\boldsymbol{u}(t)]\,\mathrm{d}t} \tag{6.4.12}$$

since we can obtain a simple solution in this case consisting of a linear feedback. F and $Q(t)$ are positive semi-definite matrices, while we assume that $R(t)$ is positive definite.

The Hamiltonian for this problem is

$$H = \tfrac{1}{2}\boldsymbol{x}^{\mathrm{T}}Q(t)\boldsymbol{x} + \tfrac{1}{2}\boldsymbol{u}^{\mathrm{T}}R(t)\boldsymbol{u} + \boldsymbol{\lambda}^{\mathrm{T}}A(t)\boldsymbol{x} + \boldsymbol{\lambda}^{\mathrm{T}}B(t)\boldsymbol{u}$$

The costate vector $\boldsymbol{\lambda}$ therefore satisfies the equation

$$\begin{aligned}\dot{\boldsymbol{\lambda}} &= \frac{\partial H}{\partial \boldsymbol{x}} \\ &= -Q(t)\boldsymbol{x} - A^{\mathrm{T}}(t)\boldsymbol{\lambda}\end{aligned} \tag{6.4.13}$$

Along an optimal trajectory we have $\partial H/\partial \boldsymbol{u} = \boldsymbol{0}$ and so

$$\boldsymbol{u} = -R^{-1}(t)B^{\mathrm{T}}(t)\boldsymbol{\lambda}$$

and using (6.4.11) we have

$$\dot{\boldsymbol{x}}(t) = A\boldsymbol{x}(t) - BR^{-1}B^{\mathrm{T}}\boldsymbol{\lambda} \tag{6.4.14}$$

From (6.4.13), and (6.4.14) we obtain the equation

$$\boxed{\begin{pmatrix}\dot{\boldsymbol{x}} \\ \dot{\boldsymbol{\lambda}}\end{pmatrix} = \begin{pmatrix} A & -BR^{-1}B^{\mathrm{T}} \\ -Q & -A^{\mathrm{T}}\end{pmatrix}\begin{pmatrix}\boldsymbol{x} \\ \boldsymbol{\lambda}\end{pmatrix}} \tag{6.5.15}$$

The boundary conditions for this equation are given by

$$\left.\begin{aligned}\boldsymbol{x}(t_0) &= \boldsymbol{x}_0 \quad \text{(given)} \\ \boldsymbol{\lambda}(t_f) &= \frac{\partial\theta}{\partial\boldsymbol{x}} = F\boldsymbol{x}(t_f)\end{aligned}\right\} \tag{6.4.16}$$

(the latter follows from (6.4.1a)). This two-point boundary value problem can be converted to a non-linear initial value problem. However, we shall see later how to derive the latter equation in a different way.

So far nothing has been said about the existence of optimal controls. It turns out that one must consider more general controls than piecewise continuously differentiable ones. If the system dynamics are defined by the equation

$$\dot{\boldsymbol{x}}(t) = \boldsymbol{f}(t, \boldsymbol{x}(t), \boldsymbol{u}(t))$$

then we consider the set

$$F(t, \boldsymbol{x}) = \{\boldsymbol{f}(t, \boldsymbol{x}, \boldsymbol{u}) : \boldsymbol{u} \in U\}$$

In the case where U is compact and $F(t, \boldsymbol{x})$ is convex for each $(t, \boldsymbol{x}) \in \mathbb{R}^{n+1}$ we can obtain sufficient conditions for the existence of an optimal control under fairly mild conditions on $\boldsymbol{f}$ (see Fleming and Rishel, 1975). However, if $F(t, \boldsymbol{x})$ is not convex then the existence theory is much more difficult and requires the introduction of even more general controls. In fact a generic control is assumed to have probability measures as values. Such controls are said to be **chattering** or **relaxed** and the reader is referred to Young (1969), Lee and Markus (1967) and Warga (1972) for a discussion of this theory.

6.4.3 Singular Control Theory

In deriving necessary conditions for the optimization of a cost functional subject to the states satisfying dynamical system equations we must examine the Hamiltonian $H(\boldsymbol{x}, \boldsymbol{u}, \boldsymbol{\lambda}, t)$ and determine the value(s) of $\boldsymbol{u}$ which make H a minimum. If $\boldsymbol{u}$ is unconstrained then we require

$$H_{\boldsymbol{u}} = \frac{\partial H}{\partial \boldsymbol{u}} = \boldsymbol{0} \tag{6.4.17}$$

while if $\boldsymbol{u} \in \Omega$ (a bounded closed set) then we must satisfy condition (6.4.4a). However, it may happen that neither of these conditions give any information about the control $\boldsymbol{u}$. In particular, if H is linear in $\boldsymbol{u}$ then we have

$$H = h(\boldsymbol{x}, \boldsymbol{\lambda}, t) + \boldsymbol{g}^{\mathrm{T}}(\boldsymbol{x}, \boldsymbol{\lambda}, t)\boldsymbol{u} \tag{6.4.18}$$

for some functions h and $\boldsymbol{g}$, and if $\boldsymbol{u}$ is unbounded then, for a minimum, we must have

$$H_{\boldsymbol{u}} = \boldsymbol{g}(\boldsymbol{x}, \boldsymbol{\lambda}, t) = \boldsymbol{0} \tag{6.4.19}$$

Since g is independent of $\boldsymbol{u}$ this gives no value for $\boldsymbol{u}$ on an extremal. Furthermore if $\boldsymbol{u}$ is constrained by $||\boldsymbol{u}|| \leqslant 1$, and if (6.4.19), together with the other necessary conditions, has a solution then we cannot take

$$u_i = -\operatorname{sgn} \frac{g_i(\boldsymbol{x}, \boldsymbol{\lambda}, t)}{||\boldsymbol{g}(\boldsymbol{x}, \boldsymbol{\lambda}, t)||} \tag{6.4.20}$$

as we would if g were non-zero. Any extremal arc satisfying (6.4.19) in the unconstrained case is said to be **singular**. Note that $H_{uu} = 0$ along such an arc. In the case where $\boldsymbol{u} \in \Omega$ is constrained we have the 'bang-bang' solution (6.4.20) when $H_u = \boldsymbol{0}$ and singular arcs when (6.4.17) is satisfied.

Since the Pontryagin necessary conditions give no information on a singular arc, other types of necessary conditions must be sought. Note that we already know that a necessary condition for a minimum is that

$$\boxed{H_{uu} \geqslant 0} \tag{6.4.21}$$

(i.e. this matrix is non-negative definite). This is known as the **Legendre–Clebsch condition**, and, of course, gives no further information in the singular control problem. However, by considering special (non-weak) control variations it can be shown that the following generalized

Legendre–Clebsch necessary conditions hold for singular arcs (see Bell and Jacobson, 1975)

$$(-1)^k \frac{\partial}{\partial u}\left[\left(\frac{\mathrm{d}}{\mathrm{d}t}\right)^{2k} H_u\right] \geqslant 0, \quad k = 0, 1, 2, \ldots \tag{6.4.22}$$

It turns out that these conditions usually do give extra information on the control u.

Example 6.4.4 (Johnson and Gibson, 1963; Bryson and Ho, 1969) Consider the problem of minimizing

$$J = \tfrac{1}{2} \int_0^\infty x_1^2 \, \mathrm{d}t$$

subject to

$$\dot{x}_1 = x_2 + u, \quad \dot{x}_2 = -u$$

$$x_1(0), x_2(0) \text{ given}, \quad x_1(\infty) = x_2(\infty) = 0$$

Then

$$H = \lambda_1(x_2 + u) + \lambda_2(-u) + \tfrac{1}{2} x_1^2$$

and so

$$\dot{\lambda}_1 = -x_1, \qquad \dot{\lambda}_2 = -\lambda_1$$

On singular arcs

$$\lambda_1 - \lambda_2 = 0$$

Since the latter condition must hold on some time interval, it follows that

$$\begin{aligned} 0 &= \dot{\lambda}_1 - \dot{\lambda}_2 \\ &= -x_1 + \lambda_1 \end{aligned}$$

on a singular arc. It can also be shown (section 6.7, exercise 9) that if H does not depend explicitly on time t, then H is constant along an optimal trajectory. Hence

$$H = \tfrac{1}{2} x_1^2 + \lambda_1 x_2 + (\lambda_1 - \lambda_2) u = \text{const} = 0$$

since $t_f = \infty$, and on a singular arc we have

$$H = \tfrac{1}{2} x_1^2 + x_1 x_2 = 0$$

The singular arcs are therefore the lines $x_1 = 0, x_1 + 2x_2 = 0$. Given any $x_1(0), x_2(0)$ (off the singular arc) we can use impulsive control (i.e. $u = K\,\delta(t)$ for some constant K) to drive the system to the singular arc and

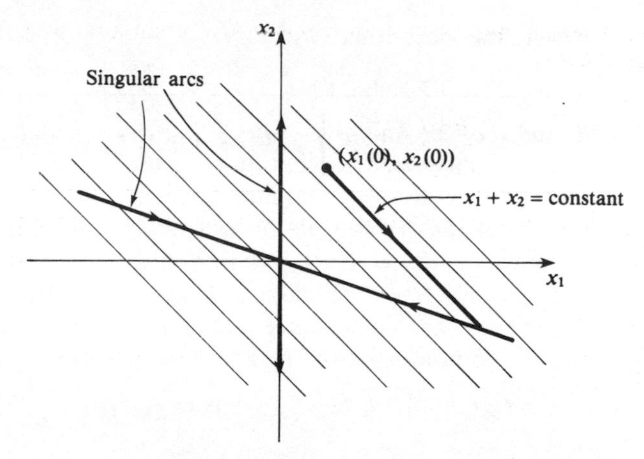

Fig. 6.7 Singular arcs.

then we follow the singular arc to the orign (see fig. 6.7). Note that

$$\frac{d^2}{dt^2}\left(\frac{\partial H}{\partial u}\right) = -\dot{x}_1 + \dot{\lambda}_1 = 0$$

and so

$$u = -(x_1 + x_2)$$

is the control along the singular arc. If u is bounded, then we can use bang-bang controls off the singular arc.

The difficulties which arise in singular control have led many authors to study the second variation, defined in section 6.3.5. Recall that the second variation is given by

$$\delta^2 J(x_0; y) = \int_{t_0}^{t_f} (y^T L_{xx} y + 2\dot{y}^T L_{\dot{x}x} y + \dot{y}^T L_{\dot{x}\dot{x}} \dot{y})\, dt \qquad (6.4.23)$$

If the state $\boldsymbol{x}$ satisfies the equations

$$\dot{\boldsymbol{x}} = \boldsymbol{f}(\boldsymbol{x}, \boldsymbol{u}, t)$$

$$\boldsymbol{x}(t_0) = \boldsymbol{x}_0, \quad \boldsymbol{N}(\boldsymbol{x}(t_f), t_f) = \boldsymbol{0}$$

then the variation $\boldsymbol{y}$ must satisfy (to first order) the equations

$$\dot{y} = f_x y + f_u v \qquad (6.4.24)$$

$$y(t_0) = 0, \quad N_{t_f}\tau + N_{x_f}(\dot{x}\tau + y)_{t=t_f} = 0 \qquad (6.4.25)$$

for control and final time variations $\boldsymbol{v}$ and τ respectively. By (6.3.20) we know that $\delta^2 J \geqslant 0$ is a necessary condition for an extremum of the original problem. However, the variations

$$\mathbf{y}=\mathbf{0}, \quad \mathbf{v}=\mathbf{0}, \quad \tau=0$$

satisfy (6.4.24) and (6.4.25) and give $\delta^2 J = 0$. We are thus led to consider the minimization of the second variation (6.4.23) subject to the constraints (6.4.24) and (6.4.25) (the **accessory minimization problem**). However, by writing $\mathbf{w}=\dot{\mathbf{v}}$, Goh (1966a, 1966b) has shown that the singular accessory minimum problem can be transformed to a non-singular one and has derived the conditions (6.4.22) from the classical Legendre–Clebsch condition. A further distinct necessary condition (Jacobson, 1970; Gabasov, 1968, 1969) states that $\delta^2 J$ is non-negative for all admissible controls only if

$$H_{ux}f_u + f_u^{\mathrm{T}} W f_u \geqslant 0 \quad \text{for all } t \in [t_0, t_f]$$

where

$$-\dot{W} = H_{xx} + f_x^{\mathrm{T}} W + W f_x$$

$$W(t_f) = (\theta + N^{\mathrm{T}}\boldsymbol{\nu})_{xx}\,|_{t=t_f}$$

Since the accessory minimization problem is so useful in singular control, it is important to search for sufficient conditions for the non-negativity of $\delta^2 J$. Before stating these conditions note that if $\delta^2 J > 0$ then the extremal of the original problem is a weak local minimum. In the non-singular case (where $H_{uu} > 0$ for all t) it can be shown (Bell and Jacobson, 1975) that a necessary and sufficient condition for $\delta^2 J(v(\cdot)) \geqslant k\,||\,v(\cdot)\,||^2$ for some $k > 0$ and some suitable norm on the control space is that there exists a function $S(\cdot)$ for all $t \in [t_0, t_f]$ which satisfies the **Riccati equation**

$$\begin{aligned} -\dot{S} &= H_{xx} + S f_x + f_x^{\mathrm{T}} S - (H_{ux} + f_u^{\mathrm{T}} S)^{\mathrm{T}} H_{uu}^{-1} (H_{ux} + f_u^{\mathrm{T}} S) \\ S(t_f) &= (\theta + N^{\mathrm{T}}\boldsymbol{\nu})_{xx}\,|_{t=t_f} \end{aligned} \tag{6.4.26}$$

provided (6.4.24) is completely controllable (see chapter 4). Note that the condition $\delta^2 J(v(\cdot)) \geqslant k\,||\,v(\cdot)\,||^2$ is stronger than positive definiteness.

In the singular case where H_{uu}^{-1} does not exist for all t, Jacobson (1971) has shown that for the non-negativity of $\delta^2 J$ it is sufficient that for all $t \in [t_0, t_f]$ there exists a continuously differentiable, symmetric matrix function P such that

$$\boxed{\begin{pmatrix} \dot{P} + H_{xx} + P f_x + f_x^{\mathrm{T}} P & (H_{ux} + f_u^{\mathrm{T}} P)^{\mathrm{T}} \\ H_{ux} + f_u^{\mathrm{T}} P & H_{uu} \end{pmatrix} \geqslant 0} \tag{6.4.27}$$

for all $t \in [t_0, t_f]$ and

$$\boxed{(\theta + N^{\mathrm{T}}\boldsymbol{\nu})_{xx}\,|_{t=t_f} - P(t_f) \geqslant 0} \tag{6.4.28}$$

Further results on singular linear quadratic problems by singular perturbation have been derived by O'Malley and Jameson (1975, 1977).

6.5 DYNAMIC PROGRAMMING

6.5.1 The Principle of Optimality (Bellman, 1957)

Consider the problem of minimizing the cost functional

$$J = \int_{t_0}^{t_f} \phi(x, u)\, dt$$

with $x(t_0) = x_0$ fixed and $x(t_f)$ free. Let γ be an optimal trajectory shown in fig. 6.8(a), and let γ_1 be the trajectory which begins on x_1 and coincides with γ between x_1 and $x(t_f)$. Then dynamic programming is based on the **principle of optimality** which may be stated in one of (at least) two equivalent forms:

(a) The optimal policy does not depend on the previous history of the state of the system, but only on the state at the time instant considered.
(b) The second section γ_1 of an optimal trajectory γ is itself an optimal trajectory.

It should be noted, however, that if γ_0 denotes the subtrajectory of γ between x_0 and x_1 then γ_0 is not necessarily optimal. However, if x_0 and $x(t_f)$ are fixed then it is clear that any section of an optimal trajectory is optimal. For if x_1 and x_2 are fixed on an optimal trajectory $\gamma = \gamma_0 \cup \gamma_1 \cup \gamma_2$ (fig. 6.8(b)) and γ_1 is not optimal between x_1 and x_2 then clearly if γ_1' is an optimal trajectory between x_1 and x_2, $\gamma' = \gamma_0 \cup \gamma_1' \cup \gamma_2$ has lower cost than γ and so γ would not be optimal.

6.5.2 Discrete Dynamic Programming

In order to understand dynamic programming it is simpler to consider first the optimization of a discrete cost functional

$$J = \sum_{k=0}^{N-1} \phi(x_k, u_k) + \theta(x_N) \tag{6.5.1}$$

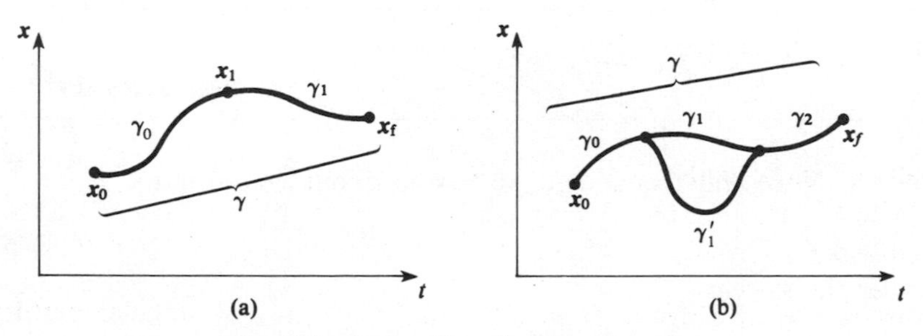

Fig. 6.8 Optimality of sub-trajectories.

subject to the discrete dynamical system (i.e. difference equation)

$$x_{k+1} = x_k + f(x_k, u_k), \quad x_0 \text{ fixed} \tag{6.5.2}$$

and $u \in \Omega$ (a constraint set). Then we wish to find a sequence u_0, $u_1, \ldots, u_{N-1}$ which minimizes (6.5.1) subject to (6.5.2) and $u \in \Omega$. Consider the stage $N-1$. By optimality, u_{N-1} does not depend on the past history but only the current state x_{N-1}. Also, u_{N-1} only affects those terms in (6.5.2) relating to the 'times' $N-1$ and N. Denote these terms by Q_{N-1}. Then

$$Q_{N-1} = \phi(x_{N-1}, u_{N-1}) + \theta(x_N)$$

However,

$$x_N = x_{N-1} + f(x_{N-1}, u_{N-1})$$

and so x_N also depends on u_{N-1}. Now let V_{N-1} denote the minimum of Q_{N-1} over all $u_{N-1} \in \Omega$. Then V_{N-1} depends on x_{N-1} and so we have

$$\begin{aligned} V_{N-1}(x_{N-1}) &= \min_{u_{n-1} \in \Omega} Q_{N-1} \\ &= \min_{u_{N-1} \in \Omega} \{\phi(x_{N-1}, u_{N-1}) + \theta[x_{N-1} + f(x_{N-1}, u_{N-1})]\} \end{aligned}$$

Next consider the time $N-2$ and define

$$Q_{N-2} = \phi(x_{N-2}, u_{N-2}) + Q_{N-1}$$

Then we let $V_{N-2}(x_{N-2})$ be given by

$$\begin{aligned} \min_{\substack{u_{N-2} \\ u_{N-1}} \in \Omega} Q_{N-2} &= \min_{u_{N-2} \in \Omega} [\phi(x_{N-2}, u_{N-2}) + \min_{u_{N-1} \in \Omega} Q_{N-1}] \\ &= \min_{u_{N-1} \in \Omega} [\phi(x_{N-2}, u_{N-2}) + V_{N-1}(x_{N-1})] \end{aligned}$$

since u_{N-1} has no effect before time $N-1$. Hence,

$$V_{N-2}(x_{N-2}) = \min_{u_{N-2} \in \Omega} \{\phi(x_{N-2}, u_{N-2}) + V_{N-1}[x_{N-2} + f(x_{N-2}, u_{N-2})]\}.$$

By induction we have

$$\boxed{V_{N-K}(x_{N-K}) = \min_{u_{N-K} \in \Omega} [\phi(x_{N-K}, u_{N-K}) + V_{N-K+1}[x_{N-K} + f(x_{N-K}, u_{N-K})]\}} \tag{6.5.3}$$

When $K = N$ we know x_0 and so we have a minimization in u_0.

Example 6.5.1
Consider the system

$$\dot{x} = Ax + Bu, \quad x \in \mathbb{R}^n, u \in \mathbb{R}^m$$

with $\boldsymbol{u}$ unconstrained and suppose that the cost J is quadratic

$$J = \int_0^{t_f} (\boldsymbol{x}^{\mathrm{T}} Q \boldsymbol{x} + \boldsymbol{u}^{\mathrm{T}} R \boldsymbol{u})\, \mathrm{d}t$$

We divide $(0, t_f)$ into N time intervals of length $\Delta = t_f/N$. If $\boldsymbol{u}$ is constant over a time interval of length Δ, then

$$\boldsymbol{x}(t) = \exp(At)\boldsymbol{x}_0 + \int_0^t \exp[A(t-\tau)] B\boldsymbol{u}(\tau)\, \mathrm{d}\tau$$

and, for the first time interval, let $\boldsymbol{u} = \text{const} = \boldsymbol{u}(0)$. Then,

$$\begin{aligned} \boldsymbol{x}(\Delta) &= \exp(A\Delta)\boldsymbol{x}(0) + \int_0^{\Delta} \exp[A(\Delta-\tau)] B\boldsymbol{u}(0)\, \mathrm{d}\tau \\ &= M\boldsymbol{x}(0) + N\boldsymbol{u}(0) \end{aligned}$$

for some matrices M and N, and by time invariance

$$\boldsymbol{x}_{k+1} = M\boldsymbol{x}_k + N\boldsymbol{u}_k$$

where $\boldsymbol{x}_k = \boldsymbol{x}(k\Delta)$, $\boldsymbol{u}_k = \boldsymbol{u}(k\Delta)$. We can also write the cost functional J in the approximate form

$$\frac{J}{\Delta} \approx \sum_{k=0}^{N-1} (\boldsymbol{x}_k^{\mathrm{T}} Q \boldsymbol{x}_k + \boldsymbol{u}_k^{\mathrm{T}} R \boldsymbol{u}_k)$$

From the general theory

$$V_{N-k}(\boldsymbol{x}_{N-k}) = \min_{\boldsymbol{u}_{N-k}} [\boldsymbol{x}_{N-k}^{\mathrm{T}} Q \boldsymbol{x}_{N-k} + \boldsymbol{u}_{N-k}^{\mathrm{T}} R \boldsymbol{u}_{N-k} + V_{N-k-1}(\boldsymbol{x}_{N-k+1})]$$

Since $\boldsymbol{x}_{N-k+1} = M\boldsymbol{x}_{N-k} + N\boldsymbol{u}_{N-k}$ it is easy to see (by induction) that V_{N-k} is a quadratic form in $\boldsymbol{x}_{N-k}$, and so we may write

$$V_{N-k}(\boldsymbol{x}_{N-k}) = \boldsymbol{x}_{N-k}^{\mathrm{T}} E_{N-k} \boldsymbol{x}_{N-k}$$

for some symmetric matrix E_{N-k}. Thus

$$\begin{aligned} \boldsymbol{x}_{N-k}^{\mathrm{T}} E_{N-k} \boldsymbol{x}_{N-k} = \min_{\boldsymbol{u}_{N-k}} [&\boldsymbol{x}_{N-k}^{\mathrm{T}} Q \boldsymbol{x}_{N-k} + \boldsymbol{u}_{N-k}^{\mathrm{T}} R \boldsymbol{u}_{N-k} + \\ &(M\boldsymbol{x}_{N-k} + N\boldsymbol{u}_{N-k})^{\mathrm{T}} E_{N-k+1} \times (M\boldsymbol{x}_{N-k} + N\boldsymbol{u}_{N-k})] \end{aligned} \tag{6.5.4}$$

Since $\boldsymbol{u}$ is unconstrained we obtain the classical optimum

$$\boldsymbol{u}_{N-k}^{\mathrm{T}} R = -(M\boldsymbol{x}_{N-k} + N\boldsymbol{u}_{N-k})^{\mathrm{T}} E_{N-k+1} N$$

and so

$$\boldsymbol{u}_{N-k} = -G_{N-k}\boldsymbol{x}_{N-k}$$

where

$$G_{N-k} = -(R + N^{\mathrm{T}} E_{N-k+1} N)^{-1} N^{\mathrm{T}} E_{N-k+1} M \tag{6.5.5}$$

Substituting the optimal $\boldsymbol{u}$ in (6.5.4) gives

$$E_{N-k} = (Q + G_{N-k}^{\mathrm{T}} R G_{N-k}) + (M - N G_{N-k})^{\mathrm{T}} E_{N-k+1} (M - N G_{N-k}) \tag{6.5.6}$$

We can use (6.5.5) and (6.5.6) to solve for G_{N-k}, E_{N-k} inductively from $k = 1$. The initial values G_{N-1}, E_{N-1} depend on the boundary conditions. For example, if $\boldsymbol{x}_N$ is unspecified then $\boldsymbol{x}_{N-1}^{\mathrm{T}} Q \boldsymbol{x}_{N-1} + \boldsymbol{u}_{N-1}^{\mathrm{T}} R \boldsymbol{u}_{N-1}$ is minimized by taking $\boldsymbol{u}_{N-1} = \boldsymbol{0}$. Then $G_{N-1} = 0$ and $E_{N-1} = Q$.

6.5.3 Continuous Dynamic Programming

Consider now the continuous system

$$\dot{\boldsymbol{x}} = \boldsymbol{f}(\boldsymbol{x}, \boldsymbol{u}, t)$$

and the minimization of the cost functional

$$J = \int_{t_0}^{t_f} \phi(\boldsymbol{x}, \boldsymbol{u}, t)\, \mathrm{d}t \quad (t_0, t_f \text{ fixed})$$

Suppose that an optimal trajectory from $\boldsymbol{x}_0$ to $\boldsymbol{x}_f$ exists. Denote by $V(\boldsymbol{x}_0, t_0)$ the minimum value of J along the trajectory. If t is any time in the interval (t_0, t_f) and $\boldsymbol{x}(t)$ is the state of the optimal trajectory at time t then by the principle of optimality the portion of the optimal trajectory between $\boldsymbol{x}(t)$ and $\boldsymbol{x}_f$ is itself optimal with cost $V(\boldsymbol{x}(t), t)$. Consider a neighbouring point $(\boldsymbol{x}', t')$ on the trajectory where $\boldsymbol{x}' = \boldsymbol{x}(t + \Delta t)$, $t' = t + \Delta t$. Then

$$\begin{aligned} V(\boldsymbol{x}, t) &= \min_{\boldsymbol{u}(\tau) \in \Omega} \int_t^{t_f} \phi(\boldsymbol{x}, \boldsymbol{u}, \tau)\, \mathrm{d}\tau \\ &= \min_{\boldsymbol{u}(\tau) \in \Omega} \left[\phi(\boldsymbol{x}, \boldsymbol{u}, t) \Delta t + \int_{t'}^{t_f} \phi(\boldsymbol{x}, \boldsymbol{u}, \nu)\, \mathrm{d}\nu \right] + O_1(\Delta t) \\ &= \min_{\boldsymbol{u}(\tau) \in \Omega} \left[\phi(\boldsymbol{x}, \boldsymbol{u}, t) \Delta t + \min_{\boldsymbol{u}(\nu) \in \Omega} \int_{t'}^{t_f} \phi(\boldsymbol{x}, \boldsymbol{u}, \nu)\, \mathrm{d}\nu \right] + O_1(\Delta t) \end{aligned}$$

since the first term depends only on $\boldsymbol{u}(t)$ at t and is unaffected by $\boldsymbol{u}(\nu)$ for $\nu \in [t', t_f]$. (In the following derivation $O_i(\Delta t)$, $i = 1, 2, \ldots$ will denote a generic quantity of the form $C_1 t^2 + C_2 t^3 + \ldots$) Hence

$$V(\boldsymbol{x}, t) = \min_{\boldsymbol{u}(t) \in \Omega} [\phi(\boldsymbol{x}, \boldsymbol{u}, t)\, \Delta t + V(\boldsymbol{x}', t')] + O_1(\Delta t)$$

Also

$$\boldsymbol{x}' = \boldsymbol{x}(t) + \boldsymbol{f}(\boldsymbol{x}(t), \boldsymbol{u}(t), t) t + O_2(\Delta t)$$

Now assuming that V has partial derivatives with respect to x_i and t, then

$$V(\boldsymbol{x}', t') = V(\boldsymbol{x}(t), t) + \left(\dot{\boldsymbol{x}}^{\mathrm{T}} \frac{\partial V}{\partial \boldsymbol{x}} + \frac{\partial V}{\partial t}\right)\Delta t + O_3(\Delta t)$$

$$= V(\boldsymbol{x}(t), t) + \left(\boldsymbol{f}^{\mathrm{T}} \frac{\partial V}{\partial \boldsymbol{x}} + \frac{\partial V}{\partial t}\right)\Delta t + O_3(\Delta t)$$

Hence

$$V(\boldsymbol{x}, t) = \min_{\boldsymbol{u}(t) \in \Omega} \left[\phi(\boldsymbol{x}, \boldsymbol{u}, t)\Delta t + V(\boldsymbol{x}, t) + \boldsymbol{f}^{\mathrm{T}} \frac{\partial V}{\partial \boldsymbol{x}}\Delta t + O_3(\Delta t)\right] + O_1(\Delta t)$$

$$= V(\boldsymbol{x}, t) + \frac{\partial V}{\partial t}\Delta t + \min_{\boldsymbol{u}(t) \in \Omega} \left[\phi(\boldsymbol{x}, \boldsymbol{u}, t)\Delta t + \boldsymbol{f}^{\mathrm{T}} \frac{\partial V}{\partial \boldsymbol{x}}\Delta t\right] + O_4(\Delta t)$$

(of course, V and $\partial V/\partial t$ are independent of $\boldsymbol{u}$). Hence

$$\boxed{-\frac{\partial V}{\partial t} = \min_{\boldsymbol{u}(t) \in \Omega} \left[\phi(\boldsymbol{x}, \boldsymbol{u}, t) + \boldsymbol{f}^{\mathrm{T}} \frac{\partial V}{\partial \boldsymbol{x}}\right]} \tag{6.5.7}$$

This is **Bellman's dynamic programming equation** and is a partial differential equation for V. In the case of a fixed end point the boundary condition is $V(\boldsymbol{x}(t_f), t_f) = 0$.

To see how dynamic programming relates to the maximum principle we shall 'prove' the latter assuming (6.5.7). This proof is restrictive, of course, because it places the strong condition on V that it be differentiable. However, it does provide geometrical insight and so is useful from a pedagogical point of view. Using the previous notation we replace t by a new state variable x_{n+1}. Then

$$\dot{x}_{n+1} = f_{n+1} = 1, \quad x_{n+1} = 0, \text{ when } t = 0$$

and

$$\frac{\partial V}{\partial t} \equiv \frac{\partial V}{\partial x_{n+1}}$$

Also we define the state variable x_0 by

$$\dot{x}_0 = f_0 = \phi(\boldsymbol{x}, \boldsymbol{u}, t) = \phi(\boldsymbol{x}, \boldsymbol{u}, x_{n+1}), \quad x_0 = 0 \text{ when } t = 0$$

Then we wish to minimize $x_0|_{t=t_f}$. Let $\tilde{\boldsymbol{x}} = (x_0, x_1, \ldots, x_n, x_{n+1})^{\mathrm{T}}$ be the generalized state vector and put

$$\tilde{\boldsymbol{f}} = (f_0, f_1, \ldots, f_{n+1})^{\mathrm{T}}$$

$$\tilde{\boldsymbol{\lambda}} = \left(1, \frac{\partial V}{\partial x_1}, \ldots, \frac{\partial V}{\partial x_n}, \frac{\partial V}{\partial x_{n+1}}\right)^{\mathrm{T}}$$

and

$$\tilde{V} = x_0 + V(x_1, \ldots, x_{n+1})$$

We can write Bellman's equation in the form

$$0 = \min_{u(t)\in\Omega}\left[\phi(x, u, t) + f^{\mathrm{T}}(x, u, x_{n+1})\frac{\partial V}{\partial x} + \frac{\partial V}{\partial x_{n+1}}\right]$$

$$= \min_{u(t)\in\Omega} (\tilde{\lambda}^{\mathrm{T}}\tilde{f})$$

$$= \min_{u(t)\in\Omega} \tilde{H}$$

where H is the Hamiltonian

$$\tilde{\lambda}^{\mathrm{T}}\tilde{f} = \sum_{i=0}^{n+1} \tilde{\lambda}_i \tilde{f}_i$$

Hence $u(t)$ is chosen to minimize the Hamiltonian $\tilde{H}$ subject to $u(t) \in \Omega$ for each t. Note that at every point of an optimal trajectory the minimum value is the constant 0.

Consider the effect of varying $\tilde{\lambda}$ with respect to a representative point $x(t)$ along the optimal trajectory. Then $\tilde{\lambda} = \tilde{\lambda}(x(t))$ and so

$$\frac{\mathrm{d}\tilde{\lambda}_i}{\mathrm{d}t} = \frac{\mathrm{d}}{\mathrm{d}t}\left(\frac{\partial \tilde{V}}{\partial \tilde{x}_i}\right)$$

$$= \sum_{j=0}^{n+1} \frac{\partial}{\partial \tilde{x}_j}\left(\frac{\partial \tilde{V}}{\partial \tilde{x}_i}\right)\frac{\mathrm{d}\tilde{x}_j}{\mathrm{d}t} \tag{6.5.7}$$

$$= \sum_{j=0}^{n+1} \frac{\partial^2 \tilde{V}}{\partial \tilde{x}_j \partial \tilde{x}_i}\tilde{f}_j, \quad i = 1, \ldots, n+1$$

and since $\tilde{\lambda}_0 = 1$ we also have $\mathrm{d}\tilde{\lambda}_0/\mathrm{d}t = 0$. Along the optimal trajectory we have

$$\tilde{H}_{\min} = (\tilde{\lambda}^{\mathrm{T}}\tilde{f})_{\min}$$

$$= \sum_{j=0}^{n+1} \frac{\partial \tilde{V}}{\partial \tilde{x}_j}\tilde{f}_j = 0$$

Clearly, at a fixed time t, $\tilde{H}$ is a minimum in x for $u(t)$ fixed at the optimal value, since any point not on the optimal trajectory will give a larger value of $\tilde{H}$. Hence

$$\frac{\partial \tilde{H}}{\partial \tilde{x}_i} = \frac{\partial}{\partial \tilde{x}_i}\left(\sum_{j=0}^{n+1} \frac{\partial \tilde{V}}{\partial \tilde{x}_j}\tilde{f}_j\right)$$

$$= \sum_{j=0}^{n+1} \frac{\partial^2 \tilde{V}}{\partial \tilde{x}_i\, \partial \tilde{x}_j}\tilde{f}_j + \sum_{j=0}^{n+1} \frac{\partial \tilde{V}}{\partial \tilde{x}_j}\, \frac{\partial \tilde{f}_j}{\partial \tilde{x}_i} = 0$$

Hence

$$\frac{d\tilde{\lambda}_i}{dt} = -\sum_{j=0}^{n+1} \tilde{\lambda}_j \frac{\partial \tilde{f}_j}{\partial \tilde{x}_i}$$

$$= -\frac{\partial \tilde{H}}{\partial \tilde{x}_i}$$

by using (6.5.7) and the fact that in the expression for $\tilde{H}$ only $\tilde{f}_i$ depends explicitly on $\boldsymbol{x}$. We have therefore derived the canonical equations and established the maximum principle. We emphasize again, however, that we have assumed that the second partial derivatives of $\tilde{V}$ exist.

6.6 THE LINEAR REGULATOR AND GENERALIZATIONS

6.6.1 The Linear Regulator

In example 6.4.3 we considered the linear–quadratic regulator problem–namely the minimization of the cost functional

$$J = \boldsymbol{x}^{\mathrm{T}}(t_f)F\boldsymbol{x}(t_f) + \int_{t_0}^{t_f} [\boldsymbol{x}^{\mathrm{T}}(t)Q(t)\boldsymbol{x}(t) + \boldsymbol{u}^{\mathrm{T}}(t)R(t)\boldsymbol{u}(t)]\, dt$$

subject to the linear dynamics

$$\dot{\boldsymbol{x}}(t) = A(t)\boldsymbol{x}(t) + B(t)\boldsymbol{u}(t) \qquad (6.6.1)$$

and we obtained a two-point boundary value problem by applying the maximum principle. In this section we shall obtain the solution in a different form by using dynamic programming. Note that $R(t)$ is assumed to be positive definite so that there are no singular controls.

From Bellman's equation (6.5.7) we obtain

$$\min_{u \in \mathbb{R}^m} [\boldsymbol{x}^{\mathrm{T}}Q\boldsymbol{x} + \boldsymbol{u}^{\mathrm{T}}R\boldsymbol{u} + V_t + V_x^{\mathrm{T}}(A\boldsymbol{x} + B\boldsymbol{u})] = 0$$

and

$$V(\boldsymbol{x}, t_f) = \boldsymbol{x}^{\mathrm{T}}F\boldsymbol{x}$$

Hence

$$\boldsymbol{x}^{\mathrm{T}}Q\boldsymbol{x} + V_t + V_x^{\mathrm{T}}A\boldsymbol{x} + \min_{u}(\boldsymbol{u}^{\mathrm{T}}R\boldsymbol{u} + V_x^{\mathrm{T}}B\boldsymbol{u}) = 0$$

Writing $C = V_x^{\mathrm{T}}B$ we have

$$\boldsymbol{u}^{\mathrm{T}}R\boldsymbol{u} + C\boldsymbol{u} = (\boldsymbol{u} + \tfrac{1}{2}R^{-1}C^{\mathrm{T}})^{\mathrm{T}}R(\boldsymbol{u} + \tfrac{1}{2}R^{-1}C^{\mathrm{T}}) - \tfrac{1}{4}CR^{-1}C^{\mathrm{T}}$$

and since R is positive definite (by assumption – no singular control) the minimum value is $-\tfrac{1}{4}CR^{-1}C^{\mathrm{T}}$, which is taken on when

$$\boldsymbol{u} = -\tfrac{1}{2}R^{-1}C^{\mathrm{T}}$$

$$= -\tfrac{1}{2}R^{-1}B^{\mathrm{T}}V_x$$

The Bellman equation is thus

$$V_t + \boldsymbol{x}^{\mathrm{T}}Q\boldsymbol{x} + V_x^{\mathrm{T}}A\boldsymbol{x} - \tfrac{1}{4}V_x^{\mathrm{T}}BR^{-1}B^{\mathrm{T}}V_x = 0$$

$$V(\boldsymbol{x}, t_f) = \boldsymbol{x}^{\mathrm{T}}F\boldsymbol{x}$$

As we saw in the discrete case it is reasonable to search for a symmetric quadratic form

$$V(\boldsymbol{x}, t_f) = \boldsymbol{x}^{\mathrm{T}}P(t)\boldsymbol{x}$$

which will satisfy this equation. By direct substitution we have

$$\boldsymbol{x}^{\mathrm{T}}Q\boldsymbol{x} + \boldsymbol{x}^{\mathrm{T}}\dot{P}\boldsymbol{x} + 2\boldsymbol{x}^{\mathrm{T}}PA\boldsymbol{x} - \boldsymbol{x}^{\mathrm{T}}PBR^{-1}P^{\mathrm{T}}\boldsymbol{x} = 0$$

or

$$\boldsymbol{x}^{\mathrm{T}}(\dot{P} + Q + PA + A^{\mathrm{T}}P - PBR^{-1}B^{\mathrm{T}}P)\boldsymbol{x} = 0$$

since $\boldsymbol{x}^{\mathrm{T}}PA\boldsymbol{x} = \boldsymbol{x}^{\mathrm{T}}A^{\mathrm{T}}P\boldsymbol{x}$. Hence we require that P satisfy the **Riccati equation**

$$\begin{aligned} -\dot{P} &= Q + PA + A^{\mathrm{T}}P - PBR^{-1}B^{\mathrm{T}}P \\ P(t_f) &= F \end{aligned} \tag{6.6.2}$$

The optimal control is

$$\boldsymbol{u}(t) = -R^{-1}B^{\mathrm{T}}P\boldsymbol{x}(t) \tag{6.6.3}$$

and it is easy to see that the optimal cost is

$$J^* = \boldsymbol{x}^{\mathrm{T}}(t_0)P(t_0)\boldsymbol{x}(t_0) \tag{6.6.4}$$

In the regulator problem we wish to keep $\boldsymbol{x}(t)$ 'close' to the origin while using the minimum control. Suppose that we now require instead to make $\boldsymbol{x}(t)$ follow the input $\boldsymbol{r}(t)$. Then we consider the cost functional

$$J = [\boldsymbol{x}^{\mathrm{T}}(t_f) - \boldsymbol{r}^{\mathrm{T}}(t_f)]F[\boldsymbol{x}(t_f) - \boldsymbol{r}(t_f)] + \int_{t_0}^{t_f} \{[\boldsymbol{x}^{\mathrm{T}}(t) - \boldsymbol{r}^{\mathrm{T}}(t)]Q(t)[\boldsymbol{x}(t) - \boldsymbol{r}(t)] + \boldsymbol{u}^{\mathrm{T}}R(t)\boldsymbol{u}\}\,\mathrm{d}t$$

This functional together with the linear dynamics (6.6.1) constitutes the **tracking problem** and it is easy to generalize the above results to solve this problem. In fact we obtain the optimal control

$$\boldsymbol{u}(t) = -R^{-1}B^{\mathrm{T}}P\boldsymbol{x}(t) - R^{-1}B^{\mathrm{T}}\boldsymbol{s}(t) \tag{6.6.5}$$

where the first term is as before with P given by (6.6.2) and $\boldsymbol{s}(t)$ is a feed-forward term given by the unique solution of the equation

$$\frac{\mathrm{d}}{\mathrm{d}t}\boldsymbol{s}(t) = -(A + BR^{-1}B^{\mathrm{T}}P)^{\mathrm{T}}\boldsymbol{s}(t) + Q\boldsymbol{r}(t) \tag{6.6.6}$$

with

$$\boxed{s(t_f) = Fr(t_f)} \tag{6.6.7}$$

If we require the infinite time problem

$$J = \int_{t_0}^{\infty} [x^{\mathrm{T}}(t)Qx(t) + u^{\mathrm{T}}(t)Ru(t)] \, \mathrm{d}t$$

(time invariant Q and R) with the time-invariant dynamics (6.6.1), i.e. $A(t) = A$, $B(t) = B$, then it is more difficult to obtain an optimal control since it is necessary to prove that the solution $P(t)$ of the Riccati equation (6.6.2) (which depends on t_f) tends to a finite matrix as $t_f \to \infty$. However, it can be shown (Anderson and Moore, 1971; Russell, 1979; Kwakernaak and Sivan, 1972; Curtain and Pritchard, 1978) that if $(A^{\mathrm{T}}, Q^{1/2})$ and (A, B) are stabilizable, then the optimal control is given by

$$u(t) = -R^{-1}B^{\mathrm{T}}Px(t)$$

as before, where P is now the solution (which exists) of the matrix Riccati equation

$$\boxed{Q + PA + A^{\mathrm{T}}P = PBR^{-1}B^{\mathrm{T}}P} \tag{6.6.8}$$

(Note that a pair (A, B) is called **stabilizable** if there exists a matrix F such that $A + BF$ is a stable matrix.)

6.6.2 Receding Horizon Control

One of the drawbacks of the linear–quadratic optimal control is that, being a linear feedback, it responds in the same way to large and small signals. The difference signal produced by such a feedback may be due to noise disturbances which do not represent true signals in the input–output error and so the control should ignore small error signals as much as possible while responding optimally to large errors. An approximation to this type of behaviour is produced by **receding horizon control** which we now discuss, following Shaw (1979). Consider again the linear plant

$$\dot{x}(t) = Ax(t) + Bu(t), \quad x(t_0) = x_0$$

together with the performance criterion

$$J = \int_{t_0}^{t_0+T} u^{\mathrm{T}}(\tau)Ru(\tau) \, \mathrm{d}\tau \tag{6.6.9}$$

subject to the constraint

$$x(t_0 + T) = 0 \tag{6.6.10}$$

The constraint (6.6.10) forces the trajectory to the origin in time T. Since we require exact control to the origin we must assume that (A, B) is a controllable pair. In this case it is easy to see, using the above methods, that the optimal control is

$$u(t) = -R^{-1}B^{T}\exp[-A^{T}(t-t_0)]\,W^{-1}(T)x_0, \quad t_0 \leqslant t \leqslant t_0 + T \qquad (6.6.11)$$

where $W(T)$ satisfies the **Lyapunov equation**

$$\frac{dW}{dT} = BR^{-1}B^{T} - AW - WA^{T}, \; W(0) = 0 \qquad (6.6.12)$$

However, the control u given by (6.6.11) is open loop and so is not stable to variations in the system parameters. Hence we seek a suboptimal feedback control which resembles (6.6.11). The receding horizon philosophy states that we should apply the control (6.6.11) as if, at each time t, we were beginning a new optimization interval of length T. In this case we put $t_0 = t$ in (6.6.11) and obtain the control

$$u(t) = -R^{-1}B^{T}W^{-1}(T)x(t) \qquad (6.6.13)$$

This feedback control law can be made more effective by allowing T to vary (with the state $x(t)$). We then obtain a non-linear feedback which can be shown to have the desired effects stated above. The choice of T can be made in many different ways (see Shaw, 1979), but in some cases the choice is clear and we shall give a simple example in the next section. Note that the receding horizon principle can be generalized to infinite dimensional systems (see Banks, 1983).

If the choice of T is made implicitly via the relation

$$V(x, T) = F(T) \qquad (6.6.14)$$

where F is monotonically decreasing in T and

$$V(x, T) = x^{T}(t)W^{-1}(T)x(t) \qquad (6.6.15)$$

then we can show that the control (6.6.13) is stabilizing if

$$\frac{DF}{dT} < \frac{dV}{dT} \qquad (6.6.16)$$

In fact we have

$$W(T) = \int_0^T \exp(-A\tau)BR^{-1}B^{T}\exp(-A^{T}\tau)\,d\tau$$

and so dW/dT is non-negative definite and

$$\frac{dW^{-1}}{dT}(T) = -W^{-1}\frac{dW}{dT}W^{-1} \leqslant 0$$

Now

$$\dot{V} = -x^{\mathrm{T}}W^{-1}(T)\left(BR^{-1}B^{\mathrm{T}} + \frac{\mathrm{d}W}{\mathrm{d}T}\right)W^{-1}(T)x + \frac{\partial V}{\partial T}\frac{\mathrm{d}T}{\mathrm{d}T}$$

and by (6.6.14), $T = F^{-1}(V)$. Hence

$$\frac{\mathrm{d}T}{\mathrm{d}t} = \frac{\mathrm{d}F^{-1}}{\mathrm{d}V}\frac{\mathrm{d}V}{\mathrm{d}t}$$

and so

$$\dot{V}\left(1 - \frac{\mathrm{d}F^{-1}}{\mathrm{d}V}\frac{\partial V}{\partial T}\right) = -x^{\mathrm{T}}W^{-1}\left(BR^{-1}B^{\mathrm{T}} + \frac{\mathrm{d}W}{\mathrm{d}T}\right)W^{-1}x$$

It follows that, under the condition (6.6.16) the control is stabilizing.

Example 6.6.1 (Banks, 1980)
We shall apply the receding horizon principle to the case of the pursuer–evader problem. Suppose that a pursuer P is chasing an evader E, measured in some inertial reference frame as in fig. 6.9(a). If the 'look angle' $\sigma(t)$ is constant and the relative velocity along the line of sight PE is negative, it is clear from fig. 6.9(b) that capture is inevitable.

In classical proportional navigation we make the pursuer's flight path turning rate proportional to $\dot{\sigma}$, i.e.

$$\dot{\theta}_{\mathrm{P}}(t) = K(t)\dot{\sigma}(t), \quad K(t) > 1 \tag{6.6.17}$$

The equations of motion can be derived from

$$r\dot{\sigma} = v_{\mathrm{E}} \sin \phi_{\mathrm{E}} - v_{\mathrm{P}} \sin \phi_{\mathrm{P}}$$

$$\dot{r} = v_{\mathrm{E}} \cos \phi_{\mathrm{E}} - v_{\mathrm{P}} \cos \phi_{\mathrm{P}}$$

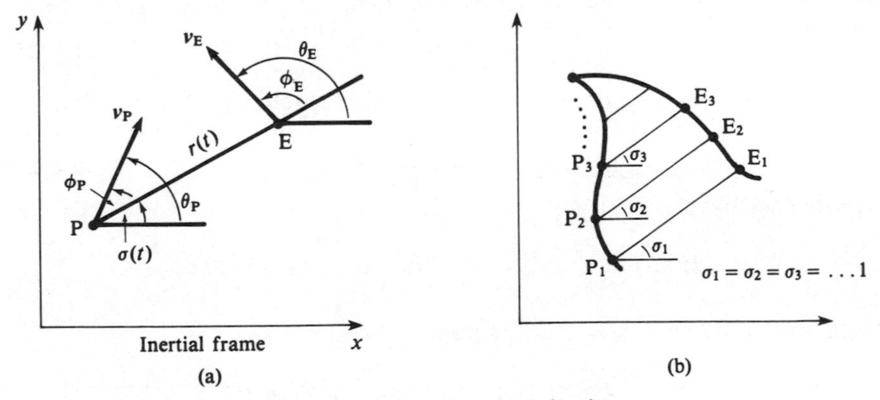

Fig. 6.9 Proportional navigation.

that is

$$r\ddot{\sigma} + 2\dot{r}\dot{\sigma} = -v_P \cos\phi_P \dot{\theta}_P - \dot{v}_P \sin\phi_P + v_E \cos\phi_E \dot{\theta}_E + \dot{v}_E \sin\phi_E$$

Using (6.6.17) we have

$$r\ddot{\sigma} + (2\dot{r} + v_P \cos\phi_P K)\dot{\sigma} = -v_P \sin\phi_P + v_E \cos\phi_E \dot{v}_E + \dot{v}_E \sin\phi_E$$

Assuming $\dot{v}_P = \dot{\theta}_E = \dot{v}_E = 0$ (i.e. a pursuer of constant speed and a non-manoeuvering evader), we have

$$r\ddot{\sigma} - \dot{r}(\Lambda_n - 2)\dot{\sigma} = 0$$

where Λ_n is given by

$$K(t) = -\frac{\Lambda_n \dot{r}}{(v_P \cos\phi_P)}$$

If $\dot{r}(t) < 0$ then the substitution $\rho = -\ln r/r_0$ gives

$$\frac{d\dot{\sigma}}{d\rho} + (\Lambda_n - 2)\dot{\sigma} = 0$$

and so

$$\dot{\sigma} = \dot{\sigma}_0 \left(\frac{r}{r_0}\right)^{\Lambda_n - 2} \to 0 \quad \text{if } \Lambda_n > 3,\ \dot{r} < 0$$

Now $\dot{\theta}_P v_P \cos\phi_P$ is the normal acceleration and so the proportional navigation (PN) law may be written in the form†

$$a_P^{\perp} = -\Lambda_n \dot{r}\dot{\sigma}$$

which is how it is normally applied.

We shall now show how, by using receding horizon control, we can embed PN in a whole class of navigation laws. Consider the geometry of fig. 6.10 and assume the evader has zero acceleration. Then the equations of motion are

$$\dot{x}_1 = x_3 = \text{relative velocity along } y$$

$$\dot{x}_2 = x_4 = \text{relative velocity along } z$$

$$\dot{x}_3 = -a_{P_y}$$

$$\dot{x}_4 = -a_{P_z}$$

If $-a_{P_y}$, $-a_{P_z}$ are controls u_1, u_2 then we have

$$\dot{x}_1 = x_3, \quad \dot{x}_2 = x_4, \quad \dot{x}_3 = u_1, \quad \dot{x}_4 = u_2$$

or

$$\dot{x}(t) = Ax(t) + Bu(t), \quad x(t_0) = x_0$$

† $a_P^{\perp}$ refers to the acceleration perpendicular to the pursuer.

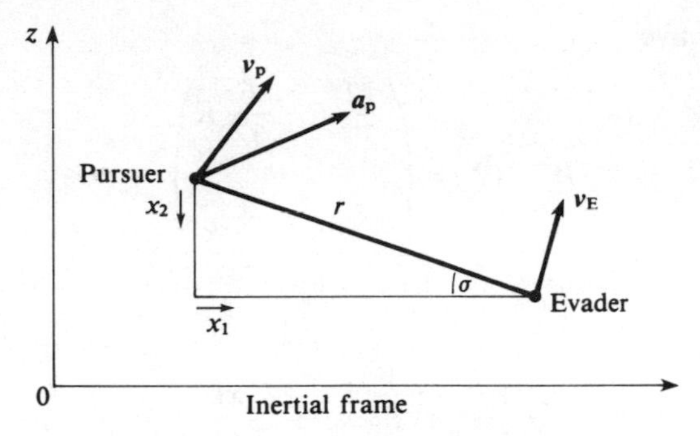

Fig. 6.10 Pursuer and evader coordinates.

where

$$A = \begin{pmatrix} 0 & I \\ 0 & 0 \end{pmatrix}, \quad B = \begin{pmatrix} 0 \\ I \end{pmatrix}$$

We seek to minimize the cost function

$$J = \int_{t_0}^{t_0+T} \boldsymbol{u}^{\mathrm{T}}(\tau) R \boldsymbol{u}(\tau)\, \mathrm{d}\tau$$

with the terminal constraint

$$\boldsymbol{x}(t_0 + T) = \mathbf{0}$$

Since (A, B) is a controllable pair we can use the receding horizon control

$$\boldsymbol{u}(t) = -R^{-1}B^{\mathrm{T}}W^{-1}(T)\boldsymbol{x}(t)$$

Now, W is given by

$$W(T) = \int_0^T \exp(-A\tau)BR^{-1}B^{\mathrm{T}} \exp(-A^{\mathrm{T}}\tau)\, \mathrm{d}\tau$$

where

$$BR^{-1}B^{\mathrm{T}} = \begin{pmatrix} 0 \\ I \end{pmatrix} R^{-1}(0 \quad I) = \begin{pmatrix} 0 & 0 \\ 0 & R^{-1} \end{pmatrix}$$

Clearly, $A^2 = 0$. Hence

$$\exp(-A\tau) = \begin{pmatrix} I & -\tau I \\ 0 & I \end{pmatrix}$$

and it is easy to see that

$$W(T) = \begin{pmatrix} \dfrac{T^3}{3}R^{-1} & -\dfrac{T^2}{2}R^{-1} \\ -\dfrac{T^2}{2}R^{-1} & TR^{-1} \end{pmatrix}$$

Moreover we have

$$W^{-1}(T)=\begin{pmatrix}\frac{12}{T^3}R & \frac{6}{T^2}R\\ \frac{6}{T^2}R & \frac{4}{T}R\end{pmatrix}$$

and so

$$u(t)=-\left(\frac{6}{T^2}I \quad \frac{4}{T}I\right)x(t)$$

(Note that u is independent of R). In terms of r and σ, we have

$$x_1=r\cos\sigma$$

$$x_2=r\sin\sigma$$

$$x_3=-r\dot\sigma\sin\sigma+\dot r\cos\sigma$$

$$x_4=r\dot\sigma\cos\sigma+\dot r\sin\sigma$$

so that

$$u(t)=\begin{Bmatrix}-T^{-2}[6r\cos\sigma+4T(-r\dot\sigma\sin\sigma+\dot r\cos\sigma)]\\ -T^{-2}[6r\sin\sigma+4T(r\dot\sigma\cos\sigma+\dot r\sin\sigma)]\end{Bmatrix}$$

If we can only control the lateral acceleration, then $u_1=0$ and if σ is small we obtain

$$u_2(t)=-\frac{2}{T^2}[3r\sigma+2(r\dot\sigma+\dot r\sigma)T]$$

We can now choose T to be dependent on the states. In fact, we can take T to be the 'time to go' t_g, i.e. $T=-r/\dot r=t_g$. Then

$$\boxed{u_2(t)=-\frac{2\dot r^2}{r}\sigma+4\dot r\dot\sigma}$$

This is PN with $\Lambda_n=4$ and a correction $-2\dot r^2\sigma/r$. Moreover, if we take $T=(-r/\dot r)/\alpha$ for some integer α then we have

$$u_2(t)=-(6\alpha^2-4\alpha)\frac{\dot r^2\sigma}{r}+4\alpha\dot r\dot\sigma$$

For example if $\alpha=2$ so that $T=t_g/2$, then

$$\boxed{u_2(t)=-16\frac{\dot r^2\sigma}{r}+8\dot r\dot\sigma}$$

6.7 EXERCISES

1. Find the maxima and minima of the function $f(x_1, x_2) = x_1^2 + x_2^2$ subject to the constraint $g(x_1, x_1) = ax_1 + bx_2 + c$ and interpret the result graphically.

2. Minimize $f = (x_1 + 1)(x_2 - 2)$ over the region $0 \leqslant x_1 \leqslant 2$, $0 \leqslant x_2 \leqslant 1$ by using the Kuhn–Tucker necessary conditions.

3. Let the function $f: C^2[0,1] \to C^2[0,1]$ be defined by $f(\phi) = \phi^2$ for $\phi \in C^2[0,1]$ and let $\psi \in C^2[0,1]$. Determine the directional derivative of f at ϕ in the direction ψ.

4. Show that $C^2[0,1]$ is not a finite-dimensional vector space.

5. Prove theorem 6.3.2.

6. Find the function $x(t) \in C^2[0, t_f]$ which passes through the points $x(0) = x_0 > 0$ and $x(t_f) = 0$ and which minimizes the integral

$$J = \int_0^{t_f} (x^2 + T^2\dot{x}^2)\,\mathrm{d}t$$

where T^2 is constant.

7. As in example 6.3.2, minimize the functional (6.3.16) with fixed initial and terminal constraints $x(t_0) = x_0$, $x(t_f) = d$ with the dynamics

$$\dot{x} = ax + u$$

8. Discuss the time optimal control problem of driving the state of the system

$$\dot{x}_1 = x_2$$

$$\dot{x}_2 = -x_1 + u$$

to the origin in minimum time, where $|u| \leqslant 1$.

9. Show that if the Hamiltonian does not depend explicitly on time t, then H is constant along an optimal trajectory. (Hint: Show that we may write

$$\frac{\mathrm{d}H}{\mathrm{d}t} = \frac{\partial \phi}{\partial t} + \lambda^{\mathrm{T}} \frac{\partial f}{\partial t} = \dot{u}^{\mathrm{T}} \frac{\partial H}{\partial u}$$

where $H = \phi + \lambda^{\mathrm{T}} f$.)

10. Use the principles of discrete dynamic programming to find the shortest path from A to B in fig. 6.11. You may only move along diagonals from left to right and the numbers represent the length of the path joining the corresponding two nodes.

11. Show that the optimal cost in the linear regulator problem is given by (6.6.4).

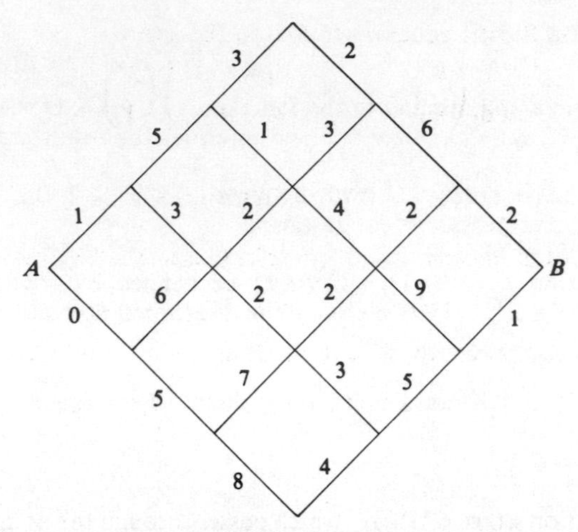

Fig. 6.11 An example of discrete dynamic programming.

12. Write down the Riccati equation for the optimal control problem

$$\frac{d^3x}{dt^3}+3\frac{d^2x}{dt^2}-2\frac{dx}{dt}+x=u$$

$$\min_u J$$

where

$$J=x^2(1)+\int_0^1 u^2\,dt$$

13. Apply the maximum principle to the bilinear control problem

$$\dot{x}=\begin{pmatrix}0 & 1\\ 1 & 1\end{pmatrix}x+u\begin{pmatrix}1 & 1\\ 1 & 2\end{pmatrix}x$$

$$\min_u J$$

where

$$J=\mathbf{x}(1)\mathbf{x}(1)+\int_0^1 u^2\,dt$$

and show that the optimal control is constant. Investigate the general system

$$\dot{x}=Ax+uBx$$

with the same functional J, when $AB\neq BA$.

14. Show that, if $X(t)$ is a matrix function of t which is non-singular for each t, then (if the derivative exists)

$$\frac{dX^{-1}(t)}{dt}=X^{-1}(t)X(t)X^{-1}(t)$$

Hence show that if we put

$$P(t)=Y(t)X^{-1}(t)$$

we may write the Ricatti equation (6.6.2) in the form

$$\frac{\mathrm{d}}{\mathrm{d}t}\begin{pmatrix} X \\ Y \end{pmatrix} = \begin{pmatrix} -A^{\mathrm{T}} & BR^{-1}B^{\mathrm{T}} \\ Q & A \end{pmatrix}\begin{pmatrix} X \\ Y \end{pmatrix}$$

with

$$Y(0) = FX(0)$$

15. Taking $Q = 0$ in the linear regulator problem show that we have

$$P(t)\boldsymbol{x}(t) = \exp[-A^{\mathrm{T}}(t - t_0)]P(t_0)\boldsymbol{x}_0$$

and hence show that we may write the optimal control in the form

$$\boldsymbol{u}(t) = -R^{-1}B^{\mathrm{T}}\exp[-A^{\mathrm{T}}(t - t_0)]P(t_0)\boldsymbol{x}_0$$

(Hint: use

$$\boldsymbol{x} = A\boldsymbol{x} - BR^{-1}B^{\mathrm{T}}P\boldsymbol{x}$$

and find

$$\frac{\mathrm{d}}{\mathrm{d}t}(P\boldsymbol{x}).)$$

16. Show that (6.6.13) is the optimal control for the cost (6.6.9) by using exercise 15 and by proving that if $F = \alpha I$ and W is given by

$$\dot{W}(t) = AW + WA^{\mathrm{T}} - BR^{-1}B^{\mathrm{T}}, \quad W(t_f) = \frac{I}{\alpha}$$

then $PW = I$ for any α. Then let $\alpha \to \infty$, thus forcing the final state to zero.

REFERENCES

Anderson, B. D. O and Moore, J. B. (1971) *Linear Optimal Control*, Prentice-Hall.

Banks, S. P. (1980) Guidance laws and receding horizon control, *Conf. on Math. Systems Theory*, Warwick University.

Banks, S. P. (1983) The receding horizon principle for distributed systems, *J. Franklin Inst.*, 315, 435–51.

Bell, D. J. and Jacobson, D. H. (1975) *Singular Optimal Control Problems*, Academic Press.

Bellman, R. (1957) *Dynamic Programming*, Princeton University Press.

Bellman, R. (1964) *Invariant Imbedding and Time-Dependent Transport Processes*, Elsevier.

Bryson, A. E. and Ho, Y-C (1969) *Applied Optimal Control*, Blaisdell.

Curtain, R. F. and Pritchard, A. J. (1978) *Infinite Dimensional Linear Systems Theory*, Springer-Verlag.

Fleming, W. H. and Rishel, R. W. (1975) *Deterministic and Stochastic Optimal Control*, Springer-Verlag.

Gabasov, R. (1968) Necessary conditions for optimality of singular control. *Engng Cybern.*, No. 5, 23–37.

Gabasov, R. (1969) On the theory of necessary optimality conditions governing special controls, *Sov. Phys. Dokl.*, 13, 1094–5.

Girsanov, I. V. (1972) *Lectures on Mathematical Theory of Extremum Problems, Lecture Notes in Economics and Systems*, Springer-Verlag.

Goh, B. S. (1966a) The second variation for the singular Bolza problem, *SIAM J. Control*, 4, 309–25.

Goh, B. S. (1966b) Necessary conditions for singular extremals involving multiple control variables, *SIAM. J. Control*, 4, 716–31.

Hadley, G. (1964) *Nonlinear and Dynamic Programming*, Addison-Wesley.

Jacobson, D. H. (1970) On conditions of optimality for singular control problems, *IEEE Trans. Aut. Control*, 15, 109–10.

Jacobson, D. H. (1971) A general sufficiency theorem for the second variation, *J. Math. Anal. Applic.*, 34, 578–89.

Johnson, C. D. and Gibson, J. E. (1963) Singular solutions in problems of optimal control, *IEEE Trans. Aut. Control*. 8, 4–15.

Kwakernaak, H. and Sivan, R. (1972) *Linear Optimal Control Systems*, Wiley-Interscience.

Lee, E. B. and Markus, L. (1967) *Foundations of Optimal Control Theory*, Wiley.

O'Malley, R. E. and Jameson, A. (1975) Singular perturbations and singular arcs, Part I, *IEEE Trans. Aut. Control*, 20, 218–26.

O'Malley, R. E. and Jameson, A. (1977) Singular perturbations and singular arcs, Part II, *IEEE Trans. Aut. Control*, 22, 328–37.

Pontryagin, L. S., Boltyanskii, V. G., Gamkrelidze, R. V. and Mishchenko, E. F. (1964) *The Mathematical Theory of Optimal Processes*, Pergamon.

Rosenbrock, H. H. and McMorran, P. D. (1971) Good, bad or optimal? *IEEE Trans. Aut. Control.*, 16, 552–4.

Russell, D. (1979) *Mathematics of Finite-Dimensional Control Systems*, Dekker.

Shaw, L. (1979) Nonlinear control of linear multivariable systems via state-dependent feedback gains, *Trans. IEEE Aut. Control*, 24, 108–12.

Warga, J. (1972) *Optimal Control of Differential and Functional Equations*, Academic Press.

Young, L. C. (1969) *Lectures on Calculus of Variations and Optimal Control Theory*, Saunders.

7 STOCHASTIC SYSTEMS

7.1 INTRODUCTION

In the previous chapters on control we have considered systems which are completely deterministic in the sense that knowledge of the initial states of the system completely determines the time evolution of the states. However, in all real systems the states are not entirely deterministic, but are affected, to a greater or lesser extent, by random noise the origin of which is discussed in chapter 1. Moreover, in any model of given physical system it is likely that the parameters of the model (for example, the elements of the 'A' matrix in a linear system) will be unknown or uncertain and must themselves be modelled by some form of random process. Finally, one must be able to deal with the problem of controlling such noisy systems.

It is these problems which we shall consider in this chapter and we shall assume a familiarity with the basic theory of probability and stochastic processes. The reader unfamiliar with these concepts should consult the literature cited in appendix 2 before proceeding with this material. (The concepts of more advanced probability theory, required for continuous filtering, are given in appendix 2.) The general problem of determining various aspects of a random system is called **estimation** and may be loosely divided into **state estimation** and **parameter estimation** and we shall discuss these problems along with stochastic control in the following sections.

We shall begin by considering discrete linear systems, since these present the simplest problem for estimation theory, and develop the celebrated Kalman filter for such systems. The generalizations necessary to deal with continuous filtering will then be considered along with the basic theory of the control of linear stochastic systems with quadratic cost and Guassian noise. The numerical difficulties of the implementation of the Kalman filter will be discussed since these are often extremely important in practical situations. We shall also present a very brief introduction to non-linear filtering and finally the problems of parameter estimation and system identification will be introduced.

The material for this chapter is necessarily fairly mathematical and we shall use the concepts from appendices 1 and 2 without further comment.

7.2 DISCRETE ESTIMATION AND FILTERING†

7.2.1 Non-recursive Estimation

Let x be a scalar random variable with mean $E(x) = x_0$ and variance $\sigma_x^2 = E[(x - x_0)^2]$, and suppose we obtain m measurements $y(k)$, $1 \leqslant k \leqslant m$ of x corrupted by additive noise $v(k)$, $1 \leqslant k \leqslant m$, where $v(k)$ has zero mean and variance σ_v^2 independent of k. Then we have

$$y(k) = x + v(k), \quad 1 \leqslant k \leqslant m \tag{7.2.1}$$

We shall frequently regard the data $y(k)$, $1 \leqslant k \leqslant m$, as a vector $y = [y(1), \ldots, y(m)]^{\mathrm{T}}$.

A **linear estimator** of x is, by definition, an expression of the form

$$\hat{x} = \sum_{i=1}^{m} h_i y(i) \tag{7.2.2}$$

for some constants h_i. Define the estimation error e by $e = E(x - \hat{x})$. The optimal mean square estimator is given by minimizing the expectation $p_e \triangleq E(x - \hat{x})^2$. Then

$$\frac{\mathrm{d}p_e}{\mathrm{d}h_j} = -2E\left\{\left[x - \sum_{i=1}^{m} h_i y(i)\right] y(j)\right\} = 0 \tag{7.2.3}$$

Hence

$$\sum_{i=1}^{m} h_i E[y(i)y(j)] = E[xy(j)], \quad j = 1, \ldots, m$$

or

$$P_y h = p_{xy} \tag{7.2.4}$$

where P_y is the $m \times m$ **data autocorrelation matrix**

$$\boxed{P_y = \{E[y(i)\, y(j)]\}_{1 \leqslant i, j \leqslant m}} \tag{7.2.5}$$

$h = (h_1, \ldots, h_m)^{\mathrm{T}}$ and p_{xy} is the cross-correlation of x and y

$$\boxed{p_{xy} = E(xy)} \tag{7.2.6}$$

† See also Anderson and Moore (1979) and Box and Jenkins (1970).

Note that equation (7.2.3) may be written in the form

$$\mathrm{E}[ey(j)] = 0, \quad 1 \leqslant j \leqslant m \tag{7.2.7}$$

Relation (7.2.7) is called the **orthogonality principle** for reasons which will become clear later. Now,

$$h = P_y^{-1} p_{xy} \tag{7.2.8}$$

and so by (7.2.2)

$$\boxed{\hat{x} = p_{xy}^{\mathrm{T}} P_y^{-1} y} \tag{7.2.9}$$

Equation (7.2.4) is called the **Wiener–Hopf equation** and the process of estimating x by (7.2.9) is called the **Wiener filter.** Note that the mean square error is given by

$$\begin{aligned} p_e = \mathrm{E}(e^2) &= \mathrm{E}\left\{e\left[x - \sum h_i y(i)\right]\right\} = \mathrm{E}(ex) \quad \text{(by (7.2.7))} \\ &= \mathrm{E}(x^2) - \sum_{i=1}^{m} h_i \mathrm{E}[xy(i)] \\ &= \mathrm{E}(x^2) - \sum_{i=1}^{m} h_i p_{xy}(i) \\ &= \mathrm{E}(x^2) - p_{xy}^{\mathrm{T}} P_y^{-1} p_{xy} \end{aligned}$$

by (7.2.8).

It should be noted at this point that we have not used the fact that $y(k)$ is related to x by (7.2.1). All that we have assumed up to now is that we may estimate x by a linear combination (7.2.2) of the measurements. Suppose that (7.2.1) holds and that v has zero mean and that $v(k)$ is uncorrelated with $v(j)$ $(j \neq k)$ and x. Then

$$\mathrm{E}[v(j)v(k)] = \begin{cases} 0 & j \neq k \\ \sigma_v^2 & j = k \end{cases}$$

and

$$\mathrm{E}[xv(j)] = 0$$

Assume also that $\mathrm{E}(x) = 0$ and write $\mathrm{E}(x^2) = \sigma_x^2$. Then

$$\begin{aligned} P_y(i,j) = \mathrm{E}[y(i)y(j)] &= \mathrm{E}\{[x + v(i)][x + v(j)]\} \\ &= \sigma_x^2 + \sigma_v^2 \delta_{ij} \end{aligned}$$

and

$$p_{xy}(j) = \mathrm{E}[xy(j)] = \mathrm{E}(x^2) = \sigma_x^2$$

Hence, by (7.2.4), we obtain

$$(\sigma_v^2 + m\sigma_x^2)\sum_{i=1}^{m} h_i = m\sigma_x^2$$

and since this equation is symmetric in h_i, $h_1 = h_2 = \ldots = h_m$, and we have

$$h_i = \frac{\sigma_x^2}{m\sigma_x^2 + \sigma_v^2}, \quad 1 \leqslant i \leqslant m$$

Thus,

$$\hat{x} = \frac{1}{m+\gamma}\sum_{i=1}^{m} y(i) \tag{7.2.10}$$

where $\gamma = \sigma_v^2/\sigma_x^2$. Moreover

$$p_e = \frac{\sigma_v^2}{m+\gamma}$$

Note that, for large signal/noise ratio (σ_x^2/σ_v^2), $\gamma \ll m$ and so the estimator (7.2.10) reduces to the simple mean

$$\left[\sum_{i=1}^{m} y(i)\right]\Big/ m$$

Returning to the relation (7.2.7) we shall now discuss the orthogonality principle in more detail. We introduce the vector space H which consists of all linear combinations of x and the measurements $y(j)$, i.e.

$$H = \left\{\alpha_0 x + \sum_{i=1}^{m} \alpha_i y(i) \colon \boldsymbol{\alpha} = (\alpha_0, \alpha_1, \ldots, \alpha_m) \in \mathbb{R}^{m+1}\right\} \tag{7.2.11}$$

For any pair of elements $v, w \in H$ we define their **inner product** by

$$\langle v, w\rangle = \mathrm{E}(vw)$$

(Of course, elements of H are random variables.) If $v \in H$ we call $\|v\| = (\langle v, v\rangle)^{1/2}$ the **norm** of v, and if $v, w \in H$ satisfy $\langle v, w\rangle = 0$ then v and w are said to be **orthogonal** and we write $v \perp w$. Since the elements of H are random variables, the latter is equivalent to v and w being uncorrelated. More generally, if $S \subseteq H$ is a subspace of H and $v \perp w$ for all $w \in S$ then we write $v \perp S$. Now let $H_{[y(1), \ldots, y(m)]}$ denote the subspace of H spanned by the measurements y. Then relations (7.2.2) and (7.2.7) merely state that

$$\hat{x} \in H_{[y(1), \ldots, y(m)]} \tag{7.2.12a}$$

$$(x - \hat{x}) \perp H_{[y(1), \ldots, y(m)]} \tag{7.2.12b}$$

(7.2.12b) explains the use of the term 'orthogonality principle'. It is easy to see that the inner product space (7.2.11) is isomorphic to $\mathbb{R}^q$ for some q and if the measurements are uncorrelated then $q = m$ or $m + 1$, depending on whether x is linearly dependent on the y or not. A finite-dimensional vector space with an inner product is called a Hilbert space. The generalization of the above remarks to infinite-dimensional Hilbert spaces is important in continuous filtering as we shall see later.

7.2.2 Recursive Estimation

The main drawbacks of the Wiener (non-recursive) filter given in the last section are the following:

(a) It requires previous knowledge (or stored estimates) of P_y (the autocorrelation matrix).
(b) The number of data samples must be specified in advance. If more samples become available we must recalculate the estimate.
(c) The matrix P_y has to be inverted, which, if m is large, is numerically demanding.

These objections can be overcome by rewriting the estimator in recursive form. Consider again the process

$$y(k) = x + v(k)$$

and let

$$\hat{x}(k) = \sum_{i=1}^{k} h(i, k)\, y(i)$$

(This time, we are varying the number of samples k, so that the estimate x and the estimation parameters $h(i)$ will depend on k.) From the results of the last section we have

$$h(i, k+1) = \frac{1}{k+1+\gamma}, \quad h(i, k) = \frac{1}{k+\gamma}$$

$$p_e(k+1) = \frac{\sigma_v^2}{k+1+\gamma}, \quad p_e(k) = \frac{\sigma_v^2}{k+\gamma}$$

Hence

$$\begin{aligned} \frac{p_e(k+1)}{p_e(k)} &= \frac{h(i, k+1)}{h(i, k)} \\ &= \frac{k+\gamma}{k+1+\gamma} \qquad (7.2.13) \\ &= \frac{1}{1 + p_e(k)/\sigma_v^2} \end{aligned}$$

Also

$$\hat{x}(k+1) = \frac{1}{k+1+\gamma} \sum_{i=1}^{k} y(i) + \frac{1}{k+1+\gamma}\, y(k+1)$$

$$= \frac{k+\gamma}{k+1+\gamma}\, \hat{x} + \frac{1}{k+1+\gamma}\, y(k+1) \tag{7.2.14}$$

$$= \frac{p_e(k+1)}{p_e(k)}\, \hat{x}(k) + \frac{p_e(k+1)}{\sigma_v^2}\, y(k+1)$$

Equations (7.2.13) and (7.2.14) provide a recursive form of the Wiener filter. Using initial estimates $\hat{x}(1)$ and $p_e(1)$ we find $p_e(k+1)$ from (7.2.13) and then $x(k+1)$ from (7.2.14). The initial estimates $\hat{x}(1)$ and $p_e(1)$ may be found by using non-recursive methods.

Now put

$$a(k+1) = \frac{p_e(k+1)}{p_e(k)}$$

$$b(k+1) = \frac{p_e(k+1)}{\sigma_v^2}$$

Then, by (7.2.14),

$$\hat{x}(k+1) = a(k+1)\hat{x}(k) + b(k+1)y(k+1)$$

However,

$$a(k+1) = \frac{p_e(k+1)}{p_e(k)}$$

$$= \frac{\sigma_v^2}{\sigma_v^2 + p_e(k)}$$

$$= 1 - \frac{p_e(k)}{\sigma_v^2 + p_e(k)}$$

$$= 1 - b(k+1)$$

and so we obtain

$$\boxed{\hat{x}(k+1) = \hat{x}(k) + b(k+1)[y(k+1) - \hat{x}(k)]} \tag{7.2.15}$$

We can write this relation in a slightly different form by defining $H_k = H_{[y(1), \ldots, y(k)]}$ and noting that we can write

$$\hat{x}(k+1) = S_{k+1} = T_{k+1}$$

where $S_{k+1} \in H_k$ and $T_{k+1} \perp H_k$. Also, if $\tilde{x}(k+1) = x - \hat{x}(k+1)$, then

$$x = \hat{x}(k+1) + \tilde{x}(k+1) = S_{k+1} + [T_{k+1} + \tilde{x}(k+1)]$$

where $S_{k+1} \in H_k$ and $[T_{k+1} + \tilde{x}(k+1)] \perp H_k$ by (7.2.12). Now, orthogonal decompositions are unique and therefore, since we also have

$$x = \hat{x}(k) + \tilde{x}(k)$$

where $\hat{x}(k) \in H_k$, $\tilde{x}(k) \perp H_k$ (from the general non-recursive theory) it follows that $S_{k+1} = \hat{x}(k)$. Furthermore, T_{k+1} is the projection of x onto $H_{k+1} - H_k$ and so $T_{k+1} = 0$ (if $H_{k+1} = H_k$) or

$$T_{k+1} = \frac{\langle x, \tilde{y}(k+1)\rangle}{\langle \tilde{y}(k+1), \tilde{y}(k+1)\rangle} \tilde{y}(k+1)$$

where

$$\tilde{y}(k+1) = y(k+1) - P_k y(k+1)$$

and P_k is the projection operator on H_k. Hence (7.2.15) may be written in the form

$$\boxed{\hat{x}(k+1) = \hat{x}(k) + \frac{\langle x, \tilde{y}(k+1)\rangle}{\langle \tilde{y}(k+1), \tilde{y}(k+1)\rangle} [y(k+1) - P_k y(k+1)]} \tag{7.2.16}$$

We can now see clearly that the recursive estimator has the following structure: the new estimate at 'time' $k+1$ is equal to the old estimate $\hat{x}(k)$ plus a correction term proportional to $\tilde{y}(k+1)$. Now $\tilde{y}(k+1)$ is orthogonal to H_k, i.e. $\tilde{y}(k+1)$ is orthogonal to the previous measurements and represents the 'new' information at the $(k+1)$th step. For this reason, the mutually orthogonal terms $\tilde{y}(k)$ are said to form the **innovations sequence**.

7.2.3 Discrete Kalman Filter

Up to now we have considered a single fixed random variable x with additive noise. We suppose now that the signal x itself is a dynamic process which may be modelled by an equation of the form

$$x(k) = ax(k-1) + w(k-1) \tag{7.2.17}$$

(We shall treat the scalar equation for simplicity of notation. The vector case will be a simple notational change.) Assume that $x(k) = 0$ and $w(k) = 0$ for $k < 0$ and that

$$\begin{aligned} \mathrm{E}[w(k)] &= 0 \\ \mathrm{E}[w(k)w(j)] &= \sigma_w^2 \delta_{kj} \\ \mathrm{E}[x(k)w(k)] &= 0 \end{aligned}$$

Suppose the measurement equation is

$$y(k) = cx(k) + v(k) \tag{7.2.18}$$

where $v(k)$ and $w(k)$ are uncorrelated. Then we seek a recursive estimator of the form

$$\hat{x}(k) = a(k)\hat{x}(k-1) + b(k)y(k)$$

As before we minimize $p_e(k) = \mathrm{E}[e^2(k)]$, where $e(k) = x(k) - \hat{x}(k)$. Then

$$p_e(k) = \mathrm{E}[a(k)\hat{x}(k-1) + b(k)y(k) - x(k)]^2$$

and so

$$\frac{\partial p_e(k)}{\partial a(k)} = 2\mathrm{E}\{[a(k)\hat{x}(k-1) + b(k)y(k) - x(k)]\hat{x}(k-1)\} = 0$$

$$\frac{\partial p_e(k)}{\partial b(k)} = 2\mathrm{E}\{[a(k)\hat{x}(k-1) + b(k)y(k) - x(k)]\, y(k)]\} = 0$$

Hence we obtain the orthogonality relations

$$\mathrm{E}[e(k)\hat{x}(k-1)] = 0 \tag{7.2.19}$$

$$\mathrm{E}[e(k)y(k)] = 0 \tag{7.2.20}$$

From (7.2.19) we have

$$\begin{aligned}\mathrm{E}(\{a(k)[\hat{x}(k-1) - x(k-1)] + a(k)x(k-1)\}\hat{x}(k-1))\\ = \mathrm{E}\{[x(k) - b(k)y(k)]\hat{x}(k-1)\}\end{aligned}$$

and so

$$\begin{aligned}a(k)\mathrm{E}[e(k-1)\hat{x}(k-1) + x(k-1)\hat{x}(k-1)]\\ = E(\{x(k)[1 - cb(k)] - b(k)v(k)\}\hat{x}(k-1))\end{aligned}$$

by using (7.2.18). However,

$$\begin{aligned}\mathrm{E}(e(k-1)\hat{x}(k-1)) &= \mathrm{E}\{e(k-1)[a(k-1)\hat{x}(k-2) + b(k-1)y(k-1)]\}\\ &= 0\end{aligned}$$

by (7.2.19) and (7.2.20). Also $\mathrm{E}[v(k)\hat{x}(k-1)] = 0$ since the $(k-1)$th estimate and $v(k)$ are uncorrelated. Hence,

$$a(k)\mathrm{E}[x(k-1)\hat{x}(k-1)] = [1 - cb(k)]E[x(k)\hat{x}(k-1)] \tag{7.2.21}$$

Now

$$\begin{aligned}\hat{x}(k-1) = a(k-1)\hat{x}(k-2) + acb(k-1)x(k-2)\\ + cb(k-1)w(k-2) + b(k-1)v(k-1)\end{aligned}$$

and since each term on the right is uncorrelated with $w(k-1)$ we have

$$\mathrm{E}[w(k-1)\hat{x}(k-1)] = 0$$

Hence, we have

$$\begin{aligned}\mathrm{E}[x(k)\hat{x}(k-1)] &= \mathrm{E}\{[ax(k-1) + w(k-1)]\hat{x}(k-1)\}\\ &= a\mathrm{E}[x(k-1)\hat{x}(k-1)]\end{aligned} \tag{7.2.22}$$

(using (7.2.17)). It follows from (7.2.21) and (7.2.22) that

$$a(k) = a[1 - cb(k)]$$

and so

$$\boxed{\hat{x}(k) = a\hat{x}(k-1) + b(k)[y(k) - ac\hat{x}(k-1)]} \qquad (7.2.23)$$

The first term on the right-hand side of (7.2.23) is the best estimate of $x(k)$ without any additional information at step k, while the second term is a correction based on the kth observation. We can determine the multipliers $b(k)$ as follows. Firstly we recall that

$$\begin{aligned} p_e(k) &= \mathrm{E}[e^2(k)] \\ &= \mathrm{E}\{e(k)[\hat{x}(k) - x(k)]\} \end{aligned}$$

Since

$$\hat{x}(k) = a(k)\hat{x}(k-1) + b(k)y(k)$$

the orthogonality relations imply that

$$p_e(k) = -\mathrm{E}[e(k)x(k)]$$

Moreover, (7.2.18) and (7.2.20) imply that

$$c\mathrm{E}[e(k)x(k)] = -\mathrm{E}[e(k)v(k)]$$

and so

$$\begin{aligned} p_e(k) &= \frac{1}{c}\mathrm{E}[e(k)v(k)] \\ &= \frac{1}{c}\mathrm{E}\{[\hat{x}(k) - x(k)]v(k)\} \\ &= \frac{1}{c}\mathrm{E}\{[a(k)\hat{x}(k-1) + b(k)y(k) - x(k)]v(k)\} \\ &= \frac{1}{c}b(k)\mathrm{E}[y(k)v(k)] \\ &= \frac{1}{c}b(k)\,\sigma_v^2 \end{aligned}$$

Hence

$$b(k) = \frac{cp_e(k)}{\sigma_v^2}$$

Note, furthermore, that

$$\begin{aligned} p_e(k) &= \mathrm{E}[\hat{x}(k) - x(k)]^2 \\ &= \mathrm{E}\{a\hat{x}(k-1) + b(k)[y(k) - ac\hat{x}(k-1)] - x(k)\}^2 \\ &= \mathrm{E}\{-a[1-cb(k)]e(k-1) - [1-cb(k)]w(k-1) + b(k)v(k)\}^2. \end{aligned}$$

Now the cross product terms obtained by expanding the square are zero since $e(k-1)$, $w(k-1)$, $v(k)$ are mutually uncorrelated. Hence

$$p_e(k) = a^2[1-cb(k)]^2 p_e(k-1) + [1-cb(k)]^2\sigma_w^2 + b^2(k)\,\sigma_v^2$$

and since $p_e(k) = b(k)\,\sigma_v^2/c$, we have

$$b(k)\{\sigma_v^2 + c^2[a^2 p_e(k-1) + \sigma_w^2]\} = c[a^2 p_e(k-1) + \sigma_w^2]$$

and finally

$$b(k) = \frac{c[a^2 p_e(k-1) + \sigma_w^2]}{\sigma_v^2 + c^2\sigma_w^2 + c^2 a^2 p_e(k-1)}$$

The filter can now be written in the form

$$\left.\begin{aligned} \hat{x}(k) &= a\hat{x}(k-1) + b(k)[y(k) - ac\hat{x}(k-1)] \\ b(k) &= cp_1(k)[c^2 p_1(k) + \sigma_v^2]^{-1} \\ &\text{where} \\ p_1(k) &= a^2 p(k-1) + \sigma_w^2 \\ p(k) &= p_1(k) - cb(k)p_1(k) \end{aligned}\right\} \quad (7.2.24)$$

(where we have dropped the subscript e from p).

A simple generalization of the above argument to the time-dependent vector equations

$$\left.\begin{aligned} \mathbf{x}(k+1) &= A(k)\mathbf{x}(k) + \mathbf{w}(k) \\ \mathbf{y}(k) &= C(k)\mathbf{x}(k) + \mathbf{v}(k) \end{aligned}\right\} \quad (7.2.25)$$

leads to the filter

$$\begin{aligned} \hat{\mathbf{x}}(k) &= A(k-1)\hat{\mathbf{x}}(k-1) + K(k)[\mathbf{y}(k) - C(k)A(k-1)\hat{\mathbf{x}}(k-1)] \\ K(k) &= P_1(k)C^{\mathrm{T}}(k)[C(k)P_1(k)C^{\mathrm{T}}(k) + R(k)]^{-1} \\ &\text{where} \\ P_1(k) &= A(k-1)P(k-1)A^{\mathrm{T}}(k-1) + Q(k-1) \\ P(k) &= P_1(k) - K(k)C(k)P_1(k) \end{aligned} \quad (7.2.26)$$

We shall also write $P_1(k) = P(k \mid k-1)$ for reasons which will become clear shortly. In (7.2.26) Q and R are the covariance matrices of $\mathbf{w}$ and $\mathbf{v}$ respectively. Note that we must start the recursion by defining $P(0)$. Since $P(0)$ is the error covariance and is completely unknown the filter is usually started with $P(0) = I \times 10^{\alpha}$ with α large, although numerical stability of the filter is often not achieved. We shall say more about the numerical computation of the filter later.

Example 7.2.1
Suppose we measure the position of a target (by radar, for example) which is moving radially away from the observer with an acceleration a. Then we have

$$\ddot{r} = a$$
$$y = r + v$$

where v, assumed to be zero, is mean white noise. (White noise is discussed later, its properties are not too important here.) Then put $\boldsymbol{x}(t) = [r(t), \dot{r}(t)]^{\mathrm{T}}$ and we obtain

$$\dot{\boldsymbol{x}}(t) = \begin{pmatrix} 0 & 1 \\ 0 & 0 \end{pmatrix} \boldsymbol{x}(t) + \begin{pmatrix} 0 \\ 1 \end{pmatrix} a$$
$$y(t) = (1 \quad 0)\boldsymbol{x}(t) + v(t)$$

The solution of the differential equation is

$$\boldsymbol{x}(t) = \exp[A(t-\tau)]\boldsymbol{x}(\tau) + \int_{\tau}^{t} \exp[A(t-s)] \begin{pmatrix} 0 \\ 1 \end{pmatrix} a \, \mathrm{d}s$$

where

$$A = \begin{pmatrix} 0 & 1 \\ 0 & 0 \end{pmatrix}$$

and

$$\exp[A(t-\tau)] = \begin{pmatrix} 1 & t-\tau \\ 0 & 1 \end{pmatrix}$$

Put $t = kT$, $\tau = (k-1)T$, and for simplicity assume that the sampling time $T = 1$. Then we obtain the equations

$$\boldsymbol{x}(k) = \begin{pmatrix} 1 & 1 \\ 0 & 1 \end{pmatrix} \boldsymbol{x}(k-1) + a\begin{pmatrix} \frac{1}{2} \\ 1 \end{pmatrix}$$
$$= \tilde{A}\boldsymbol{x}(k) + \boldsymbol{w}(k)$$
$$y(k) = (1 \quad 0)\boldsymbol{x}(k) + v(k)$$

where $\boldsymbol{w}(k) = a(\frac{1}{2} \quad 1)^{\mathrm{T}}$ and is deterministic, so $Q(k) = 0$.

The Kalman filter becomes

$$\hat{x}(k) = \begin{pmatrix} 1 & 1 \\ 0 & 1 \end{pmatrix} \hat{x}(k-1) + K(k)[y(k) - (1 \quad 1)\hat{x}(k-1) - \tfrac{1}{2}a]$$

$$K(k) = P_1(k)\begin{pmatrix} 1 \\ 0 \end{pmatrix} [(1 \quad 0)P_1(k)(1 \quad 0)^{\mathrm{T}} + \sigma_v^2]^{-1}$$

$$P_1(k) = \begin{pmatrix} 1 & 1 \\ 0 & 1 \end{pmatrix} P(k-1) \begin{pmatrix} 1 & 0 \\ 1 & 1 \end{pmatrix}$$

$$P(k) = P_1(k) - K(k)(1 \quad 0)P_1(k)$$

where σ_v^2 is the variance of $v(k)$ which is assumed constant. In this case we can usually obtain a fairly good initial guess of the initial state $x(0)$ and so we may take

$$P(0) = \begin{pmatrix} 5 & 0 \\ 0 & 1 \end{pmatrix}$$

for example.

7.2.4 Optimal Prediction

Consider again the system equations (7.2.25) and suppose that instead of requiring an estimate $\hat{x}(k)$ of $x(k)$ based on all the measurements $y(j)$, $1 \leqslant j \leqslant k$, we wish to predict $x(k)$ from the measurements $y(j)$, $1 \leqslant j \leqslant i$ where $i < k$. Intuitively we may guess that, without any information for the time instants $i+1, \ldots, k$ the best prediction of $x(k)$ is just

$$\boxed{\hat{x}(k \mid i) = \Phi(k, i)\hat{x}(i)} \tag{7.2.27}$$

where Φ is the transition matrix (appendix 1) of the system

$$x(k+1) = A(k)x(k)$$

$\hat{x}(i)$ is just the optimal estimate based on $y(1), \ldots, y(i)$ obtained in the last section and $\hat{x}(k \mid i)$ denotes the optimal prediction of $x(k)$ based on $y(1), \ldots, y(i)$. The relation (7.2.27) is, in fact, true and we may prove it simply as follows.

We know that the optimal estimate $\hat{x}(k)$ given $y(1), \ldots, y(k)$ is just the projection of $x(k)$ on a certain Hilbert space $H_{[y(1), \ldots, y(k)]}$. However, this projection is just the conditional expectation (appendix 2) of $x(k)$ given $y(1), \ldots, y(k)$. Hence we can write

$$\hat{x}(k) = \mathrm{E}[x(k) \mid y(1), \ldots, y(k)]$$

In just the same way it follows that

$$\hat{x}(k \mid i) = \mathrm{E}[x(k) \mid y(1), \ldots, y(i)]$$

Now, using (7.2.25) we have

$$x(k) = \Phi(k, i)x(i) + \sum_{j=i+1}^{k} \Phi(k, j)w(j-1)$$

and so

$$\begin{aligned}\hat{x}(k \mid i) &= \mathrm{E}[\Phi(k, i)x(i) \mid y(1), \ldots, y(i)] \\ &\quad + \mathrm{E}\left[\sum_{j=i+1}^{k} \Phi(k, j)w(j-1) \mid y(1), \ldots, y(i)\right] \\ &= \Phi(k, i)\mathrm{E}[x(i) \mid y(1), \ldots, y(i)] \\ &\quad + \sum_{j=i+1}^{k} \Phi(k, j)\mathrm{E}[w(j-1) \mid y(1), \ldots, y(i)] \qquad (7.2.28)\end{aligned}$$

However, the second term in the last expression is zero since $w(j-1)$(for $j > i$) is uncorrelated with, and hence independent of (for Gaussian processes), $y(1), \ldots, y(i)$. Therefore,

$$\hat{x}(k \mid i) = \Phi(k, i)\hat{x}(i)$$

as expected.

We can also obtain an expression for the error covariance

$$P(k \mid i) \triangleq [x(k) - \hat{x}(k \mid i)]^2$$

as follows. (For a vector x, x^2 means $x^{\mathrm{T}}x$.) We have

$$\begin{aligned}\tilde{x}(k \mid i) &\triangleq x(k) - \hat{x}(k \mid i) \\ &= \Phi(k, i)x(i) + \sum_{j=i+1}^{k} \Phi(k, j)w(j-1) - \Phi(k, \mathrm{i})\hat{x}(i) \\ &= \Phi(k, i)\tilde{x}(i \mid i) + \sum_{j=i+1}^{k} \Phi(k, j)w(j-1)\end{aligned}$$

Hence, using the properties of the transition matrix, we have

$$\begin{aligned}\tilde{x}(k \mid i) &= \Phi(k, k-1)\Phi(k-1, i)\tilde{x}(i \mid i) + \Phi(k, k)w(k-1) \\ &\quad + \sum_{j=i+1}^{k-1} \Phi(k, j)w(j-1) \\ &= \Phi(k, k-1)\,\Phi(k-1, i)\tilde{x}(i \mid i) \\ &\quad + w(k-1) + \Phi(k, k-1)\sum_{j=i+1}^{k-1} \Phi(k-1, j)w(j-1) \\ &= \Phi(k, k-1)\left[\Phi(k-1, i)\tilde{x}(i \mid i) + \sum_{j=i+1}^{k-1} \Phi(k-1, j)w(j-1)\right] \\ &\quad + w(k-1) \\ &= \Phi(k, k-k)\tilde{x}(k-1 \mid i) + w(k-1)\end{aligned}$$

and it follows that $\tilde{x}$ is zero mean Gaussian with the Markov property, i.e. $\tilde{x}(k \mid i)$ depends only on the most recent past for $i = k - 1$. Now, returning to (7.2.28) we see that

$$P(k \mid i) = \mathrm{E}\left[\Phi(k, i)\tilde{x}(i \mid i) + \sum_{j=i+1}^{k} \Phi(k, j)w(j-1)\right]^2$$

However, it is easy to see that $\tilde{x}(i \mid i)$ is uncorrelated with $w(j-1)$ for $j > i$. Hence

$$P(k \mid i) = \Phi(k, i)\mathrm{E}[\tilde{x}(i \mid i)^2\Phi^{\mathrm{T}}(k, i) + \sum_{j=i+1}^{k} \Phi(k, i)\mathrm{E}[w(j-1)]^2\Phi^{\mathrm{T}}(k, i)$$

since $\mathrm{E}[w(j)\omega^{\mathrm{T}}(k)] = 0$ for $j \neq k$, and finally,

$$\boxed{P(k \mid i) = \Phi(k, i)P(i \mid i)\Phi^{\mathrm{T}}(k, i) + \sum_{j=i+1}^{k} \Phi(k, i)Q(j-1)\Phi^{\mathrm{T}}(k, i)} \tag{7.2.29}$$

(Recall that $Q = \mathrm{E}(w)^2$.)

7.2.5 Optimal Smoothing

Frequently we are presented with noisy data which must be used to estimate the state of a dynamical system. In such a case we wish to estimate the state $x(k)$ of the system at time k given j measurements for $0 \leqslant k < j$. If N is a fixed integer and we estimate

$$\hat{x}(k \mid N), \quad 1 \leqslant k \leqslant N - 1 \tag{7.2.30}$$

we speak of **fixed-interval smoothing,** while the estimate

$$\hat{x}(k \mid j), \quad j = k + 1, k + 2, \ldots \tag{7.2.31}$$

is called **fixed-point smoothing.** Finally if we determine

$$\hat{x}(k \mid k + N), \quad k = 0, 1, \ldots, N(N \text{ fixed}) \tag{7.2.32}$$

then this constitutes **fixed-lag smoothing,** and represents the situation where we estimate the state $x(k)$ N steps behind the measurements.

Consider first single stage optimal smoothing. We have

$$\hat{x}(k \mid k + 1) = \mathrm{E}[x(k) \mid y(1), \ldots, y(k), \tilde{y}(k + 1 \mid k)] \tag{7.2.33}$$

where

$$\begin{aligned}\tilde{y}(k + 1 \mid k) &= y(k + 1) - \hat{y}(k + 1 \mid k)\\ &= y(k + 1) - C(k + 1)\hat{x}(k + 1 \mid k)\\ &= C(k + 1)\tilde{x}(k + 1 \mid k) + v(k + 1)\end{aligned}$$

(Relation (7.3.33) is obvious since $y(1), \ldots, y(k)$, $\tilde{y}(k+1 \mid k)$ clearly span the same Hilbert space as $y(1), \ldots, y(k+1)$). Hence,

$$\begin{aligned}\hat{x}(k \mid k+1) &= \mathrm{E}[x(k) \mid y(1), \ldots, y(k)] + \mathrm{E}[x(k) \mid \tilde{y}(k+1 \mid k)] \\ &= \hat{x}(k) + P_{x\tilde{y}} P_{\tilde{y}\tilde{y}}^{-1} \tilde{y}(k+1 \mid k) \qquad (7.2.34)\end{aligned}$$

where

$$\begin{aligned}P_{x\tilde{y}} &= \mathrm{E}[x(k)\tilde{y}^{\mathrm{T}}(k+1 \mid k)] \\ P_{\tilde{y}\tilde{y}} &= \mathrm{E}[\tilde{y}(k+1 \mid k)\tilde{y}^{\mathrm{T}}(k+1 \mid k)]\end{aligned}$$

since $x(k)$ and $\tilde{y}(k+1 \mid k)$ are jointly Gaussian with zero mean (appendix 2). Now

$$P_{x\tilde{y}} = \mathrm{E}[x(k)\tilde{x}^{\mathrm{T}}(k+1 \mid k)]C^{\mathrm{T}}(k+1) + \mathrm{E}x(k)v^{\mathrm{T}}(k+1)$$

The second term is zero, so using

$$\tilde{x}(k+1 \mid k) = \Phi(k+1, k)x(k) + w(k) \quad (\Phi(k+1, k) \triangleq A(k))$$

we have

$$\begin{aligned}P_{x\tilde{y}} &= \mathrm{E}[x(k)\tilde{x}^{\mathrm{T}}(k)]\Phi^{\mathrm{T}}(k+1, k)C^{\mathrm{T}}(k+1) \\ &= \mathrm{E}[\tilde{x}(k)\tilde{x}^{\mathrm{T}}(k)]\Phi^{\mathrm{T}}(k+1, k)C^{\mathrm{T}}(k+1) \qquad (7.2.35) \\ &= P(k)\Phi^{\mathrm{T}}(k+1, k)C^{\mathrm{T}}(k+1)\end{aligned}$$

where $P(k)$ is the optimal filtering error covariance, since $\hat{x}(k)$ is orthogonal to $\tilde{x}(k)$. Also,

$$\begin{aligned}P_{\tilde{y}\tilde{y}} &= \mathrm{E}[\tilde{y}(k+1 \mid k)\tilde{y}^{\mathrm{T}}(k+1 \mid k)] \\ &= C(k+1)P(k+1 \mid k)C^{\mathrm{T}}(k+1) + R(k+1)\end{aligned} \qquad (7.2.36)$$

where $P(k+1 \mid k)$ is the optimal prediction error covariance. From (7.2.34)–(7.2.36) it follows that

$$\boxed{\hat{x}(k \mid k+1) = \hat{x}(k) + F(k \mid k+1)[y(k+1) - C(k+1)\Phi(k+1, k)\hat{x}(k)]} \qquad (7.2.37)$$

where

$$\begin{aligned}F(k \mid k+1) = {} & P(k)\Phi^{\mathrm{T}}(k+1, k)C^{\mathrm{T}}(k+1) \\ & + [C(k+1)P(k+1 \mid k)C^{\mathrm{T}}(k+1) + R(k+1)]^{-1} \qquad (7.2.38)\end{aligned}$$

Note that the initial condition is $\hat{x}(0) = \tilde{x}(0) = \mathbf{0}$, and that $\hat{x}(k \mid k+1)$ requires the optimal filtered expression $\tilde{x}(k)$ plus a correction.

We can write (7.2.37) and (7.2.38) in a different form if $P(k+1 \mid k)$ is non-singular. In fact, using (7.2.26) it is clear that

$$\begin{aligned}C^{\mathrm{T}}(k+1)[C(k+1)P(k+1 \mid k)C^{\mathrm{T}}(k+1) + R(k+1)]^{-1} \\ = P^{-1}(k+1 \mid k)K(k+1)\end{aligned}$$

(Note that $P_1(k)$ in (7.2.26) is just $P(k \mid k-1)$.) Hence,

$$F(k \mid k+1) = D(k)K(k+1)$$

where

$$D(k) = P(k)\Phi^{\mathrm{T}}(k+1, k)P^{-1}(k+1 \mid k) \tag{7.2.39}$$

and so F is related to the Kalman gain K. Therefore, by (7.2.37)

$$\hat{x}(k \mid k+1) = \hat{x}(k) + D(k)K(k+1)[y(k+1) - C(k+1)\Phi(k+1, k)\hat{x}(k)]$$

Hence, using (7.2.26), we have

$$\boxed{\hat{x}(k \mid k+1) = \hat{x}(k) + D(k)[\hat{x}(k+1) - \hat{x}(k+1 \mid k)]} \tag{7.2.40}$$

We may generalize the above reasoning to obtain the optimal fixed-interval smoothing algorithm as follows. Take $\{y(1), \ldots, y(N-1), \tilde{y}(N \mid N-1)\}$ to be the measurements and then for $k < N$ we have

$$\begin{aligned} \hat{x}(k \mid N) &= \mathrm{E}[x(k) \mid y(1), \ldots, y(N-1), \tilde{y}(N \mid N-1)] \\ &= \hat{x}(k \mid N-1) + P_{x\tilde{y}}P_{\tilde{y}\tilde{y}}^{-1}\tilde{y}(N \mid N-1) \end{aligned} \tag{7.2.41}$$

where

$$P_{x\tilde{y}} = \mathrm{E}[x(k)\tilde{y}^{\mathrm{T}}(N \mid N-1)]$$

and

$$P_{\tilde{y}\tilde{y}} = C(N)P(N \mid N-1)C^{\mathrm{T}}(N) + R(N)$$

To evaluate $P_{x\tilde{y}}$ we proceed as before. In fact

$$\begin{aligned} P_{x\tilde{y}} &= \mathrm{E}[x(k)\,\tilde{x}^{T}(N \mid N-1)]\,C^{\mathrm{T}}(N) \\ &= \mathrm{E}[x(k)\tilde{x}^{\mathrm{T}}(N-1)]\,\Phi^{\mathrm{T}}(N, N-1)C^{\mathrm{T}}(N) \end{aligned}$$

It is easy to check using (7.2.25) and (7.2.26) that

$$\tilde{x}(N-1) = [I - K(N-1)C(N-1)]\,\tilde{x}(N-1 \mid N-2) - K(N-1)v(N-1)$$

and so

$$P_{x\tilde{y}} = \mathrm{E}[x(k)\tilde{x}^{\mathrm{T}}(N-1 \mid N-2)]\,[I - K(N-1)C(N-1)]^{\mathrm{T}}\Phi^{\mathrm{T}}(N, N-1)C^{\mathrm{T}}(N)$$

However,

$$[I - K(N-1)\mathrm{C}(N-1)]^{\mathrm{T}} = P^{-1}(N-1 \mid N-2)P(N-1)$$

from (7.2.26), assuming $P(N-1 \mid N-2)$ is invertible. Hence

$$\begin{aligned} P_{x\tilde{y}} = \mathrm{E}[x(k)\tilde{x}^{\mathrm{T}}(N-1 \mid N-2)]P^{-1}(N-1 \mid N-2)P(N-1) \\ \times \Phi^{\mathrm{T}}(N, N-1)C^{\mathrm{T}}(N) \end{aligned}$$

We therefore have

$$\begin{aligned}F(k \mid N) &\triangleq P_{x\tilde{y}}P_{\tilde{y}\tilde{y}}^{-1}\\ &= \mathrm{E}[x(k)\,\tilde{x}^{\mathrm{T}}(N-1 \mid N-2)]\,P^{-1}(N-1 \mid N-2)P(N-1)\\ &\quad \times \Phi^{\mathrm{T}}(N, N-1)C^{\mathrm{T}}(N)[C(N)P(N \mid N-1)C^{\mathrm{T}}(N) + R(N)]^{-1}\\ &= \mathrm{E}[x(k)\,\tilde{x}^{\mathrm{T}}(N-1 \mid N-2)]\,P^{-1}(N-1 \mid N-2)D(N-1)K(N)\end{aligned}$$

using (7.2.26) and (7.2.39). Just as above we can show that

$$\begin{aligned}\mathrm{E}[x(k)\tilde{x}^{\mathrm{T}}(N-1 \mid N-2)] &= \mathrm{E}[x(k)\tilde{x}^{\mathrm{T}}(N-2 \mid N-3)]\\ &\quad P^{-1}(N-2 \mid N-3)P(N-2)\Phi^{\mathrm{T}}(N-1, N-2)\end{aligned}$$

and so

$$\begin{aligned}F(k \mid N) &= E[x(k)\tilde{x}^{\mathrm{T}}(N-2 \mid N-3)]\\ &\quad P^{-1}(N-2 \mid N-3)D(N-2)D(N-1)K(N)\end{aligned}$$

and by induction

$$\begin{aligned}F(k \mid N) &= \mathrm{E}[x(k)\tilde{x}^{\mathrm{T}}(k+1 \mid k)]P^{-1}(k+1) \mid k)D(k+1)\ldots\\ &\quad D(N-2)D(N-1)K(N)\\ &= D(k)D(k+1)\ldots D(N-1)K(N) \qquad (7.2.42)\end{aligned}$$

using the reasoning leading to (7.2.35) and (7.2.39) again.

We now prove by induction that the generalization of (7.2.40), namely

$$\boxed{\hat{x}(k \mid N) = \hat{x}(k) + D(k)[\hat{x}(k+1 \mid N) - \hat{x}(k+1 \mid k)] \quad 0 \leqslant k \leqslant N-1} \qquad (7.2.43)$$

is true. By (7.2.40), it is true for $N = k+1$. If it is true for N replaced by $N-1$, then by (7.2.41) and (7.2.42) we have

$$\begin{aligned}\hat{x}(k \mid N) &= \hat{x}(k) - D(k)\hat{x}(k+1 \mid k) + D(k)[\hat{x}(k+1 \mid N-1)\\ &\quad + D(k+1)\ldots D(N-1)K(N)\,\tilde{y}(N \mid N-1)]\end{aligned}$$

Note, however, that

$$\hat{x}(k+1 \mid N) = \hat{x}(k+1 \mid N-1) + F(k+1 \mid N)\,\tilde{y}(N \mid N-1)$$

just as in (7.2.41) and using (7.2.42) it follows that (7.2.43) is true for all N. Note that (7.2.43) must be solved backwards in time starting with the filtered estimate $\hat{x}(N \mid N)$.

We can also evaluate the error covariance $P(k \mid N)$ for the smoothing process, although it should be noted that $\tilde{x}(k \mid N)$ is not a Gauss–Markov process so that $P(k \mid N)$ (and $\mathrm{E}[\tilde{x}(k \mid N)] = 0$) does not completely characterize this process (see Meditch, 1969). In fact, by (7.2.43),

$$\tilde{x}(k \mid N) + D(k)\hat{x}(k+1 \mid N) = \tilde{x}(k) + D(k)\hat{x}(k+1 \mid k)$$

where

$$\tilde{x}(k \mid N) = x(k) - \hat{x}(k \mid N)$$

Since

$$\mathrm{E}[\tilde{x}(k \mid N)\hat{x}^{\mathrm{T}}(k+1 \mid N)] = \mathrm{E}[\tilde{x}(k)\hat{x}^{\mathrm{T}}(k+1 \mid k)] = 0$$

we have

$$P(k \mid N) + D(k)P_{\hat{x}\hat{x}}(k+1 \mid N)D^{\mathrm{T}}(k) = P(k) + D(k)P_{\hat{x}\hat{x}}(k+1 \mid k)D^{\mathrm{T}}(k) \tag{7.2.44}$$

where

$$P_{\hat{x}\hat{x}}(k+1 \mid j) = \mathrm{E}[\hat{x}(k+1 \mid j)\hat{x}^{\mathrm{T}}(k+1 \mid j)]$$

Now

$$x(k+1) = \hat{x}(k+1 \mid k) + \tilde{x}(k+1 \mid k)$$

and so

$$P(k+1) = P_{\hat{x}\hat{x}}(k+1 \mid k) + P(k+1 \mid k)$$

Similarly

$$P(k+1) = P_{\hat{x}\hat{x}}(k+1 \mid N) + P(k+1 \mid N)$$

and so from (7.2.44), we obtain

$$\boxed{P(k \mid N) = P(k) + D(k)[P(k+1 \mid N) - P(k+1 \mid k)]D^{\mathrm{T}}(k)} \tag{7.2.45}$$

with boundary condition

$$P(k+1 \mid N) = P(N) \quad \text{when } k = N-1$$

To obtain the fixed-point smoothing algorithm note that (7.2.41) gives

$$\hat{x}(k \mid j) = \hat{x}(k \mid j-1) + F(k \mid j)\tilde{y}(j \mid j-1)$$

and so

$$\boxed{\hat{x}(k \mid j) = \hat{x}(k \mid j-1) + B(j)[\hat{x}(j) - \hat{x}(j \mid j-1)]} \tag{7.2.46}$$

using (7.2.37) and (7.2.42), where

$$B(j) = \prod_{j=k}^{N-1} D(i)$$

The estimates $\hat{x}(k \mid j)$ may be obtained by forward recursion from $\hat{x}(k)$. Of course, $B(j)$ may be obtained by recursion from

$$\boxed{B(j) = B(j-1)D(j-1)}$$

It is easy to show that the error covariance $P(k \mid j)$ satisfies

$$P(k \mid j) = P(k \mid j-1) + B(j)[P(j) - P(j \mid j-1)]B^{\mathrm{T}}(j) \tag{7.2.47}$$

(see Meditch (1969) for details).

Finally note that the fixed-lag smoothing algorithm can be obtained by similar reasoning (again see Meditch for a full discussion). We just note here the algorithm

$$\begin{aligned}\hat{x}(k+1 \mid k+1+N) = {} & \Phi(k+1, k)\hat{x}(k \mid k+N) \\ & + E(k+1+N)K(k+1+N)\tilde{y}(k+1+N \mid k+N) \\ & + U(k+1)[\hat{x}(k \mid k+N) - \hat{x}(k)], \quad k \geqslant 0\end{aligned}$$

with $\hat{x}(0 \mid N)$ as initial condition.

$$E(k+1+N) = \prod_{i=k+1}^{k+N} D(i)$$

$$D(l) = P(l)\Phi^{\mathrm{T}}(l+1, l)P^{-1}(l+1 \mid l)$$

$$U(k+1) = Q(k)\Phi^{\mathrm{T}}(k, k+1)P^{-1}(k), \quad k \geqslant 0 \tag{7.2.48}$$

The error covariance algorithm is

$$\begin{aligned}P(k+1 \mid k+1+N) = {} & P(k+1 \mid k) - E(k+1+N)K(k+1+N) \\ & C(k+1+N) \\ & \times P(k+1+N \mid k+N)E^{\mathrm{T}}(k+1+N) \\ & - D^{-1}(k)[P(k) - P(k \mid k+N)][D^{\mathrm{T}}(k)]^{-1}, \\ & k \geqslant 0\end{aligned}$$

with initial condition $P(0 \mid N)$.

$$\tag{7.2.49}$$

Note that the initial conditions for (7.2.48) and (7.2.49) must be obtained from the optimal fixed-point algorithms (7.2.46) and (7.2.47).

7.3 CONTINUOUS ESTIMATION AND FILTERING

7.3.1 Orthogonal Projections

In order to study the estimation theory of continuous systems, it is necessary to introduce a Hilbert space which corresponds to the finite dimensional space considered in section 7.2. An introduction to Hilbert spaces and stochastic processes is given in appendices 1 and 2 respectively, and we shall call on these ideas without further comment. If $(\Omega, \mathcal{B}, P)$ is a probability space then we shall consider the space H of (equivalence classes of) random variables X for which $\mathrm{E}X^2 < \infty$. Then it can be shown that H is a Hilbert space with inner product

$$\boxed{\langle X, Y\rangle = \mathrm{E}(XY)} \tag{7.3.1}$$

and norm

$$\|X\| = \mathrm{E}(X^2)$$

Note that a sequence $\{X_n\}$ converges to X in H if $\|X_n - X\| \to 0$ and we say that X_n tends in **quadratic mean** (q.m.) to X. Of course, we can ask if $X_n(\omega) \to X(\omega)$ for certain values of ω. In fact it can be seen that if $X_n \overset{\text{q.m.}}{\to} X$ then $X_n(\omega) \to X(\omega)$ for almost all ω and we write $X_n \overset{\text{a.s.}}{\to} X$ (see appendix 2). Since we often deal with signals of expectation zero, it is convenient to introduce the subspace

$$H_0 = \{X \in H\colon \mathrm{E}X = 0\}$$

The orthogonal complement $H_0^{\perp}$ of H_0 in H is clearly one dimensional and spanned by the function $1(\omega) = 1$ for all $\omega \in \Omega$. (For

$$X = (X - \mathrm{E}X) + \mathrm{E}X \times 1 \quad \text{and} \quad \langle X - \mathrm{E}X, 1\rangle = 0$$

for any $X \in H$.) It is also easy to see that if $X_n \to X$ in H then $\mathrm{E}X_n \to \mathrm{E}X$ and $\sigma(X_n) \to \sigma(X)$ where σ is the variance.

Consider now a stochastic process $\{Y_t, t \in \mathbb{R}^+\} \subseteq H_0$ on a probability space $(\Omega, \mathcal{B}, P)$ and suppose that this process is a collection of measurements of a random variable $X \in H_0$. As in the finite-dimensional case we introduce the Hilbert subspace $H^t_{\{Y\}}$ of H_0 which is the space of all linear combinations

$$\sum \alpha_i Y_{t_i} \qquad (t_i \leqslant t)$$

of the measurements. We often write $H_{\{Y\}} = H^{\infty}_{\{Y\}}$. Clearly, for any $t_1 \leqslant t_2$

$$H^{t_1}_{\{Y\}} \subseteq H^{t_2}_{\{Y\}} \subseteq H_{\{Y\}}$$

Now, since $X \in H_0$ we can write

$$X = X_1 + X_2, \quad X_1 \in H^t_{\{Y\}}, X_2 \perp H^t_{\{Y\}}$$

If P_t denotes the projection on $H^t_{\{Y\}}$, then

$$P_t X = X_1, \quad P_s X = P_s P_t X \quad \text{for } s \leqslant t$$

Since

$$\| X - P_t X \| = \min_{Z \in H^t_{\{Y\}}} \| X - Z \| \text{ (appendix 1)}$$

in any Hilbert space, it follows that $P_t X$ (i.e. the projection of X on the space spanned by the observations) is the best linear estimator (in the least squares sense) of X, just as in the finite-dimensional case. Moreover, it is easy to see that if Y_t is continuous in H_0, i.e. $Y_s \xrightarrow{\text{q.m.}} Y_t$ as $s \to t$ then if $X \in H^t_{\{Y\}}$, $P_s X \to X$ as $s \to t$ (see Davis, 1977).

7.3.2 The Covariance Function

For a zero mean stochastic process $\{Y_t\}$ we introduce the covariance function

$$\boxed{r(t,s) = \mathrm{E}(Y_t Y_s) = \langle Y_t, Y_s \rangle} \tag{7.3.2}$$

The function r determines the structure of the space $H_{\{Y\}}$ since for any pair of elements V, $W \in H_{\{Y\}}$ we can write

$$V = \lim_{n \to \infty} \sum \alpha_i^n Y_{t_i^n}$$

$$W = \lim_{n \to \infty} \sum \beta_j^n Y_{t_j^n}$$

and so

$$\langle V, W \rangle = \lim_{n \to \infty} \sum_i \sum_j \alpha_i^n \beta_j^n r(t_i^n, t_j^n)$$

Now, if $\{Y_t\}$ is continuous in H_0, then it can be shown (Doob, 1953) that there is a process $\{\tilde{Y}_t\}$ such that $Y_t(\omega) = \tilde{Y}_t(\omega)$ for almost all ω and all t and such that $\tilde{Y}_t(\omega)$ is measurable in t. Replacing $\{Y_t\}$ by $\{\tilde{Y}_t\}$ if necessary it can be shown (see Davis, 1977) that the integral

$$Z = \int_a^b g(t)\, Y_t(\omega)\, \mathrm{d}t$$

is well defined and belongs to $H_{\{Y\}}$ if g is measurable and

$\int_a^b g^2(s)\, \mathrm{d}s < \infty$. Moreover

$$\|Z\|^2 = \int_a^b \int_a^b g(t)\, g(s) r(t,s)\, \mathrm{d}s\, \mathrm{d}t$$

Since $H_{\{Y\}}$ is determined by the covariance function it is not surprising that normal processes (which, of course, are determined by their mean and covariance) have a special importance in $H_{\{Y\}}$. In fact, if $Y_n \overset{\text{q.m.}}{\to} Y$ and each Y_n is normal, then Y is normal. This means that if $\{Y_t\}$ is a normal process then the optimal estimate $\hat{X} = P_t X$ of X given above is normal along with $\tilde{X} = X - \hat{X}$. Since, also, $\hat{X} \perp \tilde{X}$ it follows that $\hat{X}$ and $\tilde{X}$ are uncorrelated, and being normal are therefore independent. Hence, using the properties of conditional expectation, if $\phi_{X|Y}(u)$ denotes the conditional characteristic function of X given $\{Y_s, s \leqslant t\}$, then

$$\begin{aligned}\phi_{X|Y}(u) &= \mathrm{E}[\exp(iuX) \mid Y_s, s \leqslant t] \\ &= \mathrm{E}[\exp(iu\hat{X}) \exp(iu\tilde{X}) \mid Y_s, s \leqslant t] \\ &= \exp(iu\hat{X})\mathrm{E}[\exp(iu\tilde{X})] \\ &= \exp(iu\hat{X} - \tfrac{1}{2}\sigma^2 u^2)\end{aligned}$$

where $\sigma^2 = \mathrm{var}(X)$.

7.3.3 Orthogonal Increments and White Noise

It is difficult to proceed much further with the continuous estimation problem without introducing more restrictions on the stochastic measurements. However, it turns out that the following assumption is satisfied by many practical systems and leads to considerable simplification in the general problem. We say that the process $\{Y_t\}$ has **orthogonal increments** (o.i.) if

$$\boxed{\langle (Y_\tau - Y_\sigma), (Y_t - Y_s) \rangle = 0} \tag{7.3.3}$$

provided $(\tau, \sigma) \cap (t, s) = \phi$. Note that the process $\{Y_t'\}$ defined by $Y_t' = Y_t - Y_0$ has orthogonal increments and so we shall assume without loss of generality that $Y_0 = 0$. Let us evaluate the covariance function for an o.i. process $\{Y_t\}$. We have, for $t > s$,

$$\begin{aligned}r(t, s) &= \mathrm{E}(Y_t Y_s) \\ &= \mathrm{E}\{[(Y_t - Y_s) + Y_s]\, Y_s\} \\ &= \mathrm{E}(Y_s^2)\end{aligned}$$

Similarly, if $s > t, r(t, s) = \mathrm{E}(Y_t^2)$. Hence

$$r(t, s) = \mathrm{E}(Y_{t\wedge s}^2) \quad \text{where } t \wedge s = \min(t, s)$$

In addition, we say that $\{Y_t\}$ has **stationary increments** if

$$\mathrm{E}(Y_t - Y_s)^2 = \mathrm{E}(Y_{t+r} - Y_{s+r})^2 \quad \text{for all } r, s, t$$

Then $Y_t = (Y_t - Y_s) + Y_s$ and so

$$\begin{aligned}\mathrm{E}(Y_t^2) &= \mathrm{E}(Y_t - Y_s)^2 + \mathrm{E}\,Y_s^2 \\ &= \mathrm{E}(Y_{t-s})^2 + \mathrm{E}\,Y_s^2\end{aligned} \tag{7.3.4}$$

Since $\mathrm{E}(Y_0^2) = 0$, the only possible continuous solution of (7.3.4) is

$$\mathrm{E}(Y_t^2) = \sigma^2 t$$

where $\sigma^2 = \mathrm{E}(Y_1^2)$. Hence, for a stationary o.i. process we have

$$\boxed{r(t,s) = \sigma^2 t \wedge s} \tag{7.3.5}$$

Finally we say that $\{Y_t\}$ has **independent increments** (i.i.) if

$$\mathrm{E}(Y_t - Y_s \mid Y_\tau - Y_\sigma) = \mathrm{E}(Y_t - Y_\sigma) \quad \text{if } (t,s) \cap (\tau, \sigma) = \phi$$

If $\mathrm{E}(Y_t) < \infty$ for all t, then an i.i. process has orthogonal increments.

We shall now consider the two most important stochastic processes the first of which has zero mean and stationary **normal** independent increments and is called **Brownian motion**. If $W_0 = 0$ and $\mathrm{E}W_1^2 = 1$ we say that $\{W_t\}$ is a **standard** Brownian motion. From (7.3.5) we have $r(t,s) = \sigma^2 t \wedge s$ for Brownian motion, where $\sigma^2 = \mathrm{E}W_1^2$ and so the process $\{(1/\sigma)W_t\}$ is standard. It is easy to show that Brownian motion exists; in fact, if $\{\phi_n\}_{n \geqslant 0}$ is a basis of the Hilbert space $L^2[0,1]$ (see appendix 1), and $\{\xi_n\}_{n \geqslant 0}$ is a sequence of independent identically distributed random variables each of which is $N(0,1)$ (normal with mean zero, variance 1), then the sequence

$$W_t^n = \sum_{i=0}^{n} \xi_i \int_0^t \phi_i(s)\,\mathrm{d}s, \quad t \in [0,1], \quad n \geqslant 0 \tag{7.3.6}$$

is easily seen to converge in H to a standard Brownian motion. Let $W_t = \lim W_t^n$. Then, if $\{W_t^n\}$, $n \geqslant 1$ is a sequence of standard independent Brownian motions defined on $[0,1]$ (which clearly exists)

$$\begin{aligned} W_t &= W_t^1 & t &\in [0,1] \\ W_t &= W_n + W_{t-n}^{n+1} & t &\in (n, n+1] \end{aligned} \tag{7.3.7}$$

defines a standard Brownian motion on $[0,\infty)$. Clearly, from (7.3.6), sample paths of each W_t^n are continuous (i.e. $W_t^n(\omega)$ is continuous in t for each $\omega \in \Omega$), and it can be shown that the same is true of W_t. However W_t is not differentiable, for

$$\mathrm{var}(\dot{W}_t^n) = \sum_{i=0}^{n} [\phi_i(t)]^2 \to \infty \quad \text{as } n \to \infty$$

so $\dot{W}_t^n$ does not converge in H.

The second important process which we require is the **white noise** process $\{\zeta_t\}$, which is normal and satisfies $\mathrm{cov}(\zeta_t, \zeta_s) = 0$ for $t \neq s$. We would like to define this process as the derivative of Brownian motion; however, as noted above, this process does not exist (in H) and any such process would have to have infinite variance, i.e. $\mathrm{var}(\zeta_t) = \infty$. Since we cannot differentiate Brownian motion, perhaps we can integrate white noise.

In fact, if

$$W_t = \int_0^t \zeta_s \, \mathrm{d}s \tag{7.3.8}$$

then, for $t > s$,

$$\begin{aligned} \operatorname{cov}(W_t, W_s) &= \mathrm{E}\int_0^s \int_0^t \mathrm{E}(\zeta_u \zeta_v)\, \mathrm{d}u\, \mathrm{d}v \\ &= \int_0^s \int_0^t \delta(u-v)\, \mathrm{d}u\ v \\ &= s \end{aligned}$$

Hence (7.3.8) defines a standard Brownian motion and so we may define white noise indirectly by this integral.

7.3.4 Wiener Integrals

In order to proceed any further with the estimation theory of continuous systems it is necessary to have a more explicit description of the spaces $H^t_{\{Y\}}$, $0 \leqslant t \leqslant \infty$. First denote by $L^2[0,1]$ the standard Hilbert space of (equivalence classes of) measurable functions f defined on $[0,1]$ such that

$$\int_0^1 f^2(t)\, \mathrm{d}t < \infty$$

and with the inner product

$$\langle \mathrm{f}, g\rangle = \int_0^1 f(t)\, g(t)\, \mathrm{d}t$$

Then, if $g \in L^2[0,1]$ we define the **Wiener integral of g with respect to the process $\{Y_t\}$**, denoted by

$$\int_0^\infty g(s)\, \mathrm{d}Y_s$$

by

$$\begin{aligned} W(g) &\triangleq \int_0^\infty g(s)\, \mathrm{d}Y_s \\ &= \lim_{l\to\infty} \sum_{i=0}^{n_l - 1} g_{li}(Y_{t^l_{i+1}} - Y_{t^l_i}) \end{aligned} \tag{7.3.9}$$

where g_l is a sequence of step functions converging to g in $L^2[0,1]$, such that the value of g_l on $[t_i^l, t_{i+1}^l]$ is g_{li}. It is then easy to show that

$$\begin{aligned} \mathrm{E}[W(g)]^2 &= \int_0^\infty g^2(t)\,\mathrm{d}t \\ &= \|g\|^2 \end{aligned}$$

Moreover, for any $t \in [0,\infty]$, we have

$$\boxed{H^t_{\{Y\}} = \left\{\int_0^\infty g(s)\,\mathrm{d}Y_s \colon g \in L^2[0,t]\right\}} \tag{7.3.10}$$

i.e. $H^t_{\{Y\}}$ is isomorphic to $l^2[0,t]$. Hence, if any $t \in \mathbb{R}^+$ and $Z_t \in H^t_{\{Y\}}$ is any q.m. continuous process (where $\{Y_t\}$ is stationary and has orthogonal increments) then $(Z_t - Z_s) \perp H^s_{\{Y\}}$ for $s \leqslant t$ implies that we can find $g \in L^2[0,t]$ such that

$$Z_t = \int_0^t g(s)\,\mathrm{d}Y_s$$

Hence the isomorphism (7.3.10) is important in proving that any q.m. continuous process in $H^t_{\{Y\}}$ with the orthogonality property above can be expanded as a Wiener integral.

We have shown above that the best linear estimator of X given $\{Y_t\}$ is P_tX. When $\{Y_t\}$ has stationary orthogonal increments we are in a position to calculate this projection. In fact, since $P_tX \in H^t_{\{Y\}}$ we can write

$$P_tX = \int_0^t g(s)\,\mathrm{d}Y_s, \quad g \in L^2[0,t]$$

Since $(X - P_tX) \perp H^t_{\{Y\}}$ we have

$$(X - P_tX) \perp \int_0^t h(s)\,\mathrm{d}Y_s, \quad \text{for all } h \in L^2[0,t]$$

that is

$$\begin{aligned} \mathrm{E}\left[X\int_0^t h(s)\,\mathrm{d}Y_s\right] &= \mathrm{E}\left[\int_0^t g(s)\,\mathrm{d}Y_s\right]\left[\int_0^t h(s)\,\mathrm{d}Y_s\right] \\ &= \int_0^t g(s)h(s)\,\mathrm{d}s \end{aligned}$$

Choosing, in particular

$$h(s) = \begin{cases} 1 & \text{if } s \in [0,r] \quad (r \leqslant t) \\ 0 & \text{otherwise} \end{cases}$$

we have

$$\mathrm{E}(XY_r) = \int_0^r g(s)\,\mathrm{d}s$$

which means that $E(XY_r)$ is differentiable and so

$$\boxed{P_tX = \int_0^t \left\{\frac{d}{ds} E[XY_s]\right\} dY_s} \tag{7.3.11}$$

7.3.5 Estimation in Continuous Systems

We shall now consider the theory of estimation for continuous time systems. Since we are concerned generally with multi-dimensional systems we must consider vector processes $\{x_t\}$. The subspace $H^t_{\{x\}}$ of H is then the space spanned by all the components x^i_t of x for any finite collection of times. Moreover, the covariance of an o.i. process $\{x_t\}$ generalizes from (7.3.5) to

$$R(t,s) = \Gamma(t \wedge s), \quad \Gamma = \text{cov}(x_t)$$

where Γ is the covariance matrix of x_t with the ijth element $\text{cov}(x^i_1, x^j_1)$. The isomorphism (7.3.10) also holds in the vector case. If x_t and y_t are given as Wiener integrals (with w_t an o.i. process with covariance $I_n t \wedge s$†)

$$x_t = \int_0^t B(s)\, dw_s$$

$$y_t = \int_0^t C(s)\, dw_s$$

for matrices B, C with $b_{ij}, c_{ij} \in L^2[0,t]$ for all i, j then it is easy to see that

$$Ex_t = Ey_t = 0$$

and

$$\text{cov}(x_t, y_s) = \int_0^{t \wedge s} B(u)\, C^T(u)\, du \tag{7.3.12}$$

Suppose now that we consider a linear system corrupted by white noise in the form

$$\dot{x} = A(t)x_t + C(t)\,\zeta(t)$$

Since the process $\zeta(t)$ is not well defined we integrate this equation and replace $\zeta(t)$ formally by (dw_t/dt) where w_t is a Brownian motion. Then

$$x_t - x_0 = \int_0^t A(s)x_s\, ds + \int_0^t C(s)\, dw_s \tag{7.3.13}$$

---

† I_n is the $n \times n$ unit matrix and $t \wedge s = \min(t, s)$.

We often write this in the 'differential form'

$$\boxed{dx_t = A(t)x_t\,dt + C(t)\,dw_t, \quad x_0 = x} \tag{7.3.14}$$

but this must always be interpreted as in (7.3.13). It is then easy to see that the stochastic differential equation (7.3.14) has the unique solution

$$x_t = \Phi(t,0)x + \int_0^t \Phi(t,s)C(s)\,dw_s \tag{7.3.15}$$

where Φ is the transition matrix corresponding to $A(t)$. (In fact this is true for any stationary o.i. process $\{w_t\}$.) If x is orthogonal to $H_{\{w\}}$, and $m_0 = \mathrm{E}x$, $Q_0 = \mathrm{cov}(x)$, then using (7.3.12) we can show that $m(t) = \mathrm{E}x_t$ and $Q(t) = \mathrm{cov}(x_t)$ satisfy

$$\dot{m} = A(t)m, \quad m(0) = m_0 \tag{7.3.16}$$

and

$$\dot{Q} = A(t)Q + QA^{\mathrm{T}}(t) + CC^{\mathrm{T}}, \quad Q(0) = Q_0 \tag{7.3.17}$$

Suppose now that we have a continuous process $\{z_t\} \in H_0$ and a stationary o.i. process $\{w_t\}$ and let

$$y_t = \int_0^t z_s\,ds + \int_0^t G(s)\,dw_s, \quad \text{(with } G(s)\,G^{\mathrm{T}}(s) \text{ positive definite for all } s\text{)}$$

be an observation. (We are thinking of y_t as being given by

$$\dot{y}_t = z_t + G(t)\zeta_t$$

for a white noise process ζ_t.) Then, if P_t denotes the projection on $H^t_{\{y\}}$ we define

$$\hat{z}_t = P_t z_t, \quad \tilde{z}_t = z_t - \hat{z}_t$$

and the **innovations process** (see Kailath, 1968; Kailath and Frost, 1968)

$$\boxed{\nu_t = y_t - \int_0^t \hat{z}_s\,ds} \tag{7.3.18}$$

It can be shown that the integral on the right is well defined. Then, if $(w_t - w_s) \perp H^s_{\{w,z\}}$ for all $t > s$ and any s it can be shown (Davis, 1977) that ν_t has orthogonal increments

$$\mathrm{cov}\,\nu_t = \int_0^t GG^{\mathrm{T}}\,ds$$

and

$$H^t_{\{y\}} = H^t_{\{\nu\}}, \quad \text{for all } t \tag{7.3.19}$$

i.e. the innovations generate the same subspace of H as $\{y\}$.

We are now in a position to derive the Kalman filter for the system

$$\begin{aligned} \mathrm{d}\boldsymbol{x}_t &= A(t)\boldsymbol{x}_t\,\mathrm{d}t + C(t)\,\mathrm{d}\boldsymbol{v}_t, \quad \boldsymbol{x}_0 = \boldsymbol{x} \\ \mathrm{d}\boldsymbol{y}_t &= H(t)\boldsymbol{x}_t\,\mathrm{d}t + G(t)\,\mathrm{d}\boldsymbol{w}_t, \quad \boldsymbol{y}_0 = \boldsymbol{0} \end{aligned} \tag{7.3.20}$$

where $\{\boldsymbol{v}_t\}$, $\{\boldsymbol{w}_t\}$ are o.i. processes with $H^t_{\{w\}} \perp H^t_{\{v\}}$ and $\boldsymbol{x} \perp H_{\{w,v\}}$. Let

$$\hat{\boldsymbol{x}}_t = P_t\boldsymbol{x}_t$$

and

$$\tilde{\boldsymbol{x}}_t = \boldsymbol{x}_t - \hat{\boldsymbol{x}}_t$$

We shall assume without loss of generality that $\mathrm{E}\boldsymbol{x}_0 = \boldsymbol{0}$. The innovations process is

$$\begin{aligned} \mathrm{d}\boldsymbol{\nu}_t &= \mathrm{d}\boldsymbol{y}_t - \mathrm{H}\hat{\boldsymbol{x}}_t\,\mathrm{d}t \\ &= \mathrm{H}\tilde{\boldsymbol{x}}_t\,\mathrm{d}t + G\,\mathrm{d}\boldsymbol{w}_t \end{aligned} \tag{7.3.21}$$

Note that $H^t_{\{x\}} \perp H^t_{\{w\}}$ so that $\{\boldsymbol{\nu}_t\}$ is an o.i. process with

$$\mathrm{cov}(\boldsymbol{\nu}_t) = \int_0^t GG^{\mathrm{T}}\,\mathrm{d}s$$

Let $D(t) = (GG^{\mathrm{T}})^{-1/2}$ and define

$$\pi_t = \int_0^t D(s)\,\mathrm{d}\boldsymbol{\nu}_s$$

Then by (7.3.12)

$$\begin{aligned} \mathrm{cov}(\boldsymbol{\pi}_t) &= \int_0^t DGG^{\mathrm{T}}D^{\mathrm{T}}\,\mathrm{d}s \\ &= It \end{aligned}$$

and so π_t has stationary orthogonal increments. Now $H^t_{\{\pi\}} = H^t_{\{\nu\}} = H^t_{\{y\}}$ and $\hat{\boldsymbol{x}}_t \in H^t_{\{y\}}$ and so by (7.3.11) we have

$$\hat{\boldsymbol{x}}_t = \int_0^t \frac{\mathrm{d}}{\mathrm{d}s}\mathrm{E}(\boldsymbol{x}_t\,\pi_s^{\mathrm{T}})\,\mathrm{d}\pi_s \tag{7.3.22}$$

It is easy to see that the process

$$\boldsymbol{q}_t = \hat{\boldsymbol{x}}_t - \hat{\boldsymbol{x}}_0 - \int_0^t A\hat{\boldsymbol{x}}_s\,\mathrm{d}s$$

has orthogonal increments with respect to $H^t_{\{y\}} = H^t_{\{\pi\}}$, i.e. $P_s(\boldsymbol{q}_t - \boldsymbol{q}_s) = \boldsymbol{0}$ if $t > s$ and so $\boldsymbol{q}_t$ must be a Wiener integral of $\boldsymbol{\pi}_s$; hence

$$\boldsymbol{q}_t = \int_0^t \Xi(s)\,\mathrm{d}\boldsymbol{\pi}_s$$

for some matrix Ξ with elements $\xi_{ij} \in L^2[0, t]$. Thus,

$$\mathrm{d}\hat{x} = A\hat{x}_t \, \mathrm{d}t + \Xi(t) \, \mathrm{d}\pi_t, \quad \hat{x}_0 = 0 \tag{7.3.23}$$

which has solution

$$\hat{x}_t = \int_0^t \Phi(t, s)\Xi(s) \, \mathrm{d}\pi_s$$

By (7.3.22) it follows that

$$\Xi(t) = \frac{\mathrm{d}}{\mathrm{d}s} E(x_t \pi_s^{\mathrm{T}}) \,|_{s=t} \tag{7.3.24}$$

Now to calculate the error covariance $P(t) = \mathrm{E}(\tilde{x}_t \tilde{x}_t^{\mathrm{T}})$ we note that

$$\pi_s = \int_0^s DH\tilde{x}_u \, \mathrm{d}u + \int_0^s DG \, \mathrm{d}w_u \quad \text{(from (7.3.21))}$$

Since $x_t \perp H^s_{\{w\}}$ we have

$$\mathrm{E}(x_t \pi_s^{\mathrm{T}}) = \int_0^s \mathrm{E}(x_t \tilde{x}_u^{\mathrm{T}}) H^{\mathrm{T}} D^T \, \mathrm{d}u$$

However,

$$x_t = \Phi(t, u)x_u + \int_u^t \Phi(t, \tau)C \, \mathrm{d}v_\tau$$

and so

$$\mathrm{E}(x_t \tilde{x}_u^{\mathrm{T}}) = \Phi(t, u)P(u)$$

since $\mathrm{d}v_\tau \perp H^u_{\{v, w\}}$, $\tau \geqslant u$, and $\hat{x}_u \perp \tilde{x}_u$. Hence

$$\mathrm{E}(x_t \pi_s^{\mathrm{T}}) = \int_0^s \Phi(t, u)P(u)H^{\mathrm{T}}(u)D^{\mathrm{T}}(u) \, \mathrm{d}u$$

and so

$$\begin{aligned}
\Xi(t) &= \frac{\mathrm{d}}{\mathrm{d}s} \mathrm{E}(x_t \pi_s^{\mathrm{T}}) \,|_{s=t} \\
&= \Phi(t, s)P(s)H^T(s)D^{\mathrm{T}}(s) \,|_{s=t} \\
&= P(t)H^{\mathrm{T}}(t)D^{\mathrm{T}}(t)
\end{aligned}$$

Substituting this in (7.3.23) gives

$$\boxed{\begin{aligned}
\mathrm{d}\hat{x}_t &= (A - PH^{\mathrm{T}}(GG^{\mathrm{T}})^{-1}H)\hat{x}_t \, \mathrm{d}t + PH^{\mathrm{T}}(GG^{\mathrm{T}})^{-1} \, \mathrm{d}y_t \\
&= A\hat{x}_t \, \mathrm{d}t + PH^{\mathrm{T}}(GG^{\mathrm{T}})^{-1} \, \mathrm{d}\nu_t
\end{aligned}} \tag{7.3.25}$$

We can also write (7.3.25) in the form

$$\mathrm{d}\hat{x}_t = A\hat{x}_t \, \mathrm{d}t + PH^{\mathrm{T}}D^{\mathrm{T}}D(H\tilde{x}_t \, \mathrm{d}t + G\mathrm{d}w_t)$$

and so, from (7.3.20),

$$\mathrm{d}\tilde{x}_t = (A - PH^{\mathrm{T}}D^{\mathrm{T}}DH)\tilde{x}_t\,\mathrm{d}t + C\,\mathrm{d}v_t - PH^{\mathrm{T}}D^{\mathrm{T}}DG\,\mathrm{d}w_t \tag{7.3.26}$$

Let $\Psi(t,s)$ be the transition matrix which corresponds to $A - PH^{\mathrm{T}}D^{\mathrm{T}}DH$. Then the solution of (7.3.26) is

$$\tilde{x}_t = \int_0^t \Psi(t,s)\,C(s)\,\mathrm{d}v_s + \int_0^t \Psi(t,s)\,PH^{\mathrm{T}}D^{\mathrm{T}}DG\,\mathrm{d}w_s$$

Since the terms on the right are mutually orthogonal it is now easy to show that $P(t)$ satisfies

$$\boxed{\begin{aligned}\dot{P}(t) = C(t)\,C^{\mathrm{T}}(t) - P(t)\,H^{\mathrm{T}}(t)[G(t)\,G^{\mathrm{T}}(t)]^{-1}\,H(t)\,P(t) + A(t)\,P(t)\\ + P(t)\,A^{\mathrm{T}}(t)\end{aligned}} \tag{7.3.27}$$

with $P(0) = 0$. (If $\mathrm{E}x_0 = \boldsymbol{x} \neq \mathbf{0}$, then $P(0) = \mathrm{cov}(\boldsymbol{x})$.) It can be shown that (7.3.27) has a unique solution $P(t)$ which is bounded and non-negative on $[0, t]$.

As in the discrete case we can consider the problems of prediction and smoothing. In the former case it is easy to see that the optimal prediction $\hat{x}_{t|s}$ of x_t given measurements up to time $s(<t)$ is given by

$$\boxed{\hat{x}_{t|s} = \Phi(t,s)\hat{x}_s} \tag{7.3.28}$$

Moreover, with the notation as before, the optimal smoothed estimate $\hat{x}_{t|s}(s > t)$ is given by

$$\boxed{\frac{\mathrm{d}}{\mathrm{d}t}\hat{x}_{t|s} = A(t)\hat{x}_{t|s} + C(t)C^{\mathrm{T}}(t)P^{-1}(t)(\hat{x}_{t|s} - \hat{x}_s)} \tag{7.3.29}$$

with the terminal condition

$$\hat{x}_{s|s} = \hat{x}_s \tag{7.3.30}$$

7.4 LINEAR STOCHASTIC CONTROL†

7.4.1 Complete Observations

Consider now the optimal problem defined by the quadratic cost functional

$$J = \mathrm{E}\left[\int_0^{t_f} (x_t^{\mathrm{T}}\,Qx_t + u_t^{\mathrm{T}}Ru_t)\,\mathrm{d}t + x_{t_f}^{\mathrm{T}}Fx_{t_f}\right] \tag{7.4.1}$$

† See also Astrom (1970) and Astrom and Wittenmark (1984).

subject to the stochastic equation

$$\left.\begin{aligned} \mathrm{d}\boldsymbol{x}_t &= A\boldsymbol{x}_t\,\mathrm{d}t + B\boldsymbol{u}_t + C\,\mathrm{d}\boldsymbol{v}_t \\ \boldsymbol{x}_0 &= \boldsymbol{\xi} \end{aligned}\right\} \tag{7.4.2}$$

where $\boldsymbol{\xi}$ is a normal vector random variable with mean $\boldsymbol{m}_0$, variance P_0 and $\boldsymbol{\xi} \perp H_{\{v\}}$. Following the development of the deterministic control of such a system we shall search for the optimal control in the class of linear feedback controls of the form

$$\boldsymbol{u}(t, \boldsymbol{x}_t) = K(t)\boldsymbol{x}_t \tag{7.4.3}$$

where K is an $n \times m$ matrix with piecewise continuous elements. Substituting (7.4.3) into (7.4.1) and (7.4.2), we obtain

$$J = \mathrm{E}\left[\int_0^{t_f} \boldsymbol{x}_t^{\mathrm{T}} M_K \boldsymbol{x}_t\,\mathrm{d}t + \boldsymbol{x}_{t_f}^{\mathrm{T}} F \boldsymbol{x}_{t_f}\right] \tag{7.4.4}$$

and

$$\mathrm{d}\boldsymbol{x}_t = \tilde{A}\boldsymbol{x}_t\,\mathrm{d}t + C\,\mathrm{d}\boldsymbol{v}_t \tag{7.4.5}$$

where

$$\begin{aligned} M_K(t) &= Q(t) + K^{\mathrm{T}}(t)R(t)K(t) \\ \tilde{A}(t) &= A(t) + B(t)K(t) \end{aligned}$$

As in the deterministic case we introduce the 'value' function

$$V_K(t, \boldsymbol{x}) = \mathrm{E}\left[\int_t^{t_f} \boldsymbol{x}_s^{\mathrm{T}} M_K(s)\boldsymbol{x}_s\,\mathrm{d}s + \boldsymbol{x}_{t_f}^{\mathrm{T}} F \boldsymbol{x}_{t_f}\right]$$

where $\boldsymbol{x}_s$ satisfies (7.4.5) for $s \in [t, t_f]$ and $\boldsymbol{x}_t = \boldsymbol{x}$. Note that, if $\boldsymbol{\eta}$ is any vector random variable, then

$$\begin{aligned} \mathrm{E}(\boldsymbol{\eta}^{\mathrm{T}} M_K(s)\boldsymbol{\eta}) &= \mathrm{tr}\,[M_K(s)Q] \\ &= \mathrm{tr}\,[Q M_K(s)] \end{aligned}$$

where $Q = \mathrm{var}(\boldsymbol{\eta})$. Therefore, since

$$\boldsymbol{x}_s = \tilde{\Phi}(s, t)\boldsymbol{x} + \int_t^s \tilde{\Phi}(s, u)C\,\mathrm{d}\boldsymbol{v}_u$$

it follows that

$$\begin{aligned} V_K(t, \boldsymbol{x}) = {} & \boldsymbol{x}^{\mathrm{T}}\left[\int_t^{t_f} \tilde{\Phi}^{\mathrm{T}}(s, t)\, M_K(s)\tilde{\Phi}(s, t)\,\mathrm{d}s + \tilde{\Phi}^{\mathrm{T}}(t_f, t)\, F\tilde{\Phi}(t_f, t)\right]\boldsymbol{x} \\ & + \int_t^{t_f} \mathrm{tr}\left[M_K(s) \int_t^s \tilde{\Phi}(s, u)C(u)C^{\mathrm{T}}(u)\tilde{\Phi}^{\mathrm{T}}(s, u)\,\mathrm{d}u\right]\mathrm{d}s \\ & + \mathrm{tr}\left[\mathrm{F}\int_t^{t_f} \tilde{\Phi}(t_f, s)CC^{\mathrm{T}}\tilde{\Phi}^{\mathrm{T}}(t_f, s)\,\mathrm{d}s\right] \end{aligned}$$

where $\tilde{\Phi}$ is the transition matrix of $\tilde{A}$. If $x^{\mathrm{T}}N(t)x$ is the first term, then differentiating N easily gives

$$\dot{N} = -M_K - \tilde{A}^{\mathrm{T}}N - N\tilde{A}, \quad \text{with } \tilde{N}(t_f) = F \tag{7.4.6}$$

Simple manipulation of the other terms shows that

$$V_K(t, x) = x^{\mathrm{T}}N(t)x + \int_t^{t_f} \operatorname{tr}(C^{\mathrm{T}}NC)\, \mathrm{d}s \tag{7.4.7}$$

where N satisfies (7.4.6).

It is easy to check that V satisfies the equation

$$L_K V(t, x) = -x^{\mathrm{T}}M_K x$$

where L_K is the differential operator defined by

$$(L_K f)(t, x) = f_t + f_x^{\mathrm{T}}\tilde{A}x + \tfrac{1}{2}\operatorname{tr}\{C^{\mathrm{T}}(t) f_{xx} C(t)\} \tag{7.4.8}$$

(provided, of course, that V is sufficiently differentiable). Hence we have

$$V_K(t, x) = -\mathrm{E}\left[\int_t^{t_f} L_K V_K(s, x_s)\, \mathrm{d}s - x_{t_f}^{\mathrm{T}} F x_{t_f}\right] \tag{7.4.9}$$

L_K can be looked upon as the total derivative of V_K along the trajectories of (7.4.5). However, in contrast to the deterministic differential we have the extra term $\frac{1}{2}\operatorname{tr}\{C^{\mathrm{T}}(t)V_{K,xx}C(t)\}$. (This is a special case of **Ito's differential rule**†.) Moreover, it is clear that (7.4.9) applies to any function f of t and x for which $f(t_f, x) = x^{\mathrm{T}}Fx$ and not just V_K.

Now suppose that there is a piecewise continuous K_0 such that

$$L_K V_{K_0}(t, x) + x^{\mathrm{T}}M_K x \geqslant L_{K_0} V_{K_0}(t, x) + x^{\mathrm{T}}M_{K_0}x (=0) \tag{7.4.10}$$

for any (t, x) and all piecewise continuous K. Then

$$V_{K_0}(t, x) = -\mathrm{E}\left[\int_t^{t_f} (L_K V_{K_0}(s, x_s)\, \mathrm{d}s - x_{t_f}^{T} F x_{t_f}\right]$$

where x_t satisfies (7.4.5) for general K. Then

$$V_{K_0}(t, x) \leqslant -\mathrm{E}\left[\int_t^{t_f} (-x_s^{\mathrm{T}} M_K x_s)\, \mathrm{d}s - x_{t_f}^{\mathrm{T}} F x_{t_f}\right]$$

$$= V_K(t, x)$$

and so any K_0 satisfying (7.4.10) is optimal (for, $\mathrm{E}V_{K_0}(0, \xi) = J(K_0) \leqslant J(K)$). It follows that K_0, $V_{K_0}(t, x)$ satisfy

$$\min_k \,[V_t + \tfrac{1}{2}\operatorname{tr}(C^{\mathrm{T}}V_{xx}C) + V_x(A - BK)x + x^{\mathrm{T}}(Q + K^{\mathrm{T}}RK)x] = 0 \tag{7.4.11}$$

which is just the Bellman equation (chapter 6)

$$V_t + \tfrac{1}{2}\operatorname{tr}(C^{\mathrm{T}}V_{xx}C) + \min_u (V_x f + h) = 0$$

† See section 7.5.1.

with

$$u = Kx$$
$$f = Ax + Bu$$
$$h = x^{\mathrm{T}}Qx + u^{\mathrm{T}}Ru$$

Note the extra term $\frac{1}{2}\mathrm{tr}(C^{\mathrm{T}}V_{xx}C)$. If we try a quadratic solution $V(t, x) = x^{\mathrm{T}}S(t)x + s(t)$ then clearly we must have

$$\boxed{\begin{aligned} \dot{S} + SA + A^{\mathrm{T}}S + Q - S^{\mathrm{T}}BR^{-1}B^{\mathrm{T}}S = 0, \quad S(t_f) = F \\ \dot{s} = -\mathrm{tr}(C^{\mathrm{T}}SC), \quad s(t_f) = 0 \end{aligned}} \tag{7.4.12}$$

and the optimal control is

$$\boxed{u = -R^{-1}B^{\mathrm{T}}S(t)x} \tag{7.4.13}$$

as in the deterministic case. We have therefore shown that with complete observations the control is just the same as if there were no noise in the system.

Note that we have only shown optimality for linear feedback controls. However, by considering general non-linear theory, it can be shown that the control (7.4.13) is also optimal for a much wider class of controls.

7.4.2 Partial Observations

If we do not observe the state x_t directly, but only have the observation y_t where

$$\mathrm{d}y_t = Hx_t + \mathrm{d}t + G\,\mathrm{d}w_t \tag{7.4.14}$$

then the solution of the optimal control problem of minimizing (7.4.1) subject to (7.4.14) and (7.4.2) is not so easy as in the case of complete observations. In fact, it requires a closer examination of stochastic differential equations which is beyond the scope of this book. We shall be content merely to state the results and refer the reader to Davis (1977) and Lipster and Shiryayev (1977) for a complete discussion.

It turns out that the optimal control in the class of linear feedback controls is just

$$\boxed{u_t = -R^{-1}B^{\mathrm{T}}S(t)\hat{x}_t} \tag{7.4.15}$$

where S is as in (7.4.12) and $\hat{x}_t$ is just the filtered estimate of x_t obtained from the Kalman filter. The optimal cost can be shown to be

$$J(u^*) = m_0^T S(0) m_0 + \operatorname{tr}[P(t_f) F] + \int_0^{t_f} \{\operatorname{tr}[(GG^T)^{-1} HPSPH^T (GG^T)^{-1}] + PQ\} \, dt \tag{7.4.16}$$

where P is given by (7.3.27) and $m_0 = E(x_0)$.

The importance of (7.4.15) is that it implies that, for the linear–quadratic stochastic control problem, we may separate the estimation of the state x_t from the evaluation of the feedback control. This is known as the **separation principle**. Note that the equation (7.3.25) for the generation of the estimate should be replaced by

$$\boxed{d\hat{x}_t = A\hat{x}_t \, dt + Bu(t) \, dt + PH^T (GG^T)^{-1} \, d\nu_t} \tag{7.4.17}$$

7.4.3 Infinite Time Problems

As in the deterministic case we may consider infinite time problems in which (7.4.1) is replaced by

$$J = E\left[\int_0^\infty (x_t^T Q x_t + u_t^T R u_t) \, dt\right]$$

Unfortunately, it turns out that $J \to \infty$ as $t_f \to \infty$ with the dynamical constraint (7.4.2). We can obviate this difficulty by using the average cost

$$J = \lim_{t_f \to \infty} E\left[\frac{1}{t_f} \int_0^{t_f} (x_t^T Q x_t + u_t^T R u_t) \, dt\right] \tag{7.4.18}$$

The results are then similar to the deterministic case. In fact, for complete observations it can be shown that the optimal control which minimizes (7.4.18) is

$$\boxed{u(t) = -R^{-1} B^T S x(t)}$$

where $S(t)$ satisfies the algebraic Riccati equation

$$SA + A^T S + Q - SBR^{-1} B^T S = 0$$

provided (A, B) and $(A^T, (Q^T)^{1/2})$ are stabilizable.

In the case of partial observations given by (7.4.14) the optimal control is now

$$\boxed{u(t) = -R^{-1} B^T S z_t}$$

where

$$dz_t = (A - BR^{-1}B^{\mathrm{T}}S)z_t\,dt + P_\infty H^{\mathrm{T}}(GG^{\mathrm{T}})^{-1}(dy_t - Hz_t\,dt)$$

Here,

$$P_\infty = \lim_{t_f \to \infty} P(t)$$

where $P(t)$ is given by (7.3.27) provided (A, C), $(A^{\mathrm{T}}, H^{\mathrm{T}})$ are stabilizable. The extension of the above results of estimation and control theory to distributed systems is discussed by Curtain and Pritchard (1978).

7.5 NON-LINEAR ESTIMATION

7.5.1 Non-linear Systems and Stochastic Equations

In this section we shall give a very brief introduction to non-linear filtering and control. The mathematical apparatus required for a rigorous study of non-linear stochastic systems is beyond the scope of this book and so we shall merely state the main results, leaving the interested reader to consult the literature (see, for example, Fleming and Rishel, 1975; Fujisaki *et al.*, 1972; Gikhman and Skorohod, 1969; Ito, 1951; Jazwinski, 1966, 1970; Kallianpur, 1980).

When we consider non-linear systems with additive noise in the form

$$dx(t) = f(x, t)\,dt + \sigma(x, t)\,dw(t) \tag{7.5.1}$$

where $w(t)$ is an independent increments process, then such an equation is to be interpreted in the integrated form as we saw for linear systems. However, since σ depends on x (which is stochastic, of course), the Wiener integral is no longer sufficiently general; we must write (7.5.1) in the form

$$x(t) = x(t_0) + \int_{t_0}^{t} f(x, s)\,ds + \int_{t_0}^{t} \sigma(x, s)\,dw(s) \tag{7.5.2}$$

where the second integral is called the **Ito integral** and is defined in a similar way to the Wiener integral, except that it has a random integrand. An equation of the form (7.5.1) is called a **stochastic differential equation.**

It can be shown that if $G(x(t), t)$ is differentiable then

$$dG = \left(\frac{\partial G}{\partial x}\right)^{\mathrm{T}} dx(t) + \frac{\partial G}{\partial t}\,dt + \tfrac{1}{2}\,\sigma^{\mathrm{T}}(x, t)H(G)\sigma(x, t)\,dt \tag{7.5.3}$$

where

$$H(G) = \left(\frac{\partial^2 G}{\partial x_i \, \partial x_j}\right)$$

is the Hessian matrix of G, and $\boldsymbol{x}$ satisfies (7.5.1). (We assume that $\operatorname{cov}[\boldsymbol{w}(t)] = I$.) The formula (7.5.3) is called the **Ito differential rule** and generalizes (7.4.8)(see McGarty, 1974).

7.5.2 The System Propagation Equations

Consider again the stochastic process defined by equation (7.5.1) and let $p_{\boldsymbol{x}}(\boldsymbol{u}, t \mid \boldsymbol{x}_0, t_0)(t > t_0)$ denote the probability density function of $\boldsymbol{x}(t)$ given that $\boldsymbol{x}(t_0) = \boldsymbol{x}_0$. Then, since it can be shown that $\boldsymbol{x}(t)$ is a Markov process, we have the **Chapmann–Kolmogorov equation**

$$\boxed{p_{\boldsymbol{x}}(\boldsymbol{u}, t \mid \boldsymbol{x}_0, t_0) = \int p_{\boldsymbol{x}}(\boldsymbol{u}, t \mid \boldsymbol{v}, s)\, p_{\boldsymbol{x}}(\boldsymbol{v}, s \mid \boldsymbol{x}_0, t_0)\, \mathrm{d}\boldsymbol{v}} \tag{7.5.4}$$

where $p_{\boldsymbol{x}}(\boldsymbol{u}, t \mid \boldsymbol{v}, s)$ is the conditional probability density of $\boldsymbol{x}(t)$ given $\boldsymbol{x}(s) = \boldsymbol{v}$. The latter conditional density therefore specifies the transition of the density $p_{\boldsymbol{x}}(\boldsymbol{u}, \tau \mid \boldsymbol{x}_0, t_0)$ from one time $\tau = s$ to another $\tau = t$ and corresponds to the transition matrix $\Phi(t, t_0)$ for linear deterministic systems.

Therefore, if we can obtain an equation for the density function $p_{\boldsymbol{x}}(\boldsymbol{u}, t \mid \boldsymbol{x}_0, t_0)$ of $\boldsymbol{x}$ we shall have a complete description of the process. In the simplified case where $\boldsymbol{x}$ satisfies

$$\mathrm{d}\boldsymbol{x}(t) = \boldsymbol{f}(\boldsymbol{x}, t)\, \mathrm{d}t + \mathrm{d}\boldsymbol{w}(t)$$

where $\boldsymbol{w}$ is normal Brownian motion with $\mathrm{E}(\boldsymbol{w}\boldsymbol{w}^{\mathrm{T}}) = Qt$ then we have that $p_{\boldsymbol{x}}(\boldsymbol{u}, t \mid \boldsymbol{x}(s))$ (arbitrary s) satisfies the Fokker–Planck equation

$$\boxed{\frac{\partial p}{\partial t} = L^+ p} \tag{7.5.5}$$

where L^+ is the (forward) Fokker–Planck operator defined by

$$L^+ = -\frac{\partial}{\partial \boldsymbol{u}}(\boldsymbol{f}\cdot) + \frac{1}{2}\sum_{i=1}^{n}\sum_{j=1}^{n} Q_{ij}\frac{\partial^2(\cdot)}{\partial u_i\, \partial u_j} \tag{7.5.6}$$

For example consider the system

$$\dot{x}(t) = -ax(t) + \zeta(t)$$

where $\int \zeta(t)$ is standard Brownian motion ($Q = 1$). Then the Fokker–Planck equation is

$$\frac{\partial p}{\partial t} = -\frac{\partial(-aup)}{\partial u} + \frac{\partial^2 p}{\partial u^2} \quad [p = p_x(u, t \mid x_0, t_0)] \tag{7.5.7}$$

If $x(t_0)$ is deterministic then $p_x(u, t_0) = \delta[u - x(t_0)]$ and it is easily seen that (7.5.7) has the solution

$$p_x(u, t \mid x(t_0)) = \frac{1}{(2\pi\sigma^2)^{1/2}} \exp\left[-\frac{1}{2}\frac{(u-\bar{x})^2}{\sigma^2}\right] \tag{7.5.8}$$

where

$$\bar{x} = x(t_0)\exp[-a(t - t_0)]$$
$$\sigma^2 = 1 - \exp[-2a(t - t_0)]$$

Of course, this is a linear example and the application of the equation (7.5.5) is much more difficult for non-linear systems.

In the case where we have a measurement process

$$\mathrm{d}y(t) = \boldsymbol{h}(\boldsymbol{x}, t)\,\mathrm{d}t + \mathrm{d}\boldsymbol{w}(t) \tag{7.5.9}$$

in addition to the dynamics (7.5.1) it can be shown that the conditional density of $\boldsymbol{x}$ given the measurements, $p_x(u, t \mid y(t))$, satisfies the equation

$$\boxed{\frac{\partial p}{\partial t} = L^+ p + \left\{\frac{\mathrm{d}y}{\mathrm{d}t} - \mathrm{E}[\boldsymbol{h}(\boldsymbol{x}, t)]\right\}^{\mathrm{T}} R^{-1}(t)\left\{\boldsymbol{h}(\boldsymbol{x}, t) - \mathrm{E}[\boldsymbol{h}(\boldsymbol{x}, t)]\right\} p} \tag{7.5.10}$$

where

$$\mathrm{E}[\boldsymbol{h}(\boldsymbol{x}, t)] = \int \boldsymbol{h}(u, t)\, p_x(u, t \mid y(t))\,\mathrm{d}u$$

This is the **Kushner–Stratonovich equation** and its application to many real systems presents formidable numerical difficulties.

7.5.3 Linearization

Since the general Kushner–Stratonovich equation is so difficult to solve we often linearize the system and measurement equations. This involves a simple application of the Taylor series (assuming that the functions concerned have a convergent Taylor series expansion). Depending on the number of terms which are retained we obtain approximations of various degrees of accuracy. We shall mention here only the first-order approximation which leads to the so-called extended Kalman filter. Consider then the equation

$$\dot{\boldsymbol{x}}(t) = \boldsymbol{f}(\boldsymbol{x}, t) + \boldsymbol{\zeta}(t)$$

together with the measurement

$$\mathrm{d}y(t) = \boldsymbol{h}(\boldsymbol{x}, t)\,\mathrm{d}t + \mathrm{d}\boldsymbol{w}(t)$$

and suppose that $\mathrm{E}[\boldsymbol{w}(t)\,\boldsymbol{w}^{\mathrm{T}}(t)] = R(t)$. Expand $\boldsymbol{f}$ and $\boldsymbol{h}$ in Taylor series to first order. Then

$$\boldsymbol{f}(\boldsymbol{x}, t) \approx \boldsymbol{f}(\boldsymbol{x}_1, t) + A(\boldsymbol{x}_1, t)(\boldsymbol{x} - \boldsymbol{x}_1)$$

and

$$\boldsymbol{h}(\boldsymbol{x}, t) \approx \boldsymbol{h}(\boldsymbol{x}_1, t) + C(\boldsymbol{x}_1, t)(\boldsymbol{x} - \boldsymbol{x}_1)$$

for some nominal trajectory $\boldsymbol{x}_1(t)$. Then if

$$\boldsymbol{r}(t) \triangleq \boldsymbol{f}(\boldsymbol{x}_1, t) - A(\boldsymbol{x}_1, t)\,\boldsymbol{x}_1(t)$$
$$\boldsymbol{s}(t) \triangleq \boldsymbol{h}(\boldsymbol{x}_1, t) - C(\boldsymbol{x}_1, t)\,\boldsymbol{x}_1(t)$$

the equations become

$$\frac{\mathrm{d}\boldsymbol{x}}{\mathrm{d}t} = A(\boldsymbol{x}_1, t)\,\boldsymbol{x}(t) + \boldsymbol{r}(t) + \boldsymbol{\zeta}(t)$$
$$\mathrm{d}y(t) = C(\boldsymbol{x}_1, t)\,\boldsymbol{x}(t)\,\mathrm{d}t + \boldsymbol{s}(t)\,\mathrm{d}t + \mathrm{d}\boldsymbol{w}(t)$$

These equations are now linear and so we may apply the Kalman filter equations to obtain the estimation equations

$$\boxed{\begin{aligned} \frac{\mathrm{d}\hat{\boldsymbol{x}}(t)}{\mathrm{d}t} &= A(\boldsymbol{x}_1, t)\hat{\boldsymbol{x}}(t) + \boldsymbol{r}(t) + P(t)\,C^{\mathrm{T}}(\boldsymbol{x}_1, t)\,R^{-1}(t) \\ &\qquad \times [\boldsymbol{y}(t) - \boldsymbol{s}(t) - C(\boldsymbol{x}_1, t)\hat{\boldsymbol{x}}(t)] \\ \frac{\mathrm{d}P}{\mathrm{d}t}(t) &= A(\boldsymbol{x}_1, t)\,P(t) + P(t)\,A^{\mathrm{T}}(\boldsymbol{x}_1, t) + Q(t) \\ &\qquad - P(t)\,C^{\mathrm{T}}(\boldsymbol{x}_1, t)\,R^{-1}(t)\,C(\boldsymbol{x}_1, t)\,P(t) \end{aligned}} \tag{7.5.11}$$

The only problem remaining is to choose $\boldsymbol{x}_1(t)$. We could use the Fokker–Planck equation (7.5.5) to find an a priori estimate of the mean or we could actually let $\boldsymbol{x}_1(t) = \hat{\boldsymbol{x}}(t)$, thus obtaining the equations (7.5.11) with $\boldsymbol{x}_1$ replaced by $\hat{\boldsymbol{x}}(t)$. The first equation then becomes a non-linear equation in $\hat{\boldsymbol{x}}(t)$, and the two equations are also coupled, so that the equation for P cannot be solved independently of the first equation. The linearized estimation equations (7.5.11) with $\boldsymbol{x}_1 = \hat{\boldsymbol{x}}$ constitute the **extended Kalman filter.**

7.6 COMPUTATIONAL ASPECTS OF LINEAR FILTERING

7.6.1 Information Filter

We shall now return to the discrete Kalman filter (7.2.26) and discuss some aspects of the numerical implementation of this filter (see also Kaminski *et al.*, 1971). The first thing to notice is that the covariance matrix $P(k)$ is theoretically positive semidefinite; however, numerical inaccuracies due to finite word length may give numerical solutions to (7.2.26) which are not positive semidefinite and which lead to numerical instability. This happens in particular when certain (linear combinations) of states are observable with little noise, while others are virtually unobservable.

The standard form of the filter in (7.2.26) is also called the covariance filter since it propagates the amount of error in the estimation. If it is used when the initial state estimate is very poor we must take $P(0)$ to be large, although there is no general indication of just how large it should be. We can obviate this difficulty by writing the equations in terms of $P^{-1}(k)$ which is then the 'information content' of the estimate. Before giving these equations we first note that the filter (7.2.26) assumes that we can calculate the estimate $\hat{x}(k)$ as soon as the measurement $y(k)$ becomes available. In fact, when computing the solution numerically it must take a finite time to evaluate the estimate when the observation appears. For short sampling periods it is likely that the computation of $\hat{x}(k)$ will take up most of the time interval from $k-1$ to k and so we can only base our estimate on the observation $y(k-1)$, i.e. we must start computing $\hat{x}(k)$ before $y(k)$ becomes available. For this reason, in order to obtain the estimate at time k we actually compute the estimate at time $k-1$ and then use the one-step prediction of $\hat{x}(k)$ which is just $A(k-1)\,\hat{x}(k-1)$. (Of course, if very fast sampling is necessary and it is impossible to perform the computation in one time interval we must use longer prediction intervals.) Hence, replacing $A(k-1)\,\hat{x}(k-1)$ by $\hat{x}(k)$ (which we should write, more correctly, as $\hat{x}(k \mid k-1)$) we obtain the numerically implementable version of (7.2.26)

$$
\begin{aligned}
&\hat{x}(k+1) = A(k)\,\hat{x}_+(k)\\
&P_1(k+1) = A(k)\,P(k)\,A^{\mathrm{T}}(k) + Q(k)\\
&\text{where}\\
&\qquad \hat{x}_+(k) = \hat{x}(k) + K(k)[y(k) - C(k)\,\hat{x}(k)]\\
&\qquad P(k) = [I - K(k)\,C(k)]\,P_1(k)\\
&\qquad K(k) = P_1(k)\,C^{\mathrm{T}}(k)[C(k)\,P_1(k)\,C^{\mathrm{T}}(k) + R(k)]^{-1}
\end{aligned}
\tag{7.6.1}
$$

It is then easy to see that if we write $d(k) = P_1^{-1}(k)\hat{x}(k)$ and

$d_+(k) = P^{-1}(k)\hat{x}_+(k)$, we obtain the information filter

$$
\begin{aligned}
d(k+1) &= [I - L(k)]\,(A^{\mathrm{T}})^{-1}(k)\, d_+(k) \\
P_1^{-1}(k+1) &= [I - L(k)]\, F(k) \\
F(k) &= (A^{\mathrm{T}})^{-1}(k) P^{-1}(k) A^{-1}(k) \\
L(k) &= F(k)[Q^{-1}(k) + F(k)]^{-1} \\
\text{where} & \\
d_+(k) &= d(k) + C^{\mathrm{T}}(k)\, R^{-1}(k)\, y(k) \\
P^{-1}(k) &= P_1^{-1}(k) + C^{\mathrm{T}}(k)\, R^{-1}(k)\, C(k)
\end{aligned}
\tag{7.6.2}
$$

However, it should be noted that we must assume that the system matrix $A(k)$ is invertible. In applying (7.5.2) when the initial estimate is unknown we can simply set $P_1^{-1}(0) = 0$ and continue until P_1^{-1} is invertible, at which time we can switch to (7.6.1).

7.6.2 Square Root Filtering

Square root filtering is based on the fact that for any positive semidefinite matrix X we may write $X = \Xi\Xi^{\mathrm{T}}$ where Ξ is lower triangular. This is called the Cholesky decomposition of X and although a recursive computation of X may lead to a non-positive definite matrix, the product $\Xi\Xi^{\mathrm{T}}$ is always positive semidefinite, regardless of the numerical errors. Hence if we work entirely with Ξ we do not have the numerical ill-conditioning mentioned above, and the effective precision of the calculation can be shown to be doubled.

We shall consider the square root algorithm in the simple case of the covariance filter with scalar measurements. (For more general cases and the information square root algorithm see Kaminski *et al.* (1971).) We introduce the Cholesky decompositions

$$P_1 = S_1 S_1^{\mathrm{T}}, \quad P = SS^{\mathrm{T}}, \quad R = VV^{\mathrm{T}}$$

Then the fourth equation in (7.6.1) may be written as

$$P = SS^{\mathrm{T}} = P_1 - KCP_1 = S_1[I - aEE^{\mathrm{T}}]S_1^{\mathrm{T}}$$

where

$$E = S_1^{\mathrm{T}} C^{\mathrm{T}}$$

$$\frac{1}{a} = E^{\mathrm{T}}E + Q$$

Then writing

$$[I - aEE^{\mathrm{T}}] = [I - a\gamma EE^{\mathrm{T}}]\,[I - a\gamma EE^{\mathrm{T}}] \tag{7.6.3}$$

with

$$\gamma = \frac{1}{1 + (aQ)^{1/2}}$$

we obtain the relation

$$S = S_1 - a\gamma S_1 E E^{\mathrm{T}}$$

between the square roots of P and P_1. Hence we may replace (7.6.1) in the scalar measurement case by the square root algorithm

$$\boxed{\begin{aligned} \hat{x}_+(k) &= \hat{x}(k) + K[y(k) - c(k)\hat{x}(k)] \\ S(k) &= S_1(k) - \gamma \boldsymbol{K}\boldsymbol{E}^{\mathrm{T}} \\ \boldsymbol{K} &= aS_1(k)\boldsymbol{E} \\ \boldsymbol{E} &= aS_1^{\mathrm{T}}(k)\boldsymbol{c}^{\mathrm{T}}(k) \\ \frac{1}{a} &= \boldsymbol{E}^{\mathrm{T}}\boldsymbol{E} + Q(k) \\ &= \frac{1}{1 + [aQ(k)]^{1/2}} \end{aligned}} \tag{7.6.4}$$

7.7 SYSTEM IDENTIFICATION

7.7.1 Cross-correlation

In the final sections of this chapter we shall give a short introduction to linear system identification. The purpose of identification is to obtain a model of a given physical plant without resorting to extensive mathematical modelling, which may be unnecessary. In fact, if the system is linear to a good approximation, then an easy way to determine the frequency response is to connect a sweep oscillator to the input and measure the output amplitude and phase shift as functions of frequency. However, in many cases of process control, removing the plant from operation to inject sinusoidal signals is undesirable or even impossible and so other methods must be used.

A simple method is to inject an additive white noise signal into the input of the plant as shown in fig. 7.1. Then the output measurement $y(t)$ is given by

$$y(t) = \int_{-\infty}^{\infty} h(s)[x(t-s) + u(t-s)]\, \mathrm{d}s + \eta(t)$$

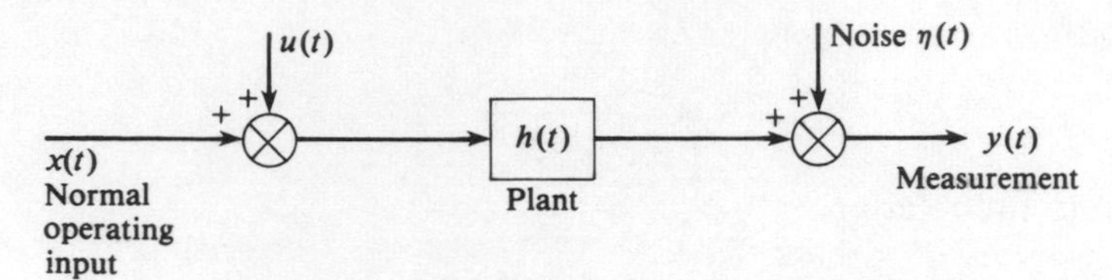

Fig. 7.1 A practical input-output system.

and so

$$\mathrm{E}[u(t)\, y(t+\tau)] = \mathrm{E}\left\{u(t) \int_{-\infty}^{\infty} h(s)[x(t+\tau-s) + u(t+\tau-s)\,\mathrm{d}s] + \mathrm{E}[u(t)\,\eta(t+\tau)]\right\}$$

For any two random processes $v(t)$, $w(t)$ we define the cross-correlation ϕ_{vw} by

$$\boxed{\phi_{vw}(\tau) = \int_{-\infty}^{\infty} v(t)\, w(t+\tau)\,\mathrm{d}t} \tag{7.7.1}$$

or by

$$\boxed{\phi_{vw}(\tau) = \lim_{T\to\infty} \frac{1}{2T} \int_{-T}^{T} v(t)\, w(t+\tau)\,\mathrm{d}t} \tag{7.7.2}$$

if the integral in (7.7.1) does not exist. Now the random process $v(t, \omega)$ is called **ergodic** if

(a) $\mathrm{E}[v(t, \omega)]$ is constant in t

(b) $\mathrm{E}[v(t, \omega)] = \int_{-\infty}^{\infty} v(t, \omega)\,\mathrm{d}t$

$$\left(\text{or } \mathrm{E}[v(t, \omega)] = \lim_{T\to\infty} \frac{1}{2T} \int_{-T}^{T} v(t, \omega)\,\mathrm{d}t\right)$$

If v and w are ergodic it follows from (7.7.1) (or (7.7.2)) that

$$\phi_{vw}(\tau) = \mathrm{E}[v(t)\, w(t+\tau)]$$

Hence, returning to the system above, we have

$$\phi_{uy} = \int_{-\infty}^{\infty} h(s)[\phi_{ux}(\tau-s) + \phi_{uu}(\tau-s)]\,\mathrm{d}s + \phi_{u\eta}(\tau)$$

if all the processes are ergodic. If, moreover, u, x and u, η are uncorrelated we have

$$\boxed{\phi_{uy} = \int_{-\infty}^{\infty} h(s)\phi_{uu}(\tau - s)\,\mathrm{d}s + \phi_{u\eta}(\tau)} \tag{7.7.3}$$

This is the **Wiener–Hopf equation** and since u was taken to be white noise we have

$$\phi_{uu}(\tau - s) = k\,\delta(\tau - s) \tag{7.7.4}$$

and so from (7.7.3) it follows that

$$\phi_{uy}(\tau) = kh(\tau) \tag{7.7.5}$$

Hence to determine the impulse response h of a plant we inject white noise at the input and cross-correlate the input and output. The constant k should be chosen so that the noise drives all the modes of the plant in the linear region.

7.7.2 Pseudo-random Binary Sequences

The above cross-correlation method assumes that a perfect white noise signal is available. However, we have seen earlier that white noise is a fictitious derivative of Brownian motion which does not exist. Hence we must find an approximation to white noise which has autocorrelation close to the delta function as in (7.7.4). Such an approximation is provided by pseudo-random binary sequences (PRBS) which can be generated by shift registers of length n in which an exclusive OR of the mth and nth bits of the register is fed back to the input, as in fig. 7.2. The shift register should never contain all zeros and so the maximum number of distinct binary numbers of length n (the length of the shift register) which can be generated is $N = 2^n - 1$. A given feedback configuration may lead to fewer than N distinct numbers and so it is important to design the system to give the maximum number

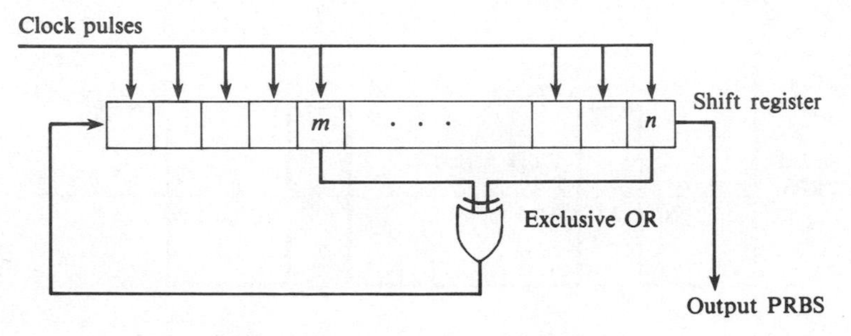

Fig. 7.2 Generating a PRBS.

N. We then say that the output is a **maximum period** PRBS. This can be achieved if m and n are chosen so that $1+x^m+x^n$ is an irreducible polynomial in the Galois field $\{0, 1\}$. (For a number of specific designs, see Davies (1970).)

Having generated a PRBS signal we usually scale it so that the '1' outputs have value a while the '0' outputs have value $-a$ (fig. 7.3). It can be shown that the autocorrelation function of such a scaled PRBS is

$$\phi_{xx}(\tau) = a^2\left[1 - \frac{|\tau|(N+1)}{N\Delta t}\right], \quad 0 \leqslant |\tau| \leqslant \Delta t$$

$$= -\frac{a^2}{N} \qquad\qquad , \quad \tau \geqslant \Delta t$$

Note that ϕ_{xx} approaches a δ function (roughly speaking) as $N \to \infty$ ($\Delta t \to 0$). Next we define the spectral density of a signal as the Fourier transform of its autocorrelation function, i.e.

$$S_{xx}(\omega) = \int_{-\infty}^{\infty} \phi_{xx}(\tau) \exp(-j\omega\tau)\, d\tau$$

For large N, the graph of $S_{xx}(\omega)$ approaches that shown in fig. 7.4. If the process spectrum has a form similar to that in fig. 7.4. then (between 0 and $2\pi/\Delta t$) the PRBS signal will resemble white noise quite closely.

Now returning to the Wiener–Hopf equation we have

$$\phi_{uy}(\tau) = \int_0^T h(s)\phi_{uu}(\tau - s)\, ds$$

and since

$$\phi_{uu}(\tau) \approx a^2 \frac{(N+1)}{N} \Delta t\, \delta(\tau) - \frac{a^2}{N}$$

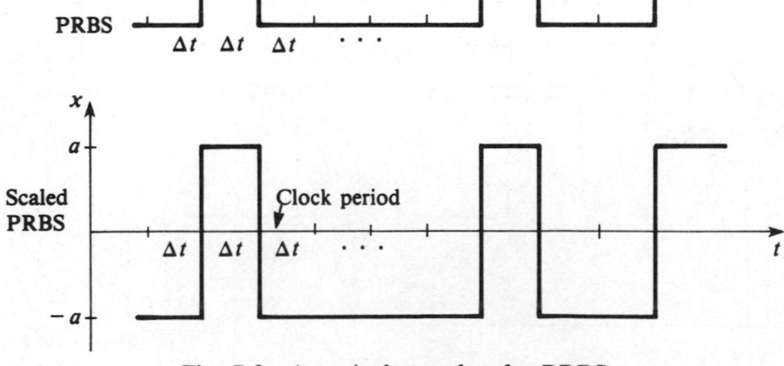

Fig. 7.3 A typical sample of a PRBS.

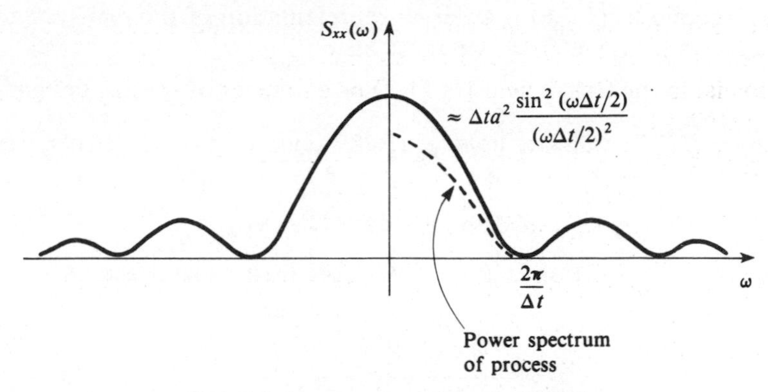

Fig. 7.4 Power spectrum of a PRBS.

we have

$$\phi_{uy}(\tau) = \frac{a^2(N+1)\,\Delta t h(\tau)}{N} - \frac{a^2}{N}\int_0^T h(s)\,\mathrm{d}s$$

For large τ the latter term is a d.c. bias which is small if N is large. The desired impulse response $h(t)$ is then found as before.

7.7.3 Parameter Identification

Another aproach to system identification is to assume that the system can be modelled by a given set of equations with unknown parameters, which are determined from a finite set of input–output measurements. Suppose, therefore, that we have a discrete system modelled by the equations

$$y(k) = \sum_{i=1}^{n} a_i y(k-i) + \sum_{j=1}^{n} b_j u(k-j) \tag{7.7.6}$$

and suppose that we obtain the (noise free) measurements $y(0), y(1), \ldots, y(N)$ from the real system when we input the sequence $u(0), u(1), \ldots, u(N) (N > n)$. Define the vectors

$$\theta(a_1, \ldots, a_n, b_1, \ldots, b_n)^{\mathrm{T}}$$
$$Y(N) = [y(n), \ldots, y(N)]^{\mathrm{T}}$$

and the matrix

$$\Psi(N) = \begin{bmatrix} y(n-1) & y(n-2) & \ldots & y(0) & u(n-1) & \ldots & u(0) \\ y(n) & y(n-1) & \ldots & y(1) & u(n) & \ldots & u(1) \\ y(n+1) & y(n) & \ldots & y(2) & u(n+1) & \ldots & u(2) \\ \vdots & \vdots & & \vdots & \vdots & & \vdots \\ y(N-1) & y(N-2) & \ldots & y(N-n) & u(N-1) & \ldots & u(N-n) \end{bmatrix}$$

Then if the model (7.7.6) is an exact representation of the real system for a given set of parameters $\boldsymbol{\theta}$, we must have

$$Y = \Psi\boldsymbol{\theta} \tag{7.7.7}$$

However, (7.7.7) will not hold generally and so for a particular set of measurements we choose the parameters $\boldsymbol{\theta}$ to minimize

$$J(\boldsymbol{\theta}) = (Y - \Psi\boldsymbol{\theta})^{\mathrm{T}}(Y - \Psi\boldsymbol{\theta}) \tag{7.7.8}$$

Minimizing this function in $\boldsymbol{\theta}$ we obtain the **least squares estimate**

$$\boxed{\hat{\boldsymbol{\theta}} = (\Psi^{\mathrm{T}}\Psi)^{-1}\Psi^{\mathrm{T}}Y} \tag{7.7.9}$$

provided the inverse matrix exists. We may call the system **identifiable** if there exists a set of inputs so that $(\Psi^{\mathrm{T}}\Psi)$ is invertible. If it is not invertible we can resort to a generalized inverse (see chapter 8), but will then not obtain a unique solution for $\hat{\boldsymbol{\theta}}$. We can also put a weighting matrix in (7.7.8), i.e.

$$J(\boldsymbol{\theta}) = (Y - \Psi\boldsymbol{\theta})^{\mathrm{T}}W(Y - \Psi\boldsymbol{\theta})$$

where W is usually diagonal and may be chosen to weight later measurements more than earlier ones. Then (7.7.9) becomes

$$\boxed{\hat{\boldsymbol{\theta}} = (\Psi^{\mathrm{T}}W\Psi)^{-1}\Psi^{\mathrm{T}}WY} \tag{7.7.10}$$

Returning to (7.7.9), as we have seen with linear estimation it is usually better to have a recursive algorithm. This may be obtained quite easily as follows. Suppose we have processed the $N+1$ measurements in $Y(N)$ as above and that we then make another observation $y(N+1)$. Then we may write

$$\Psi^{\mathrm{T}}(N+1)\Psi(N+1) = \Psi^{\mathrm{T}}(N)\Psi(N) + \psi(N+1)\psi^{\mathrm{T}}(N+1) \tag{7.7.11}$$

where

$$\psi(N+1) = [y(N)\ldots y(N-n+1)u(N)\ldots u(N-n+1)]^{\mathrm{T}}$$

Hence, if $P(N+1) = \{\Psi^{\mathrm{T}}(N+1)\Psi(N+1)\}^{-1}$ we have

$$P(N+1) = [P^{-1}(N) + \psi(N+1)\psi^{\mathrm{T}}(N+1]^{-1}$$

Using the simple formula (the **matrix inversion formula**)

$$(A + BCD)^{-1} = A^{-1} - A^{-1}B(C^{-1} + DA^{-1}B)^{-1}DA^{-1} \tag{7.7.12}$$

we have

$$\boxed{\begin{aligned}P(N+1) = P(N) - P(N)\psi(N+1)[1 + \psi^{\mathrm{T}}(N+1)P(N)\psi(N+1)]^{-1}\\ \psi^{\mathrm{T}}(N+1)P(N)\end{aligned}}$$

Also

$$\boldsymbol{\Psi}^{\mathrm{T}}(N+1)\boldsymbol{Y}(N+1) = \boldsymbol{\Psi}^{\mathrm{T}}(N)\boldsymbol{Y}(N) + \boldsymbol{\psi}(N+1)y(N+1)$$

Now by (7.79) we have

$$\hat{\boldsymbol{\theta}}(N+1) = \{P(N) - P(N)\boldsymbol{\psi}(N+1)[1 + \boldsymbol{\psi}^{\mathrm{T}}(N+1)P(N)\boldsymbol{\psi}(N+1)]^{-1} \boldsymbol{\psi}^{\mathrm{T}}(N+1)P(N)\} \times [\boldsymbol{\Psi}^{\mathrm{T}}(N)\boldsymbol{Y}(N) + \boldsymbol{\psi}(N+1)y(N+1)]$$

and so

$$\boxed{\hat{\boldsymbol{\theta}}(N+1) = \hat{\boldsymbol{\theta}}(N) + \boldsymbol{Q}(N+1)[y(N+1) - \boldsymbol{\psi}^{\mathrm{T}}\hat{\boldsymbol{\theta}}(N)]} \tag{7.7.13}$$

where

$$\boxed{\boldsymbol{Q}(N+1) = P(N)\boldsymbol{\psi}(N+1)[1 + \boldsymbol{\psi}^{\mathrm{T}}(N+1)P(N)\boldsymbol{\psi}(N+1)]^{-1}}$$

Note that in the recursive formulation we do not have to invert a matrix. A more general result can also be derived from (7.7.10) with the weighting W.

Suppose now that the measurements Y are corrupted by noise, i.e.

$$y(k) = \sum_{i=1}^{n} a_i y(k-i) + \sum_{j=1}^{n} b_j u(k-j) + v(k) \tag{7.7.14}$$

or

$$\boldsymbol{Y}(N) = \boldsymbol{\Psi}(N)\boldsymbol{\theta}^0 + \boldsymbol{v}(N)$$

where $\boldsymbol{\theta}^0$ is the true set of parameters and each element of $\boldsymbol{v}(N) = [v(n), v(n+1), \ldots, v(N)]^{\mathrm{T}}$ is assumed to be normal with mean zero and variance σ^2 (i.e. $N(0, \sigma^2 I)$). We could again use least squares to determine $\boldsymbol{\theta}$, but it turns out that the estimate is often biased in the sense that $\mathrm{E}(\hat{\boldsymbol{\theta}}) \neq \boldsymbol{\theta}^0$. We can then use other methods such as the maximum likelihood technique. Since $\boldsymbol{v}(N)$ is $N(0, \sigma^2 I)$ and $\boldsymbol{v}(N) = \boldsymbol{Y}(N) - \boldsymbol{\Psi}(N)\boldsymbol{\theta}^0$ we may write

$$f(\boldsymbol{Y} \mid \boldsymbol{\theta}^0) \triangleq (2\pi\sigma^2)^{-m/2} \exp\left[-\frac{1}{2}\frac{(\boldsymbol{Y} - \boldsymbol{\Psi}\boldsymbol{\theta}^0)^{\mathrm{T}}(\boldsymbol{Y} - \boldsymbol{\Psi}\boldsymbol{\theta}^0)}{\sigma^2}\right] \tag{7.7.15}$$

where $m = N - n + 1$, for any parameter set $\boldsymbol{\theta}$; f is the probability density of $\boldsymbol{Y}$ given $\boldsymbol{\theta}^0$. Replacing $\boldsymbol{\theta}^0$ by a general $\boldsymbol{\theta}$ in (7.7.15), we call $f(\boldsymbol{Y} \mid \boldsymbol{\theta})$ the **likelihood function.** We then seek to maximize f with respect to $\boldsymbol{\theta}$. This is equivalent to maximizing the log of f and so we define

$$-l(\boldsymbol{Y} \mid \boldsymbol{\theta}) = -\frac{m}{2}\log 2\pi - \frac{m}{2}\log \sigma^2 - \frac{1}{2}\frac{(\boldsymbol{Y} - \boldsymbol{\Psi}\boldsymbol{\theta})^{\mathrm{T}}(\boldsymbol{Y} - \boldsymbol{\Psi}\boldsymbol{\theta})}{\sigma^2}$$

Maximizing with respect to $\boldsymbol{\theta}$ and σ^2 gives

$$\boxed{\begin{aligned}\hat{\boldsymbol{\theta}} &= (\Psi^{\mathrm{T}}\Psi)^{-1}\Psi^{\mathrm{T}}Y \\ \sigma^2 &= \frac{1}{m}(Y - \Psi\hat{\boldsymbol{\theta}})^{\mathrm{T}}(Y - \Psi\hat{\boldsymbol{\theta}})\end{aligned}} \tag{7.7.16}$$

and so $\hat{\boldsymbol{\theta}}$ is just as in the least squares solution.

Consider finally the more general model

$$y(k) = \sum_{i=1}^{n} a_i y(k-i) + \sum_{j=1}^{n} b_j u(k-j) + \sum_{l=1}^{n} c_l v(k-l) + v(k) \tag{7.7.17}$$

Then the vector $\boldsymbol{v}(N)$ has covariance $R = \sigma^2 I_m$. Writing

$$\boldsymbol{\theta} = (a_1, \ldots, a_n, b_1, \ldots, b_n, c_1, \ldots, c_n)^{\mathrm{T}}$$

we can define $l(\boldsymbol{Y} \mid \boldsymbol{\theta})$ as before and $\hat{\sigma}^2$ is given by the same expression as before. However, we cannot obtain a closed form solution for $\hat{\boldsymbol{\theta}}$ and numerical techniques must be used. For a survey of such optimization techniques see Mousavi-Khalkhali (1983). The identification of non-linear systems is discussed in the survey paper by Billings (1980), while the relative merits of various identification algorithms are considered by Soderstrom *et al.* (1978).

7.8 EXERCISES

1. If three samples of a linearly increasing signal are taken at time intervals $t = 1, 2, 3$ in the presence of independent additive noise, estimate the slope x of the signal. Assume the noise samples are uncorrelated with variance σ_v^2.

2. Consider the scalar linear system

$$\dot{x}(t) = -x(t) + a + \zeta_1(t)$$
$$\dot{y}(t) = x(t) + \zeta_2(t)$$

where ζ_1 and ζ_2 are white noise processes and a is a constant, unknown forcing parameter. Derive a (continuous or discrete) Kalman filter to estimate a. (Hint: let $x_1 = x, x_2 = a, \dot{x}_2 = 0$.)

3. Write down the optimal one-step predictor for a in exercise 2.

4. Again in exercise 2, write down the equations of the mean and covariance of $(x(t), a)$ by using (7.3.16) and (7.3.17).

5. Suppose that we control the linear acceleration of a particle by an input function corrupted by white noise, i.e.

$$\ddot{x} = u + \zeta_1(t)$$

and we measure $x(t) + \zeta_2(t)$. Write down the optimal control which minimizes the cost

$$\mathrm{E}\left\{x^2(t_f) + \int_0^{t_f} [x^2(t) + u^2(t)]\ \mathrm{d}t\right\}$$

6. Verify equation (7.5.8).

7. Write down the extended Kalman filter for the system

$$\begin{aligned} \dot{x}_1 &= x_1 x_2 + \zeta_1(t) \\ \dot{x}_2 &= -x_2 + \zeta_2(t) \\ \dot{y} &= x_1 + \zeta_3(t) \end{aligned}$$

where ζ_1, ζ_2 and ζ_3 are independent.

8. Determine an algorithm to find the lower–upper triangular factorization of a matrix A, i.e. $A = LU$.

9. Evaluate the PRBS output from the system in fig. 7.5. Is it maximum length?

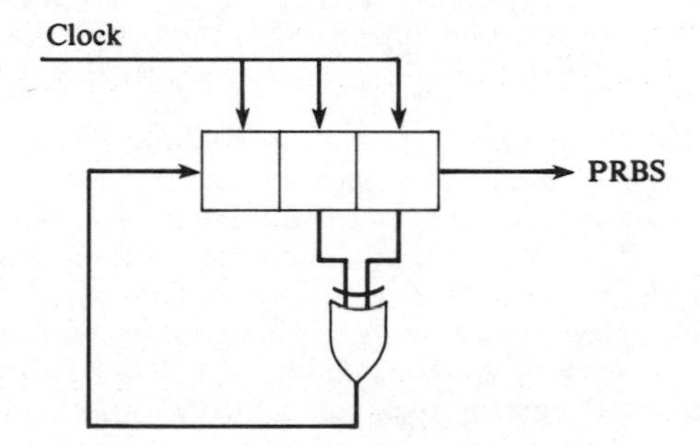

Fig. 7.5 **A simple PRBS generator.**

10. Determine the power spectral density function $S_{xx}(\omega)$ for a PRBS.

11. Prove the matrix inversion lemma (7.7.12).

12. Describe how one would implement the algorithm (7.7.13).

13. Given the measurements

k	1	2	3	4
$u(k)$	1	1	-1	-1
$y(k)$	5	4·5	4·55	-5·455

from a linear system $y(k) = b_1 y(k-1) + a_l u(k-1)$, evaluate the least square's estimate of the parameters a_1, b_1.

REFERENCES

Anderson, B. D. O. and Moore, J. B. (1979) *Optimal Control*, Prentice-Hall.

Astrom, K. (1970) *Introduciton to Stochastic Control Theory*, Academic Press.

Astrom, K. and Wittenmark, B. (1984) *Computer Controlled Systems*, Prentice-Hall.

Billings, S. A. (1980) Identification of nonlinear systems – a survey, *IEE Proc.*, 122, pt. D. 272–85.

Box, G. E. P. and Jenkins, G. M. (1970) *Time Series Analysis*, Holden Day.

Curtain, R. F. and Pritchard, A. J. (1978) *Infinite Dimensional Linear Systems Theory*, Springer-Verlag.

Davies, W. D. T. (1970) *System Identification for Self-adaptive Control*, Wiley Interscience.

Davis, M. H. A. (1977) *Linear Estimation and Stochastic Control*, Chapman and Hall.

Doob, J. L. (1953) *Stochastic Processes*, Wiley.

Fleming, W. H. and Richel, R. W. (1975) *Deterministic and Stochastic Optimal Control*, Springer-Verlag.

Fujisaki, M., Kallianpur, G. and Kunita, H. (1972) Stochastic differential equations for the nonlinear filtering problems, *Osaka J. Maths.*, 9, 19–40.

Gikhman, I. I. and Skorohod, A. V. (1969) The Theory of Stochastic Processes, Springer-Verlag.

Ito, K. (1951) On stochastic differential equations, *Mem. Am. Math. Soc.*, 4, 1–51.

Jazwinski, A. H. (1966) Filtering for nonlinear systems, *IEEE Trans. Aut. Control*, 11, 765–6.

Jazwinski, A. H. (1970) *Stochastic Processes and Filtering Theory*, Academic Press.

Kailath, T. (1968) An innovations approach to least squares estimation – Part I: linear filtering in additive white noise, *IEEE Trans. Aut. Control*, 13, 646–54.

Kailath, T. and Frost, P. (1968) An innovations approach to least squares estimation – Part II, *IEEE Trans. Aut. Control*, 13, 655–61.

Kallianpur, G. (1980) *Stochastic Filtering Theory*, Springer-Verlag.

Kaminski, P., Bryson, A. E. Jr. and Schmidt, S. F. (1971) Discrete square root filtering: a survey of current techniques, *IEEE Trans. Aut. Control*, 16, 727–36.

Lipster, R. S. and Shiryayev, A. N. (1977) *Statistics of Random Processes*, Parts I and II, Springer-Verlag.

McGarty, T. P. (1974) *Stochastic Systems and State Estimation*, Wiley.

Meditch, J. S. (1969) *Stochastic Optimal Linear Estimation and Control*, McGraw-Hill.

Mousavi-Khalkhali, S. A. (1983) *Applications of Conjugate-Gradient Methods to the Minimum Energy Control of Electric Vehicles*, Ph.D. Thesis, Sheffield University.

Soderstrom, T., Ljung, L. and Gustavsson, I. (1978) A theoretical analysis of recursive identification methods, *Automatica*, 14, 231–44.

8 ADAPTIVE CONTROL, SELF-TUNING CONTROL AND VARIABLE-STRUCTURE SYSTEMS

8.1 INTRODUCTION

In the previous chapters on theoretical control design methods, we have implicitly assumed a perfect knowledge of the system structure and parameters. By 'system structure' we mean the type of differential equations which model the real physical plant. These equations may be ordinary (linear or non-linear) differential equations, delay equations, partial differential equations, etc. Having chosen the 'correct' form of equation to represent the system we must then consider the correct choice of parameters. In the case of a linear continuous system this means choosing some or all of the elements of the 'A' matrix in the model

$$\dot{x}_m = A_m x_m + B_m u_m$$

(We shall frequently distinguish model and plant variables by using the subscripts m and p respectively.) If the matrix A_m is partially (or totally) unknown then we must design a controller which has variable parameters which can be changed as functions of the elements of A_m. Hence if we use a parameter identification algorithm (recursive least squares, for example, as discussed in chapter 7) to estimate the parameters, we can then feed these estimated values into the controller to 'tune' the control input for the particular parameter set obtained. We then obtain the so-called self-tuning controllers.

Alternatively we can fix the model parameters A_m and alter the plant control parameters B_p so that we 'force' the real system to follow the model as closely as possible. This leads to model-reference adaptive control where variations in the plant parameters A_p are adapted to by changes in the plant control parameters B_p which keeps the plant output close to the model output.

We shall first consider model-reference adaptive control in this chapter and then the more modern developments in self-tuning control. In both these methods, however, the basic dynamic structure of the model is fixed, with only the parameters changing to compensate for plant parameter variations. There seems to have been little work done on the 'best' choice of model structure, but the important case of variable structure systems in which the control law is chosen deliberately to give the system a particularly desirable structure has been extensively studied and we shall discuss the main results in the last section of this chapter.

8.2 ADAPTIVE CONTROL

8.2.1 Lyapunov Design of Model Reference Adaptive Controllers

The basic idea of model reference adaptive control (MRAC) is to choose the control $\boldsymbol{u}'$ in the real system

$$\dot{\boldsymbol{x}}_p = A_p\boldsymbol{x}_p + B_p\boldsymbol{u}' \quad (\boldsymbol{x}_p \in \mathbb{R}^n, \boldsymbol{u}' \in \mathbb{R}^m) \tag{8.2.1}$$

to be of the form $\boldsymbol{u}' = Q(\boldsymbol{u} + F\boldsymbol{x}_p)$ so that the resulting system

$$\dot{\boldsymbol{x}}_p = (A_p + B_pQF)\boldsymbol{x}_p + (B_pQ)\boldsymbol{u} \tag{8.2.2}$$

'follows' the desired model system

$$\dot{\boldsymbol{x}}_m = A_m\boldsymbol{x}_m + B_m\boldsymbol{u} \tag{8.2.3}$$

as closely as possible (see Narendra and Kudva, 1974). By 'following the model as closely as possible' we mean that the error vector $\boldsymbol{e} \triangleq \boldsymbol{x}_m - \boldsymbol{x}_p$ should tend to zero asymptotically. The structure of the basic MRAC system is shown in fig. 8.1. As can be seen, the object is to vary the parameters of the control matrices Q and F so that

$$\lim_{t\to\infty} \| \boldsymbol{e}(t) \| = 0$$

The simplest assumption which can be made in order that the adaptive property of asymptotically decaying error is possible is that the controls in the plant can be chosen so that the resulting closed-loop system exactly matches the model. In this case we can clearly find matrices Q^* and F^* so that

$$\left.\begin{aligned} B_pQ^* &= B_m \\ A_p + B_mF^* &= A_m \end{aligned}\right\} \tag{8.2.4}$$

Under the assumption of exact matching, the error vector satisfies the equation

$$\begin{aligned} \dot{\boldsymbol{e}} &= A_m\boldsymbol{e} + (A_m - A_p - B_pQF)\boldsymbol{x}_p + (B_m - B_pQ)\boldsymbol{u} \\ &= A_m\boldsymbol{e} + B_m\Phi\boldsymbol{x}_p + B_m\Psi Q(\boldsymbol{u} + F\boldsymbol{x}_p) \end{aligned} \tag{8.2.5}$$

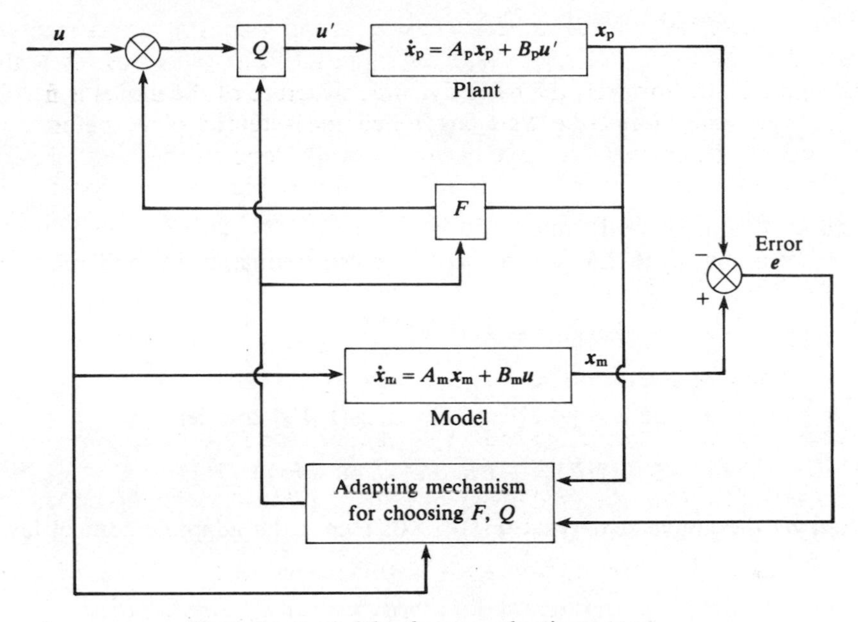

Fig. 8.1 A model-reference adaptive system.

where

$$\Phi \triangleq F^* - F(t)$$
$$\Psi \triangleq [Q^{-1}(t) - Q^{*-1}]$$

(Note that we are assuming that B_p and B_m are of full rank and so Q^* is non-singular.)

To demonstrate the asymptotic stability of (8.2.5) by a proper choice of Φ and Ψ consider first the system

$$\left.\begin{aligned} \dot{e} &= Ae + D\Xi z(t) \\ \dot{\Xi} &= -\Gamma D^{\mathrm{T}} P e z^{\mathrm{T}}(t) \end{aligned}\right\} \tag{8.2.6}$$

where $z(t)$ is a bounded vector-valued function, $\Gamma = \Gamma^{\mathrm{T}} > 0$, D is an $n \times m$ matrix of full rank and P is a positive definite matrix satisfying Lyapunov's equation

$$A^{\mathrm{T}}P + PA = -R, \quad R = R^{\mathrm{T}} > 0$$

for some matrix R. This system is asymptotically stable (i.e. $\| e(t) \| \to 0$ as $t \to \infty$) as can be seen in the following way. Let

$$V = \tfrac{1}{2}[e^{\mathrm{T}}Pe + \mathrm{tr}(\Xi^{\mathrm{T}}\Gamma^{-1}\Xi)]$$

Then V is positive definite and

$$\begin{aligned}\dot{V} &= \tfrac{1}{2}e^{\mathrm{T}}Re + e^{\mathrm{T}}PD\Xi z + \mathrm{tr}(\dot{\Xi}\Gamma^{-1}\Xi) \\ &= -\tfrac{1}{2}e^{\mathrm{T}}Re + \mathrm{tr}[(D^{\mathrm{T}}Pez^{\mathrm{T}} + \Gamma^{-1}\dot{\Xi})^{\mathrm{T}}\Xi] \\ &= -\tfrac{1}{2}e^{\mathrm{T}}Re \end{aligned} \tag{8.2.7}$$

and so $\dot{V}$ is negative definite.

Returning to (8.2.5) we can write the error equation in the form

$$\dot{e} = A_{\mathrm{m}}e + B_{\mathrm{m}}[\Phi\Psi]\begin{bmatrix} x_{\mathrm{p}} \\ Q(u + Fx_{\mathrm{p}}) \end{bmatrix}$$

and so if we define $\Xi = [\Phi\Psi]$ and $\Gamma = \mathrm{diag}[\Gamma_1\Gamma_2]$ and let

$$\dot{\Xi} = [\dot{\Phi}\dot{\Psi}] = -\Gamma B_{\mathrm{m}}^{\mathrm{T}}Pe[x_{\mathrm{p}}^{\mathrm{T}}\ (u + Fx_{\mathrm{p}})^{\mathrm{T}}Q^{\mathrm{T}}]$$

then by the above remarks, $\|e(t)\| \to 0$. Hence the adaptive control law takes the form

$$\boxed{\begin{aligned} e &= x_{\mathrm{m}} - x_{\mathrm{p}} \\ \dot{F} &= \Gamma_1 B_{\mathrm{m}}^{\mathrm{T}}Pex_{\mathrm{p}}^{\mathrm{T}} \\ \dot{Q} &= Q\Gamma_2 B_{\mathrm{m}}^{\mathrm{T}}Pe(u + Fx_{\mathrm{p}})^{\mathrm{T}}Q^{\mathrm{T}}Q \end{aligned}} \tag{8.2.8}$$

for some Γ_1, $\Gamma_2 > 0$, P satisfying $A_{\mathrm{m}}^{\mathrm{T}}P + PA_{\mathrm{m}} = -R$. The adapting mechanism is shown in fig. 8.2.

Note that we have only considered the asymptotic stability of $e(t)$ in the equation (8.2.6). The global asymptotic stability of Ξ in (8.2.6) (and hence of F and Q in (8.2.8)) does not follow from (8.2.7).

8.2.2 Identification of Parameters Using MRAC

We can use the ideas of MRAC to identify unknown parameters in a plant

$$\dot{x}_{\mathrm{p}} = A_{\mathrm{p}}x_{\mathrm{p}} + B_{\mathrm{p}}u$$

In this case the roles of plant and model in section 8.2.1 are reversed and we choose a model of the form

$$\dot{x}_{\mathrm{m}} = Cx_{\mathrm{m}} + [A_{\mathrm{m}}(t) - C]x_{\mathrm{p}} + B_{\mathrm{m}}(t)u$$

where C is a stable matrix and A_{m}, B_{m} have adjustable elements. If we can

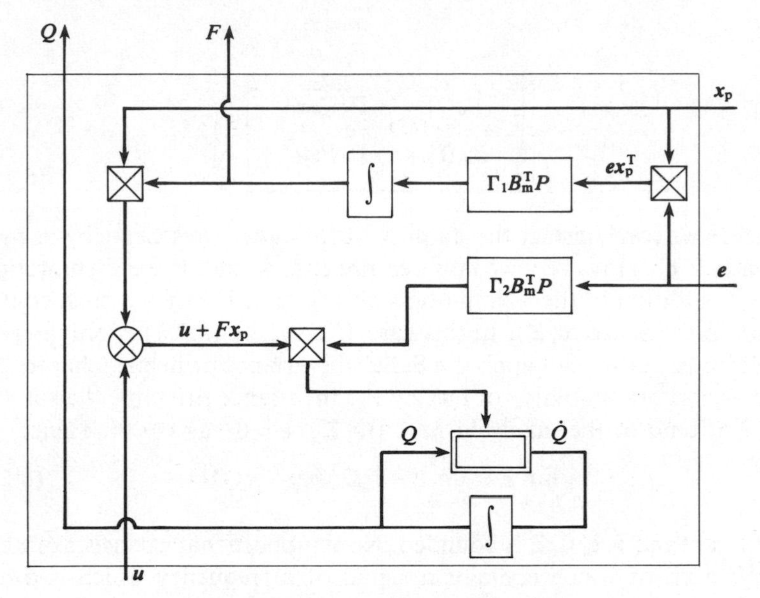

Fig. 8.2 The adapting mechanism

design a system for which

$$\lim_{t \to \infty} A_m(t) = A_p$$

$$\lim_{t \to \infty} B_m(t) = B_p$$

$$\lim_{t \to \infty} [\boldsymbol{x}_m(t) - \boldsymbol{x}_p(t)] = \lim_{t \to \infty} \boldsymbol{e}(t) = \boldsymbol{0}$$

then the model parameters will asymptotically approach those of the plant thus providing plant parameter estimates. Now, as before, we have

$$\dot{\boldsymbol{e}}(t) = C\boldsymbol{e}(t) + \Phi \boldsymbol{x}_p + \Psi \boldsymbol{u}$$

where

$$\Phi = A_m - A_p$$

$$\Psi = B_m - B_p$$

In just the same way as above we see that the adaptive laws are

$$\begin{aligned} \dot{\Phi} &= -\Gamma_1 P \boldsymbol{e} \boldsymbol{x}_p^T \\ \dot{\Psi} &= -\Gamma_2 P \boldsymbol{e} \boldsymbol{u}^T \end{aligned} \tag{8.2.9}$$

where P is positive definite and satisfies

$$\begin{aligned} C^T P + PC = -Q, \quad Q = Q^T > 0 \\ \Gamma_1 = \Gamma_1^T > 0, \qquad \Gamma_2 = \Gamma_2^T > 0 \end{aligned} \tag{8.2.10}$$

Hence

$$\boxed{\begin{aligned} \dot{A}_{\mathrm{m}}(t) &= -\Gamma_1 P e x_{\mathrm{p}}^{\mathrm{T}} \\ \dot{B}_{\mathrm{m}}(t) &= -\Gamma_2 P e u^{\mathrm{T}} \end{aligned}}$$

As before we can consider the stability of this adaptive scheme by using the system (8.2.6). However, we now require that Φ and Ψ are asymptotically stable in addition to the asymptotic stability of e. Hence we must consider (8.2.6) on the space (e, Ξ). In this case $\dot{V}$ given by (8.2.7) is only negative semidefinite, but we can apply La Salle's invariance principle (chapter 5) to prove asymptotic stability. In fact by the invariance principle the solutions of (8.2.6) tend to the manifold $M = \{(e, \Xi) : e = \mathbf{0}\}$ as $t \to \infty$. Hence

$$\lim_{t \to \infty} \dot{\Xi} = \lim_{t \to \infty} [-\Gamma D^{\mathrm{T}} P e z^{\mathrm{T}}] = 0 \tag{8.2.12}$$

Since $V > 0$ and $\dot{V} \leqslant 0$, Ξ is bounded. Now suppose that z consists of r components each of which contains a signal of a frequency which is distinct from that of the others. Then, since Ξ is bounded and (8.2.6) is a differential equation with periodic coefficients, (8.2.12) implies that $\Phi(t)$ tends to a constant matrix as $t \to \infty$. On M we have

$$D\Xi z \equiv 0 \quad (e \equiv 0)$$

and since D is invertible it follows that

$$\sum_{i=1}^{r} \xi_i z_i(t) \equiv 0$$

where ξ_i are the columns of Ξ. Since the z_i contain distinct frequency components they are linearly independent and so $\xi_i \equiv 0$ and $\Xi \equiv 0$. Asymptotic stability now follows from the invariance principle. Now to apply this result to the system (8.2.11) we have $z^{\mathrm{T}} = (x_{\mathrm{p}}^{\mathrm{T}}\ u^{\mathrm{T}})$ and so even if u is periodic then x_{p} is not necessarily periodic. However, if the plant is uniformly asymptotically stable then x_{p} is of the form $x_{\mathrm{p}} = \tilde{x}_{\mathrm{p}} + \varepsilon(t)$ where x_{p} is periodic and $\varepsilon \to 0$ as $t \to \infty$. It can then be shown (Narendra and Kudva, 1974) that if u contains $q \geqslant (n+1)/2$ distinct frequencies then $\Xi[x_{\mathrm{p}}^{\mathrm{T}}\ u] = 0$ implies $\Xi = 0$ and we again obtain asymptotic stability. Hence we can say that the system can be adaptively identified if the input excites sufficiently many modes in the system.

8.2.3 MRAC by Hyperstability

We can also design adaptive systems by using the theory of hyperstability outlined in chapter 5; in fact, following Landau and Courtiol (1974) we consider the basic system

$$\dot{x}_m = A_m x_m + B_m u_m \tag{8.2.13}$$

$$\dot{x}_p = A_p x_p + B_p u_p \tag{8.2.14}$$

consisting of a model with state $x_m \in \mathbb{R}^n$ and a plant with state $x_p \in \mathbb{R}^n$ and we determine the plant input in the form

$$u_p = -K_p x_p + K_m x_m + K_u u_m \tag{8.2.15}$$

for some variable matrices K_p, K_m, K_u to be specified by the design. As in section 8.2.1 we assume perfect model matching. Then, using (8.2.13)–(8.2.15) we have

$$0 = [A_m - A_p + B_p(K_p - K_m)]x_p + (B_m - B_p K_u)u_m$$

if $x_m = x_p$. Since this must be true for all x_p, u_m we must have

$$A_m - A_p + B_p(K_p - K_m) = 0 \tag{8.2.16}$$

$$B_m - B_p K_u = 0 \tag{8.2.17}$$

Hence for perfect model matching we must be able to solve these equations for $(K_p - K_m)$ and K_u. Consider (8.2.17) and define the Moore–Penrose pseudo-inverse $B_p^\dagger$ of B_p by (see Boulljon, 1971)

$$B_p^\dagger = (B_p^T B_p)^{-1} B_p^T$$

assuming $(B_p^T B_p)^{-1}$ exists. Then (8.2.17) implies that $K_u = B_p^\dagger B_m$. However, K_u obtained in this way may not necessarily satisfy (8.2.17). It will only do so if $(I - B_p B_p^\dagger)B_m = 0$. A similar argument applies to (8.2.16) and so sufficient conditions for exact model matching are

$$\boxed{\begin{aligned}(I - B_p B_p^\dagger)B_m &= 0\\ (I - B_p B_p^\dagger)(A_m - A_p) &= 0\end{aligned}} \tag{8.2.18}$$

Note that exact model matching is equivalent to knowing the structure of the plant dynamics (i.e. the orders of the differential operators involved). If conditions (8.2.18) do not hold, Curran (1971) shows how to modify the model to satisfy these equations.

Now suppose that in (8.2.15), $K_p = K_p(t, e)$ and $K_u = K_u(t, e)$ are chosen to depend on the error $e \triangleq x_m - x_p$ (and time). For small deviations from some nominal values K_p and K_u we may write

$$K_p(t, e) = K_p - \Delta K_p(t, e)$$

$$K_u(t, e) = K_u + \Delta K_u(t, e)$$

Then the plant input may be written

$$u_p = u_{p1} + u_{p2}$$

where

$$u_{p1} = -K_p x_p + K_m x_m + K_u u_m$$

$$u_{p2} = \Delta K_p(t, e) x_p + \Delta K_u(t, e) u_m$$

Now from (8.2.13) and (8.2.14) we obtain

$$\begin{aligned}\dot{e} &= A_m x_m - A_p x_p + B_m u_m - B_p u_p \\ &= A_m x_m - A_p x_p + A_m x_p - A_m x_p + B_m u_m - B_p(-K_p x_p + K_m x_m + K_u u_m) \\ &\quad - B_p[\Delta K_p(t, e) x_p + \Delta K_u(t, e) u_m] \\ &= A_m e + (A_m - A_p) x_p + B_m u_m - B_p K_u u_m - B_p \Delta K_u(t, e) u_m \\ &\quad + B_p K_p x_p - B_p \Delta K_p(t, e) x_p - B_p K_m x_p + B_p K_m x_p - B_p K_m x_m \\ &= (A_m - B_p K_m) e + B_p[B_p^{\dagger}(A_m - A_p) - K_m + K_p - \Delta K_p(t, e)] x_p \\ &\quad + B_p[B_p^{\dagger} B_m - K_u - \Delta K_u(t, e)] u_m\end{aligned}$$

(using (8.2.18)). We can therefore write the error equation in the form

$$\boxed{\dot{e} = (A_m - B_p K_m) e + B_p w_1} \tag{8.2.19}$$

where

$$\begin{aligned}-w_1 = {} & [\Delta K_p(t, e) - B_p^{\dagger}(A_m - A_p) + K_m - K_p] x_p \\ & + [\Delta K_u(t, e) - B_p^{\dagger} B_m + K_u] u_m\end{aligned}$$

We can also introduce the transformed error v given by

$$\boxed{v = De} \tag{8.2.20}$$

Comparing the results of chapter 5 (section 5.4) we therefore see that to apply hyperstability we require that

$$\int_0^{t_1} v^T(t) w(t)\, dt \geqslant -\gamma_0^2 \quad \text{for all } t_1 \geqslant 0 \tag{8.2.21}$$

for some constant γ_0^2 where $w = -w_1$. It is easy to see that (8.2.21) will hold if we choose

$$\Delta K_p(t, v) = \int_0^t \tilde{L} v (Q x_p)^T\, d\tau + L v (Q x_p)^T + \Delta K_p(0) \tag{8.2.22}$$

$$\Delta K_u(t, v) = \int_0^t \tilde{M} v (R u_m)^T\, d\tau + M v (R u_m)^T + \Delta K_u(0) \tag{8.2.23}$$

where $\tilde{L}$, Q, $\tilde{M}$ and R are positive definite matrices, and L and M are non-negative definite matrices.

Consider now the general linear system

$$\dot{x} = Ax + Bu$$
$$y = Cx$$

where (A, B) is a controllable pair and (C, A) is an observable pair. Then it can be shown (Anderson, 1967) that the transfer function $C(sI - A)^{-1}B$ of this system is positive real (see chapter 5) if and only if for any positive definite matrix H there exists a positive definite matrix P such that

$$PA + A^{\mathrm{T}}P = -H$$
$$B^{\mathrm{T}}P = C$$

Applying this to the system (8.2.20) and (8.2.21) and recalling that, from Popov's theorem (theorem 5.4.9), a system is hyperstable if and only if the linear part is positive real, we see that D should be chosen according to the prescription

$$\boxed{D = B_{\mathrm{p}}^{\mathrm{T}}P} \tag{8.2.24}$$

where P is the solution of the Lyapunov equation

$$\boxed{(A_{\mathrm{m}} - B_{\mathrm{p}}K_{\mathrm{m}})^{\mathrm{T}}P + P(A_{\mathrm{m}} - B_{\mathrm{p}}K_{\mathrm{m}}) = -H} \tag{8.2.25}$$

for any positive definite symmetric H. (Note that $A_{\mathrm{m}} - B_{\mathrm{p}}K_{\mathrm{m}}$ must be a stable matrix.)

It is clear that the adaptive controller (8.2.22) and (8.2.23) is of the 'proportional + integral' kind and the complete MRAC system (or **model following system** as it is called) is shown in fig. 8.3. Moreover, it should be noted that, provided (8.2.24) and (8.2.25) are satisfied, any choice of ΔK_{p}, ΔK_{u} for which (8.2.21) holds will guarantee stability. Another such choice is the 'relay + integral' control specified by the equations

$$\Delta K_{\mathrm{p}}(t, v) = \int_0^t \tilde{L}v(Qx_{\mathrm{p}})^{\mathrm{T}}\,\mathrm{d}\tau + \operatorname{sgn} v(Qx_{\mathrm{p}})^{\mathrm{T}} + \Delta K_{\mathrm{p}}(0) \tag{8.2.26}$$

$$\Delta K_{\mathrm{u}}(t, v) = \int_0^t \tilde{M}v(Ru_{\mathrm{m}})^{\mathrm{T}}\,\mathrm{d}\tau + \operatorname{sgn} v(Ru_{\mathrm{m}})^{\mathrm{T}} + \Delta K_{\mathrm{u}}(0) \tag{8.2.27}$$

where

$$\operatorname{sgn} v = (\operatorname{sgn} v_1, \operatorname{sgn} v_2, \ldots, \operatorname{sgn} v_{\mathrm{m}})$$

We can summarize the steps necessary to design the MRAC system as follows:

(a) Determine $A_{\mathrm{m}}, B_{\mathrm{m}}$ so that (8.2.18) are valid.

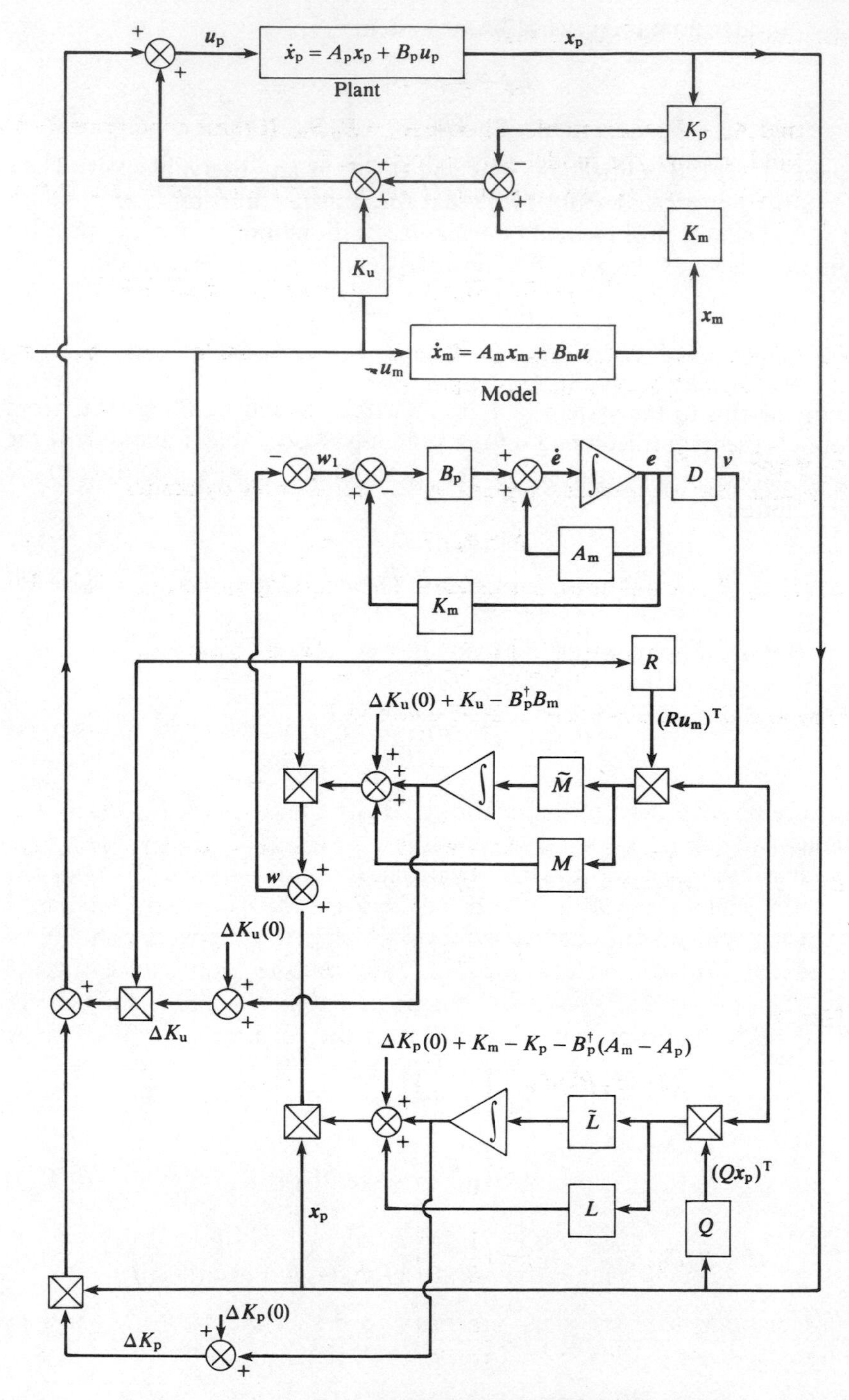

Fig. 8.3 An adaptive controller using hyperstability theory.

(b) Choose K_m and K_p so that

$$K_p - K_m = B_p^\dagger(A_p - A_m)$$

and $A_m - B_p K_m$ is stable. Choose $K_u = B_p B_m$. If these conditions do not hold, change the model parameters A_m, B_m.

(c) Choose $\Delta K_p(0)$, $\Delta K_u(0)$ and H (arbitrarily) and solve (8.2.25). Of course,

$$P = \int_0^\infty [\exp(A_m - B_p K_m)^T t] H [\exp(A_m - B_p K_m) t] \, dt$$

(d) Check that $((A_m - B_p K_m), B_p)$ is a controllable pair and that $(D, A_m - B_p K_m)$ is an observable pair.

Example 8.2.1

Suppose that we consider a plant with second-order dynamics

$$\ddot{x} + a_p \dot{x} + b_p x = u_p$$

where a_p, b_p are unknown parameters. Then putting $x_1 = x$, $x_2 = \dot{x}$, we have

$$\dot{x}_p = \frac{d}{dt}\begin{pmatrix} x_1 \\ x_2 \end{pmatrix} = \begin{pmatrix} 0 & 1 \\ -b_p & -a_p \end{pmatrix} x_p + \begin{pmatrix} 0 \\ 1 \end{pmatrix} u_p$$

One would then obviously choose the model

$$\dot{x}_m = \begin{pmatrix} 0 & 1 \\ -b_m & -a_m \end{pmatrix} x_m + \begin{pmatrix} 0 \\ 1 \end{pmatrix} u_m$$

where a_m and b_m are to be determined by conditions (a)–(d) above. Since $B_p = (0 \quad 1)^T$ we have $B_p^\dagger = (0 \quad 1)$ and so

$$I - B_p B_p^\dagger = \begin{pmatrix} 1 & 0 \\ 0 & 0 \end{pmatrix}$$

Hence

$$\begin{aligned}(I - B_p B_p^\dagger) B_m &= \begin{pmatrix} 1 & 0 \\ 0 & 0 \end{pmatrix}\begin{pmatrix} 0 \\ 1 \end{pmatrix} \\ &= \begin{pmatrix} 0 \\ 0 \end{pmatrix}\end{aligned}$$

$$\begin{aligned}(I - B_p B_p^\dagger)(A_m - A_p) &= \begin{pmatrix} 1 & 0 \\ 0 & 0 \end{pmatrix}\begin{pmatrix} 0 & 0 \\ -b_m + b_p & -a_m + a_p \end{pmatrix} \\ &= \begin{pmatrix} 0 & 0 \\ 0 & 0 \end{pmatrix}\end{aligned}$$

and so condition (8.2.18) is satisfied and exact model matching is possible (as, of course, is intuitively obvious). Now

$$K_u = (0 \quad 1)\begin{pmatrix}0\\1\end{pmatrix} = 1$$

and

$$K_p - K_m = (0 \quad 1)\begin{pmatrix}0 & 0\\ -b_p + b_m & -a_p + a_m\end{pmatrix}$$

$$= (-b_p + b_m \quad -a_p + a_m)$$

Therefore it is natural to choose $K_p = (-b_p \quad -a_p)$, $K_m = (-b_m \quad -a_m)$. Also

$$A_m - B_p K_m = \begin{pmatrix}0 & 1\\ -b_m & -a_m\end{pmatrix} - \begin{pmatrix}0\\1\end{pmatrix}(-b_m \quad -a_m)$$

$$= \begin{pmatrix}0 & 1\\0 & 0\end{pmatrix}$$

However, such a choice of K_p and K_m will not produce a stable matrix $A_m - B_p K_m$ and so we modify K_p and K_m so that

$$K_m = (-b_m - \beta, \quad -a_m - \alpha)$$

$$K_p = (-b_p - \beta, \quad -a_p - \alpha)$$

Then

$$\tilde{A} \triangleq A_m - B_p K_m$$

$$= \begin{pmatrix}0 & 1\\ \beta & \alpha\end{pmatrix}$$

This is a stable matrix for any $\alpha, \beta < 0$. Moreover

$$(B_p \tilde{A}^T B_p) = \begin{pmatrix}0 & \beta\\ 1 & \alpha\end{pmatrix}$$

and so $(\tilde{A}, B_p)$ is a controllable pair for any $\beta \neq 0$. Finally, if for simplicity we let $H = I$, and $\alpha = -3$, $\beta = -2$, then

$$P = \int_0^\infty \exp(\tilde{A}^T t)\exp(\tilde{A} t)\, dt$$

where

$$\exp(\tilde{A} t) = Q\begin{pmatrix}\exp(\lambda_1 t) & 0\\ 0 & \exp(\lambda_2 t)\end{pmatrix}Q^{-1}$$

and

$$\lambda_1 = \frac{\alpha + (\alpha^2 + 4\beta)^{1/2}}{2} = -1$$
$$\lambda_2 = \frac{\alpha - (\alpha^2 + 4\beta)^{1/2}}{2} = -2$$

with

$$Q = \begin{pmatrix} \beta/\lambda_1 & \beta/\lambda_2 \\ 1 & 1 \end{pmatrix}$$

$$Q^{-1} = \begin{pmatrix} \dfrac{\lambda_1\lambda_2}{\beta(\lambda_2 - \lambda_1)} & \dfrac{-\lambda_1}{\lambda_2 - \lambda_1} \\[2ex] \dfrac{-\lambda_1\lambda_2}{\beta(\lambda_2 - \lambda_1)} & \dfrac{\lambda_2}{\lambda_2 - \lambda_1} \end{pmatrix}$$

(see section 8.5, exercise 4). A simple calculation shows that

$$P = \begin{pmatrix} 1 & -\frac{1}{2} \\ -\frac{1}{2} & \frac{1}{2} \end{pmatrix}$$

and then $D = (-\frac{1}{2} \;\; \frac{1}{2})$. $(\tilde{A}^{\mathrm{T}}, D^{\mathrm{T}})$ is a controllable pair and so we have completed the design.

Note finally that a comparison between Lyapunov and hyperstability design of MRAC systems has been given by Narendra and Valavani (1980) and it is found that, for uniformly bounded plant signals, both methods yield exactly the same results. This does not imply, however, that one method will not be more convenient than another in any particular application.

8.2.4 Adaptive Observation

In the above MRAC designs we have assumed that all the plant states $\boldsymbol{x}_{\mathrm{p}}$ are available for the adaptive controller. This will not be the case when we just have input–output measurements, and then it is necessary to base the adaptive design on state estimates obtained from some form of observer. Such designs now abound in the literature and are typically of the forms given by Carroll and Lindorff (1973), Carroll (1974), Kudva and Narendra (1973) and Lüders and Narendra (1974). Identification schemes based on input–output data are also given by Lion (1967) and Anderson (1974). We shall follow the method of Kreisselmeier (1977) in which the Luenberger observer is written in a structurally different form which is suitable for adaptive observation.

Consider, then, the usual linear (single-input single-output) system†

$$\left.\begin{aligned} \dot{\boldsymbol{x}}(t) &= A\boldsymbol{x}(t) + \boldsymbol{b}u(t), \quad \boldsymbol{x}(0) = \boldsymbol{x}_0 \\ y(t) &= \boldsymbol{c}^{\mathrm{T}}\boldsymbol{x}(t) \end{aligned}\right\} \tag{8.2.28}$$

where the state dimension n of a minimal representation is assumed to be known. Since A, $\boldsymbol{b}$ and $\boldsymbol{c}$ are unknown and the representation is minimal, by a change of coordinates we can assume that A, $\boldsymbol{b}$ and $\boldsymbol{c}$ have the canonical representations

$$A = \begin{pmatrix} -a_1 & 1 & 0 & \dots & 0 \\ -a_2 & 0 & 1 & \dots & 0 \\ \vdots & \vdots & & & \vdots \\ -a_{n-1} & 0 & & \dots & 1 \\ -a_n & 0 & & \dots & 0 \end{pmatrix}, \quad \boldsymbol{b} = \begin{pmatrix} b_1 \\ b_2 \\ \vdots \\ \\ b_n \end{pmatrix}, \quad \boldsymbol{c} = \begin{pmatrix} 1 \\ 0 \\ \vdots \\ \\ 0 \end{pmatrix} \tag{8.2.29}$$

Recall now from chapter 4 that the standard form of the Luenberger observer may be written

$$\dot{\hat{\boldsymbol{x}}} = F\hat{\boldsymbol{x}}(t) + \boldsymbol{g}y(t) + \boldsymbol{h}u(t), \quad \hat{\boldsymbol{x}}(0) = \hat{\boldsymbol{x}}_0 \tag{8.2.30}$$

where

$$F = \begin{pmatrix} -f_1 & 1 & 0 & & \dots & 0 \\ -f_2 & 0 & 1 & & 0 \dots & 0 \\ \vdots & \vdots & & & & \vdots \\ -f_{n-1} & 0 & & & \dots & 1 \\ -f_n & 0 & & & \dots & 0 \end{pmatrix}, \quad \boldsymbol{g} = \begin{pmatrix} g_1 \\ g_2 \\ \vdots \\ \\ g_n \end{pmatrix}, \quad \boldsymbol{h} = \begin{pmatrix} h_1 \\ h_2 \\ \vdots \\ \\ h_n \end{pmatrix}$$

and the f_i are chosen so that F has a desired set of eigenvalues with negative real parts less than $-\sigma(\sigma > 0)$. We define, as usual, the state observation error

$$\boldsymbol{e}(t) = \hat{\boldsymbol{x}}(t) - x(t)$$

Then $\boldsymbol{e}$ decays exponentially if $\boldsymbol{g}$ and $\boldsymbol{h}$ satisfy

$$\boldsymbol{g}\boldsymbol{c}^{\mathrm{T}} = A - F, \quad \boldsymbol{h} = \boldsymbol{b} \tag{8.2.31}$$

In this case

$$\boldsymbol{e}(t) = \exp(Ft)\boldsymbol{e}_0$$

The relations (8.2.31) imply that we should choose $\boldsymbol{g}$ and $\boldsymbol{h}$ according to

$$g_i = f_i - a_i, \quad h_i = b_i \tag{8.2.32}$$

However, since a_i and b_i are unknown we require to determine an adaptive

†For the multi-input, multi-output case, see Narendra and Kudva (1974).

observer so that the parameters $\boldsymbol{g}$ and $\boldsymbol{h}$ converge exponentially to those given by (8.2.32). The true (unknown) values of $\boldsymbol{g}$ and $\boldsymbol{h}$ which satisfy (8.2.32) will be denoted by $\boldsymbol{g}^*$ and $\boldsymbol{h}^*$.

To write the observer in a different form, define the $2n$ systems

$$\begin{aligned} \dot{\boldsymbol{\xi}}_i(t) &= F\boldsymbol{\xi}_i(t) + \boldsymbol{e}_i y(t), \quad \boldsymbol{\xi}_i(0) = \boldsymbol{0} \\ \dot{\boldsymbol{\xi}}_{i+n}(t) &= F\boldsymbol{\xi}_{i+n}(t) + \boldsymbol{e}_i u(t), \quad \boldsymbol{\xi}_{i+n}(0) = \boldsymbol{0} \end{aligned} \tag{8.2.33}$$

where $\boldsymbol{e}_i = (0, 0, \ldots, 1, 0, \ldots, 0)$ is the ith unit vector. It follows from (8.2.30) and (8.2.33) that

$$\hat{\boldsymbol{x}}(t) = [\boldsymbol{\xi}_1(t)\,\boldsymbol{\xi}_2(t)\,\ldots\,\boldsymbol{\xi}_{2n}(t)]\boldsymbol{p} + \exp(Ft)\hat{\boldsymbol{x}}_0 \tag{8.2.34}$$

where $\boldsymbol{p} = [\boldsymbol{g}^{\mathrm{T}}\boldsymbol{h}^{\mathrm{T}}]^{\mathrm{T}}$. Now

$$(sI - F^{\mathrm{T}})^{-1}\boldsymbol{e}_1 = \det{}^{-1}(sI - F)[s^{n-1}\,s^{n-2}\,\ldots\,s\,1]^{\mathrm{T}}$$

However,

$$(sI - F^{\mathrm{T}})^{-1}\boldsymbol{e}_i = \det{}^{-1}(sI - F)P_i(s)$$

for some polynomial matrix P_i of order less than n. Hence

$$P_i(s) = T_i[s^{n-1}\,s^{n-2}\,\ldots\,s\,1]^{\mathrm{T}}$$

for some constant matrix T_i, and so

$$(sI - F)^{-1}\boldsymbol{e}_i = T_i(sI - F^{\mathrm{T}})^{-1}\boldsymbol{e}_1, \quad 1 \leqslant i \leqslant n$$

The equations (8.2.33) then reduce to pair of equations

$$\boxed{\begin{aligned} \dot{\boldsymbol{\zeta}}_1(t) &= F^{\mathrm{T}}\boldsymbol{\zeta}_1(t) + \boldsymbol{e}_1 y(t), \quad \boldsymbol{\zeta}_1(0) = \boldsymbol{0} \\ \dot{\boldsymbol{\zeta}}_2(t) &= F^{\mathrm{T}}\boldsymbol{\zeta}_2(t) + \boldsymbol{e}_1 u(t), \quad \boldsymbol{\zeta}_2(0) = \boldsymbol{0} \end{aligned}} \tag{8.2.35}$$

with

$$\boldsymbol{\xi}_i(t) = T_i\boldsymbol{\zeta}_1(t), \quad \boldsymbol{\xi}_{i+n}(t) = T_i\boldsymbol{\zeta}_2(t), \quad 1 \leqslant i \leqslant n$$

Note that with the correct parameters $\boldsymbol{p}^*$, (8.2.34) implies that

$$\boldsymbol{x}(t) = [\boldsymbol{\xi}_1(t)\,\ldots\,\boldsymbol{\xi}_{2n}(t)]\boldsymbol{p}^* + \exp(Ft)\boldsymbol{x}_0 \tag{8.2.36}$$

since then, $\boldsymbol{x}(t) = \hat{\boldsymbol{x}}(t) - \exp(Ft)\boldsymbol{e}_0$. Hence, from (8.2.34) and (8.2.36) we obtain

$$\boxed{\boldsymbol{e}(t) = [\boldsymbol{\xi}_1(t)\,\ldots\,\boldsymbol{\xi}_{2n}(t)]\,(\boldsymbol{p} - \boldsymbol{p}^*) + \exp(Ft)\boldsymbol{e}_0} \tag{8.2.37}$$

Relation (8.2.37) shows that the observation error can be separated into the part which follows from parameter mismatch and that due to initial condition mismatch.

Now define

$$\boldsymbol{z}(t) = [\boldsymbol{\zeta}_1^{\mathrm{T}}(t)\,\boldsymbol{\zeta}_2^{\mathrm{T}}(t)]^{\mathrm{T}}$$

From the identity $c^{\mathrm{T}}(sI-F)^{-1}e_i = e_i^{\mathrm{T}}(sI-F^{\mathrm{T}})^{-1}e_1$ together with (8.2.33) and (8.2.35) it follows that $c^{\mathrm{T}}[\xi_1(t) \ldots \xi_{2n}(t)] = z^{\mathrm{T}}(t)$ and so from (8.2.34) the observer output is

$$\hat{y}(t) = \hat{z}(t)p + c^{\mathrm{T}}\exp(Ft)\hat{x}_0$$

and from (8.2.37) the output observation error is

$$\begin{aligned}\eta(t) &\triangleq \hat{y}(t) - y(t) \\ &= z^{\mathrm{T}}(t)(p - p^*) + c^{\mathrm{T}}\exp(Ft)e_0\end{aligned}$$

We must now choose an adapting law for p so that $p(t) \to p^*$ as $t \to \infty$. An obvious policy is to seek to minimize $\eta^2(t)$ with respect to $p(t)$. The derivative of η^2 with respect to p is $2z(t)\eta(t)$ and so if we let $\dot{p}$ satisfy

$$\dot{p}(t) = -Gz(t)[\hat{y}(t) - y(t)] \tag{8.2.38}$$

where G is a positive definite symmetric gain matrix, then $\dot{p}$ points in the steepest descent direction for the minimization of η^2. It can be shown (Kreisselmeier, 1977) that if

$$\begin{aligned}\mathbf{0} \leqslant k_1 I &\leqslant \int_t^{t+T} z(\tau)z^{\mathrm{T}}(\tau)\,\mathrm{d}\tau \\ &\leqslant k_2 I, \quad \text{for all } t \geqslant 0\end{aligned} \tag{8.2.39}$$

for some constants k_1, k_2, T then the adaptive observer is globally exponentially stable, i.e. $\tilde{p}(t) \to p^*$ and $e(t) \to \mathbf{0}$ exponentially fast. Moreover the exponential degree of stability is no less than

$$\min\left\{\sigma, \frac{k_1\lambda_{\min}(G)}{[1 + nk_2\lambda_{\max}(G)]^2}\right\}$$

where $\lambda_{\min}(G)$, $\lambda_{\max}(G)$ are the minimum and maximum eigenvalues of G. Note that the condition (8.2.39) on z (relating to k_1) is essentially a measure of the linear independence of the elements of $z(t)$. It can be seen that the functions $z_1(t), \ldots, z_{2n}(t)$ are linearly independent (as functions of t) if the input $u(t)$ contains at least n distinct frequencies. Thus we see again that, for adaptive systems, the degree of excitation of the system is crucial.

Another type of adaptive law can be obtained by minimizing the cost functional

$$\begin{aligned}J(t) &= \int_0^t [z^{\mathrm{T}}(t)p(t) + c^{\mathrm{T}}\exp(F\tau)\hat{x}_0 - y(\tau)]^2 \exp[-q(t-\tau)]\,\mathrm{d}\tau \\ &= \int_0^t [z^{\mathrm{T}}(t)\,\Delta p(t) + c^{\mathrm{T}}\exp(F\tau)e_0]^2 \exp[-q(t-\tau)]\,\mathrm{d}\tau\end{aligned}$$

(from (8.2.36)), where $\Delta p = p - p^*$. Then

$$\frac{\partial J}{\partial p} = 2[R(t)p(t) + r(t)]$$

where

$$R(t) = \int_0^t z(\tau)z^{\mathrm{T}}(\tau) \exp[-q(t-\tau)] \, \mathrm{d}\tau \tag{8.2.40}$$

and

$$r(t) = \int_0^t z(\tau)[c^{\mathrm{T}} \exp(F\tau)\hat{x}_0 - y(\tau)] \exp[-q(t-\tau)] \, \mathrm{d}\tau \tag{8.2.41}$$

As before we choose the adaptive law as follows:

$$\boxed{\dot{p}(t) = -G[R(t)p(t) + r(t)]} \tag{8.2.42}$$

for some positive definite symmetric gain G. Note that R and r may be generated from the differential equations

$$\dot{R}(t) = -qR(t) + z(t)z^{\mathrm{T}}(t), \quad R(0) = 0$$

$$\dot{r}(t) = -qr(t) + z(t)[c^{\mathrm{T}} \exp(Ft)\hat{x}_0 - y(t)], \quad r(0) = \mathbf{0} \tag{8.2.43}$$

It can then be shown that if $z(t)$ is bounded and there exist k, T such that

$$\int_t^{t+T} z(\tau)z^{\mathrm{T}}(\tau) \, \mathrm{d}\tau \geqslant kI > 0, \quad \text{for all } t \geqslant 0$$

then the adaptive observer is globally exponentially stable with rate at least

$$\min[\sigma, q, k\lambda_{\min}(G) \exp(-qT)] \tag{8.2.44}$$

The adaptive observer (8.2.42) is shown in fig. 8.4.

We can apply the above adaptive observer to the adaptive regulation problem in the following way (Kreisselmeier, 1982). The current parameter estimate (which we now denote $\hat{p}$) is given by $\hat{p} = [\hat{g}^{\mathrm{T}} \hat{h}^{\mathrm{T}}]^{\mathrm{T}}$ and we define

$$\hat{A} = F + \hat{g}c^{\mathrm{T}}, \quad \hat{b} = \hat{h}$$

and

$$\hat{Q} = [\hat{b}\hat{A}\hat{b} \ldots \hat{A}^{n-1}\hat{b}]$$

Let $P(s)$ be any nth-order monic polynomial with left half plane zeros (i.e. a **Hurwitz** polynomial). Then we define

$$\boxed{\begin{aligned} \dot{\hat{\eta}} &= -\hat{Q}[\hat{Q}^{\mathrm{T}}\hat{\eta} - e_n], \quad \hat{\eta}(0) = \hat{\eta}_0 \\ \hat{k} &= -P(\hat{A}^{\mathrm{T}})\hat{\eta} \end{aligned}} \tag{8.2.45}$$

(Note that $P(\hat{A}^{\mathrm{T}})$ is just the matrix obtained when $\hat{A}^{\mathrm{T}}$ is substituted for s in P.) Then we define the adaptive control law by

$$\boxed{u = \hat{k}^{\mathrm{T}}\hat{x} + u_t} \tag{8.2.46}$$

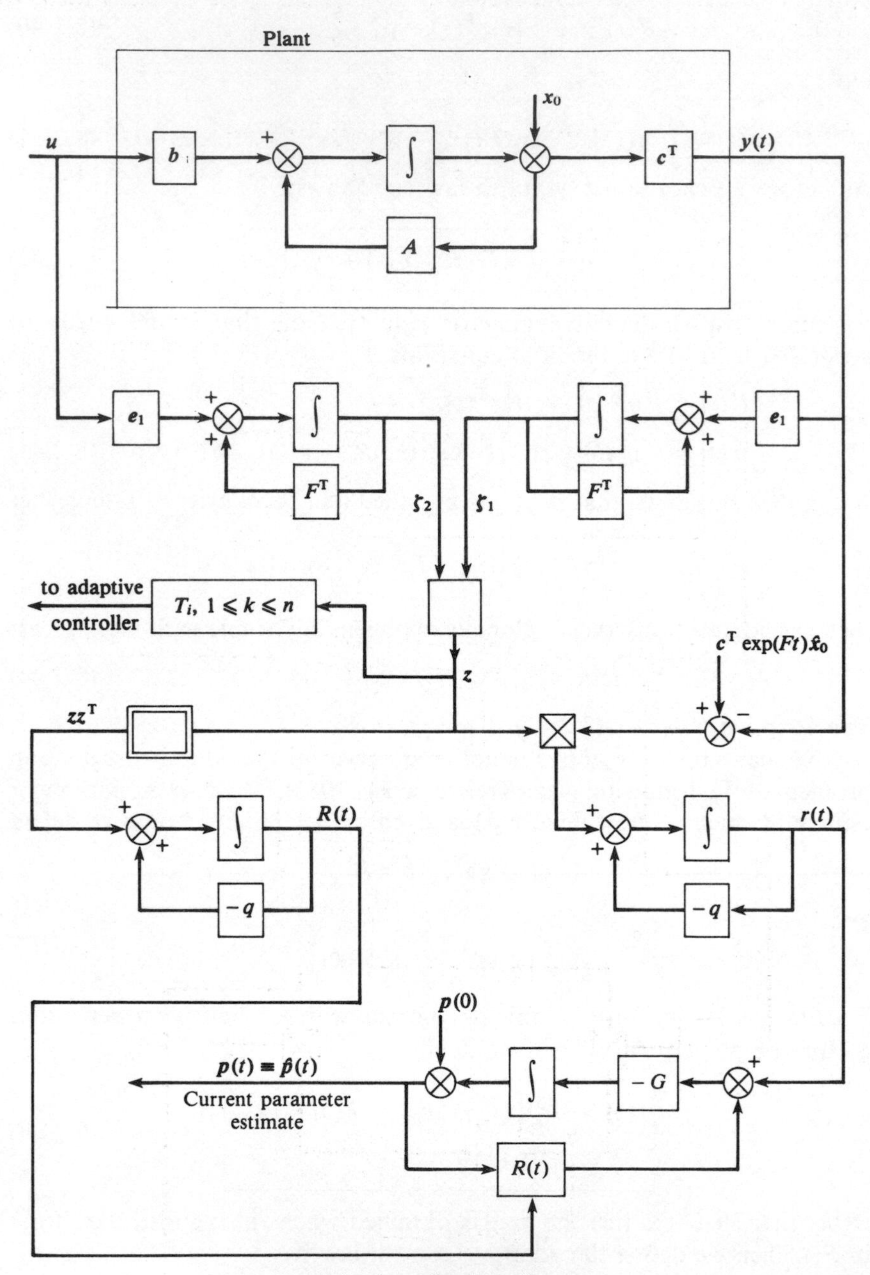

Fig. 8.4 An adaptive observer.

where u_t is an external signal which must provide sufficient excitation to the plant to ensure proper adaptation. If $\hat{x}_0$ is taken to be the zero then, by (8.2.34) we have

$$\hat{x}(t) = M\hat{p}$$

where $M = [\xi(t), \ldots, \xi_{2n}(t)]$ and so (8.2.46) becomes

$$u = \hat{k}^{\mathrm{T}} M\hat{p} + u_t \tag{8.2.47}$$

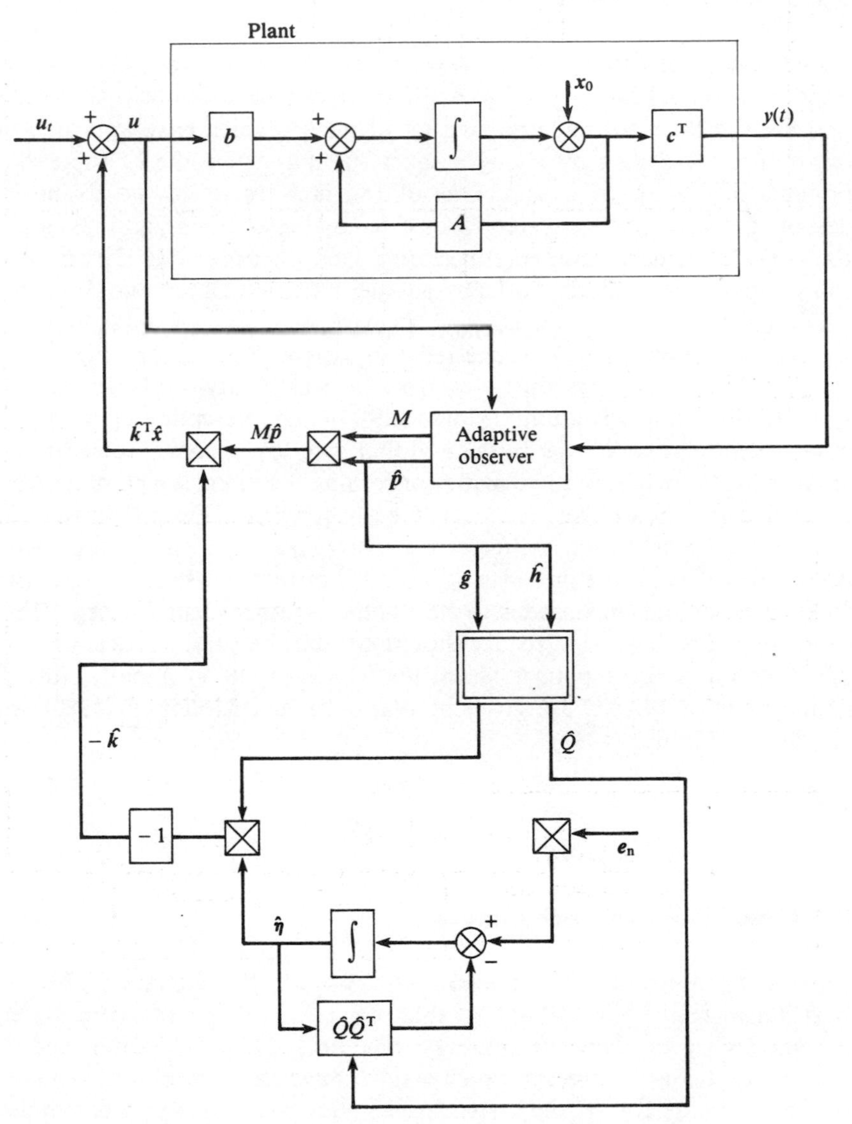

Fig. 8.5 **An adaptive control system with observer.**

This control law can be shown to be locally stable if $u_t = 0$ and globally stable (i.e. $\hat{p} \to p^*$, $x \to 0$) if u_t is sufficiently exciting. Moreover, Kreisselmeier (1982) shows that this control is effective in the presence of certain time-varying parameters and external disturbances. The complete adaptive system now appears as in fig. 8.5.

8.2.5 Bibliographical Notes

Adaptive control system design is still an active research area and so it seems important to mention at least some of the recent work to help the reader find his way through the large number of papers which have appeared on this subject. We start by giving a short summary of some of the earlier research into adaptive design. One of the first results in the Lyapunov design of MRAC control systems was given by Parks (1966), and Hang and Parks (1973) demonstrated its superiority over other methods. The method found applications in an optical tracking system (Gilbart and Winston, 1974) and in autopilots for ships (Van Amerongen and Udink ten Cate, 1973). Monopoli (1974) suggested an augmented error signal for input–output MRAC systems. However, global stability results for MRAC systems seemed elusive until Morse (1980) and Narendra *et al.* (1980) independently supplied the missing global stability proofs. However, the assumptions which have to be made concerning the unknown plant are very restrictive and are unlikely to be satisfied in practice. Hence Peterson and Narendra (1982) and Kreisselmeier and Narendra (1982) have studied the design of MRAC in the presence of bounded disturbances, allowing the error between plant and model to be bounded rather than asymptotically decreasing. The interested reader should consult the survey article of Landau (1974) and also his more recent text (Landau, 1979). Finally, further applications of MRAC are given by Narendra and Monopoli (1980) and Harris and Billings (1981).

8.3 SELF-TUNING CONTROL

8.3.1 Minimum Variance Control

Self-tuning control was introduced by Astrom and Wittenmark (1973) and is distinguished from MRAC in that we now assume that the system parameters are constant but unknown. The objective of self-tuning control is to design a control strategy which would converge to the optimal strategy that could be derived if the system parameters were known. The simplest type of optimality criterion is to choose the control u to minimize the output variance $E(y^2)$ of the system. This has applications to quality control where we require a product to have uniform characteristics.

Consider the system in fig. 8.6 where the disturbance is assumed to be filtered white noise and we suppose that all the roots of $C(z)$ are inside the unit circle. Taking Z-transforms we have

$$\begin{aligned} G(z^{-1}) &= z^{-k}\frac{(b_0 + b_1 z^{-1} + \ldots + b_n z^{-n})}{1 + a_1 z^{-1} + \ldots + a_n z^{-n}} \\ &= z^{-k}\frac{B(z^{-1})}{A(z^{-1})} \end{aligned} \tag{8.3.1}$$

$$\begin{aligned} H(z^{-1}) &= \frac{1 + c_1 z^{-1} + \ldots + c_n z^{-n}}{1 + a_1 z^{-1} + \ldots + a_n z^{-n}} \\ &= \frac{C(z^{-1})}{A(z^{-1})} \end{aligned} \tag{8.3.2}$$

where $k \geqslant 1$ and $b_0 \neq 0$. Let the variance of the noise sequence be σ^2 (constant). The system then has the input–output relation

$$y(t) = \frac{B(z^{-1})}{A(z^{-1})} u(t-k) + \frac{C(z^{-1})}{A(z^{-1})} \zeta(t)$$

or

$$\boxed{Ay(t+k) = Bu(t) + C\zeta(t+k)} \tag{8.3.3}$$

where we have omitted reference to the backward difference operator z^{-1}. We wish to choose the control u to minimize the 'cost'

$$I = E[y^2(t+k)]$$

From (8.3.2) we see that

$$\begin{aligned} \frac{C(z^{-1})}{A(z^{-1})}\zeta(t+k) &= [\zeta(t+k) + n_1\zeta(t+k-1) + \ldots + n_{k-1}\zeta(t+1)] \\ &\quad + [n_k\zeta(t) + n_{k+1}\zeta(t-1) + \ldots] \\ &\triangleq \tilde{n}(t+k) + \hat{n}(t+k) \end{aligned} \tag{8.3.4}$$

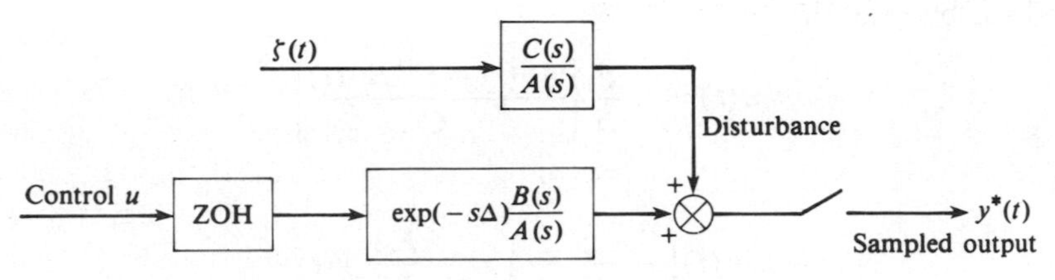

Fig. 8.6 A simple discrete system with disturbance.

for example, for some constants n_i. (This is obtained by long division by $A(z^{-1})$.) Hence

$$I = \mathrm{E}\left\{\left[\frac{B}{A}u(t) + \hat{n}(t+k) + \tilde{n}(t+k)\right]^2\right\}$$

$$= \mathrm{E}\left\{\left[\frac{B}{A}u(t) + \hat{n}(t+k)\right]^2 + n^2(t+k)\right\}$$

since $\tilde{n}(t+k)$ and $[u(t), \hat{n}(t+k)]$ are uncorrelated. Now

$$\frac{\mathrm{d}I}{\mathrm{d}u(t)} = 2\mathrm{E}\left\{\left[\frac{B}{A}u(t) + \hat{n}(t+k)\right]b_0\right\} = 0$$

and so the minimum variance control is

$$u(t) = -\frac{A}{B}\hat{n}(t+k) \tag{8.3.5}$$

Using (8.3.3)–(8.3.5) we see that the output using this control is

$$y(t+k) = \tilde{n}(t+k)$$

Note that $\hat{n}$ is the predicted disturbance and the controller cancels this part of the disturbance leaving the error $\tilde{n}(t+k)$.

To write the control (8.3.5) in an implementable form we define the polynomials E and F in z^{-1} by the identity

$$\frac{C}{A} = E + z^{-k}\frac{F}{A} \tag{8.3.6}$$

where

$$E(z^{-1}) = 1 + e_1 z^{-1} + \ldots + e_{k-1} z^{-k+1}$$

$$F(z^{-1}) = f_0 + f_1 z^{-1} + \ldots + f_{n-1} z^{-n+1}$$

Then

$$\hat{n}(t+k) = \frac{F}{A}\zeta(t)$$

$$= \frac{F}{A}\left[\frac{Ay(t) - z^{-k}Bu(t)}{C}\right]$$

and so, by (8.3.5),

$$u(t) = -\frac{F}{B}\left[\frac{Ay(t) - z^{-k}Bu(t)}{C}\right]$$

or

$$u(t)\left(\frac{C - z^{-k}F}{C}\right) = -\frac{FA}{BC}y(t)$$

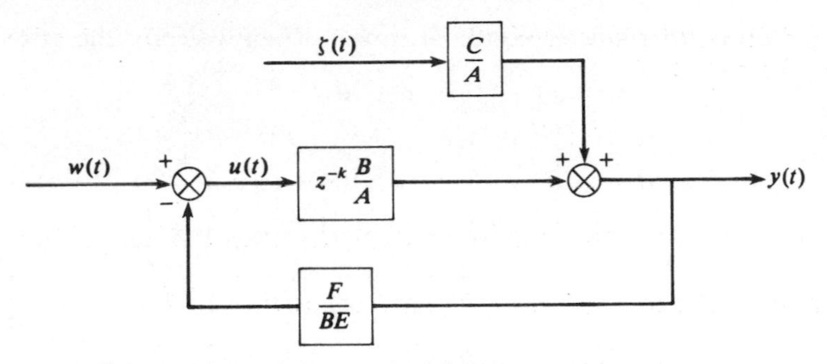

Fig. 8.7 Minimum variance feedback control system.

However, from (8.3.6), we have

$$C - z^{-k}F = AE$$

and so we obtain, finally,

$$\boxed{u(t) = -\frac{F}{BE}y(t)} \tag{8.3.7}$$

which is the feedback form of the minimum variance controller. Given an input demand $w(t)$ we can then form the feedback system shown in fig. 8.7. The closed-loop transfer function is

$$\begin{aligned} y(t) &= z^{-k}\frac{B}{A}\left[w(t) - \frac{F}{BE}y(t)\right] + \frac{C}{A}\zeta(t) \\ &= z^{-k}\frac{BE}{C}w(t) + E\zeta(t) \end{aligned}$$

from (8.3.6). Note that we have assumed that the system is minimum phase and known exactly in order to implement the control. When the system parameters are not known we must first identify the process model using a least squares algorithm. This leads to the self-tuning controller.

8.3.2 Self-tuning Regulator Based on Minimum Variance

Consider again the process (8.3.3) where we shall assume, initially, that $C = 1$, i.e.

$$Ay(t) = Bu(t-k) + \zeta(t) \tag{8.3.8}$$

If the parameters $a_1, \ldots, a_n, b_1, \ldots, b_n$ are unknown we must estimate them using, for example, the recursive least squares algorithm given in the last chapter. Let

$$\beta(t) = [\hat{a}_1(t), \ldots, \hat{a}_n(t), \hat{b}_1(t), \ldots, \hat{b}_n(t)]^{\mathrm{T}}$$

be the estimated parameter vector at time t. Then we form the error

$$\varepsilon(t) = y(t) - z^{\mathrm{T}}(t)\beta(t-1)$$

where

$$z^{\mathrm{T}}(t) = [-y(t-1), \ldots, -y(t-n), u(t-k), \ldots, u(t-k-n)]$$

Minimizing $\mathrm{E}(\varepsilon^2)$ then leads to the recursive least squares algorithm for the parameters. At each time step we can then solve the identity (cf. (8.3.6))

$$1 = \hat{A}E + z^{-k}F \tag{8.3.9}$$

for E and F which can then be substituted in the control (8.3.7). This gives rise to the **explicit** self-tuning algorithm where the control is based on the process model parameter estimates.

The explicit algorithm has the disadvantage of requiring the solution of the identity (8.3.9) at each time step. We can also obtain an **implicit** algorithm which estimates the control parameters directly as follows. From (8.3.8) we have

$$EAy(t) = EBu(t-k) + E\zeta(t)$$

and using the identity $1 = EA + z^{-k}F$ we obtain

$$y(t) = Fy(t-k) + Gu(t-k) + E\zeta(t)$$

where $G(z^{-1}) = E(z^{-1})B(z^{-1})$. Hence,

$$\begin{aligned} y(t) &= (f_0 + f_1 z^{-1} + \ldots + f_{n-1} z^{-n+1}) y(t-k) \\ &\quad + (g_0 + g_1 z^{-1} + \ldots + g_{n+k-1} z^{-n-k+1}) u(t-k) + E(z^{-1})\zeta(t) \\ &= z^{\mathrm{T}}(t)\beta(t-1) + \tilde{n}(t) \end{aligned}$$

where

$$\begin{aligned} z^{\mathrm{T}}(t) = [&y(t-k), y(t-k-1), \ldots, y(t-k-n+1), \\ &u(t-k), \ldots, u(t-2k-n+1)] \end{aligned}$$

$$\beta(t-1) = (f_0, f_1, \ldots, f_{n-1}, g_0, g_1, \ldots, g_{n+k-1})^{\mathrm{T}}$$

$$\tilde{n}(t) = \varepsilon(t) = \zeta(t) + e_1\zeta(t-1) + \ldots + e_{k-1}\zeta(t-k+1)$$

The process $e(t)$ is independent of the data vector $z(t)$ and so the recursive least squares algorithm provides unbiased estimates for F and G which can be used at each time step in the control (8.3.7). Thus

$$u(t) = -\frac{F}{BE}y(t) = -\frac{F}{G}y(t)$$

that is

$$\boxed{Gu(t) + Fy(t) = 0} \tag{8.3.10}$$

If $C \neq 1$ then it turns out that the recursive least squares algorithm still gives unbiased estimates and the same reasoning as before leads to the implicit algorithm (8.3.10).

It is important to note that the parameter estimates in the above algorithms are not unique; for example, consider the first order system

$$y(t) = ay(t-1) + bu(t-1) + \varepsilon(t)$$

with feedback control $u(t) = ky(t)$. Then, for any λ, we have

$$y(t) = (a - k\lambda)y(t-1) + (b+\lambda)u(t-1) + \varepsilon(t)$$

and so the parameters $(a - k\lambda, b + \lambda)$ give the same value for the loss function $\sum \varepsilon^2(t)$ for arbitrary λ. As in our discussion of MRAC, unique parameter estimates can be obtained if the system is subjected to a persistently exciting input.

8.3.3 Generalized Minimum Variance

In the above minimum variance controller we have not included the system input $w(t)$ or weighted the control $u(t)$ in the cost functional I. This can be overcome quite easily by considering the cost

$$I = \mathrm{E}\{[Py(t+k) - Rw(t)]^2 + [Qu(t)]^2\} \tag{8.3.11}$$

where P, Q and R are transfer functions in z^{-1} which may be specified by the designer and P_0 may be taken to be 1. As before we suppose that our system model is

$$Ay(t) = Bu(t-k) + C\zeta(t)$$

Then, using $C = EA + z^{-k}F$ we obtain

$$Cy(t) - Fy(t-k) = Gu(t-k) + EC\zeta(t)$$

with the same notation as before. Hence

$$y(t) = \frac{Fy(t-k) + Gu(t-k)}{C} + E\zeta(t)$$

$$\triangleq \hat{y}(t) + \tilde{n}(t)$$

Substituting $y(t)$ into the cost functional (8.3.11), we obtain

$$I = \mathrm{E}(\{P[\hat{y}(t+k) + \tilde{n}(t+k)] - Rw(t)\}^2 + [Qu(t)]^2)$$

$$= [P\hat{y}(t+k) - Rw(t)]^2 + [Qu(t)]^2 + \alpha^2$$

where $\alpha^2 = \mathrm{E}\{[P\tilde{n}(t+k)]^2\}$, since $\hat{y}(t+k)$, $w(t)$ and $u(t)$ are known and deterministic and $P\tilde{n}(t+k)$ is uncorrelated with these quantities. Now

$$\frac{\mathrm{d}I}{\mathrm{d}u(t)} = 2[P\hat{y}(t+k) - Rw(t)]b_0 + 2q_0Qu(t) = 0 \tag{8.3.12}$$

(where $Q = q_0 + q_1 z^{-1} + \ldots$). Note that the dependence of $\hat{y}(t+k)$ on $u(t)$ is given by

$$\frac{G}{C} u(t) = \frac{BE}{C} u(t)$$
$$= (b_0 + \alpha_1 z^{-1} + \alpha_2 z^{-2} + \ldots) u(t)$$

for some numbers α_i. Define

$$\boxed{\phi^*(t+k) \triangleq P\hat{y}(t+k) + \frac{q_0 Q}{b_0} u(t) - Rw(t)} \tag{8.3.13}$$

Then the control law, given by (8.3.12) is

$$\phi^*(t+k) = 0$$

If we also define

$$\phi(t+k) = Py(t+k) + \frac{q_0 Q}{b_0} u(t) - Rw(t)$$

then

$$\phi(t+k) = P(\hat{y}(t+k) + \tilde{n}(t+k)) + \frac{q_0 Q}{b_0} u(t) - Rw(t)$$
$$= \phi^*(t+k) + \varepsilon(t+k)$$

where $\varepsilon(t) = P\tilde{n}(t)$. Clearly, $\varepsilon(t+k)$ is uncorrelated with $\phi^*(t+k)$, which is the least squares optimal predictor of $\phi(t+k)$.

To implement the controller note that

$$\phi^*(t+k) = P\left[\frac{Fy(t) + Gu(t)}{C}\right] + \frac{q_0 Q}{b_0} u(t) - Rw(t) = 0$$

or

$$C\phi^*(t+k) = F'y(t) + G'u(t) + H'w(t) = 0$$

where

$$F' = PF$$
$$G' = PG + CQ'$$
$$H' = -CR$$
$$Q' = \frac{q_0 Q}{b_0}$$

(see fig. 8.8). It is easy to see that the closed loop transfer function is given

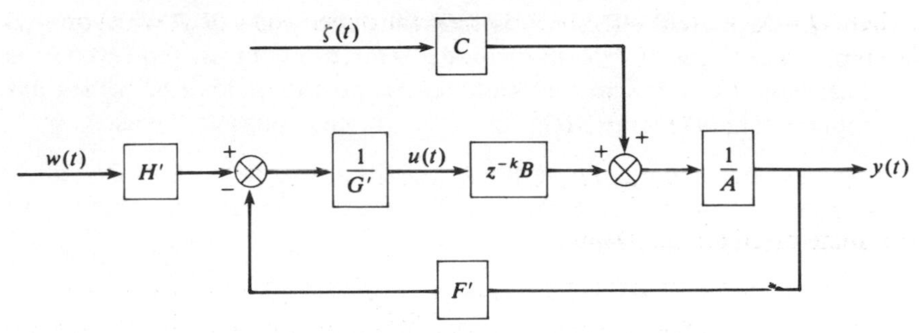

Fig. 8.8 Generalised minimum variance control system.

by

$$y(t) = z^{-k}\frac{B}{PB + Q'A}Rw(t) + \frac{PEB + Q'C}{PB + Q'A}\zeta(t)$$

which has the characteristic equation

$$\boxed{PB + Q'A = 0} \tag{8.3.14}$$

This equation can be altered by appropriate choice of P and Q', which also give rise to different controls since u is given by

$$\boxed{u(t) = \frac{CR}{PBE + CQ'}w(t) - \frac{PF}{PBE + CQ'}y(t)} \tag{8.3.15}$$

We consider three cases:

(a) If $P = 1$, $R = Q = 0$, then (8.3.15) reduces to the ordinary minimum variance control (8.3.7).
(b) If $P = 1$, $R = 1$, $Q' = \lambda$, then the transfer function is

$$y(t) = \frac{EB + \lambda C}{B + \lambda A}\zeta(t) + \frac{z^{-k}B}{B + \lambda A}w(t)$$

For $\lambda = 0$ the controller tries to cancel the plant which results in an unstable system if the open-loop system is non-minimum phase. However, if the open-loop system is stable (i.e. roots of $A(z^{-1}) = 0$ are inside the unit circle) then by choosing (or **tuning**) an appropriate value of λ we can cause the closed-loop poles to move arbitrarily close to those of the open-loop system. Note, however, that large values of λ weight the control in the cost functional at the expense of the quality of the set point tracking.
(c) If $Q = 0$ we obtain the closed-loop input–output relation

$$y(t) = E\zeta(t) + z^{-k}\frac{R}{P}w(t)$$

and so the system effectively is following the model R/P with delayed input and so this is a form of model reference control. However, it is clear from (8.3.15) that the characteristic polynomial is PBE which has roots outside the unit circle for non-minimum phase systems.

The extended minimum variance algorithm can be used to develop a self-tuning controller (see Clarke and Gawthrop, 1975, 1979). As above we obtain the control u given by

$$F'y(t) + G'u(t) + H'w(t) = 0$$

However, the parameters F', G' and H' are unknown and so, as in the ordinary minimum variance case, we can use recursive least squares to obtain estimates of these parameters. If $C = 1$ then we have

$$\begin{aligned}\phi(t+k) &= \phi^*(t+k) + \varepsilon(t+k)\\ &= F'y(t) + G'u(t) + H'w(t) + \varepsilon(t+k)\\ &= z^{\mathrm{T}}(t)\boldsymbol{\beta}(t-1) + \varepsilon(t+k)\end{aligned}$$

where

$$z^{\mathrm{T}}(t) = \{y(t), y(t-1), \ldots, u(t), u(t-1), \ldots, w(t), w(t-1), \ldots\}$$

and

$$\boldsymbol{\beta}^{\mathrm{T}}(t-1) = \{f_0', f_1', \ldots, g_0', g_1', \ldots, h_0', h_1', \ldots\}$$

We then minimize $\varepsilon^2(t+k)$ to obtain the recursive least squares estimate of $\boldsymbol{\beta}$ (where f_i', g_i' and h_i' are the respective coefficients of the polynomials F', G' and H'). Again in the case when $C \neq 1$ we can use the same algorithm to produce a self-tuner if the parameter estimates $\hat{\boldsymbol{\beta}}$ converge to the true value $\boldsymbol{\beta}$ (see Ljung, 1977).

We conclude this short introduction to self-tuning control by noting that self-tuning controllers may be designed on the basis of the pole-placement regulator in which the control is chosen so that the closed-loop transfer function has a desired set of poles. Consider, for example, the model

$$Ay(t) = z^{-k}Bu(t) + C\zeta(t)$$

where A, B and C are polynomials in z^{-1} of degrees n_A, n_B and n_C respectively, and a delay of k steps has been explicitly included. Then, using a feedback control

$$G(z^{-1})u(t) = -F(z^{-1})y(t)$$

we obtain

$$(AG + z^{-k}BF)y(t) = CG\zeta(t)$$

and so, if F and G are chosen to satisfy

$$AG + z^{-k}BF = CT \tag{8.3.16}$$

where T is some desired polynomial, then

$$y(t) = \frac{G}{T}\zeta(t)$$

so that the closed-loop poles are the roots of T. If

$$G = 1 + g_1 z^{-1} + \ldots + g_{n_G} z^{-n_G}$$

$$F = f_0 + f_1 z^{-1} + \ldots + f_{n_F} z^{-n_F}$$

and

$$T = 1 + t_1 z^{-1} + \ldots + t_{n_T T} z^{-n_T}$$

then we can solve (8.3.16) uniquely if

$$n_G = n_B + k - 1$$

$$n_F = n_A - 1$$

$$n_T = n_A + n_B + k - 1 - n_C$$

where the last condition is required to ensure the self-tuning condition that when the number of samples increases, the control converges to that which would have been designed on the basis of known process dynamics.

To design a self-tuning pole assignment control (with $C = 1$) we apply recursive least squares estimation to the model

$$Ay(t) - z^{-k}Bu(t) = \zeta(t)$$

to estimate the parameters in A and B, and solve (8.3.16) for a given T, at each sample interval, for G and F and then implement the control

$$u = -\frac{F}{G}y$$

Extended self-tuning control can be designed similarly if $C \neq 1$ (see, for example, Wellstead and Sanoff, 1981).

8.4 VARIABLE-STRUCTURE SYSTEMS

8.4.1 The Nature of Variable Structure Systems

In the above sections we have discussed two methods – model reference adaptive control and self-tuning control – in which the system model has a fixed structure and the parameters are changed in accordance with some design criterion. A rather different approach to control system design is to allow the controller structure (as a function of the states) to change in order to achieve certain desired objectives. This type of control structure is inherent, of course, in the bang-bang law which follows from the maximum principle since different non-linear feedback functions must be imposed on opposing sides of the switching surface. In the theory of variable structure systems we try to formalize the different types of behaviour which can be obtained by such structural changes. The following short discussion is based on the review article of Utkin (1977) and the monograph of Itkis (1976).

The basic idea of variable structure systems can be illustrated by considering the second-order system with state feedback kx

$$\ddot{x} + a\dot{x} + (b+k)x = 0 \tag{8.4.1}$$

The structure of (8.4.1) in the phase plane was considered in chapter 5 for various values of k and the trajectories can be sources, sinks, spirals, saddles, etc. If the control law kx is chosen so that k switches between different values which change the structure of the closed-loop system from one of these forms to another, we can obtain trajectories which are qualitatively different from any which can be obtained with a fixed k. For example, suppose the system has zero damping ($a = 0$) and that k is chosen to have two values

$$k = \begin{cases} \alpha - b \\ -\alpha - b \end{cases}$$

for some positive $\alpha > 1$, where the switching times will be considered later. Then the closed-loop system takes on one of the two structures

$$\ddot{x} + \alpha x = 0 \tag{8.4.2a}$$

$$\ddot{x} - \alpha x = 0 \tag{8.4.2b}$$

In the first case the trajectories of the system are ellipses while in the second case they form a saddle (fig. 8.9(a) and (b)). Note that, in the saddle trajectories, there is a unique stable subspace defined by the equation

$$\dot{x} + \sqrt{\alpha}x = 0 \tag{8.4.3}$$

We shall use this line and the line $x = 0$ to divide the phase plane into four regions as in fig. 8.9(c) and switching of k will occur on the boundaries of these regions. The control will then be chosen to be

$$u(t) = \begin{cases} (\alpha - b)x & \text{in regions I, III} \\ (-\alpha - b)x & \text{in regions II, IV} \end{cases}$$

It is easy then to see that the closed-loop trajectories take the form shown in fig. 8.10, and the resulting system is globally asymptotically stable, even though the two feedback systems from which it was formed are critically

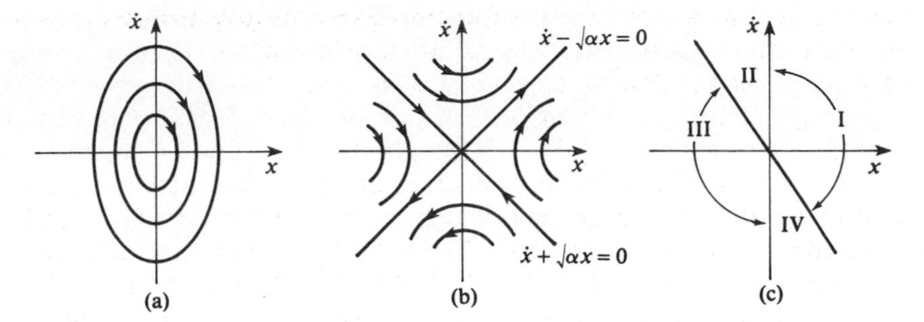

Fig. 8.9 The variable structure system (8.4.2).

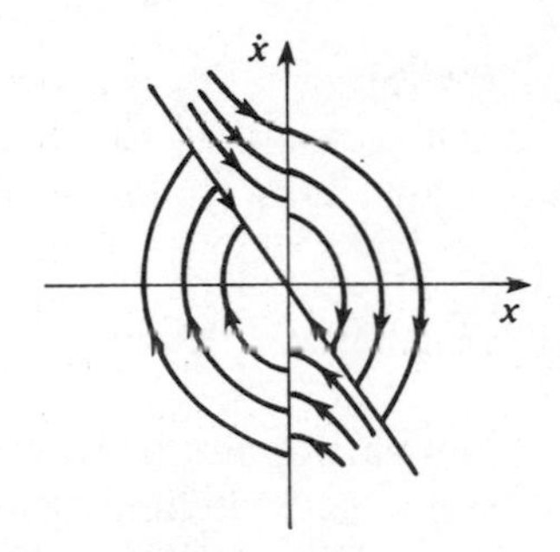

Fig. 8.10 Closed-loop trajectories for system (8.4.2).

stable and unstable, respectively. Typical waveforms of $x(t)$ for initial conditions in regions I and II are shown in fig. 8.11.

In the above example the trajectories of the closed-loop switching system are composed of pieces of the trajectories of the non-switching feedback systems and so each closed-loop trajectory resembles a non-switching trajectory on finite time intervals. It is possible, however, to obtain motions in the switching system which have no relation to those of the non-switching systems from which it is formed. For example, in the above system we do not have to choose to switch on the line defined by (8.4.3); we could force the switching to occur on the line

$$\dot{x} + cx = 0 \tag{8.4.4}$$

where $0 < c < \sqrt{\alpha}$. Then the closed-loop trajectories are as shown in fig. 8.12. A strange phenomenon now occurs on the switching line (8.4.4). In order to understand the behaviour of the system on the switching line, it is convenient to introduce two switching lines as in fig. 8.13 so that the system with elliptical trajectories switches on the line l_1 with $c = c_1$ while the 'saddle' system switches on the line l_2 with $c = c_2$ ($c_1 > c_2$). Consider an initial condition with $\dot{x}(0) > 0$, $x(0) > 0$. Then the trajectory is (part of) an

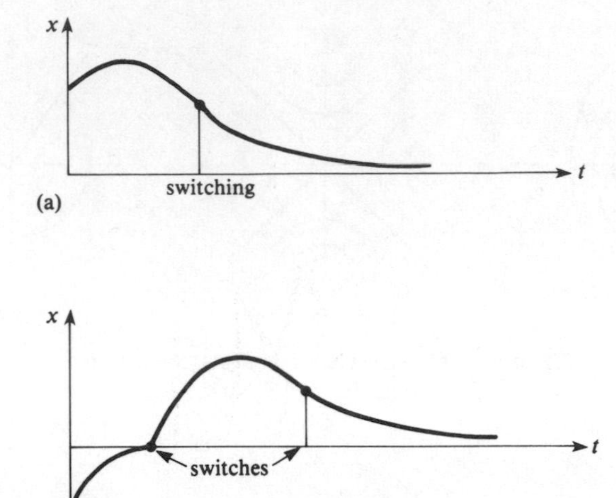

Fig. 8.11 Closed-loop trajectories as a funciton of t. (a) Initial condition in region I; (b) initial condition in region II.

ellipse until it hits the line l_1 at which point it switches to the saddle trajectory which is directed to the line l_2. Switching again occurs on l_2 and the trajectory is again elliptical until it reaches l_1. The resulting trajectory therefore remains in the shaded region R in fig. 8.13 and zig-zags to the origin. If we now let c_1, c_2 tend to c then the trajectory switches with infinite frequency on the line $\dot{x} + cx = 0$ resulting in the zig-zag line in fig. 8.12. This rapid switching will occur in a real system since there is always a physical delay in switching; however, we may idealize and say that the trajectory is forced to remain on the switching line since the trajectories on either side of the line point towards the line. For this reason the system moves stably

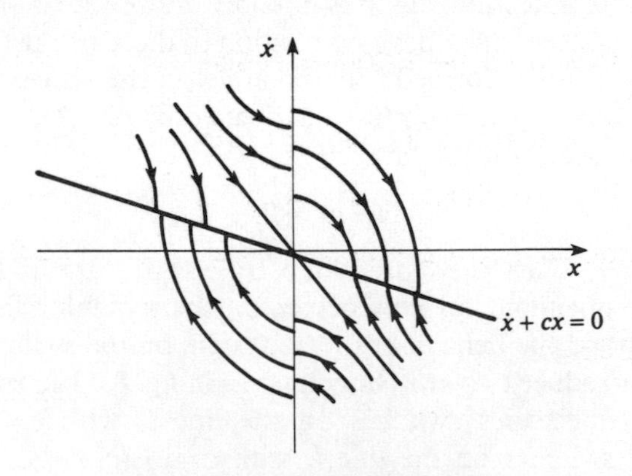

Fig. 8.12 A sliding mode.

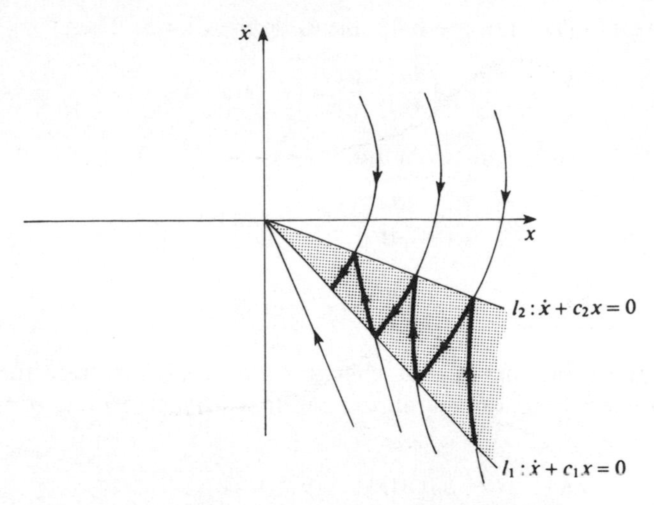

Fig. 8.13 Demonstrating the sliding condition.

down the switching line to the origin and is therefore said to be in the **sliding mode**. Note that this mode is not composed of subtrajectories of the two non-switching systems, it is a completely new type of motion.

The motion along the switching line (i.e. in the sliding mode) is specified by the equation

$$\frac{dx}{dt} + cx = 0 \tag{8.4.5}$$

and is independent of the original system parameter b (provided the sliding mode continues to exist). Sliding modes are therefore very desirable forms of motion of the composite system since we can choose c independently (to a large extent) of the system parameters, giving an inherently 'robust' system.

8.4.2 General Condition for a Sliding Mode

Suppose that a general system is specified by the non-linear equation

$$\dot{x}_i = f_i(x_1, \ldots, x_n, t), \quad 1 \leqslant i \leqslant n$$

where the functions f_i are discontinuous on a hypersurface

$$S: \sigma(x_1, \ldots, x_n) = 0 \tag{8.4.6}$$

assume that the limits

$$f_i^-(x_1, \ldots, x_n, t) = \lim_{x \to H^-} f_i(x_1, \ldots, x_n, t)$$

$$f_i^+(x_1, \ldots, x_n, t) = \lim_{x \to H^+} f_i(x_1, \ldots, x_n, t)$$

exist, where H^+, H^- are the half-spaces defined by S. Then

$$\frac{\mathrm{d}\sigma}{\mathrm{d}t} = \left(\frac{\partial \sigma}{\partial x}\right)^{\mathrm{T}} \frac{\mathrm{d}x}{\mathrm{d}t} = (f \cdot \operatorname{grad} \sigma)$$

with $f = (f_1, \ldots, f_n)$, and so the limits

$$\lim_{x \to H^-} \frac{\mathrm{d}\sigma}{\mathrm{d}t} = (f^- \cdot \operatorname{grad} \sigma)$$

$$\lim_{x \to H^+} \frac{\mathrm{d}\sigma}{\mathrm{d}t} = (f^+ \cdot \operatorname{grad} \sigma)$$

exist and represent the rate of change of σ along the trajectories. The function σ generates a one-parameter family of hypersurfaces parallel to S given by

$$S_\sigma : \sigma(x_1, \ldots, x_n) = \sigma$$

We associate H^+, H^- with the sets

$$\bigcup_{\sigma > 0} S_\sigma, \quad \bigcup_{\sigma < 0} S_\sigma$$

respectively. Then for a sliding mode we clearly require

$$\boxed{\lim_{x \to H^+} \frac{\mathrm{d}\sigma}{\mathrm{d}t} \leqslant 0 \leqslant \lim_{x \to H^-} \frac{\mathrm{d}\sigma}{\mathrm{d}t} \quad \text{or} \quad \lim_{\sigma \to 0} \sigma \frac{\mathrm{d}\sigma}{\mathrm{d}t} \leqslant 0} \tag{8.4.7}$$

(fig. 8.14(a)). Note that if either equality in (8.4.7) occurs then we may only have asymptotic approach to the switching hypersurface. In the sliding mode the trajectory remains on the hypersurface S for all times after hitting S and so in the sliding mode we require

$$\boxed{\begin{aligned} &\frac{\mathrm{d}\sigma}{\mathrm{d}t} = \left(\frac{\mathrm{d}\sigma}{\mathrm{d}x}\right)^{\mathrm{T}} \cdot f = 0 \\ &\sigma(x_1, \ldots, x_n) = 0 \end{aligned}} \tag{8.4.8}$$

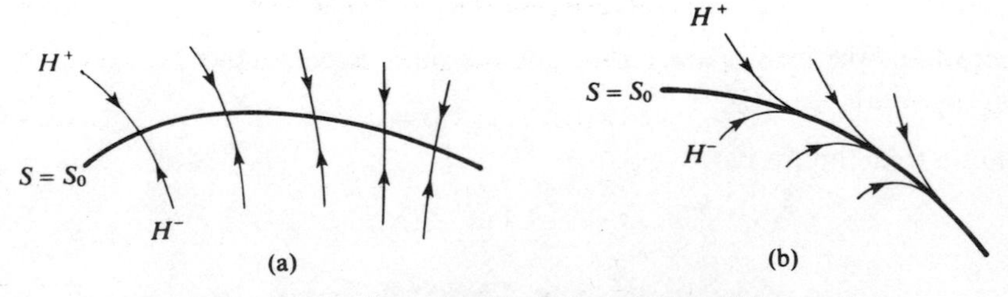

Fig. 8.14 Discontinuous trajectories on the sliding surface.

8.4.3 *n*th-order Plant with Primary Feedback

Consider the general system

$$\left.\begin{aligned} \frac{dx_i}{dt} &= x_{i+1}, \quad 1 \leqslant i \leqslant n-1 \\ \frac{dx_n}{dt} &= -\sum_{i=1}^{n} a_i x_i - bu \end{aligned}\right\} \tag{8.4.9}$$

with feedback control (depending only on x_1)

$$u = \psi x_1$$

which is switched on the hypersurfaces $x_1 = 0$ and $S: \sigma = 0$, where

$$\sigma = \sum_{i=1}^{n} c_i x_i, \quad c_n = 1 \tag{8.4.10}$$

for some positive constants c_i. Thus, we suppose that

$$\psi = \begin{Bmatrix} \alpha_1 & \text{if } x_1\sigma > 0 \\ \beta_1 & \text{if } x_1\sigma < 0 \end{Bmatrix} \tag{8.4.11}$$

To find the conditions for a sliding mode on S note that

$$\begin{aligned} \frac{d\sigma}{dt} &= \frac{dx_n}{dt} + \sum_{i=1}^{n-1} c_i \frac{dx_i}{dt} \\ &= -\sum_{i=1}^{n} a_i x_i - b\psi x_1 + \sum_{i=1}^{n-1} c_i x_{i+1} \end{aligned}$$

However, on S we have

$$x_n = -\sum_{i=1}^{n-1} c_i x_i$$

and so

$$\begin{aligned} \left.\frac{d\sigma}{dt}\right|_S &= \sum_{i=2}^{n-1} (c_{i-1} - a_i + c_i a_n - c_i c_{n-1}) x_i \\ &\quad + (c_1 a_n - a_1 - c_1 c_{n-1} - b\psi) x_1 \end{aligned} \tag{8.4.12}$$

Recalling (8.4.7) it is clear that a necessary and sufficient condition for a sliding mode on S is

$$\frac{c_{i-1} - a_n}{c_i} = c_{n-1} - a_n, \quad 2 \leqslant i \leqslant n-1 \tag{8.4.13a}$$

$$\alpha_1 \geqslant \frac{1}{b}(-a_1 + c_1 a_n - c_1 c_{n-1}) \tag{8.4.13b}$$

$$\beta_1 \leqslant \frac{1}{b}(-a_1 + c_1 a_n - c_1 c_{n-1}) \tag{8.4.13c}$$

Having obtained conditions for the existence of a sliding mode we must ensure, of course, that such a mode is stable. In fact it can be shown that the sliding mode is stable if and only if the system (8.4.9) with the control

$$u = b^{-1}(-a_1 + c_1 a_n - c_1 c_{n-1})x_1$$

has all of its eigenvalues, except possibly $\lambda_n = c_{n-1} - a_n$, with negative real parts.

Now that we have a system for which there exists a stable sliding mode we are interested in conditions which guarantee that all trajectories starting off the switching surface will actually hit the surface in finite time. From above we have

$$\begin{aligned}\frac{d\sigma}{dt} &= \sum_{i=1}^{n-1} c_i x_{i+1} - \sum_{i=1}^{n} a_i x_i - b\psi x_1 \\ &= -(a_n - c_{n-1})\sigma + \sum_{i=2}^{n-1} (c_{i-1} - a_i - c_i c_{n-1} + a_n c_i)x_i \\ &\quad + (c_{n-1}c_1 - a_1 + a_n c_1 - b\psi)x_1\end{aligned}$$

by (8.4.10). If conditions (8.4.13a) hold then

$$\frac{d\sigma}{dt} = -(a_n - c_{n-1})\sigma + (c_{n-1}c_1 - a_1 + a_n c_1 - b\psi)x_1$$

and so if $\sigma \geqslant 0$ and $a_n \geqslant c_{n-1}$, then

$$\frac{d\sigma}{dt} \leqslant (c_{n-1}c_1 - a_1 + a_n c_1 - b\psi)x_1$$

and so

$$\sigma(t) \leqslant \sigma(0) - \int_0^t (c_{n-1}c_1 - a_n c_1 + a_1 + b\psi)x_1 \, dt \tag{8.4.14}$$

Now assume that the conditions (8.4.13b) and (8.4.13c) hold in the stronger form where the inequalities are strict, and suppose that $x_1(t) \nrightarrow 0$ as $t \to \infty$. Then, from (8.4.14), it is clear that $\sigma(t) \to -\infty$ and so the trajectories must hit the switching surface. Otherwise we have $x_1(t) \to 0$, guaranteeing asymptotic stability in either case. The above reasoning also holds if $\sigma \leqslant 0$ and so we have the following sufficient conditions for a stable sliding mode to exist and such that either

$$\lim_{t \to \infty} x_1(t) = 0$$

or the surface $\sigma = 0$ is hit by all trajectories

$$\frac{c_{i-1} - a_n}{c_i} = c_{n-1} - a_n, \quad 2 \leqslant i \leqslant n-1 \tag{8.4.15a}$$

$$\alpha_1 > \frac{1}{b}(-a_1 + c_1 a_n - c_1 c_{n-1}) \tag{8.4.15b}$$

$$\beta_1 < \frac{1}{b}(-a_1 + c_1 a_n - c_1 c_{n-1}) \tag{8.4.15c}$$

$$c_{n-1} \leqslant a_n \tag{8.4.15d}$$

In fact, it can be shown that these conditions are necessary and sufficient for every point on $\sigma = 0$ to be a sliding mode, thus implying the global asymptotic stability of the system.

Finally we note that the restrictive condition (8.4.15d) can be removed if we allow **generalized sliding modes**; these are motions which do not necessarily stay on the hypersurface $S : \sigma = 0$ at all times after hitting, but are stable with respect to S in the sense that once the trajectory is 'close' to S it remains 'close' to S. Then it can be shown that if the gain b is large enough and the conditions (8.4.15a)–(8.4.15c) are satisfied with $\alpha_1 > 0 > \beta_1$, then any trajectory hits the hypersurface $S : \sigma = 0$ and any point on S belongs to a generalized sliding mode.

8.4.4 nth-order Plant with Derivative Feedback

Consider again the system (8.4.9) with the control

$$u = \sum_{i=1}^{k} \psi_i x_i \tag{8.4.16}$$

where

$$\psi_i = \begin{cases} \alpha_i & \text{if } x_i\sigma > 0, 1 \leqslant i \leqslant k \leqslant n-1 \\ \beta_i & \text{if } x_i\sigma < 0 \end{cases}$$

and

$$\sigma = \sum_{i=1}^{n} c_i x_i, \; c_n = 1$$

We are therefore allowing the control to depend on the state x_1 and its derivatives up to order $k-1$. This has the advantage of reducing the number of conditions (8.4.15a) but has the usual problem of increasing the noise in real systems because of the requirement of the state derivatives. We have

$$\frac{d\sigma}{dt} = -(a_n(t) - c_{n-1})\sigma + \sum_{i=k+1}^{n-1} (c_{i-1} - a_i(t) + c_i a_n(t) - c_i c_{n-1})x_i$$

$$+ \sum_{i=1}^{k} (c_{i-1} - a_i(t) + c_i a_n(t) - c_i c_{n-1} - b(t)\psi_i)x_i$$

where we are allowing the coefficients a_i, b to depend explicitly on time. Hence necessary and sufficient conditions for the existence of a sliding mode are

$$\inf_t a_n(t) \geqslant c_{n-1} \tag{8.4.17a}$$

$$\frac{c_{i-1} - a_i(t)}{c_i} = c_{n-1} - a_n(t) \quad k+1 \leqslant i \leqslant n-1 \tag{8.4.17b}$$

$$\alpha_i \geqslant \sup_t \frac{1}{b(t)} [c_{i-1} - a_i(t) + c_i a_n(t) - c_i c_{n-1}] \quad 1 \leqslant i \leqslant k \tag{8.4.17c}$$

$$\beta_i \leqslant \inf_t \frac{1}{b(t)} [c_{i-1} - a_i(t) + c_i a_n(t) - c_i c_{n-1}] \quad 1 \leqslant i \leqslant k \tag{8.4.17d}$$

Since the c_i are fixed, condition (8.4.17b) places a strong restriction on the time variation of the parameters a_i and so we assume that $a_{k+1}, \ldots, a_n$ are fixed. As in the case of (8.4.15), if strict inequalities in (8.4.17c) and (8.4.17d) hold then the trajectories hit the hyperplane $\sigma = 0$ from any initial position or

$$\lim_{t \to \infty} x_i(t) = 0, \quad 1 \leqslant i \leqslant k$$

As for stability on the hypersurface, it can be shown that under the conditions (8.4.17) the sliding modes on the hyperplane $\sigma = 0$ are stable if and only if all the roots of the characteristic equation

$$\boxed{\lambda^n + \sum_{i=1}^{n} a_i \lambda^{i-1} + b \sum_{i=1}^{k} \frac{1}{b} (c_{i-1} - a_i + c_i a_n - c_i c_{n-1})\lambda^{i-1} = 0} \tag{8.4.18}$$

have negative real parts, with the possible exception of the root $\lambda = c_{n-1} - a_n$. (Here we must require that all the coefficients a_i, b are fixed in time.)

Note that, if $k = n-1$, then the conditions (8.4.17a) and (8.4.17b) are vacuous and we are left with the two conditions

$$\boxed{\begin{aligned} \alpha_i &\geqslant \sup_t \frac{1}{b(t)} [c_{i-1} - a_i(t)] \\ \beta_i &\leqslant \inf_t \frac{1}{b(t)} [c_{i-1} - a_i(t)] \end{aligned}} \tag{8.4.19}$$

which can be satisfied for any plant parameters by choosing the feedback gain b sufficiently large. Since the switching surface $\sigma = 0$ is given by

$$\sum_{i=1}^{n} c_i x_i = 0$$

the sliding mode is equivalent to the equations

$$\boxed{\begin{aligned} \frac{dx_i}{dt} &= x_{i+1}, \quad 1 \leqslant i \leqslant n-2 \\ \frac{dx_{n-1}}{dt} &= -\sum_{i=1}^{n-1} c_i x_i \end{aligned}} \qquad (8.4.20)$$

and so for stability in the case of $k = n - 1$ all that we require is to choose the coefficients c_i so that the system (8.4.20) is stable.

8.4.5 Disturbance Rejection

In the above discussion no mention has been made about possible external disturbances which may affect the plant. The main effect of such disturbances is to lead to a steady-state error in the above control system, although this error can be made arbitrarily small by increasing the gain b. It is possible, however, to design variable structure systems so that the sliding mode is completely independent of the disturbance and we shall now illustrate this with the system shown in fig. 8.15. The plant has an external disturbance f and is driven by a first-order actuator. Two switching functions ψ and θ are now included, the latter being used to switch the actuator output.

The equations decribing this system are clearly

$$\left.\begin{aligned} \frac{dx_1}{dt} &= x_2 \\ \frac{dx_2}{dt} &= -a_2 x_2 - a_1 x_1 - bu + d_1 F + d_2 \dot{F} \end{aligned}\right\} \qquad (8.4.21)$$

where

$$F = a_{11} r + a_{12} \dot{r} + f, \quad a_1 = \frac{a_{11} a_{22}}{a_{21} a_{12}}, \quad a_2 = \frac{a_{11} a_{21} + a_{22} a_{12}}{a_{21} a_{12}}$$

$$d_1 = \frac{a_{22}}{a_{21} a_{12}}, \quad d_2 = \frac{1}{a_{12}}, \quad b = \frac{1}{a_{12} a_{21}}$$

$$a_{12} > 0, \quad a_{21} > 0$$

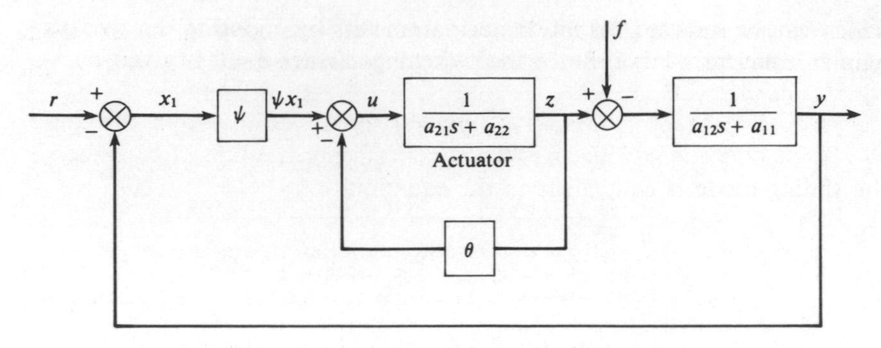

Fig. 8.15 Disturbance rejection.

The control input u is given by

$$\left.\begin{aligned} &u = \psi x_1 + \theta z \\ &\psi = \begin{cases} \alpha_1 & \text{if } x_1\sigma > 0 \\ \beta_1 & \text{if } x_1\sigma < 0 \end{cases} \\ &\theta = \begin{cases} \gamma_1 & \text{if } F\sigma > 0 \\ \rho_1 & \text{if } F\sigma < 0 \end{cases} \\ &\sigma = x_2 + cx_1 \end{aligned}\right\} \tag{8.4.22}$$

It appears from the definition of θ that we require knowledge of F to produce the appropriate switching. However, we also have the equation

$$z = F - a_{12}x_2 - a_{11}x_1 \tag{8.4.23}$$

and so F in (8.4.22) may be replaced by $z + a_{11}x_1 + a_{12}x_2$ which is known. Using (8.4.21)–(8.4.23) we have

$$\begin{aligned} \frac{dx_1}{dt} &= x_2 \\ \frac{dx_2}{dt} &= -(a_1 + a_{11}b\theta + b\psi)x_1 - (a_2 + a_{12}b\theta)x_2 + (d_1 + b\theta)F + d_2\dot{F} \end{aligned} \tag{8.4.24}$$

and so differentiating σ along the trajectories of this system we have

$$\begin{aligned} \frac{d\sigma}{dt} = &-(a_2 + a_{12}b\theta)\sigma + (c_1a_2 + c_1a_{12}b\theta - c_1^2 - a_1 - a_{11}b - b\psi)x_1 \\ &+ (d_1 + b\theta)F + d_2\dot{F} \end{aligned}$$

If we assume the fairly mild condition

$$\frac{|\dot{F}|}{|F|} \leqslant A \tag{8.4.25}$$

for some constant A, on the disturbance, then we shall obtain a sliding mode if

$$\begin{aligned}
\gamma_1 &\leqslant \frac{1}{b}(-Ad_2 - d_1) \\
\rho_1 &\geqslant \frac{1}{b}(Ad_2 - d_1) \\
\alpha_1 &\geqslant \frac{1}{b}\max_{\theta}\,[c_1(a_2 + a_{12}b\theta) - c_1^2 - a_1 - a_{11}b\theta] \\
\beta_1 &\leqslant \frac{1}{b}\min_{\theta}\,[c_2(a_2 + a_{12}b\theta) - c_1^2 - a_1 - a_{11}b\theta]
\end{aligned} \tag{8.4.26}$$

Moreover it can be shown that hitting of the switching line $\sigma = 0$ will occur if the characteristic equation (of (8.4.24) with $\psi = \alpha_1$)

$$p^2 + (a_2 + a_{12}b\theta)p + (a_1 + a_{11}b\theta + b\alpha_1)$$

has no non-negative real roots when $\theta = \gamma_1$ or $\theta = \rho_1$.

The above method of disturbance rejection may be generalized to deal with n-dimensional systems which have variable parameters; see Itkis (1976). Moreover, for a system with m zeros, we may guarantee the existence of sliding modes by switching the outputs of $m - 1$ first-order pre-filters as shown in fig. 8.16. The control switching is chosen as before and is therefore discontinuous and so the pre-filters are required to smooth the

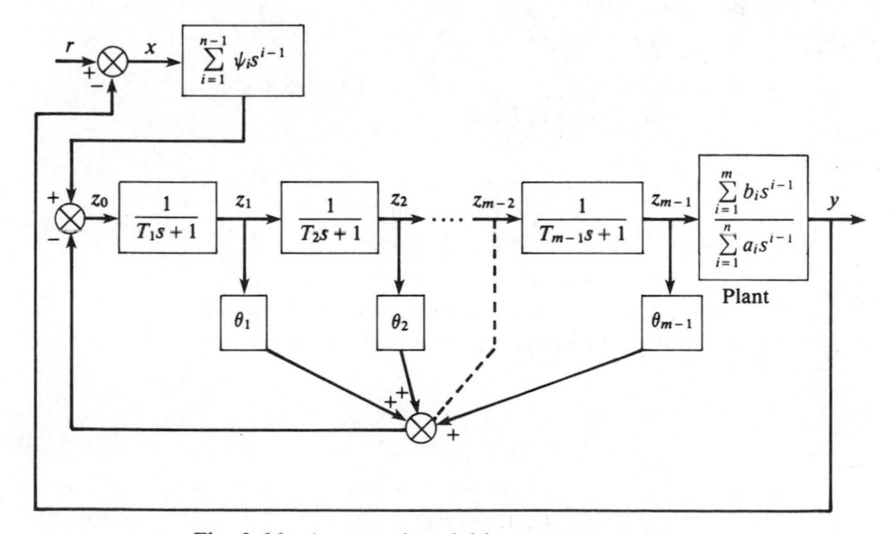

Fig. 8.16 A general variable structure system.

control signal enough so that it may be differentiated m times 'by the plant zeros'. The switching law is chosen to be

$$\boxed{v = z_0 = -\sum_{i=1}^{n-1} \psi_i x_1 - \sum_{i=1}^{m-1} \theta_i z_i} \tag{8.4.27}$$

where

$$\left.\begin{aligned} \psi_i &= \begin{cases} \alpha_i & \text{if } x_i\sigma > 0,\ 1 \leqslant i \leqslant n-1 \\ \beta_i & \text{if } x_i\sigma < 0 \end{cases} \\ i &= \begin{cases} \gamma_i & \text{if } z_i\sigma > 0,\ 1 \leqslant i \leqslant m-1 \\ \rho_i & \text{if } z_i\sigma < 0 \end{cases} \end{aligned}\right\} \tag{8.4.28}$$

8.4.6 Final Remarks

Returning to the case of the switching of $k-1$ derivatives of the state x_1, recall that the control is (cf. (8.4.16))

$$u = \sum_{i=1}^{k} \psi_i x_i$$

with

$$\psi_i = \begin{cases} \alpha_i & \text{if } x_i\sigma > 0 \\ \beta_i & \text{if } \chi_i\sigma < 0 \end{cases}$$

If we define

$$\left.\begin{aligned} \psi_i^0 &= \frac{\alpha_i + \beta_i}{2} \\ K_i &= \alpha_i - \psi_i^0 (>0) \end{aligned}\right\} \quad 1 \leqslant i \leqslant k$$

then

$$\psi_i = \psi_i^0 + \begin{cases} K_i & \text{if } x_i\sigma > 0 \\ -K_i & \text{if } x_i\sigma < 0 \end{cases}$$

Hence

$$\begin{aligned} u &= \sum_{i=1}^{k} \psi_i x_i \\ &= \sum_{i=1}^{k} \psi_i^0 x_i + \sum_{i=1}^{k} K_i \mid x_i \mid \operatorname{sgn} \sigma \end{aligned} \tag{8.4.29}$$

and so the control consists of a linear term plus a relay-like term with variable gain (a pure relay is of the form $K \operatorname{sgn} \sigma$). Lyapunov methods have been applied to relay systems by Weissenberger (1966) and to the control of bilinear systems by Longchamp (1980). The case of noisy states x_i is considered in detail by Itkis (1976), giving rise to switching surfaces which are not fixed in space since the noise will cause switching to occur off the ideal surface. Ideal sliding modes no longer exist when noise is present, but if the variance of the noise is small, similar methods to those above will predict the presence of **quasi-sliding modes**, i.e. trajectories which remain close 'on average' to the ideal switching surface. Note finally that the parameters c_i which define the switching surface are restricted by the system parameters and although they can be fixed for a fixed set of system parameters, if the latter are unknown then we must choose the c_i adaptively. We can use the methods of the previous sections to estimate the a_i which can then be used to choose appropriate c_i. Alternatively, it is possible to use specially designed adaptive systems which use information inherent in the variable structure system (see Itkis, 1976).

8.5 EXERCISES

1. Design an MRAC system for the plant

$$\frac{\mathrm{d}}{\mathrm{d}t}\begin{pmatrix} x_1^{\mathrm{p}} \\ x_2^{\mathrm{p}} \end{pmatrix} = \begin{pmatrix} 0 & 1 \\ 1 & a_{22}^{\mathrm{p}} \end{pmatrix}\begin{pmatrix} x_1^{\mathrm{p}} \\ x_2^{\mathrm{p}} \end{pmatrix} + \begin{pmatrix} 0 \\ b^{\mathrm{p}} \end{pmatrix} u, \quad a_{22}^{\mathrm{p}} < 0$$

where we have used a superscript p for convenience. (Verify condition (8.2.4) for your chosen model.)

2. Using equations (8.2.11) design an identification scheme for the system in exercise 1.

3. Find the Moore–Penrose inverse $B^{\dagger}$ of the matrix

$$B = \begin{pmatrix} 0 & 1 \\ 1 & 0 \\ 0 & 1 \end{pmatrix}$$

and determine $B^{\dagger}B$ and $BB^{\dagger}$.

4. Carry out the details of example 8.2.1 explicitly.

5. If, in the control system (8.3.3), we have $k = 3$, $n = 5$ and

$$\frac{C}{A} = \frac{1 + c_1z^{-1} + c_2z^{-2} + c_3z^{-3} + c_4z^{-4} + c_5z^{-5}}{1 + a_1z^{-1} + a_2z^{-2} + a_3z^{-3} + a_4z^{-4} + a_5z^{-5}}$$

evaluate E and F in the expression (8.3.6), i.e.

$$\frac{C}{A} = E + z^{-k}\frac{F}{A}$$

6. Design a minimum variance controller for the system

$$(1 - 0.5z^{-1})y(t) = u(t-2) + (1 - 0.3z^{-1})\zeta(t)$$

where $\zeta(t)$ is $N(0,1)$ (normally distributed with mean zero and variance 1).

7. Design a stable variable structure system of order 2 based on two linear systems with pure imaginary eigenvalues.

8. Write down the characteristic equation of the system (8.4.20).

9. Design a stable switching control system for the plant

$$\dot{x}_1 = x_2$$
$$\dot{x}_2 = x_3$$
$$\dot{x}_3 = -x_1 - 2x_2 - x_3 - bu$$

assuming that all states are measurable.

10. Determine conditions for the existence of a sliding mode for the system in fig. 8.16.

REFERENCES

Anderson, B. D. O. (1967) A system theory criterion for positive real matrices, *SIAM J. Control*, 5, 171–82.

Anderson, B. D. O. (1974) Adaptive identification of multiple-input multiple-output plants, *Proc. IEEE Conf. on Decision and Control*, 273–81.

Astrom, K. J. and Wittenmark, B. (1973) On self-tuning regulators, *Automatica,* 9, 185–99.

Boulljon, T. I. (1971) *Generalised Inverse Matrices*, Wiley-Interscience.

Carroll, R. L. (1974) A reduced adaptive observer for multivariable systems, *Proc. 15th Joint Aut. Control Conf.*, 160–3.

Carroll, R. L. and Lindorff, D. P. (1973) An adaptive observer for single-input single-output linear systems, *IEEE Trans. Aut. Control*, 18, 428–35.

Clarke, D. W. and Gawthrop, P. J. (1975) Self-tuning controller, *Proc. IEE*, 122 929–34.

Clarke, D. W. and Gawthrop, P. J. (1979) Self-tuning control, *Proc. IEE*, 126, 633–40.

Curran, R. T. (1971) Equicontrollability and its application to model following and decoupling, *Proc. 2nd IFAC Symp. on Multivariable Technical Control Systems*, Paper 1.1.3, Dusseldorf, W. Germany.

Gilbart, J. W. and Winston, G. C. (1974) Adaptive compensation for an optical tracking telescope, *Automatica*, 10, 125–31.

Hang, C. C. and Parks, P. C. (1973) Comparative studies of model reference adaptive control systems, *IEEE Trans. Aut. Control*, 18, 192–3.

Harris, C. J. and Billings, S. A. (eds.) (1981) *Self-tuning and Adaptive Control: Theory and Applications*, Peter Perigrinus.

Itkis, U. (1976) *Control Systems of Variable Structure*, Halsted Press.

Kreisselmeier, G. (1977) Adaptive observers with exponential rate of convergence, *IEEE Trans. Aut. Control.*, 22, 2–8.

Kreisselmeier, G. (1982) On adaptive state regulation, *IEEE Trans. Aut. Control*, 27, 3–17.

Kreisselmeier, G. and Narendra, K. S. (1982) Stable model reference control in the presence of bounded disturbances, *IEEE Trans. Aut. Control*, 27, 1169–75.

Kudva, P. and Narendra, K. S. (1973) Synthesis of an adaptive observer using Lyapunov's direct method, *Int. J. Control*, 18, 1201–10.

Landau, I. D. (1974) A survey of model reference adaptive techniques–theory and applications, *Automatica*, 10, 353–79.

Landau, I. D. (1979) *Adaptive Control: The Model Reference Approach*, Marcel Dekker.

Landau, I. D. and Courtiol, B. (1974) Design of multivariable adaptive model following control systems, *Automatica*, 10, 483–94.

Lion, P. M. (1967) Rapid identification of linear and nonlinear systems, *AIAA J.*, 5, 1835–42.

Ljung, L. (1977) Analysis of recursive stochastic algorithms, *IEEE Trans. Aut. Control*, 22, 551–75.

Longchamp, R. (1980) Stable feedback control of bilinear systems, *IEEE Trans. Aut. Control*, 25, 302–6.

Luders, G. and Narendra, K. S. (1974) A new canonical form for an adaptive observer, *IEEE Trans. Aut. Control*, 19, 117–19.

Monopoli, R. V. (1974) Model reference adaptive control with an augmented error signal, *IEEE Trans. Aut. Control*, 19, 474–84.

Morse, A. S. (1980) Global stability of parameter-adaptive control systems, *IEEE Trans. Aut. Control*, 25, 433–9.

Narendra, K. S. and Kudva, P. (1974) Stable adaptive schemes for system identification and control, *IEEE Trans. Systems, Man and Cyb.*, 4, Part I, 542–51; Part II, 552–60.

Narendra, K. S. and Monopoli, R. V. (1980) *Applications of Adaptive Control*, Academic Press.

Narendra, K. S. and Valavani, L. S. (1980) A comparison of Lyapunov and hyperstability approaches to adaptive control of continuous systems, *IEEE Trans. Aut. Control*, 25, 243–47.

Narendra, K. S., Lin, Y-H. and Valavani, L. S. (1980) Stable adaptive controller design, Part II: proof of stability, *IEEE Trans. Aut. Control*, 25, 440–8.

Parks, P. C. (1966) Lyapunov redesign of model reference adaptive control systems, *IEEE Trans. Aut. Control*, 11, 362–7.

Peterson, B. B. and Narendra, K. S. (1982) Bounded error adaptive control, *IEEE Trans. Aut. Control*, 27, 1161–8.

Utkin, V. I. (1977) Variable structure systems with sliding modes, *IEEE Trans. Aut. Control*, 22, 212–22.

Van Amerongen, J. and Udink ten Cate, A. J. (1973) Adaptive autopilots for ships, *Proc. IFAC/IFIP Symp. Ship Operation Automation*, Oslo, Norway.

Weissenberger, S. (1966) Stability-boundary approximations for relay-control systems via a steepest ascent construction of Lyapunov functions, *J. Basic Eng., ASME*, 88, 419–28.

Wellstead, P. E. and Sanoff, S. P. (1981) Extended self-tuning algorithm, *Int. J. Control*, 34, 433–55.

PART THREE

Microprocessor Implementation

9 INTRODUCTION TO THE Z80 MICROPROCESSOR AND ASSEMBLY PROGRAMMING

9.1 INTRODUCTION

In the first two sections of this book we have presented a substantial part of the elementary theory of the modelling and control of real systems. However, apart from the design of analogue circuits, little has been said about the hardware (and software) implementation of the control laws which have been developed. In this final section of the book we shall discuss some aspects of the implementation of control strategies on microprocessor systems. This consists of two major parts – hardware interfacing of the real system and software design of the feedback law. The former problem will be considered in detail in chapter 10 while in this chapter we shall discuss the structure and programming of the Z80 microprocessor.

Of course, feedback control algorithms can be implemented by analogue circuitry using op-amps as discussed in chapter 2. However, such circuitry has to be designed specially for each application. The main advantage of microprocessor implementation is that the hardware design is similar in many applications and only the software needs to be changed. Moreover hardware is now very cheap; it is the software development which is the major contributor to the cost. However, once robust programs have been developed, changing a small number of parameters will often allow the same software to be applied to other systems. For this reason it is important to write **portable** programs; i.e. ones which are not position dependent in the memory. Programs of this kind will run on any system (with the same processor chip) at any place in the memory; more will be said about this aspect of programming later.

The reader may wonder why it is important to understand the assembly level programming† and hardware design of microprocessor systems. Why

†See section 9.4 for assembly programming.

not simply buy a complete system with analogue to digital converters on board together with the software packages for running high level languages (such as FORTRAN or PASCAL)? Such systems can now be obtained in the price range £3000–£20 000, which for a large process plant is a negligible proportion of the cost. The answer is that in many situations there is a restriction not only on cost but more importantly on size, especially in aerospace applications where dedicated microprocessor boards must be designed to fit in a small space. Moreover, although real time languages are becoming available, it is still much easier to control the real time running of a program at assembly level as we shall see later. It is therefore important for a control engineer not only to know and understand the theoretical algorithms of control but also to be able to design and build an operational microprocessor system interfaced to the real plant.

Finally a word should be said about our choice of representative microprocessor – namely the Z80. The main reasons for choosing this particular chip are that it is cheap and readily available, it is probably the most powerful 8-bit microprocessor existing and it is well supported with software and interface chips. Of course, there are now more powerful 16- and even 32-bit microprocessors available but these are not so well supported in either aspect as the Z80. Moreover, the principles of the operation of all microprocessors are similar and the Z80 is a good representative example. In some cases, where fast sampling is involved, it may be necessary to use the faster chips, but the introduction to microprocessor implementation given here should enable the reader (armed with the technical data on the new microprocessor) to implement any other processor quite easily.

In this chapter we shall discuss a simple microprocessor (μP) system together with the basic 'architecture' of the Z80 μP, leaving the detailed interface design to chapter 10. We shall then describe, in detail, the instruction set of the Z80 μP and consider the operation of some typical instructions. No mention will be made here about logic design of computers; the interested reader should consult Krutz (1980) and Peatman (1972). The final part of this chapter will be concerned with assembly level programming and its application to some of the control algorithms developed in Part 2 of the book.

9.2 THE Z80 MICROPROCESSOR

9.2.1. A Microcomputer System

A 'minimal' microcomputer system is shown in fig. 9.1 and consists of Z80 μP, random access memory (RAM), read only memory (ROM) and a special purpose programmable input–output chip (PIO). The memory consists of two types; ROM stores fixed program and data information and can

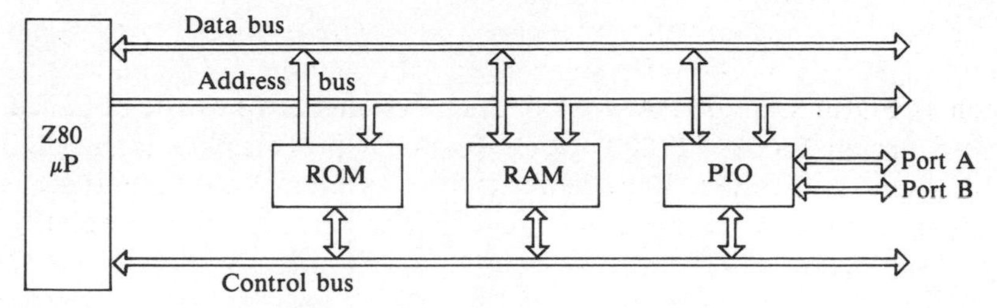

Fig. 9.1 A microcomputer system.

only be read from. RAM, on the other hand, consists of flip-flops (**static** RAM) or charge storage elements (**dynamic** RAM) and can be written into or read from. The PIO connects the microcomputer system to peripheral devices (VDUs,† keyboards, etc.) or control/data acquisition devices (digital-to-analogue converters, transducers, etc.).

Communication between the chips in the microcomputer system is implemented along three **buses**. A bus is just a parallel set of data lines grouped by function. The **address bus** is unidirectional with 16 bits; i.e. it can carry a 16-bit binary number in one direction only, namely from the Z80 chip to the rest of the system. The Z80 μP can therefore address $2^{16} = 65\,536$ different memory locations; in computer jargon $1K = 1024 = 2^{10}$ and so we may also say that the Z80 can address 64K distinct memories. The **data bus** is 8-bit and bidirectional (i.e. carries information both to and from the Z80 μP); it carries all data transfers around the system. The **control bus** carries status and control information around the system and will be discussed in more detail later.

9.2.2 Z80 Pin-out

The Z80 microprocessor is manufactured as a 40 pin DIP (dual in-line package); the pin specifications are shown in fig. 9.2(a), while the actual pin-out arrangement is shown in fig. 9.2(b). Note that some of the pin names are written with an overbar. This means that the corresponding signal is 'active low', i.e. that the voltage level at the pin is normally logic-1 and a signal is generated by pulling the line low to logic-0. It is often desirable in logic circuits to be able to disconnect (electronically) one end of a line from the other as in fig. 9.3 by means of a 'tri-state' buffer. When the $\bar{E}$ input to the buffer is low (active low!) the input logic state is transferred to the output as if the buffer were replaced by a short circuit.‡ However

† Visual display units.
‡ Note that the buffer may provide extra current drive so that it may act like a short circuit with current drive.

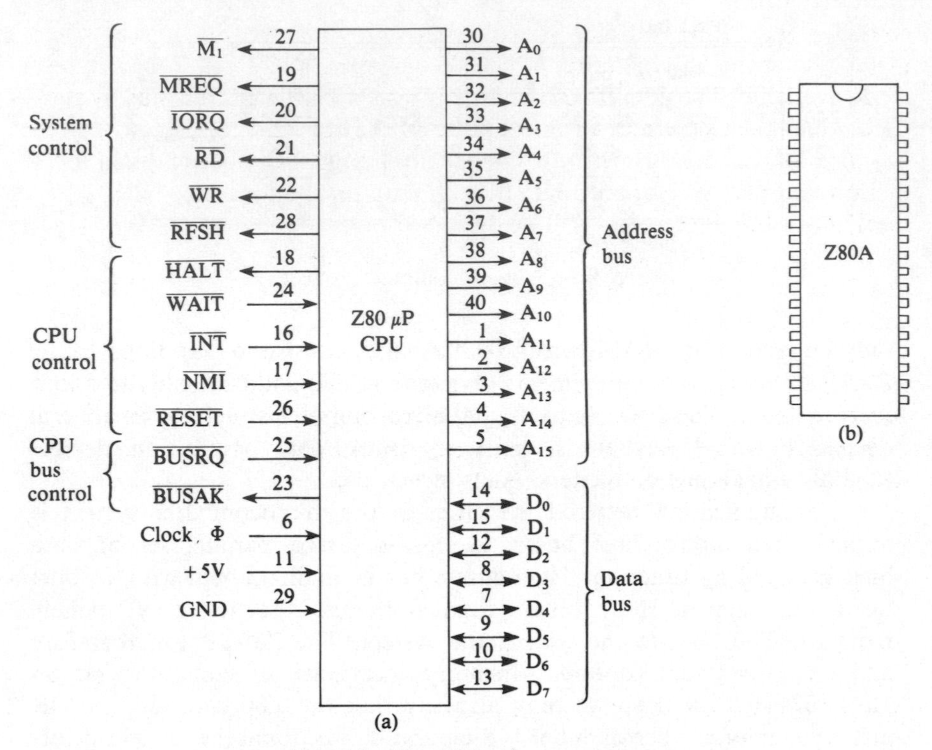

Fig. 9.2 The Z80 microprocessor.

when $\bar{E}$ is high the buffer is switched off (or **disabled**) and appears as a high impedance to logic circuit B, effectively disconnecting the two logic circuits. Such tri-state buffers are provided internally in the Z80 μP chip on the address and data buses and on the control lines $\overline{MREQ}$, $\overline{IORQ}$, $\overline{RD}$ and $\overline{WR}$. The purpose of this tri-stating will be discussed in chapter 10 in connection with interfacing techniques.

In this chapter we are mainly concerned with the software structure of the Z80 μP and the only pins which need concern us for the moment are A_0-A_{15}, D_0-D_7 forming the address and data buses, respectively, and the

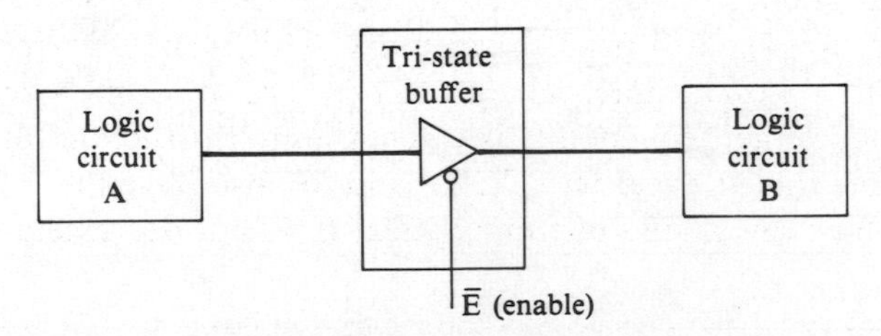

Fig. 9.3 Using tri-state buffers.

control signals $\overline{RD}$, $\overline{WR}$ and $\overline{M1}$. $\overline{RD}$ (active low) is used to 'inform' the memory that the CPU requires a byte (i.e. eight parallel bits) of data to be made available. The address of the data is placed on the address bus by the CPU which then generates a control pulse on the line $\overline{RD}$ (i.e. the logic level is pulled low for a short period). The memory must then decode this information and place the appropriate byte of data on the data bus. The exact mechanism for this process will be discussed in detail in chapter 10. When outputting a byte of data to the memory, similar remarks apply except that the CPU this time pulls $\overline{WR}$ low to indicate a memory write. The $\overline{M1}$ line will be discussed later.

The only other line that we need to mention now is the clock input (pin 6). This is just a 5-volt square wave which synchronizes the operation of the entire system and has frequency 2.5 MHz in the case of the Z80; however, a faster chip exists, the Z80A, which can be driven at 4 MHz. It is the Z80A which we shall consider in the sequel.

9.2.3 Internal Architecture

Although a complete understanding of the operation of the Z80 μP is not necessary for writing assembly level software, the programmer must be familiar with the overall architecture of the chip. This is shown in diagrammatic form in fig. 9.4. The first thing to notice is that the three system buses are continued internally on the chip and are isolated from the external buses by three buffers. The address bus is created internally by the 16-bit registers IX, IY, SP and PC. IX and IY are the **index registers** and are used to store base addresses for a sequence of memory locations, while SP is the **stack pointer** and holds the current address of the lowest occupied memory location in a specially reserved area of memory called the **stack**. More will be said about the software applications of these registers shortly. PC is the **program counter** and contains the address of the next byte of a program to be processed. Programs are stored sequentially in the random access memory (or the read only memory) starting at some memory location say 4000H for example.† (Memory locations will be written as 4-bit hexadecimal numbers from 0000H to FFFFH.) When a program is started, PC is loaded with 4000H and is automatically incremented (by one) as each byte of the program is fetched from the memory. This is shown symbolically in fig. 9.4 by the ± 1 on the internal address bus. The only time when program bytes are not fetched sequentially is when a jump instruction (or a HALT) is found. Note that the registers IX, IY and SP also have the ± 1 input on the address bus; this allows us to increment (INC) or decrement (DEC) the

†See chapter 3 for the hexadecimal number system.

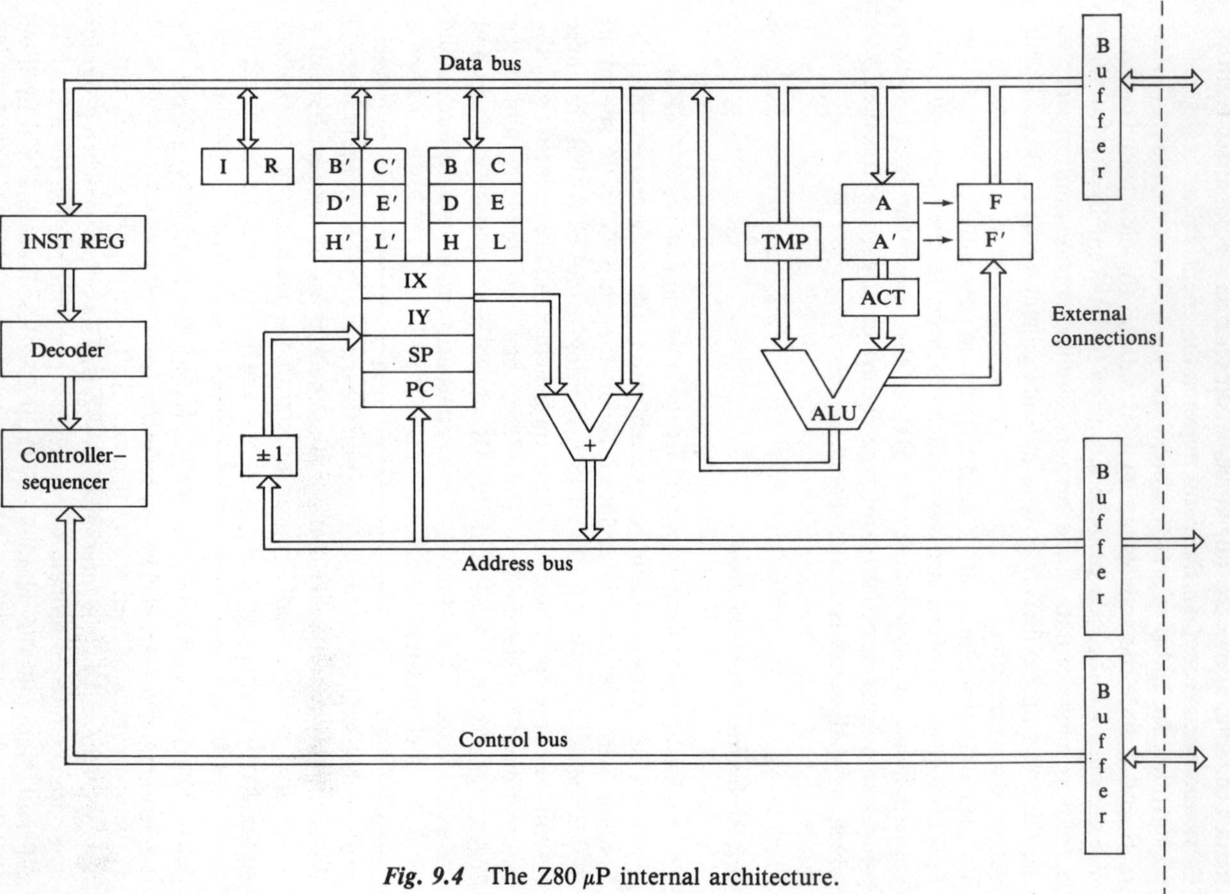

Fig. 9.4 The Z80 μP internal architecture.

contents of these registers under software control. Finally the IX and IY registers feed into a V-shaped summer, the other input of which is from the data bus. This allows the (algebraic) addition of an 8-bit offset from the base addresses stored on the index registers.

Above the 16-bit address registers are the 8-bit data registers B,C,D, E,H and L and a corresponding set B′,..., L′. These communicate with the data bus and act just like random access memories; they are designed to give a small amount of fast storage space on the chip. In fact, in some small applications one may be able to dispense with external RAM altogether and just use the internal 8-bit registers. The two complementary sets cannot be used together but a particular set can be selected by using a special exchange instruction, as we shall see later. These registers can also communicate with the address bus, but since they are only 8-bits wide we must put two together to form a 16-bit address. Hence the register pairs BC, DE and HL can each hold an address, the leftmost register holding the high byte while the other holds the low byte (in fact this accounts for the name of the HL pair). Thus if we require to store the address 34FAH in HL then we place 34H in H and FAH in L.

Moving to the right of fig. 9.4 we see a V-shaped block labelled ALU. This is the **arithmetic–logic unit** and is where all the computation of arithmetic and logical operations takes place. The ALU has inputs TMP and ACT which are just temporary registers for buffering purposes. Above ACT is a pair of registers A and A′ called the accumulators which are again complementary like the BB′, CC′, ... register pairs and can be switched with a special instruction. The accumulator (whichever of the two is selected) holds the results of arithmetic and logical operations and communicates with the data bus. The output of the ALU feeds back onto the data bus and also into another pair of special purpose registers F and F′. These are selected with the appropriate accumulator (i.e. F with A and F′ with A′). The F (and F′) register contains status information and is called the **flag register**. Each flag register has the form

S	Z	X	H	X	P/V	N	C

where X indicates a 'don't care' bit and is not used by the programmer. The six remaining bits are defined as follows:

Bit 0, C — This is the carry flag and is the carry from bit 7 of the accumulator.

Bit 1, N — The Z80 has the ability to do BCD (binary coded decimal) operations and the N flag is used to store the type of the last operation – it is set for a subtraction.

Bit 2, P/V — This is used in the P (parity) mode for logical operations, being set if the result of an operation is of even parity. For arithmetic operations it is in the V (overflow) mode and is set when an overflow occurs in signed arithmetic.

Bit 4, H This is again used in BCD in arithmetic to store a carry from the low order decimal digits (H = half-carry flag).

Bit 6, Z The zero flag is set if the accumulator is loaded with zero in certain operations.

Bit 7, S Negative sign flag. This is used in signed arithmetic and is set if the result is negative. It is, in fact, bit 7 of the accumulator.

Note that the flags C,P/V,Z,S are testable by the programmer, whereas H and N are not. The latter are used by the microprocessor and are of little interest to the programmer.

On the left of fig. 9.4 we see the main control and decode section of the Z80 μP. This part is of little concern to the programmer, although it is important to realize that when an instruction operation code (OP code) is fetched from the memory, each byte of the OP code is placed on the data bus and the internal gating on the Z80 chip places the OP code bytes successively into the instruction register (INST REG) where they are decoded. The decoded instruction is then 'executed' by the controller–sequencer, which works on an even lower level of programming than the machine (or assembly) language, called **microprogramming**. However, this topic is beyond the scope of this book (see, for example, Hill and Peterson, 1978) and is, in any case, of no significance at the assembly level. Let us also note finally the existence of two other registers – namely I and R shown in fig. 9.4. The R register stores the low order byte of an address and is used in refreshing dynamic memories, which takes place as each instruction is being decoded. It is automatically incremented after each instruction fetch. Although the R register is available to the programmer it is only used for testing purposes. The I register is the 'interrupt vector' and is used in interrupt processing to be discussed in detail in chapter 10.

9.2.4 Operation of Typical Instructions

Although not essential to the programmer, it is useful to be aware of the main features of the hardware implementation of some typical instructions. We shall therefore discuss some simple operations in this section and show how the microprogrammed control sequencer opens the correct gates for the implementation of these operations. First recall that the system clock governs the overall operation of the μP and determines the smallest time period in the system as one clock cycle, of length $T = 1/f$, where f is the frequency (in Hz). The Z80 works in **machine cycles** which are variable in length, depending on the operation, and are 3 to 5 clock cycles long. The clock cycles are numbered consecutively in each machine cycle so that a machine cycle may consist of, say, the clock cycles T1,T2,T3,T4. Each operation requires a number of machine cycles, which again varies with the operation, but the number is at most 5.

Instructions are performed in three phases, namely fetch, decode and execute. In the fetch–decode phases the OP code is brought from the memory, placed in the instruction register and decoded. The fetch–decode phases are the same for each instruction and require (part of) one machine cycle consisting of clock periods T1,T2,T3. During T1, the program counter (PC) contents are placed on the address bus (the controller enables the gates as in fig. 9.5). In T2, the program counter is incremented and in T3 the instruction is placed in the instruction register (fig. 9.6). (The process of decoding the address bus to select the correct memory contents will be considered in chapter 10.)

Once the decoding of the instruction has been performed the operation is executed. This depends, of course, on the operation so we first consider

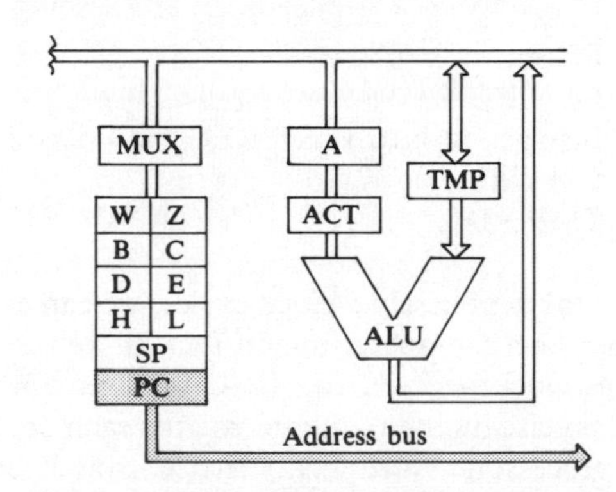

Fig. 9.5 Putting PC onto the address bus.

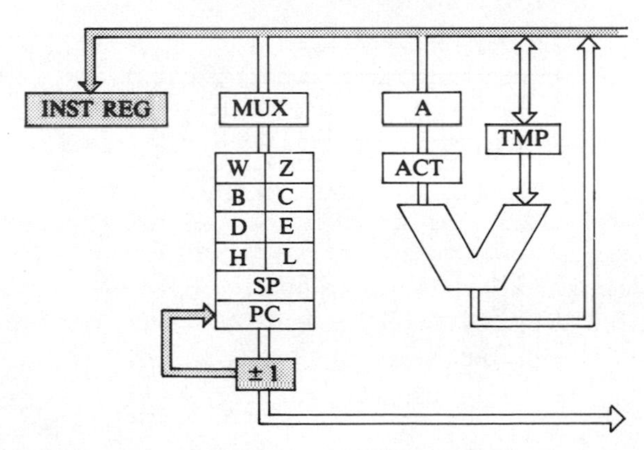

Fig. 9.6 Loading the institution into INST REG and incrementing PC.

the simple operation

LD D, C

This instruction consists of placing the byte of data which is currently in the C register into the D register. Note that the format of this operation is in **assembly** level mnemonics. LD is the assembly operation code mnemonic for 'load', while D and C are the operands. It is important to remember that in any assembly level operation (which has two operands) the source of the data is always placed after the destination. Hence in the case of the operation LD D, C the source of the data is C while its destination is D. We can now complete our discussion of the operation: in T4 the contents of C are placed in TMP (fig. 9.7) and in T5 the contents of TMP (which originated from C) are loaded into D (fig. 9.8). Hence the operation LD D, C requires 1 machine cycle lasting 5 clock cycles and we may summarize the operation symbolically as follows:

T1 PC out
T2 PC = PC + 1
T3 Instruction into INST REG (and decode)
T4 C→TMP
T5 TMP→D

Since this operation takes precisely 5 clock cycles, we can evaluate precisely the total time to perform the operation. In fact, if we use a 2 MHz clock (which has a 500 ns clock period), LD D, C requires 5×500 ns = 2.5 μs. The ability to have precise timing requirements for each operation is important for real time processing. Note finally that C cannot be transferred to D directly, but must be placed temporarily in TMP. This is simply because no gating exists on the chip for performing the direct transfer.

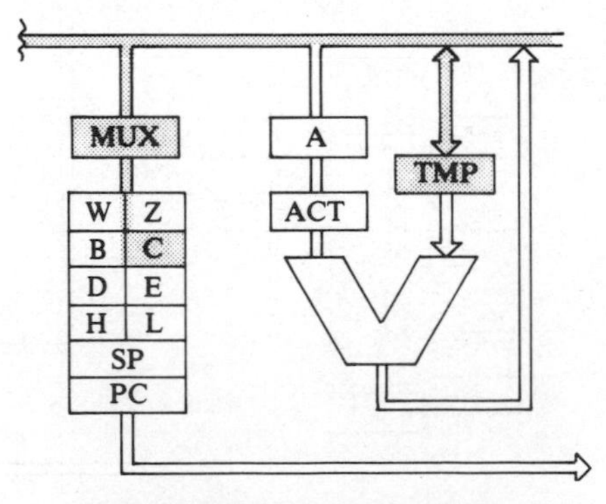

Fig. 9.7 Moving C to TMP.

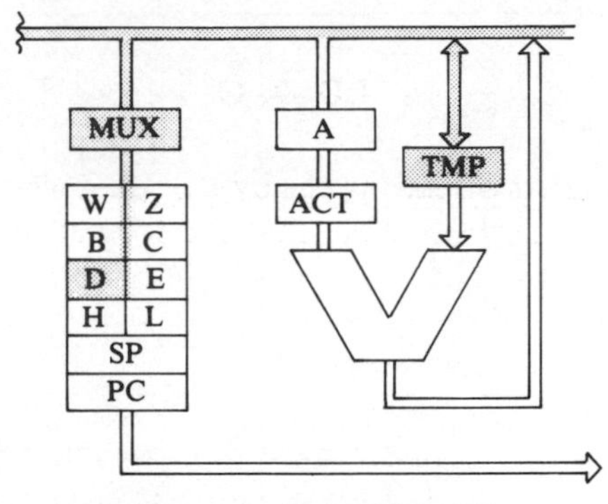

Fig. 9.8 Moving TMP to D.

We shall next consider an arithmetic operation; namely

ADD A, B

Again, ADD is the assembly mnemonic for 'add' and the operation consists of adding the contents of B to A and storing the result in A (symbolically $A + B \rightarrow A$ or $A \leftarrow A + B$). As stated above, the fetch–decode cycles are the same for each instruction (although the output of the decoder is different, of course) and so cycles T1, T2, T3 are as before. Now in T4 the contents of B are placed in TMP while those of A are placed in the temporary accumulator ACT (fig. 9.9). Now something curious happens. The addition operation is not performed immediately (in T5), but machine cycle 1 is complete with T1–T4. In T1 of the next machine cycle the fetch phase of the next instruction begins, even though the present instruction ADD A, B has not been completed. In T2 of this machine cycle PC = PC + 1 occurs since an instruction fetch has been performed. It is in T2 of this machine

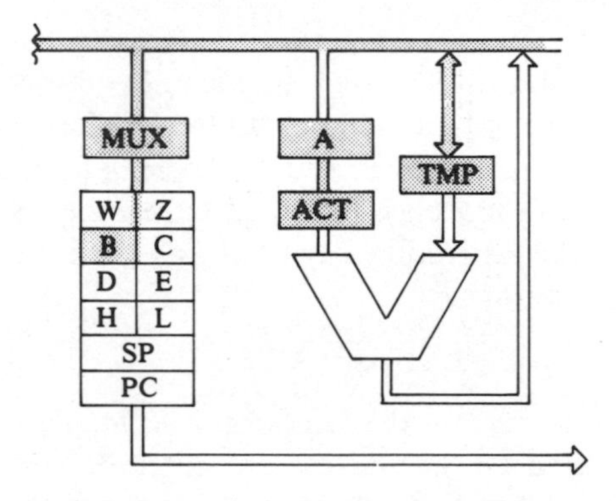

Fig. 9.9 Moving B to TMP and A to ACT.

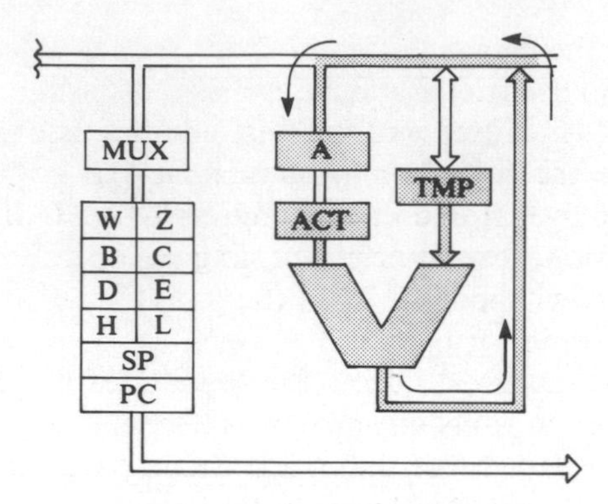

Fig. 9.10 Performing A + B→A.

cycle that the addition in the operation ADD A, B takes place (fig. 9.10). (Note that it cannot occur in T1 of the next machine cycle since the data bus is required for status information in any T1 period.) We can show this 'instruction overlap' symbolically as follows:

Instruction n	T1	T2	T3	T4	T1	T2		
Instruction $n+1$					T1	T2	T3	. . .

The reason for this overlap should be clear; if the operation had been completed in T5 before the start of the next instruction, the operation would have required 5 clock cycles. By using the fact that the ALU and the data bus are free during T2 we have saved 20 per cent of the time required to execute the instruction. Such savings in time can be extremely important.

The above arithmetic operation has both operand registers A and B on the CPU. We shall now consider an operation which requires a call to memory to obtain the second operand, namely

ADD A, (HL)

The parentheses around the HL register pair now imply that the contents of HL are to be regarded as a memory address rather than data. We may translate the operation as follows: add to A the byte of data stored in the memory location whose address is in HL and place the result back in A. The implementation of this operation begins in M1 with T1, T2, T3 cycles as before. Then in T4 we have

A→ACT

i.e. accumulator contents into the temporary accumulator. In M2 the contents of HL are placed on the address bus where they are decoded by the external circuitry and the memory puts the appropriate data on the data bus. In T3 of M2 the data is put in the TMP register. Hence by the end of M2, both sides of the ALU are 'conditioned'. As in the above operation, the clock cycle T4 is omitted and the final addition is overlapped in T2 of M3. To illustrate the operation suppose that HL contains the address 3FFFH, while the byte stored at this address is A3H. If A contains 17H before the instruction, then after performing the operation ADD A, (HL), the accumulator A will contain BAH (fig. 9.11).

The three operations considered above all require one byte of memory for storage. In other words, during the running of a program containing any of these operations, only one memory fetch will be required to perform the operation. This is because the operands are fixed and hence 'implied' by the operation code (OP). An operation which does not have a fixed source operand is

LD A, (*nn*)

Again the presence of the parentheses should immediately suggest memory rather than data. The source operand is specified by the programmer and is a 16-bit address *nn*. It should be clear that the instruction means LOAD A with the contents of memory *nn*. Before discussing the implementation of this instruction we consider how such an instruction is stored in memory as part of a program. Obviously the OP code itself requires one byte of memory and since the destination is A, which is fixed, this can be implied by the OP code byte. However, the source of the data (*nn*) is defined by a 16-bit binary number and so requires two 8-bit memories for its storage. Any 16-bit number *nn* can be written in the form

$$nn = nh \times 256 + nl$$

where *nh* is the 'high byte' of *nn* while *nl* is the 'low byte' of *nn*. It is a convention that *nl* is stored before *nh*. As we shall see later, the HEX code for

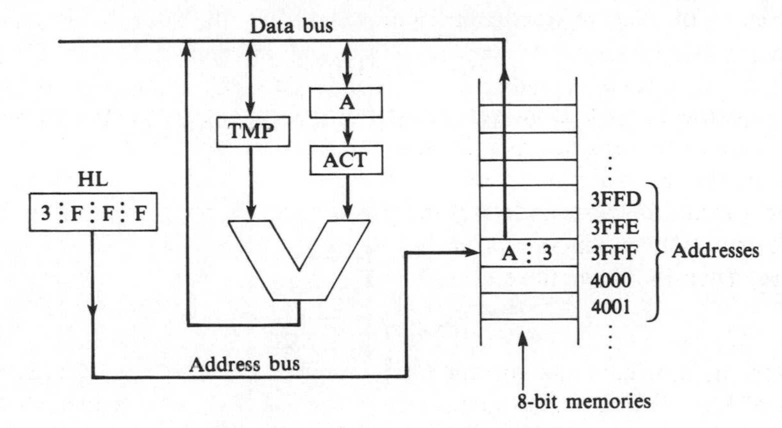

Fig. 9.11 Illustrating LD A, (HL).

the instruction LD A, (*nn*) is 3A; it follows that this operation requires three consecutive bytes of RAM for its storage (fig. 9.12). The implementation of the instruction now proceeds as follows. As before the instruction byte 3A is fetched and decoded in M1. Once the instruction has been decoded the μP 'knows' that two further memory fetch cycles are required to determine the address *nn*. This address must somehow be placed on the address bus and so it has to be stored in an address register. However, it cannot be stored in any of the registers BC, DE, HL, SP, PC, IX or IY since this will destroy the information already held there. For this reason an extra address register pair WZ is provided for temporary storage of the address *nn* in such an instruction. Since this register pair is not directly available to the programmer, it is not shown in fig. 9.4. In M2 the byte *nl* is fetched from memory and stored in the Z register (PC is, of course, automatically incremented after any memory fetch). In M3 the byte *nh* is put into the W register and finally, in M4, the contents of WZ are placed on the address bus and the data is brought from the memory location *nn* and placed in A. Note that PC is not incremented after the last memory fetch – it is important to realize that it is only incremented when fetching instruction bytes (OP codes or data), i.e. when PC is placed on the address bus.

We shall finally consider the implementation of a simple jump instruction, namely

JP *nn*

This is an unconditional jump and tells the μP to fetch the next instruction byte OP code from address *nn* rather than the one occurring sequentially in memory after the (3-byte) instruction JP *nn*. Its implementation is somewhat similar to that of the last instruction. In fact, in M1–M3 the address *nn* is fetched and placed in WZ as with LD A, (*nn*). However, at this point the instruction is effectively complete and so no machine cycle M4 is present. Instead M1 of the next instruction continues, with the difference that in T1 the WZ register contents are placed on the address bus, rather than the PC contents. Moreover, PC is not incremented after T1, but instead WZ is incremented and then loaded into PC. Hence we can show the effects of a jump type instruction and a non-jump type instruction on

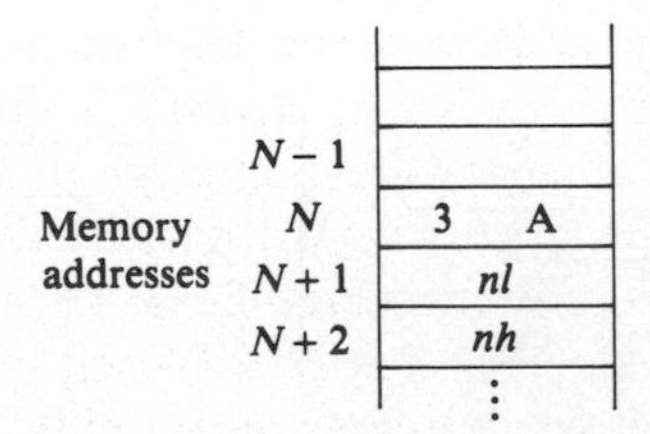

Fig. 9.12 Memory storage of words.

the implementation of the next instruction as follows:

After a jump instruction	*After a non-jump instruction*
T1: WZ→address bus	T1: PC→address bus
T2: WZ + 1→PC	T2: PC + 1→PC

This completes our discussion of the implementation of some typical instructions. Other instructions work in similar ways and the reader should now have a general idea of what type of gating is required to perform all the Z80 instructions.

9.2.5 Addressing Methods

We are now almost ready to present the complete instruction set of the Z80 μP. However, before doing so we must formally define the different types of addressing which are available on this μP. Since the ideas are common to most instructions, we shall illustrate the addressing methods by considering the LD (load) instruction. The simplest method of addressing is called **register addressing** since the operands are both registers. For example, we have the instruction

LD A, B (load A from B)

The next type is **register indirect** where the source operand contains not the data to be operated on but its address as in the example

LD A, (HL) (load A from the address stored in HL)

A similar type is **indexed addressing** as in the instruction

LD A, (IX + *d*)

The difference is that the address of the data is not stored directly in IX but is that stored in IX plus the offset *d*. Hence this instruction means 'load A with the contents of the memory location given by adding *d* to the contents of IX'. Note, however, that *d* is regarded as a **two's complement** number in order to allow forward or backward offsets. Recall that the two's complement of an 8-bit binary number B is 2^8-B. When storing signed numbers in *q* byte of memory, negative numbers are stored in two's complement form and are identified by having a '1' in the 8th bit (2^7 position) while positive numbers are stored in the usual form with a 0 in the 8th bit. Hence bit 8 is a 'sign bit' and bits 1 to 7 (or 2^0 to 2^6 positions) contain the absolute value of the number. Since 7 binary bits can store 128 different numbers we can store the numbers $-128, \ldots, -1, 0, 1, \ldots, 127$ in a byte of memory. Hence, for example, suppose that IX contains the address AB00H; then

LD A, (IX + 3)

means that A should be loaded from the memory AB03H while

LD A, (IX − 2)

means that A should be loaded from memory AAFEH since FEH = −2 as a signed two's complement number. (Note that the latter instruction is stored in memory as the 3 byte OP code DD 7E FE.) For the convenience of the reader, we give the two's complement hex and decimal equivalents of the signed numbers −128 to 127 in table 9.1. Note that the decimal equivalent of negative signed numbers is easy to remember, since it is just 256 minus the absolute value of the signed number.

So far we have seen instructions which point to memory via a 16-bit memory register. We can also point directly to memory using **extended addressing** as in the example

LD A, (F1A3H)

which means load A from the address F1A3H. (Note that when we discuss assembly programming we shall see that actual hexadecimal memory addresses are rarely used, but, rather labels, as in higher level languages.)

All the above instructions actually require the data to be brought from memory. Of course, we can load registers directly with a fixed number as in the example

LD A, 7AH

Table 9.1

signed number	hex	decimal	signed number	hex	decimal
0	0	0	−1	FF	255
1	1	1	−2	FE	254
2	2	2	−3	FD	253
⋮	⋮	⋮	⋮	⋮	⋮
125	7D	125	−126	82	130
126	7E	126	−127	81	129
127	7F	127	−128	80	128

in which A is loaded with the hex number 7A. This is called **immediate addressing.**†

Next consider another type of jump instruction – the relative jump. This uses a type of addressing **(relative addressing)** not of data but of the next instruction. A typical example is

JR e

which merely adds e to the contents of PC so that the address of the next instruction is $PC + e$ (e is a signed number to allow forward or backward jumps). However, we should note that there is a slight complication here. When this assembly language instruction is translated (by an **assembler**) to machine language, 2 is subtracted from e. To understand this we must consider the operation of this instruction in more detail. Suppose, in particular, we have the instruction JR 6. The OP code for the unconditional relative jump is 18 and we replace 6 by $6 - 2 = 4$ as noted above. The stored program containing this operation therefore has the form

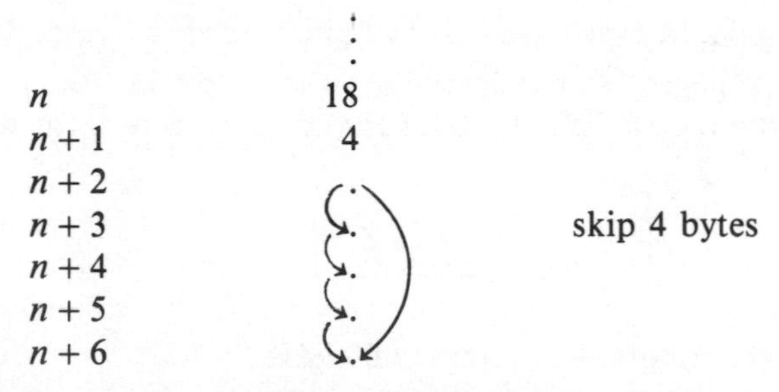

When the OP code 18 is fetched and decoded the CPU recognizes that it must fetch the next byte, namely 4, specifying the size of the jump. When this byte has been fetched, PC contains $n + 2$ and since at the assembly level, e refers to the contents of PC at the start of the instruction (i.e. n) we must subtract 2 from e in going from assembly to machine language. Similarly if $e - 2 > 127$ it is regarded as a two's complement negative number and is interpreted as a backward jump. Hence JR -2 appears in memory as

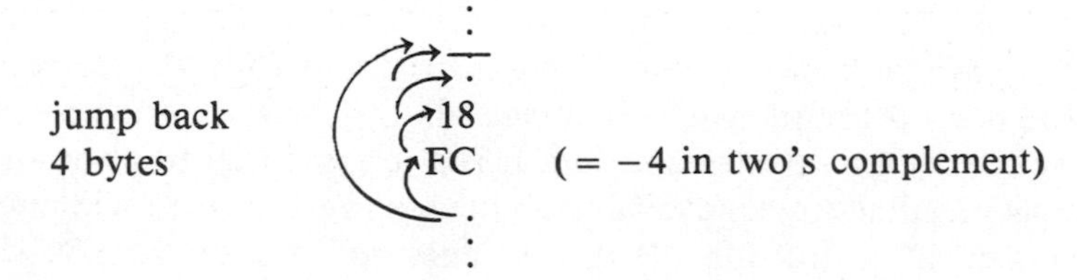

† Loading a 16-bit register pair with a 4-bit hex number is called **immediate extended addressing** as in the example LD BC, F13CH

since FC = -4 in two's complement and $e-2=-2-2=-4$. Note that $-128 \leqslant e-2 \leqslant +127$ so that $-126 \leqslant e \leqslant +129$. Having given this detailed explanation we now note (as we shall see later) that we do not use an actual number for e but a statement lable, again as we do in high level languages. The assignment of the number e to the label and the adjustment $e-2 \rightarrow e$ is all done by the assembler.

Finally, three other types of addressing which should be mentioned are (a) **bit addressing**, for example

BIT 7, A

where the 7th bit of the accumulator is tested and the zero flag is set if the bit is zero; (b) **implied addressing** as in the example

EXX

where no registers or memories are specified – they are implied by the mnemonic; (c) **page**† **zero addressing** where the instruction refers to memories in the region 0000H – 00FFH (i.e. **page zero**), for example

RST 8

which causes the program to jump to 0008H.

9.3 THE Z80 INSTRUCTION SET

9.3.1 Introduction

For completely successful assembly programming on a particular microprocessor, the programmer must be totally familiar with all the instructions that are available on the chip. These instructions form the **instruction set** of the microprocessor which is divided into groups of instructions, each group having a particular function. It is also important to be aware of the instructions that are *not* available on the microprocessor; for example, the (hypothetical) instruction LD B, (*nn*) does not exist on the Z80 μP. Only the accumulator (A register) can be loaded directly from memory. To imple-

---

† A **page** is 256 words of memory.

ment this hypothetical instruction one must perform the two (valid) instructions

LD A, (*nn*)

LD B, A

This may seem pedantic to the reader, but experience has shown that a great deal of time can be wasted by a lack of familiarity with the complete instruction set. The initial investment in time spent in memorizing the instruction set will almost certainly pay off in the long run.

The three major facets of any instruction of which the programmer must be aware are:

(a) The effect of the instruction,
(b) its timing, in terms of the number of clock cycles taken to perform the instruction,
(c) the flags which are affected.

Using the information in (b) it is clearly possible to obtain a precise† timing for any assembly program in terms of the micro clock rate. We shall now proceed to discuss the instruction set of the Z80 giving each group in a functional table.

9.3.2 The 8-bit Load group

The 8-bit load group is shown in table 9.2. Each box refers to a single instruction and contains the machine language equivalent and its timing in clock cycles (in the bottom right-hand corner of each box). Note that the machine language equivalent may contain one to four bytes, and some instructions have more than one byte in the OP code – the indexed instructions, for example. Hence the instruction LD A, (IX + *d*) has the machine language equivalent

DD
7E
d

in which DD and 7E together form the OP code. Recall that the assembly mnemonic has the form

LD destination operand, source operand

so that LD A, B has the OP code 78 (and *not* 47, which is LD B, A). Any empty box implies that the corresponding instruction does not exist.

† Precise to within the accuracy of the crystal oscillator. This timing assumes that no external interrupts occur during the execution of the program (see chapter 10).

Table 9.2 LD group: 8-bit

		Source														
		Implied		*Register*							*Reg. indirect*			*Indexed*		*Ext. Addr.*	*Imm.*
		I	R	A	B	C	D	E	H	L	(HL)	(BC)	(DE)	(IX + *d*) †	(IY + *d*)	(*nn*)	*n*
Destination Register	A	ED 57 9	ED 5F 9	7F 4	78 4	79 4	7A 4	7B 4	7C 4	7D 4	7E 7	0A 7	1A 7	DD 7E *d* 19	FD 7E *d* 19	3A *n* *n* 13	3E *n* 7
	B			47 4	40 4	41 4	42 4	43 4	44 4	45 4	46 7			DD 46 *d* 19	FD 46 *d* 19		06 *n* 7
	C			4F 4	48 4	49 4	4A 4	4B 4	4C 4	4D 4	4E 7			DD 4E *d* 19	FD 4E *d* 19		0E *n* 7
	D			57 4	50 4	51 4	52 4	53 4	54 4	55 4	56 7			DD 56 *d* 19	FD 56 *d* 19		16 *n* 7
	E			5F 4	58 4	59 4	5A 4	5B 4	5C 4	5D 4	5E 7			DD 5E *d* 19	FD 5E *d* 19		1E *n* 7
	H			67 4	60 4	61 4	62 4	63 4	64 4	65 4	66 7			DD 66 *d* 19	FD 66 *d* 19		26 *n* 7
	L			6F 4	68 4	69 4	6A 4	6B 4	6C 4	6D 4	6E 7			DD 6E *d* 19	FD 6E *d* 19		2E *n* 7

Reg. Indirect	(HL)			77 7	70 7	71 7	72 7	73 7	74 7	75 7							36 *n* 10
	(BC)			02 7													
	(DE)			12 7													
Indexed	(IX + *d*)			DD 77 *d* 19	DD 70 *d* 19	DD 71 *d* 19	DD 72 *d* 19	DD 73 *d* 19	DD 74 *d* 19	DD 75 *d* 19							DD 36 *d* *n* 19
	(IY + *d*)			FD 77 *d* 19	FD 70 *d* 19	FD 71 *d* 19	FD 72 *d* 19	FD 73 *d* 19	FD 74 *d* 19	FD 75 *d* 19							FD 36 *d* *n* 19
Ext addr.	(*nn*)			32 *n* *n* 13													
Implied	I			ED 47 9													
	R			ED 4F 9													

Destination

† $-128 \leqslant d \leqslant 127$.

Concerning the flags (i.e. F register status after the operation) we note:

Apart from instructions depending on the I and R registers, no flags are affected by the 8-bit LD group

This is extremely important since it is sometimes wrongly assumed that all instructions control the flags. Thus if we perform the instruction LD A, (*nn*) and the memory location *nn* contains 00H, then the zero flag is *not* changed by this instruction (it may already be set, of course).

9.3.3 The 16-bit Load Group

This group of instructions is shown in table 9.3. Note first that the only direct register transfers are LD SP, HL; LD SP, IX and LD SP, IY. Also the register pairs BC, DE, HL, SP, IX, IY can all be loaded directly with a given 16-bit number (immediate **extended** addressing); for example

LD HL, FA13H

loads HL with FA13 (i.e. H with FA and L with 13). The extended addressing operations now require some comment. Consider the example

LD DE, (*nn*)

This assembly mnemonic seems to be contradictory since it appears to be saying 'load the pair DE with the contents (i.e. 8 bits) of memory location *nn*'. However, since DE is a 16-bit register pair, the correct interpretation of the instruction is to load the low byte of the pair DE , namely E, from *nn* while the high byte D is loaded from $nn + 1$, the latter address being implicit in the instruction. Symbolically,

$$(nn) \rightarrow \text{E}, \quad (nn + 1) \rightarrow \text{D}$$

Similar remarks apply to the other extended address instructions.

Finally there are two types of register indirect instructions available. These involve the stack pointer SP and are of the form LD *dd*, (SP) and LD (SP), *dd* where *dd* represents one of the register pairs AF, BC, DE, HL, IX or IY. These two types of operation are usually abbreviated to POP *dd* and PUSH *dd* respectively. Consider the typical instruction PUSH IX. The operation may be written symbolically as

$$(\text{SP} - 1) \leftarrow \text{IX}_{\text{H}}, \quad (\text{SP} - 2) \leftarrow \text{IX}_{\text{L}}$$

and can be explained as follows. Firstly the stack pointer is decremented (by 1) and the contents of the stack pointer are used as the address into which the high byte of IX is loaded. Similarly SP is decremented again and used as a pointer to the low byte of IX. (Note that if SP, or any other 16-bit register, contains 0000H, then after decrementing it will contain FFFFH.)

Table 9.3 LD group: 16-bit

		Source									
		Register							*Imm. ext.*	*Ext. addr.*	*Reg. indir.*
		AF	BC	DE	HL	SP	IX	IY	*nn*	(*nn*)	(SP)
Destination Register	AF										F1 10
	BC								01 *n* *n* 10	ED 4B *n* *n* 20	C1 10
	DE								11 *n* *n* 10	ED 5B *n* *n* 20	D1 10
	HL								21 *n* *n* 10	2A *n* *n* 16	E1 10
	SP				F9 6		DD F9 10	FD F9 10	31 *n* *n* 10	ED 7B *n* *n* 20	
	IX								DD 21 *n* *n* 14	DD 2A *n* *n* 20	DD E1 14
	IY								FD 21 *n* *n* 14	FD 2A *n* *n* 20	FD E1 14
Ext addr	(*nn*)		ED 43 *n* *n* 20	ED 53 *n* *n* 20	22 *n* *n* 16	ED 73 *n* *n* 20	DD 22 *n* *n* 20	FD 22 *n* *n* 20			
Reg indir	(SP)	F5 11	C5 11	D5 11	E5 11		DD E5 15	FD E5 15			

←PUSH (row "Reg indir (SP)")

↑ POP (column "Reg. indir. (SP)")

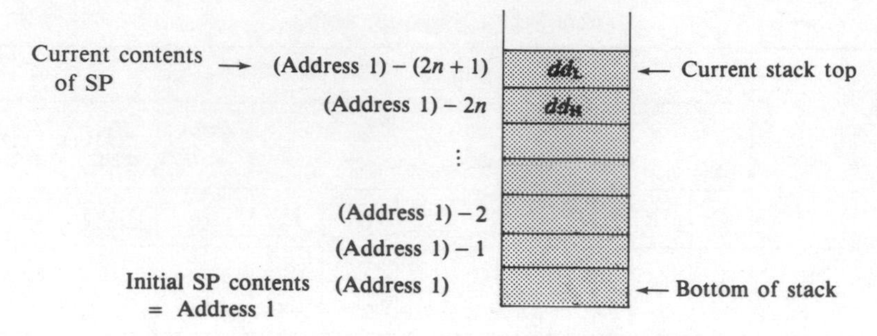

Fig. 9.13 Illustrating the stack.

The **stack** is therefore used to store register information temporarily and grows *downward* in memory. The stack pointer always points to the lowest filled byte of the stack, and should always be initialized by the programmer at the start of a program, making sure to leave enough room in memory for the complete stack. Failure to do this will result in the stack spilling over into the program or data areas usually with disastrous results. Consider now the dual instruction

POP IX

which has the symbolic operation

$$IX_L \leftarrow (SP), \quad IX_H \leftarrow (SP+1)$$

In this case the low byte of IX is loaded from the address contained in SP, then SP is incremented and the high byte of IX is loaded from the current stack pointer memory address. Finally SP is incremented again. Note that the loading and incrementing are performed in the reverse order from the decrementing and loading of the PUSH instruction. This is because the stack has a LAST-IN, FIRST-OUT structure (fig. 9.13).

No flags are affected by the 16-bit load group

9.3.4 The Exchange Group

The alternate register pairs AF′, BC′, etc. can be used by first performing an exchange operation. The exchange instructions existing on the Z80 are shown in table 9.4 and have the mnemonics EX AF, AF′; EXX; EX DE, HL; EX (SP), *dd* where *dd* = HL, IX or IY. If we wish to load the B′ register with 0BH, say, then we write

```
EXX
LD B, 0BH
```

Table 9.4 Exchanges

		Implied addressing				
		AF′	BC′, DE′, HL′	HL	IX	IY
Implied	AF	08 4				
	BC DE HL		(EXX) D9 4			
	DE			EB 4		
Reg. indirect	(SP)			E3 19	DD E3 23	FD E3 23

Note that there is no instruction of the form LD B′, 0BH, so we must always first perform the exchange. The instructions EX (SP), *dd* allow us to exchange the top two bytes of the stack with the contents of *dd*.

No flags are affected by the exchange group

9.3.5 The Block Transfer Group

In addition to the single byte or address transfers, we can also perform block transfers on the Z80; these instructions appear in table 9.5. They allow a sequence of bytes with start address in HL to be transfered to a sequence of memories with start address in DE. The instructions are explained in the table and the only point to note is that the timing of the repeat instructions (LDIR, LDDR) depends, of course, on the initial contents of the BC register pair. However, each cycle requires 21 clock cycles apart from the last (when BC = 0) which requires only 16.

Flags affected:	All instructions; N, H flags reset
	LDI, LDD; P/V flag reset if BC = 0 after operation; otherwise P/V set
	LDIR, LDDR; P/V reset
	All other flags unaffected

Table 9.5 Block transfer group

		Source		
		Reg. indirect		
		(HL)		
Destination Reg. indirect	(DE)	ED AO	'LDI'–load (DE)←(HL) inc HL, DE, dec BC	16
		ED B0	'LDIR–load (DE)←(HL) inc HL, DE, dec BC, repeat until BC = 0	*
		ED A8	'LDD'–load (DE)←(HL) dec HL, DE, dec BC	16
		ED B8	'LDDR'–load (DE)←(HL) dec HL, DE, dec BC, repeat until BC = 0	*

* = 21 if BC ≠ 0, 16 if BC = 0.

9.3.6 The Block Search Group

In many cases we wish to compare the contents of the accumulator A with those of specific memory locations. This can be accomplished by the block search group (table 9.6). The basic compare (CP) operation subtracts (HL) from A and sets certain flags. Note, however, that the A register is not affected by the operation (the subtraction A – (HL) is not performed in the accumulator). The single compare instructions CPI, CPD merely set the flags while the repeat instructions CPIR, CPDR will continue until either BC = 0 or A = (HL), i.e. a match is found between the contents of A and (HL). Again the timing is given by $(n - 1) \times 21 + 16$ where n is the number of compares performed before BC = 0 or a match is found.

Flags affected: All instructions; N flag set, C flag unaffected Z flag set if A = (HL) P/V flag reset if BC = 0 after operation; otherwise set All other flags dro†

† dro means depend on result of operation.

Table 9.6 Block search group

Search location		
Reg. indirect		
(HL)		
ED A1	'CPI' A-(HL), inc HL, dec BC	16
ED B1	'CPIR' A-(HL), inc HL, dec BC, repeat until BC = 0 or find match	*
ED A9	'CPD' A-(HL), dec HL, dec BC	16
ED B9	'CPDR' A-(HL), dec HL, dec BC, repeat until BC = 0 or find match	*

* = 21 if BC ≠ 0 and A ≠ (HL), or 16 if BC = 0 or A = (HL).

9.3.7 8-bit Arithmetic and Logic Group

Table 9.7 shows all the 8-bit arithmetic and logical operations. In all cases (apart from INC, DEC) the operation is performed on the accumulator although again the CP instructions do not affect the contents of A. Note, however, that in the assembly mnemonics for ADD, ADC, SBC the A register is explicitly included, whereas it is not in SUB, AND, XOR, OR and CP. Hence we have typical instructions

ADD A, C	A←A + C
ADC A, (HL)	A←A + (HL) + (contents of carry flag)
SUB 10H	A←A − 10H
SBC A, (IX + 3)	A←A − (IX + 3) − (contents of carry flag)
AND (IY − 4)	A←A∧(IY − 4)
XOR E	A←A⊕E
OR L	A←A∨L
CP B	A − B (*not* A←A − B), set flags

The operations of these instructions are shown on the right. The timings in table 9.7 have only been shown for ADD and INC. The other arithmetic and logical operations have identical timings to the corresponding ADD operation while DEC is similar to INC.

Table 9.7 8-bit arithmetic and logic group

	Source										
	Reg. address							*Reg indir.*	*Indexed*		*Immed.*
	A	B	C	D	E	H	L	(HL)	(IX + *d*)	(IY + *d*)	*n*
'ADD'	87 4	80 4	81 4	82 4	83 4	84 4	85 4	86 7	DD 86 *d*	FD 86 *d* 19	C6 *n* 7
Add with carry 'ADC'	8F	88	89	8A	8B	8C	8D	8E	DD 8E *d*	FD 8E *d*	CE *n*
Subtract 'SUB'	97	90	91	92	93	94	95	96	DD 96 *d*	FD 96 *d*	D6 *n*
Subtract with carry 'SBC'	9F	98	99	9A	9B	9C	9D	9E	DD 9E *d*	FD 9E *d*	DE *n*
'AND'	A7	A0	A1	A2	A3	A4	A5	A6	DD A6 *d*	FD A6 *d*	E6 *n*
'XOR'	AF	A8	A9	AA	AB	AC	AD	AE	DD AE *d*	FD AE *d*	EE *n*
'OR'	B7	B0	B1	B2	B3	B4	B5	B6	DD B6 *d*	FD B6 *d*	F6 *n*
Compare 'CP'	BF	B8	B9	BA	BB	BC	BD	BE	DD BE *d*	FD BE *d*	FE *n*
Increment 'INC'	3C 4	04 4	0C 4	14 4	1C 4	24 4	2C 4	34 11	DD 34 *d* 23	FD 34 *d* 23	
Decrement 'DEC'	3D	05	0D	15	1D	25	2D	35	DD 35 *d*	FD 35 *d*	

Flags affected: ADD, ADC; N reset, others dro
SUB, SBC; N set, others dro
AND, OR, XOR; C, N reset, others dro
CP; N set, others dro
INC; N reset, C unaffected, others dro
DEC; N set, C unaffected, others dro
Note: in logic operations, P/V is a parity flag; otherwise it is an overflow flag

9.3.8 16-bit Arithmetic Group

Certain arithmetic operations can be performed on the 16-bit registers and are shown in table 9.8. These are self-explanatory and require no further comment.

Table 9.8 16-bit Arithmetic and Logic Group

Destination		*Source*					
		BC	DE	HL	SP	IX	IY
'ADD'	HL	09 11	19 11	29 11	39 11		
	IX	DD 09 15	DD 19 15		DD 39 15	DD 29 15	
	IY	FD 09 15	FD 19 15		FD 39 15		FD 29 15
Add with carry and set flags 'ADC'	HL	ED 4A 15	ED 5A 15	ED 6A 15	ED 7A 15		
Sub. with carry and set flags 'SBC'	HL	ED 42 15	ED 52 15	ED 62 15	ED 72 15		
'INC'		03 6	13 6	23 6	33 6	DD 23 10	FD 23 10
'DEC'		0B 6	1B 6	2B 6	3B 6	DD 2B 10	FD 2B 10

Flags affected: ADD; C dro, Z, P/V, S unaffected, N reset
H unknown
ADC; C, Z, P/V, S dro, N reset, H unknown
SBC; C, Z, P/V, S dro, N set, H unknown
INC, DEC; all flags unaffected

9.3.9 Rotate and Shift Group

It is often useful to perform shift and rotate operations on registers and memories and such a group of operations is provided on the Z80 (table 9.9). The timings in the main block are again shown only on the first row, the others being similar. The operations of these instructions are shown symbolically in fig. 9.14; for example, RLC B shifts the bits of the B register one place towards bit 7 which 'falls' into the carry flag and is also returned to bit 0. RLD is most simply explained by an example; if A contains ABH and (HL) contains CDH then, after RLD, A will contain ACH and (HL) will contain DBH. Note that RLCA is identical to the operation RLC A, except in the behaviour of the flags, and similarly for RRCA, RLA and RRA.

Flags affected: RLCA, RLA, RRCA, RRA; Z, P/V, S unaffected
N, H reset, C dro
RLC, RL, RRC, RR, SLA, SRA, SRL; N, H reset
others dro
RLD, RRD; C unaffected, N, H reset,
Z, P/V, S dro
(P/V is a parity flag)

9.3.10 Bit Manipulation Group

Since it is important to be able to test and alter individual bits in a register, a group of such instructions exists on the Z80 (table 9.10). The BIT operations test a bit of the appropriate register or memory and alter the zero flag accordingly. Hence

BIT 6, p ($p = $ A, B, C, D, E, H, L, (HL), (IX + d), (IY + d))

will take the complement of bit 6 of register (or memory) p and place it in the Z flag; thus if the bit 6 of p is zero the Z flag will be set. Symbolically,

$$Z \leftarrow \bar{p}_6$$

Table 9.9 Rotates and shifts

	Source and destination											
	A	B	C	D	E	H	L	(HL)	(IX + *d*)	(IY + *d*)		A
'RLC'	CB 07 8	CB 00 8	CB 01 8	CB 02 8	CB 03 8	CB 04 8	CB 05 8	CB 06 15	DD CB *d* 06 23	FD CB *d* 06 23	RLCA	07 4
'RRC'	CB 0F	CB 08	CB 09	CB 0A	CB 0B	CB 0C	CB 0D	CB 0E	DD CB *d* 0E	FD CB *d* 0E	RRCA	0F 4
'RL'	CB 17	CB 10	CB 11	CB 12	CB 13	CB 14	CB 15	CB 16	DD CB *d* 16	FD CB *d* 16	RLA	17 4
'RR'	CB 1F	CB 18	CB 19	CB 1A	CB 1B	CB 1C	CB 1D	CB 1E	DD CB *d* 1E	FD CB *d* 1E	RRA	1F 4
'SLA'	CB 27	CB 20	CB 21	CB 22	CB 23	CB 24	CB 25	CB 26	DD CB *d* 26	FD CB *d* 26		
'SRA'	CB 2F	CB 28	CB 29	CB 2A	CB 2B	CB 2C	CB 2D	CB 2E	DD CB *d* 2E	FD CB *d* 2E		
'SRL'	CB 3F	CB 38	CB 39	CB 3A	CB 3B	CB 3C	CB 3D	CB 3E	DD CB *d* 3E	FD CB *d* 3E		
'RLD'								ED 6F 18				
'RRD'								ED 67 18				

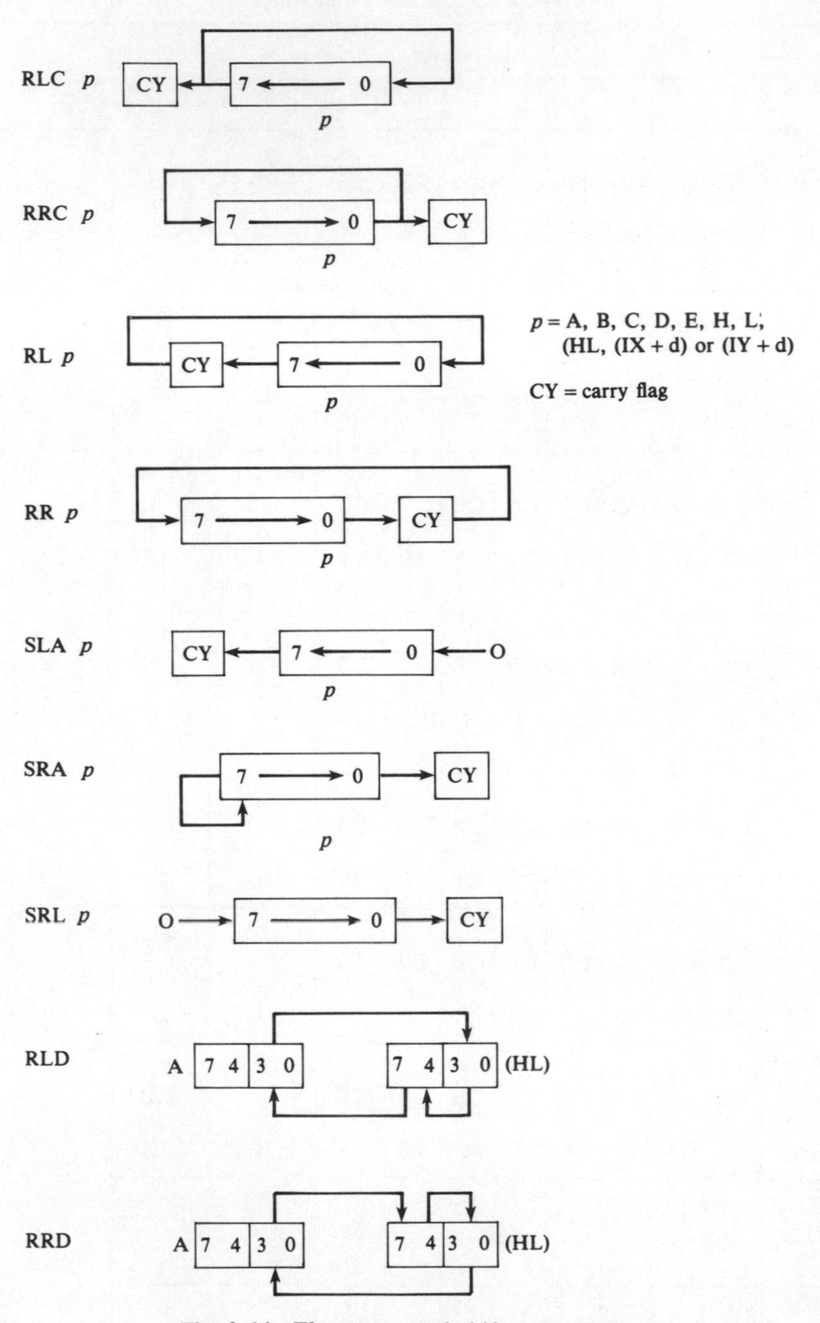

Fig. 9.14 The rotate and shift operations.

Table 9.10 Bit manipulation group

		Register addressing							*Reg. indir.*	*Indexed*	
	Bit	A	B	C	D	E	H	L	(HL)	(IX + *d*)	(IY + *d*)
Test 'BIT'	0	·47 $_{8}$	·40 $_{8}$	·41 $_{8}$	·42 $_{8}$	·43 $_{8}$	·44 $_{8}$	·45 $_{8}$	·46 $_{12}$	··46 $_{20}$	···46 $_{20}$
	1	·4F	·48	·49	·4A	·4B	·4C	·4D	·4E	··4E	···4E
	2	·57	·50	·51	·52	·53	·54	·55	·56	··56	···56
	3	·5F	·58	·59	·5A	·5B	·5C	·5D	·5E	··5E	···5E
	4	·67	·60	·61	·62	·63	·64	·65	·66	··66	···66
	5	·6F	·68	·69	·6A	·6B	·6C	·6D	·6E	··6E	···6E
	6	·77	·70	·71	·72	·73	·74	·75	·76	··76	···76
	7	·7F	·78	·79	·7A	·7B	·7C	·7D	·7E	··7E	···7E
Reset bit 'RES'	0	·87 $_{8}$	·80 $_{8}$	·81 $_{8}$	·82 $_{8}$	·83 $_{8}$	·84 $_{8}$	·85 $_{8}$	·86 $_{15}$	··86 $_{23}$	···86 $_{23}$
	1	·8F	·88	·89	·8A	·8B	·8C	·8D	·8E	··8E	···8E
	2	·97	·90	·91	·92	·93	·94	·95	·96	··96	···96
	3	·9F	·98	·99	·9A	·9B	·9C	·9D	·9E	··9E	···9E
	4	·A7	·A0	·A1	·A2	·A3	·A4	·A5	·A6	··A6	···A6
	5	·AF	·A8	·A9	·AA	·AB	·AC	·AD	·AE	··AE	···AE
	6	·B7	·B0	·B1	·B2	·B3	·B4	·B5	·B6	··B6	···B6
	7	·BF	·B8	·B9	·BA	·BB	·BC	·BD	·BE	··BE	···BE
Set bit 'SET'	0	·C7 $_{8}$	·C0 $_{8}$	·C1 $_{8}$	·C2 $_{8}$	·C3 $_{8}$	·C4 $_{8}$	·C5 $_{8}$	·C6 $_{15}$	··C6 $_{23}$	···C6 $_{23}$
	1	·CF	·C8	·C9	·CA	·CB	·CC	·CD	·CE	··CE	···CE
	2	·D7	·D0	·D1	·D2	·D3	·D4	·D5	·D6	··D6	···D6
	3	·DF	·D8	·D9	·DA	·DB	·DC	·DD	·DE	··DE	···DE
	4	·E7	·E0	·E1	·E2	·E3	·E4	·E5	·E6	··E6	···E6
	5	·EF	·E8	·E9	·EA	·EB	·EC	·ED	·EE	··EE	···EE
	6	·F7	·F0	·F1	·F2	·F3	·F4	·F5	·F6	··F6	···F6
	7	·FF	·F8	·F9	·FA	·FB	·FC	·FD	·FE	··FE	···FE

· ≡ CB ·· ≡ DD CB *d* ··· ≡ FD CB *d*

The SET and RESET operations merely set or reset the appropriate bit of the operand; for example SET 7, (HL) performs the symbolic operation

$$(HL)_7 \leftarrow 1$$

Flags affected:	BIT; C unaffected, Z dro
	P/V, S unknown, N reset, H set
	RES, SET; all flags unaffected

9.3.11 General Purpose AF Operations

There are five special operations which can be performed on the A and F registers and these appear in table 9.11. The first, DAA, means 'decimal adjust' and is used in BCD arithmetic (see section 9.6, exercise 7). The accumulator can be (one's) complemented with CPL (i.e. 0s replaced by 1s and vice versa) or two's complemented with NEG. Note that two's comp. = one's comp. + 1. The carry flag can also be complemented (CCF) or set (SCF). There is no 'clear carry flag' but this can be accomplished by AND A since the carry flat is reset by this operation and A is unaffected.

Flags affected:	DAA; C, Z, P/V, S, H, dro (P/V is parity flag)
	N unaffected
	CPL; C, Z, P/V, S unaffected, N, H set
	NEG; C, Z, P/V, S, H dro (P/V is overflow flag)
	N set
	CCF; C dro, Z, P/V, S unaffected,
	N reset, H unknown
	SCF; C set, Z, P/V, S unaffected, N, H reset

Table 9.11 General purpose AF operations

Operation	Cycles	Code
Decimal adjust, 'DAA'	4	27
Complement acc. 'CPL'	4	2F
Neg. acc 'NEG' (two's comp.)	8	ED 44
Complement carry flag, 'CCF'	4	3F
Set carry flag 'SCF'	4	37

9.3.12 Jump Instructions

The group of instructions which gives the microprocessor much of its power is the jump group shown in table 9.12. Consider first the JP and JR instructions. For example, the unconditional jump

JP *nn*

merely loads *nn* into the program counter for the next instruction thus changing the normal sequence of operations. As we have said before, the actual memory address *nn* is not used at assembly level but is replaced by a LABEL or name. Hence we may use the instruction JP in the following way:

```
JP NAME
  .
  .
  .
NAME:LD A, B
```

The label NAME (plus the colon) is translated by the assembler into the correct address *nn* (which will depend on the address of the first instruction of the program). We can also use conditional jumps which only take effect if a particular flag is set (or reset). Thus, for example, the instruction

JP NC, NAME

will have the same effect as the unconditional jump if the carry flag is reset and otherwise will essentially be ignored by the CPU. Other unconditional jumps include JP (HL), JP (IX) and JP (IY) which merely load the program counter from the appropriate address register HL, IX or IY.

We have already met the relative jump instruction JR which adds (algebraically) an offset to the PC. Again we use labels rather than the actual offset value; note, however, that conditional relative jumps are only available using the C or Z flags. Another useful jump instruction is DJNZ which means 'decrement the B register and jump if the result is non-zero'. Hence we can use this instruction as a count in a loop; first load B with the count and then place the instructions which are to be repeated in the loop terminating with DJNZ. Thus, in the program

```
LD B, OAH
NEXT: ...
  .
  .
  .
DJNZ NEXT
```

the instructions represented by dots will be repeated 10 times.

Table 9.12 Jump instructions

			Condition									
			Uncond.	Carry C	Non-carry NC	Zero Z	Non-zero NZ	Parity even PE	Parity odd PO	Sign neg. M	Sign pos. P	Reg $B \neq 0$
Jump 'JP'	Immed. ext.	*nn*	C3 *n* *n* 10	DA *n* *n* 10	D2 *n* *n* 10	CA *n* *n* 10	C2 *n* *n* 10	EA *n* *n* 10	E2 *n* *n* 10	FA *n* *n* 10	F2 *n* *n* 10	
Jump 'JR'	Relative	$PC + e$	18 $e - 2$ 12	38 $e - 2$ 7 (12)	30 $e - 2$ 7 (12)	28 $e - 2$ 7 (12)	20 $e - 2$ 7 (12)					
Jump 'JP'	Reg. indir.	(HL)	E9 4									
Jump 'JP'		(IX)	DD E9 8									
Jump 'JP'		(IY)	FD E9 8									

'CALL'	Immed. ext.	nn	CD n n 17	DC n n 10 (17)	D4 n n 10 (17)	CC n n 10 (17)	C4 n n 10 (17)	EC n n 10 (17)	E4 n n 10 (17)	FC n n 10 (17)	F4 n n 10 (17)	
Dec. B, jump 1F ≠ 0, 'DJNZ'	Relative	PC + e										10 $e-2$ 8 (13)
Return 'RET'	Reg. indir.	(SP) (SP + 1)	C9 10	D8 5 (11)	D0 5 (11)	C8 5 (11)	C0 5 (11)	E8 5 (11)	E0 5 (11)	F8 5 (11)	F0 5 (11)	
Return from int 'RETI'	Reg. indir.	(SP) (SP + 1)	ED 4D 14									
Return from non-maskable int 'RETN'	Reg. indir.	(SP) (SP + 1)	ED 45 14									

Timings in parentheses hold if condition is true; those above hold if condition is false

As in high level languages the ability to use the same set of instructions repeatedly in the form of a 'subroutine' is important. This facility is provided on the Z80 by the CALL and RET instructions. Since we must return to the main program after performing the subroutine a return address must be stored when implementing a CALL. In fact the contents of PC are PUSHed onto the stack so that CALL *nn* is like JP *nn* except that PUSH PC is also (effectively) performed. The last statement of the subroutine should be a RET instruction which is effectively a POP PC, returning to the program after the CALL instruction. Relative CALLs and RETs also exist and are similar to those for JP. The instructions RETI, RETN are return from interrupt and return from non-maskable interrupt respectively and must be used instead of RET after an interrupt service routine.

Finally a special purpose set of 'zero page' jump instructions called the restart group are shown in table 9.13. These are of the form

$$\text{RST } p \quad (p = 0, 8, 16, \ldots, 56)$$

and load PC with 0 in the high byte and p in the low byte and so have the effect of jumping to one of the 'page zero' addresses 0, 8, 16, . . . , 56. These instructions are particularly useful in interrupt processing as we shall see later. Note that RST also performs a PUSH PC so that a return address is stored on the stack.

No flags are affected by the JUMP and restart groups

Table 9.13 Restart group

Call address	OP CODE	
0000_H	$C7_{11}$	'RST 0'
0008_H	CF_{11}	'RST 8'
0010_H	$D7_{11}$	'RST 16'
0018_H	DF_{11}	'RST 24'
0020_H	$E7_{11}$	'RST 32'
0028_H	EF_{11}	'RST 40'
0030_H	$F7_{11}$	'RST 48'
0038_H	FF_{11}	'RST 56'

9.3.13 Input and Output Group

A microprocessor must be able to communicate with the 'outside world' which consists, as far as the μP is concerned, of various input–output devices. Special instructions are provided for this purpose and form the input and output groups shown in tables 9.14, 9.15. These will be discussed in more detail in chapter 10 since a complete understanding of these instructions requires a knowledge of interfacing techniques. However, we note that the instruction IN A, (*n*), for example, has the following effect. The Z80 can 'address' 256 different input or output devices and so each one has a unique 8-bit 'address'. Each device is connected onto the data bus and when IN A, (*n*) is executed the 8-bit address *n* is placed on the low 8-bits of the address bus and a control signal is generated instructing the peripheral device with address *n* to place a byte on the data bus. This byte is loaded into A. (Note that the accumulator contents at the start of the execution are placed on the upper 8 bits of the address bus for possible use by the system.) The instructions IN *r*, (C) where *rt* = A, B, C, D, E, H, or L are similar to the above instruction except that the address of the peripheral device is first placed in the C register. Block input commands are also available as can be seen from table 9.14.

The output group is similar in operation but the direction of the data is from the Z80 to the peripheral device. Hence, OUT (*n*), A outputs the byte in A (along the data bus) to the device with address *n*. (As before the accumulator contents are placed on the upper 8 bits of the address bus.)

Flags affected: OUT (*n*), A, OUT (C), *r*, IN A, (*n*); no flags affected
In *r*, (C); C unaffected, Z, P/V, S, H dro, N reset
OUTI, OUTD, INI, IND; C unaffected, P/V, S, H unknown, N set, Z set if B changes to 0
OTIR, OTDR, INIR, INDR; C unaffected, P/V, S, H unknown, N set, Z set

9.3.14 Miscellaneous Group

The final group of instructions available on the Z80 is a miscellaneous CPU control group and is shown in table 9.16. The first instruction 'NOP' means literally no operation and has no effect on the CPU registers or the memory. The next operation 'HALT' terminates execution of the current program;

Table 9.14 Input group

Input destination			Source port address: Immed. (n)	Source port address: Reg. indir. (c)	
Input 'IN'	Reg. addressing	A	D8 n 10	ED 78 11	
		B		ED 40 11	
		C		ED 48 11	
		D		ED 50 11	
		E		ED 58 11	
		H		ED 60 11	
		L		ED 68 11	
'INI' – input and inc, HL, dec. B	Reg. indir.	(HL)		ED A2 15	Block input commands
'INIR' – input, inc. HL dec. B, repeat if B ≠ 0	Reg. indir.	(HL)		ED B2 x	Block input commands
'IND' – input and dec, HL, dec. B	Reg. indir.	(HL)		ED AA 15	Block input commands
'INDR' – input, dec HL, dec. B, repeat if B ≠ 0	Reg. indir.	(HL)		ED BA x	Block input commands

Timing $x =$ 20 if $B \neq 0$
15 if $B = 0$

Table 9.15 Output Group

			Source							
			Register							*Reg. ind.*
			A	B	C	D	E	H	L	(HL)
'OUT'	Immed.	(*n*)	D3 *n* 11							
	Reg. ind.	(C)	ED 79 12	ED 41 12	ED 49 12	ED 51 12	ED 59 12	ED 61 12	ED 69 12	
'OUTI' – output inc. HL, dec. B	Reg. ind.	(C)								ED A3 15
'OUTI' – output, inc. HL, dec. B until B = 0	Reg. ind.	(C)								ED B3 *x*
'OUTD' – output, dec. HL and B	Reg. ind.	(C)								ED AB 15
'OTDR' – output, dec. HL and B until B = 0	Reg. ind.	(C)								ED BB *x*
	Port destination address									

('OUTI', 'OUTI', 'OUTD', 'OTDR' rows: Block output commands)

Timing $x =$ 20 if $B \neq 0$
15 if $B = 0$

CSE-Q*

Table 9.16 Miscellaneous CPU control

'NOP'	00 4	
'HALT'	76 4	
Disable int. 'DI'	F3 4	
Enable int. 'EI'	FB 4	
Set int. mode 0 'IM0'	ED 46 8	
Set int. mode 1 'IM1'	ED 56 8	Call to 0038_H
Set int. mode 2 'IM2'	Ed 5E 8	

after a HALT instruction has been executed the CPU performs a sequence of NOPs until it is interrupted by an external signal. The purpose of the NOP instructions is to keep the CPU 'ticking over' so to speak while continuing with the refreshing of dynamic memories.

The next two instructions DI and EI are **disable** and **enable** interrupts respectively. The Z80 can be stopped from executing a program by a signal, from an external device, called an **interrupt**. Two types of interrupt are provided on the Z80 – maskable and non-maskable. The latter type will stop the current program execution under all conditions, but the former type can be controlled by the software. Hence after performing a DI operation the CPU will ignore all maskable interrupts, while it will respond to maskable interrupts after executing EI. A more complete discussion of interrupts will be given in chapter 10 when the appropriate interfaces have been introduced.

The Z80 can respond to an external (maskable) interrupt in one of three ways, which are again set under software control by the last three instructions IM0, IM1, IM2 (interrupt modes 0, 1 and 2). In interrupt mode 0, when the external device generates an interrupt the CPU will complete the current instruction and then take the next instruction from the interrupting device rather that the memory. The external device will therefore be expected to supply at least one byte on the data bus which is interpreted by the CPU as an instruction. Usually we input a RESTART instruction since this requires only a single byte to be input. If, for example, a RES 08 instruction is input, the CPU will jump to memory location 0008H and execute the subroutine beginning at that address. On completion of this subroutine, the CPU will return to the instruction after the one which it was executing prior to the interrupt. Performing the subroutine at 0008H is referred to as **servicing** the interrupt. Note that IM0 is automatically assumed after power is initially applied to the CPU (or after a RESET).

If interrupt mode 1 is selected by executing an IM1 instruction the CPU will execute a restart to location 0038H when an interrupt occurs (provided maskable interrupts are enabled). This is similar to its response to a non-maskable interrupt, except in this case the CPU jumps to 0066H.

Finally interrupt mode 2 is selected by the IM2 instruction. To understand the operation of a mode 2 interrupt we first note that any interrupt is serviced by executing a subroutine. Hence the interrupting device must not only generate an interrupt, but also provide the CPU with the start address of the subroutine. In mode 2 interrupts the external device supplies a byte which is interpreted by the CPU as the **low** byte A_l of an address. (Note that bit 0 of A_l should always be 0.) The **high** byte A_h of this address must be loaded into the I register prior to the receipt of an interrupt by using the operation LD I, A. The CPU will then use the address A_hA_l not as the start address of the interrupt service routine, but as a pointer to this start address. Hence the start address of all the interrupt service routines (mode 2) can be stored together in a sequential table while the address A_hA_l supplied by I and the external device is used as a pointer into this table. If all memory locations in the table have the same high byte then I need only be loaded once with this byte and the different service routines are then selected by the external device via A_l. In mode 2 interrupts the CPU pushes PC onto the stack since the first instruction of the service routine may not be a CALL or RST. Note that a service routine start address is often referred to as an **interrupt vector**.

No flags are affected by the miscellaneous CPU group

9.4 ASSEMBLY PROGRAMMING

9.4.1 Assembly Language and the Assembler

In the previous section we have given a complete list of the Z80 microprocessor instructions from which our assembly level programs must be written. An assembly program consists of a sequence of these instructions written one underneath the other. As an example consider the trivial program

```
LD A, 00H
LD C, 01H
ADD A, C
HALT
```

We can add comments after any instruction separating them by a semicolon. Thus our program may be written

```
LD A, 00H; LOAD A WITH 00H
LD C, 01H; LOAD C WITH 01H
ADD A, C ; ADD C TO A, RESULT IN A
HALT
```

The comments are purely for documentation purposes and are ignored by the **assembler**. An asssembler, as we have already said, is a program which translates the assembly program into machine (or object) code. We must provide the assembler with a start address for the program and this can be done with the 'pseudo-op' ORG. Our simple program will then appear with its assembled object code as follows (omitting the comments):

```
               ORG 1000H
1000 3E00      LD A, 00H
1002 0E01      LD C, 01H
1004 91        ADD A, C
1005 76        HALT
```

The number in the first column refers to the address of the first byte of the instruction, and it is increased on each line by the number of bytes in the instruction. The numbers in the second column are the machine language equivalents of the instructions and can be checked from the tables in section 9.3.

Suppose that we now wish to include a jump statement in the program. As we have stated above, we do not refer to addresses directly but use labels in much the same way as program lines are numberd in high level languages (FORTRAN, BASIC etc.). Consider then the following program which adds the first ten natural numbers:

```
               ORG 1000H
               LD B, 0AH    ; INITIALIZE COUNT
               LD A, 00H    ; INITIALIZE SUM
               LD C, 01H    ; INITIALIZE ADDEND
          SUM: ADD A, C     ; PERFORM ADD
               INC C        ; NEXT ADDEND
or DJNZ      { DEC B        ; DECREMENT COUNT
             { JR NZ, SUM   ; JUMP BACK
               HALT
```

This program uses a standard method for performing a loop. First the number of repetitions of the instructions in the loop is placed in B as a count, then the instructions in the loop are performed, the count is decreased by 1, and a jump back to the start of the loop is performed if the count is non-zero. Note that the two bracketed instructions are equivalent to the single instruction DJNZ. Note also that the initialization statements are *outside* the loop; a common error is to put them inside the loop, often leading to an infinite loop.

If we now wish to initialize the sum (in A) to a number contained in memory rather than zero we could replace the instruction LD A, 00H by LD A, (*nn*) where *nn* is the appropriate memory address. Since we do not refer directly to absolute addresses we again use a label in place of *nn*. For example, the initial sum may represent a voltage and so we could use the label VOLT for *nn*. However, we must refer to the labels which appear in the program using another type of pseudo-op which is a kind of label declarator. These pseudo-ops are

DEFB, DEFL, DEFM, DEFS, DEFW, EQU

and are inserted after the HALT instruction in the formats

(a) LABEL:DEFB *n* $(0 \leqslant n \leqslant 256)$
(b) LABEL:DEFL *nn* $(0000 \leqslant nn \leqslant \text{FFFF})$
(c) LABEL:DEFM '*s*' (s = string)
(d) LABEL:DEFS *nn* $(0 \leqslant nn \leqslant 65535)$
(e) LABEL:DEFW *nn* $(nn = n_h n_l)$
(f) LABEL:EQU *nn* $(0000 \leqslant nn \leqslant \text{FFFF})$

In (a) the memory called LABEL is loaded with the byte *n* and can be used to initialize LABEL. The instruction (b) gives LABEL the value *nn*, i.e. specifies the address called LABEL. In (c) *s* is a string containing up to 63 characters and the instruction places the ASCII code of each character in successive memories starting at LABEL. DEFS *nn* reserves *nn* consecutive bytes of memory starting at LABEL, while DEFW *nn* places n_l in memory address LABEL and n_h in LABEL + 1. Finally in (f), LABEL is given the value *nn*.

Using VOLT as a label we can now change our program as follows:

```
      ORG 1000H
      LD B, 0AH
      LD A, (VOLT)
      LD C, 01H
  SUM:ADD A, C
      INC C
      DJNZ SUM
      HALT
 VOLT:DEFS 1
```

When the assembler 'sees' LD A, (VOLT) it will search for a label declaration after the HALT command. (If such a declaration is not included, an error message will result; this helps the programmer to eliminate any discrepancies between the labels which may occur due to mistyping.) In the example above the assembler will interpret the instruction VOLT: DEFS 1 by assigning a specific memory location to the variable VOLT. The programmer does not need to know the actual memory address 'VOLT' in this case. However, if the programmer wishes to specify the address VOLT he may replace DEFS by DEFL or EQU; for example, the last instruction may be replaced by

```
VOLT: DEFL 2000H
```

Alternatively we may specify the initial contents of VOLT using DEFB, for example

```
VOLT: DEFB 2BH
```

DEFW is useful when using 16-bit operations, as in the program

```
.
.
.
LD HL, (ADDR1)
.
.
.
HALT
ADDR1:DEFW ADDR2
ADDR2:DEFS 1
```

9.4.2 Elementary Arithmetic Software

One of the first problems in developing a microprocessor system for control applications is writing the basic arithmetic software. Since any real number R can be written in the form

$$R = \pm d_n d_{n-1} \ldots d_0 \cdot d_{-1} d_{-2} \ldots$$

(in some base) the number representations in the computer must take account of the position of the decimal point. Essentially two distinct ways of representing R in the memory exist and are called **fixed-point** or **floating-point** representations. In fixed-point arithmetic we first decide how many of the digits d_i are going to be stored (this must be finite, of course), say N. Then the numbers are of the form

$$R = \pm d_n d_{n-1} \ldots d_0 \cdot d_{-1} d_{-2} \ldots d_{-N+(n+1)}$$

If each d_i is a decimal digit then we could store two digits per byte and R

would require $(1 + N/2)$ bytes (assuming N even), as shown below:

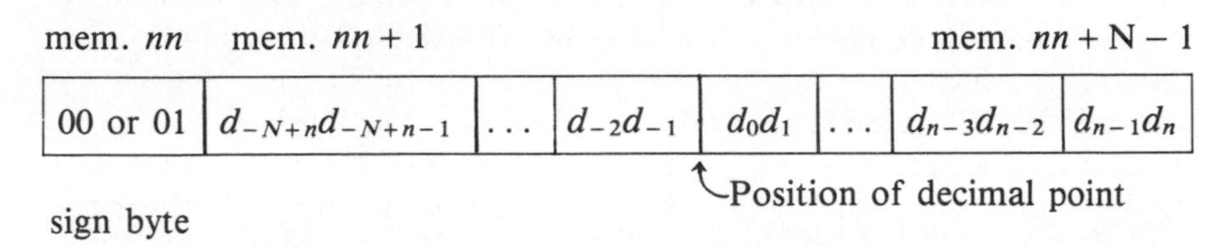

(We have also assumed that $n + 1$ is even and it is convenient to store the 'mantissa' low digits first.) It is seen that the decimal point is not included as a character since we are assuming it is fixed between the bytes indicated. Moreover the sign of the number is stored in a separate byte as 00H for a positive number and 01H for a negative number. This is clearly wasteful of space but is convenient and so this method will be used in future. If we use binary representations then we would choose $n + 1$ and N to be divisible by 8 and then we require $1 + N/8$ successive bytes to store each signed number.

In this section we shall write fixed-point arithmetic software using binary representations. To be specific we shall choose our numbers to be 32 bits long requiring 4 bytes + 1 byte for the sign; the programs can easily be altered for greater (or less) accuracy. Numbers will therefore be stored in the format

sign	m	a n	t i	s s a
mem. nn	$nn + 1$	$nn + 2$	$nn + 3$	$nn + 4$

Consider first the addition of two numbers (with the same sign) with start addresses in the DE and HL register pairs. The sum will be placed in memory starting at (IX). The following subroutine will perform this operation:

```
FPADD:LD A, (DE)
      LD (IX + 0), A  ; STORE SIGN OF SUM
      LD B, 04H       ; LOAD B WITH COUNT
      AND A           ; CLEAR CARRY
      INC DE          ; POINT TO
      INC HL          ; NEXT
      INC IX          ; BYTE
  SUM:LD A, (DE)
      ADC A, (HL)     ; ADD BYTES (WITH CARRY)
      LD (IX + 0), A  ; STORE RESULT
      INC DE
      INC HL
      INC IX
```

```
        DJNZ SUM      ; JUMP BACK
        JR C, OVERFL
        LD A, 00H     ; FLAG NO OVERFLOW
        RET
OVERFL:LD A, 01H      ; FLAG OVERFLOW
        RET
```

Since we are using fixed-point arithmetic, if the sum of the two numbers requires more than four bytes of storage then the most significant bits will 'spill over' and be lost giving an incorrect answer. We must therefore return from this subroutine with an overflow check in the A register. Checking with the timings given in the previous tables it is seen that this subroutine takes 350 clock cycles (assuming no overflow) or 87.5 μs for a 4 MHz clock.

Suppose next that we consider the subtraction of two numbers pointed to by the DE and HL register pairs as before and assume for the moment that the positive number starts at (DE). The subroutine MIN below will subtract the second mantissa from the first:

```
   MIN:LD B, 04H
       AND A; CLEAR CARRY
       INC DE
       INC HL
       INC IX
SUBTR:LD A, (DE)
       SBC A, (HL)
       LD (IX + 0), A
       INC DE
       INC HL
       INC IX
       DJNZ SUBTR
       RET
```

If the carry flag is set on return from this subroutine the second mantissa must have been larger than the first and so the answer will be incorrect. Hence we must interchange DE and HL and repeat the subtraction; then the complete fixed point subtraction subroutine FPSUB is as follows:

```
FPSUB:PUSH DE
       PUSH HL
       PUSH IX        ; SAVE ADDRESSES
       CALL MIN
       JR NC, TRUE    ; NO CARRY, ANS. CORRECT
       POP IX         ; IF CARRY FLAG SET
       POP HL         ; RESTORE ADDRESSES AND
       POP DE         ; MAKE DE POINT TO
       EX DE, HL      ; LARGEST MANTISSA
```

```
        PUSH IX
        CALL MIN      ; RECALCULATE DIFF.
        POP IX
        LD A, 01H
        LD (IX+0), A  ; ANS NEG
        RET
   TRUE:POP IX
        LD A, 00H     ; ANS POS
        LD (IX+0), A
        POP HL
        POP DE
        RET
```

This subroutine lasts for 446 or 801 clock cycles depending on the JR NC, TRUE command; i.e. it lasts for 111.5 μs or 200 μs with a 4 MHz clock.

Consider now the algebraic sum of the numbers in (DE) and (HL). If their signs are the same we call FPADD while if their signs are different we must first ensure that the positive number is in (DE) and then call FPSUB. Hence we have the subroutine

```
ALGSUM:LD A, (DE)      ; FIND SIGNS
       CP (HL)         ; OF NUMBERS AND COMPARE
       JR Z, SAMES     ; SIGNS SAME?
       CP 01H          ; NEG NO. IN (DE)?
       JR NZ, OK
       EX DE, HL
    OK:CALL FPSUB
       RET
 SAMES:CALL FPADD
       RET
```

In the worst case this subroutine runs in 217 μs.

Coming now to multiplication consider first the following simple example of binary multiplication

```
B1        101101
B2          1011     B2
          ------     --
          101101      1    B2_0
         101101       1    B2_1
        000000        0    B2_2
       101101         1    B2_3
       ---------
B3     111101111
```

The product B3 is obtained in the following way: first initialize B3 to zero, then successively add to B3 the number $B1^i \times B2_i$ where $B2_i$ is the ith bit of B2 (starting at the LSB) and $B1^i$ represents the number B1 shifted i places to the left ($0 \leqslant i \leqslant 3$). Hence we can obtain B3 by testing each successive bit of B2 and adding B1 (shifted by the appropriate number of bits) to B3 if and only if the bit is 1.

We shall now assume that our four byte binary numbers have the binary point between bytes 2 and 3. Since we are multiplying two four byte numbers together we must perform the multiplication initially in eight bytes to retain accuracy, as shown below

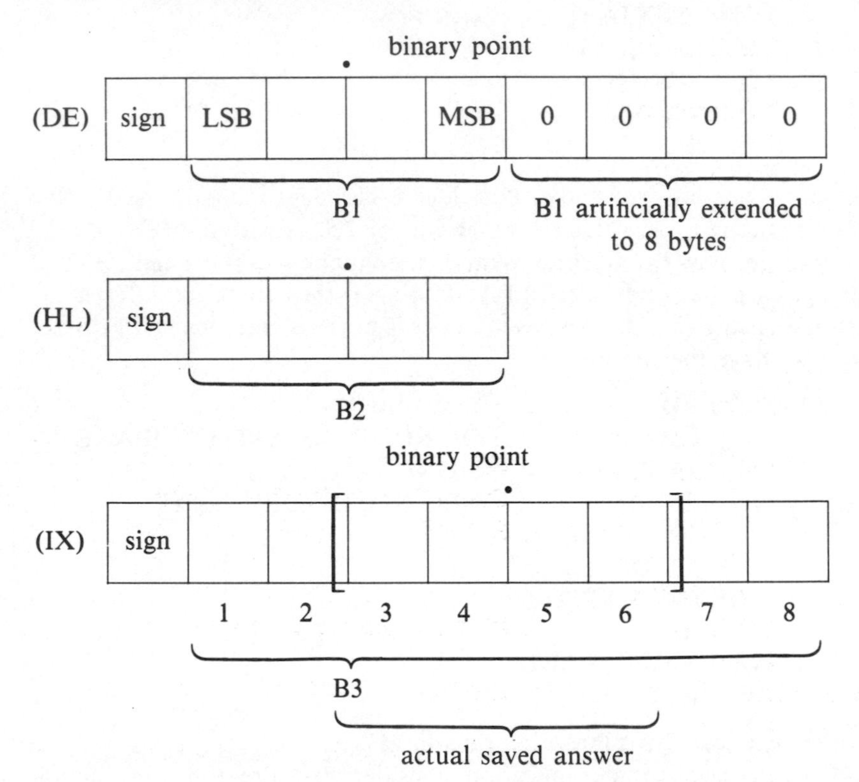

As we can see, the binary point of the answer appears between bytes 4 and 5 and since we are working in fixed-point arithmetic with only 4 bytes the actual answer which we retain is contained in bytes 3–6. Moreover, if the bytes 7 or 8 contain non-zero bits then the multiplication has 'overflowed' giving an error. Using these ideas we can now write the fixed-point 4-byte multiplication subroutine as follows:

```
FPMULT:LD A, (DE)        ; SIGN OF ANSWER IS
       XOR (HL)          ; EXCL. OR OF SIGNS OF
       LD (IX + 0), A    ; B1 AND B2
       INC DE
       INC HL
       INC IX            ; POINT TO FIRST BYTE OF MANTISSA
```

```
        PUSH DE             ; SAVE DE
        LD B, 04H           ; COUNT
        LD IY, TEMP1
NEXT1:LD A, (DE)            ; WE MUST STORE
        LD (IY + 0), A      ; B1 IN TEMP MEMORIES
        INC IY              ; SINCE B1 WILL BE
        INC DE              ; SHIFTED
        DJNZ NEXT1
        LD B, 04H
  INI1:LD A, 00H
        LD (DE), A          ; INITIALIZE 4 ARTIFICIAL
        INC DE              ; BYTES OF B1
        DJNZ INI1
        PUSH IX
        LD B, 08H
  INI2:LD (IX + 0), 00H     ; INITIALIZE ANSWER
        INC IX              ; B3 TO ZERO
        DJNZ INI2
        POP IX
        POP DE
        LD (TEMP 2), HL
        LD IY, (TEMP 2)     ; PUT HL IN IY
        LD B, 04H           ; COUNT FOR OUTER LOOP
OLOOP:LD C, B               ; SAVE B
        EXX
        LD A, (IY + 0)
        LD L, A             ; PUT BYTES OF B2 IN L
        LD B, 08H
 ILOOP:RLC L                ; CHECK BITS OF B2
        EXX
        JR NC, NOADD        ; IF BIT = 0 JUMP AROUND SUM
        PUSH DE
        PUSH HL
        PUSH IX
        LD (TEMP2), IX
        LD HL, (TEMP2)
        LD B, 08H
        AND A
        CALL SUM            ; CALL SUM SUBROUTINE IN FPADD
        POP IX
        POP HL
        POP DE
NOADD:PUSH DE
        LD B, 08H
```

```
        CALL SHIFTR      ; SHIFT B1 RIGHT (TOWARDS MSB)
        POP DE
        EXX
        DJNZ ILOOP
        EXX
        INC IY
        LD B, C          ; RESTORE OUTER LOOP COUNT
        DJNZ OLOOP
        LD B, 04H        ; RESTORE ORIGINAL
        LD IY, TEMP1     ; NUMBER B1
 NEXT2:LD A, (IY + 0)    ; TO (DE) – (DE + 3)
        LD (DE), A
        INC IY
        INC DE
        DJNZ NEXT2
        LD A, (IX – 1)   ; NOW REMOVE MIDDLE
        LD (IX + 1), A   ; 4 BYTES OF ANS. B3
        INC IX           ; AND SIGN BYTE
        LD A, (IX + 5)   ; CHECK FOR
        CP 00H           ; OVERFLOW
        JR Z, LBYTE
        LD A, 01H        ; FLAG OVERFLOW
        RET
 LBYTE:LD A, (IX + 6)
        CP 00H
        JR Z, NOOVF
        LD A, 01H        ; FLAG OVERFLOW
        RET
 NOOVF:LD A, 00H         ; FLAG NO OVERFLOW
        RET
SHIFTR:AND A             ; SHIFT B1 RIGHT
  ROTR:LD A, (DE)        ; SUBROUTINE
        RR A
        LD (DE), A
        INC DE
        DJNZ ROTR
        RET
 TEMP1:DEFS 04
 TEMP2:DEFS 02
```

It can be verified that this subroutine runs in at most 9 ms for a 4 MHz clock.

Finally, we must consider the division B3 = B1/B2 of two 4 byte numbers B1, B2 with first byte stored in (DE) and (HL) respectively. As with multiplication, in order to retain accuracy we must perform the operation in 8 bytes of memory and so we shall assume that enough memories have been assigned so that the numbers appear as below:

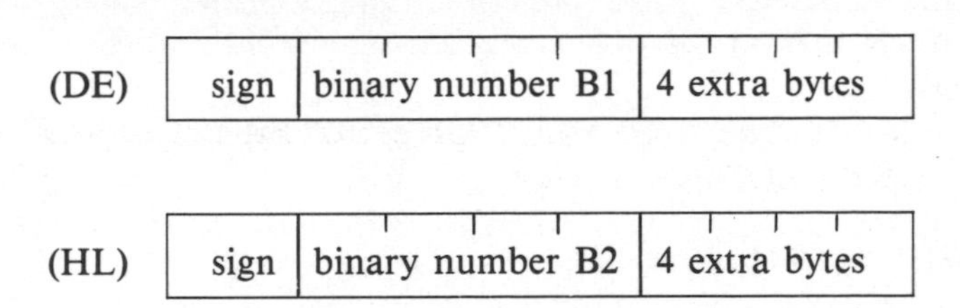

Since we are storing the numbers low bit first it will be convenient to shift the numbers to the right into the spare bytes and place zeros in the original memories:

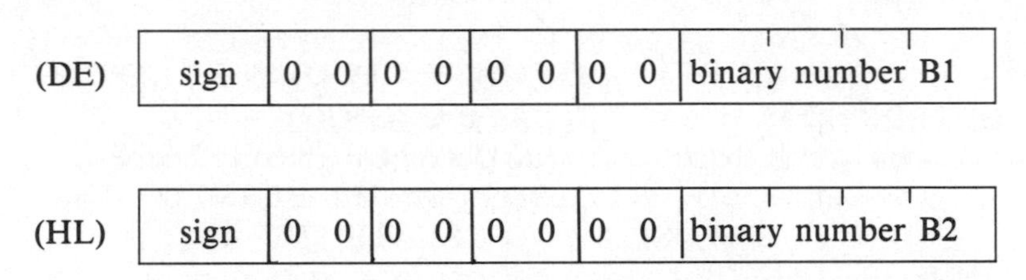

Of course, we must check if either number B1 or B2 is zero (in the latter case we have a 'divide-by-zero' error). If both numbers are non-zero then we can count the number of zeros in each stored number from the right and then shift both numbers to the right so that the first 1 appears in bit 0 of byte 8. The numbers then appear 'normalized' as follows:

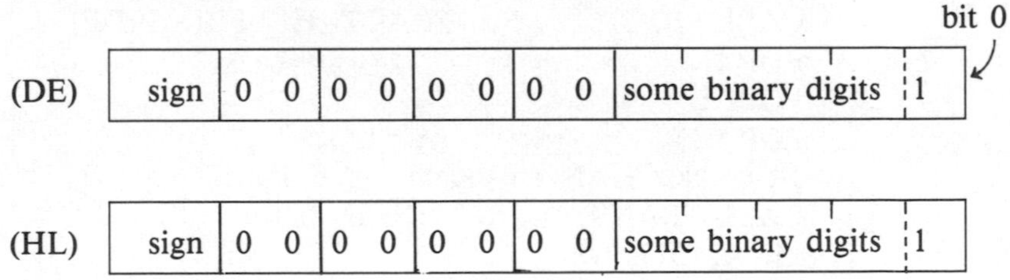

The procedure for dividing two such numbers is as follows:

(a) Subtract [(HL + 1) to (HL + 8)] from [(DE + 1) to (DE + 8)] and put result in [(IX + 1) to (IX + 8)]
(b) (i) If answer is positive, put 1 in appropriate bit of [(IY + 1) to (IY + 8)] and replace [(DE + 1) to (DE + 8)] by [(IX + 1) to (IX + 8)].
(ii) If answer is negative put 0 in appropriate bit of [(IY + 1) to (IY + 8)].
(c) Shift [(HL + 1) to (HL + 8)] one bit to the left and return to (a).

(Note that we are placing the result B3 = B1/B2 in [(IY) to (IY + 8)].) By the 'appropriate bit' in (b) we mean bit 0 of (IY + 8) the first time then bit 1 of (IY + 8) the second time, etc. until we have repeated (a)–(c) 32 times (not 64 times; the result will contain only bits from the highest four bytes (IY + 4) to (IY + 8) so that we only need to shift B2 'half-way' along the eight bytes).

The first subroutine NORMDIV will perform the division of two normalized numbers and is gven below:

```
NORMDIV:PUSH HL        ; PUT HL + 7
        EXX            ; IN HL'
        POP HL         ; THIS IS
        LD B, 07H      ; USEFUL IN
   INCR:INC HL         ; SHIFL
        DJNZ INCR      ; SUBROUTINE
        EXX
        LD B, 04H      ; COUNT FOR OUTER LOOP
 OUTLOP:LD C, B        ; SAVE COUNT IN C
        LD B, 08H      ; COUNT FOR INNER LOOP
        LD A, 01H      ; BITPOS WILL BE USED
        LD (BITPOS), A ; TO CREATE THE ANS.
 INNLOP:PUSH DE
        PUSH HL
        PUSH IX
        LD A, B        ; SAVE INNER LOOP
        LD (COUNT), A  ; COUNT
        LD, B, 08H
        CALL SUBTR     ; SUBTRACT (HL) FROM (DE)
        POP IX
        POPHL
        POP DE
        JR C, NOCHAN   ; IF ANS. -VE JUMP
        CALL TRANS     ; PUT (IX) IN (DE)
        LD A, (BITPOS)
        OR (IY + 7)    ; PLACE 1 IN APPROP.
        LD (IY + 7), A ; BIT OF RES.
 NOCHAN:LD A, (BITPOS)
        AND A
        SLA A          ; SHIFT LEFT BITPOS
        LD (BITPOS), A ; AND RESTORE
        CALL SHIFL     ; SHIFT (HL) LEFT
        LD A, (COUNT)
        LD B, A
        DJNZ INNLOP
```

```
        DEC IY
        LD B, C          ;RESTORE OUTER LOOP COUNT
        DJNZ OUTLOP
        RET
BITPOS:DEFS 01
 COUNT:DEFS 01
 TRANS:PUSH DE
        PUSH IX
        EXX              ;USE B' FOR
        LD B, 08H        ; COUNT
  NEXT:EXX
        LD A, (DE)
        LD (IX+0), A
        INC DE
        INC IX
        EXX
        DJNZ NEXT
        EXX
        POP IX
        POP DE
        RET
 SHIFL:EXX               ; USE B', HL'
        AND A            ; REGISTERS
        PUSH HL          ; NOTE THAT HL'
        LD B, 08H        ; POINTS TO RIGHT
  ROTL:LD A, (HL)        ; HAND END OF B2
        RL
        DEC HL
        DJNZ ROTL
        POP HL
        EXX
        RET
```

Having written the normalized division subroutine we must now consider the general case; for this we must calculate the effective position of the binary point and then select the correct bits for the fixed point answer. Consider first a subroutine to find the number of zeros in the MSBs of the number B1. (This routine can also be used for B2.)

```
 ZEROS:PUSH DE
        LD B, 04H
        LD C, 00H        ; C WILL CONTAIN NO. OF ZEROS
        INC DE
        INC DE
        INC DE
```

```
        INC DE
  LOOP2:DEC DE
        LD A, (DE)
        EXX
        LD B, 08H
  LOOP1:EXX
        SRL A
        JR C, END
        INC C
        EXX
        DJNZ LOOP1
        EXX
        DJNZ LOOP2
    END:POP DE
        RET
```

The complete fixed-point division subroutine now follows:

```
 FPDIV:LD A, (DE)
       XOR (HL)
       LD (IY + 0), A      ; FIND SIGN OF ANS.
       INC DE
       INC HL
       INC IY
       INC IX
       CALL ANSZER         ; INITIALIZE RESULT TO ZERO
       PUSH DE
       PUSH HL
       PUSH IX
       PUSH IY
       LD B, 04H
       LD IX, TEMP3
       LD IY, TEMP4
 NEXT5:LD A, (DE)
       LD (IX + 0), A
       LD A, (HL)
       LD (IY + 0), A
       INC DE
       INC HL
       INC IX              ; SAVE NUMBERS
       INC IY              ; B1, B2 IN
       DJNZ NEXT5          ; TEMP3, TEMP4, . . .
       POP IY
       POP IX
       POP HL
```

```
        POP DE
        CALL ZEROS
        LD B, C
NORM1:PUSH BC
        PUSH DE
        CALL SHIFTR       ; SHIFT FIRST 1 TO
        POP DE            ; RIGHT HAND END
        POP BC
        DJNZ NORM1
        PUSH DE
        PUSH HL
        LD D, C           ; STORE NO. OF ZEROS IN B1
        EX DE, HL         ; POINT TO B2
        CALL ZEROS
        LD B, C
NORM2:PUSH BC
        PUSH DE
        CALL SHIFTR
        POP DE
        POP BC
        DJNZ NORM2
        EX DE, HL
        LD A, C           ; TEST IF B2 = 0
        CP 32D
        JR Z, DZERR
        LD A, D
        CP 32D
        JR Z, ZERO
        JR NONZ
DZERR:LD A, 01H
        POP HL
        POP DE
        RET
 ZERO:LD A, 00H
        POP HL
        POP DE
        RET
 NONZ:LD A, D           ; FIND DIFF IN NO. OF ZEROS
        SUB C             ; TO GIVE RELATIVE POSITION OF
        LD (TEMP7), A     ; BINARY POINT AND STORE
        POP HL
        POP DE
        PUSH DE           ; PUT HL + 4
        PUSH HL           ; AND DE + 4
```

```
        EXX                 ; IN HL'
        POP HL              ; AND DE'
        POP DE
        LD B, 04H
INCRE:INC DE
        INC HL
        DJNZ INCRE
        EXX
        PUSH DE             ; THIS SECTION
        PUSH HL             ; MOVES B1 AND
        LD B, 04H           ; B2 ALONG
  MOV:LD A, (DE)            ; 4 BYTES AND
        LD (TEMP5), A       ; PUTS ZEROS IN
        LD A, 00H           ; FIRST 4 BYTES
        LD (DE), A
        LD A, (HL)
        LD (TEMP6), A
        LD A, 00H
        LD (HL), A
        EXX
        LD A, (TEMP5)
        LD (DE), A
        LD A, (TEMP6)
        LD (HL), A
        EXX
        INC HL
        INC DE
        DJNZ MOV
        POP HL
        POP DE
        CALL NORMDIV        ; NUMBERS ARE NORMALIZED
                            ; SO WE CAN DIVIDE
        LD A, (TEMP7)       ; NO CALCULATE
        BIT 7, A            ; POSITION OF BINARY
        JR NZ, NEG          ; POINT. CALCULATE
        CP 17D              ; NO. OF LEFT SHIFTS
        JP M, MINI          ; REQUIRED TO BRING
        CALL ANSZER         ; ANS. TO FIRST
        JR NOSHIF           ; FOUR BYTES
 MIN1:ADD A, 15D
        LD B, A
        JR SHIFT
  NEG:NEG                   ; CHECK FOR OVERFLOW
        CP 16D
```

```
        JP M, MIN2
        LD A, 01H          ; FLAG OVERFLOW
        RET
   MIN2:SUB 15D
        JR Z, NOSHIF
        NEG
        LD B, A
  SHIFT:EXX
        AND A
        PUSH IY
        LD B, 08H
   ROT1:LD A, (IY)
        RL
        DEC IY
        DJNZ ROT1
        EXX
        POP IY
        DJNZ SHIFT
 NOSHIF:LD B, 04H          ; RESTORE
        PUSH IY            ; ORIGINAL
        LD IX, TEMP3       ; NUMBERS
        LD IY, TEMP4
  NEXT6:LD A, (IX + 0)
        LD (DE), A
        LD A, (IY + 0)
        LD (HL), A
        INC DE
        INC HL
        INC IX
        INC IY
        DJNZ NEXT6
        POP IY
        LD A, (IY – 1)
        LD (IY + 3), A
        INC IY
        INC IY
        INC IY
        INC IY
        LD A, 00H          ; FLAG OVERFLOW
        RET
ANSZER:PUSH IY
        LD A, 00H
        LD B, 08H
   INIT:LD (IY + 0), A
```

```
        INC IY
        DJNZ INIT
        POP IY
        RET
TEMP3:DEFS 04
TEMP4:DEFS 04
TEMP5:DEFS 04
TEMP6:DEFS 04
TEMP7:DEFS 01
```

This subroutine takes about 19 ms and completes our basic fixed point software which will be used shortly. Note that we have written these subroutines with clarity in mind. They are by no means 'optimal' in terms of memory space or execution time.

9.4.3 Floating-Point Arithmetic

The main drawback of fixed-point arithmetic is its inability to deal with very large and very small numbers simultaneously. This can be overcome by using floating-point arithmetic where we store not only the mantissa and its sign, but also the exponent indicating the effective position of the point. For example, we could store our numbers in the following form:

exp	sign	m	a n t	i s	s a

The point is assumed to be at a fixed place in the mantissa, say at the right hand end, and the exponent is usually stored in two's complement. In this way we can store numbers of the form

$$\pm 0.b_{31}b_{30}\ldots b_0 x2^{-128} \quad \text{to} \quad \pm 0.b'_{31}b'_{30}\ldots b'_0 x2^{+127}$$

Although we shall not discuss complete floating point software, the above ideas can be extended fairly easily to the more general case.

9.4.4 Examples

We can now use the above fixed-point software to develop some simple control algorithms for computer control. Note that in many cases much simpler arithmetic software will suffice – we are using the above programs mainly for pedagogical reasons. Consider first the second-order digital filter

$$y(n) = b_1 y(n-1) + b_0 y(n-2) + a_0 x(n), \quad y(-1) = y(-2) = 0$$

which may be derived from a low pass continuous filter or it may represent the equation of a classical discrete compensator for feedback control, considered in chapter 4. We shall assume that the constants b_0, b_1 and a_0 and the signals $x(n)$ and $y(n)$ are all in the range of the fixed-point arithmetic discussed earlier, and that the input signal $x(n)$ is available at each 'time' n in the memory locations INSIG to INSIG + 4. The acquisition of external data will be considered in more detail in chapter 10. We may then implement this simple filter on a Z80 μP system using the following subroutine:

```
 FILTER:LD IX, YM2      ; INITIALIZE Y(-1)
        LD B, 05H       ;Y(-2) TO 0
NEXTB1:LD (IX + 0), 00H
        INC IX
        DJNZ NEXTB1
        LD IX, YM1
        LD B, 05H
NEXTB2:LD (IX + 0), 00H
        INC IX
        DJNZ NEXTB2
   NEWY:LD DE, AZERO    ; MULT A0 BY X(N)
        LD HL, INSIG
        LD IX, PROD1
        CALL FPMULT
        LD DE, BZERO    ; MULT B0 BY Y(N - 2)
        LD HL, YM2
        LD IX, PROD2
        CALL FPMULT
        LD DE, BONE     ; MULT B1 BY Y(N - 1)
        LD HL, YM1
        LD IX, PROD3
        CALL FPMULT
        LD IX, YM2      ; PUT Y(N - 1)
        LD IY, YM1      ; IN Y(N - 2)
        LD B, 05H
NEXTB3:LD A, (IY + 0)
        LD (IX + 0), A
        INC IX
        INC IY
        DJNZ NEXTB3
        LD DE, PROD1    ; ADD A0*X(N)
        LD HL, PROD2    ; AND B0*Y(N - 2)
        LD IX, SUM1     ; TO GIVE SUM1
        CALL ALGSUM
        LD DE, SUM1     ; ADD SUM1 TO B1*Y(N - 1)
```

```
        LD HL, PROD3
        LD IX, YM1
        CALL ALGSUM
        JR NEWY
AZERO:DEFS 09
 INSIG:DEFS 09
PROD1:DEFS 09
BZERO:DEFS 09
   YM2:DEFS 09
PROD2:DEFS 09
 BONE:DEFS 09
   YM1:DEFS 09
PROD3:DEFS 09
  SUM1:DEFS 09
```

The calculation of the update of $y(n)$ takes about 30 ms. For a system with a slow sampling period $\gg$ 30 ms this algorithm will provide the true solution (modulo the rounding errors) to the original difference equation. However, consider the effect of a sampling period just larger than 30 ms. Then (neglecting delays in the acquisition of $x(n)$) we shall not be able to calculate $y(n)$ using $x(n)$ until the end of the sampling period. Hence we will be effectively solving the equation

$$y(n) = b_1 y(n-1) + b_0 y(n-2) + a_0 x(n-1)$$

If the difference equation is the implementation of a digital compensator then this will introduce a delay into the forward path, which may have to be accounted for in the design.

As a second example we shall consider the on-line application of the scalar Kalman filter given by equations (7.2.24), which are repeated below for convenience:

$$\hat{x}(k) = a\hat{x}(k-1) + b(k)[y(k) - ac\hat{x}(k-1)] \tag{9.4.1}$$

$$b(k) = cp_1(k)[c^2 p_1(k) + \sigma_v^2]^{-1} \tag{9.4.2}$$

$$p_1(k) = a^2 p(k-1) + \sigma_w^2 \tag{9.4.3}$$

$$p(k) = p_1(k) - cb(k)p_1(k) \tag{9.4.4}$$

We shall assume that the constants a, c, σ_v^2, σ_w^2, $\hat{x}(0)$ and $p(0)$ have been initialized in memory. (These may be input from an external device using interrupts as discussed in chapter 10, or if the filter is run under similar circumstances each time then these constants may be stored at fixed places in ROM.) Moreover, we shall assume that the measurement $y(k)$ is available at time k and is input to memories YK to YK + 5 using an external interrupt and A/D converter. In the subroutine below we first calculate $p_1(k)$ from 9.4.3 and then $b(k)$ from 9.4.2 which allows the new state estimate to be obtained from 9.4.1. Finally, $p(k)$ is updated using 9.4.4 and we return to 9.4.3.

```
KALMAN:LD DE, AA
       LD HL, AA
       LD IX, ASQU    ; FIND A^2
       CALL FPMULT
       LD DE, ASQU    ; FIND P1 (K - 1)*A^2
       LD HL, PKM1
       LD IX, PROD
       CALL FPMULT
       LD DE, PROD    ; ADD σw^2
       LD HL, SIGW2
       LD IX, PIK
       CALL ALGSUM
       LD DE, CC      ; FIND C*P1(K)
       LD HL, PIK
       LD IX, PIKC
       CALL FPMULT
       LD DE, PIKC    ; FIND C^2*P1(K)
       LD HL, CC
       LD IX, PROD
       CALL FPMULT
       LD DE, PROD    ; ADDσv^2
       LD HL, SIGV2
       LD IX, SUM
       CALL ALGSUM
       LD DE, PIKC    ; FIND C*P1(K)*[C^2P1(K) + σv^2]^-1
       LD HL, SUM     ;  = B(K)
       LD IY, BK
       CALL FPDIV
       LD DE, XK      ; FIND A*X̂(K - 1)
       LD HL,AA
       LD IX,PROD
       CALL FPMULT
       LD DE,PROD;    FIND A*C*X̂(K - 1)
       LD HL, CC
       LD IX, PROD1
       CALL FPMULT
       LD A, (PROD1)  ; CHANGE SIGN
       XOR 01H        ; OF A*C*X̂(K - 1)
       LD (PROD1), A
       LD DE, PROD1   ; FIND Y(K) - A*C*X̂(K - 1)
       LD HL, YK
```

```
        LD IX, SUM
        CALL ALGSUM
        LD DE, SUM     ; FIND B(K)*[Y(K) - A*C*X̂(K - 1)]
        LD HL, BK
        LD IX, PROD1
        CALL FPMULT
        LD DE, PROD    ; FIND X̂(K)
        LD HL, PROD1
        LD IX, XK
        CALL ALGSUM
        LD DE, PIKC    ; FIND C*B(K)*P1(K)
        LD HL, BK
        LD IX, PROD
        CALL FPMULT
        LD A, (PROD)   ; CHANGE SIGN
        X0R 01H
        LD (PROD), A
        LD DE, PROD    ; FIND P(K)
        LD HL, PIK
        LD IX, PKM1
        CALL ALGSUM
        JR KALMAN
     AA:DEFS 09
   ASQU:DEFS 09
   PKM1:DEFS 09
   PROD:DEFS 09
  SIGW2:DEFS 09
    PIK:DEFS 09
     CC:DEFS 09
   PIKC:DEFS 09
    SUM:DEFS 09
     BK:DEFS 09
  PROD1:DEFS 09
     XK:DEFS 09
     YK:DEFS 09
```

This subroutine runs in about 110 ms and so our sampling period must be somewhat greater than 0.1 s. Clearly for matrix systems the calculation time of the state estimate is going to increase rapidly with dimension and so these algorithms would not be practical for large order systems with fast sampling. Other methods could then be used; for example, hardware arithmetic, parallel processing, etc.

This completes our discussion of the software structure of the Z80 μP and assembly language programming. As we have seen, the main advantage of assembly programming is the ability to obtain precise timing calculations which are so important in on-line processing. In the final chapter we shall consider the interfacing methods of the Z80 chip and external devices.

9.5 OTHER MICROPROCESSORS

In this section of the book, we are considering a specific microprocessor, namely the Z80. It is important to mention some other currently available processors which can be used to build a microcomputer system

Other 8-bit processors include the Intel 8080, 8085 (which are compatible with the Z80), the Motorola 6502 and 6800 series, of which the 6809 has hardware 8-bit multiplication, and the 8088 which has 16-bit internal architecture. True 16-bit microprocessors include the Texas Instruments 9900 series, the Zilog Z8000 and the Intel 8086. Finally some 32-bit chips are now available, including the National Semiconductor 16000 chip which has versions with 8- or 16-bit data buses, and the Motorola 68000 series.

9.6 EXERCISES

1. Explain the differences between ROM and RAM and how these differences affect their use.

2. How does a microprocessor 'know' the difference between instructions and data in a stored program?

3. Describe, in detail, the operation of the stack and its associated instructions.

4. What is the difference between the instructions

   LD BC, FF1A and LD BC, (FF1A)?

5. Explain what is meant by 'position independent code' in assembly programming.

6. Show that the overflow flag V is given by $C_6 \oplus C_7$ where C_6, C_7 are the carry bits from bit 6 to bit 7 and out from bit 7 in a two's complement addition. Give some examples of such addition.

7. The instruction DAA (decimal adjust) is used in decimal calculations and automatically changes a binary number into BCD (binary coded decimal) form. Consider the decimal sum and its BCD equivalent:

```
   73 (D)        0111 0011   (=73H)
 + 84 (D)      + 1000 0100   (=84H)
 --------      -----------
  157 (D)        1111 0111
```

When added in ordinary binary the sum on the right is not a BCD number. DAA converts 11110111 to 01010111 and sets the carry flag indicating an overflow. Write a program to add two six-digit decimal numbers obtaining the answer in decimal, each number requiring 3 bytes of memory.

8. Devise a method for converting decimal numbers to binary.

9. Calculate the total memory space required for the arithmetic software developed in section 9.4.2.

10. Write a subroutine to implement a two-dimensional Kalman filter and determine its running time per step.

REFERENCES

Hill, F. J. and Peterson, G. R. (1978) *Digital Systems: Hardware Organisation and Design*, Wiley.

Krutz, R. L. (1980) *Microprocessors and Logic Design*, Wiley.

Peatman, J. B (1972) *The Design of Digital Systems,* McGraw-Hill, Kogakusha.

10 MICROPROCESSOR INTERFACING TECHNIQUES

10.1 INTRODUCTION

In the last chapter we considered the software structure of the Z80 μP in isolation and we assumed that any external data which was required was already in the memory. This final chapter will be concerned with the hardware interfacing techniques and the associated software necessary to design a digital control system (see also Cluley, 1984). Interfacing a microprocessor to memory and external devices is made particularly simple by the existence of special 'support chips' designed precisely for this purpose. This means that nothing more than a superficial understanding of electronics is necessary to be able to design an operational microcomputer system.

We shall see that the Z80 microcomputer system is dominated by the three bus architecture and we shall show how to decode the address bus to select a particular external chip. Signals from the control bus can then be used to enable or 'turn on' the chip. The interfacing of both ROM and RAM will be discussed in detail together with the notion of 'memory map' which is concerned with the designation of various parts of the available memory to specific functions.

Next we shall discuss the implementation of input–output ports and, in particular, we shall introduce the Z80 PIO chip which is a powerful programmable input–output device specially designed for the Z80 μP. The most important method for inputting and outputting data will then be considered – this is called interrupt processing, and is designed to give the external devices 'service' only when required. The interfacing of peripherals such as keyboards, LED displays, etc., and A/D, D/A conversion and data acquisition will be discussed before some examples of real system interfaces are given. In particular we shall consider motor speed control, stepper motors and robotics, and simple process control.

10.2 THREE BUS ARCHITECTURE AND TIMING

10.2.1 The Address Bus

We have already discussed the software aspects of the three-bus architecture in chapter 9, and we shall now consider the hardware implementation of the buses. Of course, the buses are already available at the output pins of the Z80 chip, but these buses have only a small drive capability. Hence it is usual to 'buffer' the buses to increase the current sourcing and sinking available on the microprocessor buses. All the devices needed for a microcomputer system are manufactured on a variety of chips, the most widely used being the 7400 TTL series, including the low power 74LS00 series. In particular three of the hex buffers 74LS367 (fig. 10.1(a)) can be used to buffer the address buses as shown in fig. 10.1(b). The output drive capability required from any device must be calculated and the appropriate chip can then be chosen with reference to the manufacturers data sheet. For most small systems the chips which we shall use will give sufficient bus drive and so we shall not go into the actual numerical values of the output capabilities of each chip.

In fig. 10.1(a) we see that the 74LS367 device essentially places a non-inverting current amplifier in each bus line. However, note that this device

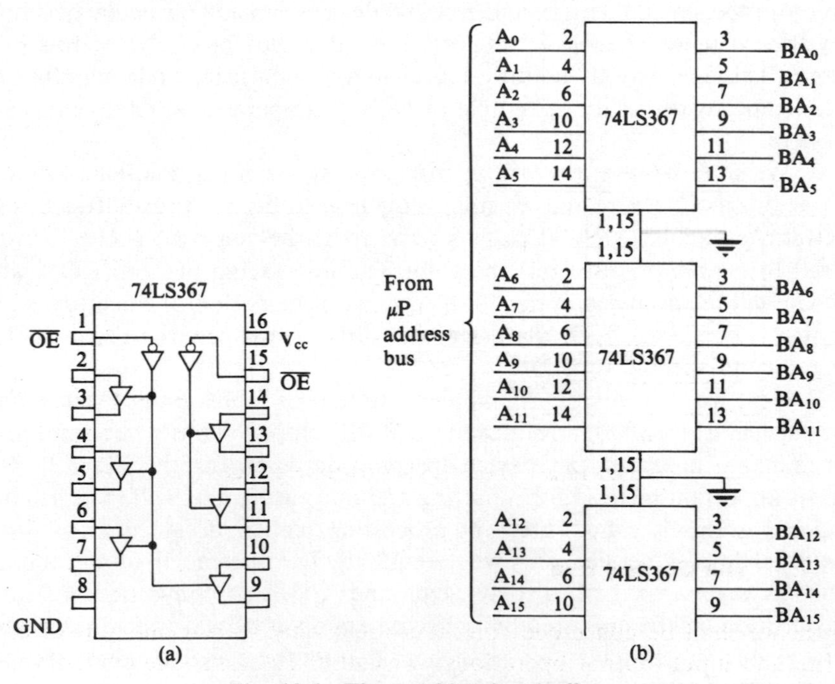

Fig. 10.1 The 74LS367 bus buffer.

is 'tri-state' which means that the amplifiers can be disabled, effectively disconnecting the buffered address bus BA_0–BA_{15} from the microprocessor address bus. Two enable lines are provided on pins 1 and 15 which are 'active low', i.e. 0 volts placed on these pins will connect the two sides of the bus. However, when logic 1 (5 volts) is put on these pins the chip is disabled and the external bus is independent of the internal μP bus. This is useful when external devices require control of the bus, as we shall see later with direct memory access. For now we have shown the buffers permanently enabled by grounding the pins 1 and 15.

10.2.2 The Data Bus

The data bus may be buffered in a similar way to the address bus. However, we must remember that the data bus is bidirectional and so we choose the bidirectional 8-bit driver 74LS245 for this purpose. This is shown in fig. 10.2(a) together with its use in the data bus in fig. 10.2(b). The chip is again shown permanently enabled via pin 19. The data direction line (pin 1) is connected to the control bus (discussed below) and must be logic 0 for data input (read) and logic 1 for data output (write).

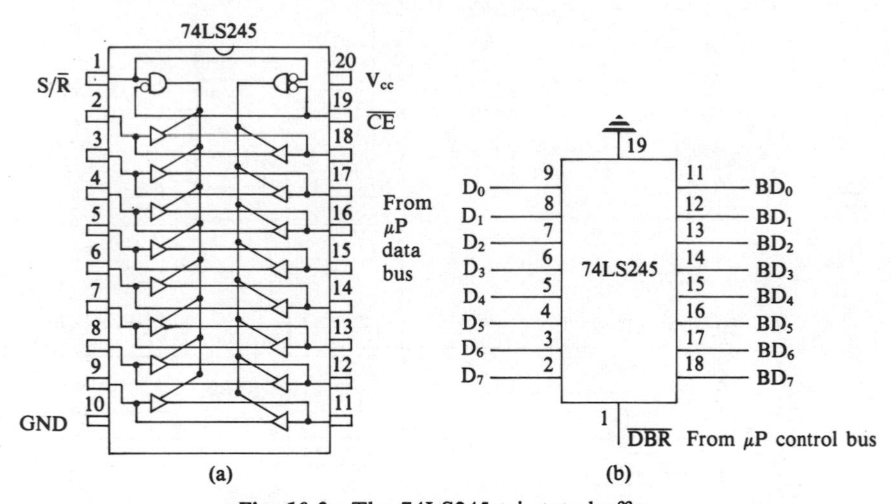

Fig. 10.2 The 74LS245 tri-state buffer.

10.2.3 The Control Bus

Since we shall be concerned for the present only with memory read and write and input–output operations we shall discuss only the control lines $\overline{WR}$, $\overline{RD}$, $\overline{IORQ}$ and $\overline{MREQ}$ (write, read, input–output request and memory request respectively). When a memory write cycle occurs we must enable the appropriate memory chip only when the $\overline{\text{WR}}$ and $\overline{\text{MREQ}}$ lines are low. Hence we form the buffered memory write request signal as

$$\overline{\text{BMEMW}} = \overline{\text{WR.MREQ}}$$

Similarly we form the buffered memory read request, input read and output write signals as

$$\overline{\text{BMEMR}} = \overline{\text{RD. MREQ}}$$

$$\overline{\text{BIOR}} \quad = \overline{\text{RD.IORQ}}$$

$$\overline{\text{BIOW}} \quad = \overline{\text{WR.IORQ}}$$

These control signals can be implemented using the 7404 inverter and the 7400 NAND gate as shown in fig. 10.3. Note that the buffered $\overline{\text{RD}}$ signal is denoted by $\overline{\text{DBR}}$ and is fed to the data bus buffer. It will enable the tri-state buffers in the 74LS245 chip for read (input to the μP) when $\overline{\text{RD}}$ is low.

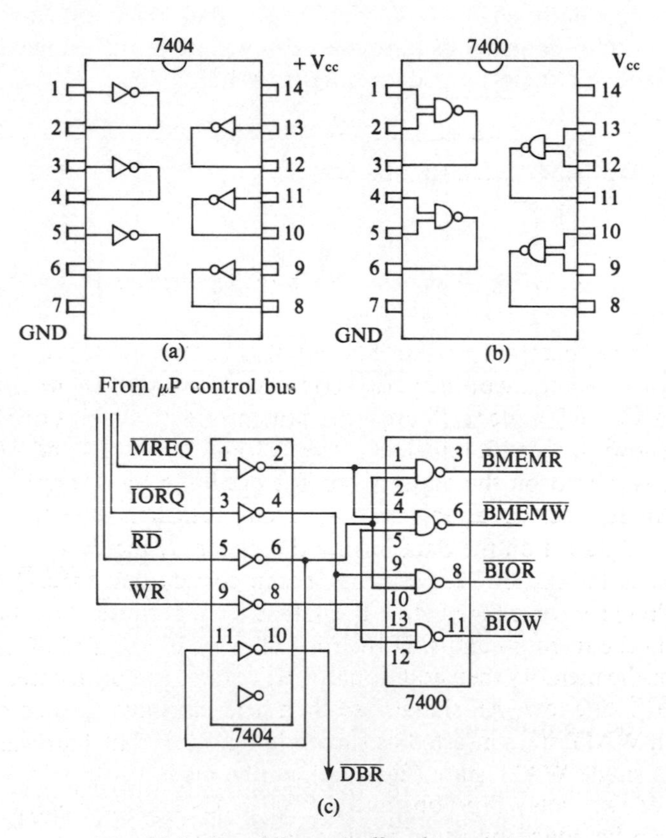

Fig. 10.3 Providing a buffered control bus.

10.2.4 Timing Considerations

It is vitally important when interfacing to ensure that the μP buses all put out the correct information at the correct times. The overall timing of the system is controlled by the Z80 CPU, but it should be remembered that buffers, latches, etc. all have propagation delays and these must be taken into account when designing the system.

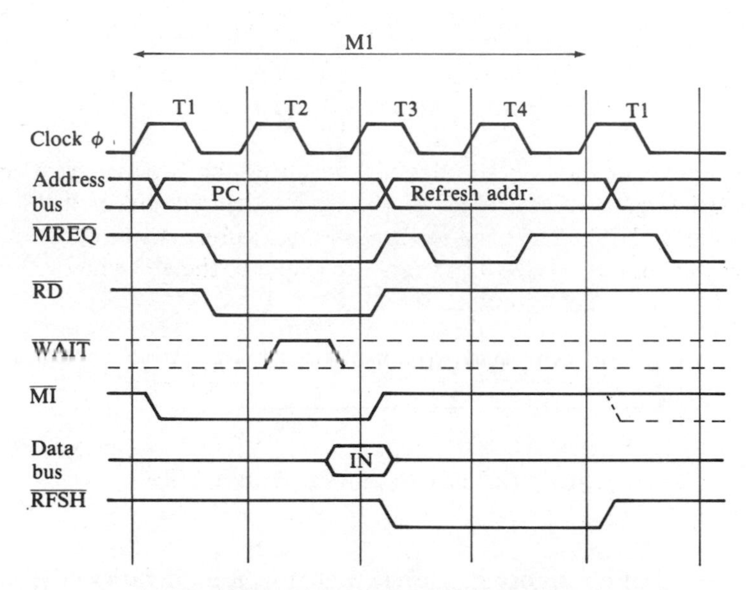

Fig. 10.4 The Z80 μp timing diagram.

A typical diagram (for an M1 cycle) is shown in fig. 10.4 and shows the times of switching of the various signals with respect to the clock. Recall that in an M1 OP code fetch cycle the contents of PC are put on the address bus as shown in fig. 10.4 in clock cycles T1 and T2. In T3 and T4 a refresh address is placed on the address bus for decoding by dynamic memories. Then $\overline{MREQ}$ and $\overline{RD}$ go low and, assuming sufficient time elapses for the data to be placed on the data bus by the memory, the data is read off the data bus at the end of T2. Note that it is important that $\overline{MREQ}$ and $\overline{RD}$ be active (low) for some time before the data is read, because of the propagation delays in the memory chips. If the time allowed by the Z80 μP is not sufficient for the memory then additional WAIT states can be inserted by pulling the $\overline{WAIT}$ line low. All signals are then held constant for one clock cycle for each WAIT state inserted as shown in fig. 10.5. The hardware for providing a single WAIT state (fig. 10.6) can be made using a 7474 chip consisting of two delay flip-flops and a 7400 NAND gate. Additional WAIT states can be added by using further 7474 chips.

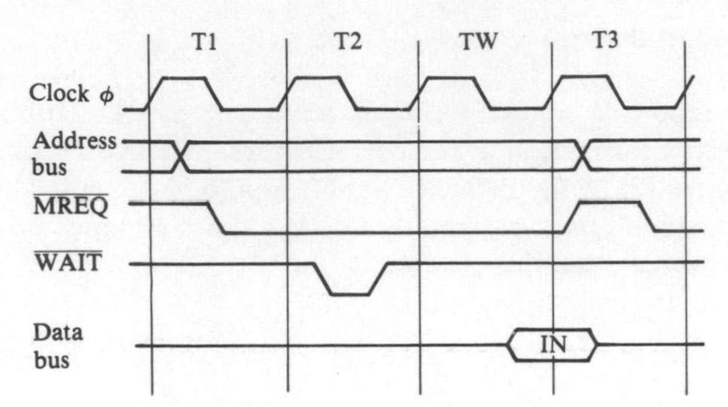

Fig. 10.5 Adding a WAIT state.

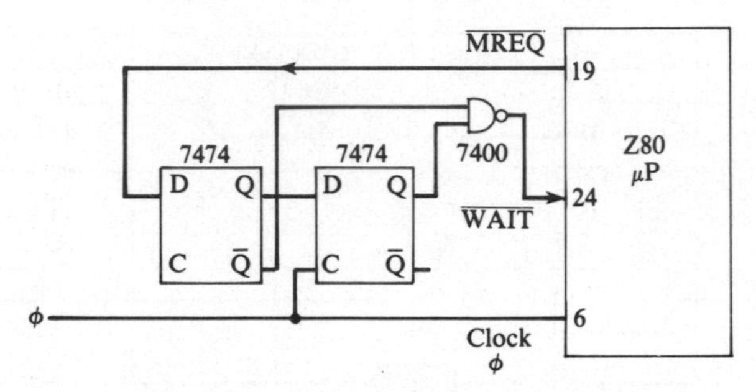

Fig. 10.6 Circuitry to activate the WAIT line.

If instead of a memory read cycle we consider an input cycle from an input port, then we obtain a similar timing diagram to that in fig. 10.4 except that $\overline{IORQ}$ is pulled low rather than $\overline{MREQ}$ during T1 and T2. When an output or write operation is performed, the timing diagram has one important difference from that of a read operation, namely the $\overline{MREQ}$ (or $\overline{IORQ}$) and data bus lines are valid *before* the $\overline{WR}$ line is pulled low in order that the data has stabilized before being latched by the peripheral device or the memory (fig. 10.7).

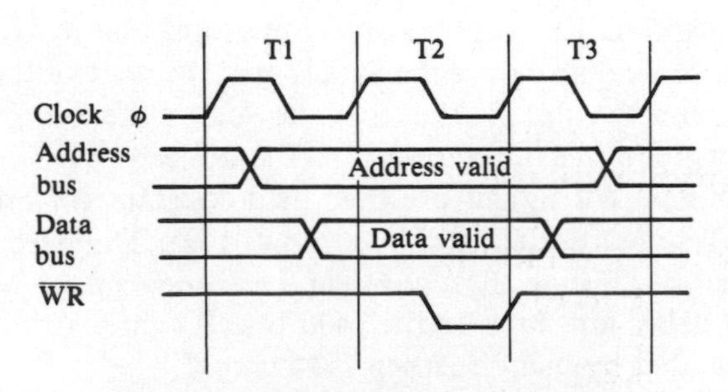

Fig. 10.7 Timing for a write cycle.

10.2.5 The System Clock and Reset Circuitry

The system clock which controls the entire operation of the microprocessor and associated chips is shown in fig. 10.8. The frequency is controlled by a 4 MHz crystal and the RC combination should provide enough phase shift for oscillation. Since we require $RC \geqslant 4 \times 10^{-6}$ and $R_1 = 330$ we must choose $C \geqslant 4 \times 10^{-6}/330 \approx 12\,000$ pF.

When power is initially connected to the Z80 chip, it enters a reset state and begins running the program starting at address 0000H (assuming one exists). A similar reset state can be entered by pulling the $\overline{\text{RESET}}$ line (pin 26) low at any time, and a simple circuit for this purpose is shown in fig. 10.9. When power is applied to R the capacitor charges up towards 5 V and $\overline{\text{RESET}}$ is disabled, at which time the program at 0000H starts to run. To reset at any successive time the switch can be closed, discharging C.

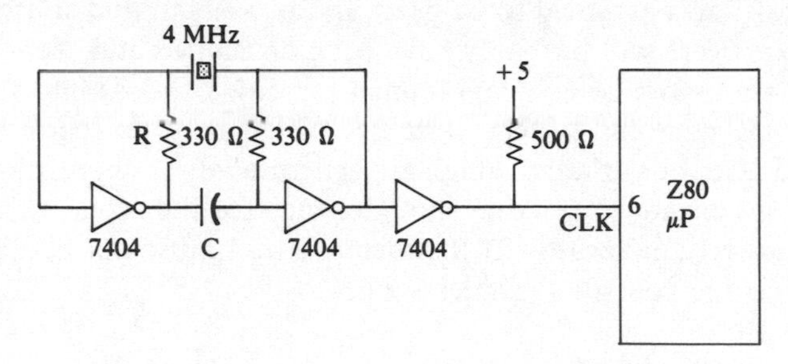

Fig. 10.8 The system clock.

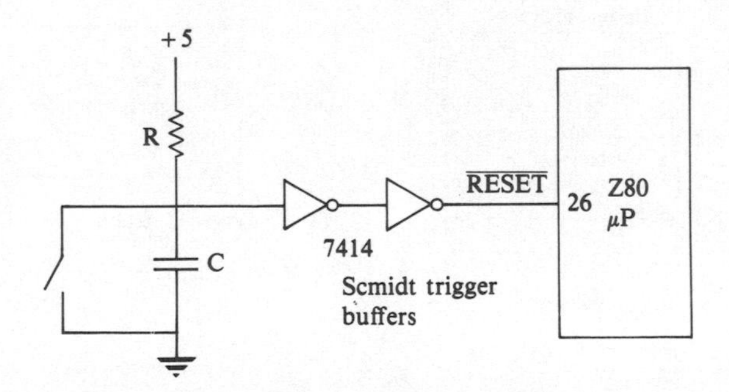

Fig. 10.9 The RESET circuit.

10.3 INTERFACING MEMORY

10.3.1 The Memory Map

The first decision which has to be made when designing a microcomputer system is what type of memory is required and how much will be needed.

The Z80 μP can address up to 64K of memory, but a small system may not require all this memory. Since the reset address is 0000H and the restart addresses are between 0000H and 00FFH it is usual to put the ROM at the bottom of the memory and any RAM above it. To be definite we shall design a system with 8K of ROM in the addresses 0000–1FFFH and 4K of RAM in the addresses 2000–2FFFH.

Having decided on the size and layout of the memory, we must assign various parts of the memory to the different software functions required by the system. This assignment is called the **memory map** and is typically of the form shown in fig. 10.10 for a digital control system. The reset address 0000H contains a JP command to the monitor program which controls the overall operation of the system. Hence when power is applied to the system the CPU will run the monitor program which contains communication software for interaction with an external operator. The ROM also contains the interrupt vector table and service routines which will be discussed later. The control algorithm is assumed to be fixed, in our system, and so is placed in ROM; this software will probably have been developed and 'debugged' on a larger system before being programmed (or downloaded) into the ROM.

The RAM contains system variables which may be required by the monitor and also the variables which are generated by the control software. Finally it is important to provide enough space for the stack which, as we have seen, grows downwards in the memory and must not be allowed to spill over into the control variables space.

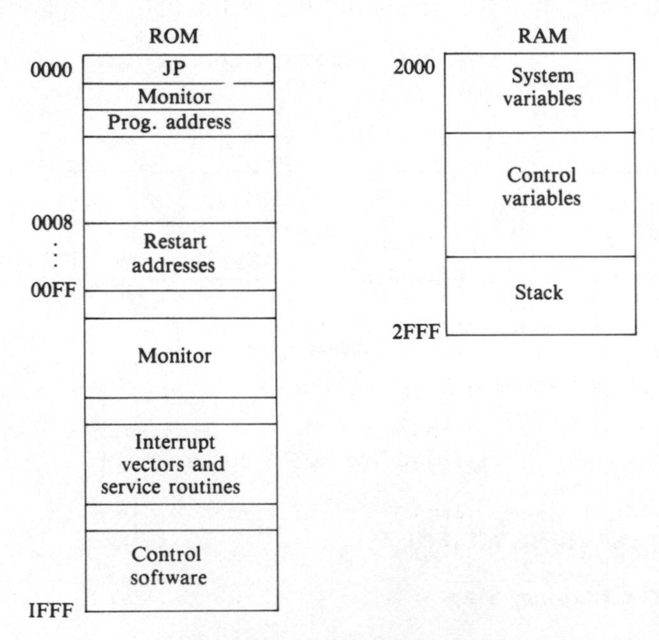

Fig. 10.10 The memory map.

10.3.2 Interfacing ROM

Once the memory map has been specified, we must select the most appropriate memory chips for our purposes. In our case we require 8K of ROM and so we shall use eight 2708 EPROMs (electrically programmable ROM). Each 2708 chip is organized into 1024 8-bit words and has ten input address lines; the pin configuration is shown in fig. 10.11(a). In the read mode the power supply requirements are shown in fig. 10.11(a); in the program mode we must apply the data to the data pins, the required address to the address pins, + 12 V to the $\overline{\text{CS}}$/WE input (pin 20) and then pulse the program pin with + 26 V for 0.5 ms two hundred times. (See Coffron (1981) for a complete description of EPROM programming hardware and software.)

The bottom 8K of memory is selected by the address lines A0–A12 with A13–A15 being logic 0. The lines A0–A9 are decoded internally by the 2708 chip and so we must decode the lines A10–A12 to provide chip select pulses for the appropriate ROM chip. The 74LS138 decoder/demultiplexer chip is ideal for our purposes and is shown in fig. 10.11(b). It has three enable inputs, two of which are logic 0 and one logic 1 – these inputs are ANDed as shown and so must all carry the appropriate voltage to enable the chip. The ABC inputs are for address lines and if we regard CBA as a binary number, with decimal equivalent *n*, then the output $\bar{n}$ goes low when we apply CBA to the input terminals. The complete ROM decoding and interface is shown in fig. 10.12 where just one of the eight 2708 chips is drawn

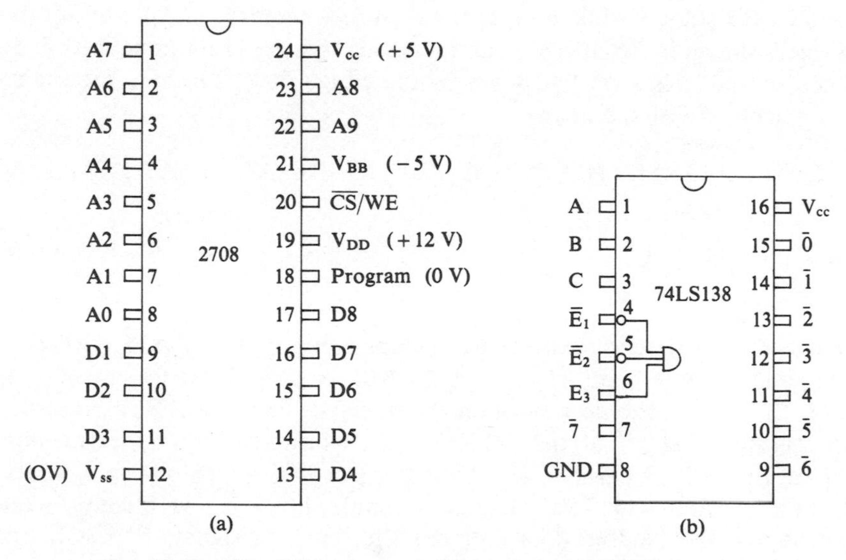

Fig. 10.11 The 2708 memory chip and the 74LS138 decoder.

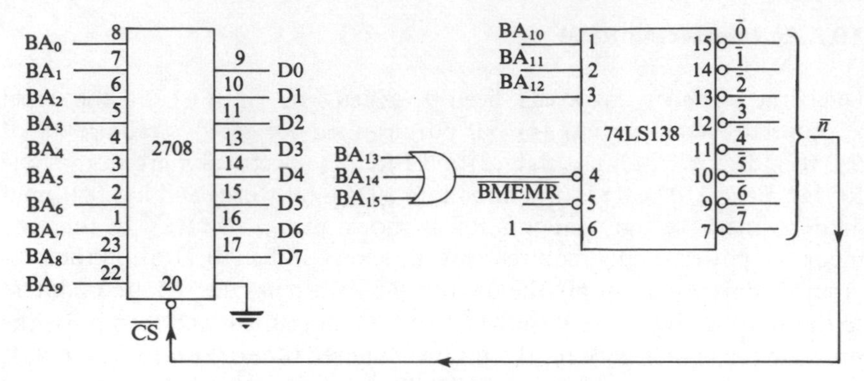

Fig. 10.12 Interfacing the 2708 memory.

for convenience. Note that we must decode address lines A13–A15 so that the ROM chips are selected only when the address is less than 2000H.

10.3.3 Interfacing Static RAM

Random access memory comes in two forms – static and dynamic. Static memory consists of flip-flops which retain their state as long as power is supplied, but dynamic RAMs lose charge and must be refreshed every few milliseconds. Static RAMs are easier to interface than dynamic RAMs but are larger and more expensive for the same memory size. For small systems, however, static RAMs are probably the best choice. We shall choose two 6116 2K × 8 static RAMs to make the required memory. The pin-out of this device is shown in fig. 10.13(a), and it is seen to have 11 input address lines. Since the addresses of RAM are in the range 2000–2FFFH, the binary equivalents arc of thc form

A15	A14	A13	A12	A11	A10	A9	A8	A7	A6	A5	A4	A3	A2	A1	A0
0	0	1	0	0 or 1	*	*	*	*	*	*	*	*	*	*	*

Hence we can select one of the 6116 chips using A13 A12 A11 = 100 (= 4 dec.) and the other with A13 A12 A11 = 101 (= 5 dec.). Hence, again using a 74LS138 decoder chip we obtain the interface in fig. 10.13(b). Note that the outputs $\overline{6}$ and $\overline{7}$ of the 74LS138 could be used for further memory expansion to 8K of RAM. To use the complete 64K of memory we must also decode A14 and A15. The principle is simple, however; we merely decode the appropriate address bits and use the decoded signal, $\overline{\text{BMEMR}}$ and $\overline{\text{BMEMW}}$ to select the correct chip.

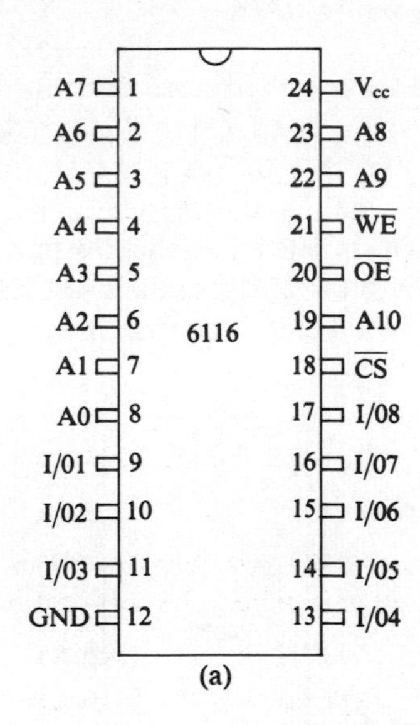

(a)

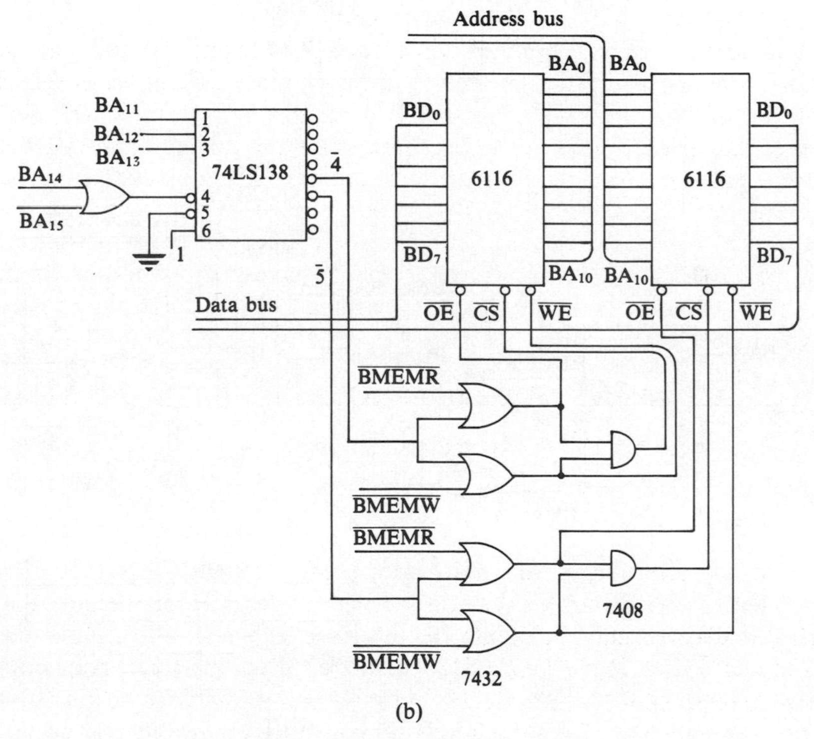

(b)

Fig. 10.13 The 6116 chip and its interface.

10.3.4 Interfacing Dynamic RAM

Interfacing dynamic RAMs is rather more difficult than static RAMs since each chip requires refreshing every few milliseconds to restore the charge which leaks away. The Z80 chip makes this task simpler by supplying the user with the R register which is automatically incremented at the end of each M1 cycle and contains in bits 0 to 6 the low part of the refresh address.

We shall use eight 4K-bit 4027 dynamic RAM chips to provide 8K bytes of RAM. The pin-out of each chip is shown in fig. 10.14(a) where it is seen

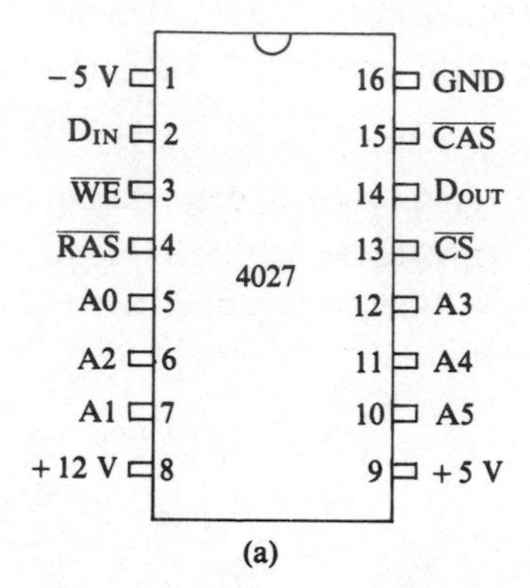

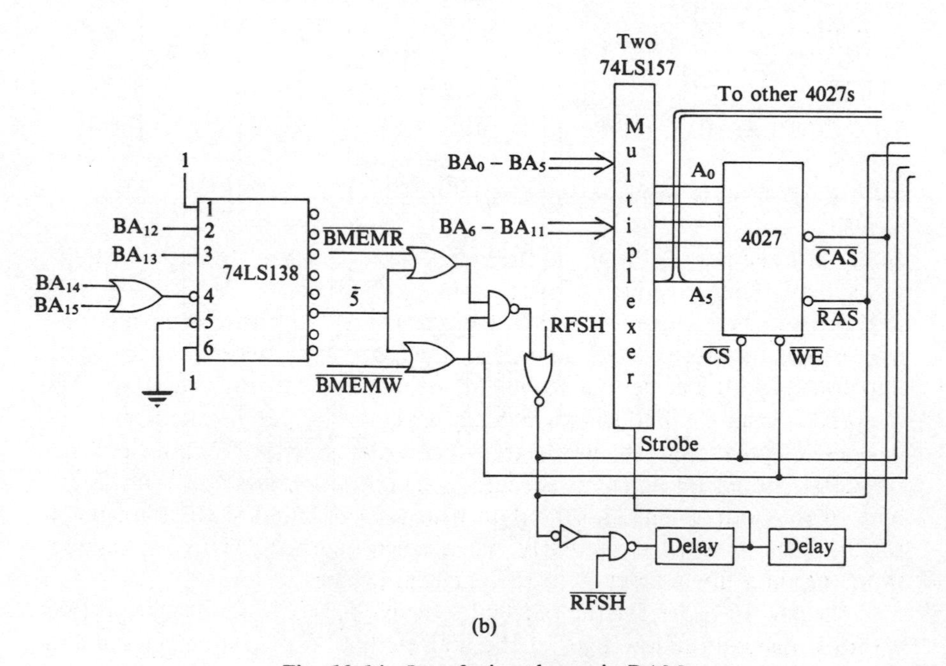

Fig. 10.14 Interfacing dynamic RAM.

to have only six address input lines, rather than the twelve which are required to address 4K of memory. The reason for this is that the address bits A0 to A11 are multiplexed into two groups of six – A0 to A5 followed by A6 to A11. The first group is interpreted as a row address while the second is a column address. By multiplexing the address bus a whole row of memory can be refreshed by a single memory read as we shall see below.

The address decoding for the chip select and $\overline{WE}$ signals can be similar to that used in fig. 10.13, except that the 1-input to the 74LS138 chip is tied to 1 rather than A11, since A11 is now input to the 4027 chips. The complete interface of the RAM chips is given in fig. 10.14(b), where for simplicity only a single 4027 chip is shown. The multiplexer (formed by two 74LS157 quad 2-input data selector chips) is designed to output bits A0–A5 when the strobe input is logic-1 and bits A6–A11 when the strobe input is logic-0. Hence when RFSH is low (i.e. not in a refresh cycle) the $\overline{RAS}$ (row address select) line goes low when $\overline{CS}$ goes low, but the strobe input to the multiplexer remains high because of the delay, and so the 4027 chips interpret the address inputs as the low bits giving a row address. After the delay the strobe input to the multiplexer goes low and A6–A11 are output to the memory chips. Then there is a further delay before the $\overline{CAS}$ (column address select) line goes low causing the 4027 chips to input the column address.

Note finally that when the RFSH signal goes low the $\overline{RAS}$ line goes low independently of the address bits A12–A15 while the strobe input to the multiplexer is held at logic-1 so that the $\overline{CAS}$ line is never enabled, allowing a complete row to be refreshed simultaneously.

10.4 INPUT–OUTPUT PORTS, INTERRUPTS AND THE PIO CHIP

10.4.1 A Simple Input–output Port

In the last chapter we discussed the complete instruction set of the Z80 μP; in particular, we mentioned instructions of the form IN A, (*n*) and OUT (*n*), A (together with more general instructions). We now consider these operations in more detail and also the hardware necessary for their implementation. Let us first recall the meaning of these instructions. The operand *n* is an 8-bit address which can label one of 256 ports (or external devices – a **port** is just the hardware which isolates the external device from the CPU. When IN A, (*n*) is executed, the address *n* is placed on the low byte of the address bus and the data which is presented to the data bus is read into the A register. Similarly, when executing the OUT (*n*), A instruction, the data flows from A to the external device.

Clearly, in order for the particular device (*n*) to be uniquely specified we must decode the low byte of the address bus to provide a port enable signal. However, since the address bus is also interfaced to the memory we must also use the $\overline{BIOR}$ and $\overline{BIOW}$ control signals so that the ports are only enabled during input–output operations. (Some microprocessors do not distinguish between memory and input–output ports and so the addresses of the latter are actually part of the memory map – hence we call this type of input–output **memory mapping**.) On the input side of the port we can use a simple tri-state buffer on the data bus such as a pair of 74LS367 hex buffer/driver chips. However, on the output side we usually wish to hold the data, which appears on the data bus, indefinitely. Since this data exists on the data bus for a very short time we must use a tri-state latch containing

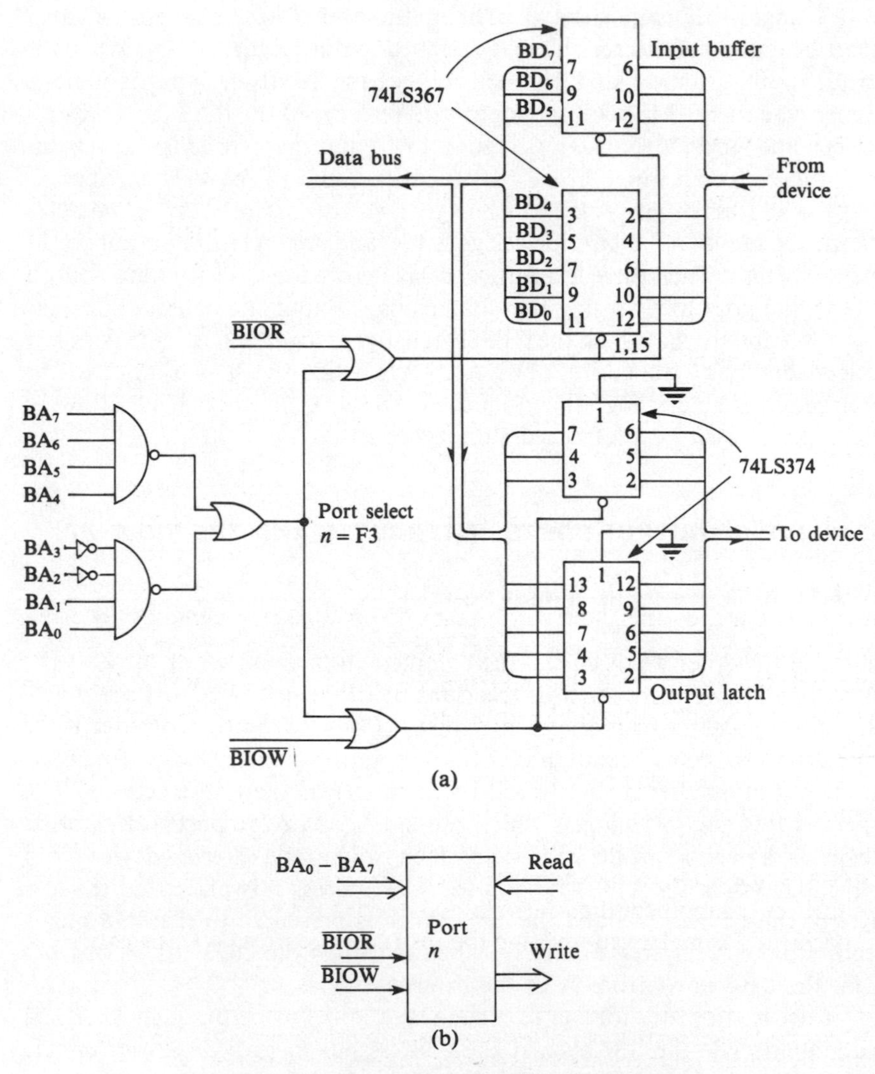

Fig. 10.15 Generating a bi-directional port.

D type flip-flops (such as the 74LS374 chip). The enable inputs to these chips will be derived from the address and control buses as described above.

We can now design a simple input–output port as in fig. 10.15(a), where we have chosen the port address $n = \text{F3H}$ (by changing the pattern of inverters on the address bus we can clearly select any port address). For simplicity we shall draw this port diagrammatically as in fig. 10.15(b), and use it to interface certain peripheral devices in the next section. Before considering the actual external devices which may be connected to the μP we shall consider the special purpose PIO chip which incorporates two input–output ports similar to that designed above, but which is also programmable by the Z80μP and so provides a very flexible input–output capability. First, however, we must consider interrupts.

10.4.2 Interrupts

As we mentioned briefly in chapter 9 the Z80 μP has two interrupt pins labelled $\overline{\text{INT}}$ and $\overline{\text{NMI}}$ which allow the operator or external circuitry to interrupt the normal program sequence. When either of these lines is pulled low, the microprocessor finishes the current operation and then executes a special interrupt subroutine to service a particular external device. The method by which the μP determines the start address of this subroutine will be seen shortly. Note first that the two types of interrupt corresponding to the $\overline{\text{INT}}$ and $\overline{\text{NMI}}$ lines are called **maskable** and **non-maskable** interrupts respectively. In the former case the μP will only respond to external interrupts on the $\overline{\text{INT}}$ line if the enable interrupts EI operation has been executed in software prior to $\overline{\text{INT}}$ going low. Interrupts can be disabled by using the DI instruction in a program and it should be noted that interrupts are disabled automatically on RESET or after the execution of a non-maskable interrupt. On the other hand, non-maskable interrupts cannot be affected by the software and whenever the $\overline{\text{NMI}}$ line is pulled low the μP will acknowledge and service the interrupt whatever the current status of the system. Also, non-maskable interrupts have priority over maskable interrupts.

Consider next the hardware necessary for generating interrupts. The simplest such circuitry is shown in fig. 10.16 where we have just used the output of a 'debounced' switch.† Suppose that at any time (i.e. asynchronous operation) the switch is toggled (i.e. closed and then released) and assume for the moment that the start address of the interrupt service routine is specified. Then after completing the instruction currently being executed,

---

† When a mechanical switch is closed, perfect contact is not maintained for a short time after. Instead, the switch contact jumps between the open and closed states for a short time until the final contact is achieved.

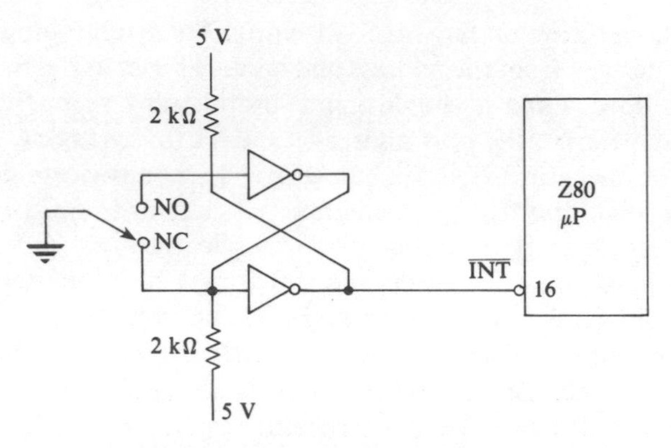

Fig. 10.16 Hardware interrupt signal circuit.

the CPU will perform a jump to the interrupt subroutine which will consist of some lines of programming terminated (usually) by

EI
RETI

Since interrupts are disabled after the acknowledgement and execution of an interrupt, it is usual to enable interrupts at the end of the service routine. (Note the use of RETI rather than RET for an ordinary subroutine.) Now suppose that the interrupt service routine is executed very quickly – quicker in fact than the length of time the switch in fig. 10.16 is held in the NO position. This is quite likely since the μP operates in microseconds while a human operator will hold the switch in the NO position for several milliseconds. In this case the interrupt line is still held low when EI is executed which would mean that the μP would respond to a spurious interrupt request. Moreoever, if the μP were to respond to interrupts immediately after execution of the EI instruction the service routine would be called again before RETI is performed. The stack may then build up and it is quite possible for it to overwrite the whole of the memory – for this reason the Z80 is designed to respond to interrupts only after the second instruction after EI.

In order to overcome the problem of the debounced switch being held down for too long, we can use it to drive a D type flip-flop as in fig. 10.17, which is reset in the first machine cycle after the start of the interrupt service CALL. The resetting signal $\overline{\text{INTA}}$ can be generated since, in this machine cycle, the CPU pulls $\overline{\text{M1}}$ and $\overline{\text{IORQ}}$ low together in clock cycle T4. (We must use $\overline{\text{M1}}$ since we must ensure that interrupts are only serviced starting in an M1 cycle.)

Having considered the types of interrupt and the hardware required to generate them, there remains the question of how the Z80 determines the

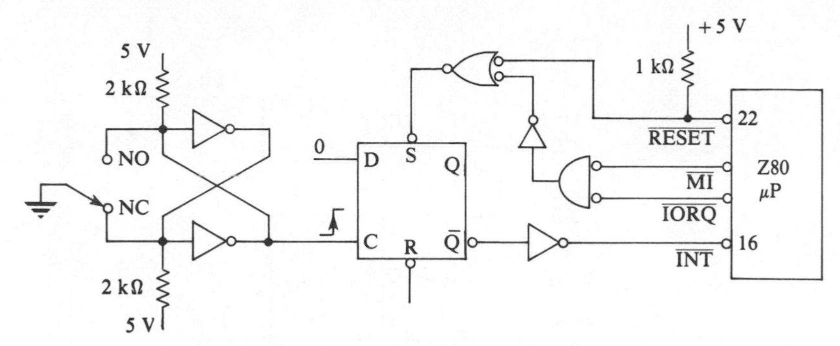

Fig. 10.17 General interrupt signal circuit.

start address of the service routine. In the case of non-maskable interrupts this is easy – on receipt of a low level signal on $\overline{\text{NMI}}$ the CPU jumps unconditionally to location 0066H. Hence if the system designer intends to use non-maskable interrupts he must ensure that a service routine starts at this location. Often the first instruction will be a jump to another part of the memory where more space is available in the memory map. The application of maskable interrupts is more difficult and the Z80 μP provides three modes of interrupt in this case. These are interrupt modes 0, 1 and 2 and are selected under software control by the respective instructions IM0, IM1, IM2.

Consider first mode 0 interrupts. If mode 0 interrupts have been selected by executing the instruction IM0, then when an interrupt occurs the CPU will finish the current instruction and then execute any instruction placed on the data bus by an external device. The instruction placed on the data bus may be any instruction and so the device would have to supply, successively, one to four bytes of the instruction. Since supplying more than one byte complicates the interface, the external device usually places a single byte RST instruction on the data bus. A simple circuit, strobed by the $\overline{\text{INTA}}$ signal, such as that shown in fig. 10.18 will place the instruction E7 on the data bus (which is the machine code for RST 32). The program will then jump to memory 0020H which could contain the first byte of a subroutine CALL. Note that the instruction placed on the data bus by the peripheral device requires two more clock cycles than is usual for that particular instruction, since the Z80 automatically inserts two WAIT states to allow time for 'daisy-chain' priority sequences (see below).

Interrupt mode 1 is very similar to non-maskable interrupts in that the start address of the service routine is fixed – this time at 0038H rather than 0066H. However, the number of clock cycles to perform this restart instruction is two more than normal.

Mode 2 interrupts are the most powerful since they allow the service routine start addresses to be dynamically altered under software control. To

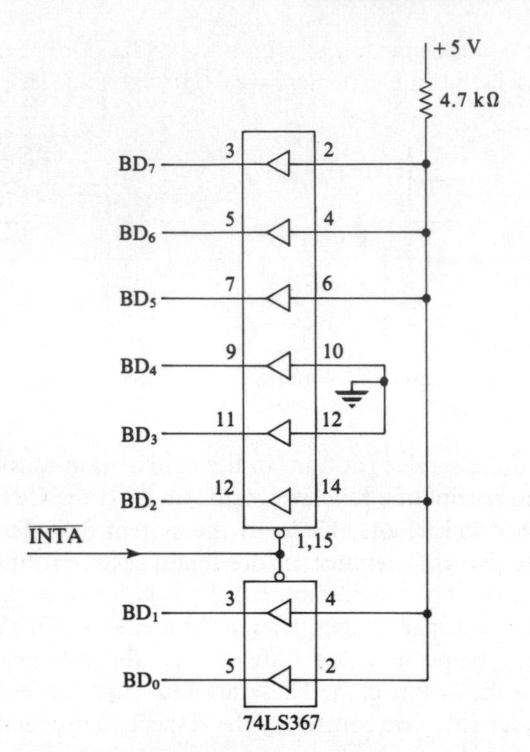

Fig. 10.18 Generating a RESET address.

understand mode 2 interrupts, note first that the designer must initially decide on the required number (say N) of interrupt service routines which can be placed anywhere in the memory (within the limits of the memory map). Having written and stored these subroutines, their start addresses are then stored in a table at some other part of the memory. Each address requires two bytes of memory, of course, and so the service routine start addresses will be stored in $2N$ successive locations, in the usual way with low byte followed by high byte. The table must start on an even-numbered memory location. When a mode 2 interrupt occurs the external device must point to (i.e. supply the address of) one of the addresses in the table. In fact the high byte of the table address is provided by the I register and is loaded before the occurrence of the interrupt by using a LD I, A instruction. Hence the external device needs to supply only the low byte of the table address which will be of the form

7 bits from peripheral	0

since the addresses in the table start on even numbered locations. To clarify this general discussion somewhat, suppose that some external device has an interrupt service routine starting at memory 10FFH and suppose that this address is stored in the table in memories 20A0H and 20A1H as shown below.

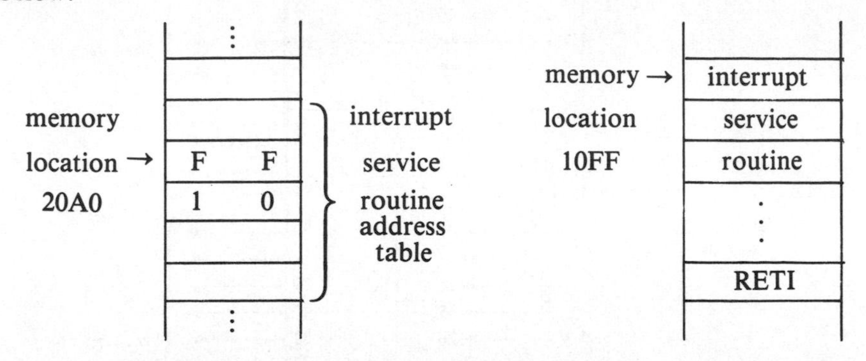

Hence when a mode 2 interrupt occurs from this device we must ensure that we have loaded I with 20H and the device must supply (along the data bus) the byte A0H to give the complete address table pointer 20A0H.

10.4.3 The PIO Chip

Many of the hardware and software interfacing techniques described above can be implemented simply using the special purpose PIO (programmable input–output) chip. The pin-out of this chip is shown in fig. 10.19(a) and it is seen to contain two input–output ports (A and B) each with **handshake** lines ARDY, $\overline{\text{ASTB}}$, BRDY and $\overline{\text{BSTB}}$ which will be discussed later. The Z80 CPU (buffered) data bus is connected directly to the PIO, as are the CPU control lines $\overline{\text{IORQ}}$, $\overline{\text{RD}}$ and $\overline{\text{M1}}$ signals. The PIO is also capable of generating interrupts and has an output line $\overline{\text{INT}}$ which can be connected directly to the similar pin on the Z80 μP. Interrupt priority is governed by the interrupt enable in and out pins which are used for **daisy chain** connection with other PIOs (see below).

Each of the ports A and B can be programmed to be in the data or control modes and so there are effectively four ports rather than two, thus we require four port addresses to select each port uniquely. This can be done in a similar way to the port select in fig. 10.15 except that we only decode BA_2–BA_7 to generate the $\overline{\text{CE}}$ (chip enable) signal. The two remaining address lines BA_0 and BA_1 are used to select port A or port B and control or data respectively. Hence, if we choose the addresses

00 Port A data
01 Port B data
02 Port A control
03 Port B control

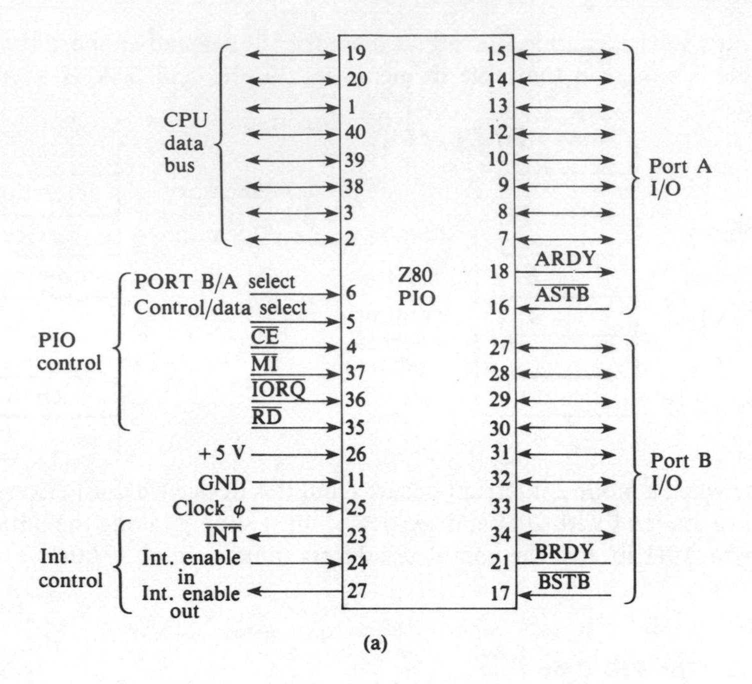

(a)

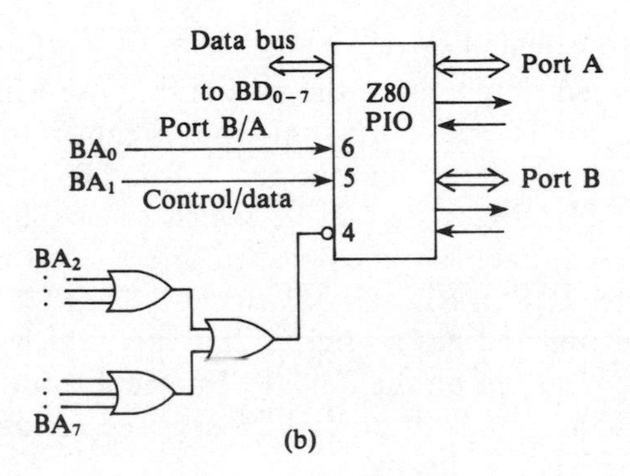

(b)

Fig. 10.19 The Z80 PIO.

then we have interface shown in fig. 10.19(b). (Note that a logic 0 on pin 6 selects port A while a 0 on pin 5 selects data.)

The PIO has four operating modes:

Mode 0 (byte output)
Mode 1 (byte input)
Mode 2 (byte bidirectional – port A only)
Mode 3 (control)

These are selected by outputting the byte D7 D6 0 0 1 1 1 1 to the desired control port, where

$$D7\ D6 = \begin{cases} 00 & \text{for mode 0} \\ 01 & \text{for mode 1} \\ 10 & \text{for mode 2} \\ 11 & \text{for mode 3} \end{cases}$$

10.4.4 PIO Mode 0 Operation

This is the simplest mode of operation of the PIO and is used to output a data byte to a peripheral device. The following simple program illustrates mode 0 operation:

```
LD A, 0FH     ; PROGRAM PIO TO MODE 0 BY
OUT (02H), A  ; OUTPUTTING BYTE 0F TO PORT 02
LD A, 26H     ; OUTPUT BYTE 26H TO
OUT (00H), A  ; PORT A
HALT
```

Usually this simple method of outputting a byte to a peripheral device will not be acceptable since the device may not be ready for the data, or the CPU may be outputting bytes too quickly for the peripheral. To overcome this difficulty we must use interrupt processing so that the peripheral can 'request' a data byte and then the CPU will only output the byte when the device is ready for it. The PIO 'handshake' lines $\overline{\text{ASTB}}$ and ARDY (or $\overline{\text{BSTB}}$ and BRDY for port B) are used for this purpose, as shown in fig. 10.20. The peripheral requests a byte on the port A by pulling the $\overline{\text{ASTB}}$ line low. When the PIO senses the active $\overline{\text{ASTB}}$ signal it 'passes on' the interrupt request to the Z80 CPU by activating the $\overline{\text{INT}}$ line. The Z80 PIO supports mode 2 interrupts and so when the CPU receives the interrupt request it will expect the PIO to supply the low byte of the interrupt vector table address. The interrupt service routine can then consist of outputting a byte to the PIO port A. On receiving the output byte, the PIO will activate the ARDY line (logic 1) to inform the peripheral that the data byte is available on the port A bus and can be read.

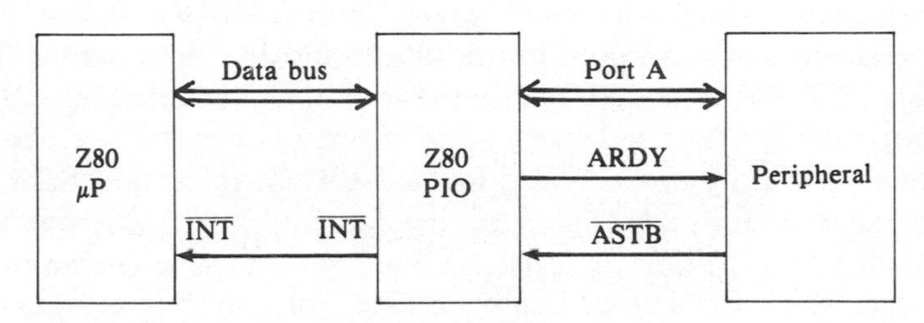

Fig. 10.20 Interfacing the PIO chip.

Consider the following assembly program:

```
INITL1:IM2              ; SET Z80 INTERRUPT MODE 2
       LD HL, VTAB      ; START ADDRESS OF VECTOR TABLE
       LD A, H
       LD I, A          ; SET I REG TO HIGH BYTE OF VTAB
       LD IY, SERV1     ; SET IN SERVICE ROUTINE START
       LD (VTAB + 04H), IY ; ADDRESS IN VECTOR TABLE
       LD A, 04H        ; OUTPUT INTERRUPT VECTOR
       OUT (02H), A     ; TO PIO CONTROL PORT A
       LD A, 0FH        ; SET PIO MODE 0
       OUT (02H), A
       LD A, 87H        ; ENABLE PIO INTERRUPTS
       OUT (02H), A
       JP MAIN1
MAIN1:EI                ; ENABLE INTERRUPTS
       :
       JP MAIN1
SERV1:PUSH AF           ; SAVE CPU REGISTERS
       LD A, (PERIPH)   ; OUTPUT BYTE IN
       OUT (00H), A     ; PERIPH TO PIO
       POP AF
       EI               ; RE-ENABLE INTERRUPTS
       RETI
```

The following points may be noted. Firstly, we must initialize the system using INITL1. This routine defines the start address of the interrupt vector table and then loads the address of the byte output service routine into the table as the third address entry. The interrupt vector must then be programmed into the PIO and we must remember to enable the PIO interrupts so that the PIO will pass on interrupt requests to the CPU. This is done by outputting the interrupt control word 87H to the control port A of the PIO. (More will be said about this interrupt control word later, but for now it is sufficient to know that the byte 87H will enable the PIO interrupts.) Following the initialization routine we suppose that some main processing loop MAIN1 has been written. This may calculate the output bytes to the peripheral or it may contain processing which has no connection with the peripheral device. In the interrupt service routine SERV1 we first save any CPU registers which are used in the subroutine and then output the byte stored in PERIPH to the PIO. The PIO will then activate the ARDY line and the peripheral can use this signal to strobe in the data byte. (Note that as soon as the $\overline{\text{ASTB}}$ line is pulled low the PIO disables the ARDY line so that the external device does not read the data bus until the output byte has been loaded into the port A register.) Finally we must re-enable the interrupts since the CPU automatically disables interrupts immediately after acknowledging an existing interrupt.

It is clear from the above discussion that the peripheral device must have the appropriate logic to interpret and implement the handshake signals ARDY and $\overline{\text{ASTB}}$. This means that it must be capable of generating interrupts on the $\overline{\text{ASTB}}$ line, but it must only do so after the ARDY line has been enabled and the present data byte has been read. If it were to generate another interrupt before the ARDY line has been activated by the PIO, then the present data byte being output may be lost. The PIO output port is a latch and so the output byte will be held indefinitely until the peripheral is ready to read it and the port will only be overwritten by a new output byte after the next interrupt is generated.

10.4.5 PIO Mode 1 Operation

PIO mode 1 operation is used to input a byte to the CPU from a peripheral device and is similar to mode 0 operation apart from the data direction and the interpretation of the handshake signals. In this case (referring again to the hardware of fig. 10.20) the peripheral pulls the $\overline{\text{ASTB}}$ line low, requesting an interrupt, when it has a data byte waiting to be read by the CPU. The ARDY line is activated by the PIO when the port A input register is empty and ready to receive the data.

The software for mode 1 operation is similar to that given above which may be modified as follows:

```
INITL2:IM2
       LD HL, VTAB
       LD A, H
       LD I, A
       LD IY, SERV2
       LD (VTAB + 06H), IY
       LD A, 06H
       OUT (02H), A
       LD A, 4F              ; SET PIO MODE 1
       OUT (02H), A
       LD A, 87H
       OUT (02H), A
       JP MAIN2
MAIN2:EI
       :
       JP MAIN2
 SERV2:PUSH AF
       IN A, (00H)
       LD (PERIPH), A
       POP AF
       EI
       RETI
```

The service routine SERV2 will often contain some extra lines of programming to process the input byte in PERIPH, but it must be ensured that there is enough time to do this processing before the next byte is available at the terminal.

10.4.6 PIO Mode 2 Operation

If we wish to use the same PIO port for input and output operations we can program the PIO in mode 2. This facility is only available on port A, however, since both sets of handshake lines must be used. The port A handshake lines $\overline{ASTB}$ and ARDY are used for output synchronization while port B lines $\overline{BSTB}$ and BRDY are used for input operations. Moreover, while port A is programmed in mode 2, port B must be programmed in mode 3.

The two service routines SERV1 and SERV2 can be used in mode 2 operation, but one major difference between mode 2 byte output and mode 0 operation should be noted. In mode 0 operation the output port behaves like a latch and the data byte can be read at any time, independently of the $\overline{ASTB}$ signal. However, in mode 2 output the data is only valid on the port data bus, and can therefore only be read by the peripheral, while $\overline{ASTB}$ is active. Hence when the peripheral pulls the $\overline{ASTB}$ line low it must read the current data on the port data bus. Since the PIO does not generate an interrupt until the rising edge of $\overline{ASTB}$, the data read by the peripheral is that which was output during the previous interrupt request–service cycle. Thus the peripheral is always one step behind the CPU in its received data.

The hardware for mode 2 operation is similar to that in fig. 10.20, but now the peripheral must be capable of handling all four handshake lines and be able to generate interrupts on both $\overline{ASTB}$ and $\overline{BSTB}$ for output and input respectively. The only change in the software requirement is a new initialization routine as follows:

```
INITL3:IM2
        LD HL, VTAB
        LD A, H
        LD I, A
        LD IY, SERV1
        LD (VTAB + 04H), IY
        LD IY, SERV2
        LD (VTAB + 06H), IY
        LD A, 04H          ; PLACE OUTPUT INTERRUPT
        OUT (02H), A       ; VECTOR IN PORT A
        LD A, 06H          ; PLACE INPUT INTERRUPT
        OUT (03H), A       ; VECTOR IN PORT B
        LD A, 8FH          ; SET PIO MODE 2
        OUT (02H), A       ; IN PORT A
```

```
LD A, 0CFH          ; SET PIO MODE 3
OUT (03H), A        ; IN PORT B
LD A, 0FFH          ; SET MASK BYTE FOR
OUT (03H), A        ; FOR PORT B
LD A, 87H           ; ENABLE PIO INTERRUPTS
OUT (02H), A        ; FOR PORT A
OUT (03H), A        ; FOR PORT B
JP MAIN3
```

Here, again, MAIN3 is some routine which is executed while interrupts are not in progress. The only comment which needs to be made about INITL3 is the mask byte 0FFH which is output to port B. This is a requirement after setting mode 3 operation even though it is of no significance here. The mask byte will be discussed in detail in the next section.

10.4.7 PIO Mode 3 Operation

In PIO mode 3 operation the data direction of each individual bit of the port data bus can be programmed independently of the rest. Moreover, in this mode no handshake lines are used; instead the PIO monitors certain bits of the port data bus for interrupt generation. Which of the bits are monitored can again be programmed and if more than one data line is used for interrupt control we can program the PIO to respond to the appropriate logic level on any chosen data line or only on all such lines. Hence the PIO can be programmed to perform an OR or an AND function on the interrupt data lines. The lines chosen for interrupt monitoring are said to be **unmasked** data lines while the remaining data lines are **masked**.

In programming mode 3 operation of a PIO we must therefore output certain bytes to the desired control port in addition to those which are used in the other modes. First we program mode 3 operation by outputting the byte CFH. Since we can select the data direction on each data line we next output a 'mask' byte which controls this choice. If the byte contains a 1 in the *n*th place ($0 \leqslant n \leqslant 7$) then the *n*th line of the PIO data bus is chosen as an input line, while if the *n*th bit is 0 then it is an output line. The next byte output to the control port is the interrupt control word and corresponds to the byte used in the other modes for enabling the interrupts. However, instead of just using the 7th bit, all four bits d_4–d_7 now carry information as follows:

Interrupt control word:	d7	d6	d5	d4	d3	d2	d1	d0
	enable interrupt	AND/ OR	HIGH/ LOW	mask byte follows	0	1	1	1

(d6–d4: mode 3 only)

As before, if d7 is logic 1 then interrupts are enabled. As stated above the lines which are chosen to be monitored for interrupts can be ANDed or ORed. If d6 is logic 1 then they are ANDed together, so that all must be at the appropriate logic level to generate an interrupt. Bit 5 is used to select which logic level is to be considered active for interrupts, if d5 = 0 then low logic is chosen, i.e. if d6 = 1 then all lines monitored for interrupts must be at logic 0 for an interrupt to be generated. Finally if d4 = 1 then the PIO is programmed to interpret the next byte output to the control port as a mask byte specifying which data lines are to be monitored for interrupts. If a bit of this output byte is logic 1 then the corresponding data line is masked, i.e is ignored. Only lines corresponding to bits at logic 0 will be monitored for interrupt generation.

We can now write the following initialization routine for mode 3 interrupts:

```
INITL4:IM2
        LD HL, VTAB
        LD A, H
        LD I, A
        LD IY, SERV3
        LD (VTAB + 08H), IY
        LD A, 08H
        OUT (02H), A
        LD A, 0CFH            ; SET MODE 3 FOR
        OUT (02H), A          ; PORT A
        LD A, 0FH             ; DEFINE INPUT LINES
        OUT (02H), A          ; FOR PORT A
        LD A, 97H             ; SET INTERRUPT
        OUT (02H), A          ; CONTROL WORD
        LD A, 0F8H            ; MONITOR D0, D1, D2
        OUT (02H), A
        JP MAIN4
```

MAIN4 is again a routine which is run between interrupts. When an interrupt occurs the service subroutine SERV3 is executed. SERV3 may be of the following form:

```
SERV3:PUSH AF
      IN A, (00H)     ; INPUT BYTE FROM PORT A
      AND 0FH
      :
      LD A, (TEMP)
      OUT (00H), A
      POP AF
      EI
      RETI
```

Note that we selected d0–d3 as port input lines in INITL4 and so when the IN A, (00H) instruction is performed, we must remove any spurious bits from d4–d7 using AND 0FH.

10.4.8 Interrupt Priorities

Since the PIO has two ports which can be interrupt driven simultaneously (apart from in mode 2) there may exist a conflict when interrupts appear on both ports. Suppose that both ports are interfaced to separate peripherals which are capable of generating interrupts and that, in the interrupt service routines, we enable interrupts (EI instruction) at the start of the routines. Hence each service routine can itself be interrupted while being executed. The PIO responds differently, however, to interrupts appearing at port A than to those appearing at port B. In fact, port A has higher **priority** than port B so that if an interrupt has occurred at port B and the appropriate service routine is being executed, then an interrupt appearing at port A will be acknowledged immediately and the service routine for port A will be executed. The port B service routine will only be completed after the routine for port A is finished. On the other hand, an interrupt occurring at port B while the service routine for port A is being executed will be held **pending** and will only be serviced after the completion of the port A service routine.

If we require several ports in our system, each having the programming facilities of the PIO, then we can use a number of PIOs each with its interrupt line connected (together) onto the CPU interrupt line. We have seen above how the PIO resolves interrupt priorities internally between ports A and B, but how can we resolve such priorities between the different PIOs? The PIO itself simplifies this task by providing the interrupt enable lines IEI, IEO (pins 24, 27; fig. 10.19(a)). These lines can be connected as in fig. 10.21 where we have used three PIOs (more could be added, of course). Such a connection is called a **daisy chain** and can easily be understood from the following rules for the IEI and IEO logic levels:

(a) A PIO can request an interrupt only if IEI is high.
(b) If a PIO requests an interrupt IEO is pulled low.
(c) If IEI is low, then IEO is low.

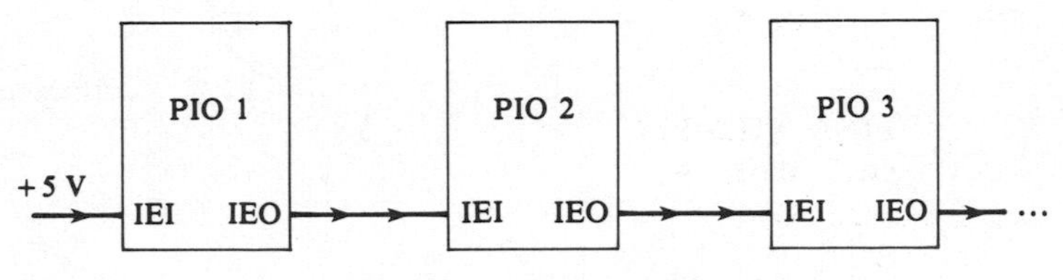

Fig. 10.21 Daisy chaining.

Hence we see that PIO1 can always respond to an interrupt since IEI is always high, and that the priorities are given by

$$\text{PIO1 port A} > \text{PIO1 port B} > \text{PIO2 port A} > \text{PIO2 port B} > \text{PIO3 port A} > \text{PIO3 port B}$$

where > means the port on the left has higher priority than that on the right.

10.5 INTERFACING PERIPHERALS

10.5.1 Simple Output Devices

It is often important to be able to use an output port to drive various types of loads. For example, each data line on the output port can drive an LED as shown in fig. 10.22. The 7406 buffer may not be required if the port is contained in a PIO, but it is always good practice to check on the port output drive and current sink capabilities before connecting any peripheral devices.

When interfacing heavier loads it is usually necessary to use the buffered output line to drive the base of a transistor or a Darlington pair (fig. 10.23). One can also connect a relay to the port, but it is important to

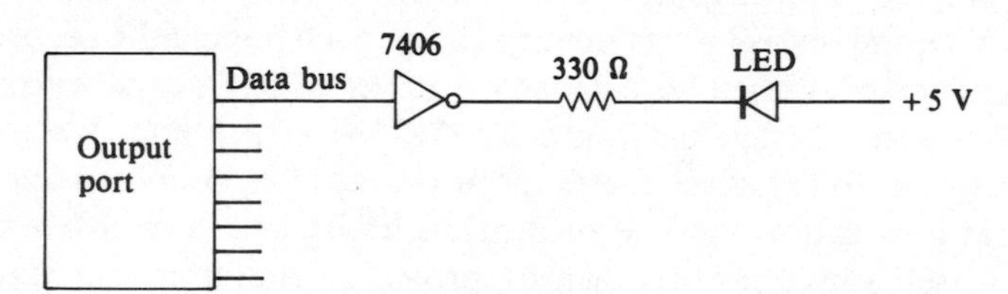

Fig. 10.22 Interfacing an LED.

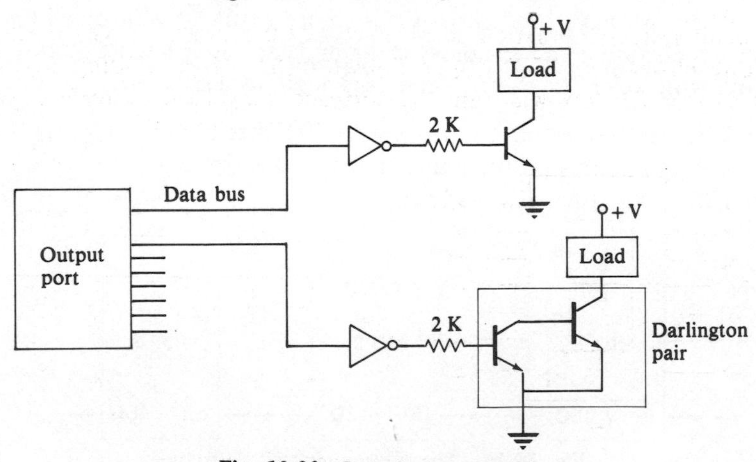

Fig. 10.23 Interfacing a load.

remember the protection diode across the inductor to prevent high voltages (generated by by back e.m.f due to rapid changes in voltage) appearing at the buffer (fig. 10.24). Finally, when driving very heavy loads it is best to use an opto-isolator of the type shown in fig. 10.25. The port line is connected to an LED as in fig. 10.22 which drives the base of an opto-transistor giving complete isolation of the output circuit from the port. In this way there is no risk of high voltages appearing on the chip pins.

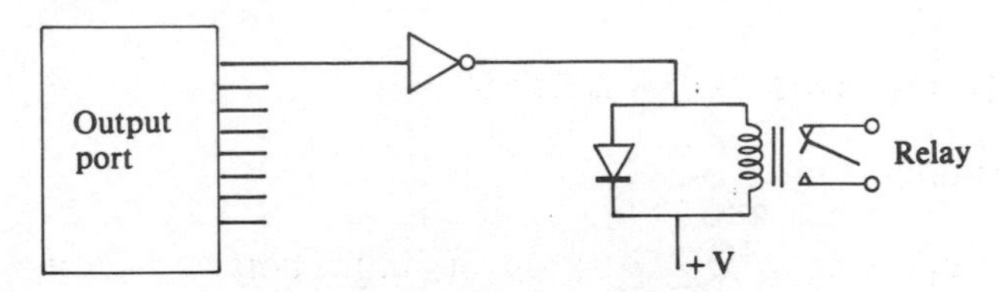

Fig. 10.24 Interfacing a relay.

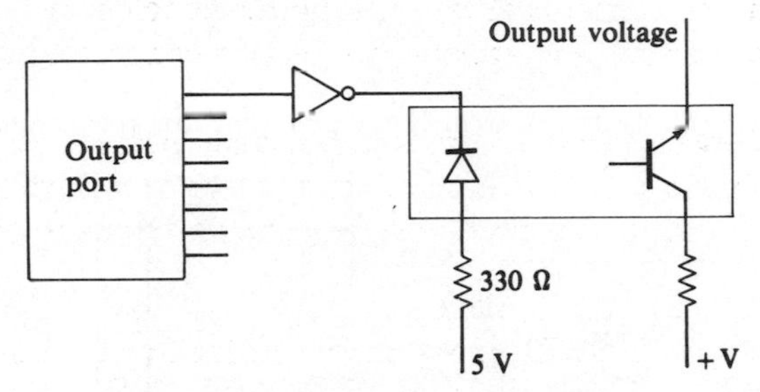

Fig. 10.25 Using an opto-isolator.

10.5.2 Interfacing Seven Segment Displays

For small systems numerical displays can be useful and a typical example is the seven-segment LED display shown in fig. 10.26. Pulling any of the lines a–g or DP low will light the corresponding LED. Interfacing these

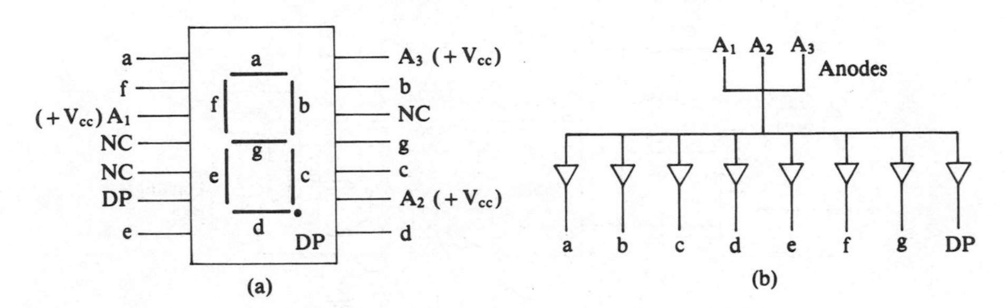

Fig. 10.26 The 7-segment LED display.

displays is made particularly easy by the 7447 BCD to seven-segment decoder chip (fig. 10.27). We can therefore connect two such displays to a port as in fig. 10.28 and we can output the number 65 (say) using

```
LD A, 65H
OUT (n), A
```

where n is the port address. Similar interfaces can be designed using the lower power liquid crystal devices.

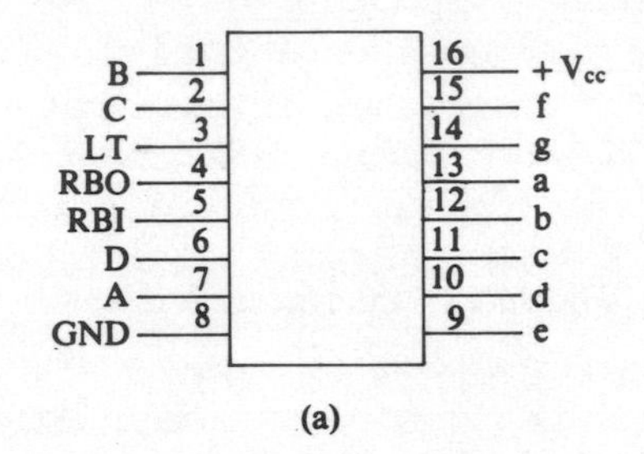

(a)

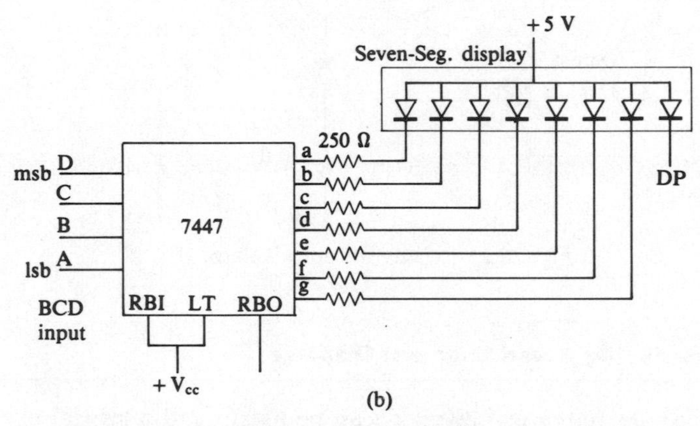

(b)

Fig. 10.27 A BCD decoder for the 7-segment display.

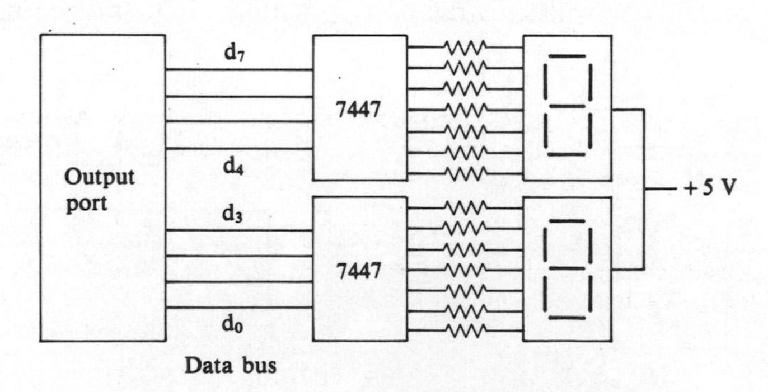

Fig. 10.28 Interfacing two 7-segment displays.

10.5.3 Interfacing a Simple Keyboard

We can construct a simple 64 character keyboard interface as in fig. 10.29. Note that the keyboard is interfaced to an input–output port with address FFH which consists of a tri-state buffer on the input side and a latch on the output side. Since this is a 'dumb' keyboard we must 'poll' it periodically to see if it requires service, i.e. if any key is being pressed. The simplest way to do this is to provide an oscillator which will generate an interrupt every 10 ms (say). If we tie this to the $\overline{\text{NMI}}$ line of the Z80 μP then we can put the instruction JP KEYB at location 0066H, where KEYB is the keyboard service routine.

To understand the operation of the keyboard, suppose that we output the byte 11111110 to the port output latch. This will put zero volts onto row R_0. If any key in this row is pressed, the corresponding column line will go low. Hence if we input the byte $C_0C_1 \ldots C_7$ appearing at the input buffer, any pressed keys will give a zero in the appropriate bit. We can repeat this by outputting a zero successively on each row line R_i and then input the byte $C_0C_1 \ldots C_7$.

For simplicity we shall assume that only a single key should be pressed at any time. (Most keyboards have shift keys, but this can easily be accommodated by a simple change in the following program.) We can now write a subroutine to detect a single key pressed and determine its value in the following way (note that D is returned with 00H unless exactly one key is pressed, in which case D is returned with the BCD number D_1D_2 where D_1 = row number and D_2 = column number):

```
SINGK:LD B, 08H        ; SET ROW COUNT
      LD D, 00H        ; INITIALIZE D
      LD C, 7FH        ; SET ROW OUTPUT BYTE TO 01111111
NEXTR:LD A,C
      LD L, D          ; SAVE CURRENT D IN L
      LD H, B          ; SAVE ROW COUNT
      OUT (FF), A      ; OUTPUT ROW BYTE
      IN A, (FF)       ; INPUT COLUMN BYTE
      CP FFH           ; TEST AND JUMP IF
      JR Z, NOKEY      ; NO KEY PRESSED
      LD D, H          ; IF KEY PRESSED
      SLA D            ; PUSH VALUE (DECIMAL)
      SLA D            ; TO HIGH NIBBLE
      SLA D
      SLA D
      LD E, 00H        ; SET NO. OF KEYS PRESSED IN ROW TO 0
      LD B, 08H        ; SET COLUMN COUNT
NEXTK:RLA              ; PUSH KEY BIT TO CARRY FLAG
      JR C, KEYOP      ;JUMP IF KEY OPEN
      PUSH AF          ; IF CLOSED
```

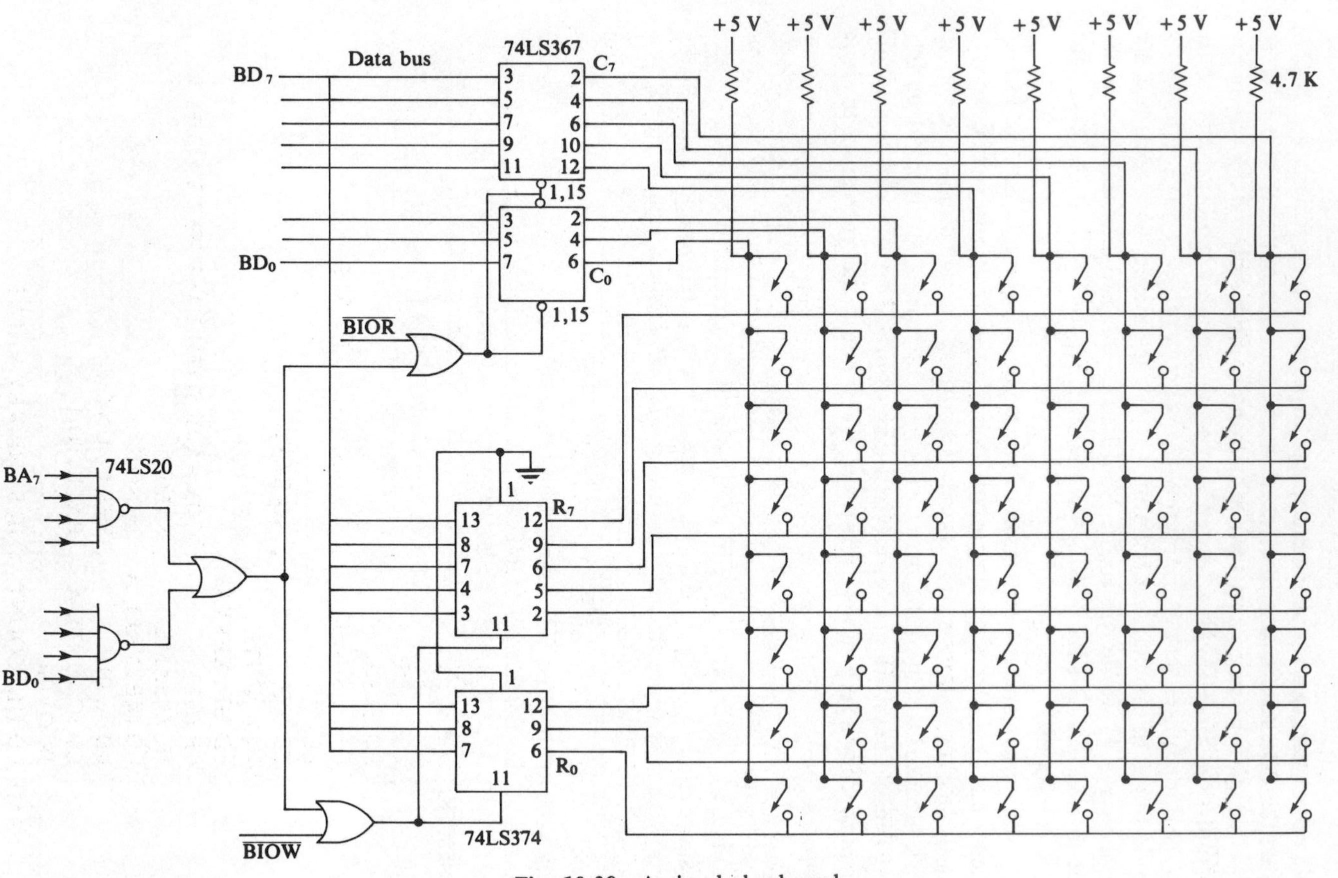

Fig. 10.29 A simple keyboard.

```
        LD A, D         ; PUT COL.
        OR B            ; COUNT
        LD D, A         ; IN D
        POP AF
        INC E           ; INC NO. OF KEYS PRESSED
KEYOP:DJNZ NEXTK
        LD A, E         ; TEST IF MORE THAN
        CP 01H          ; ONE KEY PRESSED
        JP P, MOKEY
        LD A, L         ; TEST IF ANY KEYS PRESSED
        CP 00H          ; ON PREVIOUS ROWS
        JR NZ, MOKEY
        RRC C           ; SHIFT ROW OUTPUT BYTE
        LD B, H         ; RESTORE COUNT
NOKEY:DJNZ NEXTR
        RET
MOKEY:LD D, 00H         ; IF MORE THAN ONE KEY
        RET             ; PRESSED SET D=0
```

The above subroutine will detect a single key closure; however, there is a problem with the simple keyboard we are considering – namely, the keys are not 'debounced' which means that high and low logic levels will appear on the input port pin connected to a pressed key. These values will oscillate until a stable low logic level exists. We could debounce the keys with hardware as in fig. 10.16, but we shall now show how to implement a software debounce. This can be done simply by insisting that for (say) ten consecutive calls to keyboard we detect the same key being pressed. This means that the subroutine SINGK must return the same (non-zero) key value ten times in the D register. Our keyboard service routine therefore becomes:

```
KEYB:PUSH AF            ; SAVE REGISTERS
     PUSH BC
     PUSH DE
     PUSH HL
     CALL SINGK         ; FIND SINGLE KEY STROKE
     LD A, D
     CP 00H             ; NO KEY?
     JR Z, INITL
     LD HL, KEYVAL
     CP (HL)            ; COMPARE WITH LAST KEY VALUE
     JR Z, SAME         ; SAME?
     LD (KEYVAL), A     ; IF NOT
     LD A, 01H          ; SET COUNT
     LD (COUNT), A      ; BACK TO 1
     JR OVER
SAME:LD HL, COUNT       ; IF SAME KEY VALUE
     INC (HL)           ; AS IN LAST CALL TO KEYB
```

```
         LD A, 0AH          ; INC COUNT AND COMPARE
         CP (HL)            ; WITH 10
         JR Z, KEYOK
   OVER:POP HL
         POP DE
         POP BC
         POP AF
         RETN
  KEYOK:CALL KEYINT         ; IF KEY PRESSED FOR TEN CALLS
                            ; JUMP TO MONITOR PROG.
  INITL:LD A, 00H           ; INITIALIZE COUNT
         LD (COUNT), A      ; AND
         LD (KEYVAL), A     ; KEYVAL
         JR OVER
  COUNT:DEFS 1
 KEYVAL:DEFS 1
```

The subroutine KEYINT used above will be some monitor routine which accepts a new key value and interprets it depending on the context. (KEYINT may, for example, display the character corresponding to the keyvalue on a VDU.)

10.5.4 Interfacing a VDU

A special purpose microprocessor controller may require interfacing to some visual display unit, usually a monitor. Without going into detail we shall now discuss the general method for implementing such an interface. First recall that a video monitor uses a scanning electron beam which is modulated with the displaying information. If we require, say, 80 characters on a line then we must modulate the beam on each line scan with 80 successive memory bytes; moreover, if each character is an 8×8 dot matrix then eight successive line scans will produce a complete row of characters.

Since the video output is scanned rapidly and a phosphor screen has no 'memory' the information display must be stored in RAM. This will mean, for high resolution graphics, that up to 16K of memory will be used to store the screen.† (This storage requirement can be reduced by storing only character values and using a character generator chip.) In the simple system which we shall describe, we shall assume that the Z80 μP only writes

†For example, suppose each character is formed by an 8×8 set of pixels. Then if there are 24 rows and 64 columns of characters on the screen, the memory requirement will be $24 \times 64 \times 8 = 12$K.

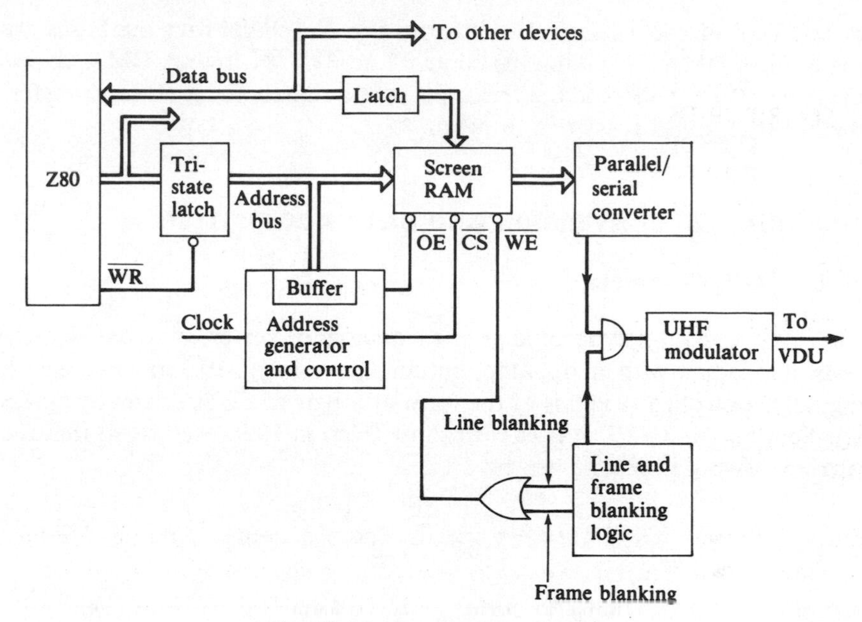

Fig. 10.30 Interfacing a VDU.

to the screen RAM while the screen logic only reads from it. A schematic diagram of a typical video interface is shown in fig. 10.30. To read the memory to the screen we must provide an address generator and control which outputs the appropriate bytes at the correct times. Note that the screen refresh is performed independently of the microprocessor. Each byte is converted to serial data which is then mixed with the line and frame blanking signals. Writing into the screen RAM from the CPU must be restricted to the line and frame blanking periods so that 'glitches' do not appear on the screen and so the address and data buses must be latched and the monitor software must ensure that data is output to the memory at the appropriate speed. This can be done via an interrupt service routine.

10.5.5 Direct Memory Access (DMA)

The only control signals on the Z80 chip which we have not yet mentioned are $\overline{\text{BUSRQ}}$ and $\overline{\text{BUSAK}}$. These signals are used to allow external devices to obtain fast access to the memory, i.e. data is transferred directly between the memory and the peripheral without passing through the CPU. When a device requires a DMA transfer, it pulls the line $\overline{\text{BUSRQ}}$ low. The CPU responds by putting all three of its internal bus drivers into the high impedance state, thus effectively disconnecting the Z80 from the external buses. The microprocessor then informs the peripheral that the buses are ready for a DMA by pulling the control line $\overline{\text{BUSAK}}$ low. A DMA chip is available which will generate the appropriate signals for a DMA transfer, but we shall not discuss this in detail.

10.6 D/A, A/D CONVERSION AND DATA ACQUISITION

10.6.1 D/A Conversion

The most common type of digital to analogue converter is based on a resistance ladder with an op. amp. output shown in fig. 10.31(a) where each digitally controlled switch is of the form of a pair of FETs driven by an RS flip-flop (fig. 10.31(b)). It is easy to show (section 10.9, exercise 4) that the current flowing in R_f is given by

$$i_N = -\frac{1}{6R}\sum_{i=0}^{N-1}\frac{V_i}{2^{N-1-i}} \tag{10.6.1}$$

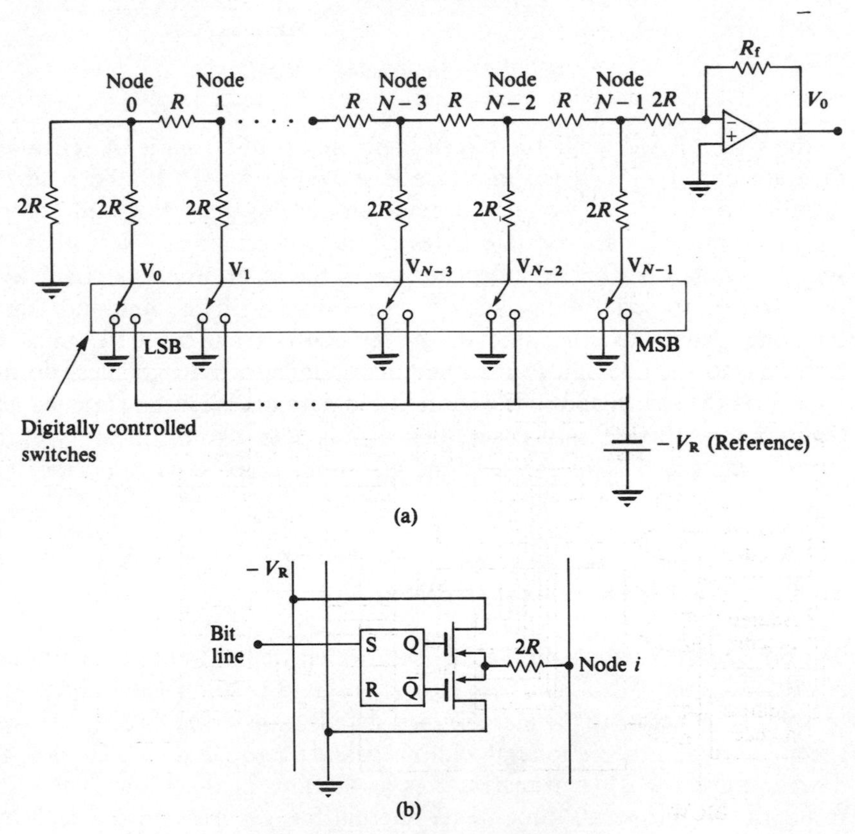

Fig. 10.31 A D/A converter.

and so the output voltage is

$$V_0 = -\frac{R_f}{6R}\sum_{i=0}^{N-1}\frac{V_i}{2^{N-1-i}}$$

$$= -\frac{R_f}{6R2^{N-1}}\sum_{i=0}^{N-1}2^iV_i$$

However $V_i = 0$ or $-V_R$ depending on the binary input and so

$$V_0 = \frac{R_f V_R}{6R2^{N-1}} \times \text{binary input}$$

Appropriate values of V_0 can be set by choosing suitable values for R_f, V_R and R, assuming N has been fixed.

Suppose we choose V_0 to have a maximum value of 10 V. Then we must decide on the resolution which is required for a particular application. For example, an 8-bit D/A converter will resolve 256 different levels between 0 V and 10 V, giving a minimum resolution interval of about 40 mV, while a 16-bit D/A converter will resolve 65536 levels with an interval of 0.15 mV. D/A converters are available on DIP chips with 8-, 10-, 12- or 16-bit resolutions (with or without the output op. amp.). Interfacing an 8-bit D/A to the Z80 is particularly easy – in fact, it can be connected directly to a buffered output port (such as a PIO). Interfacing a 16-bit D/A to an 8-bit μP chip is not so easy and two multiplexed output bytes are required. A simple circuit for such an interface is given in fig. 10.32. Three address decoders of the form shown in fig. 10.15 are required which output logic-0 when BA_0–BA_7 (low byte of the address bus) contains E0, E1 or E2. When a 16-bit number is to be output to the D/A converter, the low byte

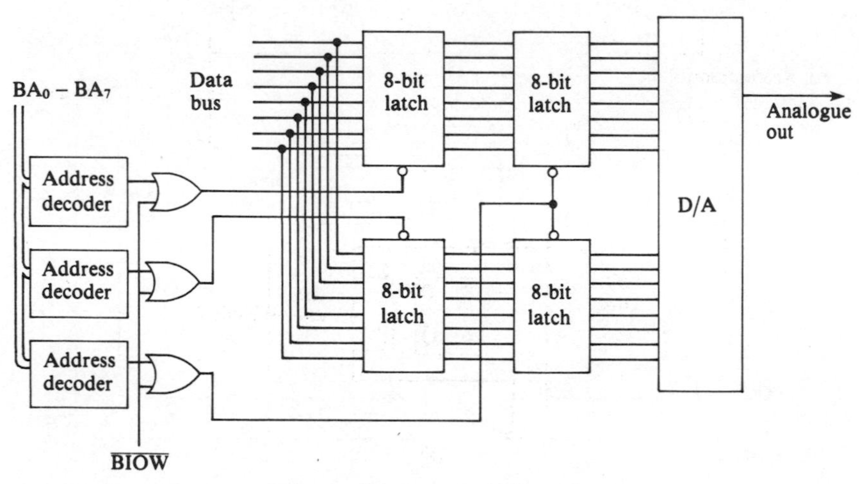

Fig. 10.32 Interfacing a D/A.

is output first to the port E0, then the high byte to the port E1 and finally the complete 16-bit number is output to the port E2 and hence to the converter.

10.6.2 A/D Conversion

Analogue signals from the 'real world' must be converted into a form which can be accepted by a microprocessor system, which means 8-bit binary numbers in the case of the Z80 μP. A typical A/D converter is shown in fig. 10.33 and consists of a binary counter, the output of which is converted to an equivalent analogue signal by a D/A converter. The output of the D/A is then compared to the analogue input and when they are equal the comparator output stops the counter. The binary number b stored by the counter then satisifies the relation

$$|b - E_{in}| < \frac{E_{max}}{2^n - 1}$$

where E_{in} is the analogue input signal, E_{max} is the maximum analogue input and $n = 8$ for an 8-bit converter. ($E_{max}/(2^n - 1)$ is called the **discretization error.**)

Since computer control systems are usually designed with fixed sampling intervals of, say, T seconds, an A/D converter may be interfaced to the Z80 μP using an interrupt driven system. The control logic of the A/D converter can accept a start pulse which resets the counter and initiates a conversion sequence. When the comparator goes low, indicating a match between the analogue input and counter output, the control logic generates an end of count (EOC) pulse which can be used by the CPU to input the

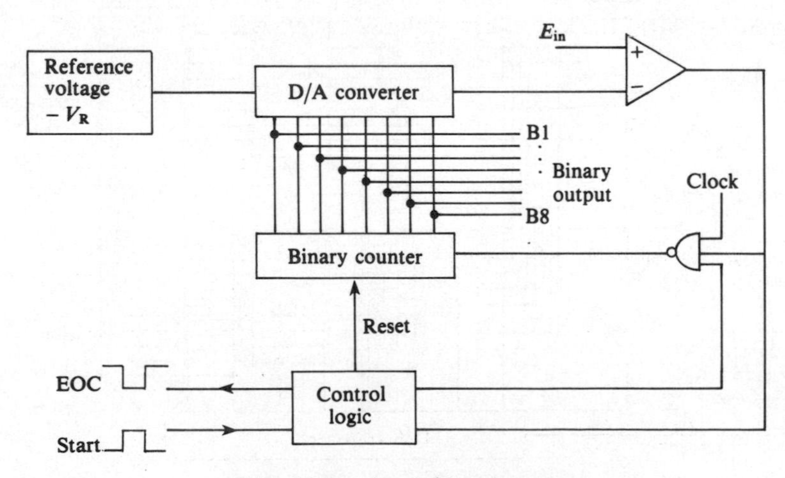

Fig. 10.33 An A/D converter.

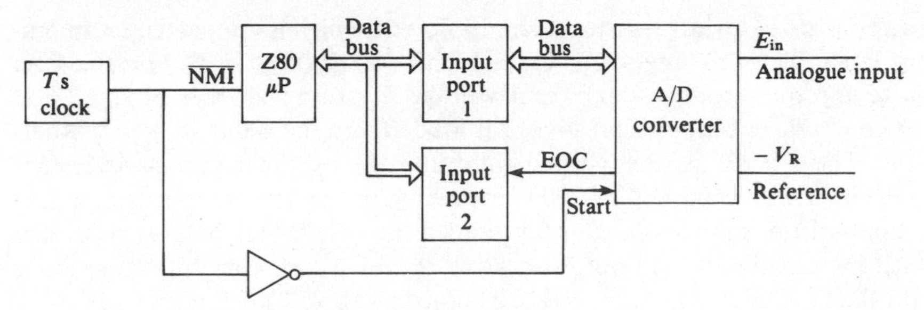

Fig. 10.34 Interfacing an A/D.

A/D output byte. A possible interface is shown in fig. 10.34, where only a single bit of input port 2 is used (say D_0) to input the EOC signal. When the T second clock generates an interrupt the counter starts and simultaneously the Z80 jumps to the interrupt service routine which first loops around an IN operation for port 2 until an EOC signal is detected. The data byte can then be input from port 1 and processed in some way (for example, the system may be running an on-line Kalman filter). Of course, the sampling interval T must be long enough to allow the interrupt service routine to be completed and so we now see why the ability to evaluate precise timings for all the operations is essential.

The above A/D converter has the drawback of being fairly slow in that a particular conversion cycle can take up to 2^n clock periods. A faster method is to use successive approximations where the analogue input E_{in} is compared with $E_{max}/2$, to determine if E_{in} is in the upper or lower half of the full scale deflection. If $E_{in} < E_{max}/2$, then E_{in} is next compared with $E_{max}/2^2$ while if $E_{in} > E_{max}/2$, then E_{in} is compared with $(E_{max}/2) + (E_{max}/2^2)$. Continuing in this way we divide the search interval in two until the difference is within the discretization error. A hardware implementation of the successive approximation A/D is given in fig. 10.35

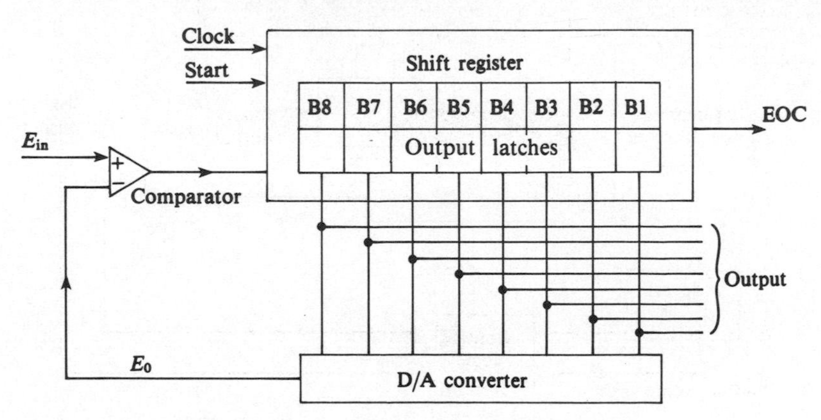

Fig. 10.35 A shift register A/D converter.

and consists of a shift register each flip-flop of which is connected to an output latch. The shift register is loaded with 10000000 (1 in B8) on the start pulse and this produces an output voltage E_0 from the D/A of $E_{max}/2$. If the comparator output is positive, 1 is loaded into the B8 latch, and 0 otherwise. The 1 in B8 is then shifted right and the output from the latches is then either 11000000 or 01000000 depending on the first comparison. Again a comparison is made and 1 or 0 is loaded into the B7 latch. This continues until the comparator output is negative at which time the shift pulse stops and the EOC signal is produced. The conversion will take at most 16 clock cycles. Interfacing this A/D is just as for the previous one.

10.6.3 Data Acquisition Systems

Since the above types of A/D converters compare the analogue input with an internally generated analogue signal, we must ensure that the analogue input does not change during the conversion cycle. For this reason, we usually pass the analogue input through a sample and hold circuit before feeding it into the A/D converter. Moreover, the analogue signal may come from a variety of systems or transducers and so it must also be 'conditioned' into a form suitable for the A/D converter. It may, in fact, vary from a few millivolts to several volts and so must first be passed through an instrumentation amplifier, preferably with variable gain. Finally in a microprocessor controlled system we may have several analogue inputs and so a multiplexer usually precedes the amplifier to switch between the different channels.

The combination of a multiplexer, programmable amplifier, sample/hold circuit and A/D converter is called a **data acquisition system** and is shown in fig. 10.36. The input channels can be selected from an output port;

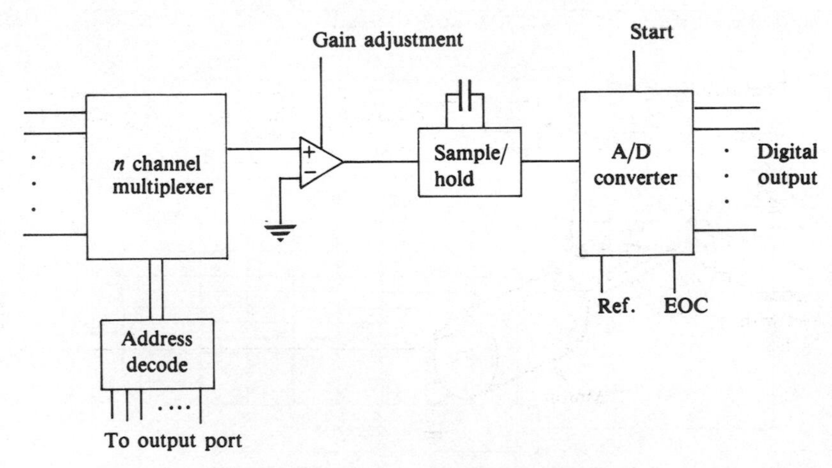

Fig. 10.36 A data acquisition system.

for example, a sixteen channel multiplexer can be addressed from a 4-bit binary number. The instrumentation amplifier gain can also be selected from an output port. Data acquisition systems of this type are readily available on chips which can be interfaced directly to a microprocessor.

10.7 EXAMPLES OF REAL SYSTEM INTERFACES

10.7.1 Microprocessor Speed Control of Motors by Phase-locked Loops

In this section we shall discuss the microcomputer control of a d.c. motor. We shall first consider a standard analogue system (see also Sen, 1981; Geiger, 1981) and then compare this with a computer control design.

The basic idea of phase-lock loop speed control of a d.c. motor is to generate a square wave with frequency proportional to the angular velocity of the output shaft and compare this with a reference frequency which represents the desired speed. If the phases of the two signals are 'locked', i.e. have a constant difference, then the signal frequencies and hence the actual and desired speeds will be equal. The basic control system is shown in fig. 10.37 and consists of a d.c. motor with an optical tachogenerator on the motor shaft, the output of which is transformed to a square wave to produce a wave with frequency proportional to the motor speed. This is compared with a reference signal and the phase lock system produces the appropriate feedback.

If we wish to drive the motor in both directions, then we must place two sensors on the disk so that their output frequencies are out of phase by (say) 90° as in fig. 10.38. As shown, the output Q of the flip-flop can be

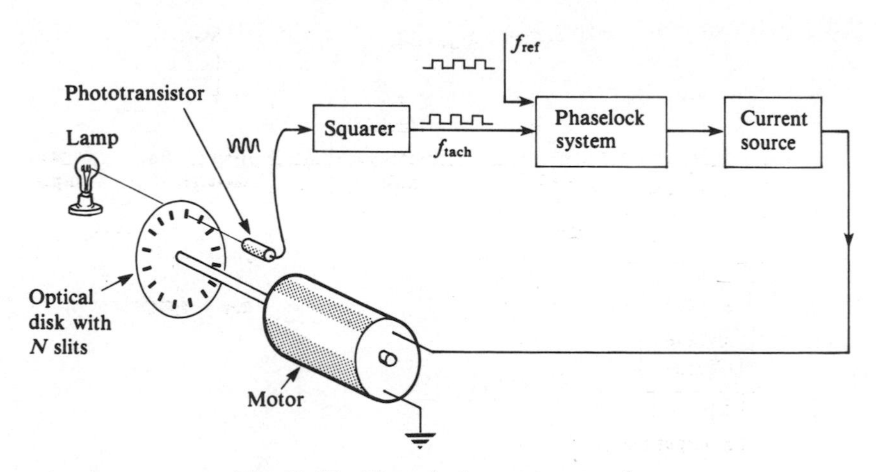

Fig. 10.37 Phase-lock motor control.

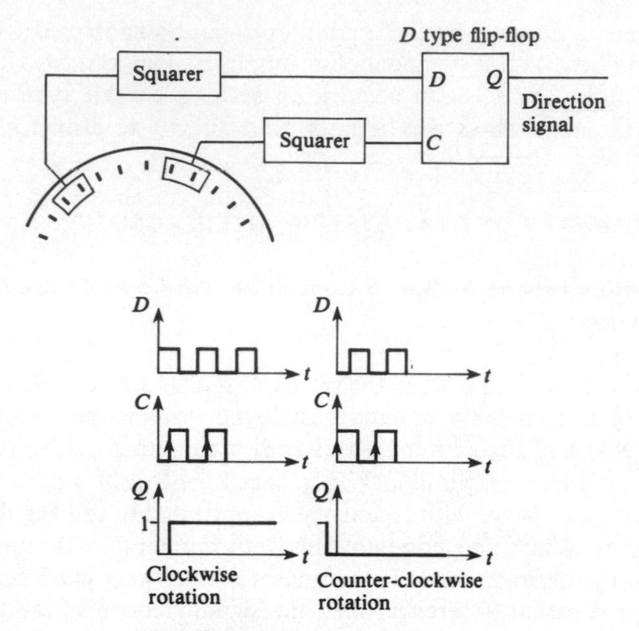

Fig. 10.38 Direction indicator.

arranged to be 1 for clockwise rotation and 0 for counter-clockwise rotation. In fig. 10.39 we show how to obtain a voltage proportional to the frequency of the square-wave optical sensor output signal. If τ is the monostable pulse time, then

$$\begin{aligned} V &= \text{d.c. component of monostable output} \\ &= \tau f V_{cc} \end{aligned} \tag{10.7.1}$$

and we choose τ so that

$$\tau = \frac{1}{f_{max}} \tag{10.7.2}$$

where f_{max} is the maximum input frequency. The circuits in figs. 10.38 and

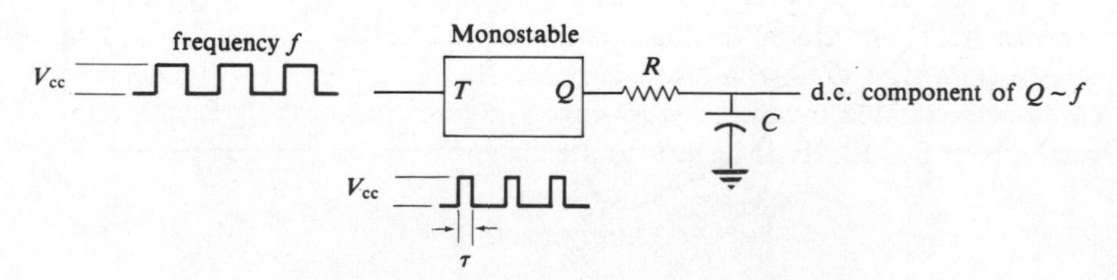

Fig. 10.39 Extracting the d.c. component of a square wave.

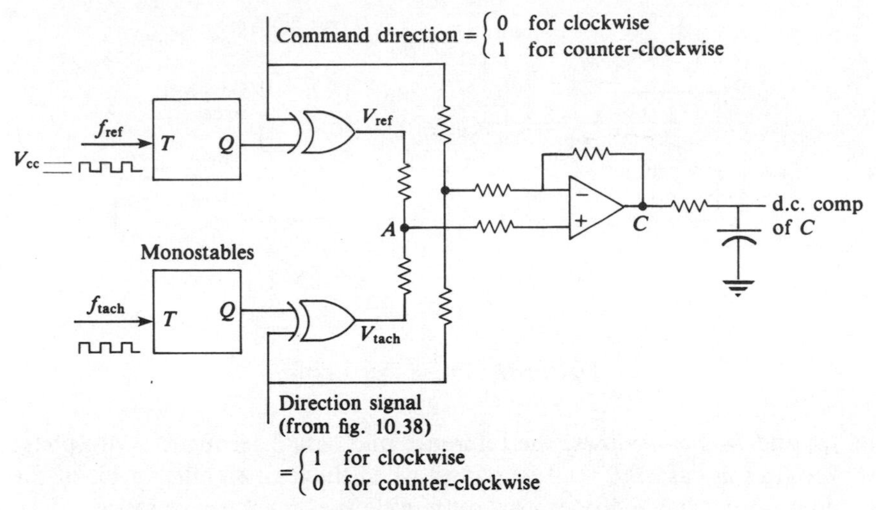

Fig. 10.40 The frequency to voltage converter.

10.39 can be combined to form a four-quadrant velocity error detector as in fig. 10.40. It is easy to see (section 10.9, exercise 5) that the d.c. output of this circuit is

$$\frac{\pm V_{cc}}{2}\tau(f_{ref} - f_{tach}) \tag{10.7.3a}$$

if the command and actual directions are the same and

$$\frac{\pm V_{cc}}{2}\tau(f_{ref} + f_{tach}) \tag{10.7.3b}$$

if they are different (+ sign for clockwise and – sign for counter-clockwise rotation). If the gain of this frequency-to-voltage converter is denoted by $k_{f/V}$, then the transfer function of the speed sensing system is

$$\frac{V}{\omega} = \frac{k_{f/V}}{s\tau_{RC} + 1}$$

where τ_{RC} = RC filter time constant.

The simple frequency-to-voltage converter in fig. 10.40 can be used to produce the proportional feedback control (see also Clayton (1979, p. 336) for an improved d.c. separation circuit). An integrated error signal can easily be generated by a binary up–down counter followed by a D/A converter as in fig. 10.41. It is easy to see that

$$V_0 = k_i \int (f_{ref} - f_{tach})\, dt + C \tag{10.7.4}$$

for some constants k_i and C. Note that we can also provide phase lock

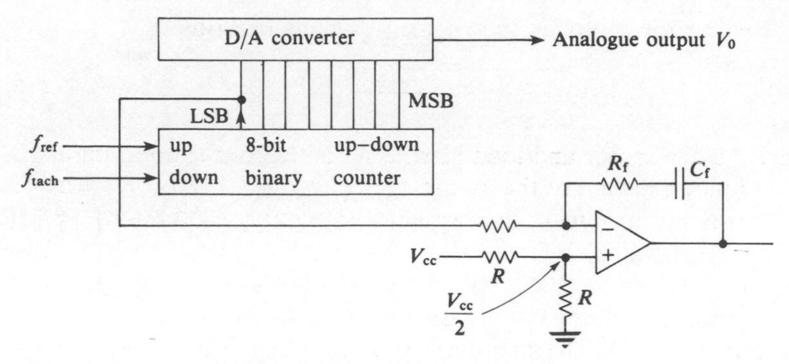

Fig. 10.41 Phase lock circuit.

of f_{ref} and f_{tach} as follows: the reference and actual frequencies are phase locked (in this case 180° out of phase) when the least significant bit of the counter output has a 50 per cent duty cycle (i.e. mark/space ratio† of 1). Hence we can produce another feedback signal by integrating the LSB of the binary counter. The op. amp. output in fig. 10.41 is 0 when the LSB has a 50 per cent duty cycle. The complete feedback system is shown in fig.

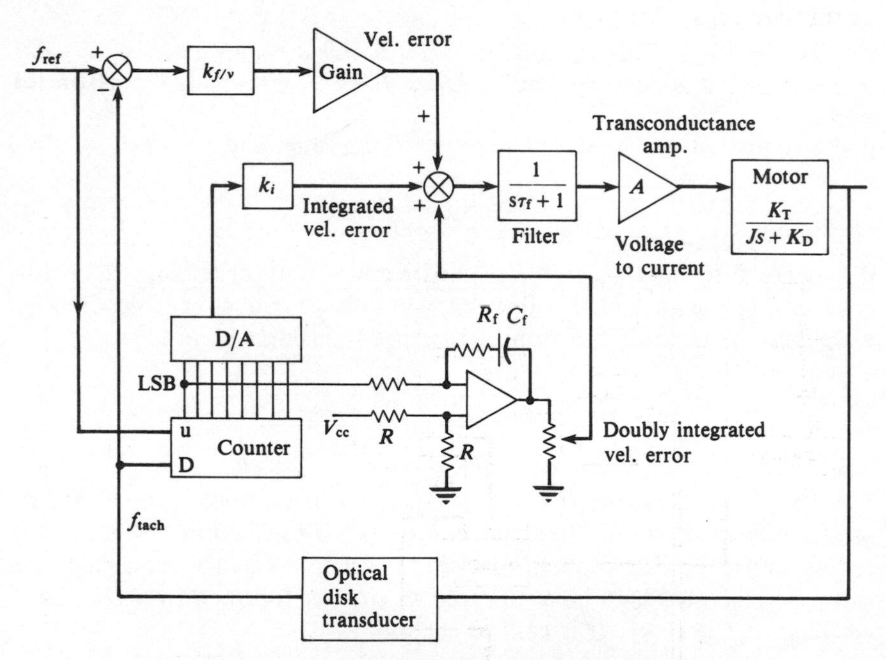

Fig. 10.42 The complete phase lock motor control loop.

†The mark/space ratio of a signal is the ratio of the time for which the signal is logic-1 to the time for which it is logic-0.

10.42 where the motor is shown with transfer function

$$\frac{K_T}{J_s + K_D} \tag{10.7.5}$$

where J is the motor and load inertia, K_T is the torque constant and K_D is the damping constant of the motor. (Total torque $T = K_T \times I$, where I is the motor winding current.) The classical design of the filters using Bode plots is given by Geiger (1981) and is a simple application of the techniques in chapter 4.

A possible system for the microprocessor control of a d.c. motor is shown in fig. 10.43. The information input to the μP consists of the voltage V_{tach} produced by the optical disk transducer (modulated by the direction signal) and the reference voltage which is proportional to the desired speed (again modulated by the desired direction). For example we may choose V_{tach} and V_{ref} so that voltages in the range $[0, V_{max}/2]$ correspond to counter-clockwise rotation while those in the range $[V_{max}/2, V_{max}]$ correspond to clockwise rotation. We need two separate ports to input this data and these are shown in fig. 10.43 to be interrupt driven with a clock which generates interrupts at the correct intervals (related, of course, to the A/D transfer times and the appropriate service routine execution times). We have chosen the $\overline{\text{NMI}}$ line for the feedback signal and the $\overline{\text{INT}}$ line for the reference voltage. This is, however, completely arbitrary. Feedback gains and integration is done by the CPU and the control signal is placed on the output port.

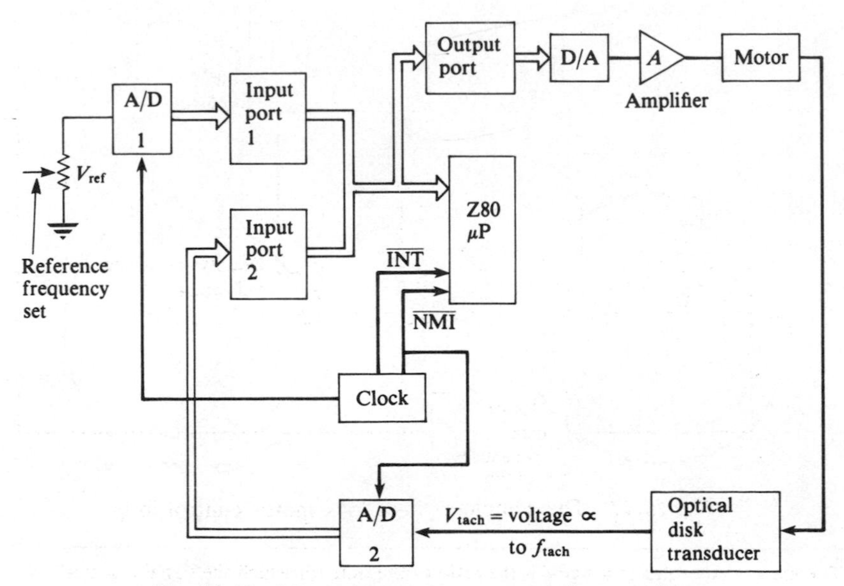

Fig. 10.43 Microprocessor motor speed controller.

10.7.2 Stepper Motors and Simple Robotics

The use of motors in control actuation is clearly very important. We shall consider in this section the stepper motor and its application to a simple robot arm. Note that one could also use d.c. servo motors, but stepper motors have the advantage of not requiring A/D and D/A converters. However they have the disadvantage that the position of the shaft may depend on the load inertia and so this may have to be compensated for. For simplicity, we shall ignore the load inertia effects (see Fitzgerald and Kingsley, 1961).

A typical motor consists, as with any other motor, of a stator and a rotor. Both of these are toothed, and each tooth on the stator is an electromagnet with two independent (oppositely oriented) coils so that it can be given a north or south polarity. The rotor consists of two toothed

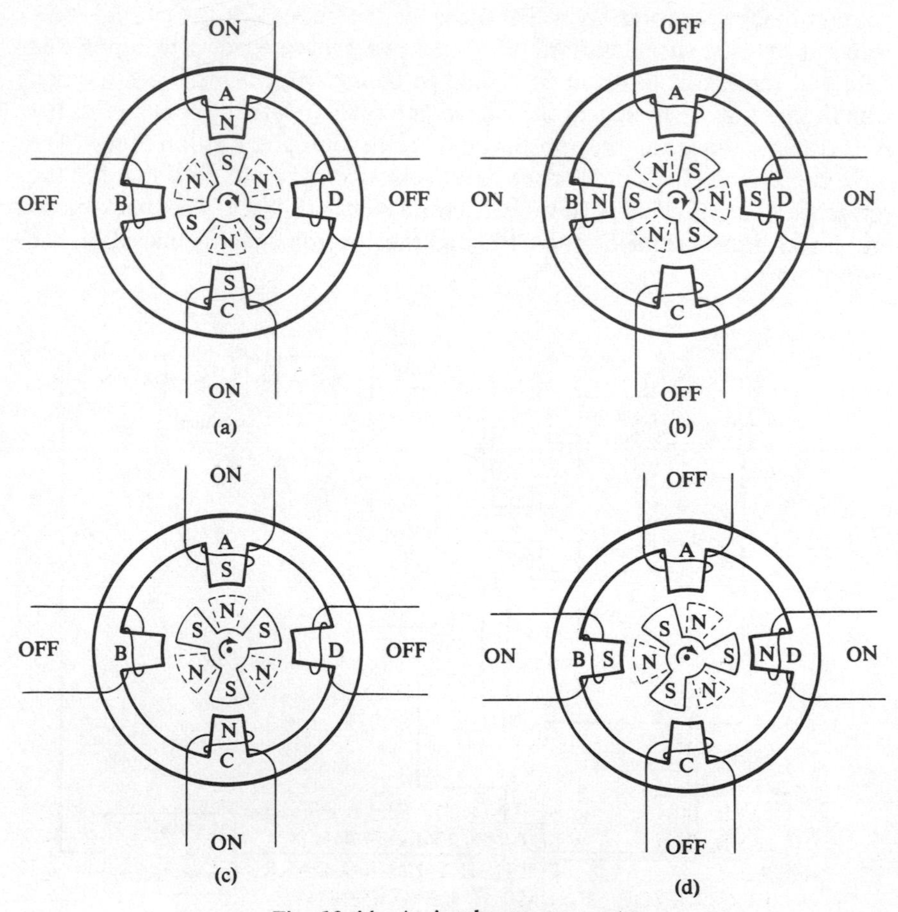

Fig. 10.44 A simple stepper motor.

cylinders, one of which has permanent magnetic south poles and the other north poles on each tooth. The two cylinders are fixed together at a half-tooth spacing. Finally the rotor has more teeth than the stator. An example is shown in fig. 10.44. The stator teeth windings are connected in series into four distinct circuits as follows; first number the stator teeth, say clockwise, 1 to n (n even) and label the corresponding windings s_1, n_1, s_2, n_2, . . . where s_i (n_i) is the south (north) winding of tooth i. Then the windings are connected in series in the form

$$s_1 \text{ to } n_3 \text{ to } s_5 \text{ to } n_7 \ldots$$
$$n_1 \text{ to } s_3 \text{ to } n_5 \text{ to } s_7 \ldots$$
$$s_2 \text{ to } n_4 \text{ to } s_6 \text{ to } n_8 \ldots$$
$$n_2 \text{ to } s_4 \text{ to } n_6 \text{ to } s_8 \ldots$$

The sequence of diagrams in fig. 10.44 shows how the stepper motor can be rotated **(single phase)** by creating the appropriate poles on the stator.† Note that making adjacent teeth have the same polarity will produce more

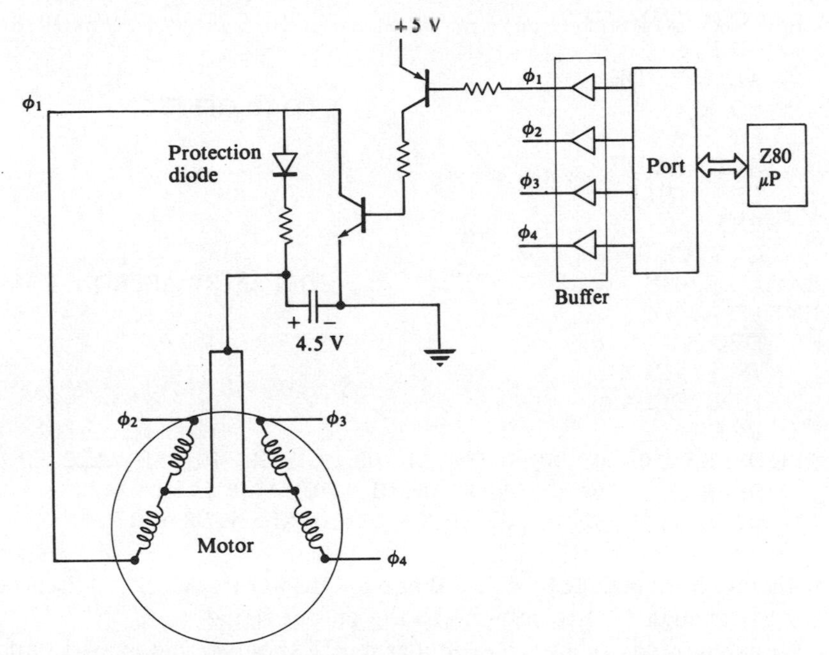

Fig. 10.45 Interfacing a stepper motor to a microprocessor port.

†By making A and C into north and south poles respectively, and switching off B and D, the rotor takes the position in fig. 10.44(a). A and C are then switched off and B and D are made into north and south poles. The north pole on the rotor nearest to D and the south pole nearest to B are then attracted to D and B respectively, as shown, turning the rotor through 30°.

torque (this is the **dual phase** method), while by alternating single and dual phase methods the stator can be made to turn by a half-step (see Giacomo, 1979).

The four series winding circuits described above can be connected to an output port of a Z80 μP as shown in fig. 10.45 (where only a single drive circuit is shown, the others being similar). The appropriate phases of the winding circuits can be driven by outputting the correct byte on the output port. We can store these bytes as a set of tables in the memory, one for each mode of operation; i.e. single phase, dual phase or half-step. The following subroutine will then drive the motor one step (or a half-step); the C′ register is assumed to contain the correct table offset and will be incremented or decremented on each call to the subroutine, C contains the port address and the memory PHASE contains the offset for the desired stepping method – 0 for single phase, 8 for dual phase and 16 for half-step.

```
STEPM:EXX
      LD HL, TABLE
      LD D, 00H
      LD A, (PHASE)
      LD E, A
      ADD HL, DE
      LD A, C                                ; LOAD OFFSET
      LD E, A
      ADD HL, DE
      LD A, (HL)
      EXX
      OUT (C), A
DELAY:LD B, 08H                              ; DELAY BY APPROX. 4 MS
 NEXT:LD A, 7D
      DEC A
      JR NZ, NEXT
      DJNZ DELAY
      RET
TABLE:DEFB 07, 0E, 0D, 0B, 07, 0E, 0D, 0B  ; SINGLE-PHASE VALS.
      DEFB 06, 0C, 09, 03, 06, 0C, 09, 03  ; DUAL-PHASE VALS.
      DEFB 06, 0E, 0C, 0D, 09, 0B, 03, 07  ; HALF-STEP VALS.
```

Note that we have included a delay of about 4 ms (4 MHz clock) so that the motor has enough time to respond to the output signal.

We can now design a simple robot arm with shoulder, elbow, wrist and hand as in fig. 10.46. Each joint has a stepper motor and a corresponding output port. Suppose that a given movement of the arm is stored in memory as a sequence of bytes starting at MOVE. Each byte will be assumed to be of the form

$$000m_4 00m_1 m_0$$

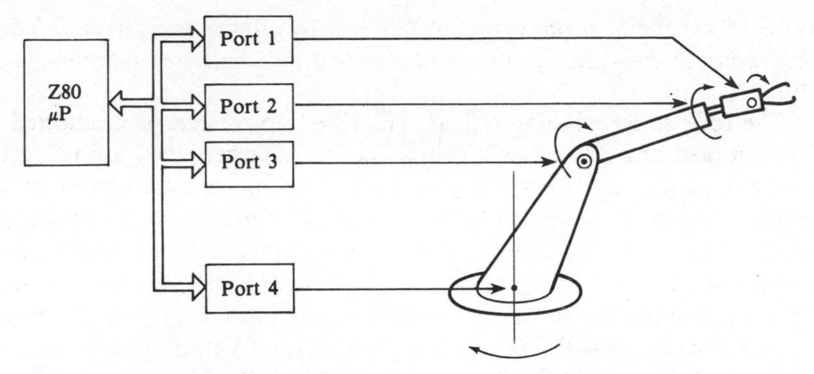

Fig. 10.46 A microprocessor controlled robot arm.

where m_4 equals 1 for a clockwise motion and 0 for a counter-clockwise motion of the motor number m_1m_0 (i.e. $m_1m_0 = 00$ for motor 1, 01 for motor 2, 10 for motor 3 and 11 for motor 4). Then the next subroutine will drive the arm through the sequence of moves stored in the memory:

```
   GO:LD A, 00H
      LD (PHASE), A    ; SELECT MODE, SAY SINGLE-
                       ; PHASE
      LD A, 00H        ; INITIALIZE TABLE
      LD (STAT), A     ; OFFSET FOR EACH MOTOR
      LD (STAT + 1), A ; ASSUME ALL ARE 00H
      LD (STAT + 2), A
      LD (STAT + 3), A
      LD IY, STAT
      LD IX, MOVE
 NMOV:LD A, (IX + 0)   ; INPUT MOVE
      LD E, A
      LD C, F0H        ; PORT ADD. OF MOTOR 1
      PUSH IY
      AND 0FH
      JR Z, NOUP
      LD B, A
   UP:INC IY           ; POINT TO CORRECT OFFSET
      INC C            ; AND PORT ADDRESS
      DJNZ UP
 NOUP:LD A, (IY + 0)
      LD D, A
      LD A, E
      AND F0H
      CP 00H           ; CLOCKWISE?
      JR Z, CLOCKW
      DEC D
```

```
        JR DONE
 CLOCK:INC D
  DONE:LD A, D
        AND 7           ; KEEP OFFSET
                          BETWEEN 0 AND 7
        LD (IY + 0), A  ; STORE NEW OFFSET
        POP IY
        EXX
        LD C, A
        EXX
        CALL STEPM
        LD A, (NMOVES)
        DEC A
        LD (NMOVES), A
        INC IX
        JR NZ, NMOV
        HALT
  STAT:DEFS 4
  MOVE:DEFS 30000
NMOVES:DEFS 1
```

We have assumed here that the port addresses are F0, F1, F2 and F3 and have also reserved room for 30 000 moves.

10.7.3 Process Control

In this final section we shall consider briefly a typical process control system which may be a chemical plant, nuclear reactor, milling machine, water treatment plant, etc. The inputs and outputs of the process will typically be temperatures, pressures, concentrations, flow rates, etc., and are related by some equation which is usually linearized about a nominal trajectory or operating point, giving rise to a matrix transfer function

$$G(s) = [G_{ij}(s)]$$

(For specific examples of real processes see Rosenbrock (1974) and Patel and Munro (1982)). The microcomputer control system (often referred to as DDC or direct digital control) is generally of the form shown in fig. 10.47 and consists of the plant, data acquisition system, microcomputer, D/A converter, analogue comparator and transducers. Using the multivariable control theory given in chapter 4 we can determine a digital compensator

$$G_c(z)$$

which can be implemented on the computer using the software developed in the last chapter. For a detailed discussion of process control the reader

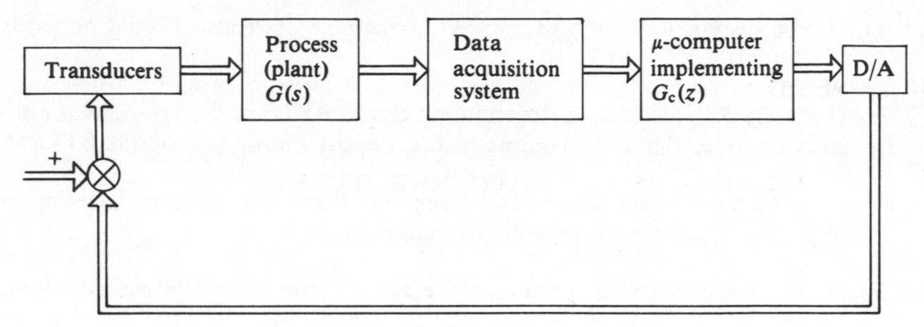

Fig. 10.47 A general DCC system.

should consult Chard (1983), Harrison (1978), Cassell (1983) and Katz (1981).

10.8 CONCLUSION

In this book we have considered the three main aspects of control engineering – modelling and simulation, control theory and microcomputer implementation. In Part 1 the modelling and simulation of both finite and infinite dimensional systems was discussed and this will always be the first task of a control system designer since it is desirable to have the best possible model for the system which can only usually be tested by simulations. The many different types of algorithms for controlling deterministic and stochastic systems were covered in some detail in Part 2, although in a book of this length only the basic theory could be given for most ideas. The reader should supplement his study with more specialist texts on these subjects. Finally in Part 3 we have presented the software and hardware aspects of microprocessor control system design, showing how to interface the Z80 μP to a real system.

This book is mainly introductory in nature, but it is hoped that certain ideas have been covered in greater detail giving the reader an impression of the more modern developments of the subject. Control and information sciences are now diversifying into the areas of robotics, image processing, intelligent software (including pattern recognition for robotic vision systems), expert systems (in which expert knowledge is placed on a computer data base for the use of non-specialists), non-linear system theory, large scale and hierarchical decentralized control systems and many other fields. If this book generates more interest in control and its related fields then it will have served a useful purpose.

10.9 EXERCISES

1. Discuss the significance of the memory map of a microcomputer system and describe the nature of position-independent coding.

2. Why is the system ROM usually placed at memory locations 0000H onwards?

3. Design an output port with address A0H and write a program to output a sequence of 255 bytes of data stored in memories BYTE to BYTE + 254 at about 1 s intervals. (Use the delay routine in the stepper motor subroutine STEPM.)

4. Interface the port in exercise 3 to two 7-segment displays so that the bytes output by the program are shown as 2-bit hex equivalents.

5. Discuss the three interrupt modes on the Z80 μP and their relative usefulness.

6. Describe, in detail, the daisy chain operation of the PIO.

7. Prove equation (10.6.1).

8. Show that the velocity error detector output (in fig. 10.40) is given by the expressions (10.7.3a) and (10.7.3b).

REFERENCES

Cassell, D. A. (1983) *Microcomputers and Modern Control Engineering,* Reston.
Chard, A. (1983) *Software Concepts in Process Control,* National Computing Centre.
Clayton, G. (1979) *Operational Amplifiers,* 2nd edn, Butterworth.
Cluley, J. (1984) *Interfacing to Microprocessors,* MacMillan.
Coffron, J. W. (1981) *Practical Hardware Details for 8080, 8085, Z80 and 6800 Microprocessor Systems,* Prentice-Hall.
Fitzgerald, A. E. and Kingsley, C. (1961) *Electric Machinery,* McGraw-Hill.
Geiger, D. F. (1981) *Phaselock Loops for DC. Motor Speed Control,* Wiley.
Giacomo, P. (1979) A stepper motor primer, *Byte,* 4(2), 90; 4(3), 142.
Harrison, T. J. (ed.)(1978) *Minicomputers in Industrial Control,* Instrument Society of America.
Katz, P. (1981) *Digital Control Using Microprocessors,* Prentice-Hall.
Patel, R. V. and Munro, N. (1982) *Multivariable System Theory and Design,* Pergamon.
Rosenbrock, H. H. (1974) *Computer Aided Control System Design,* Academic Press.
Sen, P. C. (1981) *Thyristor DC Drives,* Wiley.

APPENDIX 1

SUMMARY OF MATHEMATICAL IDEAS USED IN THE BOOK

In this appendix we shall list the mathematical results which are used in this book. All theorems and formulae will be presented without proofs, which can be found in the cited references. For general engineering mathematics, see Jeffrey (1974).

A1.1 VECTOR ANALYSIS (See Spain, 1965)

In the discussion of electromagnetism and fluid dynamics given in chapter 1 we used certain vector identities which are contained in the following list. The reader may be interested in proving them or referring to the text cited above.

If f, g are scalar functions of (x, y, z) and $\boldsymbol{F} = (F_1, F_2, F_3)$, $\boldsymbol{G} = (G_1, G_2, G_3)$ and $\boldsymbol{H} = (H_1, H_2, H_3)$ are vector valued functions, then we have the following identities:

$$\nabla \cdot (\nabla \wedge \boldsymbol{F}) = 0$$

$$\nabla \cdot \nabla f = \nabla^2 f$$

$$\nabla \wedge \nabla f = 0$$

$$\nabla \cdot (f\boldsymbol{G}) = \boldsymbol{G} \cdot \nabla f + f \nabla \cdot \boldsymbol{G}$$

$$\nabla \wedge (f\boldsymbol{G}) = \nabla f \wedge \boldsymbol{G} + f(\nabla \wedge \boldsymbol{G})$$

$$\nabla \wedge (\nabla \wedge \boldsymbol{F}) = \nabla(\nabla \cdot \boldsymbol{F}) - \nabla^2 \boldsymbol{F}$$

$$\boldsymbol{F} \cdot (\boldsymbol{G} \wedge \boldsymbol{H}) = \boldsymbol{G} \cdot (\boldsymbol{H} \wedge \boldsymbol{F}) = \boldsymbol{H} \cdot (\boldsymbol{F} \wedge \boldsymbol{G})$$

$$\nabla \wedge (\boldsymbol{F} \wedge \boldsymbol{G}) = \boldsymbol{F}(\nabla \cdot \boldsymbol{G}) - \boldsymbol{G}(\nabla \cdot \boldsymbol{F}) + (\boldsymbol{G} \cdot \nabla)\boldsymbol{F} - (\boldsymbol{F} \cdot \nabla)\boldsymbol{G}$$

where

$$\nabla = \left(\frac{\partial}{\partial x}, \frac{\partial}{\partial y}, \frac{\partial}{\partial z}\right)$$

$$F \cdot G = F_1G_1 + F_2G_2 + F_3G_3$$

$$F \wedge G = \begin{vmatrix} \hat{\imath} & \hat{\jmath} & \hat{k} \\ F_1 & F_2 & F_3 \\ G_1 & G_2 & G_3 \end{vmatrix}$$

and

$$\hat{\imath} = (1, 0, 0), \quad \hat{\jmath} = (0, 1, 0), \quad \hat{k} = (0, 0, 1)$$

Moreover, if S is a closed surface bounding the volume V, then we have the **divergence (or Gauss') theorem**

$$\oint_S F \cdot \mathrm{d}S = \int_V \nabla \cdot F \, \mathrm{d}V$$

and if l is a closed curve bounding an open surface S, then we have **Stokes' theorem**

$$\oint_l F \cdot \mathrm{d}l = \int_S (\nabla \wedge F) \cdot \mathrm{d}S$$

Note that we have also used the **stress tensor** in the text. All we need to know here is that a second-rank Cartesian tensor T has a matrix representation T_{ij} $(1 \leqslant i, j \leqslant 3)$. A particularly important tensor is the **Kronecker delta** defined by

$$\delta_{ij} = \begin{cases} 1 & \text{if } i = j \\ 0 & \text{if } i \neq j \end{cases}$$

A1.2 SET THEORY (See Dieudonné, 1969)

We require only the most elementary set theory in this book. In axiomatic set theory the concept of 'set' and the symbol '$\in$' are undefined, but for our purposes we can think of a set as a collection of objects and we can read $\in$ as 'belongs to'. Small finite sets can be specified by writing the elements in braces; for example

$$S = \{a, b, c\}$$

while, in general, we specify the elements belonging to a set as those satisfying some property $P(\cdot)$. In this case we write

$$S = \{x : P(x) \text{ is true}\}$$

and in particular we define the set

$$\mathbb{R} = \{x : x \text{ is a real number}\}$$

A **real number** is, of course, an expression of the form

$$d_n d_{n-1} \ldots d_0 \cdot d_{-1} d_{-2} \ldots$$

written in some base. $\mathbb{R}^+$ will denote the set $\{x \in \mathbb{R} : x \geqslant 0\}$, and $[a, b]$, $[a, b)$, $(a, b]$, (a, b) denote the sets of real numbers x such that $a \leqslant x \leqslant b$, $a \leqslant x < b$, $a < x \leqslant b$, $a < x < b$, respectively, for any real numbers a, b. The set of n-dimensional vectors is defined by

$$\mathbb{R}^n = \{x = (x_1, \ldots, x_n) : x_i \in \mathbb{R}, 1 \leqslant i < n\}$$

If $\{S_i\}_{i \in I}$ is a collection of sets (indexed by a set I), we denote the **intersection** and **union** of this family by

$$\bigcap_{i \in I} S_i, \quad \bigcup_{i \in I} S_i$$

In particular, if S_1, S_2 are sets, we write $S_1 \cap S_2$, $S_1 \cup S_2$ for their intersection and union. We define

$$S_1 \backslash S_2 = \{x \in S_1 : x \notin S_2\}$$

and call this the **set difference** of S_1 and S_2. Moreover

$$S_1 \times S_2 \triangleq \{(x_1, x_2) : x_1 \in S_1, x_2 \in S_2\}$$

is called the **Cartesian product** of S_1 and S_2 ($\triangleq$ means 'equal by definition').

Finally, a **function** from S_1 to S_2 is a subset F of $S_1 \times S_2$ such that

$$(x_1, x_2), (x_1, x_3) \in F \Rightarrow x_2 = x_3$$

If (x_1, y), $(x_2, y) \in F$ imply that $x_1 = x_2$, then we say that F is **one-to-one**. The set

$$D = \{x \in S_1 : \text{there exists } y \in S_2 \text{ such that } (x, y) \in F\}$$

is the **domain** of F and

$$R = \{y \in S_2 : \text{there exists } x \in S_1 \text{ such that } (x, y) \in F\}$$

is the **range** of F. If $R = S_2$ we say that F is **onto**. A function F is usually written $F : S_1 \rightarrow S_2$ or $F : D \rightarrow S_2$.

A1.3 LINEAR ALGEBRA

(See Lang, 1970; Fletcher, 1972; Barnett, 1971; Strang, 1980)

A **vector space over a field** K (where $K = \mathbb{R}$ or $\mathbb{C}$, the set of complex numbers) is a set V such that there exist two functions

$\cdot : K \times V \to V, + : V \times V \to V$ called **scalar multiplication** and **vector addition** respectively, which satisfy the axioms

(a) (i) $(v_1 + v_2) + v_3 = v_1 + (v_2 + v_3)$
  (ii) $v_1 + v_2 = v_2 + v_1$
  (iii) there exists $0 \in V$ such that $0 + v = v + 0 = v$ for all $v \in V$
  (iv) there exists $-v \in V$ such that $v + (-v) = -v + v = 0$ for all $v \in v$

(b) (i) $\alpha \cdot (v_1 + v_2) = \alpha \cdot v_1 + \alpha \cdot v_2$, for all $\alpha \in K$, $v_1, v_2 \in V$
  (ii) $(\alpha + \beta) \cdot v = \alpha \cdot v + \beta \cdot v$, for all $\alpha, \beta \in K$, $v \in V$
  (iii) $\alpha \cdot (\beta \cdot v) = (\alpha\beta) \cdot v$, for all $\alpha, \beta \in K$, $v \in V$
  (iv) $1 \cdot v = v$, $0 \cdot v = 0$, for all $v \in V$

We usually omit the '$\cdot$' in scalar multiplication and write simply αv.

Examples
(1) $\mathbb{R}^n$ is a vector space over $\mathbb{R}$ under the usual elementwise operations.
(2) If $\mathbb{C}$ denotes the field of complex numbers, then

$$\mathbb{C}^n = \{(z_1, \ldots, z_n) : z_i \in \mathbb{C}\}$$

is a vector space over $\mathbb{C}$.

Definition
Vectors $v_1, \ldots, v_m \in V$ are **linearly independent** if the equation

$$\alpha_1 v_1 + \ldots + \alpha_m v_m = 0 \quad (\alpha_i \in K) \tag{A1.1}$$

implies that $\alpha_i = 0$, $1 \leqslant i \leqslant m$. Otherwise $v_1, \ldots, v_m$ are **linearly dependent** which means that there exist scalars $\alpha_1, \ldots, \alpha_m$, not all zero, such that (A1.1) holds.

A **basis** of a vector space V is a linearly independent set contained in V which is not contained in any larger linearly independent set. If V has a finite basis with n vectors, we can identify V with the vector space K^n of n dimensional vectors with elements in K.

A **matrix of order $m \times n$ over K** is an array of scalars (from K) of the form

$$\begin{pmatrix} k_{11} & k_{12} & \ldots & k_{1n} \\ \vdots & \vdots & & \vdots \\ k_{m1} & k_{m2} & \ldots & k_{mn} \end{pmatrix} \in K^{m \times n}$$

and a **linear equation** is an expression of the form

$$Ax = b \tag{A1.2}$$

where A is an $m \times n$ matrix, $x \in K^n$, $b \in K^m$. (A1.2) can also be written

$$\sum_{i=1}^{n} a_i x_i = b$$

where a_i is the ith column vector of A. Hence (A1.2) states that the vectors $a_1, \ldots, a_n, b$ are linearly dependent.

One of the central results of linear algebra is the reduction of a matrix to a 'canonical' form. First we define the eigenvalues of an $n \times n$ matrix A over $\mathbb{C}$ to be the solutions of the polynomial equation

$$\det(A - \lambda I_n) = 0$$

where I_n is the identity matrix given by $I_n = (\delta_{ij})_{1 \leqslant i,j \leqslant n}$. If A and B are matrices which are related by

$$B = PAP^{-1} \tag{A1.3}$$

for some invertible matrix P, we say that A and B are **similar** or that B is obtained from A by a similarity transformation. It can then be shown that any $n \times n$ matrix A over $\mathbb{C}$ is similar to a matrix of the form

$$\begin{pmatrix} J_1 & 0 & & \ldots & & 0 \\ 0 & J_2 & 0 & \ldots & & 0 \\ \vdots & \vdots & \vdots & & & \vdots \\ 0 & 0 & 0 & \ldots & 0 & J_k \end{pmatrix}$$

(Jordan canonical form)

where

$$\sum_{i=1}^{k} \dim J_i = n$$

and each J_i is of the form

$$J_i = \begin{pmatrix} \lambda_i & 0 & & \ldots & & & 0 \\ 1 & \lambda_i & 0 & & & \ldots & 0 \\ 0 & 1 & \lambda_i & & 0 & \ldots & 0 \\ \vdots & \vdots & & & \vdots & & \vdots \\ 0 & 0 & & \ldots & 1 & & \lambda_i \end{pmatrix} \text{ or } J_i = (\lambda_i)$$

in which λ_i is an eigenvalue of A. (Note that λ_i may equal λ_j for some $i \neq j$ and that all the eigenvalues of A appear in J).

In general, if $A \in \mathbb{C}^{n^2}$ then the P matrix in (A1.3) which reduces A to Jordan form will be complex. If $A \in \mathbb{R}^{n^2}$ then there exists a similarity transformation with $P \in \mathbb{R}^{n^2}$ which reduces A to the **real normal form**

$$\begin{pmatrix} R_1 & 0 & & \ldots & & 0 \\ 0 & R_2 & 0 & \ldots & & 0 \\ \vdots & & & & & \vdots \\ 0 & & & \ldots & 0 & R_l \end{pmatrix}$$

where each R_i is of the form

$$\begin{pmatrix} D_i & 0 & & \cdots & & 0 \\ I_2 & D_i & 0 & \cdots & & 0 \\ \vdots & & & & & \vdots \\ 0 & & & \cdots & 0 \quad I_2 & D_i \end{pmatrix}$$

where

$$D_i = \begin{pmatrix} a_i & -b_i \\ b_i & a_i \end{pmatrix}$$

and $a_i \pm jb_i$ are eigenvalues of A, or R_i is a Jordan block if λ_i is a real eigenvalue.

A1.4 REAL ANALYSIS (See Apostol, 1965)

$\mathbb{R}^n$ is important as the canonical example of a vector space of dimension n over $\mathbb{R}$. However, for the purposes of the analysis of functions defined on $\mathbb{R}^n$ it is necessary to have some notion of 'nearness' in $\mathbb{R}^n$, which can be specified by a norm.

Definition
A **norm** on a vector space V (of finite or infinite dimension) is a function $\|\cdot\| : V \to \mathbb{R}^+$ such that

$$\left.\begin{array}{ll} \text{(a)} & \|x\| = 0 \text{ if and only if } x = 0 \\ \text{(b)} & \|\alpha x\| = |\alpha| \, \|x\| \text{ if } \alpha \in K, x \in V \\ \text{(c)} & \|x+y\| \leqslant \|x\| + \|y\|, \; x, y \in V \end{array}\right\} \tag{A1.4}$$

The **Euclidean norm** on $\mathbb{R}^n$ is given by

$$\|x\| = \left(\sum_{i=1}^{n} x_i^2\right)^{\frac{1}{2}}, \quad x = (x_i) \in \mathbb{R}^n$$

The set

$$B_\varepsilon(x_0) = \{x \in V : \|x - x_0\| \leqslant \varepsilon\}$$

is called the (closed) ball of radius ε with centre x_0. A set $S \subseteq V$ is **open** if each point $s \in S$ is contained in some ball contained in S. The set $\mathcal{O}$ of all open sets in V is said to form the **topology** of V. For general notions of topology the reader may consult Kelly (1955).

Perhaps the most important theorem of real analysis is Taylor's theorem which may be stated as follows.

Taylor's Theorem

If $f: \mathbb{R}^n \to \mathbb{R}$ has continuous partial derivatives of order m at each point, then for any $\boldsymbol{a} \neq \boldsymbol{b}$ in $\mathbb{R}^n$ we have

$$f(\boldsymbol{b}) - f(\boldsymbol{a}) = \sum_{k=1}^{m-1} \frac{1}{k!} \mathrm{d}^k f(\boldsymbol{a}; \boldsymbol{b} - \boldsymbol{a}) + \frac{1}{m!} \mathrm{d}^m f(\boldsymbol{z}; \boldsymbol{b} - \boldsymbol{a})$$

where $\boldsymbol{z} = \lambda \boldsymbol{a} + (1 - \lambda)\boldsymbol{b}$ for some $\lambda \in [0, 1]$ and

$$\mathrm{d}^k f(\boldsymbol{x}; \boldsymbol{t}) = \sum_{i_1=1}^{n} \sum_{i_2=1}^{n} \cdots \sum_{i_k=1}^{n} \mathrm{D}_{i_1 i_2 \ldots i_k} f(\boldsymbol{x}) t_{i_1} t_{i_2} \ldots t_{i_k}$$

with $\boldsymbol{t} = (t_1, \ldots, t_n)$ and

$$\mathrm{D}_{i_1 i_2 \ldots i_k} f(\boldsymbol{x}) = \frac{\partial^{i_1 \ldots i_k}}{\partial x_{i_1} \ldots \partial x_{i_k}} f(\boldsymbol{x})$$

Of course, $\partial / \partial x_{i_j}$ denotes the partial derivative with respect to x_{i_j}.

This theorem is often applied with $\boldsymbol{b} = \boldsymbol{x}$, and $\boldsymbol{a} = \boldsymbol{0}$ and $m = 2$, in which case we have

$$f(\boldsymbol{x}) = f(\boldsymbol{0}) + \{\text{grad } f(\boldsymbol{0})\}\boldsymbol{x} + \text{higher order terms}$$

where

$$\text{grad } f(\boldsymbol{0}) = \left(\frac{\partial f}{\partial x_1}(0), \ldots, \frac{\partial f}{\partial x_n}(0) \right)$$

Any function $\boldsymbol{f}: \mathbb{R}^n \to \mathbb{R}^n$ may be expressed in the form

$$\boldsymbol{f} = (f_1, \ldots, f_n), \quad \text{where } f_i : \mathbb{R}^n \to \mathbb{R}$$

and then Taylor's theorem with $m = 2$ may be applied to each f_i to give

$$\boldsymbol{f}(\boldsymbol{x}) = \boldsymbol{f}(\boldsymbol{0}) + J_{\boldsymbol{f}}(\boldsymbol{0})\boldsymbol{x} + \text{higher order terms}$$

where

$$J_{\boldsymbol{f}}(\boldsymbol{0}) = \left(\frac{\partial f_i}{\partial x_j} \right)_{1 \leqslant i, j \leqslant n}$$

is the ***Jacobian matrix*** of $\boldsymbol{f}$ at $\boldsymbol{0}$.

A1.5 LINEAR DIFFERENTIAL EQUATIONS AND THE TRANSITION MATRIX (See Coddington and Levinson, 1955)

Consider the linear homogeneous differential equation

$$\dot{\boldsymbol{x}}(t) = A(t)\boldsymbol{x}(t), \quad \boldsymbol{x}(t_0) = \boldsymbol{x}_0 \tag{A1.5}$$

where $x(t) \in \mathbb{R}^n$ and $A(t)$ is a continuous matrix-valued function. Then it is easy to show that this equation has a unique solution of the form

$$x(t) = \Phi(t, t_0)x_0$$

where the matrix function $\Phi(t, t_0)$ satisfies

(a) $\Phi(t_2, t_1)\Phi(t_1, t_0) = \Phi(t_2, t_0)$, for all t_1, t_2

(b) $\Phi(t, t_0)$ is invertible for all t, t_0 and

$$\Phi^{-1}(t, t_0) = \Phi(t_0, t)$$

(c) $\dfrac{d\Phi}{dt}(t, t_0) = A(t)\Phi(t, t_0)$

(d) $\dfrac{d\Phi^T}{dt}(t_0, t) = -A^T(t)\Phi^T(t_0, t)$

Φ is called the **transition matrix** of the equation (A1.5).

The inhomogeneous equation

$$\dot{x}(t) = A(t)x(t) + b(t), \quad x(t_0) = x_0$$

has the solution

$$x(t) = \Phi(t, t_0)x_0 + \int_{t_0}^{t} \Phi(t, s)b(s)\, ds \tag{A1.6}$$

as can be seen by direct substitution.

In the case of the autonomous system

$$\dot{x}(t) = Ax(t)$$

the transition matrix clearly takes the form

$$\Phi(t, t_0) = \exp[A(t - t_0)]$$

A1.6 COMPLEX FUNCTION THEORY (See Saff and Snider, 1976)

We shall assume that readers are familiar with the elementary theory of complex functions and merely remind them of the following wellknown results.

Cauchy's Theorem

If Γ is a simple closed positively oriented curve and f is analytic inside and on Γ except at the points $z_1, \ldots, z_n$ inside Γ, then

$$\int_{\Gamma} f(z)\, dz = 2\pi j \sum_{k=1}^{n} \mathrm{Res}(z_k)$$

Recall that if z_0 is an isolated singularity† of f, then f may be written in a neighbourhood of z_0 as a Larent series

$$f(z) = \sum_{k=-\infty}^{\infty} a_k(z - z_0)$$

where

$$a_k = \frac{1}{2\pi j} \int_\gamma \frac{f(z)}{(z - z_0)^{k+1}} \mathrm{d}z$$

for any simple closed curve γ surrounding z_0. Then we define the residue of f at z_0 as

$$\mathrm{Res}(z) = a_{-1}$$

If $a_{-j} = 0$, $j \geqslant k$ for some $k > 1$, then z_0 is called a **simple pole**.

Principle of the Argument

If f is analytic and non-zero at each point of a simple closed positively oriented curve Γ and is meromorphic (i.e. has only simple poles) inside Γ, then

$$\frac{1}{2\pi i} \int_\Gamma \frac{f'(z)}{f(z)} \mathrm{d}z = N_0(f) - N_p(f) \tag{A1.7}$$

where $N_0(f)$ and $N_p(f)$ are the number of zeros and poles of f inside Γ (counting multiplicities). Moreover it is easy to show that the left-hand side of (A1.7) is just the number of times the contour $f(\Gamma)$ winds around the origin (in the positive direction) in the $w = f(z)$ plane.

Finally we have used the notion of conformal map in the book, and we shall now give a definition of this concept. An analytic function $w = f(z)$ of the complex variable z is **conformal** at z_0 if $f'(z_0) \neq 0$. It is easy to show that such a function preserves the angle between two smooth (oriented) curves through z_0.

A1.7 HILBERT SPACES (See Taylor, 1958; Helmberg, 1969)

Definition

A sequence $\{x_n\}$ belonging to a normed vector space V is a **Cauchy sequence**, if for any given ε

$$\| x_n - x_m \| \leqslant \varepsilon, \quad \text{when } n, m \geqslant N$$

†That is f is analytic in a neighbourhood of z_0, but not at z_0.

for some N (depending on ε). Note that a convergent sequence is a Cauchy sequence.

A normed vector space is **complete** if every Cauchy sequence converges.

Definition
An **inner product space** V (over $K = \mathbb{R}$ or $\mathbb{C}$) is a vector space together with a function $\langle .,. \rangle : V \times V \to K$ which is linear in both variables and satisfies

$$\langle x, y \rangle = \overline{\langle y, x \rangle}, \quad x, y \in V$$

A **Hilbert space** is an inner product space which is complete under the norm

$$\| x \| = \{\langle x, x \rangle\}^{1/2}$$

Examples
(1) $\mathbb{R}^n = \{(x_1, \ldots, x_n) : x_i \in \mathbb{R}\}$ with inner product

$$\langle x, y \rangle = \sum_{i=1}^{n} x_i y_i$$

(2) $\mathbb{C}^n$ with inner product

$$\langle x, y \rangle = \sum_{i=1}^{n} x_i \bar{y}_i$$

(3) The space $C[0, 1]$ of (real or complex valued) continuous functions defined on the interval $[0, 1]$ is an inner product space under the inner product

$$\langle f, g \rangle = \int_0^1 f(t)\bar{g}(t)\, dt$$

However, $C[0, 1]$ is not complete under this inner product. By adding to $C[0, 1]$ the limits of Cauchy sequences which are not already in $C[0, 1]$ (these are called measurable functions) we obtain the space $L^2[0, 1]$ which is a Hilbert space under the above inner product. Note that $C[0, 1]$ is **dense** in $L^2[0, 1]$ in the sense that any element of $L^2[0, 1]$ is arbitrarily close to an element of $C[0, 1]$ in the norm defined by the inner product.

Definition
A subset S of a Hibert space H is a **subspace** of H if it is itself a Hilbert space under the same operations.

We say that two vectors $h_1, h_2 \in H$ are **orthogonal** if

$$\langle h_1, h_2 \rangle = 0$$

It can then be shown that for any subspace S in H and any vector $h \in H$ we can write

$$h = h_S + h_S^{\perp}$$

where $h_S \in S$ and $\langle h_S^{\perp}, g \rangle = 0$ for all $g \in S$. h_S is called the **orthogonal projection** of h on S.

Definition

An (orthonormal) **basis** of a Hilbert space H is a set of vectors $B = \{b_\alpha\}_{\alpha \in A}$

indexed by some set A such that

$$\langle b_\alpha, b_\beta \rangle = 0, \quad \alpha \neq \beta$$
$$\| b_\alpha \| = 1, \quad \alpha \in A$$

and any $h \in H$ can be written

$$h = \sum_{\alpha \in A} h_\alpha b_\alpha$$

for some scalars h_α. Note that $h_\alpha = \langle h, b_\alpha \rangle$ and is called the αth **Fourier coefficient** of h with respect to the basis B. H is said to be **separable** if we can take A to be the set of integers.

Note that $L^2[0, 1]$ is separable with basis

$$\{\cos n\pi x, \sin n\pi x\}_{n \geqslant 0}$$

(This gives the usual Fourier coefficients of a continuous function.)

Finally we define the Fourier transform of an element $f \in L^2[0, 1]$ by

$$\hat{f}(\omega) \triangleq (Ff)(\omega) = \frac{1}{2\pi} \int_{-\infty}^{\infty} f(t)\exp(-j\omega t)\,dt$$

It can be shown that F maps an element of $L^2(-\infty,\infty)$ into $L^2(-\infty,\infty)$ and in fact, F is one-to-one and onto with inverse

$$(F^{-1}\hat{f})(t) = \int_{-\infty}^{\infty} \hat{f}(\omega)\exp(j\omega t)\,d\omega$$

Moreover

$$\frac{1}{2\pi} \int_{-\infty}^{\infty} \hat{f}^2(\omega)\,d\omega = \int_{-\infty}^{\infty} f^2(t)\,dt$$

that is

$$\frac{1}{2\pi} \|\hat{f}\| = \|f\|$$

REFERENCES

Apostol, T. M. (1965) *Mathematical Analysis*, Addison-Wesley.
Barnett, S. (1971) *Matrices in Control Theory*, Van Nostrand Reinhold.
Coddington, E. A. and Levinson, N. (1955) *The Theory of Ordinary Differential Equations*, McGraw-Hill.
Dieudonné, J. (1969) *Foundations of Modern Analysis*, Academic Press.
Fletcher, T. J. (1972), *Linear Algebra*, Van Nostand Reinhold.
Helmberg, G. (1969) *Introduction to Spectral Theory in Hilbert Space*, North-Holland.
Jeffrey, A. (1974) *Basic Mathematics for Engineers and Technologists*, Nelson.
Kelley, J. L. (1955) *General Topology*, Van Nostrand.
Lang, S. (1970) *Algebra*, Addison-Wesley.
Saff, E. B. and Snider, A. D. (1976) *Fundamentals of Complex Analysis*, Prentice-Hall.
Spain, B. (1965) *Vector Analysis*, Van Nostrand.
Strang, G. (1980), *Linear Algebra and its Applications*, Academic Press.
Taylor, A. E. (1958) *Introduction to Functional Analysis*, Wiley.

APPENDIX 2

ELEMENTARY PROBABILITY THEORY

(See also Feller, 1957; Burrill, 1972; Halmos, 1950; Doob, 1953: Box and Jenkins, 1970; Soong, 1981.)

A2.1 EVENTS AND PROBABILITY MEASURES

In the classical theory of probability we consider random experiments in which the number of outcomes is finite. If

$$\Omega = \{\omega_1, \ldots, \omega_N\}$$

denotes the set of outcomes of the experiment we assign a probability $p(\omega_i) \geqslant 0$ to each outcome so that

$$\sum_{i=1}^{N} p(\omega_i) = 1$$

Example
Let Ω be the outcomes of tossing an (unbiased) coin. ω_1 = 'heads', ω_2 = 'tails', then $p(\omega_1) = p(\omega_2) = \frac{1}{2}$.

More generally we consider an arbitrary set Ω as the set of outcomes of an experiment. Subsets of Ω may also be regarded as the outcomes of experiments although not all subsets of Ω may be meaningful outcomes. Hence we consider a certain class of subsets $\mathfrak{S}$ of Ω called **events**† on which a numerical function P is defined, with properties to be specified below. It

†An event is just a possible outcome of a random experiment

is clear that the set $\mathfrak{S}$ should have the following properties

$$A \in \mathfrak{S} \Rightarrow A' = \Omega \backslash A \in \mathfrak{S} \tag{A2.1}$$

$$A, B \in \mathfrak{S} \Rightarrow A \cup B \in \mathfrak{S} \tag{A2.2}$$

since the complement of an event should be an event, and the union of two events, which occurs if either A or B (or both) occur, should be an event. Moreover, since we also consider limiting procedures, we also often require the stronger condition that

$$A_i \in \mathfrak{S}, \quad 1 \leqslant i < \infty \Rightarrow \bigcup_{i=1}^{\infty} A_i \in \mathfrak{S} \tag{A2.3}$$

A set $\mathfrak{S}$ satisfying (A2.1) and (A2.2) is called an **algebra** (of sets) while one satisfying (A2.1) and (A2.4) is called a **σ-algebra**. The pair $(\Omega, \mathfrak{S})$ where $\mathfrak{S}$ is a σ-algebra on Ω is called a **measurable space**.

A **probability measure** P defined on an algebra $\mathscr{A}$ is a set function such that

$$0 \leqslant P(A) \leqslant 1, \quad P(\Omega) = 1, \quad A \in \mathscr{A} \tag{A2.4}$$

$$P\left(\bigcup_{n=1}^{\infty} A_n\right) = \sum_{n=1}^{\infty} P(A_n), \quad A_i \in \mathscr{A} \quad (\sigma\textbf{-additivity}) \tag{A2.5}$$

whenever

$$\bigcup_{n=1}^{\infty} A_n \in \mathscr{A}$$

and the A_i are pairwise disjoint (i.e. $A_i \cap A_j = \phi$ if $i \neq j$).

It is easy to show that a probability measure defined on an algebra $\mathscr{A}$ may be uniquely extended to one defined on the σ-algebra $\sigma(\mathscr{A})$ generated by $\mathscr{A}$ (i.e. the σ-algebra which is the intersection of all σ-algebras containing $\mathscr{A}$). For example, consider $\Omega = [0, 1)$ and let

$$\mathscr{A} = \{[a, b) : 0 \leqslant a \leqslant b \leqslant 1\}$$

where we set $[a, b) = \phi$ if $a = b$. Clearly, $\mathscr{A}$ is an algebra and we define

$$P([a, b)) = b - a$$

It can be shown that P is σ-additive and so extends to a probability measure on $\mathfrak{S} = \sigma(\mathscr{A})$. We denote the extension also by P and call a set $N \in \mathfrak{S}$ such that $P(N) = 0$ a **null** set.† We then define

$$\mathfrak{S}_L = \sigma(\mathscr{A} \cup \mathscr{N}) \quad \text{(the } \sigma\text{-algebra generated by } \mathscr{A} \text{ and } \mathscr{N}\text{)}$$

where $\mathscr{N}$ is the set of all null sets in $\mathfrak{S}$. Again P extends to a probability measure on $\mathfrak{S}_L$ called the **Lebesgue measure**. This construction can be

---

† To make $\mathfrak{S}_L$ 'complete' we must add in the events with zero probability.

generalized to a measure on a set Ω. Note, however, that a general measure μ on Ω does not satisfy $\mu(\Omega) = 1$. If $\mu(\Omega) < \infty$, μ is called a **finite** measure. The triple $(\Omega, \mathfrak{S}, \mu)$ is called a **measure space**.

A2.2 PROBABILITY DISTRIBUTION FUNCTIONS AND RANDOM VARIABLES

Generalizing the above example on $[0, 1)$ we consider the set $\mathbb{R}^n$ and let $\mathscr{R}^n$ denote the set of all rectangles in $\mathbb{R}^n$, where a **rectangle** is a set of the form

$$\{x \in \mathbb{R}^n : x_i \in A_i,\ A_i \text{ an interval},\ 1 \leqslant i \leqslant n\} = \prod_{i=1}^{n} A_i$$

The elements of the σ-algebra $\mathscr{B} \triangleq \sigma(\mathscr{R}^n)$ are called the **Borel sets** of $\mathbb{R}^n$.

Let μ be a finite non-negative measure on $\mathscr{B}$ and define the function D on $\mathbb{R}^n$ by

$$D(x) = \mu(\{y : -\infty < y_i < x_i,\ 1 \leqslant i \leqslant n\}) \tag{A2.6}$$

Clearly

$$D(x) \to 0 \text{ as } x_i \to -\infty \quad \text{(for each } i\text{)}$$

and

$$D(x) \to \mu(\mathbb{R}^n) \text{ as } (x_1, \ldots, x_n) \to (\infty, \ldots, \infty)$$

D is called the **distribution function** corresponding to the measure μ, and if $\mu(\mathbb{R}^n) = 1$ it is called the **probability distribution function**.

Now consider two measurable spaces $(\Omega_1, \mathfrak{S}_1)$, $(\Omega_2, \mathfrak{S}_2)$. A function $f: \Omega_1 \to \Omega_2$ is called **measurable** if

$$A \in \mathfrak{S}_2 \Rightarrow f^{-1}(A) \in \mathfrak{S}_1$$

A measurable function $X : (\Omega, \mathfrak{S}) \to (\mathbb{R}, \mathscr{R})$, where $(\Omega, \mathfrak{S})$ has a probability measure P defined on it, is called a (real) **random variable**. Clearly, sets of the form $\{\omega \in \Omega : X(\omega) < a\}$ for any real a are events and so we can define the function

$$P_X(a) = P(\{\omega : X(\omega) < a\})$$

P_X is called the **probability distribution function** of X, i.e. the probability that the random variable X is less than a. Note that if we define

$$P_X(A) = P(\{\omega : X(\omega) \in A\}), \quad A \in \mathscr{R}$$

then $(\mathbb{R}, \mathscr{R}, P_X)$ becomes a probability measure space.

More generally, if $X = (X_1, \ldots, X_n)$ are n real random variables then $X : (\Omega, \mathfrak{S}) \to (\mathbb{R}^n, \mathscr{R}^n)$ is measurable and the function

$$P_X(a) = P(\{\omega : X_i(\omega) < a_i,\ 1 \leqslant i \leqslant n\}), \quad a \in \mathbb{R}^n \tag{A2.7}$$

is the **joint probability distribution function** of X. Also,

$$P_X(A) = P(\{\omega : X(\omega) \in A\}), \quad A \in \mathscr{R}^n$$

defines a probability measure on $(\mathbb{R}^n, \mathscr{R}^n)$.

A2.3 INTEGRATION

Let $(\Omega, \mathfrak{S}, \mu)$ be a measure space and let $f : \Omega \to \mathbb{R}$ be a non-negative measurable function. We wish to define the integral

$$\int_\Omega f(\omega)\, d\mu$$

Consider first a **simple function** defined on Ω by

$$S = \sum_{i=1}^{k} s_i I_{A_i}$$

where

$$I_{A_i}(\omega) = \begin{cases} 1 & \omega \in A_i \\ 0 & \omega \in \Omega \backslash A_i \end{cases}$$

is called the **indicator function** of A_i. If the A_i are disjoint (i.e. $A_i \cap A_j = \phi$ if $i \neq j$) then S takes the constant value s_i on A_i. Then we define

$$\int_\Omega S(\omega)\, d\mu = \sum_{i=1}^{k} s_i \mu(A_i)$$

It can be shown that any non-negative measurable function f is the limit of an increasing sequence of simple functions S_f^i, $1 \leqslant i < \infty$. (i.e. $S_f^{i+1}(\omega) \geqslant S_f^i(\omega)$ for each i, ω). We then define

$$\int_\Omega f(\omega)\, d\mu = \lim_{i \to \infty} \int_\Omega S_f^i(\omega)\, d\mu$$

The limit on the right always exists (but may be ∞). If f is a general measurable function we define

$$\int_\Omega f(\omega)\, d\mu = \int_\Omega f^+(\omega)\, d\mu - \int_\Omega f^-(\omega)\, d\mu \tag{A2.8}$$

where

$$f^+ = \max(f, 0)$$
$$f^- = -\min(f, 0)$$

provided at least one term on the right of (A2.8) is finite. Finally, if

$$\int_{\Omega} |f(\omega)| \, d\mu < \infty$$

we say that f is integrable.

Examples

(1) If $(\Omega, \mathfrak{S}, P)$ is a probability measure space and X is an integrable random variable, we define the **expectation** (or **mean**) of X by

$$EX = \int_{\Omega} X(\omega) \, dP \tag{A2.9}$$

(Note that (A2.9) is often written $\int_{\Omega} X(\omega) P(d\omega)$. We also define the **variance** VX of X by $E[(X - EX)^2]$.

(2) If $X = (X_1, \ldots, X_n)$ is a vector of random variables defined on $(\Omega, \mathfrak{S}, P)$ and $f: (\mathbb{R}^n, \mathscr{R}^n) \to (\mathbb{R}, \mathscr{R})$ is a measurable function, then

$$f[X(\cdot)]$$

is a random variable on $(\Omega, \mathfrak{S}, P)$ and f is a random variable defined on $(\mathbb{R}^n, \mathscr{R}^n, P_X)$, where P_X is the probability distribution of X. Also we clearly have

$$\int_{\Omega} f[X(\omega)] \, dP = \int_{\mathbb{R}^n} f(x) \, dP_X \tag{A2.10}$$

The value on the right is called the **Lebesgue-Stiltjes integral** of f with respect to P_X.

A2.4 PROBABILITY DENSITY FUNCTIONS

Let $(\Omega, \mathfrak{S}, \mu_1)$, $(\Omega, \mathfrak{S}, \mu_2)$ be measure spaces. We say that μ_1 is **absolutely continuous** with respect to μ_2 if

$$\mu_2(A) = 0 \Rightarrow \mu_1(A) = 0 \quad (A \in \mathfrak{S})$$

If μ_1 and μ_2 are **σ-finite** in the sense that Ω can be written as

$$\bigcup_{i=1}^{\infty} A_i, \quad A_i \in \mathfrak{S}$$

where $\mu_1(A_i)$, $\mu_2(A_i) < \infty$ for each i, then the following **Radon–Nikodym theorem** can be proved:

If μ_1 is absolutely continuous with respect to μ_2 (μ_1, μ_2 σ-finite) then there exists a (unique up to sets of μ_2 measure 0) finite-valued measurable function f on Ω such that

$$\mu_1(A) = \int_A f \, d\mu_2, \quad A \in \mathfrak{S}$$

f is called the **Radon–Nikodym derivative** of μ_1 with respect to μ_2 and is written $d\mu_1/d\mu_2$.

In particular, if the probability distribution function P_X of a (vector) random variable X is absolutely continuous with respect to Lebesgue measure, then we can find a non-negative measurable function $p_X(\boldsymbol{x})$, $\boldsymbol{x} \in \mathbb{R}^n$ such that

$$P_X(A) = \int_A p_X(\boldsymbol{x})\, d\boldsymbol{x}, \quad A \in \mathbb{R}^n$$

p_X is called the (joint) probability density function of $X = (X_1, \ldots, X_n)$.

A2.5 THE GAUSSIAN DISTRIBUTION

An important example of a density function is given by

$$p_X(x) = \frac{1}{\sigma(2\pi)^{1/2}} \exp\left[-\frac{1}{2}\left(\frac{x-\mu}{\sigma}\right)^2 \right]$$

for some constants σ, μ. It is easy to see that

$$EX = \mu, \quad VX = \sigma$$

X is called a **normal** or **Gaussian** random variable. The distribution function is just

$$P_x(x) = \int_{-\infty}^{x} p_x(x)\, dx$$

More generally, we say that the vector $(X_1, \ldots, X_n)$ of random variables is jointly Gaussian or normally distributed if it has the joint density function

$$p_X(\boldsymbol{x}) = \frac{1}{(2\pi)^{n/2}(\det \Sigma)^{1/2}} \exp\left[-\frac{1}{2}(\boldsymbol{x}-\boldsymbol{\mu})^T \Sigma^{-1} (\boldsymbol{x}-\boldsymbol{\mu}) \right]$$

where

$$\Sigma = [E(X_i X_j)]$$

is called the **covariance matrix** of X.

A2.6 CONVERGENCE OF RANDOM VARIABLES

If $\{X_n\}$ is a sequence of random variables on $(\Omega, \mathfrak{S}, P)$, we say that

(a) X_n **converges almost surely** (a.s.) to the random variable X if there exists $A \in \mathfrak{S}$ with $P(A) = 0$ such that

$$\lim_{n\to\infty} |X_n(\omega) - X(\omega)| = 0 \quad \text{for } \omega \notin A$$

i.e. X_n converges to X pointwise, apart from on a set of probability zero.

(b) X_n **converges in probability** to X if for every $\varepsilon > 0$,

$$P(|X_n - X| \geqslant \varepsilon) \to 0 \text{ as } n \to \infty$$

(c) X_n **converges in νth mean** ($\nu > 0$) to X if

$$\mathrm{E}|X_n - X|^{\nu} \to 0 \text{ as } n \to \infty$$

It can be shown that a.s. convergence and convergence in νth mean both imply convergence in probability (to the same limit). (If $\nu = 2$ we also say that X_n converges in **quadratic** mean.)

The set of all random variables with $\mathrm{E}X^2 < \infty$ forms a Hilbert space $L^2(\mathfrak{S})$ with inner product

$$\langle X, Y \rangle = \mathrm{E}(XY)$$

A2.7 CONDITIONAL EXPECTATION

If $(\Omega, \mathfrak{S}, P)$ is a probability space and $P(D) \neq 0$ for some D we define the **conditional probability** of A given D by

$$P(A \mid D) = \frac{P(A \cap D)}{P(D)} \text{ for any } A \in \mathfrak{S}$$

The set function P_D defined by

$$P_D(A) = P(A \mid D)$$

is clearly a probability measure on Ω and so $(\Omega, \mathfrak{S}, P_D)$ is also a probability space. If X is a random variable on Ω then we can define the **conditional expectation** of X given D by

$$\mathrm{E}(X \mid D) = \int_{\Omega} X \, \mathrm{d}P_D$$

Note that if X is the indicator function I_A of A, then

$$\begin{aligned} \mathrm{E}(I_A \mid D) &= \int_{\Omega} I_A \, \mathrm{d}P_D \\ &= P_D(A) \\ &= P(A \mid D) \end{aligned}$$

If $P(D) = 0$ we cannot use the above approach and so we proceed as follows. For any integrable random variable on Ω let ϕ be the function defined on a σ-algebra $\mathscr{E} \subseteq \mathfrak{S}$ by

$$\phi(D) = \int_D X \, \mathrm{d}P \quad \text{for } D \in \mathscr{E}$$

Then ϕ is clearly a measure on $\mathscr{E}$ and is absolutely continuous with respect to P and so by the Radon–Nikodym theorem there is an $\mathscr{E}$-measurable function g such that

$$\phi(D) = \int_D g \, dP|_{\mathscr{E}}$$

where $P|_{\mathscr{E}}$ means the probability measure P restricted to $\mathscr{E}$. We write

$$E(X \mid \mathscr{E}) = g$$

If $\{Y_1, Y_2, \ldots\}$ is a set of random variables we define the σ-algebra $\mathscr{G}$ **generated** by this set to be the smallest σ-algebra containing the sets $Y_i^{-1}(\mathscr{R})$ for each i. Then we write

$$E(X \mid Y_1, Y_2, \ldots) = E(X \mid \mathscr{G})$$

It is easy to see that this reduces to the usual classical definition for $P(X < x \mid Y_1 = y_1, Y_2 = y_2, \ldots)$ when $P(Y_1 = y_1, Y_2 = y_2, \ldots) > 0$.

Note that we can define the conditional expectation also as a projection in $L^2(\mathfrak{S})$. In fact, if X is a random variable in $L^2(\mathfrak{S})$ and $\mathscr{E} \subseteq \mathfrak{S}$ is a sub-σ-algebra of $\mathfrak{S}$, then the set of $\mathscr{E}$-measurable functions in $L^2(\mathfrak{S})$ forms the closed subspace $L^2(\mathscr{E})$ and so we can define

$$E(X \mid \mathscr{E}) = \text{orthogonal projection of } X \text{ on } L^2(\mathscr{E})$$

A2.8 STOCHASTIC PROCESSES

A **stochastic process** $\{X_t, t \in T\}$ is a family of random variables defined on a common probability space $(\Omega, \mathfrak{S}, P)$ with $T \subseteq \mathbb{R}$. Usually we take T to be an interval or a discrete set. Since each X_t is a function defined on Ω we often write $X_t(\omega)$ and for a fixed ω we say that the function

$$t \to X_t(\omega)$$

is a **sample path** of the process.

Although we can define the joint probability functions of any countable collection of the X_t we cannot generally define such a joint probability for the whole set since the set

$$\{\omega : X_t(\omega) \geqslant 0, \text{ for all } t \in T\} = \bigcap_{t \in T} \{\omega : X_t(\omega) \geqslant 0\}$$

may not be an event. Since we require to define integrals of the form

$$\int_a^b X_t(\omega) \, dt$$

we must overcome the above difficulty. This can be done as follows. The process $\{X_t, t \in T\}$ is **separable** if there is a countable set $S \subseteq T$ and a null set N such that for any closed set $K \subseteq [-\infty, \infty]$ and any open interval I we have

$$\{\omega : X_t(\omega) \in K,\ t \in I \cap T\} \backslash \{\omega : X_t(\omega) \in K,\ t \in I \cap S\} \subseteq N$$

and

$$\{\omega : X_t(\omega) \in K,\ t \in I \cap S\} \backslash \{\omega : X_t(\omega) \in K,\ t \in I \cap T\} \subseteq N$$

i.e. these two sets differ by a null set.

It can then be shown that for every process $\{X_t, t \in T\}$, there exists a separable process $\{\bar{X}_t, t \in T\}$ such that

$$P(X_t = \bar{X}_t) = 1 \quad \text{for all } t \in T$$

A process $\{X_t, t \in T\}$ is said to be **Markov** if, for any increasing collection $t_1, \ldots, t_n \in T$ we have

$$P(X_{t_n} \leqslant x_n \mid X_{t_i} = x_i,\ 1 \leqslant i \leqslant n-1) = P(X_{t_n} \leqslant x_n \mid X_{t_{n-1}} = x_{n-1})$$

i.e. X depends only on the immediate past.

REFERENCES

Box, G. E. P. and Jenkins, G. M. (1970) *Time Series Analysis*, Holden Day.
Burrill, C. W. (1972) *Measure, Integration and Probability*, McGraw-Hill.
Doob, J. L. (1953) *Stochastic Processes*, Wiley.
Feller, W. (1957) *An Introduction to Probability Theory and its Applications, Vol. I*, Wiley.
Halmos, P. R. (1950) *Measure Theory*, Van Nostrand.
Soong, T. T. (1981) *Probabilistic Modelling and Analysis in Science and Engineering*, Wiley.

INDEX

Abscissa of convergence, 188
Absolute maximum, 322
Absolute stability, 290
Acceleration error, 204
Accessory minimization problem, 354
Action integral, 171
Adaptive control, 423, 424
  observation, 435
Adder, 79, 108
Addressing methods, 485
  bit, 488
  extended, 486
  immediate, 487
  implied, 488
  indexed, 485
  register, 485
  zero page, 488
ADI, 158
Admittance matrix, 28
Aerodynamic derivatives, 22
Algebraic equations, 143
Algorithm 138, 143, 532
  explicit, 156
  implicit, 155
Aliasing, 223
Alternating direction implicit method, 158
Amplifier, 78, 114
  design of, 116
Amplitude, 68, 105
Angle of attack, 21, 54
Artificial dither, 307
Assembly language, 488, 513
  programming, 513
Augmented Euler equations, 337
Auto correlation matrix, 374
Autonomous system, 35

Base (of number system), 125
Bellman's equation, 355, 359
Bernoulli's theorem, 45
Bilinear system, 312
  transform, 231
Binary coded decimal, 129
Binary numbers, 126
Blasius' theorem, 52
Block diagram, 61, 226
Bode diagram, 119, 204
Bolza problem, 333
Boundary condition, 37, 150
Brownian motion, 395
Bus, 473, 538
  address, 473, 538
  control, 473, 539
  data, 473, 539

Calculus of variations, 332
Canonical form, 593
  controllable, 245
  observable, 245
Capacitor, 24
Cauchy–Riemann equations, 49
Causal system, 191, 222
Centre, 269
Chapman–Kolmogorov equation, 408
Characteristic, 159
  curves, 152, 159
  directions, 151
  gains, 252
  locus, 252
  polynomial, 234
Chattering controls, 351
Cholesky decomposition, 412
Circle criterion, 290, 297
Circulation, 48
Classical compensation, 212
Codes, 128
  ASCII, 132
  BCD, 129
  Gray, 130
Comoving derivative, 45
Complete observations, 402
Complex potential, 49
Complex velocity, 49

Computer, 124
analogue, 77
digital, 124
Constrained optimization, 324
Constraint qualification, 331
Constraints, 324
Continuity equation, 43
Controllability, 243
Convergence, 598
in probability, 607
in quadratic mean, 392
Converter (A/D, D/A), 572
Convolution, 189
Covariance function, 393
Cross correlation, 413
Cubic spline, 147

Damping ratio, 194
Data acquisition, 576
Degrees of freedom, 5, 172
Describing function, 301
Desensitivity, 117
Diagonal dominance, 250
Difference equation, 222
Differential equation, 192
delay, 54
ordinary, 34
partial, 36
Differentiator, 106
Diode, 90
Zener, 98
Directional derivative, 333
Discrete dynamic programming, 355
Discrete Kalman filter, 379
Discrete systems, 219
Discretization error, 574
Discriminant, 271
Displacement
real, 7
virtual, 7
Distributed systems, 35
Disturbance rejection, 461
Division controller, 317
Drift stabilization, 113
Dual-input describing function, 305
Dynamic programming, 355

Eigenvalues, 70
Eigenvectors, 71
Electric field, 22
Elliptic equation, 152, 162
Energy, 6, 10
Equilibrium
point, 259
solution, 261
Ergodic process, 414
Error, 111
discretization, 123
quantizing, 574
rounding, 131
Error analysis, 111
Error constant, 204
Estimation, 373
parameter, 373
state, 373
Euler's angles, 15
Euler's equations, 17
Euler's method, 138
Extended Kalman filter, 410

Feedback, 239
Finite-differences, 153
Finite element method, 170
Fluid dynamics, 41
Fokker–Planck equation, 408
Force, 4
aerodynamic, 21
conservative, 9
dissipative, 10
frictional, 10
generalized, 9
viscous, 10
FORTRAN, 140
Fourier's law, 40
Frequency response, 68, 105, 199
Function, 133
generator, 103

Gain, 84
Gain margin, 202
Gateaux derivative, 334
Gaussian elimination, 156
Gaussian random variable, 606
Gauss' theorem, 590
Generalized coordinates, 5
Generalized minimum variance, 447
Generalized Nyquist criterion, 253
Gershgorin band, 250
Global maximum, 322
Gray code, 129
Gyroscope, 17

Hamiltonian function, 343
Hamilton's principle, 335
Harmonic function, 48
Heat diffusion equation, 39
Hexadecimal number system, 126
Hidden oscillations, 228
Hyperbolic equation, 152, 158
Hyperstability, 297

Identification, 66
parameter, 417
Impedance matrix, 27
Improper node, 270
Independent increments, 395

Inductor, 24
Inertia, 12
  ellipsoid of, 13
  moment of, 12
  products of, 13
Inertial reference frame, 4
Infinite time problem, 363, 406
Information filter, 411
Innovations, 379
  process, 399
  sequence, 379
Integrator, 79, 107
Interpolation, 145
Interrupt, 512
  vector 513
Invariant zero, 249
Inverse function, 101
Inverse Nyquist array, 250
Inverting terminal, 78
Ito integral, 407
Ito's differential rule, 404, 408

Jacobi, 341
  condition, 341
  method, 166
Jacobian matrix, 325
Jordan form, 593
Jury's criterion, 231

Kalman filter, 379
Keyboard, 567
Kinetic energy, 6, 7
Kirchhoff's laws, 24
Krasovskii's method, 287
Kuhn–Tucker theory, 329
Kushner–Stratonovich equation, 409

Lagrange interpolation formula, 145
Lagrange multiplier, 326
Lagrange problem, 333
Lagrange's equations, 9
Lagrangian dynamics, 7
Lagrangian function, 326
Laplace operator, 39
Laplace transformation, 187
La Salle's invariance principle, 281
Lead–lag compensation, 212
Least squares estimation, 418
Legendre–Clebsch condition, 351
Lift coefficient, 54
Limit cycle, 265, 303
Linear estimator, 374
Linearization, 35, 259, 409
Linear quadratic problem, 349, 361
Linear system, 34, 81, 85
Local maximum, 322
Lyapunov
  equation, 364
  function, 278
  stability, 274
  theorem, 278

Machine cycle, 478
Machine language, 133
Magnetic field, 22
Mass–spring–damper system, 11
Maximum likelihood function, 419
Maxwell's equations, 22
Mayer problem, 333
McMillan form, 249
Mean, 605
Memoryless system, 221
Memory map, 543
Microprocessor, 471
  Z80, 473
  68000, 535
Microprogramming, 478
Minimal system, 238
Minimum variance control, 442
Model following system, 431
Modelling error, 3
Model reference control, 424
Modified Z-transform, 228
Motor, 32
  d.c., 32, 33
  stepper, 582

Natural frequency, 194
Navigation law, 365
Newton's equations, 4
Newton's laws, 4
Nichol's chart, 206
Node, 24, 268
Noise, 58
Non-linear estimation, 407
Non-linearity, 258
  backlash, 302
  deadzone, 98, 302
  hysteresis, 98, 302
Non-minimum phase system, 196
Normal modes, 38
Norton's theorem, 26, 29
Number systems, 125
Nyquist, 199
  criterion, 199
  diagram, 200

Objective function, 328
Observability, 243
Observer, 246
Octal numbers, 126
Operational amplifier, 78
Orthogonality principle, 375
Ostrowski bands 250
Overflow 478
Overshoot 194

Parabolic equation, 152
Parity, 130
Parseval's theorem, 293
Partial fraction expansion, 190
Partial observations, 405
Phase margin, 202
Phase response, 105
Phase plane, 263
Phase shift, 105
PIO, 555
Poincaré–Bendixson theorem, 266
Pole, 192, 249
Pontryagin's maximum principle, 342, 344
Popov criterion, 291
Port, 549
Position error, 204
Potential difference, 23
Potentiometer, 102
PRBS, 415
Predator–prey equations, 55
Predictor–corrector method, 142
Principle of optimality, 355
Principle of the argument, 201
Process control, 586

Quasilinear equation, 150
Quasi-sliding mode, 465

RAM 473
  dynamic, 473, 548
  static, 473, 546
Random number generation, 179
Random variable, 603
Random vector, 606
Rate of deformation tensor, 46
Realization, 236
Receding horizon control, 363
Recursive estimation, 377
Register, 475
Regulator problem, 361
Relative left primeness, 241
Relaxed control, 351
Riccati equation, 354, 362
Rigid body, 13
ROM, 473, 545
Root locus, 207
Routh's criterion, 197
Runge–Kutta method, 139

Saddle, 268
Sample path, 60
Sampling theorem, 224
Scaling factors, 82
Secondary minimum problem, 342
Second variation, 341
Self-tuning control, 423
Sensitivity, 69
Separation principle, 406
Shape functions, 173
Simplex, 172
Simulation
  analogue, 77
  digital, 123
  language, 148
Simultaneous displacement method, 166
Singular control, 351
Singularity, 264
Sink, 268
Sliding mode, 455
Smith form, 249
Smith predictor, 235
Source, 270
Spectrum, 224
Square root filtering, 412
Stability, 196
Stabilizability, 244, 363
Stack, 475
State variable, 233
Stationary increments, 394
Steady state error, 203
Stochastic equation, 407
Stochastic process, 59
Stoke's theorem, 590
Strange attractor, 181
Stream function, 49
Stress tensor, 44
Strict equivalence, 242
Successive displacement method, 167
Successive overrelaxation method, 167
Superfluous coordinates, 5
Sylvestor's criterion, 286
Synthesis, 218
System
  delay, 54
  distributed, 35
  finite-dimensional, 33
  infinite-dimensional, 36
  lumped, 35
  non-linear, 88, 407
  stochastic, 58, 373
System zero, 195

Taylor series, 137
Taylor's theorem, 595
Terminal manifold, 336
Thévenin's theorem, 26
Time constant, 193
Time-invariant system, 191
Timing, 541
Top, 17
Torque, 18
Tracking problem, 362
Transfer function, 61, 87, 108, 192, 223, 315
Transistor, 564
Transmission zero, 249

Transversality conditions, 335
Tridiagonal matrix, 154, 182
Trim conditions, 21
Two-point boundary value problem, 338
Two's complement, 128

Unconstrained optimization, 322, 333
Uniform distribution, 179
Unimodular matrix, 242
Unit impulse 221
Unit step, 188

Variable gradient method, 288
Variable structure system, 451
Variance, 605
Variational, 334
VDU, 570
Vector field, 263
Velocity, 6
Velocity error, 204
Virtual work, 7
Volterra series, 315

Wave equation, 36
Weierstrass–Erdmann corner conditions, 339
White noise, 395
Wiener filter, 375
Wiener–Hopf equation, 375, 415
Wiener integral, 396
Work, 6

Zero, 192
  decoupling, 243
  system, 247
  transmission, 249
Zero order hold, 225
Z-transform, 219